ELEMENTS OF
POWER ELECTRONICS

THE OXFORD SERIES IN ELECTRICAL AND COMPUTER ENGINEERING

M.E. Van Valkenburg, *Senior Consulting Editor*
Adel S. Sedra, *Series Editor, Electrical Engineering*
Michael R. Lightner, *Series Editor, Computer Engineering*

Allen and Holberg, *CMOS Analog Circuit Design*
Bobrow, *Elementary Linear Circuit Analysis, 2nd Ed.*
Bobrow, *Fundamentals of Electrical Engineering, 2nd Ed.*
Campbell, *The Science and Engineering of Microelectronic Fabrication*
Chen, *Linear System Theory and Design*
Chen, *System and Signal Analysis, 2nd Ed.*
Comer, *Digital Logic and State Machine Design, 3rd Ed.*
Cooper and McGillem, *Probabilistic Methods of Signal and System Analysis, 2nd Ed.*
Franco, *Electric Circuits Fundamentals*
Houts, *Signal Analysis in Linear Systems*
Jones, *Introduction to Optical Fiber Communication Systems*
Krein, *Elements of Power Electronics*
Kuo, *Digital Control Systems, 3rd Ed.*
Lathi, *Modern Digital and Analog Communications Systems, 2nd Ed.*
McGillem and Cooper, *Continuous and Discrete Signal and System Analysis, 3rd Ed.*
Miner, *Lines and Electromagnetic Fields for Engineers*
Roberts, *SPICE, 2nd Ed.*
Santina, Stubberud and Hostetter, *Digital Control System Design, 2/E*
Schwarz, *Electromagnetics for Engineers*
Schwarz and Oldham, *Electrical Engineering: An Introduction, 2nd Ed.*
Sedra and Smith, *Microelectronic Circuits, 4th Ed.*
Stefani, Savant, and Hostetter, *Design of Feedback Control Systems, 3rd Ed.*
Van Valkenburg, *Analog Filter Design*
Warner and Grung, *Semiconductor Device Electronics*
Wolovich, *Automatic Control Systems*
Yariv, *Optical Electronics in Modern Communications, 5th Ed.*

ELEMENTS OF POWER ELECTRONICS

Philip T. Krein
University of Illinois at Urbana-Champaign

New York   Oxford
OXFORD UNIVERSITY PRESS
1998

OXFORD UNIVERSITY PRESS

Oxford   New York
Athens   Auckland   Bangkok   Bogota   Bombay   Buenos Aires
Calcutta   Cape Town   Dar es Salaam   Delhi   Florence   Hong Kong
Istanbul   Karachi   Kuala Lumpur   Madras   Madrid   Melbourne
Mexico City   Nairobi   Paris   Singapore   Taipei   Tokyo   Toronto   Warsaw
and associated companies in
Berlin   Ibadan

Published by Oxford University Press, Inc.,
198 Madison Avenue, New York, New York, 10016
http://www.oup-usa.org

Library of Congress Cataloging-in-Publication Data
Krein, Philip T., 1956–
Elements of power electronics / Philip T. Krein.
p.   cm. — (The Oxford series in
electrical and computer engineering)
Includes bibliographical references and index.
ISBN 0-19-511701-8 (cloth)
1. Power electronics.   I. Title.   II. Series.
TK7881.15.K74   1997
621.31'7—dc21   96-37809
                            CIP

Printing (last digit): 9 8 7 6 5 4
Printed in the United States of America
on acid-free paper

In memory of
THOMAS P. FITZGERALD,
1953–1996

CONTENTS

PREFACE

INTRODUCTION

Power electronics is one of the broadest growth areas in electrical technology. Today, electronic energy processing circuits are needed for every computer system, every digital product, industrial systems of all types, automobiles, home appliances, lamps and lighting equipment, motor controllers, and just about every possible application of electricity. At one time, the growth was pushed by energy conservation goals. Today, there are many more benefits in terms of reliable, lightweight power processors. A host of new applications is made possible by improvements in semiconductors and by better understanding of power electronics. Motors with integrated electronic controls will soon be the norm. Portable telephones and communication devices demand tightly optimized power management. Advanced microprocessors need special techniques to supply their power. Utilities worry about the quality of their product, and about how to use electronics for more effective power delivery.

This text presents modern power electronics in its many facets. But it is not a loose collection of information. Rather, the intent is to lay down a firm conceptual base from which engineers can examine the field and practice its unusual and challenging design problems. What makes the treatment different? First, a sound scientific framework is established, then students are encouraged to observe how the many converter types and methods branch out naturally from this framework. Second, the treatment is structured for aspiring student engineers. It is written to help students synthesize their electrical engineering study, as they finish an education and begin a career or advanced study. Third, it covers a great deal of support material, such as models for passive components and basic design strategies for magnetics, that is rarely taught but is ubiquitous for the practicing designer.

With a few important exceptions, past treatments of power electronics begin with devices, then develop specific application circuits case by case. In such a broad field, students with little experience are hard pressed to find the deep commonality. A few hours of Web browsing confirms how much misinformation exists about power electronics and design of conversion circuits. Modern devices have reached the point at which they no longer limit the applications. Imaginative designers have found a huge variety of solutions to many types of power electronics problems. It is essential to develop a system-level understanding of the needs and techniques, since a device focus can be unnecessarily constraining. Even so, books continue to be published following the practice of past treatments. Notable exceptions include the 1981 book by Peter Wood, *Switching Power Converters* (Van Nostrand), where a switching function approach was introduced as the first unifying framework for power electronics. More recently, the text by Kassakian, Schlecht, and Verghese, *Principles of Power Electronics* (Addison-Wesley, 1991), builds on Wood's framework with many extensions. Unfortunately, these two books tend to be best suited for students pursuing advanced degrees. This new book owes its roots to the Wood text, and shares the philosophy of the Kassakian text. However, from the outset it was planned for undergraduate students or other engineers with no prior power electronics background.

Why study power electronics? First, because it is fun. Power electronic circuits and systems are the basic energy blocks needed for things that move, light up, cook a meal, fire a combustion cylinder, or display information on a video monitor. Second, because it makes use of all a student's knowledge of electrical engineering, and aims at a new level of understanding. To most students, circuit laws are lifeless mathematical equations. To the power electronics engineer, Kirchoff's laws are the beacon that guides a design—and the snare that catches the unwary or careless. A power electronics engineer needs a working understanding of circuits, semiconductor devices, digital and analog design techniques, electromagnetics as it affects layout and device action, power systems and machines, and the inner action of major applications. Third, because of the challenge. Since power processing is needed just about everywhere, there are few areas with more variety of design tasks. A power electronics expert might work on a 10 MW backup system one day, and on a 1 W system for battery processing the next. Fourth, because of the opportunity. The next personal computer you buy will have a power supply as big as the rest of its electronics combined. It will place extreme performance demands on the supply, and will require total reliability. The power supply will be a significant fraction of the cost to build the computer. Yet the computer manufacturer employs dozens of hardware-software engineers for every power electronics engineer. The need is there, and will grow.

ORGANIZATION AND USE

The book is organized into five parts. Four are here in your hands. Part V, the laboratory supplement, is available through a World Wide Web site. In Part I, the framework for power electronics is established. The three chapters in Part I offer a historical perspective, and establish key framework concepts such as switching functions, equivalent methods for filter design, diode circuit analysis, and regulation. Part II covers all the major converter classes—dc–dc, ac–dc, dc–ac, ac–ac, and resonant converters—in considerable depth. Students are often surprised to learn that they can become effective designers of useful converters by the time they are through Chapter 4. Chapter 8 presents perhaps the first undergraduate text material on the emerging subject of resonant converters. Part III covers the issues of components, from models for sources and loads to power semiconductors to the circuits that drive them. Unique features include the fundamental approach to magnetics design, coverage of wire sizing and parasitic resistance effects, and extensive examples. Part IV introduces control methods, again at the undergraduate level. Chapters 15 and 16 discuss general control issues and develop the popular frequency domain design approach. Chapter 17 provides a new perspective on an especially simple approach to large-signal control.

The book is big because of the breadth of the field. The general layout supports a first course at the senior level, based on Part I and Part II. A second course would cover Part III and Part IV. At the University of Illinois, there is just one course at present. We attempt to cover Chapters 1–6, 11, and 12 in detail, with briefer treatments of 7–10 and 13–14. The chapters are relatively independent, so a variety of course arrangements can be supported. As prerequisites, students should recognize that all their basic course work in electrical engineering will be brought to bear for the study of power electronics. Prior courses in circuits, in electronics, in systems, and in electromagnetics are essential. Prior courses in electromechanics, analog or digital circuit and filter design, and power systems can be helpful, but are not vital.

A few things are not here. Space and time do not permit detailed coverage of individual applications. Motor control and telecommunications power are two examples. It is not possible to provide adequate coverage for dc or ac motor control, or for telecommunications power system design. Beyond the introduction in Chapter 6, the motor control application is left to books from others. The telecommunications application is left to a number of dc–dc converter examples.

The book makes extensive use of computer tools, and students are encouraged to follow this lead. However, no floppy disk is included because few readers find time to learn the programs on such disks. Instead, several example listings are given in the Appendix. Also, students may visit the Web site http://power.ece.uiuc.edu/krein_text to find copies of programs for downloading and to obtain updates to programs or to course materials. A group of industry-based students developed extensive Mathcad® applications. Some of these can be found on the site. Additional problems will be posted as well. Instructors can request Web access to problem solutions through the publisher.

A few words on chapter problems: In this book, a great many of the problems have a design orientation. This means the problems are open-ended, and not always completely specified. Students are encouraged to think about the *context* of a problem, and fill in information when necessary. There are no tricks here. In general, each problem attempts to describe a real system.

UNITS, STANDARDS, AND SIGNIFICANT DIGITS

In general, the International System (SI) of Units is used throughout the text, consistent with IEEE standards. There are some exceptions in magnetics and capacitive components, in which the centimeter is common as the unit of length. If units are not listed explicitly, SI units should be assumed. Appendix B provides a review of some of the unit issues. When possible, graphics symbols are taken from *IEEE Standard 315-1975* (reaffirmed, 1993). The standard gives procedures for creating combination symbols. Unusual symbols, such as that for an ideal ac current source, attempt to follow the procedures.

There are dozens of numerical examples and hundreds of numerical problems in this text. It is important to be aware of significant digit issues. A real circuit application commonly has only about two significant digits. Tolerances on capacitors, inductors, and timing elements are wide. However, in power electronics we are often interested in small differences for efficiency measurements or other detailed information. It is important that small differences not be lost to round-off error in repeated calculations. In examples here, digits are carried through, and round-off is performed only as the last step in the computation.

ACKNOWLEDGMENTS

I am indebted to the many students who have given their insight and their active help in bringing this text about. Richard Bass, now at Georgia Institute of Technology, shared many suggestions on the lab material and as we began to use these notes in place of a published text. Pallab Midya, now at Motorola Corporation, along with Dr. Bass, provided some of the insights now present in Chapter 17 and in the lab sections of Part V. Christopher Nekolny, now at Commonwealth Edison, assisted in first-round production of Part I. A special debt is owed to Rüdiger Munzert, who spent many hours proofreading during his exchange visit from Technische Hochschule Darmstadt. He helped me simplify the English throughout the text, and made personal contributions to Chapter 17.

Final round preparation could not have been completed without the efforts of four current students. Matthew Greuel created most of the lab figures. Daniel Logue found a way to build camera-ready computer output for simulation figures. Richard Muyshondt created several simulations and assisted with proofreading. Luis Amaya became the resident SPICE expert for circuit simulation. I am very grateful for all of this assistance.

My colleagues in power electronics have made many constructive suggestions. The ideas from David Torrey of Rensselaer Polytechnic Institute have been particularly useful, and I appreciate the encouragement. The comments of Thomas Sloane of Alpha Technologies were especially significant and challenging. Others provided encouraging comments too numerous to be mentioned here.

I am grateful for the support of my wife, Sheila Fitzgerald Krein, and family in this project. They have tolerated the long hours and extra workload with grace. It has been a difficult time because of some family tragedies, and I deeply appreciate their forbearance.

CREDITS AND CAVEATS

Many power conversion circuits and control techniques are the subject of active patent protection. The author cannot guarantee that specific circuits or methods described in the text are available for general use. This is especially true of resonant conversion material in Chapter 8.

Power electronics by its nature is an excellent subject for laboratory study. However, it brings many more hazards than more familiar areas of electronics. Readers who plan experimental work in the field should take proper safety precautions in the laboratory.

Mathcad is a registered trademark of Mathsoft, Inc. *Mathematica* is a registered trademark of Wolfram Research, Inc. *PSPICE* is a registered trademark of MicroSim Corporation. Matlab is a registered trademark of The MathWorks, Inc. Xantrex, Lambda, Kyosan, Magnetek, Semikron, Vicor, Tektronix, and Motorola are registered trademarks of their respective companies.

NOMENCLATURE

Symbol	Meaning
α	Phase delay angle
β	Turn-off angle
δ	Difference angle, for relative phase control
ϵ	Electric permittivity
η	Efficiency, P_{out}/P_{in}
λ	Flux linkage, Wb-turns
μ	Magnetic permeability
ξ	Time constant ratio, t/T
ρ	Resistivity
σ	Electrical conductivity; Stefan–Boltzman constant
τ	Time constant, L/R or RC
ϕ	Flux; phase angle
ω	Radian frequency; radian shaft rotational speed
$\mathscr{P}$	Permeance
$\mathscr{R}$	Reluctance
A	Area
B	Magnetic flux density
C	Capacitance
D	Duty ratio
E	Electric field
F	Force
G	Open-loop transfer function
H	Magnetic field intensity; feedback transfer function
I	Current
J	Current density
K	Closed-loop transfer function
L	Inductance
M	Modulating function

N	Number of turns
P	Power
Q	Reactive power; quality factor
R	Resistance
S	Apparent power
T	Period; temperature
V	Voltage
W	Work; energy
X	Reactance
Y	Admittance
Z	Impedance
a	Turns ratio
b	Fourier sine coefficient
c	Constant (in general)
d	Time-varying duty ratio
e	Control error
f	Frequency
g	Gap length
h	Heat transfer coefficient
i	Time-varying current
j	$\sqrt{-1}$
k	Modulation index; gain
l	Length
m	Integer index
n	Integer index
p	Integer index; instantaneous power
q	Switching function
s	Laplace operator
t	Time
u	System input; Heaviside's step function
v	Time-varying voltage
x	State variable
y	Output variable

Special Circuit Symbols

	Ideal dc voltage source		Ideal ac voltage source
	Ideal dc current source		Ideal ac current source
	Voltage source (ac or dc)		Generic transistor (BJT, FET, IGBT)

PART I

PRINCIPLES

CHAPTER 1

Figure 1.1 The magnificent energy of Niagara Falls readily converts to electricity for transport to users far away. (Top: Niagara Falls. Bottom: Marimbondo Hydroelectric Power Plant, courtesy of Furnas Centrais Eléctricas, Brazil.)

BACKGROUND

1.1 THE ENERGY BASIS OF ELECTRICAL ENGINEERING

In 1748, Benjamin Franklin used his newly invented electric motor to roast a turkey for a riverbank party. As we commemorate 250 years since this invention and first practical use, electricity has become the energy of choice for modern life. Electrical energy consumption continues to grow, especially as a fraction of all energy used. Let us take a few moments to consider why this medium is so important.

Energy in electrical form has major advantages:

- It is easy to transport. A few metal wires can transmit the energy needs of an entire city.
- In contrast to heat energy, it can be converted back and forth to mechanical energy or other energy forms with high efficiency.
- It is atomic in character and thus comes in small packages.
- It is intense. Virtually any energy need or power rate, from tiny communication circuits to gigawatt lasers, can be met electrically.

What about some of the alternatives? Moving water has been used as an energy source for centuries. The Niagara River in North America and the Nile River in Africa have enough flow and sufficient elevation change to provide many thousands of megawatts to surrounding cities. The simplest way to transport this kind of power by water is obvious: pipe a river to an energy destination point! A much cheaper method is to convert a river's energy in a hydroelectric plant, then move it over an electrical transmission line. All of the awe-inspiring energy represented by Niagara Falls (the American side is shown in Figure 1.1 at the top) can be converted in a plant similar to the Brazilian facility at the bottom of Figure 1.1, then transported over a single conventional electric circuit.

Heat is also common as a basic energy form. In many large cities, low-pressure steam from power plants is piped to buildings as a source of space heat or process steam. The low temperature of any reasonable heat source precludes wide use of energy in this form. A one-volt battery, for example, has potential energy that corresponds to a thermal source at 50,000°C.

TABLE 1.1
Examples of Electrical Energy Forms

Purpose	Typical Form of Electricity	Expected "Ideal Form"
Very high-power generation	Three-phase ac, 50–60 Hz, 10–30 kV	Polyphase ac
Bulk energy transport	Three-phase ac, up to 765 kV	dc, 500 kV and up
Domestic wiring	Split single-phase ac, 120 V (U.S.), 220 V (Europe)	Possibly low-voltage dc
Electric motors	Single-phase ac, three-phase ac, dc	Polyphase ac
Digital electronics	+5 V dc	Low-voltage dc (2 V and below are discussed)
Analog electronics	+12 V, ±12 V, ±15 V, lower levels	Bipolar dc
Fluorescent lighting	Single-phase ac, approximately 220 V	High-frequency ac
Storage battery applications	Load dependent	Fixed dc current
Medical and industrial magnetic devices	Depends on available sources	High-current dc
Photovoltaic source conversion	(No typical practice)	Matched to peak power transfer impedance
Mobile power systems	+12 V dc (automotive), +28 V dc (aircraft), 400 Hz ac (aircraft, marine)	dc at 48 V, 150 V and higher levels
Telephone systems	+48 V dc and other dc levels	Application dependent
Underground power cable	Polyphase ac	Bipolar dc
Superconducting systems	Very high dc currents	High-current dc
Portable equipment	1.5 V to 12 V battery levels	Highest possible efficiency, with dc voltage levels for digital, analog, radio-frequency, and display electronics

Waste heat cannot come close to the broad applications of any electrical source. Energy conversion from heat is another critical issue. All conversion is limited fundamentally by the Second Law of Thermodynamics and therefore by the Carnot cycle efficiency. In the case of thermal conversion, the Carnot limit is very restrictive. Even a large steam-based power plant has conversion efficiency below 40%. The Carnot limit is much less important in electrical systems. Electric motors with 98% conversion efficiency or better have been constructed.

This is not to say that energy in electrical form does not come without problems. Consider that:

- Energy storage in electrical form is difficult.
- Electricity takes a wide range of different forms. The form best suited for large-scale energy transport, for instance, is not the best form for digital logic circuits. Table 1.1 gives some examples.

Storage remains one of the key limiting factors in the application of electricity. The only device which actually stores electrons is the capacitor. The stored energy is given by $\frac{1}{2}CV^2$. The following example illustrates the feeble energy capabilities of typical capacitors.

Example 1.1.1   Capacitor energy[1]
A 2700 μF capacitor like the one shown in Figure 1.2 is connected across a 120 V dc source. How much energy does this capacitor store? How long will this energy support a 60 W light bulb?

---

[1]Appendix B lists many of the important SI units.

A 2700 μ F capacitor in a 120 V circuit stores $\frac{1}{2}CV^2$, or 19.4 J of energy. A 60 W bulb uses 60 J/s. The energy stored in this capacitor will keep the bulb burning for (19.4 J)/(60 J/s) = 0.32 s. At higher voltage, perhaps 375 V, the capacitor can maintain the bulb for 3.2 s.

A one-liter capacitor can store a thousand joules or so. In contrast, a liter of gasoline stores about 10 MJ as burned in a typical engine. Electrical energy storage will be difficult for years to come.

We see electrical energy in dc batteries or ac line supplies, single-phase circuits or three-phase circuits, 5 V or 3.3 V logic levels, bipolar 12 V levels, neighborhood distribution wires at perhaps 12,000 V ac, transmission systems rated at almost a million volts, and a variety of frequencies. Each of these myriad forms has its own purpose, and any application is best matched to a particular type of source.

Electrical engineering differs from other engineering disciplines in part because it works with the intangibles of atomic physics. In the strictest sense, people do not use electricity. As in Figure 1.3, they use mechanical force and work, light, information, heat, and other more tangible results of energy consumption. Electrical engineering is concerned with conversion among these many forms.

1.2 WHAT IS POWER ELECTRONICS?

The goal of an energy conversion process is to best serve the needs of a consumer, while supporting highly efficient generation and transport of energy. Consider a few subdisciplines of electrical engineering, with their conversion roles:

Figure 1.2 The capacitor is the only device capable of direct electrical energy storage. The device here, rated at 2700 μF at 375 V, can keep a 60 W light bulb burning for only 3.2 s.

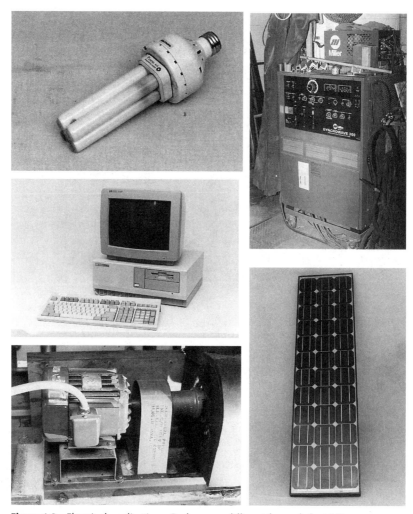

Figure 1.3   Electrical applications: Each uses a different form of electricity.

- *Signal processing* engineers consider conversions between information and electricity.
- *Antenna* engineers consider the most efficient, reliable ways to convert the energy in electromagnetic waves into circuit currents and voltages.
- *Electromechanics* engineers consider conversions between mechanical and electrical energy forms.
- *Quantum electronics* engineers often concern themselves with the interaction of optical signals with electric circuits.

There are many other examples, but a basic issue has been left out: What about conversion, with emphasis on energy flow, among the various forms of electrical sources? This is the framework of power electronics. Power electronics engineers design circuits to convert electrical energy among its many useful forms. The discipline is defined in terms of

Figure 1.4 Control, energy, and power electronics are interrelated.

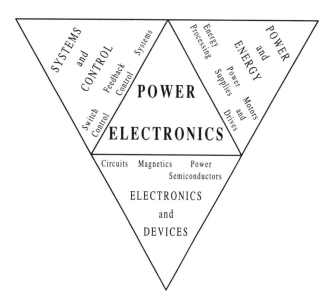

electrical energy conversion, applications, and electronic devices. This suggests that power electronics is one of the broadest subdisciplines of electrical engineering. Many students find that the study of power electronics shows them how electrical engineering fits together as a cohesive discipline.

It is helpful to have a more specific definition.

> **Definition:** *Power electronics involves the study of electronic circuits intended to control the flow of electrical energy. These circuits handle power flow at levels much higher than the individual device ratings.*

As shown in Figure 1.4, the subject represents a median point combining topics in energy systems, electronics, and control.[2] In the study of electrical engineering, power electronics must be listed on a par with digital electronics, analog electronics, and radio-frequency electronics to reflect its distinct nature. Figure 1.5 reflects this parity.

Although power electronics stands distinct from other electronics areas, most students and engineers have some familiarity with a few of the major applications. Rectifiers, or circuits for ac to dc conversion, are well-known examples of energy conversion circuits. Inverters, or dc to ac converter circuits, have been important for nearly a century. Since the 1960s, new semiconductor technologies have dramatically broadened practical possibilities for conversion circuits. A universal characteristic of power electronic circuits is that they manage the flow of electrical energy between some sort of source and a load. The parts in a circuit must direct electrical flows, not impede them. Small components able to manipulate heavy energy flows are of interest, analogous to valves in a plumbing system. Let us consider some examples in an attempt to better understand the definition.

---

[2]Figure 1.4 is based on material in J. Motto, ed., *Introduction to Solid State Power Electronics.* Youngwood, PA: Westinghouse, 1977.

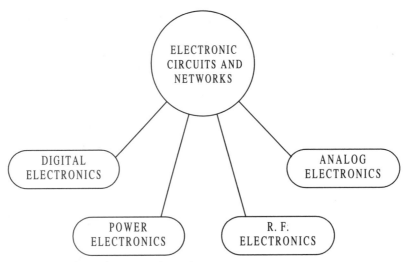

Figure 1.5 Advanced topics in electronics.

Example 1.2.1 An audio amplifier is certainly an electronic circuit, and often it handles considerable energy levels. In North America, a typical stereo receiver converts 60 Hz ac energy and very low-power FM electromagnetic signals into substantial power levels at audio frequency. Is this power electronics?

Maybe, but most amplifier circuits do not really handle high relative energy levels, and the most common types are not usually considered examples of power electronics. A commercial 100 W amplifier usually is designed with transistors big enough to dissipate the full 100 W. The devices are used primarily to reconstruct the audio information rather than to manipulate energy flows. In this situation, the ratio of energy handled to the device energy consumption is about 1 : 1. Conventional audio amplifiers are rarely more than 50% efficient in use. On the other hand, more specialized power electronics amplifiers do exist. They are common in portable communications products, automotive systems, and telephone products. A power electronic design for the same 100 W output might use transistors rated for only 10 W and can readily exceed 80% efficiency.

Example 1.2.2 A half-wave rectifier circuit is built with a standard 1N4004 diode and a capacitor, as shown in Figure 1.6. This device is specified for peak reverse voltage of 400 V, average forward current of 1 A, and power dissipation of 1 W. The circuit input is 60 Hz, 120 V ac RMS, and the output is 170 V dc at up to 1 A. Is this a power electronic circuit?

Yes. The diode is rated for 1 W, yet is controlling 170 W at the circuit output. The circuit controls more than 100 times as much energy as its devices consume. Rectifiers are classical examples of power electronic circuits.

Example 1.2.3 One popular small bipolar transistor is the 2N2222. This device has a rated collector-emitter breakdown voltage of 30 V, a maximum collector current of 0.8 A, and can dissipate 0.5 W. In a conventional analog circuit, it usually handles energy within its 0.5 W power dissipation rating. In principle, the device can manipulate the flow of 0.8 A in a 30 V circuit, so a power electronics engineer would list its power handling level at 24 W. This

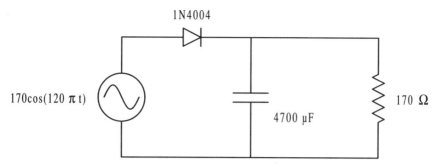

Figure 1.6 Half-wave rectifier for Example 1.2.2.

24 W control capability makes the device quite common in power supplies and other energy conversion circuits.

Example 1.2.4 The MTH15N40 is a metal-oxide-semiconductor field-effect transistor. Its manufacturer reports that it has a maximum continuous drain current rating of 15 A, maximum drain-source breakdown voltage of 400 V, and rated power dissipation of 150 W. In power electronic applications, this device can be used to control up to 15 A × 400 V = 6000 W. This transistor is rated for 150 W, yet it can control the flow in a 6 kW circuit. Several manufacturers have developed power electronic controllers for domestic refrigerators, air conditioners, and even electric vehicles based on this device and its relatives.

Power electronics designers tend to look mainly at voltage and current ratings of a device. Semiconductors of modest size can handle the energy levels in a typical home or commercial establishment. Semiconductor circuits for lighting, industrial motors, and even locomotives are in regular use.

High-power applications lead to some interesting issues. For example, in an inverter, the semiconductors often manipulate 20 times their rated power or more. A small design error or minor change could alter this somewhat, perhaps to a factor of 25. This small change might put large additional stresses on the devices, leading to quick and catastrophic failure. The flip term "smoke test" has a clear and direct meaning when the circuit under test is a power converter. The associated excitement leads to a black magic view of the subject. Many engineers find out the hard way that power semiconductors make fast, but expensive, fuses.

1.3 THE NEED FOR ELECTRICAL CONVERSION

In the 1880s, two major inventions determined the course of electric power systems. George Stanley built a practical transformer, and so allowed convenient conversion among various voltage levels for ac. Nikola Tesla invented the polyphase ac system, and with it showed an easy method for converting electrical energy into mechanical energy. The advantages of low-frequency ac were compelling to designers as early as the 1890s. Such systems form the basis of power systems worldwide.

However, many applications require direct current for proper operation. Electrochemical processes and most electronic circuits are among these. It turns out that dc, at very high voltage levels (HVDC), is the most economical way to transmit electricity over long distances. Until recently, dc motors were the choice for motion control applications. All this means that rectification has been necessary ever since Thomas Edison's original dc system of the early 1880s began to be supplanted by ac. Rectification has become a more acute issue with the rise of electronics and computers. A personal computer often uses dc power at five or more different voltages, ranging from the internal 3 V battery for the real-time clock to 15 kV or more for the accelerating voltage in the cathode-ray tube display. The next generation may well see a revolution in superconducting materials, which will bring new conversion requirements. Modern conversion needs reach well beyond rectifiers, and include:

- Voltage level conversion (for dc as well as ac). A typical system not only needs multiple levels, but it often requires them to be mutually isolated so that their loads remain separate.
- Frequency conversion. Rectification is one example (e.g., 60 Hz input, 0 Hz output). Another example is conversion between the 50 Hz system used in about half the world and the 60 Hz system used in the other half. Mobile systems such as aircraft often use higher frequencies, with 400 Hz serving as one standard value. Much higher frequencies are used for induction heating.
- Waveshape conversion (square, sine, triangle, others). Sinusoidal waveforms for power minimize interference with frequency-multiplexed communication systems. They have other advantages in steady-state ac systems. However, sine waves are not always best for power conversion or motors. Square waves are better for rectification. Triangle or trapezoid waves are used in some motors.
- Polyphase conversions. Single-phase ac power is by far the most widely available form of electrical energy. However, polyphase sources are by far the best form when energy is to be converted and transported. This type of conversion is important for introducing speed control and motor efficiency improvements into household appliances, for instance.

How do we accomplish these conversions? Originally, the straightforward way was to link a motor and generator on the same mechanical shaft, as in Figure 1.7. For example, an ac motor powered with ac electricity could drive a dc generator and thus perform ac to dc conversion. This process converts electricity to mechanical form along the way. This method sometimes still applies when power levels are very high (beyond 1 MW or so), provided that the desired frequencies match available motors and generators. Commercial machines are generally rated for dc, for 50 Hz, and for 60 Hz. There are a few electric railway systems rated for lower frequencies such as 16.7 Hz or 25 Hz.

Some of the difficulties with this process of electromechanical conversion include:

- Limited conversion ranges and functions.
- Bulky, heavy, noisy equipment.

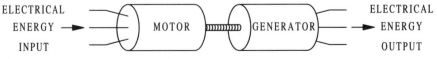

Figure 1.7 Motor-generator conversion.

- Substantial maintenance requirements.
- Slow response times and limited control capability.

In the past, the distinction was made between rotating converters based on machines, and static converters based on electronic circuits. The term *static power conversion* has given way to the more general term *power electronics*.

The general nature of the conversion issues at hand can be summarized as follows:

1. Electricity must always be converted back and forth to the energy forms of interest to people.

2. Electrical engineers are in the business of energy conversion. Whether the issue involves information processing, motors, communication systems, remote sensing, or device fabrication, electrical energy is the means to an end.

3. Electricity is easy to control. A light switch, a volume knob, or a cathode ray tube manipulate electrons with speed and precision.

4. Electrical supplies are needed in a variety of subtly different forms: ac, dc, high or low voltage, high or low current, and so on.

5. Some applications of electricity are not compatible. Voltage transformation with magnetic transformers requires ac. Many chemical processes require dc. Ac motors operate at speeds that depend on the source frequency. Dc motors have speeds that can be adjusted as a function of voltage or current.

6. Conversion of electricity among its various forms is important for a wide range of applications.

7. Compact electronic components with adequate ratings are available, so that electronic alternatives to motor-generator sets can be built.

While rectification has always been a key issue, there are many possible conversion objectives. The technological significance of many of these is growing as new applications become available.

1.4 HISTORY

1.4.1 The Early History of Rectifying Devices

In many ways, the search since the 1880s for better rectifier methods has grown into the entire field of power electronics. The basic form of the diode rectifier circuit was discussed in the nineteenth century, and the modern 50 kW rectifier in Figure 1.8 uses the same principles. What makes the early idea significant is the recognition that the underlying process is fundamentally nonlinear, and cannot be done with any combination of linear circuit elements such as resistors, capacitors, and inductors.

One familiar nonlinear device is a rectifying diode—an element that conducts differently depending on the direction of current flow. The term *diode* actually refers to a generic two-terminal electronic element, but has come to be almost interchangeable with the rectifi-

Figure 1.8   Industrial rectifier.

cation function. While the silicon P-N junction diode is the most common example today, many other technologies yield a rectifying two-terminal element. One early example is the selenium diode, used by C. T. Fritts in a rectifier circuit as early as 1883. The development of the vacuum tube diode about twenty years later was essential to practical applications, but it is interesting that semiconductor rectifiers existed well before the invention of vacuum tubes.

The vacuum diode is limited in fundamental ways by the low current density possible in a vacuum system. A major improvement came when mercury was included in rectifier tubes. The mercury arc tube opened the way to multimegawatt power levels, even at voltages as low as a few hundred volts. A 1905 paper by Charles Proteus Steinmetz considered the performance of mercury tubes for rectification. The waveforms in that paper can easily be duplicated in modern rectifier systems, and represent a broad selection of the possibilities of power rectification.

A typical rectifier system, dating from the 1940s, used mercury arc tubes to convert power from a 50 Hz 2400 V bus into 3000 V dc for a railway locomotive. Arc tubes are still used in certain specialized circumstances, such as rectification beyond 1 MV.

Since before the invention of the transistor, semiconductor diodes have dominated at all but the highest power levels. By the late 1930s, single devices formed from selenium, copper oxide, and other nonlinear materials were manufactured commercially. The P-N junction diode appeared late in the 1940s, and now is the dominant technology, although Schottky barrier diodes offer an alternative in many low-voltage situations. Modern diodes are extremely cheap—even those with ratings high enough for almost any electronic load. Devices with ratings up to about 3 A and 600 V are manufactured in huge lots. Diodes rated at more than 15 kV are readily available, and currents up to about 6000 A also can be achieved (although not both simultaneously). One figure of merit is the power handling rating—the product of voltage and current ratings. Individual diodes exist with power handling capabilities above 36 MW.

The fabrication methods for diodes have evolved rapidly. Today, hundreds of diodes with power handling ratings up to perhaps 500 W each can be fabricated on a single silicon wafer. The highest power devices use the opposite method: Individual diodes formed from complete single wafers are available today, even for 20 cm wafers and larger. One of the biggest challenges with large single devices is packaging: Making a 6000 A connection to a thin, brittle disk 20 cm across is a formidable task.

Complementing the packaging challenge is the challenge to find improved materials for higher power handling. Germanium is sometimes used, but it is more sensitive to high temperatures than silicon. Gallium arsenide power rectifiers have entered the commercial arena, and other compound materials are being examined for power rectifiers. Silicon carbide and even diamond film promise new opportunities to reach extreme power levels during the coming decades. The objective for the latter materials is to increase power handling levels for a given size semiconductor by allowing devices to operate at temperatures beyond 200°C.

1.4.2 Inverters, Controlled Rectifiers, and the SCR

In many cases, the distinction between a rectifier and an inverter is artificial: In a rectifier, energy flows from an ac source to a dc load. In an inverter, the flow is from a dc source to an ac load. An inverter thus has much the same function as a rectifier, except for the direction of energy flow. A "generic conversion circuit," connecting an ac circuit to a dc circuit, could support bidirectional energy flow. Such a circuit provides dual rectifier and inverter operation. The example in Figure 1.9 uses parallel and series elements to handle thousands of megawatts.

Although a dual use rectifier and inverter circuit is possible in principle, the rectifier diode does not support such a circuit. A diode, as a true two-terminal element, is a passive device. This means that its behavior is determined solely by terminal conditions, and there is no direct opportunity for adjustment or other control. A rectifier built with diodes cannot be "told" to begin working as an inverter, since there is no circuit input to alter the function of the devices.

One of the most important power electronic devices, the silicon-controlled rectifier or SCR, addresses this need for control. The SCR, introduced in 1957, provides the function of a diode with the addition of a third terminal for control. The conventional SCR will not conduct unless a signal is applied to this control terminal, or *gate*. Once a gate signal is present, the device operates more or less as a conventional diode. In this way, the gate permits adjustment of the conduction behavior, and leads to the concept of an adjustable diode.

The SCR was not the first technology to provide controlled rectifier function. By the 1920s, passive circuit methods were combined with vacuum diodes to create similar functions. Grid control was used with mercury arc tubes to provide controlled rectification by the 1930s. The cycloconverter—a complicated controlled rectifier adapted for ac–ac conversion—was introduced in about 1931. The SCR provides the function of a grid-controlled

Figure 1.9 A rectifier-inverter control set for the Pacific Intertie high-voltage dc line between Oregon and Southern California. Courtesy of Los Angeles Department of Water and Power. Photographer: Peter S. Garra.

arc tube, but with much better speed and efficiency. The device brought about a revolution in electronic power conversion. Through the use of the SCR, electronic converters for electrochemical processing, transportation systems, industrial dc motor controls, and electric heating and welding became practical during the 1960s. Such familiar applications as variable-speed kitchen appliances and lamp dimmers rely on the SCR and its relatives for control. It is sometimes said that power electronics began when the SCR was introduced.

Once a controlled rectifier can be built, the step to inverters is a small one. Inverters are the critical conversion method for most alternate energy resources. Sources as diverse as wind energy, solar panels, battery banks, and superconducting magnetic energy storage (SMES) rely on inverter circuits to transfer their energy to an ac power grid. The SCR remains crucial for these kinds of systems.

Very high power levels have always been an important application for inverters and controlled rectifiers. This is because dc power is the most economical form for transmission of energy over very long distances. Beyond about 1000 km or so, wavelength effects begin to bring trouble to ac power networks. Resonances and reflections can affect behavior or create failures. Dc power avoids these fundamental problems, and high-voltage dc (HVDC) power transmission remains an important application. The power levels can be extreme: A major link from the Columbia River to Southern California on the U.S. Pacific coast is rated at up to 600 kV and 6000 MW. These levels are far beyond the capacity of any individual device, and large series and parallel combinations of devices must be used to provide diode or controlled rectifier functions. In this particular case, each line terminal can act either as a rectifier or inverter (there are two sets of devices at each end) so that the line power can be adjusted for seasonal changes in energy flow. Some installations elsewhere in the world support only unidirectional power flow.

1.4.3 Inversion from dc Voltage

Two of the fastest-growing inverter applications are not as well served by the SCR. These are circuits for independent backup power and circuits for control of ac motors. Two small modern commercial units are shown in Figure 1.10. The most straightforward inverter circuits use timing information from the ac voltage source to control their operation. Backup circuits and motor controllers do not have access to this sort of time reference information. Without such timing information, inverter control can be complicated. Motor control and backup applications were difficult to build from electronic circuits until relatively recently. One early example was the Stir-Lec I, an experimental electric vehicle built by General Motors in 1968. This car used SCRs in an innovative but complicated arrangement to convert dc power from batteries for an ac motor.

Both backup power and ac motor control systems were considered to be classical applications of motor-generator sets prior to the semiconductor revolution. Diesel-driven generator systems remain the standard choice for large backup power sources. Battery backup is common in dc applications, such as telephone networks and communication equipment. Batteries are becoming more common in small backup applications. Equipment rated for the 0–1 kW range is now readily available.

The growth in low-power battery-backup inverters can be attributed in part to developments in transistor technology. In power electronics applications, bipolar transistors (BJTs),

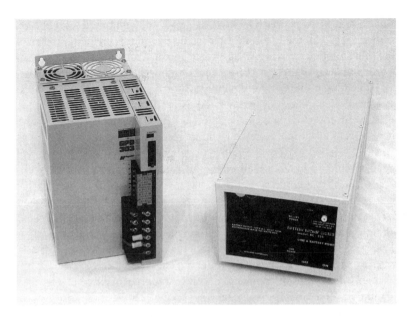

Figure 1.10 Two commercial inverters, one for backup power and the other for motor control.

field-effect transistors (FETs), and more recent insulated-gate bipolar transistors (IGBTs) are in wide use. Power BJTs were first developed for the U.S. space program. By the late 1960s, power handling capacities had reached 1 kW. These ratings were well-matched to small solar panel units. Ratings quickly reached 5 kW early in the 1970s—sufficient to meet the needs of small computers and other light commercial backup applications. By the early 1980s, an experimental device with 1 MW power handling level had been constructed on a single 7.5 cm wafer.

As power BJT technology was developed, improvements in FETs and advances in SCRs began to make inroads in these applications. FETs have certain advantages over BJTs. They are more convenient to use in many circuits. After the power FET was introduced commercially in 1975, it rapidly came into use for low-level conversion applications. This device offers an interesting contrast to other power semiconductors. While power BJTs, diodes, and SCRs are individual components with large geometries, or even full single-wafer devices, power FETs are formed using MOSFET technology as parallel combinations of thousands or even millions of devices. Recent FETs have power handling levels well beyond 10 kW, and have been being applied to inverters for computer power, and even for heavier loads such as AM broadcast transmitters and inverters for ac motors up to about 50 HP.

The IGBT, introduced in the mid-1980s, acts in many ways as a combination device. In practice, its behavior is like that of a BJT, with a power FET to supply the base current. This combination allows the IGBT to approach power handling levels of BJTs, while retaining the simplified operating characteristics of FETs. Recent IGBTs have power handling capability above 500 kW. They dominate motor control inverters.

Figure 1.11 shows just a few of the range of devices and packages used in power electronics today.

Motor control and motor drives are often considered a separate field related to power electronics. The most important long-term goal of engineers designing motor controls has been to supplant dc motors with ac machines. In a typical commercial ac motor controller, the incoming ac power is rectified to create a dc source. This dc voltage supplies an inverter virtually identical to that used in a backup power application. Control of ac motors has been an important technological objective since Tesla introduced the polyphase induction motor in the late 1880s. Dc motors are common in control applications, because their speed can be altered simply by adjusting the input dc voltage level, and their output torque can be manipulated through control of their main winding current. They have major disadvantages in cost and reliability: a true dc motor has brushes and a mechanical commutator which must be maintained. Ac motors, and especially induction motors, are inherently cheaper to build and maintain than dc machines. They have better power-to-weight ratio than dc machines, and can operate at higher speeds. Moving parts are few, and only the bearings themselves require upkeep if motor ratings are observed. However, the speed of an ac machine is tied to the input frequency, and the torque is adjusted by altering the magnetic field levels in the device.

The difficult challenge of providing adjustable magnetic field and input frequency makes ac motors hard to control. Before about 1980, the extra cost of control equipment far

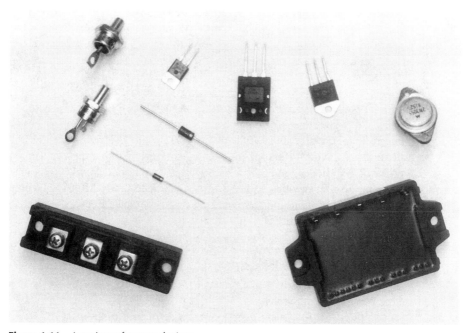

Figure 1.11   A variety of power devices.

exceeded the cost disadvantage of dc machines, and dc systems were common when control of motor speed or torque was needed in an application. In a few cases, the reliability advantages of ac machines were critically important. Motor-generator sets provided adjustable frequency for these applications. Examples include the Scherbius combination, a combined ac motor–ac generator–ac motor, and the Ward–Leonard combination, a dc or ac motor–dc generator–dc motor combination. These are summarized in Figure 1.12.

The SCR can be used with some difficulty in electronic converters for ac motor controls, as mentioned above in the example of the 1968 electric car drive. One of the first examples was the so-called static Scherbius system, in which combinations of SCRs substituted for the functions of the Scherbius motor–generator arrangement. The SCR is hard to use in such a system because its gate controls only the turn-on behavior. It is possible to alter an SCR so that current can be turned off by means of a negative gate signal. This device, called a *gate turn-off SCR*, or GTO, is used in some large motor control circuits today, particularly those on ac locomotives.

More recent electronic circuits built from power FETs or IGBTs meet the basic functional requirements of ac motor control with inverters. In the early 1990s, the cost of these electronic drives began to drop so dramatically that now the combination of a power electronic circuit and an ac motor is cheaper than the cost of an equivalent dc motor system. This development is bringing extraordinarily rapid change in manufacturing and industrial processing. Advanced ac motion control equipment has reached the cost and performance level at which almost any automation application can be addressed. Another widely anticipated device, the MOSFET-controlled GTO or MCT, will make circuits for 100 kW and beyond more convenient.

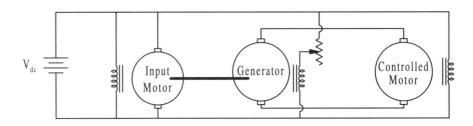

WARD-LEONARD SYSTEM FOR DC MOTOR SPEED CONTROL

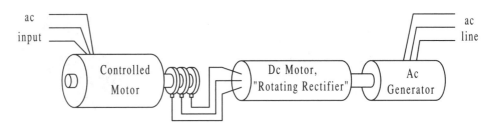

SCHERBIUS METHOD FOR AC MOTOR SPEED CONTROL

Figure 1.12 Ward–Leonard and Scherbius machine combinations.

1.4.4 Power Supplies and dc–dc Conversion

Power supply circuits for radios, computers, portable communications, television sets, and so on, are a commonplace although often neglected element of electrical equipment. This application dominates the power electronics industry in many ways, because of the huge quantities involved. Some typical supplies intended as components of larger systems are illustrated in Figure 1.13. The earliest power supplies for vacuum tube electronics were rectifiers, followed by filtering circuits to create a smooth dc output. Until quite recently, most power supplies took this same form, usually with the addition of a transformer at the ac input to provide the correct output level.

This conventional power supply style matured after about 1967, when integrated series regulator circuits were developed and routinely used at the supply output. A series regulator circuit is a form of amplifier, which provides a very tightly fixed output even from a somewhat noisy rectified signal. The combination of transformer, rectifier, and regulator is referred to as a *linear power supply* since the output filtering circuit is based on a linear amplifier (the circuit as a whole is still nonlinear). Classical rectifier power supplies in this form are among the most widely used electronic circuits, and form the power supply for hundreds of millions of small appliances and electronic accessories.

As costs of electronics decline, the power supply becomes a larger fraction of system cost and design effort. One major manufacturer estimates that power supply cost will soon reach 50% of the total cost of a typical electronic product, such as a cordless telephone or personal computer. This situation makes new technology developments in power supplies critically important. Late in the 1960s, use of dc sources in aerospace applications led to the

Figure 1.13   A typical computer power supply and two fixed-output switching supplies.

application of dc–dc conversion circuits for power supplies. The basic circuit arrangements are much older, and grew out of rectifier applications. Although the basic ideas are old, power semiconductors can be used to make these circuits inexpensive and reliable. In a typical arrangement, an ac source from a wall outlet is rectified without any transformation; the resulting high dc voltage is converted through a dc–dc circuit to the 5 V, 12 V, or other level required by the application. These *switched-mode power supplies* are rapidly supplanting linear supplies across the full spectrum of circuit applications.

A personal computer often requires three different 5 V supplies, two 12 V supplies, a −12 V supply, a 24 V supply, and perhaps a few more. This doesn't include the supplies for the video display or peripheral devices. Only a switched-mode supply can support such complex requirements without high costs. The bulk and weight of linear supplies make them infeasible for hand-held communication devices, calculators, notebook computers, and similar equipment.

Switched-mode supplies often take advantage of FET semiconductor technology. Trends toward high reliability, low cost, and miniaturization have reached the point at which a 5 V power supply sold today might last 500,000 hours (more than 50 years!), and provide 100 W of output in a package with volume of less than 20 cm^3, for a price of less than \$0.50 per watt. This type of supply brings an interesting, if mundane, dilemma: The ac line cord to plug it in actually takes up more space than the power supply itself. Innovative concepts such as integrating a power supply within a connection cable will be used in the future.

Device technology for supplies is being driven by expanding needs in the automotive industry, the telecommunications industry, and markets for portable equipment. The automotive industry uses well-defined voltage levels, but the amount and complexity of electronic hardware in a typical car continue to increase. Power conversion for this industry must be cost effective, yet rugged enough to survive the high vibration and wide temperature range to which a passenger car is exposed. Global communications is possible only when sophisticated equipment can be used almost anywhere in the world. This brings a special challenge, since electrical supplies are neither reliable nor consistent in much of the world. In North America, voltage swings in the domestic ac supply are often less than ±5% around a nominal value. In many developing nations, the swing can easily be ±25%—when power is available. Communications equipment must tolerate these swings, and must also match a wide range of possible backup sources. Given the enormous size of possible markets for telephones and consumer electronics in developing countries, the need for flexible-source equipment is clear. Portable equipment challenges designers to obtain the best possible performance from small batteries. Equipment must use as little energy as possible. The low voltages used for portable battery packs, which range from less than 2 V up to only about 10 V, make any conversion circuit difficult.

1.4.5 Power Electronics as a Practice

The field of power electronics is relatively new as a distinct discipline, even though the applications and fundamental methods have a long history. As a special electrical engineering practice, it dates from about the time the SCR was introduced. The term was coined about 1970 to describe the subject as a cohesive discipline. Engineers active in power electronics are typically involved in one of six areas:

1. Power supplies, either in general, or for specific power needs in computers, electronic instruments, telecommunications, portable equipment, or other electronic circuits.
2. Power semiconductor devices.
3. Electronic motor drives.
4. Support of energy conversion and control in large power systems. This includes control of power networks, high-voltage dc systems, superconducting magnetic energy storage, alternate energy conversion, and controllers for generators.
5. Electric transportation and mobile power, including electric cars, railways, aircraft electric power systems, and spacecraft power systems.
6. Magnetics.

Examples of industries which need experts in power electronics include:

1. *Computer* industry: need for lots of power at a variety of low dc voltages. Backup power is also important. The trend to voltage levels below 5 V will require considerable innovation.
2. *Telecommunications* industry: needs power for transmission, signal processing, and auxiliary services. Highly reliable power sources must be distributed throughout a large system. Rapid communication system growth in developing countries offers vast opportunity. There is extraordinary opportunity for portable communications technology, as cellular telephones, pagers, and a variety of personal devices grow in use.
3. *Aerospace* industry: lightweight converters for aircraft energy needs and for spacecraft electric power processing. Reliability requirements are extreme. Conversion from solar power and other alternate energy sources. The all-electric aircraft is a recent development driven partly by advances in power electronics.
4. *Electronic equipment* industry: power supplies for instruments, consumer electronics, portable and remote measurement devices, and many other products. Many of these represent major growth areas in power electronics. For example, power electronic ballasts for high-efficiency fluorescent lighting are being manufactured in high volumes.
5. *Industrial controls*: motor and motion control applications. Power sensing and measurement. Advanced power-electronic ac motor controls are now available commercially. They can be expected to supplant almost all existing dc motor systems over the next two decades.
6. *Automobile* industry: electric actuators and energy control. Electric traction systems. A typical passenger car contains dozens of electric motors—each controlled with power electronics. Air pollution concerns are forcing the issue of electric cars.
7. *Electric power* industry: emergency backup power supplies. Alternate energy source conversion. HVDC transmission. Power supply quality, and direct electronic control of utility grids. It is possible to move more energy through existing connections if power electronic controls are added. Devices suitable for the power grid are eagerly anticipated.

These examples span nearly the full range of both light and heavy industry. Power electronics today influences the design of everything from office equipment and home appliances to high-speed transportation and satellites.

1.4.6 Summary and Future Developments

Much of this historical discussion has described the special semiconductors used for power conversion circuits. The special energy processing applications addressed by power electronics have also been described. Not long ago, the devices were in many ways the limiting factors in converter design. The fast pace of change in devices has brought about entire new families of power electronic applications, and devices are less and less constraining. The important classes of power electronic components routinely reach power handling levels at least equal to a household appliance load, to a small industrial process, or to most elements of an automobile. Today's designer chooses a circuit and device because it is a good match for the application. Many alternatives are often available. It has been said that power electronics is a "device-driven" field. This is no longer the case. Now, the field has become a true "applications-driven" subject. This trend will continue into applications such as those suggested by Figure 1.14. The figure shows an industrial robot, a portable telephone, a powerful portable computer, an electric car, and a utility plant.

The chronology is summarized in Table 1.2. Among the key opportunities in the next few years are:

1. Electronic ac motor drives. These offer a revolution in energy conservation and industrial processes. It is becoming possible to use sophisticated controls for refrigerators, fans, conveyers, and even washing machine motors. Energy usage can be reduced considerably in many cases.

Figure 1.14   Some growth areas for power electronic applications.

2. Electric transportation. Power electronics is the cornerstone of modern initiatives for electric, hybrid, and alternative vehicles. Today, power electronics can run an electric car at well over 150 km/hr. Battery energy can be conserved well enough for practical cars to reach about 150 km of range with conventional batteries. Novel ideas and innovations are needed for the best designs.

3. Electrical systems for high-efficiency lighting and appliances. Modern fluorescent lights are many times more efficient than incandescent bulbs. They require high-frequency power electronics for optimum operation. Similar new developments extend to appliances such as microwave ovens and other major energy consumers.

4. Power supplies for telecommunications. There are billions of people without telephones or other means of outside communication. Many developing countries do not plan to build expensive hardwired networks. Satellite links, wireless communications, and other new technologies must be integrated with existing equipment. Each different application has unique power supply needs.

5. Power supplies in battery-based portable products. Energy conservation is crucial in battery systems. Proper use of power electronics can easily make a 50% difference or more in battery life. Converters for applications below 5 V are a special challenge.

6. Power electronic applications in electric utility networks. It is possible to extend conversion advantages to extreme energy levels. Systems with fast control of all energy flows can be imagined. This gives a tremendous range of opportunities for lower cost electricity distribution.

Novel power electronic methods are being applied to audio amplifiers, cellular telephones, and microprocessors. The list is long, and there is need for skilled engineers who can apply the methods in unconventional ways.

TABLE 1.2
Summary of Chronology of Electronic Power Conversion

Dates	Device or Technology	Conversion Technologies
1880s	Transformer, M-G sets	Electromechanical units for ac–dc conversion, voltage level shifting for ac.
1900s	Vacuum diodes	Development of major applications.
1920s	Mercury-arc tubes	Electronic rectification. Electronic circuits for ac–dc and dc–ac conversion. Basic techniques worked out for ac–ac conversion.
1930s	Selenium rectifiers, grid control	"Semiconductor" rectifier technologies in regular production.
1940s	Magnetic amplifiers	Electronic power amplifiers. Further advances in electronic conversion.
1950s	Semiconductors	Inception of electronic conversion for high-voltage dc power transmission. Growing need for small power supplies for electronic gear.
1960s	Silicon-controlled rectifiers (SCRs)	High-power semiconductor devices. These quickly replaced gas tubes, and made controllable ac–dc converters practical and cheap.
1970s	Power bipolar transistors	Substantial simplification of dc–ac and dc–dc conversion techniques. Emergence of power electronics as a separate discipline.
1980s	Power field-effect transistors	New methods for dc–dc conversion. The influence of device properties on power electronics begins to wane. Rapid expansion of markets for miniature power supplies.
1990s	IGBT	Nearly any application now possible. Emphasis on the best alternative for a given application.

1.5 GOALS AND METHODS OF ELECTRICAL CONVERSION

1.5.1 The Basic Objectives

The function of a power conversion circuit is to control the energy flow between a given electrical source and a given load, as depicted in Figure 1.15. A converter must manipulate flow, but should not consume energy. The reason is simple. Since a power converter appears between a source and a load, any energy used within the converter is lost to the overall system. To be useful, a converter should have high input-output energy efficiency, $\eta = P_{out}/P_{in}$. This is the first and primary design objective in power electronics:

EFFICIENCY TARGET $\rightarrow$ 100 %

We seek lossless processes to implement converters.

A power converter connected between a source and a load also affects system reliability. If the energy source is perfectly reliable (it is on all the time), then a failure in the converter affects the user (the load) just as if the energy source fails. An unreliable power converter creates an unreliable system. Reliability requirements are extreme in many applications. To put this in perspective, consider that a typical U.S. household loses electric power only a few minutes a year. Energy is available 99.999% of the time. A converter must be better than this to avoid degrading a system. As high efficiency becomes more routine, this second reliability objective grows in importance:

RELIABILITY TARGET $\rightarrow$ NO FAILURES OVER APPLICATION LIFETIME

Reliability is often a more difficult objective than efficiency. Imagine just trying to prove that a circuit will not fail over decades of use. Spectacular failures are a traditional property of power supplies, and other equipment is often damaged in the process.

1.5.2 The Efficiency Objective—The Switch

As simple a circuit element as a light switch like the one in Figure 1.16 gives a reminder that the extreme requirements in power electronics are not especially novel. Ideally, when a switch is on, it provides $v_{switch} = 0$, and will carry any current imposed on it. When a switch is off, it blocks the flow of current ($i_{switch} = 0$), regardless of the voltage across it. The *device power*, $v_{switch}i_{switch}$, is identically zero at all times. The switch controls energy flow with

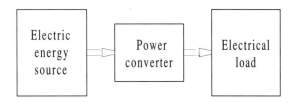

Figure 1.15 Basic electric power converter.

no loss. Reliability is high, too. Household light switches perform over decades of use and perhaps 100,000 operations. Of course, a mechanical light switch does not meet all the practical needs. A switch in a power supply often functions 100,000 times each second. Even the best mechanical switch will not last beyond a few million cycles.

A circuit built from ideal switches will be lossless. The advantages of switching circuits are so significant that many people equate power electronics with the study of switching power converters. Other lossless elements such as capacitors, inductors, and conventional transformers, might also be useful for conversion. The complete concept is shown in Figure 1.17, which illustrates a power electronic system. Such a system consists of an energy source, an electrical load, a power electronic circuit, and control functions. The power electronic circuit contains switches, lossless energy storage elements, and magnetic transformers. The controls take information from the source, load, and designer, then determine how the switches operate to achieve the desired conversion. Usually, the controls are built up with conventional low-power analog and digital electronics.

1.5.3 The Reliability Objective—Simplicity and Integration

It is well established in military electronics that the more parts there are in a system, the more likely it is to fail. As we shall see, power electronic circuits tend to have few parts, especially in the main energy flow paths. The necessary operations must be carried out through shrewd use of those parts. Often, this means that sophisticated control strategies are applied to seemingly simple conversion circuits.

One way to avoid the reliability-complexity tradeoff is to use highly integrated components. A high-end microprocessor, for example, contains more than a million parts. Since all interconnections and signals flow within a single chip, the reliability is nearly that of a single part. An important parallel trend in power electronic devices is the integrated module. Manufacturers seek ways to package several switching devices, their interconnections,

Figure 1.16 A switch and its electrical terminal values.

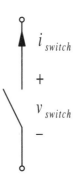
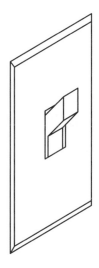

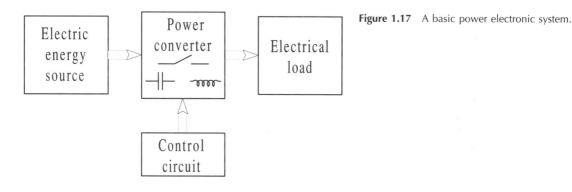

Figure 1.17 A basic power electronic system.

and protection components together as a unit. Control circuits for converters are also integrated as much as possible to keep reliability high.

1.5.4 Important Variables and Notation

In a power electronic system, several electrical quantities are of special interest. Efficiency has already been identified. Maximum values of currents and voltages will be needed to determine the necessary device ratings. Energy flow is the underlying objective, and power and energy levels in each part of the system are very important. We are most interested in energy flow over reasonable lengths of time. The power electronic circuit must work to alter the flow from source to load. The average energy flow rate, or *average power*, is therefore of particular interest. Some important quantities:

- Average power at a specified location. This represents useful energy flow over time.
- Peak values of voltages and currents. These determine device ratings.
- Average values of voltages and currents. These represent the dc values in a circuit.
- RMS voltages and currents. These represent power in resistors, and often determine the losses in a converter.
- Waveforms. Power electronic circuits often have clear graphical properties. Study of waveforms is often the most direct way to evaluate a circuit's operation.
- Device power. Switches are not quite ideal, and some residual power will be lost in them.

All these quantities are crucial to a basic understanding of power electronics and the circuits studied in it.

Notation for average and RMS values of some periodic function *v(t)* will be given as

$$\langle v \rangle = \frac{1}{T}\int_0^T v(t)\, dt, \quad V_{\text{RMS}} = \sqrt{\frac{1}{T}\int_0^T v^2(t)\, dt} \tag{1.1}$$

Table 1.3 summarizes the notational practices in this text.

1.5.5 Conversion Examples

Let's look at a few examples of power electronic systems.

Example 1.5.1   Consider the circuit shown in Figure 1.18. It contains an ac source, a switch, and a resistive load. It is therefore a simple but complete power electronic system.

Just to get an idea of how power conversion might take place, let us assign some kind of control to the switch. What if the switch is turned on whenever $V_{ac} > 0$, and turned off otherwise? The input and output voltage waveforms are shown in Figure 1.19. The input has a time average of 0, and an RMS value equal to $V_{peak}/\sqrt{2}$. The output has a nonzero average value given by

$$\langle v_{out}(t) \rangle = \frac{1}{2\pi} \left(\int_{-\pi/2}^{\pi/2} V_{peak} \cos \theta \, d\theta + \int_{\pi/2}^{3\pi/2} 0 \, d\theta \right)$$

$$= \frac{V_{peak}}{\pi} = 0.3183 V_{peak} \tag{1.2}$$

and an RMS value equal to $V_{peak}/2$ (Confirm this RMS value as an exercise). The output has some dc voltage content. The circuit can be thought of as an ac–dc converter.

The circuit in Example 1.5.1 is a half-wave rectifier with a resistive load. A diode can be substituted for the switch. The example shows that a simple switching circuit can perform power conversion functions. But, notice that a diode is not, in general, the same as the switch. A diode places restrictions on the current direction, while a true switch would not. An ideal switch allows control over whether it is on or off, while a diode's operation is constrained by circuit variables. Consider a second half-wave circuit, now with a series L-R load, shown in Figure 1.20.

Example 1.5.2   A series D-L-R circuit has ac voltage-source input. This circuit operates much differently than the half-wave rectifier with resistive load. Remember that a diode will be on if forward biased, and off if reverse biased. In this circuit, an off diode will give $i = 0$.

TABLE 1.3
Nomenclature Summary

Notation	Description
$v(t)$, $i(t)$, etc.	Instantaneous values of voltage, current, power, or other quantities. Given in lower-case notation. Time is usually shown explicitly.
$\langle v \rangle$, $\langle i \rangle$	Bracket notation for average or dc quantities. Averages are defined over some time period T in integral form.
V, I, P	Upper-case form. Used for explicit dc source values. Also used as an alternative notation for averages, especially for average power.
V_{RMS}, I_{RMS}	The true root mean square value associated with a given time function. RMS quantities are defined over a time period T in integral form. In conventional power systems practice, a given voltage or current is an RMS value unless explicitly stated otherwise.
$\bar{v}(t)$, $\bar{i}(t)$	Moving-average quantities. An average form which retains time dependence by moving the "integral window" as a function of t. A simple alternative $P(t)$ will usually be used to indicate the moving average of power.
$\tilde{V}$, $\tilde{I}$	Complex phasor quantity, with magnitude given in RMS units.
$\tilde{v}(t)$, $\tilde{i}(t)$	Small-signal perturbation or ripple. A small change around a constant level.

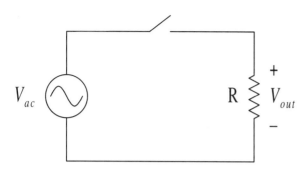

Figure 1.18   A simple power electronic system.

Whenever the diode is on, the circuit is the ac source with R-L load (Figure 1.20b). Let the ac voltage be $V_0 \cos(\omega t)$. From Kirchhoff's Voltage Law (KVL),

$$V_0 \cos(\omega t) = L\frac{di}{dt} + Ri \qquad (1.3)$$

Let us assume that the diode is initially off (this assumption is arbitrary, and we will check it out as the example is solved). If the diode is off, $i = 0$, and the voltage across the diode $v_d = v_{ac}$. The diode will become forward-biased when v_{ac} becomes positive. The diode will turn on when the input voltage makes a zero-crossing in the positive direction. This allows us to establish initial conditions for the circuit: $i(t_0) = 0$, $t_0 = -\pi/(2\omega)$. The differential equation can be solved in the conventional way[3] to give

$$i(t) = V_0 \left[\frac{\omega L}{R^2 + \omega^2 L^2} \exp\left(\frac{-t}{\tau} - \frac{\pi}{2\omega\tau}\right) + \frac{R}{R^2 + \omega^2 L^2} \cos(\omega t) + \frac{\omega L}{R^2 + \omega^2 L^2} \sin(\omega t) \right] \quad (1.4)$$

where τ is the time constant L/R. What about diode turn-off? One first guess might be that the diode turns off when the voltage becomes negative, but this is not correct. We notice from the solution that the current is not zero when the voltage first becomes negative (Check this!). If the switch attempts to turn off, it must drop the inductor current to zero instantly. The derivative of current in the inductor, di/dt, would become negative infinite. The inductor voltage $L(di/dt)$ similarly becomes negative infinite—and the devices are destroyed. What really happens is that the falling current allows the inductor to maintain forward bias on the diode. The diode will turn off only when the current reaches zero. A diode has very definite properties that determine the circuit action, and both the voltage and current are relevant.

We have considered the diode in these two example circuits so far. While the device acts as a switch, we do not have any control over its behavior. Let us consider a different way to operate the switch in the first example circuit.

Example 1.5.3   Consider again the circuit of Figure 1.18. Instead, turn the switch on whenever $V_{ac} > V_{peak}/2$, and turn it off one-half cycle later. The input and output voltage wave-

---

[3]Many of the equations in this book can be readily analyzed using symbolic computer programs such as *Mathematica*. Appendix C provides a list of computer code necessary for many equations in the text, including the one here.

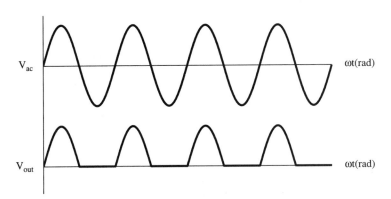

Figure 1.19 Input and output waveforms for example circuit.

forms are shown in Figure 1.21. The input has $\langle v \rangle = 0$ and $V_{\text{RMS}} = V_{\text{peak}}/\sqrt{2}$. The switch turns on as the input waveform crosses the line $V_{\text{peak}}/2$, 30° after the zero crossing point. The output average value is given by

$$\langle v_{\text{out}}(t) \rangle = \frac{1}{2\pi}\int_{-\pi/3}^{2\pi/3} V_{\text{peak}}\cos\theta \, d\theta = \frac{\sqrt{3}}{2\pi}V_{\text{peak}} = 0.2757 \, V_{\text{peak}} \qquad (1.5)$$

which is only 87% of the previous dc value. The RMS value is still $V_{\text{peak}}/2$.

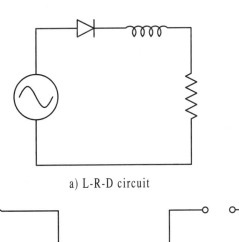

a) L-R-D circuit

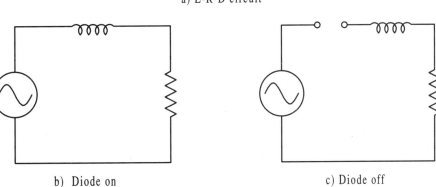

b) Diode on   c) Diode off

Figure 1.20 Half-wave rectifier with *L-R* load for Example 1.5.2.

In the Example 1.5.3 circuit, a diode cannot be used to perform the requested conversion. The circuit still performs rectification, but some more general switch will be needed to permit the necessary control. The rectifier example shows how a switch can be used to obtain a conversion function. From the examples, rectifier operation can be adjusted by manipulating switch timing or the load properties. The dc output depends on when the switch is turned on or off. However, the output was not a clean dc waveform. We need filtering to recover the dc value. Any type of low-pass filter could, in principle, allow recovery of the dc output. Filters are one way in which energy storage elements are applied in power electronics.

Storage elements also appear at intermediate points in many power converters. An example appears in Figure 1.22. In this circuit, the left switch is turned on to store energy in the inductor. The right switch sends energy from the inductor into the load. The inductor mediates energy transfer through the system, and adds flexibility to the converter. Let us consider a possible way of operating this circuit.

Example 1.5.4 The switches in the circuit of Figure 1.22 are controlled to operate in alternation: when the left switch is on, the right one is off, and so on. What does the circuit do if each switch operates half the time? The inductor and capacitor have large values.

When the left switch is on, the source voltage V_{in} appears across the inductor. When the right switch is on, the output voltage V_{out} appears across the inductor. If this circuit is to be a useful converter, we want the inductor to receive energy from the source and deliver it to the load without loss. Over time, this means that energy should not be allowed to build up in the inductor (it should flow through, instead). The power into the inductor therefore must equal the power out, at least over some reasonable period of time. The average power in should equal the average power out of the inductor. Let us denote the inductor current as i. The input is a constant voltage source. Assuming that the inductor current is also constant, since L is large, the average power into L is

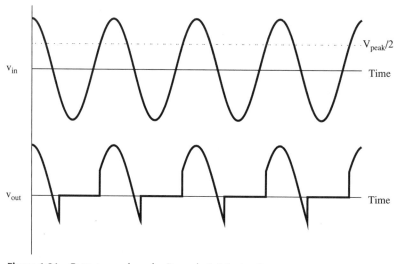

Figure 1.21 Output waveform for Example 1.5.3 circuit.

$$P_{\text{in}} = \frac{1}{T}\int_0^{T/2} V_{\text{in}}i \, dt = \frac{V_{\text{in}}i}{2} \tag{1.6}$$

For the average power out of L, we must be careful about current directions. The current out of the inductor will have a value $-i$. The average power out is

$$P_{\text{out}} = \frac{1}{T}\int_{T/2}^{T} -iV_{\text{out}} \, dt = -\frac{V_{\text{out}}i}{2} \tag{1.7}$$

Again, if this circuit is to be useful as a converter, the net energy flow should be from the source to the load over time. $P_{\text{in}} = P_{\text{out}}$ requires that $V_{\text{out}} = -V_{\text{in}}$.

The circuit of Figure 1.22, when operated as described in the example, serves as a polarity reverser. The output voltage magnitude is the same as that of the input, but the output polarity is negative with respect to the reference node. The circuit is often used to generate a negative supply for analog circuits from a single positive input level. If the inductor in the polarity reversal circuit is moved instead to the input, a step-up function is obtained. Consider the circuit of Figure 1.23 in the following example.

Example 1.5.5   The switches of Figure 1.23 are controlled in alternation. Each switch is on during half of each cycle. Determine the relationship between V_{in} and V_{out}.

The inductor's energy should not build up when the circuit is operating normally as a converter. A power balance calculation can be used to relate the input and output voltages. Again let i be the inductor current. When the left switch is on, power is injected into the inductor. Its average value is

$$P_{\text{in}} = \frac{1}{T}\int_0^{T/2} V_{\text{in}}i \, dt = \frac{V_{\text{in}}i}{2} \tag{1.8}$$

Power leaves the inductor when the right switch is on. Again we need to be careful of polarities, and remember that the current should be set negative to represent output power. The result is

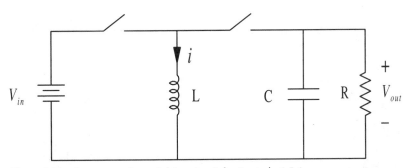

Figure 1.22   Energy transfer switching circuit for Example 1.5.4.

$$P_{\text{out}} = \frac{1}{T}\int_{T/2}^{T} -(V_{\text{in}} - V_{\text{out}})i \; dt = -\frac{V_{\text{in}}i}{2} + \frac{V_{\text{out}}i}{2} \tag{1.9}$$

Equating the input and output power,

$$\frac{V_{\text{in}}i}{2} = -\frac{V_{\text{in}}i}{2} + \frac{V_{\text{out}}i}{2} \quad \text{and} \quad 2V_{\text{in}} = V_{\text{out}} \tag{1.10}$$

and we see that indeed the output voltage is double the input. Many seasoned engineers find the dc-dc step-up function of Figure 1.23 to be surprising. Yet Figure 1.23 is just one example of such action. Others (including circuits related to Figure 1.22) are used in systems from CRT electron guns to spark ignitions for automobiles.

The circuits of the examples have few components. A commercial step-up circuit is shown in Figure 1.24. The left switch is implemented as four power MOSFETs (metal-oxide-semiconductor field effect transistor) in parallel, while the right switch is a diode. This circuit actually takes in an ac supply, then rectifies it and boosts up the result. It can supply up to 3000 W at 400 V dc from a 240 V ac source. There are extra components for control functions, but the power electronics are very much those of Figure 1.23. Another commercial circuit is shown in Figure 1.25. Although it is far more complicated than the preceding examples, its power electronic heart is the polarity reversal circuit.

The history of power electronics has tended to flow like these examples: a circuit with a particular conversion function is discovered, analyzed, and applied. As the circuit moves from a simple laboratory test to a complete commercial product, control and protection functions are added. The power portion of the circuit remains close to the original idea. The natural question arises as to whether a systematic approach to conversion is possible. Can we start with a desired function and design an appropriate converter? Are there underlying principles that apply to design and analysis? Where do all these control functions come from and what do they try to accomplish? How do the circuits work? In this text, we will begin to see how the various aspects of energy flow manipulation, sensing and control, the energy source, and the load fit together in a complete design. The goal is a systematic treatment of power electronics. Keep in mind that while many of the circuits look deceptively simple, all are nonlinear systems with unusual behavior.

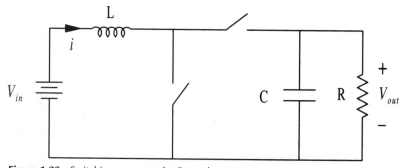

Figure 1.23   Switching converter for Example 1.5.5.

1.6 RECAP

> **Definition:** *Power electronics involves the study of electronic circuits intended to control the flow of electrical energy. These circuits handle power flow at levels much higher than the individual device ratings.*

Energy conversion is our business because people use light, mechanical work, information, heat, and other tangible results of energy—not electricity. In most electronics, devices are limited by their ability to dissipate lost energy. In power electronics, we are interested in how much energy flow a device can manipulate, and intend to keep the dissipation as low as possible. Power handling ratings (the product of ratings for voltage and current) are larger than power dissipation ratings by more than a factor of 100 for many devices.

Voltage level and frequency conversion are the most common needs for electrical energy conversion. All major industries use power converters. Significant growth is expected in ac drives, electric transportation, portable power, telecommunications applications, and utility applications over the next few years.

A power converter is positioned between a source and a load. A primary objective is 100% efficiency. Very high reliability is also important. The switch is a familiar lossless device that can manipulate energy flow. Many circuits can be analyzed through energy conservation considerations. Functions such as dc voltage conversion can be created with switch networks.

> **Definition:** *A power electronic system consists of an electrical source and load, a power electronic circuit containing switches and energy storage, and control functions. The power electronic circuit portion often has relatively few parts, and most of the components in a commercial system perform control functions.*

Figure 1.24 Photograph of commercial boost converter.

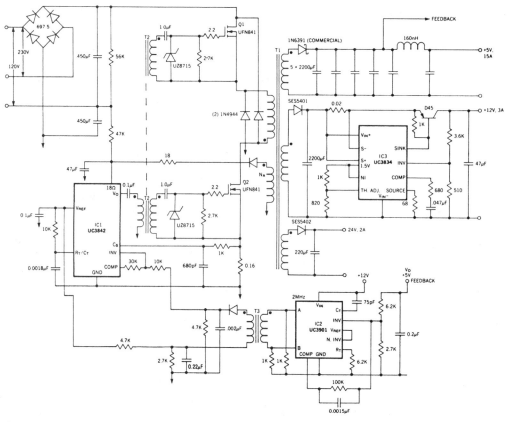

Figure 1.25 Commercial power converter (based on polarity reverser). From R. Patel, D. Reilly, and R. Adair, "150 Watt Flyback Regulator," *Unitrode Power Supply Design Seminar*. Lexington, MA: Unitrode, 1986. Reprinted with permission of Unitrode Corporation.

Power electronics today is evolving from a device-driven field to an applications-driven field. Not long ago, the possible conversion functions were limited by the types of switch technologies. Today, new applications such as ac motor controllers can be built with several types of devices. In the future, application requirements will not be limited by device capabilities.

PROBLEMS

1. List five products that you own which require electrical energy conversion.

2. Try to list the motors and their functions in a typical household (e.g., refrigerator compressor motor). Power electronic controllers can save at least 10% of the energy used by many of these motors.

3. List some possible applications of rectifiers. Which of these could benefit from the ability to adjust the dc output level with a controlled rectifier?

4. Consider the circuit of Example 1.5.3, operated more generally. Let us turn the switch on whenever $V_{ac} > V_{peak}/k$, where k is an adjustable parameter. The switch will turn off one-half cycle later. What is the output average voltage as a function of k?

5. In the polarity reversal circuit of Example 1.5.4, we used a power balance to determine the output voltage, assuming that each switch is on during half of the period T. What will the output be if the left switch is on 75% of T and the right switch is on 25% of T? Again, the input and output power must match.

6. The step-up converter of Example 1.5.5 showed doubling at the output. Instead, keep the left switch on for 95% of T, and the right switch for 5% of T. What is the ratio V_{out}/V_{in} in this case?

7. The circuit in Figure 1.26 offers another arrangement of switches and energy storage. Assume that the switches act in alternation, and that each is on for 50% of the period T. What is the ratio V_{out}/V_{in}?

8 Compute the time at which the diode turns off each cycle in a half-wave rectifier with series R-L load. See Example 1.5.2.

9. A switching power converter is designed to have efficiency of 95% when the output load is 100 W. The efficiency increases linearly up to 97% with 200 W output (i.e., efficiency is 96% at 150 W, and so on). Similarly, the efficiency decreases to 94% at 50 W output. The converter will be damaged whenever the power dissipated inside it exceeds 6 W. If the converter is intended to operate for output loads between 20 W and some maximum limit, what is the upper limit of output power which can safely be supplied?

10. A designer constructs the converter of Problem 1.9. The purchaser finds that the efficiency is slightly worse than expected: It is 94.5% at 100 W, 96.5% at 200 W, and so on. What power limit will the purchaser encounter?

11. The customers of a certain electric utility company use energy at a combined rate of 10 GW. Of this amount, 30% goes toward fluorescent lighting. The utility finds that it costs 10¢ to generate a kilowatt for an hour. A small company has developed a new switching power converter for supplying energy to fluorescent lights. The converter allows a 40% reduction in energy consumption of this type of lighting. If every utility customer installed the new converters, how much money would be saved in energy costs each year?

12. Power balance can be a delicate thing. The circuit in Figure 1.27 is a step-up converter intended to provide 24 V output from a 12 V source. The capacitor is rated for use at up to 35 V, and the inductor is ample for the job. The switches operate to give no net energy gain in the inductor. However, a problem occurs if for some reason the resistive load is disconnected: The capacitor bursts into flame. Think about the change in P_{out} when the resistor is suddenly removed, and explain the trouble. *Don't try this in the laboratory without safety glasses and a stout circuit cover.*

13. The version of the polarity reverser circuit shown in Figure 1.28 includes a model for inductor losses. The resistor R_L represents loss in the magnetic material and in the wires. The switches operate in alternation. Each is on 50% of a cycle. The inductor and capacitor values are large. What is V_{out} for this converter in terms of V_{in}, R_L, and R? (*Hint*: Conservation of energy requires that the input average power match the output average power plus the internal loss $I_L^2 R_L$.)

14. Many nonswitching power conversion methods conserve current as well as energy. In one such example, a circuit provides 10 V_{dc} output from a 20 V_{dc} source. The load draws 5 A. Input power must supply the output plus any losses, but the input and output currents must match. What is the efficiency for this converter? Comment on this approach.

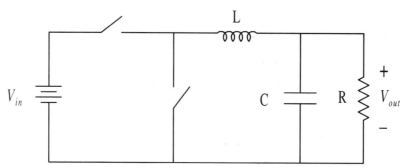

Figure 1.26 Alternative dc conversion circuit.

15. A designer wants to obtain the advantages of the polarity reverser but avoid the change of sign. One way to do this is to cascade two converters. The second "reverses" the output of the first. Draw this combination. If the left-hand switch in each converter is on 25% of a period, and the right-hand switch is on 75% of each period, what is the overall ratio V_{out}/V_{in}?

16. An inverter produces a voltage square wave at 60 Hz, with a peak value of 200 V. What is the RMS value of this voltage?

17. In the boost converter of Figure 1.27, the inductor has series resistance of 2 Ω. What voltage will be measured at the output? Keep in mind that the input power must supply this additional loss in addition to the output.

18. An inverter for an electric vehicle has typical efficiency of 98%. There are six power devices in the inverter. They dissipate power in approximately equal shares. Each can handle power dissipation of 250 W, voltage of 600 V, and current of 600 A. Other parts of the inverter consume about 200 W total. The main dc power source is 350 V_{dc}. What is the maximum power that can be achieved at the output without violating any limits?

19. What will happen over a long time interval to the circuit of Figure 1.22 if there is an initial mismatch in power flows? You might consider the start-up case, in which $i_L = v_C = 0$ initially.

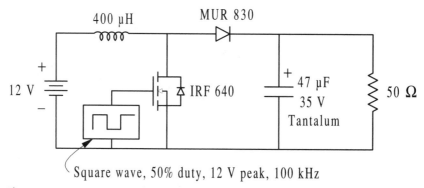

Figure 1.27 Boost converter for 12 V to 24 V conversion.

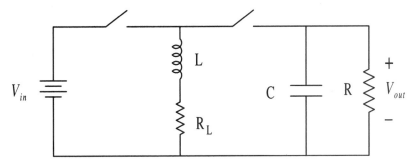

Figure 1.28 Polarity reverser with lossy inductor model.

20. Efficiency is hard to measure. In a laboratory setting, a certain converter is measured to have input power of 28.2 W and output power of 28.4 W. However, the meter accuracy on this scale is ±0.75%. With this level of accuracy, what range of power loss values is consistent with the data?

REFERENCES

H. D. Brown, J. J. Smith, "Current and voltage wave shape of mercury arc rectifiers," *Trans. AIEE*, vol. 52, p. 973, 1933.

Y. Ferguson, *Electric Railway Engineering*. London: Macdonald and Evans, Ltd., 1953.

B. Franklin, *Experiments and Observations on Electricity Made at Philadelphia in America*, I. Bernard Cohen, ed. Cambridge: Harvard Univ. Press, 1941.

C. T. Fritts, "A new form of selenium cell," *American J. Science*, vol. 26, p. 465, 1883.

L. O. Grondahl, P. H. Geiger, "The new electronic rectifier," *Trans. AIEE*, vol. 46, p. 357, 1927.

C. F. Harding, *Electric Railway Engineering*. New York: McGraw-Hill, 1926.

C. C. Herskind, "Grid controlled rectifiers and inverters," *Trans. AIEE*, vol. 53, p. 926, 1934.

C. C. Herskind, W. McMurray, "History of the Static Power Converter Committee," *IEEE Trans. Industry Appications*, vol. IA-20, no. 4, pp. 1069–1072, July 1984.

A. W. Hull, H. D. Brown, "Mercury arc rectifier research," *Trans. AIEE*, vol. 50, p. 744, 1931.

E. T. Moore, T. G. Wilson, "Basic considerations for DC–DC conversion networks," *IEEE Trans. Magnetics*, p. 620, Sept. 1966.

E. T. Moore, T. G. Wilson, "DC to DC converter using inductive-energy storage for voltage transformation and regulation," *IEEE Trans. Magnetics*, p. 18, March 1966.

E. L. Owen, "Origins of the inverter," *IEEE Industry Applications Mag.*, vol. 2, p. 64, January 1996.

E. L. Owen, M. M. Morack, C. C. Herskind, A. S. Grimes, "Ac adjustable-speed drives with electronic converters—the early days," *IEEE Trans. Industry Applications*, vol. IA-20, p. 298, March 1984.

H. Rissik, *The Fundamental Theory of Arc Converters*. London: Chapman and Hall, 1939.

H. Rissik, *Mercury-Arc Current Converters*. London: Sir Issac Pitman and Sons, 1935.

A. C. Seletzky, S. T. Shevki, "Grid and plate currents in grid-controlled mercury vapor tubes," *J. Franklin Institute*, vol. 215, March 1933.

F. G. Spreadbury, *Electronic Rectification*. London: Constable and Company, 1962.

C. P. Steinmetz, "The constant current mercury-arc rectifier," *Trans. AIEE*, vol. 24, p. 271, 1905.

CHAPTER 2

Figure 2.1 Mechanical and semiconductor switches are the main elements of all energy control circuits and systems.

ORGANIZING AND ANALYZING SWITCHES

2.1 INTRODUCTION

We have seen that circuits involving switches and energy storage can be operated to give energy conversion functions. Switches come in a huge variety; a small sample is depicted in Figure 2.1. It is important to develop a systematic way to identify appropriate power electronic circuits, and to find out how they can be controlled to give the desired performance. Consider, for example, that circuits exist to:

- Provide a very tightly fixed dc output from an uncertain input voltage source. A power supply of this type can be used anywhere in the world without adjustments. It must adapt itself to the available energy source while maintaining a precise output.

- Provide polarity reversal. This is extremely helpful in bridging digital and analog electronics.

- Step a dc voltage up or down. The step-up circuit in Chapter 1 is a kind of dc transformer.

- Rectify, invert, and otherwise facilitate energy flow between ac and dc circuits, or between ac circuits acting at different frequencies.

The most common way to achieve these conversions is to store energy in a circuit element, then deliver it to a load to alter the voltage or current. The examples in Chapter 1 offer a "case study list" for power conversion. In these circuits, the use of switches and energy storage has been rather ad hoc. A few devices were interconnected and a power electronic function was asserted for the whole. We need a more general approach to construct and operate circuits for energy conversion. The polarity reverser and the step-up circuit in Chapter 1 both operate by switching periodically with a 50% on-time. How will operation change as the on-time moves from this 50% value?

One way to identify possible conversion circuits is to just write down all possible combinations of two or three switches, along with a few storage elements. This is a real brute force approach and does not give insight into how we might make the best choice for a given function. But, there is a better way! In this chapter, we will be concerned with basic tools for organizing and operating switch circuits. Once the switches have been organized, we will consider some of the device issues that determine the behavior of converters. Diode circuits, for instance, can be analyzed by taking the device behavior into account. A key element is that switch action is periodic in time for almost all power converters. We will take advantage of this to develop tools for converter analysis.

2.2 THE SWITCH MATRIX

We already know that some switches will lie between an electrical source and a load. There might be just one switch, or perhaps a large group. In any case, there is a complexity limit: If a converter has m inputs and n outputs, even the densest possible collection of switches would have a single switch between each input line and each output line. The $m \times n$ switches in the circuit can be arranged according to their connections. The pattern suggests a matrix, as shown in Figure 2.2.

Power electronic circuits are of two types:

1. *Direct switch matrix circuits.* In these circuits, any energy storage elements are connected to the matrix only at the input and output terminals. The storage elements effectively become part of the source or load. A full-wave rectifier with external low-pass filter is an example of a direct switch matrix circuit. In the literature, these circuits are sometimes just called *matrix converters.*

2. *Indirect switch matrix circuits,* also termed *embedded converters.* These circuits, like the polarity reverser example, have energy storage elements connected *within* the matrix structure. There are usually very few storage elements in such a case, and indirect switch matrix circuits are often analyzed as a cascade connection of two direct switch matrix circuits with the storage in between.

The switch matrices in realistic applications are usually small. For example, a typical source has two connections, as does a typical load. A two-by-two switch matrix is just about as complicated as it gets in this situation. A rectifier example is given in Figure 2.3. The ac

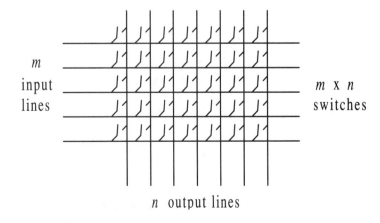

Figure 2.2   The general switch matrix.

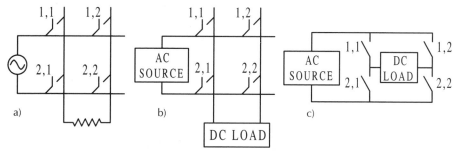

Figure 2.3   Rectifier conversion matrix.

source and the load of Figure 2.3a have two connections each. More generally, these represent a source and load with a 2×2 matrix in between, as in Figure 2.3b. The matrix is commonly drawn as the H-bridge shown in Figure 2.3c. In contrast, a half-wave rectifier uses just one input and one output line—a single switch forms its "matrix." A more complicated example is the three-phase ac–dc converter shown in Figure 2.4. There are three possible inputs, and the two terminals of the dc circuit provide outputs, to give a 3×2 switch matrix. Figure 2.5 shows a switch matrix similar to that in a personal computer power supply. In this case, there are five separate dc loads, and the switch matrix is 2×10. There are very few practical converters with more than about twenty-four switches, and most designs use fewer than twelve switches.

A switch matrix provides a clear way to organize devices for a given application. It also helps to focus the effort into three major problem areas. Each of these areas must be addressed effectively in order to produce a useful power electronic system.

- The "Hardware" Problem → Build a switch matrix.
- The "Software" Problem → Operate the matrix to achieve the desired conversion.
- The "Interface" Problem → Add energy storage elements to provide the filters or intermediate storage necessary to meet the application requirements.

Figure 2.4   Three-phase ac to dc converter switch matrix.

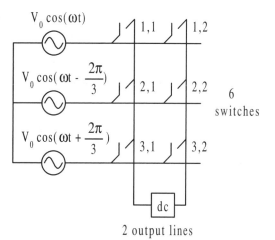

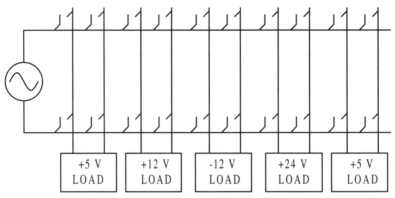

Figure 2.5   Rectifier with five separate dc outputs (ten output lines).

In a generic circuit, such as that of Figure 2.3*b*, we must decide what electronic parts to use, how to operate them, and how best to filter the output to satisfy the needs of the load.

As a first step, replace the switches in the general rectifier of Figure 2.3 with diodes. One choice of diode directions gives the full-wave bridge rectifier shown in Figure 2.6. Other devices would be used if the objective were an inverter or a controlled rectifier. We will look at consistent choices of device types throughout our study of power electronics. The diode directions must be consistent with circuit laws and must provide the intended energy flow. The choice of diodes gives one solution to the hardware problem, while the choice of diode directions in effect represents a solution of the software problem for a rectifier circuit. Let us consider the issues of circuit laws, first, and examine how they affect switch matrices.

2.3 THE REALITY OF KIRCHHOFF'S VOLTAGE AND CURRENT LAWS

Granted, a switch circuit can perform power conversion, and granted, it can be described as a switch matrix. How do we make the choices of switch operation? Consider first the simple circuit of Figure 2.7—something we might try for ac–dc conversion. This circuit has problems. Kirchhoff's Voltage Law (KVL) tells us that the "sum of voltage drops around a closed loop is zero." Imagine what will happen here if the switch is closed: the sum of volt-

Figure 2.6   Diode bridge rectifier, with switch labels.

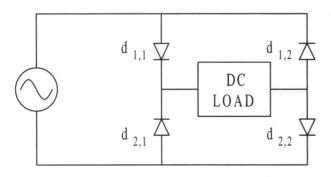

Figure 2.7 Hypothetical ac–dc converter.

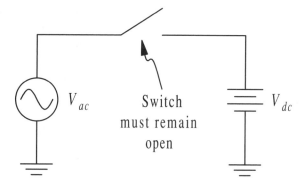

$$V_{ac} \qquad \text{Switch} \qquad V_{dc}$$
$$\text{must remain}$$
$$\text{open}$$

ages around the loop is not zero! In reality, a very large current will flow, and cause a large $I \times R$ (current $\times$ resistance) drop in the wires. KVL will be satisfied by the wire voltage drop, but a fire might result. (We would hope that fuses will blow or circuit breakers will trip, instead.) KVL should serve as a warning: "Don't connect unequal voltage sources directly."

You are familiar with KVL as an abstract rule of circuit analysis. In power electronics, KVL takes on a much more concrete meaning. There is nothing to prevent someone from building the circuit of Figure 2.7, but it will cause problems as soon as the switch is closed. The photograph in Figure 2.8 shows a dangerous combination of high-power ac and low-power dc voltages. Good sense says that the circuit should not be used. This is the reality of KVL: We can try to violate KVL, but will not succeed. As you should realize, this means

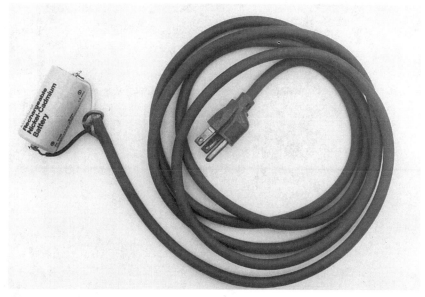

Figure 2.8 A dangerous mix of ac and dc sources. No sensible person would test this without extensive safety precautions.

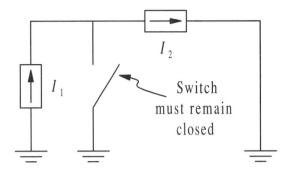

Figure 2.9 Converter involving current sources.

certain things about the switch matrix. *We must avoid switching operations that connect unequal voltage sources.* Notice that a wire, or dead short, can be thought of as a voltage source with $V = 0$, so the warning is a generalization for avoiding shorts across an individual voltage source.

A similar constraint holds for Kirchoff's Current Law (KCL). KCL states that "currents into a node must sum to zero." When current sources are present in a converter, we must avoid any attempts to violate KCL. Consider the simple circuit shown in Figure 2.9. If the current sources are different and if the switch is opened, the sum of the currents into the node will not be zero. In a real circuit, high voltages will build up and cause an arc to create another current path. This situation has real potential for damage, and a fuse will not

Figure 2.10 An expensive motor controller, destroyed by an accidental attempt to violate KVL. Notice the holes in the transistor cases. The semiconductors literally exploded through their packages.

help. If we try to violate KCL, we will not manage it. *We must avoid operating switches so that unequal current sources are connected in series.* An open circuit can be thought of as a current source with $I = 0$, so the warning applies to the problem of opening an individual current source.

In contrast to conventional circuits, in which KVL and KCL are automatically satisfied, switches do not "know" KVL or KCL. If a designer forgets to check, and accidentally shorts two voltages or breaks a current source connection, some problem will result. The photograph in Figure 2.10 shows a dc–dc converter for a motor control system. This circuit—worth about $1,000 in U.S. currency—was destroyed when an error connected the input voltage source to a reversed voltage. The circuit breakers did not have time to react.

There are some interesting implications of the current law restrictions when energy storage is included. Look at the circuits of Figure 2.11. Both represent "circuit law problems." In Figure 2.11, the voltage source will cause the inductor current to ramp up indefinitely, since $V = L\, di/dt$. We might consider this to be a "KVL problem," since the long-term effect is similar to shorting the source. In Figure 2.11*b*, the current source will cause the capacitor voltage to ramp toward infinity. This causes a "KCL problem"; eventually, an arc will form to create an additional current path, just as if the current source had been opened. Of course, these connections are not a problem if they are only temporary. However, it should be evident that an inductor will not support dc voltage, and a capacitor will not support dc current.

Circuit laws are not exclusively linked to problems. Both KVL and KCL provide very positive guidance about how to design a switch matrix. They provide excellent working rules for design. Often, all but one or two possibilities for switches and their operation can be identified simply through careful application of basic circuit laws. Let us return to the diode bridge example of Figure 2.6 to illustrate. Some of the possible connections are shown in Figure 2.12. It is easy to see that any diode combination that connects both of the left devices or both of the right devices in the same direction will attempt to violate KVL, as in Figure 2.12*a* and 2.12*b*. The combination shown in Figure 2.12*c* does not create any KVL or KCL problems, but the diodes will never allow any current flow through the resistor. Just a little more thought quickly reduces the switch arrangements to one combination: there is only one connection of the diode bridge that permits energy to flow without violating KVL. This is shown in Figure 2.12*d*.

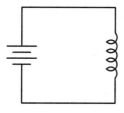

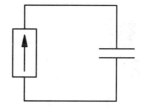

a) Inductive circuit with
   long-term KVL problem

b) Capacitive circuit with
   long-term KCL problem

Figure 2.11   KVL and KCL problems in simple energy storage connections.

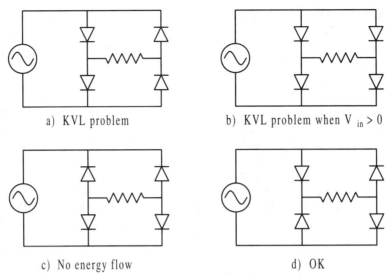

a) KVL problem   b) KVL problem when $V_{in} > 0$

c) No energy flow   d) OK

Figure 2.12 Diode bridge directions.

KCL determines many effects of current interconnections. In Chapter 1, we compared a half-wave rectifier with resistive and *R-L* loads. Consider a half-wave rectifier with a current source load, shown in Figure 2.13. KCL requires that we provide a current path for the output source. Since there is only one loop in the circuit, this path requires the diode to be on at all times. No switching action will ever occur, and the combination will not function as a power converter. (The result is quite different when a diode bridge supplies the current source, since several loops can be formed. Try this as an exercise.)

Power electronics is perhaps the only major subject in which a designer must think carefully about obeying circuit laws!

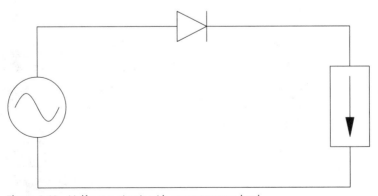

Figure 2.13 Half-wave circuit with current-source load.

2.4 THE SWITCH STATE MATRIX AND SWITCHING FUNCTIONS

With the organizing framework of the switch matrix, and now that we have the initial notion of restrictions on interconnections, it is natural to see whether we can find some organized way for studying how switches should operate. Each switch in a switch matrix is either on or off. This can be represented with a mathematical matrix, called the switch state matrix, which corresponds to the circuit. This is a matrix $\mathbf{Q}(t)$, with m rows and n columns. Each element $q_{ij}(t)$ is 1 when the corresponding switch is on and 0 when it is off. The elements of $\mathbf{Q}(t)$ are referred to as switching functions, and are important in the design of converters.

Example 2.4.1   A general example of a switch state matrix $\mathbf{Q}(t)$:

m rows (m inputs) n columns (n outputs)

$$\text{element } q_{ij} = \begin{cases} 1, & \text{if switch at location } i, j \text{ is on} \\ 0, & \text{if switch is off} \end{cases}$$

At a specific time t_0, the matrix will have a particular numerical value, such as:

$$\mathbf{Q}(t_0) = \begin{bmatrix} 1 & 0 & 0 & 0 & 0 & \cdots & 0 \\ 0 & 0 & 1 & 0 & 1 & \cdots & 0 \\ 0 & 1 & 0 & 0 & 0 & \cdots & 1 \\ 0 & 0 & 0 & 0 & 0 & \cdots & 0 \\ 1 & 0 & 0 & 1 & 0 & \cdots & 0 \\ . & \cdots \\ . & \cdots \\ . & \cdots \\ 0 & 0 & 1 & 1 & 1 & \cdots & 0 \end{bmatrix}$$

The matrix might often be as small as 2×2. The software problem can be stated as choosing the switching functions a desired operation.

Example 2.4.2   Consider a converter with three ac input lines, an input reference node, and two output lines. The restrictions imposed by KVL can be written in mathematical terms from the switch state matrix. Write an expression to represent the KVL restrictions.

The switch state matrix Q(t) for this converter has eight elements, q_{11} through q_{41}, and q_{12} through q_{42}. Assume that the inputs are three distinct voltages. To satisfy KVL, we must ensure that none of these voltages become interconnected. Notice in Figure 2.14 that this means no more than one switch in either column of the matrix can be on at any time. For the switch state matrix, either all four entries in a column should be zero, or a single entry can be one. If we add entries, such as $q_{11} + q_{21} + q_{31} + q_{41}$, the KVL restriction means that this column sum must not exceed one. The KVL restriction can be written:

$$\sum_{i=1}^{m} q_{ij} \leq 1 \quad \text{for any } j \tag{2.1}$$

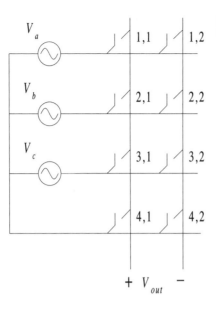

Figure 2.14   Three-phase input, two-line output switch matrix.

If this expression is satisfied, there will never be two or more switches in a column on together, and KVL will be satisfied. In this example, each column acts separately. It is fine (for KVL) to have 1,1 and 1,2 on together, for instance—or 1,1 and 2,2, and so on.

Let us consider a 2×2 matrix with both a voltage and current source to illustrate the power of the mathematical treatment.

Example 2.4.3   For the circuit shown in Figure 2.15, write separate expressions that represent the KVL and KCL restrictions. When both are taken together, what do they imply about switch operation?

The KVL restriction to avoid shorting voltages means again that the column sums of $Q(t)$ must not exceed 1. The KCL restriction requires a current path at all times. A little

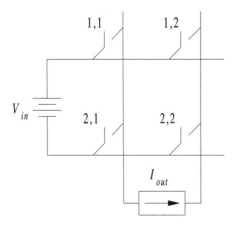

Figure 2.15   Input voltage and output current 2×2 converter.

thought shows that this path requires at least one switch in each column on at any given time. KCL would be satisfied even if two switches in each column are on, since multiple switches simply lead to multiple paths for current flow. The results to meet KVL are $q_{11} + q_{21} \leq 1$ and $q_{12} + q_{22} \leq 1$, as before. The KCL requirement for a current path can be written $q_{11} + q_{21} \geq 1$ and $q_{12} + q_{22} \geq 1$. When these two requirements are combined, the result is $q_{11} + q_{21} = 1$ and $q_{12} + q_{22} = 1$. This is a mathematical expression to represent the observation that to satisfy KVL and KCL in the 2 × 2 switch matrix of Figure 2.15, each column of the matrix must have exactly one switch on at any moment in time. Fewer switches on would break the path required by KCL, while more would short voltage sources. In any column, one and only one switch can be on at any moment.

The example gives only a flavor of the issues at hand. For instance, if the input in Figure 2.15 is a current source while the output is a voltage source, the KVL and KCL restrictions require the row sums to be exactly 1. The switch state matrix treatment narrows down the possible switch action. A disadvantage is that the mathematics tends to mask the physical meaning of the constraints. It is crucial to remember that $q_{11} + q_{21} = 1$ does not represent just the notion that "the matrix column sum is 1," but rather serves as a shorthand for the requirement that "one and only one switch among these can be on at any moment." If switch operation violates the constraints, the converter, source, and load might be damaged.

The switch state matrix also helps describe a converter's operation. Let us consider another converter with three inputs to introduce the concepts.

Example 2.4.4   A converter with three input sources, four input lines, two output lines, and a resistive load is shown in Figure 2.16. In many practical applications, it is desirable to share a common reference node at the source and the load. In this case, a simple way to accomplish the common node is to turn switch 4,2 on and leave it closed. How does this affect the switch state matrix? Write an expression for $V_{out}(t)$ with the switching functions.

Figure 2.16 Converter with three input sources, a common input node, and a resistive load for Example 2.4.4.

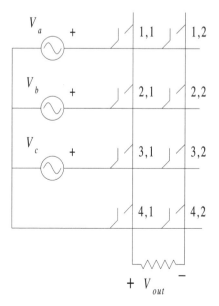

In this converter, at most one switch can be on in each column. If switch 4,2 is on all the time (i.e., $q_{42} = 1$ always), then no other switch in the second column can be closed. The switch state matrix becomes

$$\mathbf{Q}(t) = \begin{bmatrix} q_{11}(t) & 0 \\ q_{21}(t) & 0 \\ q_{31}(t) & 0 \\ q_{41}(t) & 1 \end{bmatrix}$$

The KVL restriction also requires $q_{11} + q_{21} + q_{31} + q_{41} \le 1$. With this choice of $\mathbf{Q}$, we see that the output voltage is V_a when switch 1,1 is on, V_b when switch 2,1 is on, V_c when switch 3,1 is on, and zero when switch 4,1 is on. It is permissible to have all four of these switches off, in which case no current can flow and the output is zero. Observe that this can be written

$$V_{\text{out}}(t) = q_{11}(t)V_a(t) + q_{21}(t)V_b(t) + q_{31}(t)V_c(t) \tag{2.2}$$

In matrix notation, this is a matrix multiplication $\mathbf{V}_{\text{out}} = \mathbf{V}_{\text{in}}\mathbf{Q}$, with $\mathbf{V}_{\text{in}}$ defined as the row vector $[V_a \quad V_b \quad V_c \quad 0]$, and $\mathbf{V}_{\text{out}}$ defined as the row vector $[V_{\text{out}} \quad 0]$.

We will use expressions such as (2.2) to analyze choices of switch timing as we explore converter operation.

The actual choice of switching functions depends on the desired converter operation. The proper choices in this example can create rectifiers or ac–ac converters. Let us try two different choices of switching functions just to get an idea of the waveforms and operating possibilities. For the moment, the choices are arbitrary. Later on, we will see how to make a deliberate choice of switch operation.

Example 2.4.5   Begin with the converter of Example 2.4.4. Let the inputs be 60 Hz sinusoids, $V_a(t) = V_0 \cos(120\pi t)$, $V_b(t) = V_0 \cos(120\pi t - 2\pi/3)$, $V_c(t) = V_0 \cos(120\pi t + 2\pi/3)$. Operate the switches so that $q_{11} + q_{21} + q_{31} = 1$, with each of these three on symmetrically one-third of each 60 Hz cycle. Time the switches so each turns on at the peak point of the corresponding ac input voltage. Plot $V_{\text{out}}(t)$ for this case.

Figure 2.17 shows the three voltages at the top, while the three switching functions described in the problem statement are shown in the center. The output voltage is formed from pieces of the inputs. V_{out} is V_a when $q_{11} = 1$, and so on. The waveform is shown at the bottom of Figure 2.17. This waveform has a nonzero average value. This dc content suggests that the circuit is a rectifier. The waveform is typical in an industrial controlled rectifier system. The dc value is $0.413 \, V_0$ (check this by integration).

Example 2.4.6   Consider the same converter as in Example 2.4.5, except that the switches are to be turned on symmetrically near the positive and negative peaks of the input voltages. The switching functions are

$$q_{11} = \begin{cases} 1 & \text{near positive and negative peaks of } V_a \\ 0 & \text{elsewhere} \end{cases}$$

$$q_{21} = \begin{cases} 1 & \text{near positive and negative peaks of } V_b \\ 0 & \text{elsewhere} \end{cases}$$

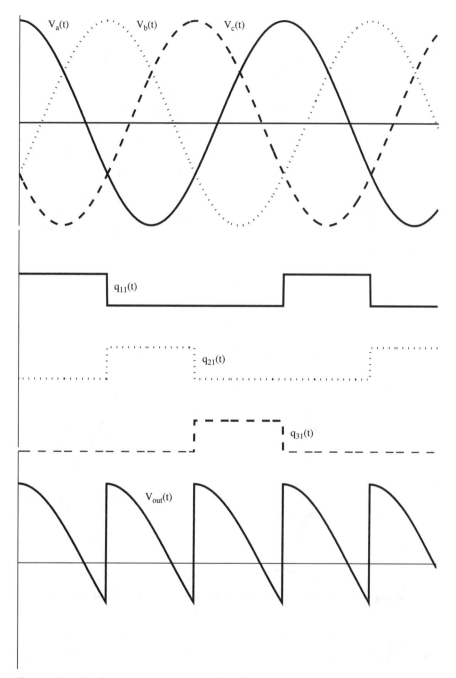

Figure 2.17 The three input voltages, switching functions, and output for Example 2.4.5.

$$q_{31} = \begin{cases} 1 & \text{near positive and negative peaks of } V_c \\ 0 & \text{elsewhere} \end{cases}$$

Plot $V_{out}(t)$ for this choice.

Input voltage and switching function waveforms corresponding to this choice are shown in Figure 2.18. The output waveform can be constructed in a straightforward manner, and is shown in Figure 2.19. It is interesting to observe that V_{out} has no average value. This choice of **Q** produces an ac–ac converter rather than a rectifier. The waveform is almost a square wave with a frequency of 180 Hz. The circuit is a reasonable 60–180 Hz converter. Other switching function choices will yield entirely different results.

Switching functions determine the output behavior of any switch matrix. Since they describe converter operation, switching functions are powerful tools for power electronics design. They serve as the basis for formulating many converter problems, and for comparing alternatives. We will take advantage of this in subsequent work.

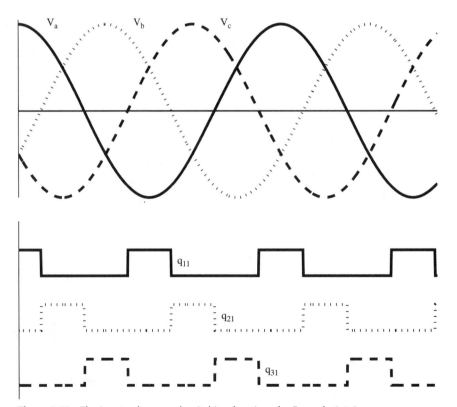

Figure 2.18 The input voltages and switching functions for Example 2.4.6.

Figure 2.19 Output voltage for Example 2.4.6.

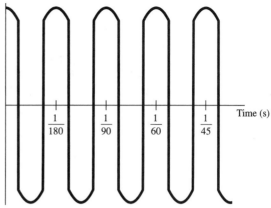

Time (s)

$\dfrac{1}{180}$ $\dfrac{1}{90}$ $\dfrac{1}{60}$ $\dfrac{1}{45}$

2.5 OVERVIEW OF SWITCHING DEVICES

A switching function has a value of either 1 or 0; either a switch is on or it is off. When on, it will carry any amount of current for any length of time and in any direction. When off, it will never carry current, no matter how much voltage is applied. This combination of conditions describes an *ideal switch*. It is entirely lossless. An ideal switch changes from its on state to its off state instantaneously.

A *real switch* is some available device that approximates an ideal switch in one way or another. Real switches such as the devices shown in Figure 2.20 differ from the ideal in characteristics such as:

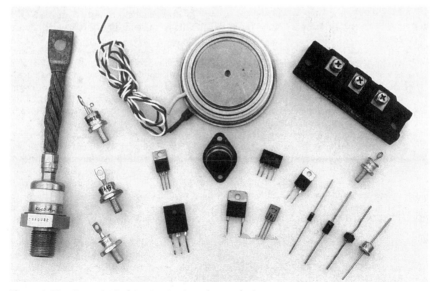

Figure 2.20 Some typical power semiconductor devices.

- Limits on the amount or direction of on-state current.
- A nonzero on-state voltage drop (such as a diode forward voltage).
- Some level of leakage current when the device is supposed to be off.
- Limitations on the voltage that can be applied when off.
- Operating speed. The time of transition between the on and off states can be important.

The degree to which properties of an ideal switch must be met by a real switch depends on the application. For example, a diode can easily be used to conduct dc current; the fact that it conducts only in one direction is often an advantage, not a weakness. Some characteristics of the most common power semiconductors are listed in Table 2.1. The table shows a wide variety of speeds and rating levels. As a rule, faster speeds apply to lower ratings. For each device type, cost tends to increase both for faster devices and for devices with higher power handling capacity.

Conducting direction and blocking behavior are fundamentally tied to the device type, and these basic characteristics constrain the choice of device for a given conversion function. Consider again a diode. This part carries current in only one direction and always blocks current in the other. Ideally, it exhibits no forward voltage drop or off-state leakage current. The ideal diode is an important switching device, yet it does not have all the characteristics of an ideal switch. It is convenient to define a special type of switch to represent this behavior: the *restricted switch*.

TABLE 2.1
Some Modern Semiconductor Switch Types and Their Basic Characteristics

Device Type	Characteristics of Power Devices
Diode	Current ratings from under 1 A to more than 5000 A. Voltage ratings from 10 V to 10 kV or more. The fastest power devices switch in less than 20 ns, while the slowest require 100 μs or more. The function applies in rectifiers and dc–dc circuits.
BJT	(Bipolar Junction Transistor) Conducts collector current (in one direction) when sufficient base current is applied. Power device current ratings from 0.5 A to 500 A or more; voltages from 30 V to 1200 V. Switching times from 0.5 μs to 100 μs. The function applies to dc–dc circuits; combinations with diodes are used in inverters.
FET	(Field Effect Transistor) Conducts drain current when sufficient gate voltage is applied. Power FETs have a parallel connected reverse diode by virtue of their construction. Ratings from about 1 A to about 100 A and 30 V up to 1000 V. Switching times are fast, from 50 ns or less up to 200 ns. The function applies to dc–dc conversion, where the FET is in wide use, and to inverters.
SCR	(Silicon Controlled Rectifier) Conducts like a diode after a gate pulse is applied. Turns off only when current becomes zero. Prevents current flow until a pulse appears. Ratings from 10 A up to more than 5000 A, and from 200 V up to 6 kV. Switching requires 1 μs to 200 μs. Widely used for controlled rectifiers.
GTO	(Gate Turn-Off SCR) An SCR which can be turned off by sending a negative pulse to its gate terminal. Can substitute for BJTs in applications where power ratings must be very high. The ratings approach those of SCRs, and the speeds are similar as well. Used in inverters rated beyond about 100 kW.
TRIAC	A semiconductor constructed to resemble two SCRs connected in reverse parallel. Ratings from 2 A to 50 A and 200 V to 800 V. Used in lamp dimmers, home appliances, and hand tools. Not as rugged as many other device types.
IGBT	(Insulated Gate Bipolar Transistor) A special type of power FET that has the function of a BJT with its base driven by an FET. Faster than a BJT of similar ratings, and easy to use. Ratings from 10 A to more than 600 A, with voltages of 600 V to 1200 V. The IGBT is popular in inverters from about 1 kW to 100 kW.
MCT	(MOSFET-controlled Thyristor) A special type of SCR that has the function of a GTO with its gate driven from an FET. Much faster than conventional GTOs, and easier to use. These devices are entering the market after several years of development.

TABLE 2.2
The Types of Restricted Switches

Action	Name	Device
Carries current in one direction, blocks in the other	Forward-conducting reverse-blocking (FCRB)	Diode
Carries or blocks current in one direction	Forward-conducting forward-blocking (FCFB)	BJT
Can carry in one direction or block in both directions	Forward-conducting bidirectional-blocking (FCBB)	GTO
Can carry in both directions, but blocks only in one direction	Bidirectional-carrying forward-blocking (BCFB)	FET
Can carry or block in both directions	Bidirectional-carrying bidirectional blocking (BCBB)	Ideal switch

> **Definition:** *A* restricted switch *is an ideal switch with the addition of restrictions on the direction of current flow and voltage polarity. The* ideal diode *is one example of a restricted switch.*

The diode always permits current flow in one direction, while blocking flow in the other. It therefore represents a *forward-conducting reverse-blocking* restricted switch. This FCRB function is automatic—the two diode terminals provide all the necessary information for switch action. Other restricted switches require a third gate terminal to determine their state. Consider the polarity possibilities, given in Table 2.2. Additional functions such as bidirectional-conducting reverse-blocking (BCRB) can be obtained simply by reverse connection of one of the five types above (in this case, the BCFB in reverse). Commercial IGBTs can be built to meet either FCBB or BCFB functions.

Symbols for restricted switches can be built up by interpreting the diode's triangle as the current-carrying direction, and the bar as the blocking direction. The five types can be drawn as in Table 2.3. Although the restricted switch symbols are not in common use, they

TABLE 2.3
Restricted Switch Symbols, with the Corresponding Device Symbols

Restricted Switch Type and Symbol	Device and Conventional Symbol
FCRB	Diode
FCFB	BJT
FCBB	GTO
BCFB	FET
BCBB	Ideal Switch

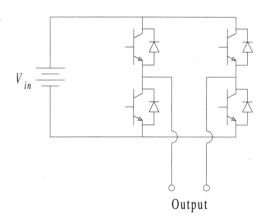

Figure 2.21 An inverter, such as this one, requires the BCFB function.

Output

show the polarity behavior of the switching devices. A circuit drawn with restricted switches represents an idealized power converter.

As an example of the utility of the restricted switch, consider an inverter intended to deliver ac load current from a dc voltage source. A switch matrix built to perform this function must be able to manipulate ac current and dc voltage. Regardless of the physical arrangement of the matrix, we would expect BCFB switches to be useful for this conversion. This is a correct result: Modern inverters operating from dc voltage sources are built with FETs or IGBTs. An IGBT example is shown in Figure 2.21. The symbol arrangement shown in Figure 2.22 points out that the transistor-diode combination shows the intended BCFB behavior. The hardware problem for this inverter consists of picking a device to meet the necessary ratings while coming as close as possible to a true BCFB. As new devices are introduced to the market, it is straightforward to determine what types of converters will use them. Table 2.4 provides a list of examples.

2.6 ANALYZING DIODE SWITCHING CIRCUITS

Many simple switching circuits behave in a straightforward fashion. The circuits often use only a few switching devices. Although the switches do not lend themselves to standard circuit analysis, the small number of possible on and off states in a typical application allows

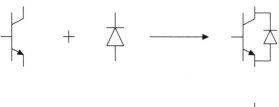

Figure 2.22 BJT-diode combination to create BCFB function.

TABLE 2.4
Choice of Restricted Switch Based on Converter Function

Converter Function	Device Type	Comments
Rectification (no control)	FCRB	Automatic operation does not allow any control.
Controlled rectifiers, from ac voltage	FCBB	Must manipulate either turn-on or turn-off to provide control. Must handle bipolar voltage.
Inverters, from dc voltage	BCFB	Must handle bipolar current.
dc–dc converters	FCFB, FCRB	Only a single polarity is needed.
ac–ac converters	BCBB	A full bidirectional switch is needed for general ac–ac conversion.

a straightforward analysis method. The idea is simple: Each switch defines two possible configurations of the complete circuit. A diode bridge rectifier, for example, contains $n = 4$ switches and generates $2^n = 16$ possible configurations of the complete circuit. Each configuration is easy to analyze. A bridge example is shown in Figure 2.23.

Given that the individual configurations are easy to study, the analysis of a circuit such as the diode bridge requires information about the circuit status: At a particular moment in time, or under certain conditions, which configuration correctly describes the circuit? A *trial*

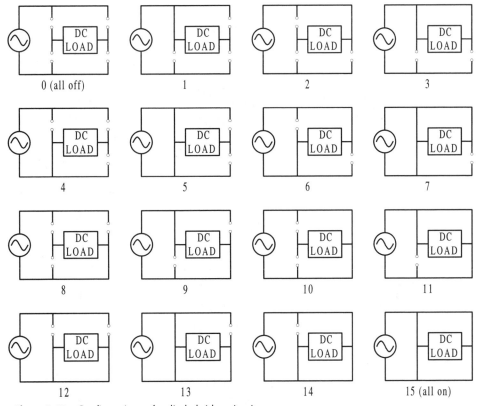

Figure 2.23   Configurations of a diode bridge circuit.

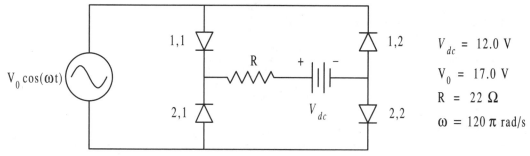

Figure 2.24 Low-cost battery charger for Example 2.6.1.

method is one widely used approach to address this question. Think about a network of ideal diodes. No matter what else happens, each diode in the network will either be on or off. If on, the device current will be positive. If off, the device forward voltage will be negative. The trial method involves an educated guess about the circuit configuration. Since the circuit must follow KVL and KCL, the correct configuration must obey these laws while properly reflecting the behavior of the diodes. The procedure is as follows:

1. Assume a state (on or off) for each switch in the circuit.
2. Solve the network equations to find currents in on switches, and voltages across off switches.
3. When diodes are used as switches, all forward currents must be positive; all forward voltages must be negative.
4. If the currents and voltages are consistent with requirements, the circuit is solved. If they are not consistent, the initial assumptions of switch state must have been in error.
5. Change the assumptions, and repeat until a consistent solution is found.
6. Assemble solutions, if necessary, to give a complete time function.

This method is easy when only a few switches appear. It lends itself to automation, and some commercial extensions to SPICE[1] analyze diode circuits in this manner. Strictly speaking, this is not a trial-and-error approach for two reasons. First, with a little practice it is not hard to pick out reasonable configurations (KVL and KCL eliminate many possibilities right away). Second, it is usually possible to test the switch action in a general way, so that the configurations can be determined without exhaustive checks.

Example 2.6.1 As shown in Figure 2.24, an ideal diode bridge supplies a dc voltage source in series with a resistance. The energy input comes from an ac source. This circuit is typical of the power stage of many inexpensive battery chargers. Assume the battery is connected properly so that $V_{dc} > 0$. Use the trial method to find and plot the resistor current and the

---

[1]SPICE (Simulation Program with Integrated Circuit Emphasis) is the most widely used circuit simulation tool. Although the base program can be obtained without charge, many commercial vendors have added special features for power electronic circuits. Often, student versions are available free or for a nominal cost.

four diode switching functions. What is the average power delivered to the dc source? What happens if the battery is connected backwards by accident?

1. Guess a state for the diodes, such as all devices off. This corresponds to configuration 0 in Figure 2.23.

2. Solve the circuit. Since the switches are off, $I = 0$. We need to check and ensure that each diode is reverse biased, consistent with the assumed off state. Let us check a KVL loop through the positive current direction. This loop obviously requires $v_{in} - V_{dc} - v_{11} - v_{22} = 0$. We require both $v_{11} < 0$ and $v_{22} < 0$. Notice that these cannot both be true unless $V_{dc} > v_{in}$. Following the other direction, we see $v_{in} + v_{12} + V_{dc} + v_{21} = 0$. In this case, if $V_{dc} > -v_{in}$, both v_{12} and v_{21} could be negative. For our guess to be correct, both $V_{dc} > v_{in}$ and $V_{dc} > -v_{in}$ must be true. In combination, this requires $V_{dc} > |v_{in}|$. That is, whenever $V_{dc} > |v_{in}|$, all four diodes will be off, and I_{out} will be zero.

3. Test for consistency. The assumption will be consistent if the condition is met. If at any moment $V_{dc} < |v_{in}|$, at least one diode will exhibit a positive bias voltage, and the assumption will be invalid.

4. Examine other possibilities when the condition is not met. For instance, any configuration with just one diode on will have zero current at all locations—inconsistent with any device being on. We might guess that both of the left diodes 1,1 and 2,1 are on, and the others are off. In this configuration, KVL is violated because of the short circuit path imposed on v_{in}. If this configuration is in place in a real circuit somehow, large current will flow in the loop. Since the diodes are in a reverse series connection, they cannot both carry positive current simultaneously—and they both will not be on together. The conditions are not consistent with diode properties, so this configuration does not occur. As an exercise, try the various possibilities to determine that the only plausible configurations here are configuration 0 (all switches off), configuration 6 (1,2 and 2,1 on), and configuration 9 (1,1 and 2,2 on).

5. Change assumptions until consistency is achieved. Consider the configuration with 1,1 and 2,2 on. In this case, the KVL loop requires $v_{in} - V_{dc} - i_{out}R = 0$. The diode operation requires $i_{out} > 0$, $v_{21} = -v_{in} < 0$, and $v_{12} = -v_{in} < 0$. The assumed configuration is consistent with the constraints provided $v_{in} > 0$ and $v_{in} > V_{dc}$. Now check the configuration with 1,2 and 2,1 on. The KVL loop requires $-v_{in} - V_{dc} - i_{out}R = 0$. The diodes require $i_{out} > 0$, $v_{11} = v_{in} < 0$, and $v_{22} = v_{in} < 0$. All three constraints are satisfied when $-v_{in} > V_{dc}$.

At this point, all configurations have been checked, and we can begin to summarize the circuit's operation. Figure 2.25 shows $v_{in}(t)$, V_{dc}, and the switching functions $q_{11}(t)$ and $q_{12}(t)$. The function q_{11} is high when $v_{in} > V_{dc}$ and low otherwise. Similarly, q_{12} is high when $-v_{in} > V_{dc}$. Notice that $q_{22} = q_{11}$ and $q_{21} = q_{12}$ in this circuit. The resistor voltage v_r is $v_{in} - V_{dc}$ when 1,1 is on, $-v_{in} - V_{dc}$ when 1,2 is on, and 0 when all switches are off, as we found when testing configurations. The waveform is given in Figure 2.26. This result, $\max(|v_{in}| - V_{dc}, 0)$, is a full-wave rectified signal with an offset of V_{dc}.

The average power delivered to the battery is the integral

$$P_{bat} = \frac{1}{T}\int_{-T/2}^{T/2} V_{dc}i_{out}(t)\,dt = \frac{1}{T}\int_{-T/2}^{T/2} V_{dc}\frac{\max(|v_{in}(t)| - V_{dc}, 0)}{R}\,dt \qquad (2.3)$$

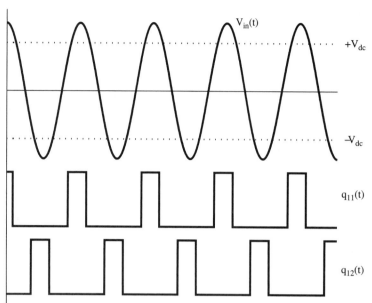

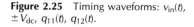

Figure 2.25 Timing waveforms: $v_{in}(t)$, $\pm V_{dc}$, $q_{11}(t)$, $q_{12}(t)$.

The maximum function can complicate the evaluation when mathematical software is used, but it is easy enough to translate the integral into simpler form by using the switching times t_{on} and t_{off}. These correspond to the times when $V_{dc} = |v_{in}|$, or $\omega t = \cos^{-1}(V_{dc}/V_0)$. For convenience, change variables to $\theta = \omega t$, and take advantage of symmetry so that only one half-pulse needs to be evaluated. The integral reduces to

$$P_{bat} = \frac{2}{\pi} \int_0^{\cos^{-1}(V_{dc}/V_0)} V_{dc} \frac{V_0 \cos \theta - V_{dc}}{R} \, d\theta \tag{2.4}$$

Figure 2.26 Resistor voltage $i_{out}R$.

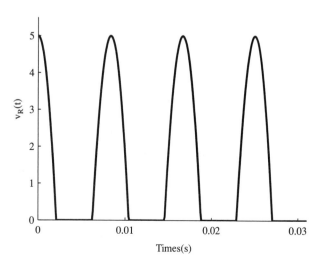

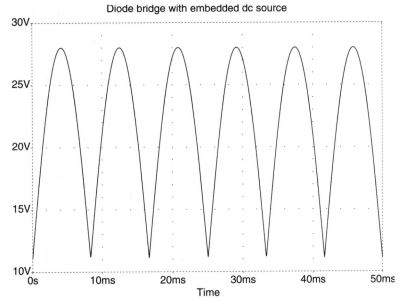

Figure 2.27   Resistor voltage when the dc source of Figure 2.24 is reversed.

For the values in Figure 2.24, this integral gives P_{bat} = 0.894 W.

What if the battery is connected backwards? The basic analysis still applies, and we can test a negative value of V_{dc} to see what will happen. The exercise is left to the reader, and the resistor current waveform for this situation is shown in Figure 2.27.

In Example 2.6.1, the trial method eliminated all but three circuit configurations out of 16. With only three cases to check, circuits can be analyzed with modest effort. Let us look at a more complicated case as a second example.

Example 2.6.2   Figure 2.28 shows a diode bridge with an ac load at the output. Dc sources are included to model the forward voltage drop of silicon power diodes. This circuit and its relatives are sometimes used to generate special waveshapes for testing or communications. Plot the resistor voltage waveform.

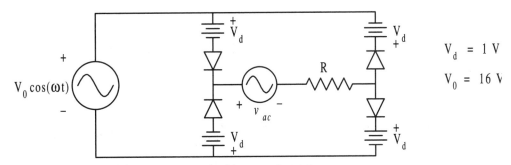

Figure 2.28   Diode bridge with ac source load for Example 2.6.2.

A quick check can be used to demonstrate that only three configurations are plausible in this circuit, just as in the previous example (try the case with both upper diodes on as one test). Let us evaluate these three.

Case 0: No switches on. Since $i = 0$, the current is consistent with the off state. The voltages are in reverse bias provided that $|v_{in}| < v_{ac} + 2V_d$.

Case 6: 1,2 and 2,1 on. Now $v_{out} = -v_{in} - 2V_d - v_{ac}$. This is valid when $-v_{in} > v_{ac} + 2V_d$.

Case 9: 1,1 and 2,2 on. In this configuration, $v_{out} = v_{in} - 2V_d - v_{ac}$. It is valid when $v_{in} > v_{ac} + 2V_d$.

Now, let $v_{ac} = (\frac{1}{2})V_0 \cos(\omega t)$. The output resistor waveform is:

$$v_{out}(t) = \begin{cases} \dfrac{V_0}{2}\cos(\omega t) - 2, & \text{1,1 and 2,2 on} \\[2mm] -\dfrac{3V_0}{2}\cos(\omega t) - 2, & \text{1,2 and 2,1 on} \\[2mm] 0, & \text{none on} \end{cases} \tag{2.5}$$

giving rise to the waveform plotted in Figure 2.29.

The examples for diode circuit analysis so far have been static, meaning that dynamic variables are not important to the treatment. Dynamics complicate the behavior, since the diode behavior will become a function of past history as well as the actual configuration. To solve the network equations in a given configuration, we will need proper initial conditions. The basic trial method will still work out if the initial conditions are considered.

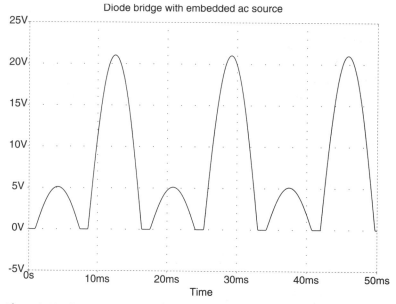

Figure 2.29 Output resistor waveform for Example 2.6.2.

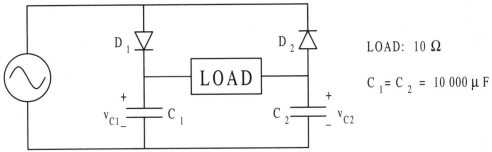

Figure 2.30   Diode-capacitor bridge for power supply input interface.

There are many important dynamic diode circuits. As an example, let us look at one diode–capacitor combination that is very common in power supplies.

Example 2.6.3   The diode–capacitor bridge shown in Figure 2.30 is widely used for power supplies intended to function over extreme input voltage ranges. Assume that the capacitors start out uncharged, and that the ac power is applied at time $t = 0$. The input voltage is $V_0 \sin(\omega t)$ for this example, to avoid excessive current in the circuit. Determine the output voltage.

In this circuit, there are only three valid configurations. When the voltage is first applied, the left diode is off, but quickly becomes forward biased. Since the capacitor voltages are initially zero, the right diode will be reverse biased initially. The operating sequence is as follows:

Time $t = 0^+$ to 1/240: left diode turns on, right diode remains off. The circuit configuration is shown in Figure 2.31a. The capacitor voltage v_{C1} is identical to v_{in}, and current

Figure 2.31   Circuit configurations and results for diode–capacitor bridge.

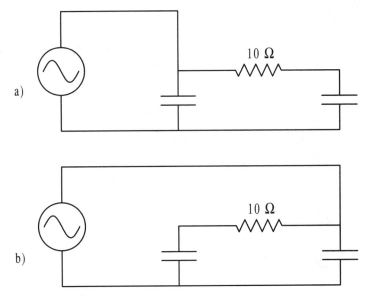

flows into C_1. Since the RC time constant is very slow compared to the sinusoidal change in v_{in}, the right capacitor will not charge up much as yet. As long as diode D_1 is on, v_{C_1} must track v_{in}. This configuration will remain valid until shortly after v_{in} has peaked. At that time, $dv_{in}/dt < 0$, so that $i_{C_1} < 0$ as well. Once the diode current tries to go negative, the configuration becomes inconsistent. The diode must switch off. In essence, the left diode–capacitor pair functions as a peak detection circuit: C_1 will charge up to V_0.

Time $t = 1/240^+$ to $1/120$: After the voltage peak, both diodes are off and reverse biased. The bias on the right diode is maintained in reverse because the voltage v_{C_2} is still close to zero. The bias on the left diode results from the slow RC time constant: while v_{C_1} falls exponentially at a slow rate, the sine wave is falling toward zero. Negative bias is therefore maintained across D_1, and the voltage v_{C_1} stays close to V_0.

Time $t = 1/120^+$ to $1/60$: The right diode–capacitor pair is just a mirror image of the left pair. During the falling portion of the negative half-cycle, v_{C_2} tracks v_{in}, and the capacitor eventually charges to $-V_0$. Just after the negative peak, the voltage begins to rise, and the current in C_2 reverses. Once the flow in D_2 tries to reverse, the right diode shuts off.

After a full cycle, C_1 maintains a voltage close to V_0 at all times, while C_2 holds a voltage close to $-V_0$. The resistor voltage is approximately $2V_0$; this circuit has voltage-doubling behavior. To confirm the details, a SPICE simulation result for this circuit is given in Figure 2.32. The command set for the run can be found in the Appendix.

The trial method shows the importance of considering the various configurations of a switching circuit. The individual configurations can be studied with conventional circuit techniques, even though the whole is often complicated. It is crucial to take an organized step-by-step approach for power electronic circuits. When diodes are involved, the configurations can be checked and tested against diode behavior. An ability to sketch out quickly and consider the operation of several circuit configurations is a helpful skill for

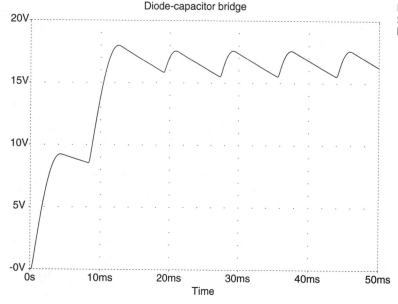

Figure 2.32 Output waveform (from SPICE simulation) for diode–capacitor bridge.

power electronics analysis. A common mistake among newcomers to the field is a reluctance to draw lots of circuit diagrams—even trivial ones—to obtain a view of converter and switch matrix action.

2.7 THE SIGNIFICANCE OF FOURIER ANALYSIS

The switch action in a conversion circuit is most often periodic. The waveforms themselves are rather choppy and sometimes discontinuous. We must form an approximate dc output of a rectifier, for instance, by piecing together "chunks" of the ac input. Given the strange waveforms and the periodic action, how can we determine whether a conversion operation is successful? In the 60 Hz to 180 Hz conversion example, the 180 Hz result was evident in the waveform. But how big was the intended 180 Hz signal? What filter would make the output sinusoidal? How can alternative approaches be compared to determine which works best? What should be done to get some specific conversion, such as 60 Hz to 400 Hz or something else?

These questions, along with such basic issues as the choice of switching frequency and the nature of switching functions, can be answered through Fourier analysis. You recall the idea: Any physically reasonable periodic signal can be expressed as a sum of sine waves with frequencies that are integer multiples of the fundamental. In a power converter, a waveform can be represented as a Fourier Series. As in the case of circuit laws, this is more than a mathematical construct. The sine waves in the series have physical reality, and we can examine the series to find out whether it contains the desired result. We can compare different waveforms to see which one gives the largest desired sinusoid while producing the smallest possible coefficients at other frequencies.

In many other electrical engineering specialties, the Fourier Transform gives helpful representations for unknown or highly variable signals. In power electronics, the waveforms are usually well known, and it is the process of generating the waveforms that is of interest. We are often concerned with a single frequency. The Fourier Series tends to be more directly applicable to the problems of power electronics than the Fourier Transform.

Fourier analysis is a very powerful tool for both analysis and design of power converters. It guides the selection of switching frequencies and helps define the switch action for a desired conversion result. It helps us pick out the salient features of pieced waveforms. It is used directly for control of many high-power inverter circuits. It is even the basis for some modern standards that define the quality of a near-sinusoidal electrical source. In the next section, a short review of Fourier Series analysis is provided. The analysis is then applied to converter waveforms and to switching functions. One crucial aspect is the ability of a converter to control the flow of energy. Fourier Series analysis identifies some fundamental requirements to make energy transfer successful.

2.8 REVIEW OF FOURIER SERIES

Nearly any periodic function of time, $f(t)$ such that $f(t) = f(t + T)$, can be written as a sum of sinusoids known as the Fourier Series,

$$f(t) = a_0 + \sum_{n=1}^{\infty} a_n \cos(n\omega t) + b_n \sin(n\omega t) \qquad (2.6)$$

where the radian frequency ω is defined as $\omega = 2\pi/T$. This result applies to signals with jump discontinuities (often encountered in power electronics). While there are a few very special periodic functions without Fourier Series representations, the properties of these exceptions are so bizarre as to be physically unrealistic. Any periodic signal that can be produced by a real circuit meets the mathematical conditions for a Fourier Series. The coefficients in the series expression are given by

$$a_0 = \frac{1}{T} \int_{\tau}^{\tau+T} f(t)\, dt$$

$$a_n = \frac{2}{T} \int_{\tau}^{\tau+T} f(t) \cos(n\omega t)\, dt \qquad (2.7)$$

$$b_n = \frac{2}{T} \int_{\tau}^{\tau+T} f(t)\sin(n\omega t)\, dt$$

The integrals can be computed beginning at any time τ, as long as the interval of integration includes a full period. We can choose any convenient value for τ, such as $\tau = 0$ or $\tau = -T/2$. Again, $\omega = 2\pi/T$.

In power electronics, it is often helpful to make the change of variables $\theta = \omega t$. The new variable θ defines an *angular time*, with units of radians, that helps emphasize the shape of a waveform rather than the explicit time. In the angular time coordinate, with $\tau = 0$, the series coefficients become

$$a_0 = \frac{1}{2\pi} \int_{0}^{2\pi} f(\theta)\, d\theta$$

$$a_n = \frac{1}{\pi} \int_{0}^{2\pi} f(\theta)\cos(n\theta)\, d\theta, \qquad (2.8)$$

$$b_n = \frac{1}{\pi} \int_{0}^{2\pi} f(\theta)\sin(n\theta)\, d\theta$$

The angular scale is very convenient for waveforms that differ in frequency but not in shape. For example, the qualitative behavior of a diode bridge is the same whether the input is 50 Hz, 60 Hz, 400 Hz, or something else. Keep in mind that angular time is formally a change of variables.

An alternate form of the Fourier Series is frequently used by electrical engineers. In this case, the sine and cosine terms are combined into phase-shifted cosine functions. The series has the form

$$f(t) = \sum_{n=0}^{\infty} c_n \cos(n\omega t + \theta_n) \qquad (2.9)$$

Here,

$$c_0 = a_0, \qquad \theta_0 = 0, \qquad c_n = \sqrt{a_n{}^2 + b_n{}^2}, \qquad \theta_n = -\tan^{-1}(b_n/a_n) \qquad (2.10)$$

In a typical power converter, the function $f(t)$ is a piecewise sinusoid: Many source waveforms are sinusoidal, and jumps are added as switch action selects the connections between inputs and outputs. Integrals for these waveforms for the Fourier coefficients are relatively easy to compute. A computer program to automate this procedure is described in Appendix C. In solving the integrals, the *orthogonality relations*

$$\int_0^\pi \sin(mx)\sin(nx)\,dx = \begin{cases} 0, & m \neq n \\ \pi/2, & m = n \end{cases} \qquad (2.11)$$

$$\int_0^\pi \cos(mx)\cos(nx)\,dx = \begin{cases} 0, & m \neq n \\ \pi/2, & m = n \end{cases} \qquad (2.12)$$

are of special importance. It is also helpful to recall some symmetry relationships: If the function $f(t)$ has *even symmetry* (i.e., symmetry about the y-axis, like $\cos\theta$), the coefficients $b_n = 0$ for all n. If the function has *odd symmetry* (i.e., symmetry about the origin, like $\sin\theta$), the coefficients $a_n = 0$ for all n.

Definition: *Each term* $c_n \cos(n\omega t + \theta_n)$ *is called a* Fourier component *or a* harmonic *of the function* f(t). *The Fourier component corresponding to* n *is called the* nth component *or the* nth harmonic. *The coefficient* c_n *is the* component amplitude, *and* θ_n *is the* component phase.

Definition: *The term* $c_0 = a_0$ *is called the* dc component *of* f(t). *The term* $c_1 \cos(\omega t + \theta_1)$ *is called the* fundamental *of* f(t). *The frequency* 1/T *is the* fundamental frequency *of* f(t).

Each component also has an RMS value, $c_{0(\mathrm{RMS})} = c_0$, $c_{n(\mathrm{RMS})} = c_n/\sqrt{2}$.

It is important to realize that, while a given waveform might have an infinite number of Fourier components, in most practical applications only one is really desired. For example, if $f(t)$ is the output voltage waveform of an ac–dc converter, we would be interested in c_0. If $f(t)$ is the output of a 60 Hz to 180 Hz ac–ac converter, the value of c_n corresponding to 180 Hz would be of interest. For the purposes of power electronics, it is convenient to define the following terms.

Definition: *The term* wanted component *refers to the one Fourier component which is desired by the user or designer in a waveform. All other Fourier components are called* unwanted components.

Given that there is only one wanted component in most situations, the others are, in effect, noise. An example of a typical rectifier will help to illustrate the ideas.

Example 2.8.1   Compute the Fourier Series of a full-wave rectified signal $|V_0 \cos(\omega_{in}t)|$. What is the fundamental frequency? Which is the wanted component, and what is its RMS value? What are the amplitude and frequency of the largest unwanted component?

The waveform in question is plotted in Figure 2.33. This waveform has period $T = \pi/\omega_{in}$. Choose $\tau = -T/2$ to take advantage of symmetry for the integrals. The radian frequency associated with the Fourier Series in this case is, by definition, $\omega = 2\pi/T = 2\omega_{in}$. The Fourier coefficients, from equation (2.7), are

$$
\begin{aligned}
a_0 &= \frac{\omega_{in}}{\pi} \int_{-\pi/(2\omega_{in})}^{\pi/(2\omega_{in})} V_0 \cos(\omega_{in}t)\, dt = \frac{2\omega_{in}}{\pi} \int_0^{\pi/(2\omega_{in})} V_0 \cos(\omega_{in}t)\, dt \\[2mm]
&= \frac{2\omega_{in}}{\pi} \left[\frac{V_0}{\omega_{in}} \sin(\omega_{in}t) \right]_0^{\pi/(2\omega_{in})} = \frac{2V_0}{\pi}
\end{aligned}
$$

$$
a_n = \frac{2\omega_{in}}{\pi} \int_{-\pi/(2\omega_{in})}^{\pi/(2\omega_{in})} V_0 \cos(\omega_{in}t) \cos(n\omega t)\, dt, \qquad \omega = 2\omega_{in}
$$

$$
b_n = \frac{2\omega_{in}}{\pi} \int_{-\pi/(2\omega_{in})}^{\pi/(2\omega_{in})} V_0 \cos(\omega_{in}t) \sin(n\omega t)\, dt, \qquad \omega = 2\omega_{in}
$$

(2.13)

The coefficients a_n and b_n can benefit from the change of variables $\theta = \omega_{in}t$. Then integral tables or computer tools can readily be applied. The results are

$$
a_n = \frac{4}{\pi} \int_0^{\pi/2} V_0 \cos(\theta) \cos(2n\theta)\, d\theta = \frac{4V_0}{\pi} \frac{\cos(n\pi)}{1 - 4n^2}
$$

$$
b_n = \frac{2}{\pi} \int_{-\pi/2}^{\pi/2} V_0 \cos(\theta) \sin(2n\theta)\, d\theta = 0 \quad \text{for all } n \geq 1
$$

(2.14)

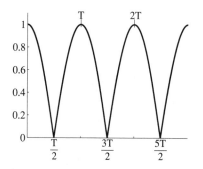

Figure 2.33   Full-wave rectified cosine function.

In summary, the Fourier Series for a full-wave rectified cosine can be written from equation (2.10) as

$$f(t) = \frac{2V_0}{\pi} + \frac{4V_0}{\pi} \sum_{n=1}^{\infty} \frac{\cos(n\pi)}{1 - 4n^2} \cos(n\omega t), \qquad \omega = 2\omega_{in} \tag{2.15}$$

The fundamental frequency is twice that of the ac input. The wanted component is the dc term, with an RMS value of $2V_0/\pi$. The amplitudes c_n decrease rapidly as n increases, and the largest unwanted component is the $n = 1$ term

$$\frac{4V_0}{3\pi} \cos(2\omega_{in}t) \tag{2.16}$$

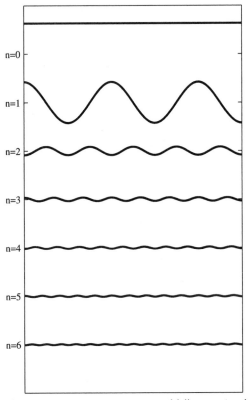

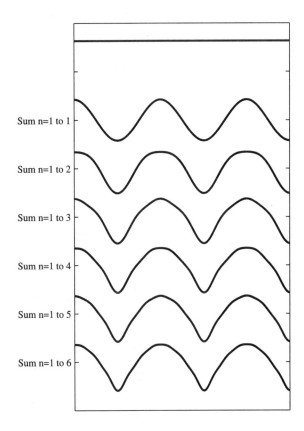

Figure 2.34 Fourier components of full-wave signal.

Look at the various components of the full-wave rectified signal to see the reconstruction of the original waveform. This is shown in Figure 2.34. Since the coefficients fall so rapidly with increasing n, only a few terms are needed to give an excellent approximation of the signal.

An inverter might switch a dc input to approximate an ac output. The simplest version would produce a square wave of a given radian frequency ω_{out}. The series for a square wave will be useful in several types of converters, so let us consider it as a second example.

Example 2.8.2   One of the simplest inverters switches between polarities of a dc input to create a square wave. This wave can be filtered to give an approximate sinusoidal output. Find the Fourier Series of a square wave of radian frequency ω. In an inverter, which component is wanted? What is its amplitude?

A square wave of amplitude V_0 and frequency ω will exhibit period $T = 2\pi/\omega$. This is plotted in Figure 2.35. The general square wave function with amplitude 1 is sometimes given the symbol $\text{sq}(\omega t)$, and is defined as follows:

$$\text{sq}(\omega t) = \text{sgn}[\cos(\omega t)] \quad \text{where } \text{sgn}(x) = \begin{cases} 1, & x > 0 \\ 0, & x = 0 \\ -1, & x < 0 \end{cases} \tag{2.17}$$

The average value of $\text{sq}(\omega t)$ is zero. As in the full-wave rectifier case, the symmetry about the y-axis is such that $b_n = 0$ for all n. With the change of variables $\theta = \omega t$, and the choice $\tau = -T/4$, the remaining coefficients a_n of $V_0\text{sq}(\theta)$ are

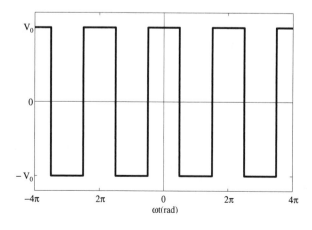

Figure 2.35   General square wave, $V_0\text{sq}(\omega t)$.

$$a_n = \frac{1}{\pi} \int_{-\pi/2}^{3\pi/2} V_0 \, \mathrm{sq}(\theta)\cos(n\theta) \, d\theta$$

$$= \frac{1}{\pi} \int_{-\pi/2}^{\pi/2} V_0 \cos(n\theta) \, d\theta - \frac{1}{\pi} \int_{\pi/2}^{3\pi/2} V_0 \cos(n\theta) \, d\theta \qquad (2.18)$$

$$= \frac{4V_0}{n\pi} \sin\left(\frac{n\pi}{2}\right)$$

Since the component amplitudes decrease as $1/n$, the $n = 1$ term is the largest. Presumably, this is the one to use in the inverter application. Therefore, the fundamental is also the wanted component. Its amplitude is $4V_0/\pi$.

In a typical power converter application, the designer identifies the wanted component, then tries to generate a waveform from a switch matrix such that the wanted component is large and the unwanted components are small. This is not a trivial task, but Fourier analysis helps to point it in the right direction.

2.9 POWER AND AVERAGE POWER IN FOURIER SERIES

In a power conversion system, the major issue is the control of energy flow between a source and a load. The usual measure of energy flow in an electrical network is the average rate at which energy is transferred, or the average power P. This can be computed from the instantaneous power, $p(t) = v(t)i(t)$, with the average integration

$$P = \frac{1}{T} \int_{\tau}^{\tau+T} v(t)i(t) \, dt \qquad (2.19)$$

Average power can also be computed by averaging the products of the Fourier Series of the voltage and current. This approach reveals some interesting properties, and we need to examine it. First, assume that the voltage $v(t)$ and the current $i(t)$ are periodic waveforms with the same period $T = 2\pi/\omega$. They have Fourier Series that can be written as

$$v(t) = \sum_{n=0}^{\infty} c_n \cos(n\omega t + \theta_n) \qquad (2.20)$$

$$i(t) = \sum_{m=0}^{\infty} d_m \cos(m\omega t + \phi_m) \qquad (2.21)$$

(notice that the coefficients and summation indices differ so we can tell them apart). The average power is the integral of the product. With the distributive law, the product can be written as a nested sum, so the integral is

$$P_{ave} = \frac{\omega}{2\pi} \int_0^{2\pi/\omega} \sum_{n=0}^{\infty} \left[\sum_{m=0}^{\infty} c_n d_m \cos(n\omega t + \theta_n)\cos(m\omega t + \phi_m) \right] dt \qquad (2.22)$$

This expression is linear, so we can integrate each term in this double series and add the result. This is expressed as

$$P_{ave} = \sum_{n=0}^{\infty} \left[\sum_{m=0}^{\infty} \frac{\omega}{2\pi} \int_0^{2\pi/\omega} c_n d_m \cos(n\omega t + \theta_n)\cos(m\omega t + \phi_m) \, dt \right] \qquad (2.23)$$

This sum of integrals is simplified considerably by applying the orthogonality relations to the terms. For a given term, the integral will be

$$\frac{\omega}{2\pi} \int_0^{2\pi/\omega} c_n d_m \cos(n\omega t + \theta_n)\cos(m\omega t + \phi_m) \, dt = \begin{cases} 0, & m \neq n \\ \dfrac{c_n d_m \cos(\theta_n - \phi_n)}{2}, & n = m \neq 0 \\ c_n d_n, & n = m = 0 \end{cases}$$

$$(2.24)$$

Only the terms in the summation for which $n = m$ will contribute to the total. The total becomes

$$P = c_0 d_0 + \sum_{n=1}^{\infty} \frac{c_n d_n}{2} \cos(\theta_n - \phi_n). \qquad (2.25)$$

But the RMS magnitude of each Fourier component is $|v_{RMS(n)}| = c_n/\sqrt{2}$ and $|i_{RMS(n)}| = d_n/\sqrt{2}$ for $n \neq 0$. This means that the average power can be written

$$P_{ave} = \sum_{n=0}^{\infty} |v_{RMS(n)}||i_{RMS(n)}| \cos(\theta_n - \phi_n) \qquad (2.26)$$

if we define phases θ_0 and $\phi_0 = 0$. Equation (2.26) states that *the average power contributed by each Fourier component is the RMS magnitude of voltage times the RMS magnitude of current times the cosine of the phase angle difference*. But it is also true that purely sinusoidal voltage and current will give this same result! None of the cross-terms, involving different frequency components of voltage and current, appear in the final expression. Two crucial facts emerge:

1. The average power for a given periodic function is the sum of average powers contributed by each Fourier component of that function.

2. Only Fourier components which appear in both the current and the voltage can contribute to the average power. Cross-frequency terms do not contribute.

The second fact can also be written as the frequency matching constraint: *Only voltage and current Fourier components of equal frequency contribute to P*. Both of the facts are extremely important. The first means that we can apply conservation of power and energy component-by-component for a signal with a Fourier Series. The second is a tight restriction on how to transfer energy. An ac–dc converter, for example, will have average output power at dc only if both the output voltage and the output current have a dc component. A dc–ac converter intended for 60 Hz output will produce average power at 60 Hz only if the output voltage and current both have a 60 Hz Fourier component.

The average power results are one of the foundations of power electronics and are the basis of all choices of switch operation. They can make analysis very direct. Consider the following examples.

Example 2.9.1   A full-wave rectified voltage waveform with peak value of 50 V is imposed on a 10 A dc current source in Figure 2.36. What is the average power in the source?

We could solve this by taking the waveforms, multiplying them, then averaging the result. This is simple enough, since the current is just a fixed value, but we have already computed the average of a full-wave signal, and should take advantage of the earlier result. The average power in the current source has been shown to be the sum of average powers of all the Fourier components. The current has only a dc component, 10 A. The voltage's dc component is $2V_0/\pi = 31.8$ V. The average power in the source is 318 W. The complications of the voltage waveform do not affect the power—only the dc component can contribute.

Example 2.9.2   A square wave with amplitude 24 V is filtered with a lossless LC filter, then applied to a resistive load, $R = 8$ Ω. The LC filter has excellent low-pass characteristics, and it can be assumed that only the fundamental component is passed. What is the average power in the resistor?

In this arrangement, the filter would typically appear as a two-port network with a com-

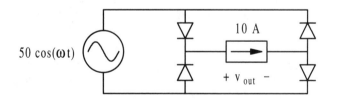

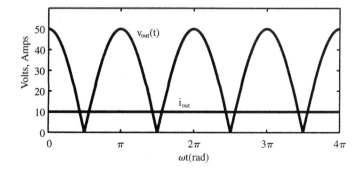

Figure 2.36   Circuit and waveforms for Example 2.9.1.

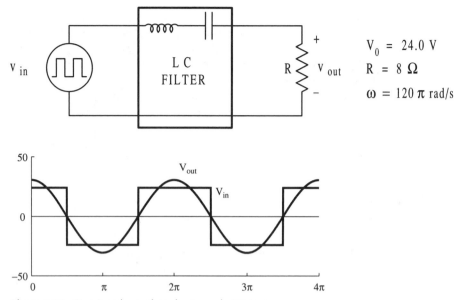

Figure 2.37 Circuit and waveform for Example 2.9.2.

mon node, as shown in Figure 2.37. Since the filter passes only the fundamental, the output voltage must be the fundamental component of the square wave, $24 \cdot 4/\pi = 30.6 \text{ V}_{\text{peak}}$. The RMS value is $30.6/\sqrt{2} = 21.6 \text{ V}$. The resistor current is proportional to the voltage, so the output voltage and current have identical waveforms. Therefore, the phase difference is zero, and the average power must be $V_{\text{RMS}}I_{\text{RMS}}$ from the fundamental component. This gives 21.6 V times (21.6 V)/(8 Ω), or 58.4 W.

Example 2.9.3   The power results become especially important when the waveforms become very complicated. Figure 2.38 shows a possible 60 Hz to 50 Hz converter, along with an output waveform produced by switching at 110 Hz. What conditions must the load exhibit for successful energy transfer? Can you say anything about the 60 Hz input?

Although the waveform is jumpy and complicated, it is just made of sinusoidal pieces. The Fourier Series can be computed relatively easily, although a computer certainly helps with the bookkeeping. The first twenty-seven components are given in Table 2.5.

This waveform has some interesting properties. Its period is 0.1 s, to give a fundamental frequency of 10 Hz. Yet the 10 Hz Fourier component is zero. It is a 10 Hz waveform with no 10 Hz content! This sort of thing is not especially unusual in power electronics, but is a bit counterintuitive. There is a nonzero 50 Hz component, however. *To attain energy transfer with this component of the voltage waveform, the load must exhibit a 50 Hz current component.* The simplest such load would be a 50 Hz current source.

On the input side, notice that the current drawn from the 60 Hz source must contain a 60 Hz frequency component to produce nonzero average flow of energy.

The requirement that the voltages and currents have some Fourier components that match in frequency must be met in order for energy to be transferred between a source and

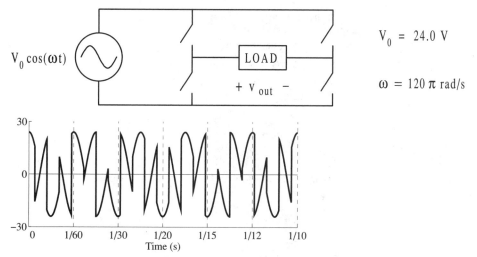

Figure 2.38 Circuit and waveform for Example 2.9.3.

a load. In many converters, this requirement is met automatically. In a resistor, for instance, the voltage and current waveforms must contain identical frequencies. In a diode converter, the switches act automatically such that frequency requirements are met. The requirements allow computation of energy flow, and will be used to define switch action in converters in which frequencies do not automatically match.

2.10 FOURIER SERIES REPRESENTATION OF SWITCHING FUNCTIONS

To take full advantage of Fourier analysis, we need to consider series representations of switching functions. The idea is to figure out the Fourier Series of a switching function in some general way, in order to avoid recomputing it multiple times in the future. Since any switching function $q(t)$ is either 0 or 1, and is normally periodic, a plot of a given $q(t)$ will be a train of square pulses. Let us take a general pulse train and figure out its Fourier Se-

TABLE 2.5
Fourier Component Values for Figure 2.38

Frequency	Amplitude	Phase
10 Hz	0	0
20 Hz	0	0
30 Hz	0	0
40 Hz	0	0
50 Hz	$0.637 V_0$	0
60–160 Hz	0	0
170 Hz	$0.637 V_0$	0
180 Hz	0	0
270 Hz	$0.212 V_0$	0

ries. A pulse train of arbitrary period T, with one pulse centered on the time $t = t_0$, is shown in Figure 2.39. The pulses in the train each have duration DT, where D is defined as the *duty cycle* or *duty ratio*. Notice that $0 \leq D \leq 1$. This pulse train has frequency $f = 1/T$ and radian frequency $\omega = 2\pi/T$.

Let us compute the Fourier components of this $q(t)$. The dc component a_0 is given by

$$a_0 = \frac{1}{T} \int_{t_0-T/2}^{t_0+T/2} q(t)\, dt \tag{2.27}$$

Only the portion of each cycle with $q(t) = 1$ will contribute to the integral. Since $q(t) = 0$ except between $t_0\text{-}DT/2$ and $t_0 + DT/2$, the integral becomes

$$a_0 = \frac{1}{T} \int_{t_0-DT/2}^{t_0+DT/2} q(t)\, dt = \frac{1}{T} \int_{t_0-DT/2}^{t_0+DT/2} 1 \, dt = \frac{DT}{T} = D \tag{2.28}$$

An inspection of the waveform in Figure 2.39 confirms that it has a time-average value of D, the duty ratio.

The values of a_n and b_n can also be found and used to compute c_n and θ_n. This is left as an exercise. The results are:

$$c_n = \frac{2}{\pi} \frac{\sin(n\pi D)}{n}, \qquad n \neq 0 \tag{2.29}$$

The value of phase is $\theta_n = -n\omega t_0$, where $\omega = 2\pi/T$. It is conventional to define a reference phase $\phi_0 = \omega t_0$, so that $\theta_n = -n\phi_0$. Thus, the Fourier Series for this very general $q(t)$ can be written

$$q(t) = D + \frac{2}{\pi} \sum_{n=1}^{\infty} \frac{\sin(n\pi D)}{n} \cos(n\omega t - n\phi_0) \tag{2.30}$$

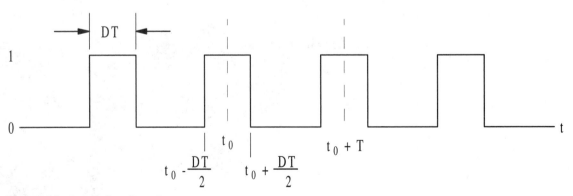

Figure 2.39 A periodic pulse train.

The Fourier representation shows that the function $q(t)$ is determined completely by just three parameters: the duty ratio D, the radian frequency $\omega = 2\pi f$ (or the period T), and the reference time t_0 (or the reference phase ϕ_0). These three numbers fully define the switching function, and switch action can always be interpreted in terms of one or more of the them.

The switch action in a power converter will usually have to adjust over time to account for any changes in the environment, the input source, or the output load. The series representation makes a few possibilities salient:

1. *Duty ratio adjustment.* The duty ratio is related to the pulse width DT. Converters that operate by adjusting the duty ratios of their switches exhibit pulse-width modulation (PWM) action.

quency adjustment. Frequency adjustment is unusual in power electronics for a very reason: The need to provide matching frequency components of voltage and current a given source or load often places tight constraints on the switching frequencies. st important exception is in dc–dc conversion. Since only the average value is of the dc case, it is possible to adjust the frequency without causing trouble. This called *frequency control.* Although true frequency modulation (FM) is rare in rsion, it is mathematically possible.

ment. One of the oldest ways to alter the behavior of a power converter is ning of the switch action. Since the wave shapes are almost always more he specific frequencies in this technique, timing adjustment is normally ular time frame $\theta = \omega t$. The term *phase control* describes the idea of action in time. Some converters vary the phase in a regular way, phase modulation (PM).

will study use either PWM or phase control to permit adjustment

2.11 S

AP

nes are the key to electronic conversion of energy. Three major aspects of design—the hardware, software, and interface problems—can be defined and studied. Consider some of the concepts so far:

- The hardware problem (build a switch matrix). Switches can be organized in a matrix. Given information about the source and load, the switch matrix dimensions (such as 2×2 or 4×3) can be determined. The sources help define the current and voltage polarities that a switch will see. This permits the selection of restricted switches to build a matrix. The restricted switch types correspond to specific kinds of devices.

- The software problem (operate the matrix to provide a desired conversion). Switching functions provide a convenient mathematical representation of switch action. Circuit laws constrain switch operation. We need to avoid switching actions that might try to violate KVL or KCL. Fourier analysis identified a frequency matching requirement to ensure that energy is transferred successfully in a circuit. The duty ratio, frequency, and phase completely define the switch action. The ultimate objective is to produce a large wanted component while ensuring that unwanted components are small.

- The interface problem (add energy storage to filter energy flow for the application requirements). This remains to be studied in depth. The frequency matching requirement helps suggest the importance of filters.

A switch matrix will either be direct, meaning that it simply interconnects the input and output, or indirect, meaning that energy storage is embedded in the matrix. Switch matrices are one of the few types of circuits in which a designer can try to breach KVL or KCL, but of course this violates the underlying physics. Accidental switch action that tries to violate circuit laws is perhaps the most common cause of failure in power converters.

- *KVL restriction*: a switch matrix must avoid interconnecting unlike voltage sources.
- *KCL restriction*: a switch matrix must avoid interconnecting unlike current sources.

Voltage sources switched across inductors, or current sources switched into capacitors, will violate KVL or KCL in the long run.

Switching functions, which are functions of time valued at 1 when the corresponding physical switch is on and 0 when it is off, offer a convenient representation of switch action. KVL and KCL restrictions, for instance, can be stated in terms of switching functions. Converter waveforms turn out to be products of switching functions and sources.

The ideal switch can carry any current, block any voltage, switch under any conditions, and change back and forth between on and off instantly. This ideal device is not yet very useful in power conversion, because it is hard to create with real devices. Semiconductor switches exhibit polarity limitations in addition to physical limits on current, voltage, and time. The polarity limitations define restricted switches. These are extremely useful in converter design. Any converter can be defined initially with restricted switches. This ideal representation is translated into corresponding semiconductor hardware once it is fully analyzed and understood. Restricted switches are defined in terms of current and voltage polarity. Five types exist:

1. Forward-conducting reverse-blocking (FCRB) switches, corresponding to ideal diodes.
2. Forward-conducting forward-blocking (FCFB) switches, corresponding to idealizations of power bipolar junction transistors.
3. Forward-conducting bidirectional-blocking (FCBB) switches, corresponding to idealized GTOs, or to series combinations of FCFB and FCRB devices.
4. Bidirectional-conducting forward-blocking (BCFB) switches, corresponding to ideal power MOSFETs.
5. Bidirectional-conducting bidirectional-blocking (BCBB) switches, corresponding to the ideal switch.

There are other important devices, such as the SCR, that add timing properties to the restricted switch behavior. These will be examined in context later on.

Diode circuits can be analyzed with a trial method. If a diode is on, its forward current must be positive. If it is off, its forward voltage must be negative. The method begins with circuit configurations, the various possible arrangements defined by switch state, then analyzes the plausible configurations to see which one is consistent with both the circuit laws and the diode properties. If a configuration's voltages or currents contradict diode properties, then switch action to change the configuration would occur in a real circuit. In a diode bridge, the trial method shows that the usual action is for devices to act in diagonal pairs.

Through Fourier analysis, we can resolve periodic signals into individual frequency components. A port in an electrical circuit will have nonzero average power only if the voltage and current at the port have a common Fourier component frequency. Furthermore, the power at a port is the sum of powers contributed by the individual frequencies. Cross-frequency terms do not contribute to average energy flow.

A real periodic function $f(t)$ can be represented with the Fourier Series

$$f(t) = \sum_{n=0}^{\infty} c_n \cos(n\omega t + \theta_n). \tag{2.31}$$

A general switching function with radian frequency ω, duty ratio D, and phase (referenced to the center of the pulse) ϕ has the Fourier Series

$$q(t) = D + \frac{2}{\pi} \sum_{n=1}^{\infty} \frac{\sin(n\pi D)}{n} \cos(n\omega t - n\phi) \tag{2.32}$$

The three parameters completely define the switching function and hence the switch action. A useful converter must permit some adjustment of its operation. Pulse-width modulation and phase control are the most common adjustment methods in power electronics.

PROBLEMS

1. Draw the switch matrix for the most general single-phase ac to dc converter. How many switches are in this matrix?

2. A dc–dc converter has a current source connected between the input terminals. Assume that some sort of electrical load appears across the output lines. Of the four switches, what combinations can be operated without violating the KCL restriction?

3. A switch matrix converts ac to dc. The input is a four-phase ac voltage source, while the output is a dc current source.

   a. Draw the converter, and label the switches.
   b. State the KVL restrictions in terms of the switching functions.
   c. State the KCL restrictions in terms of the switching functions.
   d. Naturally, both KVL and KCL restrictions must be met. State this combined requirement in terms of the switching functions.

4. An ac–ac single-phase to single-phase converter is shown in Figure 2.40. The input source has a frequency of 50 Hz. The switching functions have a frequency of 80 Hz, and a duty cycle of one-half. Plot the output voltage waveform. What is the frequency of this waveform?

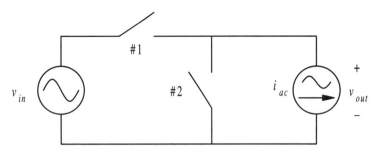

Figure 2.40 Converter of Problem 4.

5. A three-phase rectifier is shown in Figure 2.41. The current I_{out} is constant. It is proposed to turn each switch on when the corresponding voltage is positive, so switch a turns on when $V_a > 0$, and so on. Will this operating method meet requirements of KVL and KCL? If so, plot the switching function q_a and the voltage $V_{out}(t)$. If not, discuss the problem and propose a solution.

6. An ac–dc converter is shown in Figure 2.41. I_{out} has a constant, positive, value. This is a "midpoint converter" because of the use of a neutral. The switches function as ideal diodes, and are:

$$\text{ON for } I_{switch} > 0 \qquad \text{OFF for } V_{switch} < 0$$

Given: $V_a(t) = V_0 \cos(\omega t)$, $V_b = V_0 \cos(\omega t - 2\pi/3)$, $V_c = V_0 \cos(\omega t + 2\pi/3)$

a. Plot $V_{out}(t)$ on some three-phase graph paper.
b. Compute the average value of $V_{out}(t)$.
c. Sketch the switching function for switch a.

7. Find V_{out} as a function of v_{in} for the circuit of Figure 2.42. The resistors have identical values. v_{in} can have any time-varying value.

8. Find V_{out} as a function of v_{in} and V_{dc} for the circuit of Figure 2.43. The resistors are all identical.

9. A dc–dc converter has output voltage, input voltage, output current, and input current all ≥ 0.

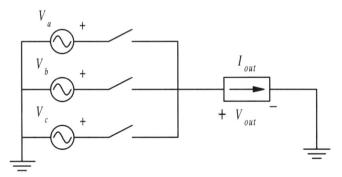

Figure 2.41 ac–dc midpoint converter (Problems 5 and 6).

Figure 2.42 Full-wave rectifier (Problem 7).

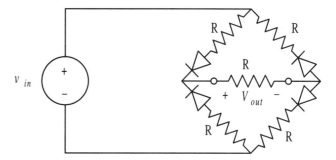

a. Draw the switch matrix for this converter.

b. What types of restricted switches can be used for this converter?

10. A certain converter requires a forward-conducting forward-blocking switch. Among the five possible restricted switch types, there are several with this capability. Which of the five types can be used to operate this converter? How did you determine this?

11. Any of the five possible restricted switches can be constructed from combinations of bipolar transistors and diodes. See if you can design ways to assemble each of the switch types solely from diode and transistor combinations. (*Hint*: The restricted switch symbols for the devices are helpful here.)

12. Derive the coefficients a_n and b_n for the general periodic pulse train of Figure 2.39. Show that the values for c_n and θ_n presented in the text are correct.

13. A single-phase ac to dc converter is operated with phase control. The input is a voltage source $V_0 \cos(\omega t)$. Each switch of the four in the matrix is on half the time. The output load is a resistor. The switches operate so that the input source is always connected to the output in some fashion.

a. Draw the switching functions and resistor voltage as a function of time when the switching function phase is $0°$.

b. Repeat for phase of $45°$.

14. Compute the Fourier Series of the output voltage for the converter of Problem 6 above.

Figure 2.43 Full-wave rectifier with internal dc source (Problem 8).

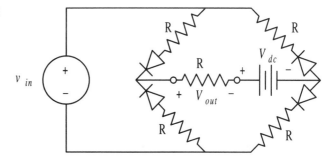

15. A converter has a single ac voltage input, of amplitude V_{in}, and two switches. When switch 1 is on, the input and output are connected. When switch 2 is on, the output is shorted. The switches operate so $q_1 + q_2 = 1$.

    a. Draw this circuit.

    b. The frequency of the input is 60 Hz, the peak value of the input voltage is V_0, and the frequency of q_1 is 10 kHz. The output average power is P_{ave}. What is the amplitude of the 60 Hz Fourier component of the input current?

16. It is proposed to convert single-phase 60 Hz voltage to single-phase 30 Hz current by a method called "integral cycle control." The circuit is shown in Figure 2.44. The method works as follows: Switch 1 will be on for one full period of $V_{in}(t)$, then it will turn off for one full period, then on for one period, and so on. This will give an output waveform like the one at the right in Figure 2.44.

    a. Plot the switching function of switch 1, $q_1(t)$.

    b. Write the Fourier Series of $q_1(t)$. (Notice that all parameters are known.) What is the value of phase, in degrees?

    c. Does this switching function provide the desired conversion? If not, plot one that does.

17. A dc–dc converter is built so that $V_{out}(t) = q_1 V_{in}(t)$. The switch operates at 50 kHz with duty ratio of 50%. This output voltage is applied to the input of a simple one-pole low-pass filter with a corner frequency at 500 Hz. What is the Fourier Series of the signal at the filter output?

18. A bridge rectifier with a current-source load is built along the lines of the one in Figure 2.36, except that SCRs substitute for diodes. The switching functions operate with a delay of 90°.

    a. Draw the output voltage waveform.

    b. What is the amplitude of the fundamental of the output voltage?

    c. Compute the average power into the load.

19. A converter has a sinusoidal ac voltage source input and an ac load. Four switches are used for the 2×2 switch matrix. The input frequency is 60 Hz. The switches operate at 40 Hz, such that $q_{1,1} = q_{2,2}$.

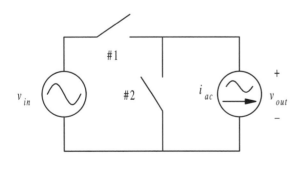
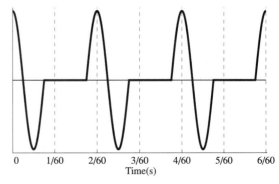

Figure 2.44 Integral cycle controller circuit and output waveform.

    a. How are the other switching functions related to $q_{1,1}$ if the load requires a current path?

    b. Plot the output voltage waveform, and determine its fundamental frequency.

    c. Assuming that the wanted output component is the one with the largest Fourier component amplitude, what frequency or frequencies are wanted at the output?

20. A three-phase current source is available. In this source, the three currents share a common reference node, since the sum of currents $I_0 \cos(\omega t) + I_0 \cos(\omega t - 2\pi/3) + I_0 \cos(\omega t + 2\pi/3) = 0$. The output is a dc load with voltage-source characteristics. Draw a switch matrix that can support conversion between these two. What types of switching devices will be needed?

REFERENCES

H. B. Dwight, *Tables of Integrals and Other Mathematical Data*. New York: MacMillan, 1961.

J. W. Motto, *Introduction to Solid State Power Electronics*. Youngwood, PA: Westinghouse, 1977.

D. L. Powers, *Boundary Value Problems*. New York: Academic Press, 1972.

P. Wood, *Switching Power Converters*. New York: Van Nostrand Reinhold, 1981.

CHAPTER 3

Figure 3.1 Commercial power supplies. These units cascade a rectifier and a dc–dc converter.

CONVERTER CONCEPTS

3.1 INTRODUCTION

In this chapter, basic concepts of the electrical energy conversion process, together with indicators of conversion quality, are introduced. We need to establish a specific set of objectives for a converter. It is important to identify performance measures to determine whether objectives are being met. How can user expectations for a power supply be translated into design goals? How do we talk about performance or quality? What should a converter try to accomplish? It is easy to demand a perfect power converter. Often, a power electronics engineer must work with a user toward reasonable performance and cost objectives (represented by the supplies in Figure 3.1). A perfect converter is impossible and a near-perfect one might involve considerable expense. How can the user's need be met with a reliable, efficient, cost-effective circuit?

Any converter design is constrained by KVL and KCL. The concept of source conversion is developed to link the circuit laws to basic converter organization. As we shall see, current sources are just as common in power electronics as voltage sources.

Distortion must be considered in any converter or power supply. How much ripple or other unwanted effects will be present at a rectifier output? How pure is the output signal of a dc–dc converter or an inverter? Distortion is an inevitable result of the nonlinear switching behavior of power electronic circuits. We must consider how to reduce unwanted effects since they cannot be eliminated outright. Performance indicators such as *total harmonic distortion* are introduced to help characterize distortion performance. *Power factor* helps to address system efficiency rather than just converter efficiency. Converter efficiency is critical, but so is the energy flow taken from the input source. Power factor measures the fraction of useable flow from the input that actually contributes to output power.

Regulation defines how closely a power supply output approaches an ideal source. A real converter shows voltage drops or other changes when the load power increases or the input voltage decreases. Commercial products usually are designed to approximate a voltage source closely, and we will examine a sample specification to gain an understanding of the concept.

Many converters construct pre-defined waveforms as part of their action. The full-wave rectified sinusoid is one example. When a waveform is fully determined, it can be treated as an *equivalent source*. It will be shown that this idea is a powerful tool for analyzing distortion waveforms and other operating characteristics of converters. Equivalent sources are especially helpful in the design of filters for converters. This brings out the *interface problem* introduced in Chapter 2. By the end of Chapter 3, the underlying issues in the development of interfaces will be considered.

3.2 SOURCE CONVERSION

One of the most basic principles of power supplies is that the user expects an ideal source. Designers of electrical equipment and electronic circuits generally assume that this type of energy supply is available. Most of us are accustomed to this idea. The electric utility grid is carefully operated to provide a sinusoidal voltage source of good quality. Student lab courses often use expensive dc supplies to ensure a fixed voltage source for projects. The idea of a device that supplies a constant voltage over a wide range of power levels is familiar.

Commercial voltage sources come quite close to ideal behavior. The typical U.S. household electrical outlet provides about 120 V from no load right up to the circuit breaker or fuse current limits. The 12 V source in a modern automobile is expected to hold firm for load currents ranging from a few milliamps to well over 100 A. Power supplies can be designed to meet very tight requirements for voltage constancy. Current sources are less familiar, but just as useful in principle. A device which supplies a fixed current over some wide voltage range is certainly a reasonable energy source.

A user of power conversion equipment would most likely expect to supply it from some source—whether the 50 Hz or 60 Hz ac mains, a battery, or some other energy source. To a power electronics designer, this expectation represents the concept of *source conversion*. According to this concept, any power converter would normally be designed so that its output resembles an ideal source. Not only does a user expect a converter to behave as an ideal source, but a typical power electronics engineer often intends the input energy source to show near-ideal behavior. A power converter usually is expected to manipulate energy flow between an ideal input and an ideal output.

> **Definition:** Source conversion *is the concept that describes the operation of most power converters: They control energy flow between near-ideal electrical sources.*

Our typical user seeks an ideal source. A reminder of the defining characteristics of ideal sources might be helpful at this point.

> **Definition:** *A* current source *supplies a given current waveform into* any *load and at* any *voltage. No external circuit or device can force a current source to change.*

> **Definition:** *A* voltage source *supplies a given voltage waveform into* any *load and at* any *current. No external device or circuit can force a voltage source to change.*

The source conversion concept creates a dilemma because of KVL and KCL restrictions. Since unlike voltage sources cannot be interconnected, and unlike current sources likewise cannot be interconnected, we can say that "voltage is not converted to voltage" and that "current is not converted to current." No switch matrix can control energy flow between two unequal voltage sources or unequal current sources. However, any voltage can be imposed on a current source and any current can be imposed on a voltage source. This means that voltage sources can be interconnected with current sources without trouble. The concept shows us that "voltage converts to current" and "current converts to voltage."

The source conversion concept is a fundamental of power electronics. In a well-designed power converter, both the input and output ought to have the characteristics of an ideal source. If the input is a voltage source, then the output should resemble a current source. If the input is a current source, then the output should have properties of a voltage source. Although voltage sources are the most familiar form for energy, current sources are quite common in power electronics. In fact, current and voltage sources are roughly equally prevalent. Most often, a current source represents the behavior of an inductor.

In some conversion applications, conversion of voltage to current or current to voltage is a problem. Perhaps the input and output are both near-ideal voltage sources. One alternative in such cases is to introduce an intermediate source, and perform a voltage–current–voltage conversion, for instance. An internal source used in this manner is called a *transfer source*.

> ***Definition:*** A transfer source *is some electrical source element, intended to consume no energy, placed in the path between an input source and an output source. It serves as a temporary location as energy is transferred between the input and output.*

When a transfer source is used, it is embedded in the converter switch structure. Recall the definition of an indirect switching converter, in which elements are positioned within the switch matrix. A transfer source is one of the most common types of such elements.

Source conversion concepts play a major role in determining the topology of switching power converters. There are converters with current input and output, voltage input and current output, and indeed all combinations, for dc–dc, dc–ac, ac–dc, and all other types of conversion functions.

Example 3.2.1   In a general sense, there are only two types of direct converters for dc–dc. Draw these two circuits and comment. What about indirect converters?

This example shows some truly crucial concepts. A direct converter interconnects an input and output with a switch matrix. For the dc–dc case, the only possibilities are current input and voltage output, or voltage input and current output, as shown in Figure 3.2. In contrast, indirect converters have no inherent limitations. As storage devices, transformers, and transfer sources are added to the matrix, the complexity goes up, and there is an unlimited number of indirect converter configurations.

3.3 DISTORTION

Since switches are always present in a useful power converter, many of the voltages and currents will be the product of a switching function and an ideal source. The nonlinear action

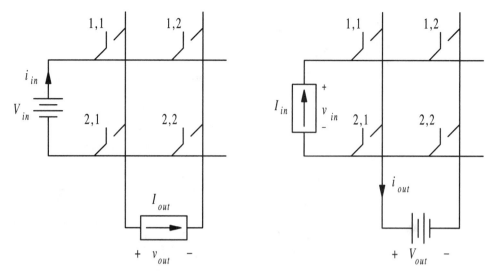

Figure 3.2   The two direct dc–dc converters.

generates an infinite sequence of Fourier components, as we saw in Chapter 2. It is very important to recognize that this behavior is fundamental: we cannot switch without generating a whole sequence of harmonics. In converter terms, there will always be unwanted components present along with the wanted one. The unwanted components can be reduced, perhaps by filters, but cannot be eliminated.

Example 3.3.1   The converter of Figure 3.3 has an ac input voltage source and a single dc output current source. The switch matrix must be operated so that (1) no input voltages are

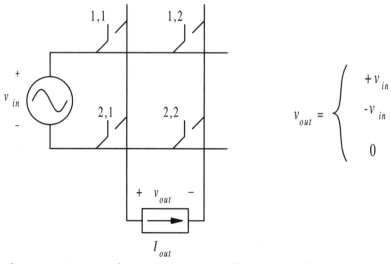

$$v_{out} = \begin{cases} +v_{in} \\ -v_{in} \\ 0 \end{cases}$$

Figure 3.3   Since a rectifier must create its output from sections of a sinusoidal waveform, the output cannot be pure dc.

shorted together, and (2) a path for the output current is always provided. What is v_{out}? Comment on the Fourier components.

If you look more closely, you will see that v_{out} must be equal to zero, to v_{in}, or to $-v_{in}$. Since the output is a dc current source, v_{out} must have a dc Fourier component in order for $P_{out(ave)}$ to be nonzero. But none of the input voltages is a nonzero dc signal at any time. *The output cannot be made to conform exactly to the purpose at hand.* More directly, we must accept the bad with the good: The output voltage will consist of pieces of the input voltages, and will contain many Fourier components along with the wanted dc component.

Example 3.3.1 holds more generally. Since the actual output waveform of any converter is constructed from the input sources, and since the desired output in general differs from the input, the output must contain unwanted components along with the wanted component.

Unwanted components are unavoidable, but do not contribute to average energy transfer if the sources are ideal. We are forced to address the problem of filtering out unwanted components if the user really expects ideal behavior. The complete collection of unwanted components defines *distortion*. The terms *harmonic distortion* or simply *harmonics* refer to this unwanted behavior. In dc applications, the term *ripple* describes the collected unwanted components. It is crucial to recognize the fundamental nature of harmonic distortion: If we intend to do energy conversion without loss, the necessary switch action must always give us distortion along with the desired waveform. If efficiency is important, it is not a question of how to avoid distortion, but rather how to reduce it through filtering to a tolerable level.

Example 3.3.2   A salesman has approached your company with a "new type of power supply." The claim is a converter able to accept either 60 Hz 120 V or 50 Hz 230 V input, and provide a low-voltage dc output at up to 1500 W. The efficiency is said to be 95%. When asked about ripple and distortion, the salesman proudly asserts that "through our advanced engineering," an output "totally free of ripple under all conditions" has been achieved. Your boss is very interested, and seeks you out as a power electronics expert for comment.

The high efficiency and flexible input certainly require switching action. Since the unit is almost certainly a switching power supply, it is physically inconsistent for it to achieve zero ripple. The claims are misinformation or hype, and the salesman will have to be encouraged to be more forthcoming about the performance of this supply.

In the initial steps of converter design, questions about distortion must be addressed. How much can the load tolerate? Are there specific frequencies that are especially bad, such that they should be avoided if possible? What kind of filtering is necessary? How do we specify distortion limits?

In most dc applications, ripple represents both a variation around the desired dc level and a possible energy effect at undesired frequencies. In a standard dc supply, it is common to specify the maximum peak-to-peak ripple, and often the ripple RMS magnitude as well. Typical numbers for peak-to-peak ripple fall in the 50–100 mV range for low dc voltages such as 5 or 12 V. A good rule of thumb is that the ripple will be about 1% of the nominal output.

While low ripple amplitude is helpful, it is not always enough. For example, the dc supplies for a computer display need to avoid frequencies that interact with the visual information. Tiny variations in dc level that happen to be synchronized to the image will appear as waviness, distorted characters, or other effects readily apparent to the human eye. High-power converters often use switching rates in the range of 1–10 kHz. The unwanted components can produce vibration in magnetic components and loads. This creates audible noise—in frequency ranges particularly annoying to people. An interesting historical note is that high-frequency ac (in the 50 kHz range) has been considered for power distribution in large spacecraft such as a space station, because the frequency range permits convenient power electronic conversion. The frequencies are not audible to humans. However, many laboratory animals can hear 50 kHz audio signals. An annoying background noise creates stress, and high-frequency ac power distribution would make it difficult to conduct biological experiments in space.

In ac applications, a different measure of harmonics is needed. In most cases, the *total harmonic distortion* (THD) gives the necessary information. THD is a ratio of the RMS value of unwanted components to the fundamental. Two definitions are in use. The first, common in the United States, is based on the fundamental. Given a signal *f(t)* with no dc component and with Fourier coefficients $c_1, c_2, c_3, ...$,

$$\text{THD} = \sqrt{\frac{\sum_{n=2}^{\infty} c_n^2}{c_1^2}} \tag{3.1}$$

When no distortion is present, the THD value will be zero. When harmonics are present, the value has no particular upper limit: if c_1 is small, the THD value can be well over 100%. For an individual Fourier component, the peak value c_n is associated with a RMS value $c_n/\sqrt{2}$. Equation (3.1) often can be treated more directly by observing that the RMS value of *f(t)* is given by

$$f_{\text{RMS}} = \sqrt{\frac{1}{2} \sum_{n=1}^{\infty} c_n^2} \tag{3.2}$$

When this is substituted into equation (3.1), the THD can be written

$$\text{THD} = \sqrt{\frac{2f_{\text{RMS}}^2 - c_1^2}{c_1^2}} = \sqrt{\frac{f_{\text{RMS}}^2 - f_{1(\text{RMS})}^2}{f_{1(\text{RMS})}^2}} \tag{3.3}$$

where $f_{1(\text{RMS})}$ is the RMS value of the fundamental component, $c_1/\sqrt{2}$.

In Europe, an alternative definition of THD, referred to here as total harmonic ratio (THR), is common. The ratio is taken relative to the total RMS value rather than just that of the fundamental. In this version, the value cannot exceed 100%:

$$\text{THR} = \sqrt{\frac{\sum_{n=2}^{\infty} c_n^2}{\sum_{n=1}^{\infty} c_n^2}} = \sqrt{\frac{f_{\text{RMS}}^2 - f_{1(\text{RMS})}^2}{f_{(\text{RMS})}^2}} \tag{3.4}$$

As in the case of THD, the THR will be zero if no harmonics are present other than the fun-

DISTORTION 91

damental. A THR value of 1 means no fundamental is present: the waveform consists entirely of distortion.

In some power conversion applications, the fundamental component of *f(t)* is not the wanted component. This is especially common in ac–ac conversion. THD and THR are not really the correct concepts in this case, since the fundamental component will not have special meaning. Let us assume that the *m*th Fourier component is the wanted one. The *total unwanted distortion* (TUD) is defined as

$$\text{TUD} = \sqrt{\frac{\sum_{n=1, n \neq m}^{\infty} c_n^2}{c_m^2}} \tag{3.5}$$

This will be small if the waveform consists almost entirely of the wanted component. That is, a small TUD value suggests that the converter is successfully performing its desired function.

Consider a few simple waveforms and their distortion measures.

Example 3.3.3 Compute the THD and THR of the general square wave sq(ωt).

First, we need RMS values and fundamentals for sq(ωt). The RMS value is trivial: sq$_{\text{RMS}}$ = 1 (check this as an exercise). The fundamental can be computed by integrating to get a_1 and b_1 for the Fourier Series. From (2.18), the series is

$$\text{sq}(\omega t) = \frac{4}{\pi} \sum_{n=1}^{\infty} \frac{\sin(n\pi/2)}{n} \cos(n\omega t) \tag{3.6}$$

The fundamental component has amplitude $4/\pi$ and RMS value $2\sqrt{2}/\pi$. The distortion values are

$$\text{THD} = \sqrt{\frac{f^2_{\text{RMS}} - f^2_{1(\text{RMS})}}{f^2_{1(\text{RMS})}}} = \sqrt{\frac{1 - \left(\frac{2\sqrt{2}}{\pi}\right)^2}{\left(\frac{2\sqrt{2}}{\pi}\right)^2}}$$

$$= \sqrt{\frac{1 - 8/\pi^2}{8/\pi^2}} = \sqrt{\frac{\pi^2}{8} - 1} = 48.3\%, \tag{3.7}$$

$$\text{THR} = \sqrt{1 - 8/\pi^2} = 43.5\%$$

In this case, the THD and THR values are not very far apart. The two measures have nearly identical values when distortion is low, but the THD rapidly becomes larger once the harmonic content reaches about 50%.

Example 3.3.4 Find the THD and THR of a symmetrical triangle waveform with peak value 1 V. This signal is sometimes given the symbol tri(*t*) (see Fig. 3.4). Notice that over the time period 0 to *T*/2, this waveform is the linear function $1 - 4t/T$.

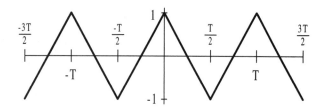

Figure 3.4 The function tri(t).

Since the wave is symmetric about the y axis, the integrals can be performed by doubling the value over the first half-period. The RMS value is

$$\text{tri}_{\text{RMS}} = \sqrt{\frac{2}{T} \int_0^{T/2} \left(1 - \frac{4t}{T}\right)^2 dt} = \frac{\sqrt{3}}{3} \tag{3.8}$$

The fundamental Fourier component is given by a_1 (the symmetry makes $b_1 = 0$),

$$c_1 = a_1 = \frac{4}{T} \int_0^{T/2} \left(1 - \frac{4t}{T}\right) \cos\left(2\pi \frac{t}{T}\right) dt = \frac{8}{\pi^2} \tag{3.9}$$

This has an RMS value of $(4\sqrt{2})/\pi^2$. The distortion measures are

$$\text{THD} = \sqrt{\frac{\frac{1}{3} - \frac{32}{\pi^4}}{\frac{32}{\pi^4}}} = \sqrt{\frac{\pi^4}{96} - 1} = 12.1\%,$$

$$\tag{3.10}$$

$$\text{THR} = \sqrt{\frac{\frac{1}{3} - \frac{32}{\pi^4}}{\frac{1}{3}}} = \sqrt{1 - \frac{96}{\pi^4}} = 12.0\%$$

Example 3.3.5 The waveform shown in Figure 3.5 is typical of the line current in a three-phase diode rectifier bridge. What is the THD of this waveform?

The time axis position does not affect the THD value. If the axis $t = 0$ is positioned to coincide with the center of a pulse, the waveform will have even symmetry, and $b_n = 0$.

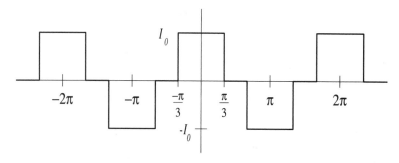

Figure 3.5 Output waveform for Example 3.3.5

The RMS value is

$$f_{RMS} = \sqrt{\frac{1}{\pi} \int_{-\pi/3}^{\pi/3} I_0^2 \, dt} = I_0 \sqrt{\frac{2}{3}} \tag{3.11}$$

The component $c_1 = a_1$ since $b_1 = 0$, and a_1 is

$$a_1 = \frac{1}{\pi} \int_{-\pi/3}^{2\pi - \pi/3} f(\theta) \cos \theta \, d\theta$$

$$= \frac{1}{\pi} \int_{-\pi/3}^{\pi/3} I_0 \cos \theta \, d\theta - \frac{1}{\pi} \int_{\pi - \pi/3}^{\pi + \pi/3} I_0 \cos \theta \, d\theta \tag{3.12}$$

$$= \frac{2\sqrt{3} \, I_0}{\pi}$$

so the RMS value associated with c_1 will be $\sqrt{6}/\pi$. The THD value is therefore

$$THD = \sqrt{\frac{\frac{2}{3} - \frac{6}{\pi^2}}{\frac{6}{\pi^2}}} = 31.08\% \tag{3.13}$$

Example 3.3.6 Find the TUD of the output waveform from the 60 Hz to 50 Hz converter of Example 2.9.3. The converter in that example switched between $+v_{in}$ and $-v_{in}$ to produce a wanted component at 50 Hz.

Since the waveform is either $+v_{in}$ or $-v_{in}$, the RMS value will be the same as the RMS value of v_{in}, or $0.707V_0$. There is no fundamental component, so the THD and THR values are not useful. The wanted component from the previous example has an amplitude of $0.636V_0 = 2V_0/\pi$. Its RMS value is $0.450V_0 = \sqrt{2}V_0/\pi$. The TUD is

$$TUD = \sqrt{\frac{f^2_{RMS} - f^2_{5(RMS)}}{f^2_{5(RMS)}}} = \sqrt{\frac{(V_0/\sqrt{2})^2 - (\sqrt{2}\,V_0/\pi)^2}{(\sqrt{2}\,V_0/\pi)^2}} \tag{3.14}$$

$$= \sqrt{\frac{1/2 - 2/\pi^2}{2/\pi^2}} = 121\%$$

This high value shows that most of the signal is composed of unwanted components. The high distortion value is not surprising given that there is an unwanted 170 Hz component just as large as the wanted component. This guarantees that the TUD is at least 100%.

3.4 REGULATION

No power supply provides an ideal source, even though this is the goal. Changes in load, input level, or even temperature can alter performance. The term *regulation* describes the ability of a power converter to compensate for these external changes and maintain the desired behavior. A supply that behaves as an ideal source would be said to have *perfect regulation*. Regulation is a key specification, and is often discussed by manufacturers. An example is given in Figure 3.6, where a typical manufacturer's data sheet is shown.

Rating	Specification Value
Voltage Rating	15V nominal with ±10% adjustment.
Current Rating	20.0 A @ 35°C, derated linearly to 4.0 A @ 75°C.
Output Voltage: Line Regulation	0.1% from minimum to maximum rated line input.
Output Voltage: Load Regulation	0.1% from 0.5 A load to full load.
Output Voltage: Ripple and Noise	25 mV$_{RMS}$, 150 mV$_{p-p}$, 20 MHz bandwidth at full load.
Output Voltage: Temperature Coefficient	0.04%/°C maximum.
Rated Input	87 to 132 V or 175 to 264 V (user selectable range), 40-440 Hz, or 125 to 175 V dc. 400 W, 700 VA max.
Efficiency	75% minimum at full load.
Overshoot	No overshoot at turn-on, turn off or power failure.
Temperature Range	Continuous operation -40°C to +75°C with derating above 35°C. Storage -55°C to +85°C.
Operating Humidity	10% to 85% RH, non-condensing.
In-Rush Limiting	The cold-start in-rush current will not exceed 40 A.
Mean Time Between Failures (MTBF)	40,000 hours per MIL-HDBK 217E.
Hold-Up Time	At full load, output remains within regulation limits for at least 16.7 ms after loss of input power.
Turn-On Delay	Output voltage reaches nominal within 0.6 s after application of input power, with no overshoot.
Stability	±0.1% for 24 hrs after power up.
Combined Regulation (source, load, temp., time)	±0.32% for 24 hrs.
EMI	Conforms to FCC Class A and MIL-STD-461A.
Fungus Proofing	Passes MIL-STD-810C, Method 508.1.
Shock	10 G, 55 Hz, all axes.
Dimensions	76 mm × 117.5 mm × 300.0 mm, 3.2 kg.
Warranty	5 years
Price	$250 (Quantity 1000)

Figure 3.6 Data sheet for a typical commercial power supply.

Let us consider some of the values shown in Figure 3.6. The *line regulation* defines the supply's ability to hold its output fixed when the input line voltage varies. This might be represented formally with the partial derivative $\partial V_{out}/\partial V_{in}$, for instance. Instead, most manufacturers specify an allowed output change over a given input range.

Definition: Line regulation *refers to the ability of a power converter to maintain the output even when the input fluctuates. The most common measure of line regulation is percentage change,*

$$\% \ Line \ reg = 100 \ \frac{V_{out(highest \ input)} - V_{out(lowest \ input)}}{V_{out(nominal)}} \qquad (3.15)$$

This quantity tells the customer directly how much change is expected when the input is uncertain. Usually, the change implies the size of the effect when the supply is operating at the rated full-load condition. Line regulation of 0.1%, for example, means that a 5 V supply, under full load, will vary by only 5 mV as the ac input changes over the full specified operating range. This 5 mV variation is much less than the ripple variation and is tested by measuring only the average values. The normalized derivative

$$\frac{\partial V_{out}/V_{out}}{\partial V_{in}/V_{in}} \qquad (3.16)$$

is also used sometimes to represent the line regulation.

The *load regulation* reflects changes seen at the output caused by altering the requested energy flow. One fundamental measure of this behavior is provided by the output impedance, $\partial V_{out}/\partial I_{out}$. There are other more common specifications, given in the definition.

Definition: Load regulation *refers to the ability of a power converter to maintain the output when the output power fluctuates. In power systems and battery-based systems, load regulation is usually referenced to the open-circuit output voltage,*

$$\% \ Load \ reg = 100 \ \frac{V_{out(no \ load)} - V_{out(full \ load)}}{V_{out(no \ load)}} \qquad (3.17)$$

In switching power supplies, the change is more commonly specified relative to nominal output voltage, as

$$\% \ Load \ reg = 100 \ \frac{V_{out(min \ load)} - V_{out(full \ load)}}{V_{out(nominal)}} \qquad (3.18)$$

Other important regulation issues include *temperature regulation*, $\partial V_{out}/\partial T$, and *long-term drift*—the variation in output over periods of hours or days.

The resistive voltage divider is useful for an illustration of regulation.

Example 3.4.1   A simple resistive voltage divider is used to step down a dc voltage. Under no-load conditions, what is the line regulation? Use the normalized variation in equation (3.16) as a measure.

A divider is shown in Figure 3.7. The ratio of output to input is well known as

$$\frac{V_{out}}{V_{in}} = \frac{R_{out}}{R_{in} + R_{out}} \tag{3.19}$$

The partial derivative $\partial V_{out}/\partial V_{in}$ gives the same ratio. Substituting into equation (3.16), we obtain

$$\frac{\partial V_{out}/V_{out}}{\partial V_{in}/V_{in}} = 1 \tag{3.20}$$

In other words, any change at the input is reflected 100% at the output. If the input changes by 5%, the output will change by 5%, and so on. This circuit exhibits no regulation, since the line variation is not attenuated. A voltage divider thus produces an unregulated dc output.

Of the various regulation issues, load regulation is perhaps the most difficult to deal with. The load, after all, is provided by the customer. Measurements and tests will depend on the customer's connections. The outcome is not always predictable. Let us consider a common example.

Example 3.4.2   A certain 5 V power supply has perfect load regulation, meaning that the output remains exactly 5 V even when the output power changes from no load to full rated load. The customer connects this supply to a load through a length of large wire, such as #12 AWG (diameter 2.0 mm), with total resistance of 0.00526 Ω. If the supply's rated output power is 100 W, what load regulation will the customer actually measure at the load terminals?

The effect here is difficult to avoid in practice: Any resistor will drop a voltage proportional to current and so alters the regulation properties of a power supply. In this exam-

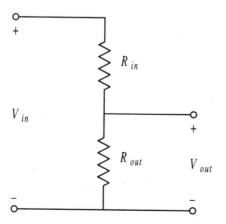

Figure 3.7   Resistive divider for dc step-down conversion.

ple circuit, the customer will measure 5.000 V at the end of the connection wires under no-load conditions. When the load actually draws 100 W, we have

$$i^2 R = 100 \text{ W}, \qquad \frac{5 \text{ V}}{R + 0.00526 \ \Omega} = i \qquad (3.21)$$

This gives $i = 20.4$ A, $R = 0.239 \ \Omega$. The voltage drop along the connection wire will be 0.108 V. The customer will measure 4.89 V at the load terminals. The load regulation at the terminals is therefore

$$\% \text{ Load reg} = 100 \ \frac{V_{\text{out(min load)}} - V_{\text{out(full load)}}}{V_{\text{out(nominal)}}} = 100 \ \frac{0.11 \text{ V}}{5 \text{ V}} = 2.2\% \qquad (3.22)$$

The regulation has been compromised from a perfect 0 to more than 2% just by connecting some wire from the power supply to the load. Compare this to the 0.1% load regulation claimed by the manufacturer in Figure 3.6. Most products must avoid this significant problem by providing sense connections to the load. Sense connections allow the power supply to react to the voltage at the point of end use, rather than just at the power supply terminals. Figure 3.8 shows both the problem circuit and the sense scheme used to correct it.

For inverters and other equipment with ac output, regulation can be specified for both the amplitude and the frequency. In many applications, the ac frequency is determined directly by switch action, and is not subject to variation with line or load. The amplitude, on the other hand, clearly is subject to voltage drops and connection effects.

The concept of regulation extends to a wide range of similar systems. In a motor controller, for instance, the speed regulation could be specified as a percent change between the speed of an unloaded motor and the speed of a motor carrying rated shaft load. In general, a *regulator* is any control system that tries to maintain a fixed output independent of possible disturbances. The more general case, in which a control system tries to follow a specific time-varying reference signal, is called a *tracking problem*.

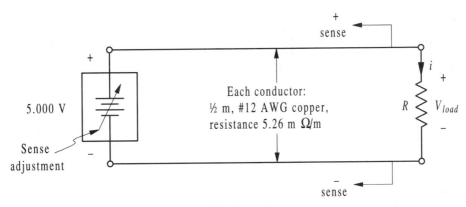

Figure 3.8 Load connection through a small resistance.

3.5 EQUIVALENT SOURCES

In many kinds of power converters, a source is processed through a switch matrix to create a well-defined, although possibly complicated, waveform. Square waves, triangular waves, rectified sine waves, and sine waves with jumps are more common than pure sine waves in a converter. Consider the action of a simple inverter circuit, shown in Figure 3.9. The switch action imposes a known square wave on the load.

When a converter produces a well-defined waveform such as this square wave, it is apparent that the load cannot distinguish it from a true square wave source. We should be able to represent the converter and its input by substituting a square waveform. This concept, called the *equivalent source concept*, is useful for many direct converters.

> **Definition:** An equivalent source *is a true voltage or current source, often nonsinusoidal, that represents the combined action of an actual source and a set of switches.*

Let us consider an example.

Example 3.5.1   A rectifier bridge, shown in Figure 3.10, supplies a series R-L load. Represent the results with an equivalent source, then find the steady-state current waveform, and find the peak-to-peak current ripple.

The rectifier bridge imposes the signal $|V_0 \cos(\omega t)|$ on the R-L circuit. Since there is no doubt about this, and since the values of R and L do not alter the diode action, we can treat $|V_0 \cos(\omega t)|$ as a nonsinusoidal voltage source. There are several ways to analyze the resulting simple network. Perhaps we might solve the differential equation

$$|V_0 \cos(\omega t)| = L\frac{di}{dt} + iR \tag{3.23}$$

Laplace transforms could be used to do this, if desired. Or we could split the source into a series combination of sources, based on the Fourier Series. This gives a set of equivalent

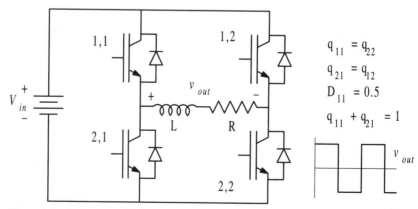

Figure 3.9   Square-wave inverter.

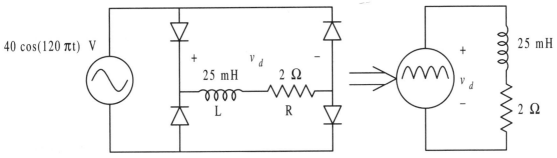

Figure 3.10 Diode bridge with *R-L* load.

sources, as in Figure 3.11. The equivalent sources provide a very strong advantage: The new circuits are linear, and avoid the nonlinearity and complication of switches. We can use superposition, Laplace transforms, or other techniques from linear network analysis to solve for *i(t)*.

Based on superposition, let us solve term-by-term for the Fourier Series of the current in this converter. Table 3.1 shows the phasor RMS voltages $c_n/\sqrt{2} \angle \theta_n$ for the various frequency terms, along with the impedance $R + j\omega L$ at each component frequency. The Fourier Series of the voltage was found in Chapter 2 to be

$$|V_0 \cos(\omega_{in}t)| = \frac{2V_0}{\pi} + \frac{4V_0}{\pi} \sum_{n=1}^{\infty} \frac{\cos(n\pi)}{1 - 4n^2} \cos(2n\omega_{in}t) \tag{3.24}$$

The current for each component frequency is *V/Z* for that frequency. The phasor voltage and current values are given in the table for the first few terms of the series. The current components drop off very quickly, since the voltage drops off roughly as the square of the frequency, while the impedance rises as the frequency. The phasor values can be multiplied by $\sqrt{2}$ and then used to write the current's Fourier Series,

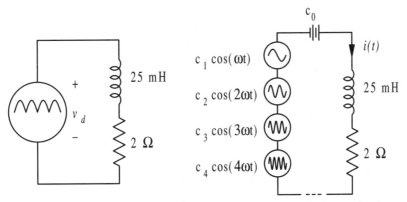

Figure 3.11 Equivalent source and Fourier Series equivalent applied to *R-L* load.

TABLE 3.1
Component-by-component Current Computation

Frequency (Hz)	Phasor Voltage (RMS)	Impedance R + jωL (Ω)	Phasor Current (RMS)
0	25.5	2	12.73
120	12.004∠0°	18.96∠83.94°	0.6333∠−83.94°
240	2.401∠180°	37.75∠86.96°	0.0636∠93.04°
360	1.029∠0°	56.58∠87.97°	0.0182∠−87.97°
480	0.572∠180°	75.42∠88.48°	0.0076∠91.52°
600	0.364∠0°	94.27∠88.78°	0.0039∠−88.78°
720	0.252∠180°	113.12∠88.99°	0.0022∠91.01°
840	0.185∠0°	131.96∠89.13°	0.0014∠−89.13°
960	0.141∠180°	150.81∠89.24°	0.0009∠90.76°

$$i(t) = 12.73 + 0.896\cos(2\omega_{in}t - 83.94°) + 0.0899\cos(4\omega_{in}t + 93.04°)$$
$$+ 0.0257\cos(6\omega_{in}t - 87.97°) + \cdots \tag{3.25}$$

The first eight Fourier harmonics give the waveforms shown in Figure 3.12. The voltage shown there is very close to the actual equivalent source. The current is even closer to the true result, since its components drop off more quickly. The waveform can be used to estimate the current ripple. Alternatively, we can use the large first harmonic as a ripple estimate. The first harmonic current of $0.6333\angle-83.94°$ A$_{RMS}$ implies peak-to-peak ripple of 1.791 A. The actual peak-to-peak ripple (obtained by solving the differential equation) is 1.778 A. The error is only 0.7%.

Figure 3.12   The voltage and current waveforms for Example 3.5.1 (based on eight Fourier components).

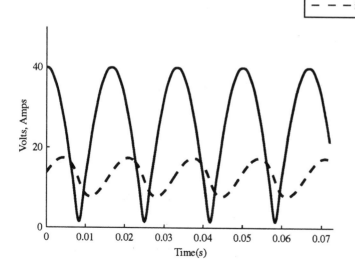

Consider an inverter circuit, such as that in Figure 3.13. If the switch action is well defined, the load is exposed to a known waveform, such as a square wave. This waveform can be treated as a source. When that step is taken, it is easy to determine how the load is affected by changes in switch action, or how the output current will change with different load choices.

3.6 INTRODUCTION TO POWER FILTERING

The source conversion concept, as well as the information about distortion and regulation, will help us define requirements for filters in power electronic circuits. This represents the third major design issue: the interface problem. Energy storage elements in a power converter normally are selected to provide the basic source characteristics desired at a particular location in a circuit. For instance, if dc current source characteristics are needed at an output port, series inductance would be added at that location. An inductor tends to maintain fixed current, and provides at least partial current source behavior. When dc voltage source characteristics are desired, parallel capacitance would be the choice, based on the capacitor's tendency to maintain fixed voltage.

In Figure 3.13, the input is a voltage source, so the output should resemble a current source. Therefore, we would expect to encounter series inductance at the load. Indeed, the circuit does not function effectively if the load is capacitive. You might review prior examples to see the choices of capacitance and inductance. When simple R-L or R-C filters are encountered in power stages, their function is nearly always consistent with the expected source properties.

When ac sources are appropriate, more complicated arrangements might be encountered. The underlying objective is to leave the wanted component intact while attenuating the unwanted components. When the wanted component is dc, the filter should show a low-pass characteristic with the proper source type. When the wanted component is ac, a band-pass filter is needed. Since only one wanted component is likely to be present, we can sometimes use resonant techniques to construct a filter. Figure 3.14 shows the inverter with a

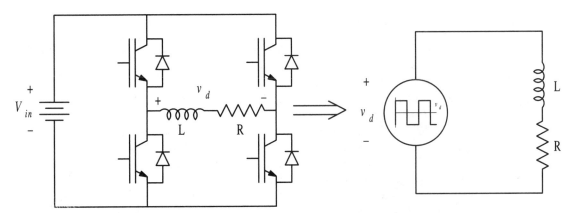

Figure 3.13   Voltage-sourced inverter and an equivalent source approach.

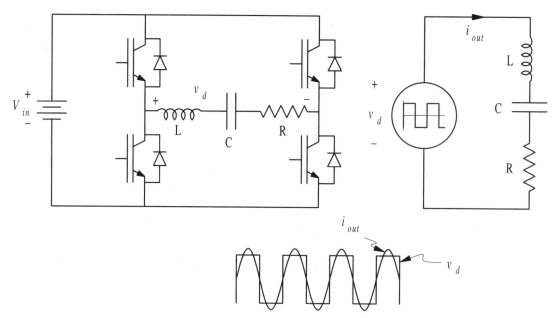

Figure 3.14   Inverter with series-resonant load.

resonant L-C-R load. Given a wanted radian frequency ω_{wanted}, notice that the choice $\omega_{\text{wanted}} = 1/\sqrt{(LC)}$ will pass the wanted frequency perfectly, while unwanted components will be attenuated.

The equivalent source concept is always extremely helpful for power filter design. Let us reexamine the circuit of Example 3.5.1 and actually develop a design process for a filter.

Example 3.6.1   A full bridge diode set, with a 60 Hz input at 40 V peak, supplies a resistive load. At full load, the output power is 325 W. The customer would like to minimize the output ripple and calls for a specification of $\pm 1\%$ of nominal output. Select and design an appropriate single-element power filter for this application. The circuit is shown in Figure 3.15.

Since the input is an ac voltage source, source conversion suggests that the output should resemble a current source. Only the dc component is wanted, so a low-pass filter is needed. It should be inductive to show current source action. From the earlier example, we saw that the first harmonic provides an excellent estimate of the ripple. The voltage waveform across the L-R pair will be the basic full-wave signal under any choice of R and L, so we can treat it as an equivalent voltage source. As in Example 3.5.1, the dc value is 25.46 V, and the first harmonic is $12.00 \angle 0°$ V_{RMS}. In a dc equivalent circuit, the inductor is represented by a short circuit. The output power is 325 W $= V^2/R$, so the resistance must be 2.00 Ω. The nominal current is $(25.46 \text{ V})/(2 \ \Omega) = 12.73$ A. The ripple must not deviate outside of the range 12.60 to 12.86 A (i.e., $\pm 1\%$). This requires the peak value of the first harmonic of current to be less than 1%, or 0.127 A, meaning an RMS value less than 0.090 A. The impedance to the first harmonic therefore must be at least 133 Ω, which is 12 V divided by 0.090 A. This impedance is far larger than the resistance, so the inductance will need to

Figure 3.15 Diode bridge with 800 W dc load.

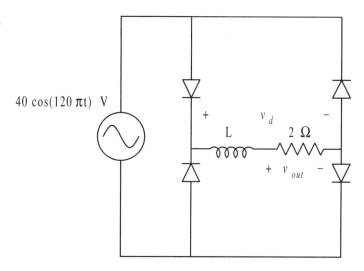

account for virtually all of this impedance. The first harmonic frequency is 120 Hz. If the reactance ωL is to be larger than 133 Ω at 120 Hz, $L > 176$ mH. The filter will require a 176 mH inductor to pass the desired dc output while attenuating the unwanted components below $\pm 1\%$ of nominal output.

The tendency in power line frequency circuits is toward very large storage devices. This is the only way to provide a high-quality interface with simple L-R or R-C circuits. When a free choice is possible, the trend is toward much higher frequencies. In Example 3.6.1, if the input had been 60 kHz instead of 60 Hz, only 176 μH would have been needed to meet the same specifications. In dc–dc supplies, in which the frequency is not constrained by power matching requirements, modern designs operate at 100 kHz and up. This is one key to miniaturization: Energy storage requirements drop as switching frequencies increase.

3.7 POWER FILTER EXAMPLES

Since a power filter is normally intended to serve a source interface function, this can be used to advantage to simplify computations in many cases. The individual elements can be checked separately, since each is supposed to maintain a constant source value. In this section, several examples of various types of converters will be studied to provide some practice with power filters.

Example 3.7.1. A buck-boost (polarity reversal) dc–dc converter is shown in Figure 3.16. The switching frequency is 100 kHz, and each switch is on 50% of a cycle. Find values of L and C to keep the inductor current ripple below $\pm 10\%$ of nominal, and the output voltage ripple below $\pm 1\%$ of nominal.

First, notice that this circuit performs voltage-to-current-to-voltage conversion. The inductor is a transfer source, and provides current source behavior for this indirect converter. From a Chapter 1 example, this circuit is known to produce an output of -12 V, so the ca-

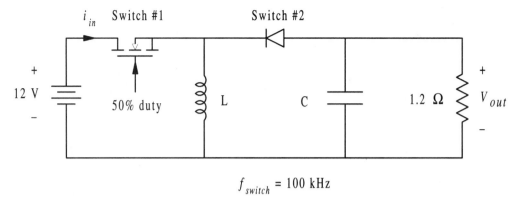

$$f_{switch} = 100 \text{ kHz}$$

Figure 3.16 Buck-boost converter for Example 3.6.2.

pacitor voltage has a nominal value of -12 V. The output load is 120 W. Since there are no losses possible in this ideal converter, the input power must be 120 W. We can use this information to find out the nominal inductor current, as follows:

- The current drawn from the input source is $q_1 I_L$. This means that the inductor current flows in the input source when switch 1 is on, and there is no flow when it is off.
- Only the dc component of this current contributes to average power, as was demonstrated in Chapter 2.
- The dc component of i_{in} must be $\frac{1}{2}I_L$. This is the average of the function $q_1 I_L$. The dc value must be 10 A to give $P_{in} = 120$ W.
- Therefore, $I_L = 20$ A here.

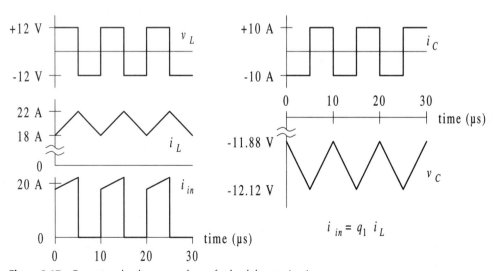

Figure 3.17 Current and voltage waveforms for buck-boost circuit.

Notice that this discussion takes advantage of the fact that I_L is nearly constant to determine that its nominal value is 20 A.

Now, let us find the component values. We are interested in the changes in I_L and V_C. The inductor current is a minimum just when switch 1 turns on, and is a maximum just as it is turned off. The voltage $v_L = L\, di/dt$ is $+12$ V while switch 1 is on, so the current changes linearly during the 5 μs on-time of switch 1. Thus $di/dt = \Delta i/\Delta t$, with $\Delta t = 5$ μs. The current should not change by more than $\pm 10\%$. This means it must not go lower than 18 A, or higher than 22 A. Therefore, the change must be less than 4 A, and

$$12 \text{ V} = L\frac{\Delta i}{\Delta t}, \qquad \frac{12 \text{ V} \times 5 \text{ } \mu\text{s}}{L} = \Delta i < 4 \text{ A} \qquad (3.26)$$

This requires $L > 15$ μH.

In the case of the capacitor, there is an R-C exponential decay while switch 1 is on. However, we need not make the problem even this difficult. The purpose of the capacitor is to keep the output voltage tightly fixed, so that the resistor current remains close to 10 A. When this is true, the current in the expression $i_C = C\, dv/dt$ will cause the voltage to change linearly, with $dv/dt = \Delta v/\Delta t$. The voltage should not change more than 0.24 V (i.e., $\pm 1\%$ of 12 V) during the 5 μs time when switch 1 is on. Thus

$$10 \text{ A} = C\frac{\Delta v}{\Delta t}, \qquad \frac{10 \text{ A} \times 5 \text{ } \mu\text{s}}{C} = \Delta v < 0.24 \text{ V} \qquad (3.27)$$

This requires $C > 208$ μF. The key waveforms appear in Figure 3.17.

Example 3.7.2   A 1 mH inductor is used in the converter shown in Figure 3.18. What is the output ripple voltage at full load?

Figure 3.18 Buck converter for Example 3.7.2.

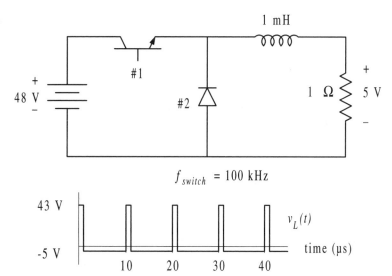

Since the output voltage and resistance are known, and the output current is very close to being constant, the output current must be 5 A. When switch 1 is on, the inductor tries to maintain a fixed 5 A value, but instead the current will rise slightly since the inductor is not infinite. The inductor voltage is $V_{in} - V_{out} = 43$ V during the switch 1 on-time. The inductor current change will be

$$v_L = L\frac{di}{dt}, \qquad (43\text{ V})\frac{\Delta t}{L} = \Delta i \qquad\qquad (3.28)$$

The duration of the switch 1 on-time will give Δt. This duration is the duty ratio 5/48 times the period 10 μs (see if you can determine why this is true). The change is therefore $\Delta i = 44.8$ mA.

Example 3.7.3   The dc–dc converter in Figure 3.19 uses an inductor to provide the necessary current-source characteristic for the switch matrix, and uses a capacitor to help ensure that the user will see voltage source behavior at the output terminals. Notice that these two can be traded off. A large capacitor, for instance, means that more inductor current ripple can be tolerated. Select values based on ±50% inductor current variation, and $\pm\frac{1}{2}$% output voltage variation. Test the result with SPICE simulation or a similar method.

Solved together, this is a fairly complicated *R-L-C* circuit analysis problem. However, there are some key simplifications: Since the output voltage is intended to be fixed, we can treat the capacitor as a fixed source for the purposes of choosing the inductor, and then can continue on.

If the output is treated as a 5 V source, its average current must be 20 A to give 100 W output. Therefore the inductor current is 20 A on average and will vary between 10 A and 30 A. While switch 1 is on, the inductor voltage is 7 V, and the current increases by 20 A. It is easy to show that switch 1 is on 5/12 of each period, or 2.08 μs. The inductor can be found as follows:

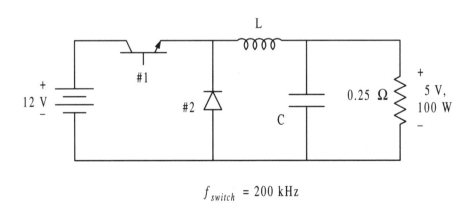

$$f_{switch} = 200\text{ kHz}$$

$$\Delta i_L = \pm 50\% \qquad \Delta v_C = \pm\frac{1}{2}\%$$

Figure 3.19   dc–dc buck converter for Example 3.7.3

$$7 \text{ V} = L\frac{\Delta i}{\Delta t}, \qquad \Delta i = 20 \text{ A}, \qquad L = \frac{7 \text{ V} \times 2.08 \text{ } \mu s}{20 \text{ A}} = 0.73 \text{ } \mu\text{H} \tag{3.29}$$

For the capacitor, we can now treat the inductor as a triangular equivalent source. The resistor current is supposed to be constant, so why not treat it as an equivalent source as well? An equivalent circuit based on these sources is given in Figure 3.20.

Given that the capacitor current follows directly from the figure, it is easy enough to find the change Δv_C over time. The current is not constant, so we cannot set dv/dt to the slope $\Delta v/\Delta t$, but i_C is a simple triangular function of time. The voltage increases whenever i_C is positive. From Figure 3.20, the current becomes positive while switch 1 is on, at time $t_0 = 1.04 \text{ } \mu s$. The current rises 10 A in 1.04 μs, which gives a slope of 9.6×10^6 A/s. The slope while switch 1 is off can be computed as -6.86×10^6 A/s. The voltage v_C will reach its minimum value at time t_0 (just after the current has been negative for a half cycle), then will climb until t_1, when the current again becomes negative. The net change $\Delta v_C = v(t_1) - v(t_0)$. Let us integrate the current during its positive half-cycle to find $v_C(t_1)$:

$$v_c(t_1) = v_c(t_0) + \frac{1}{C}\int_{t_0}^{2.08 \text{ } \mu s} 9.6 \times 10^6(t - t_0) \, dt +$$
$$\frac{1}{C}\int_{2.08 \text{ } \mu s}^{t_1} -6.86 \times 10^6(t - t_1) \, dt \tag{3.30}$$

[Of course, even doing an integral here is overkill. All we need is the area under the triangle, $\frac{1}{2}bh = \frac{1}{2}(2.5 \text{ } \mu s)(10 \text{ A}) = 12.5 \times 10^{-6}$.] The value $v(t_1) - v(t_0)$ is therefore $12.5 \times 10^{-6}/C$. To keep the ripple below $\pm\frac{1}{2}\%$, the voltage must not change by more than 0.05 V. This requires $C = 250 \text{ } \mu\text{F}$. Figure 3.21 shows a SPICE simulation of the complete circuit.

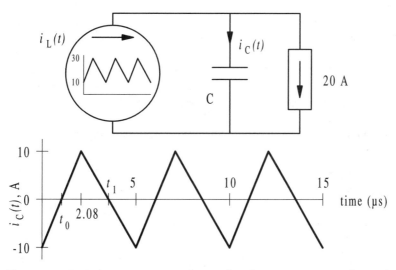

Figure 3.20 Equivalent current sources imposed on the output capacitor of Example 3.7.3.

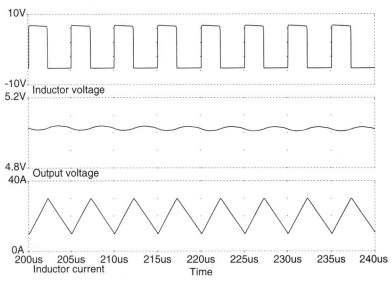

Figure 3.21 SPICE simulation of i_L and v_C for 12 V to 5 V converter, with P_{out} = 50 W, L = 0.73 μH, C = 250 μF.

3.8 POWER FACTOR

Distorted waveforms and requirements for substantial filters mean that much of the energy moving around in a power electronic circuit is not used for actual energy transfer. Unwanted components are present and create unwanted current flows in a circuit. These unwanted flows can be redirected by filtering so that they do not have a large effect at inputs or output. The flows are still present, however. They cause power loss in resistors and will require increased current ratings of switches and other devices. Distortion is an important measure of waveform quality, but we often need a representation of energy issues as well.

If a waveform has no distortion, only the wanted component is present, and this component is fully available for energy transfer. At a given location in a circuit, the power flow can be as high as $V_{RMS}I_{RMS}$. This would represent a "best case" in which no extra harmonics are present, and in which the phase angle between the current and voltage is exactly zero.

The ratio between the power flow P and the maximum possible flow $V_{RMS}I_{RMS}$ is an important figure of merit for a circuit. Define the following terms:

Definition: *The* apparent power, *symbol* S, *is the product of* V_{RMS} *and* I_{RMS} *at a given port in a network. Units of volt-amperes (VA) are used for* S *to distinguish it from an actual power flow.*

Definition: *The* power factor, *symbol* pf, *is the ratio* P/S *at a given network port. The power factor has no units and cannot exceed 1.*

Whenever $pf < 1$, the circuit carries excess currents or voltages that do not perform useful work. Utilities and large manufacturing plants are particularly concerned about power factor. For example, if the power factor drawn by a given load is 0.5, the supply system must provide twice as much current as is really necessary. Losses in the supply system are proportional to the square of the current, so supply losses increase by a factor of four if power factor drops from 1 to 0.5. This has substantial effects on the efficiency of an electric utility system. Most electric companies charge industrial customers a penalty fee if the customer draws power such that $pf < 0.8$.

Power factor in a steady-state ac network is the cosine of the phase angle between the voltage and current. In a power converter, the unwanted components also contribute to this effect. These two factors are often considered separately.

> **Definition:** *The* displacement factor, *symbol* df, *is the cosine of the phase angle between the wanted voltage and current components, given by $cos(\theta_{v(wanted)} - \theta_{i(wanted)})$. In a single-frequency sinusoidal network, displacement factor is the cosine of the phase shift between voltage and current, and power factor and displacement factor are identical.*

> **Definition:** *The* distortion factor, *symbol* k_d, *represents the effects of unwanted harmonics on the power factor. There is no consensus on a definition, but we will use* $pf = k_d df$, *or* $k_d = pf/df$, *which is one commonly used version of the distortion factor.*

Example 3.8.1   A square wave of current at 60 Hz, with amplitude 10 A, is drawn from a 120 V_{RMS}, 60 Hz voltage source by a rectifier. The square wave and the sinusoid are in phase. Find *pf*, *df*, and k_d at the input voltage source for this situation. The waveforms are shown in Figure 3.22. A circuit with this behavior is shown in Figure 3.23.

Figure 3.22   A square current waveform is not unusual for a rectifier. This presents less than ideal power factor to the input ac supply.

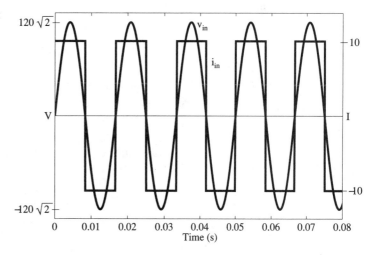

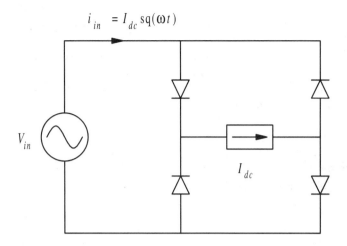

$$i_{in} = I_{dc}\,\mathrm{sq}(\omega t)$$

V_{in}

I_{dc}

Figure 3.23   Rectifier that draws a square wave input current.

The RMS voltage is 120 V here, while the RMS current is 10 A. Therefore, $S = 1200$ VA. The instantaneous power vi is the waveform in Figure 3.24, which has an average value of 1080 W. The power factor seen by the source is $pf = 0.90$. The displacement factor is 1, since there is no phase shift between the sinusoid and the wanted (60 Hz) component of the current. The distortion factor $k_d = 0.90$.

The noncontributing current flow is associated with reactive power. We take the definition from that in a single-frequency network. There is still some disagreement among experts as to the most useful definition in the context of power electronics.

> ***Definition:*** *Reactive power, symbol* Q, *is defined as the root-square difference* $\sqrt{S^2 - P^2}$. *It serves as a measure of flow in a network caused by unwanted harmonics and by interchange among energy storage elements. Reactive power contributes to internal losses, but does not contribute to net energy transport through a circuit over time.*

The goal of eliminating unwanted harmonics is similar to a goal of eliminating reactive power. Since distortion is unavoidable in a power electronic circuit, there will always be some reactive power flow. There are two major reasons to keep the power factor as close to 1 as possible:

1. System efficiency is highest when the power factor is close to 1. Low power factor means excess flows in a circuit. This generates excess i^2R loss.
2. Source effects. If power factor is low, the source will not be able to provide the maximum possible energy. Consider a conventional 120 V, 15 A wall plug. A device with $pf = 0.5$ can draw only 900 W from this plug. Any higher power will draw $I_{RMS} > 15$ A. On the other hand, a device with $pf = 1$ can draw 1800 W from the wall plug.

The source effects are beginning to be important for modern power supplies. Electric vehicles, for instance, need all the power they can obtain from the wall connection to charge quickly.

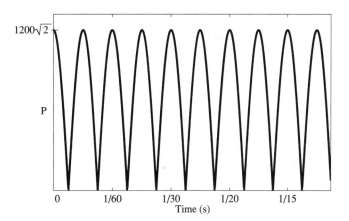

Figure 3.24 Instantaneous power p(t) for Example 3.8.1.

3.9 RECAP

The laws of circuit theory combine with practical considerations to define some of the important design aspects of power electronic systems. Most users seek an ideal source, so a typical system performs *source conversion*—controlling energy flow among near-ideal sources. In indirect converters, *transfer sources* provide a way to connect to any desired output.

Given that power converters create output waveforms by chopping an input signal, distortion cannot be eliminated entirely. Any real converter will have both wanted and unwanted Fourier components, and interfaces to maximize the wanted effects while minimizing unwanted harmonics will be necessary. The *total harmonic distortion* (THD),

$$\text{THD} = \sqrt{\frac{\sum_{n=2}^{\infty} c_n^2}{c_1^2}} \tag{3.32}$$

is a convenient measure of the distortion effect when the fundamental component is wanted.

Harmonics and ripple represent dynamic variation of output or input waveforms. Static variations caused by load changes, source changes, temperature, or other effects define the regulation properties of a converter. A power electronic system has *perfect regulation* (with conventional regulation measures equal to zero) if the output does not vary under any circumstances. It provides no regulation if variations such as the input line alter the output in a 1 : 1 correspondence. Power supply manufacturers commonly specify several regulation effects when designing and selling products. However, load regulation is dependent on the user's connections. Attention to detail is necessary to get full performance out of even a perfect power supply.

Often, the waveforms imposed on a circuit, filter or source are known and periodic. They might not be conventional dc or sinusoidal signals, but they are still well defined. This leads to the concept of an *equivalent source*. Such a source can be substituted for known switching action. When this is done, standard tools of linear analysis, such as superposition, can be used to understand the behavior of a system. The equivalent source approach is particularly helpful when designing or analyzing power filters. A power filter is an energy storage element used as an interface to create a near-ideal source.

The *power factor* provides a measure of the fraction of energy flow that actually performs useful work. Many power electronic systems display very low power factor, primarily because of the distortion they produce. Low power factor generates excess system loss and limits the energy that can be obtained from a source.

PROBLEMS

1. A full-wave diode rectifier is used to create a 60 Hz to 120 Hz converter. The dc component is removed by filtering. What is the THD of the output voltage of this converter?

2. The waveform of Figure 3.25 is generated by a certain type of inverter. What is its THD?

3. Consider the waveform of Figure 3.25 as an equivalent source. It represents an inverter driving a load, and has peak value of 100 V and average value 0 with frequency of 60 Hz. The load is a 10 Ω resistor in series with an inductor and capacitor. The *L-C* pair is resonant at 60 Hz, with $C = 220~\mu F$. What is the amplitude of the current flow in this load?

4. In the circuit of Example 3.7.1, the input source will have some limitations because of its internal resistance. The circuit is shown in Figure 3.26.

   a. If the input source resistance is 0.2 Ω, what is the output for 50% duty ratio, assuming the input capacitor C_{in} is not present?

   b. The input capacitor is to be added to help filter out input current variations. What should its value be to keep the input current variation below ±10%?

5. What is the THD of a switching function with duty ratio D, assuming that the dc component is removed by filtering?

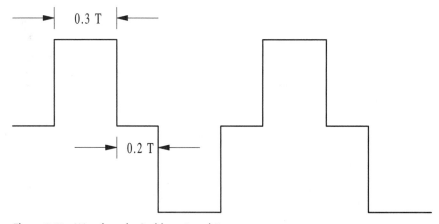

Figure 3.25 Waveform for Problems 2 and 3.

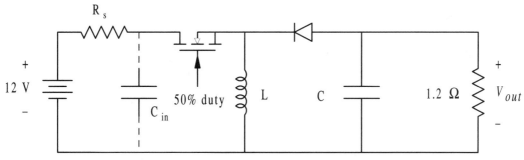

Figure 3.26   Nonideal buck-boost converter for Problem 4.

6. The circuit shown in Figure 3.27 is a type of indirect converter. Two of the energy storage elements have been blocked out. Determine whether these elements are capacitors or inductors. How can you tell?

7. Compare the following ideas for an ac voltage conversion application. The loads are purely resistive, and a switch matrix is imposed between an ac input voltage source and the load.

   • The first idea tries to adjust ac energy flow by chopping out portions of the input waveform near the zero crossings. Specifically, the switches operate to disconnect the load when the input voltage is within an angle β of the zero crossing.

   • The second idea tries to adjust ac energy flow by chopping out portions of the input waveform near the peaks. The switches disconnect the load when the voltage is within an angle γ of the peaks.

   To compare these:

   a. Plot sample waveforms for $\beta = 30°$ and for $\gamma = 30°$.

   b. Compute and plot the RMS values of load current and current THD as functions of β and γ, respectively.

   c. Compute the power factor seen by the input source with these two converters.

   d. Discuss your results.

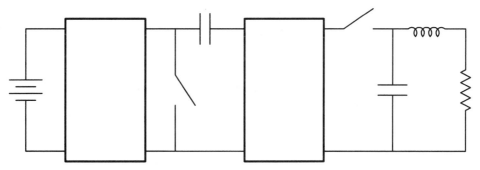

Figure 3.27   Indirect converter for Problem 6.

8. What are the THD and THR of the waveform $f(t) = \cos(\omega t) + \frac{1}{3}\cos(3\omega t) + (1/5)\cos(5\omega t)$?

9. The converter data in Figure 3.6 represents a realistic set of product capabilities. A sample set of these products is being checked for compliance with the specification. For each of the following input test scenarios, determine whether the results are within the limits set out in Figure 3.6.

   a. Load: 0.75 Ω. When the input is 125 V at 60 Hz, the output is 15.471 V. When the input is 88 V at 60 Hz, the output is 15.389 V.

   b. Load: 1.0 Ω. When the input is 130 V at 400 Hz, the output is 14.934 V. When the input is 95 V at 50 Hz, the output is 14.921 V.

   c. Load: 12.0 Ω. When the input is 124 V at 60 Hz, the output is 15.784 V. When the input is 87 V at 60 Hz, the output is 15.776 V.

10. A group of converters is advertised to meet the requirements listed in Figure 3.6. A load test is performed to check compliance. Do the tests listed fall within the allowed limits? Each was conducted with an input voltage of 120 V $\pm 1\%$ at 60 Hz.

   a. V_{out} = 15.619 V with a 30 Ω load, and V_{out} = 15.602 V with a 0.75 Ω load.

   b. V_{out} = 14.702 V with a 30 Ω load, and V_{out} = 14.700 V with a 0.75 Ω load.

   c. V_{out} = 15.011 V with a 30 Ω load, and V_{out} = 14.997 V with a 0.75 Ω load.

11. A converter claimed to meet the specifications of Figure 3.6 is plugged in and subjected to various tests over a 24-hour interval. The voltage is trimmed initially to 15.000 V. At the end of the sequence, a "final measurement" is made. If the result is 15.037 V, is this consistent with the specification? What range of final values is allowed?

12. Consider the converter data in Figure 3.6. A sample of this product is carefully adjusted to provide exactly 15 V into 50% rated load. The temperature begins to rise, partly because of internal heating. When the temperature reaches 45°C, the load current is raised to the maximum permissible value. Under worst case conditions, what output voltage would be observed after the load change?

13. A 12 V power supply is tested, and found to have average output of 12.15 V with no load and 11.98 V with rated load. What is the load regulation, expressed in conventional terms?

14. A dc–dc converter produces a 5 V supply with 2% line regulation. The output is connected to the load without sense lines, through an 0.004 Ω resistor. What is the expected total regulation (line plus load) over the full allowed input range and an output range of 10 W to 100 W?

15. Load regulation can be a serious problem in automobiles. In a certain type of car, a microprocessor in the extreme rear controls critical functions, such as the antilock brakes. The processor requires 12 V $\pm$ 2% to maintain its function. Two wires run the length of the car to supply it from a 12 V controller under the hood. The processor can draw power anywhere between 5 μW and 5 W. How low must the wire resistance be to allow sufficient load regulation?

16. Line regulation is a serious issue in international applications. The single-phase source voltage worldwide varies at least over the range 85 V_{RMS} to 250 V_{RMS}. A 5 V supply has been prepared for international applications. Over the range 95 V_{RMS} to 125 V_{RMS}, this supply output varies by

0.3%. If the variation increases proportionally as the input variation increases, what is the change in output voltage over the full input range?

17. A power electronic circuit imposes a voltage pulse train with 10% duty ratio, 10 kHz frequency, and 500 V amplitude on a series combination of 1.5 mH and 6 Ω. What is the peak-to-peak inductor current variation? If a large capacitor is added in parallel with the resistor, what is the peak-to-peak inductor current variation?

18. The output capacitor in a certain resonant converter sees net current equal to $40 \cos(50{,}000\pi t)$ A. What value of capacitance will be needed to keep the voltage variation below 5 V peak to peak?

19. The output voltage at a conventional wall plug has unusual distortion: As the voltage increases above half its peak value, in either a positive or negative direction, a square pulse 20 V high and 100 μs wide appears to be added on. The waveform without the pulse is $340 \cos(100\pi t)$ V. What are the THD and THR of this voltage waveform?

20. Some workers use power factor for dc circuits as well as ac circuits. They base the numbers directly on the definition. Consider the converter of Example 3.7.2. What is the power factor $P/(V_{RMS}I_{RMS})$ seen by the input supply?

REFERENCES

Computer Products. *Power Supply Engineering Handbook*. Pompano Beach, FL: Computer Products, 1987.

J. G. Kassakian, M. F. Schlecht, and G. C. Verghese. *Principles of Power Electronics*. Reading, MA: Addison-Wesley, 1991.

Lambda Electronics, "Series LFQ data." Melville, NY: Lambda, 1987.

R. D. Schultz, R. A. Smith, *Introduction to Electric Power Engineering*. New York: Harper & Row, 1985.

P. Wood, *Switching Power Converters*. New York: Van Nostrand Reinhold, 1981 (Reprinted by R. Kreiger, 1986).

PART II

CONVERTERS AND APPLICATIONS

CHAPTER 4

Figure 4.1 dc–dc converters are the basis of modern power supplies.

DC–DC CONVERTERS

4.1 THE IMPORTANCE OF DC–DC CONVERSION

A major reason why ac electricity was accepted as the form of choice for the modern electric power system is the magnetic transformer. The inability to conveniently change voltage levels was one of the major drawbacks of Edison's early dc system concept. While the term *transformer* normally implies a magnetic device restricted to ac signals, more strictly it describes a lossless two-port network with a certain ratio, *a*, between the input and output voltages. The *dc transformer* would be a device that, like its ac counterpart, provides lossless transfer of energy between circuits at different voltage or current levels. The only reason that the dc transformer is unfamiliar is that the magnetic technology used in practical ac transformers cannot handle direct current. In power electronics, the dc transformer is realized as a dc–dc converter. Energy is transferred between two dc circuits operating at different voltage and current levels.

dc–dc conversion is very important in electronic circuit applications, and is becoming increasingly important in a much wider range of applications. Many circuits use power at several different voltage levels. It is often more convenient to convert energy from a single source rather than trying to distribute many different supplies. This is especially true in battery operated equipment or in systems with a battery backup.

Modern fixed-output dc power supplies find their way into products ranging from home appliances to industrial controllers. At one time, most dc power supplies took energy from an ac line source, changed its level with a transformer, then rectified the result. Today, most supplies, like those in Figure 4.1, are built with dc–dc converters. The incoming ac is rectified directly with a simple diode circuit, then the high-level dc is converted to the desired levels. Modern power supplies of this type range from 2 V outputs for special logic up to 500 V or more for industrial applications. Many products are designed around 170 V inputs (the peak value of rectified 120 V ac) or 300–400 V inputs (the peak values of 230 V ac, 240 V ac, and many three-phase rectified sources). Others are intended to handle the 48 V level used in many telephone networks. Converters for 48 V to 5 V, for instance, support the logic circuits in modern telephones. Converters for 300 V to 12 V support analog power supply applications, and are also used in many commercial designs for electric automobiles.

As we explore dc–dc converters, we need to examine alternatives, then examine dc–dc switching converters based on the general switch matrix. The basic converter forms lead into a variety of special versions with input–output isolation, extreme conversion ratios, or other applications. Certain types of nonswitching circuits are commonly used for dc regulation, and some of these forms will be considered.

4.2 WHY NOT VOLTAGE DIVIDERS?

Why not just use a voltage divider for dc–dc conversion? After all, voltage dividers are often used within integrated circuits to produce low-level dc voltages. The problems with applying voltage division for power conversion are severe: a voltage divider is inherently inefficient (at best, the efficiency is the ratio V_{out}/V_{in}), and does not provide regulation. A divider cannot provide $V_{out} > V_{in}$ or change polarity.

The divider circuit in Figure 4.2 can be analyzed to find its efficiency and other power conversion properties. It supplies an output load drawing current I_{out} at voltage V_{out}. Circuit analysis shows that

$$\text{if } I_{out} = 0, \quad V_{out} = \frac{R_{out}}{R_{in} + R_{out}} V_{in}$$

$$\text{if } I_{out} \neq 0, \quad V_{out} = \frac{V_{in} - I_{out}R_{in}}{1 + R_{in}/R_{out}}$$

(4.1)

The output depends on both V_{in} and I_{out}. Furthermore, the output value cannot easily be adjusted, since the resistor values must be altered to change it. The efficiency for $I_{out} > 0$ is

$$\eta = \frac{P_{out}}{P_{in}} = \frac{V_{out}I_{out}}{V_{in}I_{in}}$$

$$= \frac{P_{out}}{P_{out} + I^2_{out}R_{in} + \dfrac{V^2_{out}}{R_{out}} + \dfrac{2P_{out}R_{in}}{R_{out}} + \dfrac{V^2_{out}R_{in}}{R^2_{out}}}$$

(4.2)

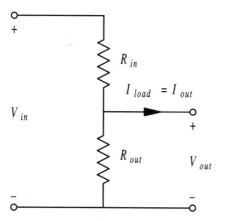

Figure 4.2 Resistive voltage divider as a dc–dc converter.

The value can be very low, especially if a large division ratio is attempted. In Chapter 3, we saw that the divider shows no line regulation in the no-load case (input changes have 100% effect at the output). The line regulation, as measured with a normalized partial derivative, actually gets worse as I_{out} increases. The load regulation measure

$$\frac{\partial V_{out}/V_{out}}{\partial I_{out}/I_{out}}\bigg|_{V_{in}=0} = 1 \tag{4.3}$$

That is, the circuit provides neither load nor line regulation. It is certainly useful as a tool for voltage sensing, but seems inappropriate for power conversion. Consider the following example.

Example 4.2.1 A voltage divider is to be built to give a 5 V output from a 12 V input. The nominal load is 5 W. Design a divider to provide 5 V ± 5% for loads ranging from 3 W to 5 W, assuming that the input line is perfectly regulated at exactly 12 V. What is the no-load output? What is the efficiency at nominal load?

In a divider, the load interacts with the resistor set to create the output. The heavier the load, the lower the output voltage. For a design, let us start with an arrangement that provides $V_{out} = 4.75$ V for a 5 W load and 5.25 V for a 3 W load. Power of 5 W at 4.75 V corresponds to a load resistor, R_{load}, of 4.51 Ω. Power of 3 W at 5.25 V corresponds to 9.19 Ω. The divider design requires

$$\frac{V_{out}}{V_{in}} = \frac{R_{out}\|R_{load}}{R_{in} + R_{out}\|R_{load}} \tag{4.4}$$

For the two loads, two equations can be developed in the unknowns R_{in} and R_{out}, then solved simultaneously. The result for 4.75 V to 5.25 V output is

$$R_{in} = 2.13 \ \Omega, \qquad R_{out} = 2.03 \ \Omega \tag{4.5}$$

For the 5 W load, this provides 4.75 V and draws 3.40 A from the 12 V source. The efficiency is 12.3%. For the 3 W load, the divider draws 3.16 A, and has efficiency of only 7.9%. While there are other (smaller) values of R_{in} and R_{out} that will meet the requirements, they will give even lower efficiencies, so these values are a "best case" choice. The no-load output will be 5.84 V.

One might argue that this converter has good load regulation, since the output varies only ±5% over the load range. However, this has come at the expense of low efficiency. In essence, the divider has been constructed to draw a much bigger current than is needed for the load, so that the load variation is swamped out by the large current draw of the divider itself. This divider dissipates more than 34 W even with no load. Since $P_{in} \gg P_{out}$, a small change in P_{out} will not alter the operation very much.

Dividers are used primarily for extremely low power levels, such as for creating multiple reference voltage levels for analog-to-digital converters. They can be applied when efficiency is not an issue.

4.3 LINEAR METHODS AND DIRECT DC–DC CONVERTERS

Switching converters for dc–dc conversion can transfer energy among differing dc voltage levels with little loss. Sometimes called *choppers*, these converters offer a way to build dc transformers. In Chapter 2, the Fourier Series results for a switching function led to duty ratio adjustment and timing adjustment as the alternatives for controlling the action of a power conversion circuit. Since the duty ratio D represents the dc component of a switching function, it will be a determining factor in the operation of dc–dc converters. Thus duty-ratio control, or *pulse width modulation*, is an important aspect of dc–dc power electronics. The switching frequency can be chosen somewhat arbitrarily in most kinds of circuits, and the duty ratio of each switch is then set to that value necessary for the desired conversion function. The Fourier Series of dc–dc converter outputs are normally scaled versions of the switching function Fourier Series, so that adjustment of the switching function average value translates immediately into an output change.

4.3.1 Linear Regulators

Several nonswitching concepts for dc transformers have been tried over the years, but it is only relatively recently that efficient, low-noise, cost-effective dc transformation has become possible. Consider one popular type of circuit, the linear regulator, the two major types of which are shown in Figure 4.3. The simple series regulator in Figure 4.3a is a transistor connected as an emitter follower (or source follower for an FET). The transistor operates in its active region rather than as a switch. If the base voltage is set to some particular value, the emitter voltage will be very close to one diode drop below that set point. The emitter voltage V_{out} becomes a function of $V_{control}$ rather than a function of the input voltage or the load

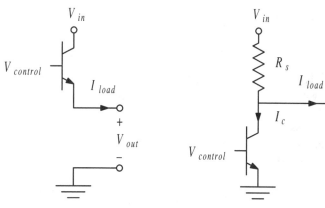

Figure 4.3 Basic circuits for linear regulation. (*a*) Series or pass regulator; (*b*) shunt regulator.

a) Series or pass regulator   b) Shunt regulator

current. This represents a "pure regulation" function: the output is independent of various disturbances. Line and load regulation are ideally perfect for this case. Since $V_{\text{out}} = V_{\text{control}} - V_{\text{be}}$, the output is a linear function of some low-power control voltage, and the term *linear regulator* is appropriate for describing the circuit.

The shunt regulator shown in Figure 4.3*b* resembles a common-emitter amplifier circuit. The transistor once again is used in its active region. The collector current will be βI_b, rather than a function of the load current. The small base current can be adjusted as necessary to produce the required output voltage. Again, the output is a linear function of the control, and the circuit is another example of a linear regulator. Zener diodes are the most common shunt regulation elements.

These linear regulators sacrifice efficiency to obtain regulation. They have the important limitation that $V_{\text{out}} \leq V_{\text{in}}$ in all situations. For a series linear regulator, the power loss and efficiency are

$$P_{\text{loss}} = (V_{\text{in}} - V_{\text{out}})I_{\text{load}}, \qquad \eta = \frac{V_{\text{out}}I_{\text{load}}}{V_{\text{in}}I_{\text{load}}} = \frac{V_{\text{out}}}{V_{\text{in}}} \tag{4.6}$$

so that high efficiency requires a step-down ratio as close as possible to 1. For the shunt regulator, the loss and efficiency are

$$P_{\text{loss}} = V_{\text{out}}I_c + (I_{\text{load}} + I_c)^2 R_s, \qquad \eta = \frac{V_{\text{out}}I_{\text{load}}}{V_{\text{in}}(I_{\text{load}} + I_c)} \tag{4.7}$$

With no load, there is a significant loss because $I_c \neq 0$. In the best case, where $I_{\text{load}} \gg I_c$, the efficiency becomes $V_{\text{out}}/V_{\text{in}}$.

Linear regulators are not really power electronic circuits, since the active device normally must have a power rating not much different from the desired output power. As converters, they have limited efficiency. However, since they bring the possibility of perfect regulation, linear regulators are often used as elements of larger conversion systems. Their function is to filter a converter output to give it a precise value. The regulation properties are used when efficiency has been addressed elsewhere in the system. In the filtering application, a switching converter produces a voltage just slightly above the desired V_{out}. This voltage has significant ripple. A linear regulator then filters this result to give a fixed V_{out}. Since the regulator's input is just slightly higher than V_{out}, the power loss is small.

4.3.2 The Buck Converter

The direct switching converter for dc voltage to dc current is shown in Figure 4.4. KVL and KCL constraints require that one and only one switch can be on in any column at a given time. This can be written $q_{1,1} + q_{2,1} = 1$ and $q_{1,2} + q_{2,2} = 1$. Only four switch combinations avoid violations of KVL or KCL:

Combination	Result
Close 1,1 and 2,2	$V_{\text{OUT}} = V_{\text{IN}}$
Close 2,1 and 2,2	$V_{\text{OUT}} = 0$

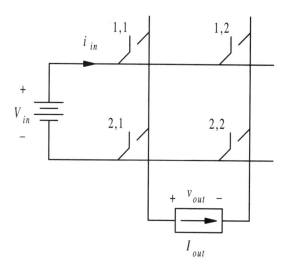

Figure 4.4   dc voltage to dc current direct converter.

Close 1,1 and 1,2 $V_{\text{OUT}} = 0$

Close 1,2 and 2,1 $V_{\text{OUT}} = -V_{\text{IN}}$

There are three possible output voltage values. Switch action selects among these three. Given that the KVL and KCL constraints are met, the output voltage can be written in terms of the switching functions in the compact form

$$
\begin{aligned}
v_{\text{out}}(t) &= q_{1,1}q_{2,2}V_{\text{in}} - q_{1,2}q_{2,1}V_{\text{in}} \\
&= q_{1,1}q_{2,2}V_{\text{in}} - (1 - q_{2,2})(1 - q_{1,1})V_{\text{in}} \qquad (4.8) \\
&= (q_{1,1} + q_{2,2} - 1)V_{\text{in}}
\end{aligned}
$$

This means that if both $q_{1,1}$ and $q_{2,2}$ are on, the output is $+V_{\text{in}}$. If neither $q_{1,1}$ nor $q_{2,2}$ is on, then both $q_{1,2}$ and $q_{2,1}$ must be on, and the output is $-V_{\text{in}}$. Otherwise, $v_{\text{out}} = 0$.

Of interest is the average value of the output, $\langle v_{\text{out}} \rangle$. When equation (4.8) is attacked with an averaging integral, constant terms such as V_{in} can be factored outside the integral, and the computation reduces to averages of the switching functions,

$$
\langle v_{\text{out}} \rangle = \frac{V_{\text{in}}}{T} \int_0^T [q_{1,1}(t) + q_{2,2}(t) - 1] \, dt = (D_{1,1} + D_{2,2} - 1)V_{\text{in}} \qquad (4.9)
$$

The magnitude of this average output will never exceed V_{in}. Since the output is always lower than the input, the name *buck converter* is used, based on older transformer terminology.

In a buck converter, switching action determines the input current as well as the output voltage. The current $i_{\text{in}}(t)$ will be

$$
i_{\text{in}}(t) = (q_{1,1} + q_{2,2} - 1)I_{\text{out}} \qquad (4.10)
$$

with average value

$$
\langle i_{\text{in}} \rangle = (D_{1,1} + D_{2,2} - 1)I_{\text{out}} \qquad (4.11)
$$

TABLE 4.1
Four-switch Buck Converter Transformation Ratio if
$q_{1,1} = q_{2,2}$

Duty Ratio of Switch 1,1	Transformation Ratio a
0	−1
0.25	−0.5
0.50	0
0.75	+0.5
1.00	+1

If a ratio $a = D_{1,1} + D_{2,2} - 1$ is defined, we see that $V_{out}/V_{in} = a$, and $I_{in}/I_{out} = a$. The circuit has the basic characteristics of a transformer, with the restriction that $|V_{out}| \le V_{in}$. One natural operating mode is to use the switches in diagonally opposite pairs, just as occurs in diode bridges. Then $q_{1,1} = q_{2,2}$ and $a = 2D_{1,1} - 1$. The converter average output is a linear function, exemplified in Table 4.1.

The generalized buck converter of Figure 4.4 becomes simpler and a little more practical if the input and output sources share a common ground reference. This can be accomplished by turning switch 2,2 on at all times, and leaving switch 1,2 off. In addition, the output current source typically consists of a large inductance in series with a load. Figure 4.5 shows the arrangement. The switches have been relabeled simply as 1 (the previous 1,1) and 2 (the previous 2,1) for simplicity. Switch 1 must be able to either carry or to prevent the flow of I_{out}, so a transistor is appropriate. Switch 2 must be able to conduct I_{out} and prevent reverse flow, a natural diode application. In practice, the term *buck converter* normally refers to this common-ground arrangement instead of the general case in Figure 4.4.

The circuit of Figure 4.5 requires $q_1 + q_2 = 1$, since a current path must be provided for the inductor whenever the transistor is off. The output is V_{in} when 1 is on and 0 when 2 is on. We can write $v_{out}(t) = q_1 V_{in}$. Thus v_{out} looks just like the switching function q_1, except that its amplitude is V_{in} instead of 1. The average value of the output is

$$\langle v_{out} \rangle = \frac{1}{T} \int_0^T q_1(t) V_{in}\, dt = \frac{V_{in}}{T} \int_0^T q_1(t)\, dt = D_1 V_{in} \tag{4.12}$$

Since the inductor cannot sustain a dc voltage drop in steady state, the resistor voltage average value must match $\langle v_{out} \rangle$. The input average current—the only component that contributes to input power flow—will be $\langle i_{in} \rangle = D_1 I_{out}$. The inductor and load act as a low-pass

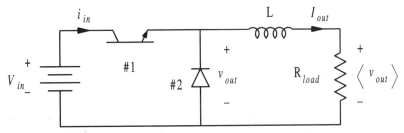

Figure 4.5   dc–dc voltage-current common-ground converter, or *buck converter*.

filter to ensure that dc is passed to the output while the unwanted ac components are attenuated. A typical output waveform appears in Figure 4.6.

The buck converter is used in many switching dc power supplies, in high-performance
dc motor controllers, and in a wide range of electronic circuit applications. This converter
type, along with some closely related circuits, is sometimes called a "buck regulator," a "step-
down" converter, or a "forward" converter. Notice the regulation properties of this circuit.
Since the output voltage computation does not involve the load current or resistance, the relationship $\langle v_{out} \rangle = D_1 V_{in}$ holds as long as the constraint $q_1 + q_2 = 1$ holds. Output is independent of load, so load regulation is perfect. On the other hand, the output is proportional
to V_{in}, so any change in input is reflected proportionally at the output. These regulation properties are exactly the same as those of an ideal transformer: Load regulation is perfect, and
line effects are completely unregulated. Some sort of control will be needed to provide line
regulation.

Some of the important relationships in the buck converter are summarized in Equation
(4.13):

$$v_{out} = q_1 V_{in}, \qquad i_{in} = q_1 I_{out}$$
$$\langle v_{out} \rangle = D_1 V_{in}, \qquad \langle i_{in} \rangle = D_1 I_{out}$$
$$p_{out}(t) = p_{in}(t) = q_1 V_{in} I_{out} \qquad\qquad (4.13)$$
$$\langle p_{out} \rangle = \langle p_{in} \rangle = D_1 V_{in} I_{out}$$
$$v_{out(RMS)} = \sqrt{\frac{1}{T} \int_0^T q_1{}^2(t) V^2{}_{in}\, dt} = V_{in}\sqrt{D_1}$$

No power is lost in either switch, so the input and output power must be equal at all times.
The switches each must carry the inductor current when on and must hold off the input voltage when off, as listed in equation (4.14), in which the subscript 1 refers to switch 1, and so
on.

$$\text{Carrying current:} \quad i_1 = q_1 I_L, \qquad i_2 = q_2 I_L$$
$$\qquad\qquad\qquad\qquad\qquad\qquad\qquad\qquad (4.14)$$
$$\text{Blocking voltage:} \quad v_1 = (1 - q_1) V_{in}, \qquad v_2 = (1 - q_2) V_{in}$$

Let us consider an application example to illustrate some of these relationships.

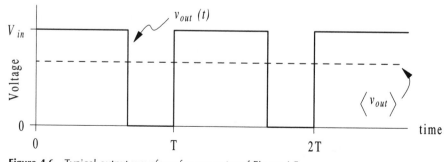

Figure 4.6 Typical output waveform for converter of Figure 4.5.

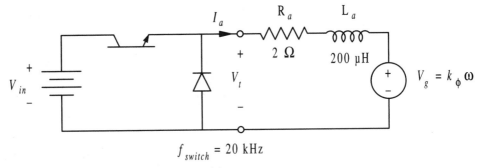

Figure 4.7 Simplified dc motor circuit model.

Example 4.3.1   A buck converter is used with a dc motor as its load. The converter duty ratio helps provide a basis for motor control. A dc motor can be associated with a circuit model, as illustrated in Figure 4.7. Sketch the motor current waveform and comment on the current-source model. What is the effect of duty ratio on the motor speed?

This circuit shows an internal voltage, or *back EMF*, proportional to shaft speed ω with units of rad/s. The circuit is inductive because of the windings and magnetic materials. The voltage applied to the inductor is either $V_{in} - I_aR_a - V_g$ or $-I_aR_a - V_g$, depending on the circuit configuration. In either case, the time-rate-of-change of inductor current is limited, and the inductor current i_L changes very little over short periods of time (such as the time it takes a switch to turn on or turn off). If there is nonzero flow in the inductor, a current path must be provided, according to KCL. The current waveform with $D_1 = 0.5$ is shown in Figure 4.8. At any point in time, the current has a positive and slowly varying value. Even though the waveform is not pure dc, current is always flowing, so a current source has validity as a model.

If the resistance R_a is small, as is the usual case, the voltage V_g should be the average value of V_t. Since V_g is proportional to motor speed, and $\langle V_t \rangle = D_1 V_{in}$, the operating speed is approximately

$$\omega_{rotor} \approx \frac{D_1 V_{in}}{k_\phi} \tag{4.15}$$

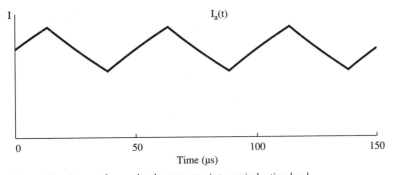

Figure 4.8 Current from a buck converter into an inductive load.

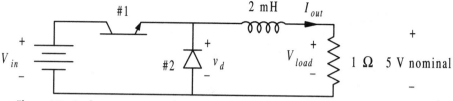

Figure 4.9 Buck converter example.

The duty ratio serves as a direct speed control, with obvious application possibilities. Many very advanced dc drive systems start out with exactly this concept, then add features such as compensation for the $I_a R_a$ voltage drop or corrections for minor speed variations caused by the mechanical load.

The buck converter has ripple in its output voltage, and also in its input current. The output L-R load serves as a filter for the ripple. A conventional buck converter supports only current flow toward its output, and always has an input voltage higher than the output. These conditions are well adapted to restricted switches.

To control the value of $\langle v_{out} \rangle$, the duty cycle of the gate control signal to switch 1 can be altered. If the devices switch quickly and have no loss, a near-perfect buck converter can be implemented. The circuit then implements a dc transformer with a step-down ratio determined by D_1. One limitation of this transformer is that it does not exhibit electrical isolation. A configuration called a *forward converter* inserts a transformer into the buck circuit for isolation. It will be discussed later in this chapter.

Here is a typical buck converter design example. It uses the previous concepts of source interfaces to demonstrate low output ripple.

Example 4.3.2   A buck converter circuit with an R-L load is shown in Figure 4.9. Let $V_{in} = 15$ V. Use a switching frequency of 50 kHz. If the output is to be 5 V nominally, what duty ratio is required? What is the actual output $V_{load}(t)$, including ripple?

Since the circuit is a buck converter, we expect $\langle v_d \rangle = D_1 V_{in}$. The duty ratio of switch 1 should be (5 V)/(15 V) = 1/3. The frequency is 50 kHz, so switch 1 should be on for 6.67 μs, then off for 13.33 μs, on so on. If the output is exactly 5 V, the inductor voltage will be $V_{in} - V_{load} = 10$ V when switch 1 is on, and $-V_{load} = -5$ V when 1 is off. The inductor current has $di/dt = (10 \text{ V})/(2 \text{ mH}) = 5000$ A/s when 1 is on. It has $di/dt = -2500$ A/s when 2 is on. During the on-time of switch 1, the circuit shows the form in Figure 4.10, and the current increases $(5 \times 10^3 \text{ A/s}) \times (6.67 \ \mu\text{s}) = 0.033$ A.

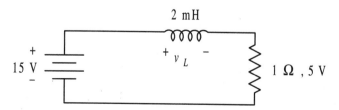

Figure 4.10 Buck converter circuit with switch 1 on.

Figure 4.11   Buck converter circuit with switch #2 on.

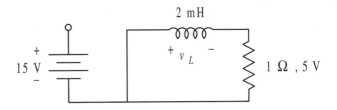

When switch 2 is on, the circuit shows an *R-L* form, as in Figure 4.11. Since the output should be close to a fixed 5.00 V, assume that the inductor voltage is -5 V whenever switch 2 is on. This assumption will be valid if the time constant L/R is much longer than the on-time of switch 2. In the circuit here, $L/R = 2$ ms, which is 150 times the switch 2 on-time. The inductor current drops $(2.5 \times 10^3$ A/s$) \times (13.33$ μs$) = 0.033$ A. The current fall must balance the rise. If it did not, the output average value would change accordingly. The current change of 0.033 A produces an output voltage change of 0.033 V for this 1 Ω load. Figure 4.12 shows the output waveform. It certainly seems reasonable to assume that $V_{\text{load}} \approx$ 5 V at all times. The output voltage is almost perfectly constant, with $V_{\text{out}} = 5 \pm 0.017$ V.

The reader is encouraged to compare the triangular waveform computed in this analysis to the actual exponential rise and fall of i_L. What is the maximum percent difference between the exponential result and the triangular prediction?

The buck converter of the above example showed almost constant output current. Since the output is so nearly a current source, the average value of V_{out} was given accurately as $D_1 V_{\text{in}}$. An interesting side note was that the average power was presented to a resistor. The resistor showed nearly constant current, but also has nearly constant voltage. The resistor cannot distinguish whether the source presented to it is a constant current source or a constant voltage source.

4.3.3 The Boost Converter

The buck converter provides dc transformation for the case of $V_{\text{out}} \leq V_{\text{in}}$. This type of converter is widely used for electronic power supplies, in which the desired output voltage is

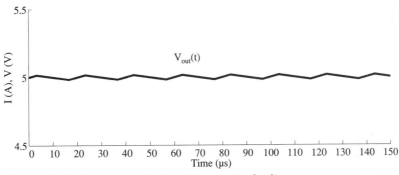

Figure 4.12   Buck converter circuit output current and voltage.

small compared to a typical ac input voltage level. The second direct dc–dc converter type has a current source input and voltage source output. Since it is reversed from the buck case, it is perhaps not surprising that $V_{out} \geq V_{in}$ for this converter, which is shown in Figure 4.13.

The relationships for this boost converter parallel those of the buck circuit. The input current and output voltage are fixed source values, while the input voltage and output current are determined by switch matrix action. For the common-ground version in Figure 4.13,

$$q_1 + q_2 = 1$$
$$v_{in}(t) = q_2 V_{out} = (1 - q_1)V_{out}$$
$$i_{out} = q_2 I_{in} = (1 - q_1)I_{in} \qquad (4.16)$$
$$\langle v_{in} \rangle = D_2 V_{out} = (1 - D_1)V_{out}$$
$$\langle i_{out} \rangle = (1 - D_1)I_{in}$$

Let $V_{in} = \langle v_{in} \rangle$ and $I_{out} = \langle i_{out} \rangle$. Then the relationships can be written

$$V_{out} = \frac{1}{1 - D_1} V_{in} \quad \text{and} \quad I_{in} = \frac{1}{1 - D_1} I_{out} \qquad (4.17)$$

The input and output power must always be equal, as must their averages. There are no places for energy to be lost in this converter.

Switch 1 carries I_{in} when on. When it is off, switch 2 must be on, and switch 1 must block V_{out}. Therefore, 1 must be a forward-conducting forward-blocking device. Switch 2 can be a diode. In the buck converter, we used a series inductor to create a current source effect. A typical boost converter uses an inductor at the input to create a current source. A capacitor would be an appropriate interface at the output to provide voltage source characteristics. Let us consider an example.

Example 4.3.3   A boost converter intended to convert a 5 V input into a 120 V output is shown in Figure 4.14. Rated output power is 50 W. The specifications call for an output voltage of 120 V ± 0.1%, and input current variation of no more than ±1%. The switching frequency is 20 kHz. Select an inductor and capacitor to meet these specifications. What should

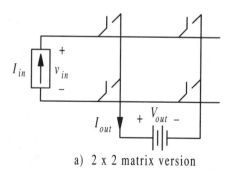

a) 2 x 2 matrix version

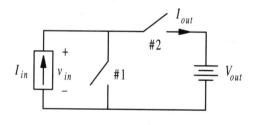

b) Common ground version

Figure 4.13   Boost dc–dc converter.

Figure 4.14 Boost converter circuit for example.

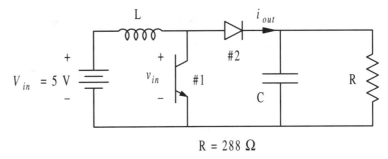

$$R = 288 \; \Omega$$

the current and voltage ratings of the switches be? If the on-time of switch 1 has uncertainty of about ± 50 ns, how much extra uncertainty will be introduced into V_{out}?

Notice that the combination of the input voltage source and the inductor effectively becomes a current source. The voltage on this source is $v_{in} = q_2 V_{out}$. However, since the inductor cannot sustain a dc voltage drop, the average voltage $\langle v_{in} \rangle = D_2 V_{out}$ must match V_{in}. With $V_{in} = 5$ V and $V_{out} = 120$ V, the duty ratio $D_2 = (5 \text{ V})/(120 \text{ V}) = 0.0417$, and switch 2 is on 2.0833 μs out of each 50 μs period. Switch 1 is on for 47.92 μs each cycle. To produce a high voltage, we need to spend a lot of time building up inductor energy, then spend a very short time discharging the energy rapidly at high potential. The output is 50 W, as is the input (no losses appear). Thus $I_{in} = 10$ A. The current is to vary by no more than $\pm 1\%$, for a total variation of 0.2 A. During the switch 1 on time, the inductor voltage is simply $V_{in} = 5$ V. The current will rise linearly as

$$5 \text{ V} = L \frac{di}{dt} = L \frac{\Delta i}{\Delta t} \tag{4.18}$$

$$\frac{(5 \text{ V})\Delta t}{L} = \Delta i \leq 0.2 \text{ A}$$

Since Δt is the on-time, 47.92 μs, the value $L > 1.20$ mH will ensure that the change in current stays below 0.2 A.

In the case of the output, the capacitor must supply 0.417 A during the long period when switch 2 is off. The voltage will fall during this interval but is not allowed to change by more than $\pm 0.1\%$, for a total change of 0.24 V. The change can be expressed approximately as

$$i_c = C \frac{dv}{dt} = C \frac{\Delta v}{\Delta t} \tag{4.19}$$

$$\frac{i_c \Delta t}{C} = \Delta v \leq 0.24 \text{ V}$$

A capacitor value $C \geq 83.3 \; \mu$F should meet the requirements. The switches will carry the inductor current of 10 A when on, and must block the full voltage V_{out} when off. Therefore, devices rated for 10 A on and 120 V off would be a minimum set of ratings. One significant practical problem in a boost converter is that the actual ratio V_{out}/V_{in} is rather limited.

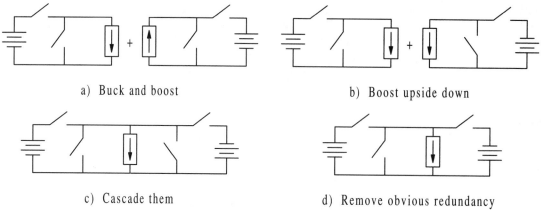

a) Buck and boost

b) Boost upside down

c) Cascade them

d) Remove obvious redundancy

Figure 4.15   Cascaded buck and boost converters.

Forward drops in real switches are one reason for this. A second reason is the difficulty in controlled very high or very low duty ratios precisely. In this circuit, ±50 ns random jitter on the switch 1 on time will place the actual on-time between 47.87 μs and 47.97 μs. This seemingly insignificant on-time variation corresponds to output uncertainty over the range 117.2 V to 123.0 V, or about ±2.5%.

4.4 INDIRECT DC–DC CONVERTERS

4.4.1 The Buck-Boost Converter

The direct dc–dc converters offer basic conversion functions. The limitation that $V_{out} \leq V_{in}$ for the buck, and $V_{out} \geq V_{in}$ for the boost are the most obvious restrictions. How can we create a more complete dc transformer function? One way is to cascade the two direct converters to form a simple indirect converter. The two converters can be adjusted independently to give any desired output ratio. This cascade arrangement is developed in Figure 4.15.

Some of the switches in the cascade are redundant and can be removed. Other switches can be eliminated by recognizing that we do not wish to consume average power in the transfer current source. In fact, only two switches are needed in the final result, as shown in Figure 4.16. This is referred to as a *buck-boost converter* or an *up-down converter* since outputs of any magnitude are possible.

Figure 4.16   Buck-boost converter.

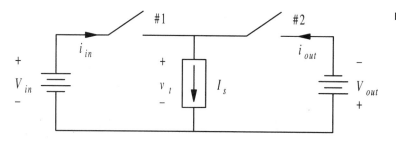

Let us examine the operation of this circuit with a focus on the transfer source, since its voltage is entirely determined by the switch action. The KVL and KCL restrictions require that one and only one switch be on at a time. The voltage across the transfer source, v_t, is V_{in} when switch 1 is on, and $-V_{out}$ when 2 is on. The transfer current source value is I_s. To meet the basic objectives of power electronics, the transfer source should not consume energy, so $\langle p_t \rangle = 0$. The relationships are

$$q_1 + q_2 = 1$$
$$v_t = q_1 V_{in} - q_2 V_{out}$$
$$p_t = v_t I_s = q_1 V_{in} I_s - q_2 V_{out} I_s$$
$$\langle p_t \rangle = D_1 V_{in} I_s - D_2 V_{out} I_s = 0$$

(4.20)

The last part of equation (4.20) can be satisfied for nonzero I_s if $D_1 V_{in} = D_2 V_{out}$. Since we expect $D_1 + D_2 = 1$, this can be written

$$V_{out} = \frac{D_1}{1 - D_1} V_{in}$$

(4.21)

The other variables determined by switch action include the input and output currents. These bring about the relationships

$$i_{in} = q_1 I_s, \qquad \langle i_{in} \rangle = D_1 I_s$$
$$i_{out} = q_2 I_s, \qquad \langle i_{out} \rangle = D_2 I_s$$
$$\langle i_{in} \rangle + \langle i_{out} \rangle = I_s$$
$$D_1 \langle i_{out} \rangle = D_2 \langle i_{in} \rangle$$

(4.22)

The cascade process required the polarity of the output boost converter to be inverted: the buck-boost converter produces a negative voltage with respect to the input, as we have seen previously in polarity reversal examples. The output in principle can range from 0 to ∞. It is zero if $D_1 = 0$ and infinite if $D_1 = 1$. When $D_1 = \frac{1}{2}$, the output magnitude is equal to the input. Except for the reversed polarity of V_{out}, the buck-boost converter provides a good transformer function.

The devices in the circuit can be determined with brief analysis. Switch 1 carries I_s when on and blocks $V_{in} + V_{out}$ when off. A forward-carrying forward blocking device is needed. The second switch can be a diode. The output load requires a capacitor to give it voltage source properties. What about the transfer current source? It must maintain constant current without power loss. There is no power dissipated in an inductor, so it is just the device needed. We do not require a special current source—only an inductor. Let us implement a converter similar to that in Example 4.3.3 to evaluate the buck-boost converter.

Example 4.4.1 Figure 4.17 shows a buck-boost converter, intended to provide -120 V output at 50 W with $+5$ V input. What are the switch duty ratios for this function? What are the switch ratings for voltage and current? What values of L and C will keep the output vari-

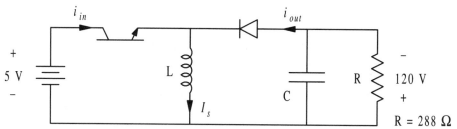

Figure 4.17 A realization of the buck-boost converter.

ation below $\pm 0.1\%$ and the transfer current variation below $\pm 1\%$? The switching frequency is 20 kHz.

Given that I_s and V_{out} are very nearly constant, we expect $D_1 V_{in} = D_2 V_{out}$. Then $D_1/D_2 = 24$ and $D_1 + D_2 = 1$. This requires $D_1 = 24/25$ while $D_2 = 1/25$. Since switch 1 is on almost the full cycle, energy in the inductor is built up over a long time interval, then released quickly into the load. The average input current must be 10 A to provide the 50 W input, while the average output current must be 0.417 A for this power level. The transfer source value $I_s = \langle i_{in} \rangle + \langle i_{out} \rangle = 10.417$ A. Each switch must carry 10.417 A when on, and must block 125 V when off. The transfer source variation is to be less than $\pm 1\%$, or ± 0.104 A. This translates to a total current change of 0.208 A. While switch 1 is on, the transfer voltage v_t is simply 5 V, and the current rises since $v_L = L\, di/dt$ is positive. The result is

$$5 \text{ V} = L\frac{di}{dt} = L\frac{\Delta i}{\Delta t}, \qquad \Delta i \le 0.208 \text{ A}$$

$$L \ge 1.15 \text{ mH} \tag{4.23}$$

For the capacitor, the allowed voltage fall is $\pm 0.1\%$, or 0.24 V. This drop will occur during the switch 1 on interval, or 24/25 of a period. The result is

$$i_C = C\frac{dv}{dt} \approx C\frac{\Delta v}{\Delta t}, \qquad \Delta v \le 0.24 \text{ V}$$

$$C \ge 83.3 \ \mu\text{F} \tag{4.24}$$

One common mistake in solving buck-boost problems involves the value of I_s. This value is not the same as either I_{in} or I_{out}, but instead is their sum. The following example illustrates this issue.

Example 4.4.2 A buck-boost converter transfers energy from a $+12$ V source to a -12 V source. The rated output load is 30 W. What are the switch duty ratios and what is the transfer current source value? What are the switch ratings?

Since $|V_{out}| = V_{in}$, the duty ratios should be $D_1 = D_2$ and $D_1 = \frac{1}{2}$. The average value of I_{in} is $P_{in}/V_{in} = 30/12 = 2.5$ A. The value of I_s is $2.5/D_1 = 5$ A. The transfer current source in this case is double the input and output average currents. The switches must carry 5 A when on, and must block $V_{in} + V_{out} = 24$ V when off.

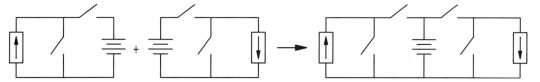

Figure 4.18 Boost converter supplying a buck converter.

4.4.2 The Boost-Buck Converter

Just as a buck converter can be cascaded with a boost converter for conversion, so a boost converter can supply a buck circuit. The idea is shown in Figure 4.18. As in the buck-boost case, many of the switches are redundant. In the final analysis, only two switches are needed. The final circuit, with energy storage elements in place, is shown in Figure 4.19. The center capacitor serves as a transfer voltage source. It takes energy in from the input side, stores it briefly, then sends it on to the load. The transfer source should exhibit $\langle p_t \rangle = 0$.

Switch action will determine the input voltage, the output voltage, and the current in the transfer source. Some of the major relationships are

$$v_{in}(t) = q_2 V_t, \qquad \langle v_{in} \rangle = D_2 V_t$$
$$v_{out}(t) = q_1 V_t, \qquad \langle v_{out} \rangle = D_1 V_t$$
$$i_t = q_2 I_{in} + q_1 I_{out}$$
$$p_t = q_2 I_{in} V_t + q_1 I_{out} V_t \qquad (4.25)$$
$$\langle p_t \rangle = 0 = D_2 I_{in} + D_1 I_{out}$$
$$D_2 I_{in} = -D_1 I_{out}$$

Like the buck-boost, the boost-buck exhibits a polarity reversal between input and output. In the literature, this arrangement is called a Ćuk converter, after the developer who patented it in the mid-1970s.

Figure 4.19 Boost-Buck converter.

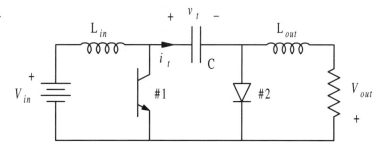

4.4.3 The Flyback Converter

The dc–dc converters examined so far all share a common reference between the input and output. Two forms were able to provide a wide range of outputs, but both reverse the polarity. These properties are often inconvenient. One possible solution is a minor modification to the buck-boost converter. Consider that the transfer current source in such a converter is an inductor. A real inductor is a device that stores energy in a magnetic field, and is commonly built as a coil wound around a piece of magnetic material. The converter operation can be considered in terms of energy: When the input-side switch is on, the input source forces current into the inductor and raises the stored energy level. When the output-side switch is on, energy is allowed to flow from the inductor into the load. Over time, the average power in the inductor must be zero, so energy does not build up in it.

There is no special reason why the inductor core cannot accommodate a second coil of wire. If this is done, one winding can be used to inject energy into the inductor, while the other can be used to remove it. The two windings would not have to share any electrical connection, and so would provide isolation between the input source and the output load. This *coupled-inductor* version of the buck-boost circuit is very popular, particularly at power levels below about 150 W. It is known as a *flyback* converter because the diode turn-on occurs when the inductor output coil voltage "flies back" as the input switch turns off. An example is shown in Figure 4.20. An important note about the operation: Although we often think of an inductor as maintaining a constant current, the action in a real device is to maintain a constant magnetic flux. With only one winding, the flux and current are proportional. With two or more windings, the flux is proportional to the total of the ampere-turn products of all windings. KCL can be satisfied by providing a current path through any winding on an individual inductor.

The flyback converter is functionally identical to the buck-boost converter. This is easy to see if the turns ratio between the windings is unity. The only difference is that a two-port inductor serves as the transfer current source. The isolation means that output connections can be reversed to cancel out the polarity change. A flyback converter has virtually all the properties of an ideal dc transformer.

The possibility of a non-unity turns ratio is a very helpful feature of the flyback converter. In the $+12$ V to -12 V converter of Example 4.4.2, insertion of a coupled inductor with 1 : 1 turns ratio will have no operational effect. A coupled inductor with 1 : 2 turns ratio would produce -24 V output for 12 V input given $D_1 = \frac{1}{2}$. A 12 : 5 turns ratio would provide -5 V output with $D_1 = \frac{1}{2}$, and so on. The 5 V to -120 V converter used in an earlier example can be built with a 5 : 120 flyback inductor to make its operation more convenient. Let us consider such a design.

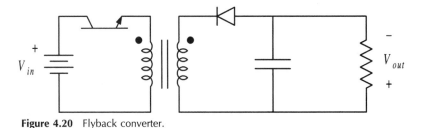

Figure 4.20 Flyback converter.

Example 4.4.3 A flyback converter for 5 V to 120 V conversion is shown in Figure 4.21. Determine the switch duty ratios and current and voltage ratings. Find L and C values to keep the output variation below $\pm 0.1\%$ and the transfer source current within $\pm 1\%$. The switching frequency is 20 kHz. What is the effect of ± 50 ns switch time uncertainty on the operation of this circuit? Notice that the output can be used for either a $+120$ V load or a -120 V load, since isolation allows either connection arrangement at the output.

As shown in Figure 4.21, a positive output can be obtained from a flyback converter by reversing the sense of the output winding polarity. The resistor draws 50 W at the output. Therefore, the average output current is 0.417 A and the average input current is 10 A. On the input side, the turns ratio means that 5 V on the inductor primary translates to 120 V on the inductor secondary. On the output side, a 5 V input will have the effect of 120 V. We can avoid the turns ratio issue by observing two facts: The input sees a circuit equivalent to a standard buck-boost with no turns ratio if the output is held at -5 V. The output sees the effect of a 120 V source as if the circuit were a straight 120 V to -120 V buck-boost converter. From the input side, then, the duty ratio should be $\frac{1}{2}$. The transfer current seen from the input side will be $(10 \text{ A})/D_1 = 20$ A, so the transistor must carry $I_s = 20$ A when on. It needs to block 10 V when off. On the output side, the duty ratio is again $\frac{1}{2}$. The diode must carry 0.833 A. It must block 240 V when off. The required $\Delta i < \pm 1\%$ of nominal translates to $\Delta i < 0.4$ A on the input side. The inductor value, measured from the input side, must satisfy the relation

$$5 \text{ V} = L\frac{\Delta i}{D_1 T}, \qquad \Delta i < 0.4 \text{ A}, \qquad D_1 T = 25 \text{ } \mu s \tag{4.26}$$

This requires $L > 313$ μH. On the output side, the capacitor must supply the load half the time, without its voltage changing more than 0.2%, or 0.24 V. Therefore.

$$0.417 \text{ A} = C\frac{\Delta v_{\text{out}}}{D_1 T}, \qquad \Delta v_{\text{out}} < 0.24 \text{ V}, \qquad D_1 T = 25 \text{ } \mu s \tag{4.27}$$

The expression requires $C > 43.4$ μF.

A random variation of 50 ns around the nominal 25 μs on time is illustrated in Figure 4.22. This variation would make the intermediate voltage vary between 4.98 V and 5.02 V.

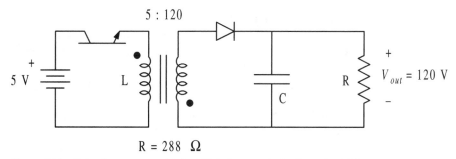

Figure 4.21 Flyback converter with 5 : 120 inductor ratio for Example 4.4.3.

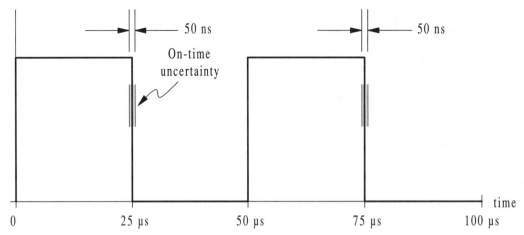

Figure 4.22   Effect of switching time variation of 50 ns.

The output would vary between 119.5 and 120.5 V, for variation of $\pm0.4\%$. This is much less than the $\pm2.5\%$ variation seen in the boost converter.

In general, flyback converters are designed to keep the nominal duty ratio close to 50%. This tends to minimize the energy storage requirements, and keeps the sensitivity to variation as low as possible. The flyback circuit is simple, and can be built with few parts. An additional advantage appears when several different dc supplies are needed: If it is possible to use two separate coils on the magnetic core of the inductor, it should be just as reasonable to use three, four, or even more coils. Each can have its own turns ratio with mutual isolation.

Figure 4.23 shows a commercial design for a 5 W flyback power supply. The ac line is rectified, then applied to the dc–dc converter. Five output windings are shown, including the one at lower left, designed to provide power for the switching function control circuitry. The dots on the coils indicate a polarity reversal from input to output—a clear indicator of a flyback circuit.

4.4.4 Other Indirect Converters

The buck-boost and boost-buck circuits are simple examples of indirect converters. They have a single transfer source and can be built with only two switches. It is possible to use several transfer sources and build up converters of increasing complexity. Only a few of the unlimited possibilities are used in practical converters. Here, we will consider one additional example.

Figure 4.24 shows the "single-ended primary inductor converter" or SEPIC circuit, which is sometimes used as an alternative to the buck-boost and boost-buck circuits when a polarity reversal is undesirable. The circuit is a boost-buck-boost cascade. Removal of redundant switches avoids the complexity implied by this extended cascade arrangement and makes the circuit relatively compact. An idealized model of the SEPIC circuit, with ideal sources substituted for the various storage elements, is shown in Figure 4.25. The switch ac-

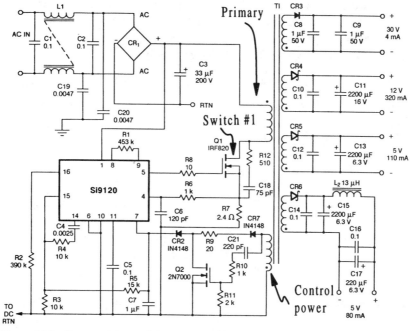

Figure 4.23 Commercial 5 W flyback converter for 170 V input and multiple outputs. From C. Varga, *Application Note AN90-2*. Santa Clara: Siliconix, 1990. Reprinted by permission.

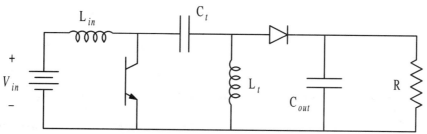

Figure 4.24 The SEPIC converter. Two transfer sources permit any input-to-output ratio without polarity reversal.

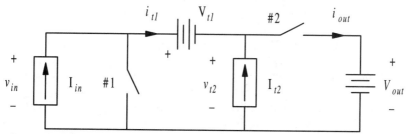

Figure 4.25 Ideal sources used to model the SEPIC circuit.

tion determines the input voltage $v_{in}(t)$, the output current $i_{out}(t)$, the current in the voltage transfer source $i_{t1}(t)$, and the voltage in the current transfer source $v_{t2}(t)$. Both transfer sources should have average power flow of zero to avoid internal losses (and to be consistent with a capacitor and an inductor as the actual devices). The relationships are

$$v_{in} = q_2(V_{out} + V_{t1}), \qquad \langle v_{in}\rangle = D_2(V_{out} + V_{t1})$$

$$i_{out} = q_2(I_{in} + I_{t2}), \qquad \langle i_{out}\rangle = D_2(I_{in} + I_{t2})$$

$$i_{t1} = -q_1 I_{t2} + q_2 I_{in}, \qquad \langle i_{t1}\rangle = 0 = -D_1 I_{t2} + D_2 I_{in} \qquad (4.28)$$

$$v_{t2} = -q_1 V_{t1} + q_2 V_{out}, \qquad \langle v_{t2}\rangle = 0 = -D_1 V_{t1} + D_2 V_{out}$$

$$q_1 + q_2 = 1, \qquad D_1 + D_2 = 1$$

Some algebra will bring out the transfer source values and input-output ratios:

$$I_{t2} = \frac{D_2}{D_1} I_{in} = \frac{1 - D_1}{D_1} I_{in}$$

$$V_{t1} = \frac{D_2}{D_1} V_{out} = \frac{1 - D_1}{D_1} V_{out} \qquad (4.29)$$

$$\langle v_{in}\rangle = D_2\left(V_{out} + \frac{1 - D_1}{D_1} V_{out}\right) = \frac{1 - D_1}{D_1} V_{out}$$

The converter has the same input-output ratio as the buck-boost converter, except that no polarity reversal occurs. As an exercise, try to determine the switch current and voltage rating requirements for this converter.

4.5 FORWARD CONVERTERS

Among the dc–dc converters examined so far, the flyback circuit comes closest to an ideal dc transformer. The coupled inductor in the flyback circuit is suggestive of a conventional magnetic transformer. Unlike a transformer, however, the coupled inductor must store energy and carries a net dc current. The buck and boost circuits lack a transfer source, so simple substitution of a coupled inductor will not give them isolation properties. In the buck and boost cases, we can seek out an ac signal within the converter, and insert a magnetic transformer at the ac location. Circuits based on this idea are called *forward converters*.

4.5.1 Basic Transformer Operation

A short overview of transformer operation is needed at this point. A magnetic transformer is normally formed by two coils on a magnetic core, as shown in Figure 4.26. The flux ϕ within the core links both coils. Define flux linkages $\lambda_1 = N_1\phi$ for coil 1 and $\lambda_2 = N_2\phi$ for coil 2. By Faraday's Law, each coil has a voltage $v = d\lambda/dt$. Therefore $v_1 = d\lambda_1/dt$ and $v_2 = d\lambda_2/dt$. The voltages follow the ratio $v_1/v_2 = N_1/N_2$, provided that $d\phi/dt \neq 0$. The latter prop-

Figure 4.26   Simple magnetic transformer.

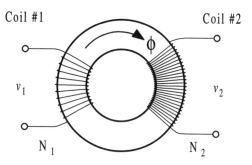

erty is very significant; it implies that only ac signals can be used. The dot location indicates consistent polarity points; that is, voltages measured with the dot taken as the high point will all have equal phase.

A real transformer exhibits several additional effects:

1. The windings exhibit resistance.
2. Some flux will leak out through the air rather than linking both coils.
3. Flux will be present even when there is no electrical load on the coils.

Wherever there is flux, it will be associated with a voltage proportional to its time-rate-of-change since $v = d\lambda/dt$. A flux is associated with an magnetomotive force (MMF) proportional to current. Thus flux implies a proportional relationship between voltage and di/dt—an inductance. Figure 4.27 shows the development of a complete transformer circuit model from the resistance and flux effects.

The magnetizing inductance L_m is perhaps the most important component of the equivalent circuit. This is the inductance seen from the input under no-load conditions. In a trans-

Figure 4.27   Real transformer effects and the implied circuit model.

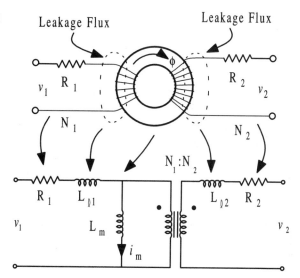

former, L_m is usually made as large as possible. This will make the associated reactance X_m large, and the magnetizing current i_m small. (As an aside, the coupled inductor of the flyback converter has relatively small L_m so that the current i_m actually dominates over other currents.) Since the inductance L_m is associated with flux in the core, it can be accessed, in some sense, from either winding. The value measured at a given winding will be proportional to the square of the number of turns. In Figure 4.27, the inductance L_m is taken at the coil 1. It could equally well be taken at the coil 2. In that case, as shown in Figure 4.28, an equivalent value

$$L'_m = L_m\left(\frac{N_2}{N_1}\right)^2 \tag{4.30}$$

represents the inductive effect of the core flux at coil 2. We will consider L_m in forward converter analysis to help determine how the transformer affects operation.

4.5.2 General Considerations in Forward Converters

Look at the buck circuit in Figure 4.29. The diode voltage is a pulse train, and it would be very convenient to insert an ac transformer at that location. However, v_{out} has a positive average value. The magnetic transformer's magnetizing inductance cannot sustain this dc voltage. Each pulse would impose positive voltage on the transformer, so that the flux would build up during each cycle. There are physical limits on flux just as there are physical limits on current or voltage. Thus simple insertion of a transformer is not feasible. There are two ways to avoid this problem and embed a transformer in a buck-type circuit:

1. Construct the converter as an ac link circuit by cascading an inverter and rectifier. This arrangement, shown in Figure 4.30, explicitly creates an ac point in the circuit, then inserts a transformer at the ac point.
2. Add a third catch winding on the coil so that flux can be conserved and a nonzero $d\phi/dt$ value can be maintained without causing other problems.

4.5.3 Catch-Winding Forward Converter

A catch-winding forward converter is shown in Figure 4.31. Tertiary coil 3 is a little like the secondary of the flyback converter's coupled inductor. While switch 1 is on, this transistor

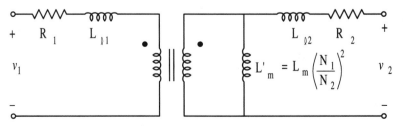

Figure 4.28   Transformer with magnetizing inductance referred to secondary.

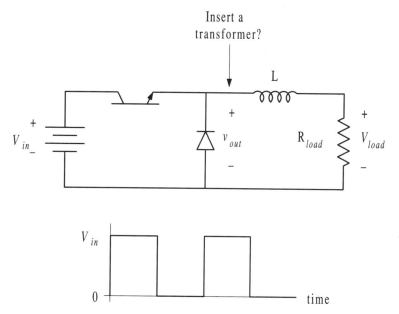

Figure 4.29 Buck converter and waveforms.

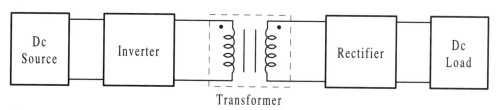

Figure 4.30 General form of an ac link converter.

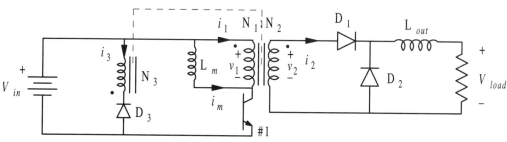

Figure 4.31 Catch-winding forward converter.

carries the primary current i_1 and also the magnetizing current i_m. When the transistor turns off, the magnetizing inductance will maintain the current flow in coil 1, such that $i_1 = -i_m$. The output diode prevents a current $i_2 = i_1(N_1/N_2)$ from flowing in coil 2 since i_1 is negative. The catch winding diode D_3 permits current $i_3 = -i_m(N_1/N_3)$ to flow. While diode D_3 is on, the voltage across coil 3 will be $-V_{in}$. Voltage $v_1 = -V_{in}(N_1/N_3)$ just after the transistor turns off and diode D_3 turns on. This negative voltage decreases the core flux. When switch 1 is turned on again, the flux can rise normally. The function of the catch winding in providing a rise and fall sequence for the flux is called *flux resetting*. The flux behavior over time is analyzed in equation (4.31), and the results are plotted in Figure 4.32. Diode D_1 will be on if and only if the transistor is on, so a simplified notation will be used in which D_1, D_2, and D_3 represent the duty ratios of the respective diodes.

$$\text{Switch 1 on:} \quad v_1 = V_{in} = \frac{d\lambda_1}{dt} = N_1\frac{d\phi}{dt}$$

$$\text{Change in } \phi: \quad \frac{d\phi}{dt} = \frac{V_{in}}{N_1} = \frac{\Delta\phi}{\Delta t}, \quad \Delta\phi = \frac{V_{in}}{N_1}D_1T \tag{4.31}$$

$$\text{Switch 3 on:} \quad v_3 = -V_{in}, \quad \frac{d\phi}{dt} = \frac{v_3}{N_3} = \frac{\Delta\phi}{\Delta t}$$

$$\text{Change in } \phi: \quad \Delta\phi = -\frac{V_{in}}{N_3}D_3T$$

Diode D_3 will not be on unless the transistor is off. Therefore, the duty ratios are such that $D_3 \leq 1 - D_1$. The flux will never be less than zero, since i_m will not be negative in this circuit. For proper operation, the flux fall must equal the flux rise. This requires

$$\frac{V_{in}}{N_1}D_1T = \frac{V_{in}}{N_3}D_3T \quad \text{or} \quad \frac{D_1}{N_1} = \frac{D_3}{N_3} \tag{4.32}$$

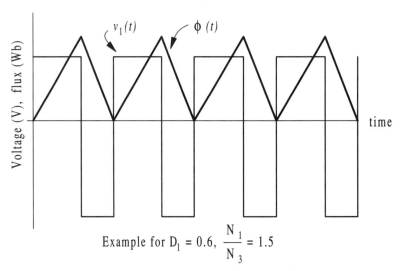

Example for $D_1 = 0.6$, $\dfrac{N_1}{N_3} = 1.5$

Figure 4.32 Flux and primary voltage vs. time for catch-winding converter.

Consider the implications of equation (4.32). Since $D_3 \leq 1 - D_1$, it is required that

$$D_1 \frac{N_3}{N_1} \leq 1 - D_1 \quad \text{or} \quad D_1 \leq \frac{N_1}{N_1 + N_3} \qquad (4.33)$$

If $N_3 = N_1$, the duty ratio of switch 1 must not exceed 50%. If the duty ratio increases beyond 50%, there will not be enough time to bring the flux down sufficiently, and the converter will not function correctly. Lower values of D_1 are no problem, since there will be time left over at the end of the switching period after the flux is brought to zero.

The secondary voltage in this converter is positive whenever the transistor is on. The diode D_1 exhibits the same switching function as the transistor, and the voltage v_{out} will be just like the output voltage of a buck converter except for the turns ratio. The output and its average value will be

$$v_{\text{out}} = q_1 V_{\text{in}} \frac{N_2}{N_1}, \qquad \langle v_{\text{out}} \rangle = D_1 V_{\text{in}} \frac{N_2}{N_1} \qquad (4.34)$$

and this forward converter is often called a *buck-derived* circuit.

4.5.4 ac Link Forward Converters

The catch-winding converter allows the transformer flux to vary, but does not make the best use of the core flux capability. Its application is limited to low power levels, and it is not a common alternative to the flyback circuit. The ac link configuration is considerably more widespread, and is very common for dc power supplies above a few hundred watts. A general example of the configuration for the buck case is shown in Figure 4.33. In this circuit, the switches in the inverter can operate to produce a voltage square wave at the transformer

Figure 4.33 Inverter-rectifier cascade to form ac link forward converter.

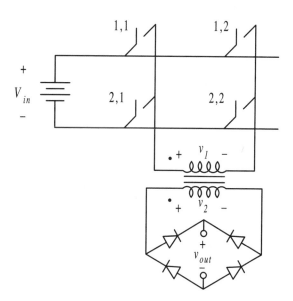

primary. The rectifier processes the square wave. This brings very considerable advantages over a conventional sine wave rectifier, since the rectified square wave produces a direct dc output without extra filtering.

A simple square wave eliminates the adjustable duty ratio of the buck converter. To remedy this, a waveform such as that shown in Figure 4.34a is more commonly used as the inverter output. The signal can be stepped up or down by the transformer, so the secondary waveform will be the same except for a turns ratio multiplier $a = N_2/N_1$. Once full-wave rectification is performed, the result will be the absolute value waveform of Figure 4.34b. This waveform is identical to the output of a buck converter. The duty ratio definition comes from Figure 4.34a; this means a factor of two will be required to account for the dual pulse arrangement at the output. The average output is

$$\langle v_{\text{out}} \rangle = 2aDV_{\text{in}} \tag{4.35}$$

This implies that no individual switch on the inverter side will be on more than 50% of each cycle.

One way to implement this concept is the bridge circuit illustrated in Figure 4.35. This circuit is called a *full-bridge forward converter* since all four switches of the inverter are present and active. The topology is common at power levels of 500 W and above. The photograph in Figure 4.36 shows a dc power supply implemented with a full-bridge forward converter. The incoming ac line is rectified directly, then the dc–dc converter provides low-voltage step-down as well as isolation. The switching function timing is adjusted to provide control, particularly to provide line regulation.

There are several other styles of buck-derived forward converters. While a detailed comparison is beyond the scope here, Figure 4.37 gives examples of the possibilities. The complexity associated with four transistors is avoided in most of these cases.

One of the more interesting topologies is the push-pull forward converter shown in Figure 4.37c. The push-pull idea takes advantage of a center-tapped transformer to reduce

Figure 4.34 Waveforms in inverter-rectifier cascade.

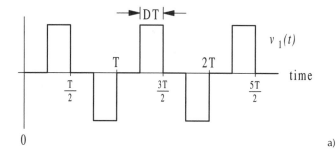

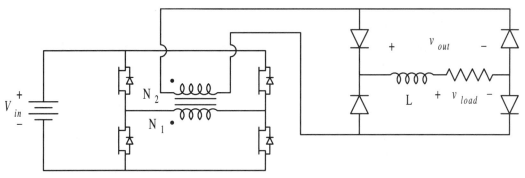

Figure 4.35 Full-bridge forward converter.

Figure 4.36 A commercial 1500 W dc power supply, based on a full-bridge forward converter.

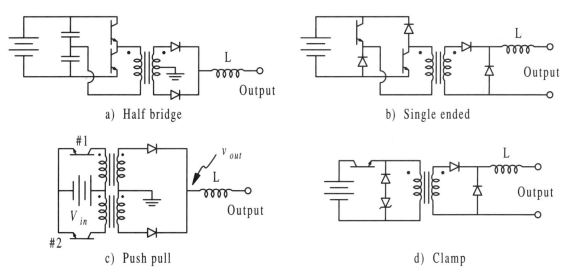

a) Half bridge

b) Single ended

c) Push pull

d) Clamp

Figure 4.37 Four alternative forward converter topologies.

to two switches. The lower switch applies a positive voltage on the primary. When it is on, the core flux increases. The upper switch acts to decrease the flux. The two switches must be controlled carefully to ensure perfect symmetry between the flux rise and fall over each switching cycle. The output rectifier also uses a center tap to eliminate two diodes. The output is

$$v_{\text{out}} = (q_1 + q_2)aV_{\text{in}}, \qquad \langle v_{\text{out}} \rangle = 2aD_1V_{\text{in}} \tag{4.36}$$

where $D_1 = D_2$. Notice that the switching functions q_1 and q_2 do not add to one in this converter. The current path is provided by the output diodes instead of through alternating action of the transistors.

4.5.5 Boost-Derived Forward Converters

Boost versions of forward converters are much less common than the buck-derived versions but are not unknown. One straightforward arrangement follows from the full-bridge converter, with a current source substituted for the input voltage source. This is illustrated in Figure 4.38. In this circuit, switches 1,1 and 2,2 need to be on to make the primary current i_{in}. Switches 1,2 and 2,1 will be turned on the make the primary current $-i_{\text{in}}$. Switch pairs 1,1 and 2,1 or 1,2 and 2,2 can be used to provide times when the primary current is zero. The secondary current will be the primary current divided by the turns ratio a. The rectifier set will see the same current as the rectifier in a conventional boost converter. The boost-derived converter's input-output relationships become

$$\langle i_{\text{out}} \rangle = \frac{2D}{a}I_{\text{in}}, \quad V_{\text{out}} = \frac{a}{2D}\langle v_{\text{in}} \rangle \tag{4.37}$$

very much along the same lines as the conventional boost converter.

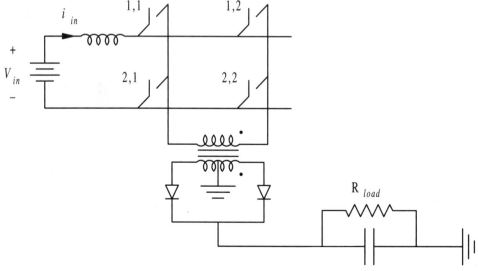

Figure 4.38 Boost-derived full-bridge forward converter.

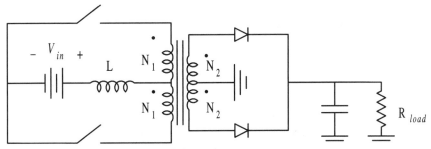

Figure 4.39   Boost-derived push-pull forward converter.

A boost-derived version of the push-pull circuit is shown in Figure 4.39. Like the full bridge circuit, it provides output equal to $aV_{in}/(2D)$.

4.6 BIDIRECTIONAL CONVERTERS

The full matrix buck converter shown in Figure 4.4 can provide both positive and negative output voltages. The negative voltage allows it to recover energy back to the input source from a true current-source output. Since power can flow in both directions, the matrix represents a *bidirectional converter*. A boost version corresponds to the matrix circuit shown in Figure 4.13a. These circuits are rarely encountered, but they do provide energy flow control when appropriate sources are available.

The buck-boost converter can give some insight into an alternative type of bidirectional converter. The output of a buck-boost is intended to be a voltage source, so bidirectional operation requires a direction change in the output current. But the output current is determined by the transfer source. This raises a natural question: Why should operation of a buck-boost converter be limited by constraints on the transfer source? In fact, the operation need not be limited in this way. The inductor can be used as a "reversible current source," and thus allows bidirectional operation, shown in Figure 4.40. The only change from the basic converter is that the restricted switch type must change to support current flow in both directions.

The inductors in real motors and other loads are not always large. For instance, for the dc motor circuit model shown in Figure 4.41, any attempt to decrease the motor terminal

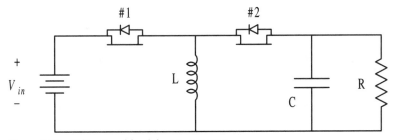

Figure 4.40   Bidirectional buck-boost converter.

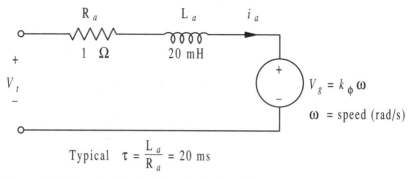

Figure 4.41 dc motor characteristics showing low inductance.

voltage with a buck converter will cause current reversal within 50 ms or so. Following the bidirectional buck-boost circuit, it might make sense to use switches capable of handling bipolar currents to account for this "reversible current source" load. The circuit, given in Figure 4.42, can either send energy to the load or recover energy after the output current reverses. The circuit can be treated as a standard buck circuit when the motor current is positive. If the current is allowed to become negative, then the motor acts as an input to a boost converter. The switches continue to operate with $q_1 + q_2 = 1$, and energy is recovered from the internal voltage of the machine.

4.7 DC–DC CONVERTER DESIGN EXAMPLES

In this section, examples of several dc–dc converters are presented. Some of these are very similar to the cases in Section 3.7, although the flavor now is more closely aimed at design rather than analysis.

Example 4.7.1 Design a converter with 12 V input and 200 V output at up to 50 W. This converter will be used as the first step in generating clean sinusoidal backup power. No isolation is needed. The output ripple should not exceed ±5%.

A boost converter seems reasonable here. It gives the most direct approach without isolation. The ratio $V_{in}/V_{out} = D_2$, so $D_2 = 0.06$ and $D_1 = 0.94$. The load current is

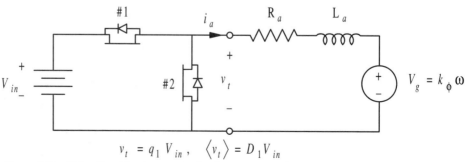

Figure 4.42 Bidirectional buck converter for dc motor control.

Figure 4.43 Candidate boost converter for Example 4.7.1.

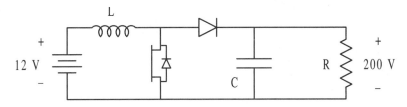

(50 W)/(200 V) = 0.25 A. The input current is (50 W)/(12 V) = 4.17 A. The inductor ripple is not specified, but it is imposed on the dc input source, so a limit should be set. Let us choose it to be ±20% (selected so that distortion power will be under 5%), for total variation of 1.67 A. The switching frequency is not given, but these current and voltage levels match well to power MOSFETs. If MOSFETs are used, a switching frequency on the order of 100 kHz would be expected.

The inductor current will increase while switch 1 is on in the sample boost converter of Figure 4.43. During this time, the inductor voltage is fixed at 12 V, so the current will change linearly with slope $\Delta i/\Delta t$. The inductor can be determined from the allowed current change,

$$v_L = 12 \text{ V} = L\frac{di}{dt} = L\frac{\Delta i}{\Delta t}, \qquad \Delta t = D_1 T, \qquad \Delta i < 1.67 \text{ A} \qquad (4.38)$$

Some algebra shows that $L > 68 \ \mu H$ should meet the current variation requirement, given a typical 100 kHz switching frequency.

The capacitor should maintain the output within a window 20 V wide. When the transistor is on, the capacitor must maintain the full 0.25 A load current. This requires

$$i_C = 0.25 \text{ A} = C\frac{\Delta v}{\Delta t}, \qquad \Delta t = D_1 T, \qquad \Delta v < 20 \text{ V} \qquad (4.39)$$

To meet the requirements, $C > 1.2 \ \mu F$ given 100 kHz switching.

Example 4.7.2   A dc–dc converter is needed for a telephone switch system. This system requires 5 V at levels between 50 W and 500 W. The available source is a nominal 48 V bus. The bus might vary by ±20%, especially because battery backup might be the only source of power at times. The input bus is referenced to earth ground. Suggest an appropriate conversion circuit for this application. Isolation is not required. Define the operating characteristics of this converter. Select component values to provide maximum output ripple of ±1%.

The buck converter should be appropriate in this case, as shown in Figure 4.44. It provides step-down and a common reference. Each switch will need to be rated for full output load current of (500 W)/(5 V) = 100 A. The highest possible line voltage is 48 V+ 20%, or about 58 V. The switches will need to block this voltage when off. The load current might seem high, but commercial power MOSFETs with ratings of 100 A and 100 V exist. Since MOSFETs are fast, we might choose a switching frequency above 20 kHz to avoid audible noise. In practice, the high current limits the maximum level, and 50 kHz is probably a good switching frequency at which to evaluate the design issues.

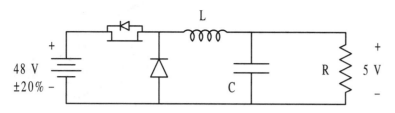

Figure 4.44 Proposed converter for Example 4.7.2.

The input voltage can vary between about 38 V and 58 V, corresponding to 48 V $\pm$ 20%. Since $V_{\text{out}} = D_1 V_{\text{in}}$, the duty ratio will need to adjust between 5/58 and 5/38 to maintain a fixed output, or $0.0868 < D_1 < 0.132$.

Let us first try to choose an inductor with $C_{\text{out}} = 0$ as a beginning. This inductor must limit the current swing to $\pm 1\%$. At full load, this implies a total current swing of 2 A. At the lightest load, the swing is only allowed to be 0.2 A. Recall that more inductance is needed to reduce ripple, so the light load situation represents the worst case. With the diode on, the inductor voltage is essentially constant at $v_L = -5$ V, and the current falls in a linear manner. We have

$$v_L = L\frac{di}{dt} \approx L\frac{\Delta i}{\Delta t}, \qquad v_L = -5 \text{ V}, \qquad \Delta t = D_2 T, \qquad \Delta i \leq 0.2 \text{ A} \qquad (4.40)$$

The period T was chosen as 20 μs. The value of D_2 falls in the range $0.869 < D_2 < 0.913$, and the inductor must be big enough to maintain $\Delta i < 0.2$ A over this full range. The result is

$$(5 \text{ V})\frac{D_2 T}{L} < 0.2 \text{ A}, \qquad L > D_2 (500\mu\text{H}) \qquad (4.41)$$

The inductor should be at least 457 μH to meet the requirements over the full converter operating range with $C_{\text{out}} = 0$. An output capacitor would allow this inductor value to be smaller. This would be helpful, since an inductor to handle 100 A is large, and inductance of 457 μH represents very considerable energy storage.

To test the capacitor issue, let us choose $L = 50$ μH, then find out how much capacitance is needed. The inductor current will vary by

$$\Delta i = (5 \text{ V})\frac{D_2 T}{L} = \frac{5 \text{ V} \times 20 \text{ } \mu\text{s}}{50 \text{ } \mu\text{H}} D_2 = (2 \text{ A})D_2 \qquad (4.42)$$

The variation will be 1.82 A for the highest value of D_2. As in Example 3.7.3, the inductor can be treated as a triangular equivalent source, while the output is a fixed source. The triangular variation is centered on the load current. The equivalent source circuit is illustrated with the capacitor current $i_C(t)$ in Figure 4.45. While i_C is positive, the voltage v_C will increase by an amount Δv_C. The current is positive half the time and has a peak value of (1 A)D_2. Integration gives

$$\frac{1}{C}\int_{v_C>0} i_C \, dt = \Delta v_C, \qquad \Delta v_C < 100 \text{ mV} \qquad (4.43)$$

Figure 4.45 Equivalent source circuit for capacitor current.

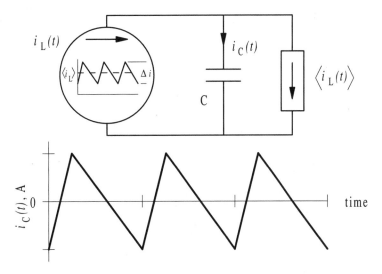

Since the integral is the triangular area $\frac{1}{2}(10\ \mu s)(i_{C(peak)})$, the capacitance value will be

$$C > \frac{5\ \mu s \times 1\ A}{0.1\ V}D_2, \qquad C > (50\ \mu F)D_2 \qquad (4.44)$$

A standard capacitor value of 47 μF will meet the specifications, with the benefit that the inductor need be only 11% of the value with $C_{out} = 0$. The combination $f_{switch} = 50$ kHz, $L = 50\ \mu$H, $C = 47\ \mu$F, and $0.0868 < D_1 < 0.132$ meets all the requirements.

Example 4.7.3 A boost converter allows inputs of 15 to 40 V_{dc}, and gives output of 120 V at 12 kW. The inductor is 10 μH. What value of f_{switch} is needed to keep the inductor current variation below $\pm20\%$ under all input voltages? The circuit is shown in Figure 4.46.

The load draws $I_{load} = (12\ kW)/(120\ V) = 100$ A in this circuit. Since there is no loss, the input current must be $(12\ kW)/V_{in}$. While switch 1 is on, the inductor voltage $v_L = V_{in}$, and the inductor current should rise by no more than 40% (the total allowed variation). This means

$$v_L = L\frac{di}{dt} = L\frac{\Delta i}{\Delta t}, \qquad \frac{V_{in}\Delta t}{L} \leq 0.4\left(\frac{12000\ W}{V_{in}}\right) \qquad (4.45)$$

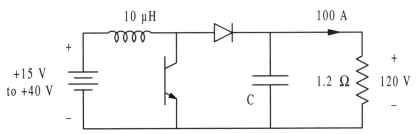

Figure 4.46 Boost converter for Example 4.7.3.

The time $\Delta t = D_1 T$ depends on the input value and the switching period. For a boost converter, $V_{out} = V_{in}/D_2$. In this case V_{out} is known, so we can find an expression for D_1:

$$V_{out} = \frac{V_{in}}{1 - D_1}, \qquad D_1 = 1 - \frac{V_{in}}{120 \text{ V}} \tag{4.46}$$

When this is substituted into equation (4.42), the result is

$$\frac{V_{in}(1 - V_{in}/120 \text{ V})}{L f_{switch}} \leq \frac{4800 \text{ W}}{V_{in}}, \qquad f_{switch}(\text{Hz}) \geq \frac{V^2_{in} - V^3_{in}/120}{4800 L} \tag{4.47}$$

The last expression must be satisfied for any allowed value of V_{in}. Since it is not linear, checking the two extreme values of 15 V and 40 V is not an adequate test. To make sure f_{switch} is large enough to meet the requirements, let us check the maximum value of the right-hand side in the final expression, then set f_{switch} larger than that maximum. We need

$$\frac{\partial}{\partial V_{in}}\left(\frac{V^2_{in} - V^3_{in}/120}{4800 L}\right) = 0 \quad \text{or} \quad 2V_{in} - \frac{3V^2_{in}}{120} = 0 \tag{4.48}$$

This indicates a relative maximum or minimum at $V_{in} = 80$ V—outside the specified range. Thus the extreme case occurs when $V_{in} = 40$ V. With this value,

$$f_{switch} \geq 22222 \text{ Hz} \tag{4.49}$$

so 22.2 kHz switching should meet the requirement.

Example 4.7.4   A dc–dc converter is to be built to meet the following specifications:

Allowed dc input voltage range: +10 V to +24 V

Output voltage: +5 V

Output voltage ripple: ±0.05 V maximum, at rated load

Rated load: 200 W, resistive

Switching frequency: 50 kHz

Common ground connection between input and output

    a. Draw a circuit that can perform this function.

    b. What range of values will be needed the switch duty ratios?

    c. Find an inductor value to meet the stated specifications.

Since the output is lower than the input and there is a common ground, a buck converter is the simplest approach. Others such as the SEPIC or the flyback could do the job, but are more complex. An appropriate buck converter is shown in Figure 4.47. If the inductor is large enough to keep the current ripple low, the duty ratio D_1 will vary between (5 V)/(10 V) and (5 V)/(24 V), while $D_2 = 1 - D_1$ will be between $\frac{1}{2}$ and 19/24. The average output current will be 40 A with a 200 W load. This could be represented with a resistance of 0.125 Ω. The load current flows in the inductor. Total load voltage variation of 0.10 V

Figure 4.47 Buck converter for Example 4.7.4

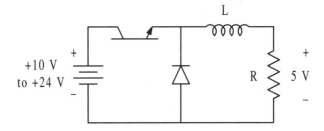

corresponds to inductor current variation of $(0.10 \text{ V})/(0.125 \ \Omega) = 0.80$ A. While the diode is on, the inductor current must not fall more than 0.80 A. The relationships are

$$\text{Diode on: } v_L = L\frac{di}{dt} \approx L\frac{\Delta i}{\Delta t}, \qquad \left| \frac{v_L D_2 T}{L} \right| = \Delta i \leq 0.80 \text{ A} \qquad (4.50)$$

The period is 20 μs for 50 kHz switching, and the inductor has 5 V applied while the diode is on. Equation (4.47) can be rewritten to establish the condition on L,

$$L \geq (125 \times 10^{-6} \text{ H})D_2 \qquad (4.51)$$

The inductor must work for any allowed value of D_2. Here the largest value of D_2 represents the worst case. Then $D_2 = 19/24$, and $L \geq 99.0 \ \mu$H. A 100 μH inductor should work in this converter.

4.8 RECAP

dc–dc converters are widely used as the basis for many types of dc power supplies. They are also common in battery-based systems and in telecommunications systems. Voltage dividers are poor substitutes for dc–dc converters, since they generate losses without operating benefits such as regulation. Linear regulators have been used as advanced output filter components for power converters. They provide very tight output control, and the converter can be designed to minimize the loss in the regulator.

Switching systems for dc–dc conversion are of primary interest to us. The chapter uses switching functions to define switch matrix action. Average values are important, and yield forms in which duty ratios substitute for switching functions. Filters need only low-pass characteristics, and the design generally follows the power filter ideas introduced in Chapter 3. Duty ratio adjustment is the only realistic method for making small adjustments in these types of systems. The general objective in designing a dc–dc converter is to prepare a circuit with the properties of an ideal dc transformer: the input and output voltages are related by the ratio a, the input and output currents are related by the inverse of a, and power is conserved as it flows through the device.

The direct voltage-to-current and current-to-voltage converters represent the buck and boost converter networks. These are very common over a wide spectrum of applications. The buck converter in particular is the basis for isolated forward converters and step-down converters for applications ranging from low-voltage portable equipment to multikilowatt dc power supplies. In the buck converter, the output average value is given by

$$V_{out} = D_1 V_{in} \tag{4.52}$$

where D_1 is the duty ratio of the active switch—normally either a BJT or an FET. For the boost converter, the opposite relationship holds:

$$V_{out} = \frac{V_{in}}{D_2} \tag{4.53}$$

where D_2 is the duty ratio of the diode in the circuit. In the cases we have studied so far, KVL and KCL require that $D_1 + D_2 = 1$. The relationships were examined based on this requirement, and are more complex for converters that do not meet it, as we will see later.

Indirect converters open up broad possibilities for various source interfaces and conversion ratios. One of the simplest represents a cascade of the buck and boost circuits. The buck-boost converter reverses polarity, and provides an average output given by

$$V_{out} = -\frac{D_1}{D_2} V_{in} \tag{4.54}$$

The inductor embedded in this converter as a transfer source carries a current equal to the sum of the average input and output current values. The boost-buck cascade is also used as a simple indirect converter. It provides the same ratio

$$V_{out} = -\frac{D_1}{D_2} V_{in} \tag{4.55}$$

The SEPIC converter is one example of a more complicated indirect circuit, with both transfer current and voltage sources. This gives the relationship

$$V_{out} = \frac{D_1}{D_2} V_{in} \tag{4.56}$$

without a polarity reversal.

The buck-boost converter is often implemented with a multiple-winding inductor. This *coupled inductor* adds isolation to the circuit, and avoids polarity problems since the output can be referenced to an arbitrary potential. The combination is called a *flyback* converter, and is one of the most important dc power supply circuits for power levels up to about 150 W. Many personal computers and consumer electronics products use flyback-based power supplies. Multiple inductor windings make it easy to create supplies with several outputs.

Buck and boost converters can be provided with isolation by configuring them as inverter-rectifier cascades. At the internal ac port, a conventional magnetic transformer can be added. Circuits based on this technique are called *forward converters*, and are commonly applied to dc–dc power supplies at levels above 200 W or so. Forward converters can be of the buck-type, if their input acts as a voltage source while the output acts as a current source.

Boost-type forward converters use an input inductor to obtain current-source behavior, and display voltage source action at their outputs.

In some applications, particularly dc motor drives, dc–dc converters need to handle energy flow in two directions. To build such *bidirectional converters*, the switching devices usually are configured to permit the output current source to be either positive or negative. The buck converter version of this concept can drive a motor with positive current to accelerate it or provide energy to a mechanical load. When the motor must be slowed, its input current can be made to reverse. Kinetic energy is recovered to the electrical source.

PROBLEMS

1. A voltage divider provides 1 V output from 5 V input. The maximum output current is 1 mA. The resistors are chosen to provide load regulation of better than 1% over the load current range of 0–1 mA. Find the resistor values necessary for this application. What is the efficiency at full load?

2. A voltage divider is used to split a 12 V battery into ±6 V levels, as shown in Figure 4.48. This divider is intended to supply up to 2 W total, split between the two output voltage levels in arbitrary ratios. Load regulation is to be better than 1%. Choose resistors to meet this need, subject to achieving the highest possible efficiency. What is the efficiency?

3. A dc–dc converter has 12 V input and 5 V output at power levels between 10 W and 60 W. The switching frequency is 120 kHz. Draw a circuit that can perform this function. Determine inductor and capacitor values necessary to keep the output ripple below ±1%.

4. A converter has 5 V input and produces 15 V output at a rated load of 4 W for an analog circuit application. The switching frequency is 250 kHz. Draw a circuit that can perform this function. Select inductor and capacitor values to keep the peak-to-peak output ripple below 200 mV.

5. A dc–dc converter is needed to provide power for a video camera pack from an automobile. In normal operation, the car electrical system varies between 10 V and 15 V. The video camera requires an accurate 10.2 V supply for proper operation and battery charging. The power level is 20 W. Isolation is recommended to avoid any issues about ground connections. At this power level, a switching frequency of 150 kHz should be feasible. Draw a circuit that can perform the desired function. What duty ratio range is needed for 10.2 V output? Choose inductor and capacitor values to limit the output ripple to 80 mV peak-to-peak.

6. A buck converter is to work from a 48 V input and supply a 15 V output at power levels from 5 W to 100 W. The design is constrained to have $L \leq 20\ \mu H$, $C \leq 22\ \mu F$, and $f_{switch} \leq 100$ kHz. What combination of these three will provide the lowest ripple for a 100 W output? What is this ripple value? What ripple will occur for the 5 W output for the same design?

Figure 4.48   Voltage divider to split a battery into bipolar sources.

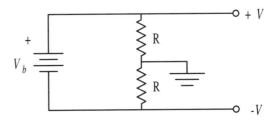

7. A dc–dc converter is needed for a control application. The load is 30 W. The input voltage ranges from 60 V to 90 V, and the output is to be 280 V ±2%. Propose a circuit to meet this need. Provide values for switching frequency and the various components. What current and voltage ratings will be needed for the semiconductors?

8. A converter for an aerospace application supplies a 48 V load from an input bus that varies between 20 V and 40 V. The load varies between 10 W and 400 W. Propose a circuit to meet the need while keeping ripple voltages and currents below ±1%. Provide values for switching frequency and the various components.

9. A dc–dc converter for an electric motor system provides an adjustable output of 100 V to 500 V from a 1500 V input. The power level ranges from 1 kW to 200 kW. The switching frequency is 8 kHz.

   a. Isolation is desired. Use a full-bridge forward converter as the basis for a design. What duty ratio range will be needed, given your choice of turns ratio?

   b. Select inductor and capacitor values to keep the output voltage ripple below ±10%.

   c. What are the current and voltage ratings needed for the semiconductors and for the magnetic transformer?

10. A dc–dc buck converter switches at ω_s rad/s. The gate of the input switch is controlled by a *pulse width control module* that automatically adjusts to maintain an output of precisely +5 V, as long as the input is higher than this. Find the switch duty ratios and the amplitude of the largest unwanted component at the output for

   a. $V_{in} = 10$ V   b. $V_{in} = 15$ V   c. $V_{in} = 7$ V.

11. A dc–dc converter is to be built to the following specifications:

   Allowed dc input voltage range: +3 V to +18 V
   Output voltage (average value): +24 V
   Output voltage ripple: ±0.05 V *maximum* at rated load
   Rated load: 120 W, resistive
   Input current ripple: ±100 mA *maximum* at rated load
   Switching frequency: 100 kHz
   Common ground between input and output

   a. Draw a circuit which can perform this function. How many switches have gates?

   b. What is the range of duty ratio values for each switch?

   c. Find circuit part (L, C, R) values which will meet the stated specifications.

12. A dc–dc converter is to be designed for +12 V output and +150 V input.

   a. Draw a circuit to perform this function.

   b. What is the duty ratio for each switch?

   c. An inductor is required in order to simulate a current source. Let $L = 2$ mH. The load is 120 Ω. Find and plot the inductor current $i_L(t)$ if the switching frequency is 50 kHz.

13. A dc–dc converter has $V_{in} = +10$ V and $V_{out} = +5 \pm 0.05$ V into rated load. The rated load is 200 W, and the switching frequency is 50 kHz.

    a. Draw a circuit which can perform this function.

    b. Find the value of L or C which will meet the specifications.

    c. Plot $I_{in}(t)$.

14. A boost converter is built as a full four-switch matrix (see Figure 4.13 for an example). It has $V_{in} = +12$ V. The input inductor and output capacitor are large. The load is resistive.

    a. If $q_{1,1}(t) = q_{2,2}(t)$, and $D_{1,1} = 0.75$, find $V_{out(ave)}$. If this condition gives $P_{out(ave)} = 120$ W, plot the output current (into the RC combination).

    b. Let $q_{1,1}(t) = q_{2,2}(t)$. The duty ratio $D_{1,1}$ drops abruptly to 0.20. What is the new value of V_{out} just after the change? Also, comment on the long-term steady behavior at the load.

15. What is $V_{out(ave)}$ as a function of load for a buck converter, given that the inductor allows some times when $i_L = 0$?

16. A dc–dc converter is shown in Figure 4.49. $R = 10$ Ω, $C = 100$ μF, L is very large, and $V_{out} = 20 \pm 0.1$ V.

    a. What are the duty ratios and frequencies of the two switching functions?

    b. What is the inductor current?

17. The "off-line" flyback converter shown in Figure 4.50 provides +12 V and +5 V outputs. Each output supplies a 20 W load. The transistor duty ratio is to be $D_T = 0.5$.

    a. What should N_5 and N_{12} be?

    b. What value of $L\mu$ will ensure a current ripple below $\pm 10\%$ with 50 kHz switching?

18. A dc–dc converter is to be built with the following specifications:

    Allowed dc input voltage range: +8 V to +18 V

    Output average voltage: +24 V

    Output voltage ripple: ± 0.06 V maximum at rated load

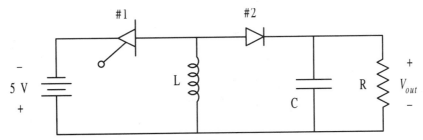

Figure 4.49   dc–dc converter for Problem 16.

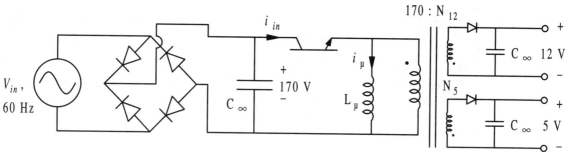

Figure 4.50 The "off-line" flyback converter.

Rated load: 300 W, resistive

Input current ripple: ±250 mA maximum at rated load

Switching frequency: 100 kHz

Common ground between input and output.

a. Draw a circuit which can perform this function.

b. What is the necessary range of duty ratio for each switch?

c. Find circuit part values (R, L, and C) which will meet the stated specifications.

19. In the converter of Figure 4.51, switch 1 operates at 200 kHz.

a. What are the duty ratios of the two switches to make $\langle v_r \rangle = 5$ V?

b. Find L such that $v_r = 5$ V $\pm 0.2\%$.

c. What is the load regulation for this circuit for small changes in R?

20. A dc–dc converter is needed to supply a 5 V load from a battery input. For flexibility, it is desired to accommodate many different battery types, so the possible input voltages range from 3.0 V to 8.0 V. The output ripple is to be less than 1% peak to peak. The load can have a power level between 10 W and 50 W.

a. Propose a conversion circuit to address these requirements. What switching frequency will be reasonable?

b. Select capacitor and inductor values for this circuit so that it will meet the requirements.

c. What voltage and current ratings will be needed for the switching devices and for the inductor?

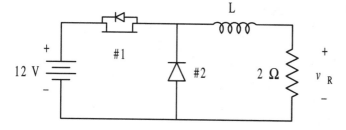

Figure 4.51 Converter circuit for Problem 19.

21. A battery charger circuit is intended to deliver a desired current into a battery, then maintain a fixed "float voltage" when full charge is reached. For a small car battery, a reasonable scenario would be to deliver 5 A to the battery until it reaches 13.8 V, then maintain the converter output at 13.8 V indefinitely.

   a. Consider an off-line flyback circuit similar to that in Figure 4.50 for this application. The duty ratio will be adjusted as necessary. Choose a turns ratio, and propose a switching frequency.

   b. What duty ratio will be needed to deliver 5 A of current when the battery is at low charge and has a terminal voltage of 11.5 V?

   c. What duty ratio will be needed when the battery voltage reaches 13.8 V?

22. A bidirectional buck-boost converter supplies the dc motor in Figure 4.41, which has $R_a = 1\ \Omega$, $L_a = 0.02$ H, and $k_\phi = 0.05$ H when speed is measured in rad/s. The motor mechanical load power is the product $V_g i_a$. A 24 V dc source is available.

   a. When the motor is driving a 100 W mechanical load at 500 rad/s, what duty ratio will be needed to supply its terminal voltage? What current will the inductor carry?

   b. When the motor is being driven as a generator, the mechanical power acts as an *input*, and $P < 0$. With a -100 W load at 400 rad/s, what duty ratio will be needed? What is the inductor current?

   c. The switching frequency is 20 kHz, and the motor is expected to have loads ± 200 W. What value of capacitance should be used at the converter output to ensure ripple in L_a of less than 2 mA$_{\text{peak-to-peak}}$?

23. You have been challenged to show why a buck converter should be used instead of a series regulator for supplying a 5 V logic load from a 12 V automotive electric system. Explain the issues, and justify through a numerical example the advantages of a buck converter in this case.

REFERENCES

G. C. Chryssis, *High-Frequency Switching Power Supplies*, 2nd ed. New York: McGraw-Hill, 1989.

J. G. Kassakian, M. F. Schlecht, and G. C. Verghese, *Principles of Power Electronics*. Reading, MA: Addison-Wesley, 1991.

D. M. Mitchell, *Dc–Dc Switching Regulator Analysis*. New York: McGraw-Hill, 1988.

Motorola, *Switchmode Designer's Guide*. Phoenix: Motorola, manual SG79/D, 1993.

A. I. Pressman, *Switching Power Supply Design*. New York: McGraw-Hill, 1991.

R. P. Severns, E. J. Bloom, *Modern Dc-to-Dc Switchmode Power Converter Circuits*. New York: Van Nostrand, 1985.

C. Varga, "Designing Low-Power Off-Line Flyback Converters Using the Si9120 Switchmode Converter IC.," *Application Note AN90-2*. Santa Clara, CA: Siliconix Inc., 1990.

G. W. Wester, R. D. Middlebrook, "Low-frequency characterization of dc–dc converters," *IEEE Trans. Aero. Elec. Sys.*, vol. AES-9, no. 3, pp. 376–385, 1973.

CHAPTER 5

Figure 5.1 Bridge rectifier designed for a dc motor drive system.

DIODE-CAPACITOR CIRCUITS AND RECTIFIERS

5.1 INTRODUCTION

The dc–dc converters of the preceding chapter were formulated directly by considering the switch matrix. The idea that voltage converts to current and current to voltage leads immediately to the buck and boost converters. Most dc–dc circuits use these building blocks. In this chapter, switching function methods and design ideas are extended to several practical diode circuits and to controlled rectifiers more generally. The focus is on circuits that take energy from a fixed ac voltage source and transfer it to a dc load. The diode bridge has been examined in previous examples, but the considerations involved in a design have been explored only in a limited fashion. The first sections of this chapter review the diode bridge rectifier, and establish a design framework for classical diode-capacitor rectifiers. Switching function methods are augmented with approximate waveforms to simplify the analysis.

Diode-capacitor combinations are not confined to rectification. Many low-power applications use diode-capacitor charge pumps for dc–dc conversion. Charge pumps use switches to create various series and parallel capacitor connections. Boosting charge-pump circuits are explored briefly.

At higher power levels, rectifier control becomes important. Diode rectifiers do not permit convenient control, and the SCR provides a solution. In later sections, the operation of controlled rectifiers is explored. Modern applications of controlled rectifiers include chemical processes, dc motor controls like the one pictured in Figure 5.1, power plant energy storage, high-voltage dc transmission, and dc power supplies of about 10 kW and higher.

The distinction between rectifiers and inverters is somewhat artificial. A rectifier takes energy from an ac source and delivers it to a dc load. In an inverter, the energy flow direction is reversed. The material in Chapter 5 thus leads naturally into the inverter discussions in Chapter 6.

5.2 RECTIFIER OVERVIEW

Rectifiers can readily be understood in terms of a switch matrix and switching functions, along the same lines as direct dc–dc converters. For example, if we wish to convert power

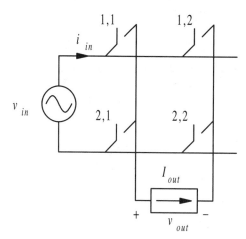

Figure 5.2 ac voltage to dc current switch matrix converter.

from an ac voltage source to a dc load, the load should have current source characteristics. The concept is shown in Figure 5.2. The switches in general must carry dc current and block ac voltage, so they should have forward-conducting bidirectional-blocking capability. The most sophisticated rectifiers use such a device, in the form of a GTO. The realities of applications do not always require the full capability of such a switch. The SCR, for instance, approximates the FCBB device, with control only over its turn-on. This is sufficient (and even preferable) for most ac voltage-based rectifiers.

Useful rectifiers can be built with uncontrolled diodes. We have examined the diode bridge circuit previously, but let us consider its operation as a rectifier and examine possible design methods. A diode bridge circuit is shown in Figure 5.3. The voltage $v_d(t)$ in the figure is the full-wave rectifier form $|V_0 \cos(\omega t)|$. The resistor voltage has the same average value as v_d, since the inductor does not support dc voltage. This means that the line poten-

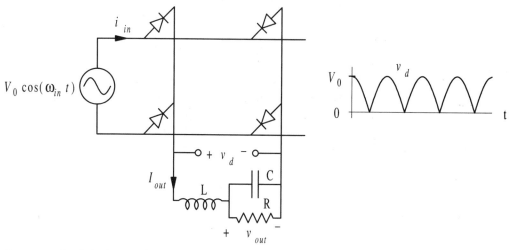

Figure 5.3 Diode bridge rectifier with output filter.

tial V_0 is linked to the output value. There is no line regulation, and the line potential is commonly stepped down with a magnetic transformer to provide the right voltage.

In many practical low-power circuits the output inductor is not used. Inductors, as magnetic components, tend to be relatively heavy and expensive. Rectifier circuits are usually constrained by ac line frequencies to operate at 50 or 60 Hz. At such low frequencies, the inductor values needed for useful filtering are high. As a result, inexpensive equipment operating below perhaps 200 W often dispenses with the inductor as a filter element. An example will help place this concern in perspective.

Example 5.2.1 A clock radio is supplied through a 120 V to 10 V, 60 Hz transformer, followed by a bridge rectifier. The appliance draws a maximum of 6 W from the rectifier. What inductor value will be needed to keep the output ripple below $\pm 1\%$, given that no filter capacitor is provided?

The ac potential is normally specified as an RMS value, so the peak voltage at the diode bridge input will be $10\sqrt{2}$ V $= 14.1$ V. If L is large enough to maintain low ripple, the resistor voltage is approximately constant. The voltage $v_d(t)$, shown in Figure 5.3, has average value of

$$\langle v_d \rangle = V_{\text{out}} = \frac{1}{\pi} \int_{-\pi/2}^{\pi/2} 10\sqrt{2} \cos\theta \, d\theta \tag{5.1}$$

This evaluates to

$$V_{\text{out}} = \frac{10\sqrt{2}}{\pi} \left[\sin\theta \right]_{-\pi/2}^{\pi/2} = (10\sqrt{2})\frac{2}{\pi} = 9.00 \text{ V} \tag{5.2}$$

Since the resistor voltage is 9.00 V, the resistor current is (6 W)/(9 V) = 0.67 A, and the resistance is 13.5 Ω. The inductor must attenuate all harmonics so that only $\pm 1\%$ variation about the 9.00 V average remains. Here, the ripple waveform will not be triangular. KVL allows us to write

$$|V_0 \cos(\omega t)| = L\frac{di_{\text{out}}}{dt} + i_{\text{out}}R \tag{5.3}$$

The circuit will operate in a periodic manner in steady-state, with an average output current of 0.67 A. There will be a small 120 Hz ripple around this average value. The equation can be rewritten to focus on this small ripple if we define a ripple current $\tilde{i}_{\text{out}}$ such that $i_{\text{out}} = \tilde{i}_{\text{out}} + \langle i_{\text{out}} \rangle$. Then we can write

$$|V_0 \cos(\omega t)| = L\frac{d\tilde{i}_{\text{out}}}{dt} + R\tilde{i}_{\text{out}} + R\langle i_{\text{out}} \rangle \tag{5.4}$$

by taking advantage of the constant average value since $d\langle i_{\text{out}} \rangle/dt = 0$ in the periodic steady state. The values $\langle i_{\text{out}} \rangle$ and $|V_0 \cos(\omega t)|$ are both known. Over the interval $t = -1/240$ to $t = +1/240$, the voltage $v_d(t)$ is $V_0 \cos(\omega t)$, and the ripple must begin and end the interval with the same value. The Mathematica commands to solve for the ripple current (over one 1/120 s period) are

vO = 10 Sqrt[2]; vave = 2 vO/Pi; pout = 6; iave = pave/vave; omega = 120 Pi;
R = vave/iave

DSolve[{vO Cos[omega*t] == L itild'[t] + R itild[t] + R iave, itild[−1/240] ==
itild[1/240]}, itild[t], t]

The ripple can also be estimated with Fourier analysis based on the fundamental. We want
the peak-to-peak value of $\tilde{i}_{out}R$ to be below $\pm 1\%$ of 9.00 V, for total variation of less than
0.18 V. A little effort with the Mathematica result yields $L \approx 1.18$ H. This is a large inductance, and represents an expensive solution for such a low power level.

Many low-power applications share this difficulty of excessive inductance. Diode-capacitor
circuits, such as the simplified bridge in Figure 5.4, were originally developed to avoid large
inductance. These circuits do not provide a means for control, but their practical importance
is hard to overstate. In the section that follows, we will consider the operation and performance of this arrangement.

5.3 THE CLASSICAL RECTIFIER—OPERATION AND ANALYSIS

When a bridge rectifier is constructed without the output inductor, the result is the classical
rectifier circuit of Figure 5.4. The simplicity of this circuit makes it almost universal for rectification below 50 W, and it is common even at higher power ratings. In many switching
supplies, the classical rectifier serves as the input section, to be followed by dc–dc conversion. The circuit is contrary to one of our power electronics design principles: It tries to interconnect two voltage sources. It might be expected to have KVL problems because of the
interaction between the ac voltage source and the capacitor. This is indeed the case, and classical rectifiers are major contributors to poor power factor and current distortion on utility
grids. Considerable research effort is in progress to avoid the problems of the circuit without resorting to high values of inductance.

The circuit output waveform is given in Figure 5.5. The waveform $|v_{in}(t)|$ is shown as
a dotted line for reference. When a given diode pair is on, the output is connected directly,
and $v_{out} = |v_{in}|$. When the diodes are off, the output is unconnected, and v_{out} decays expo-
nentially according to the RC time constant. First, the trial method will be used to consider
various possibilities for switch action. Many combinations are eliminated quickly (for instance, a single device cannot be on with all others off). Table 5.1 lists the nontrivial possi-

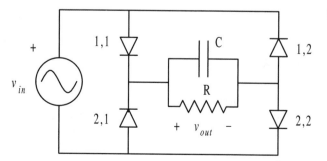

Figure 5.4   Classical diode-capacitor bridge rectifier.

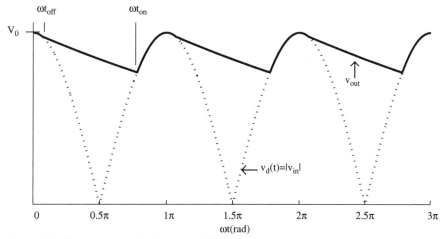

Figure 5.5 Output voltage of classical rectifier.

bilities. There are times when all switches are off. This is not a problem, since the lack of a current source allows the devices to be off without causing KCL problems.

The three allowed configurations are summarized in Figure 5.6. The major remaining part of the analysis is to determine the switching times so that the waveforms can be understood in depth. The circuit also lends itself to some powerful approximate design tools, which are illustrated in this analysis. To continue with the trial method, consider time $t = 0$, and assume that 1,1 and 2,2 are on. Since $v_{out} = v_{in}$ in this arrangement, the diode currents are $i_d = i_{in} = i_C + i_R$. The configuration will be valid provided $i_d > 0$, which means

$$C\frac{dv_{in}}{dt} + \frac{v_{in}}{R} > 0 \quad \text{or} \quad C\frac{d}{dt}[V_0 \cos(\omega t)] + \frac{V_0 \cos(\omega t)}{R} > 0 \qquad (5.5)$$

At $t = 0$, the capacitor current is zero and the resistor current is V_0/R. The sum is positive, so the configuration is valid. As time progresses, the voltage begins to fall. The voltage de-

TABLE 5.1
Diode Combinations for Classical Rectifier

On Switch

Combination	When Allowed?	Comments
1,1 and 2,1	Not allowed	KVL problem: shorts the input. Not consistent with diode directions.
1,2 and 2,2	Not allowed	KVL problem, and not consistent.
All on	Not allowed	KVL problem, and not consistent.
1,1 and 1,2	Not allowed	This forces $v_{out} = 0$, but with no output there will be no output current, so the diodes would be off.
2,1 and 2,2	Not allowed	Forces $v_{out} = 0$ and no output current will flow.
1,1 and 2,2	When $i_{in} > 0$	Voltage $v_{out} = v_{in}$ in this configuration.
2,1 and 1,2	When $i_{in} < 0$	Voltage $v_{out} = -v_{in}$ in this configuration.
All off	When $v_{out} > |v_{in}|$	Output voltage decays exponentially with time constant RC.

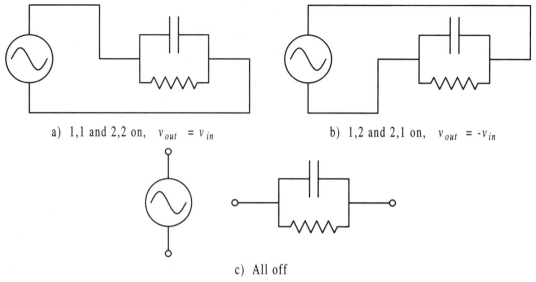

a) 1,1 and 2,2 on, $v_{out} = v_{in}$ b) 1,2 and 2,1 on, $v_{out} = -v_{in}$

c) All off

Figure 5.6 Configurations for classical rectifier.

rivative and capacitor current become negative. Eventually, the sum $i_C + i_R$ becomes zero, and the diodes shut off. This occurs at a time t_{off} that satisfies the conditions

$$C\frac{dv_{in}}{dt}\bigg|_{t_{off}} + \frac{v_{in}(t_{off})}{R} = 0 \quad \text{or} \quad -\omega C V_0 \sin(\omega t_{off}) + \frac{V_0}{R}\cos(\omega t_{off}) = 0 \qquad (5.6)$$

Equation (5.6) allows computation of the turn-off time. Dividing by V_0 and performing some algebraic manipulation,

$$\frac{1}{\omega RC} = \tan(\omega t_{off}) \qquad (5.7)$$

There is a turn-off angle $\theta_{off} = \omega t_{off}$, given by the arctangent of $1/(\omega RC)$. In many textbooks on power supplies, the turn-off angle is assumed to be 0°. This is an approximation, accurate in the limit of infinite capacitance. The actual turn-off occurs when the capacitor current and resistor current balance out exactly.

Once the diodes are off, the output decays exponentially from its initial value. The analysis is simply the solution of a first-order equation. The initial voltage will be $v_{in}(t_{off})$, and the time constant τ will be the RC product, so

$$\text{Diodes off:} \quad v_{out}(t) = V_0 \cos(\theta_{off})\exp[-(t - t_{off})/\tau] \qquad (5.8)$$

This decay will continue as long as the diodes are reverse biased, that is, $v_{out} > |v_{in}|$. When the full-wave voltage increases again during the next half-cycle, two diodes will turn on as the full-wave value crosses the decaying output. The turn-on time t_{on} in Figure 5.5 therefore must satisfy

$$V_0 \cos(\theta_{\text{off}})\exp[-(t_{\text{on}} - t_{\text{off}})/\tau] = |V_0 \cos(\omega t_{\text{on}})| \qquad (5.9)$$

This transcendental equation cannot be solved in closed form, but given numerical parameters, it can be solved iteratively.

We have expressions for the turn-off and turn-on times. The output voltage is known in considerable detail as a result. The other waveform that the switch matrix manipulates is the input current, which is $i_C + i_R$ while 1,1 and 2,2 are on, or $-(i_C + i_R)$ while 1,2 and 2,1 are on. The output voltage maximum is the peak input V_0, while the minimum output occurs at t_{on}. Thus the peak-to-peak output ripple is $V_0 - |v_{\text{in}}(t_{\text{on}})|$. Consider the implications of the ripple as we prepare an example. To guarantee small ripple, the output voltage should decay very little while the diodes are off. This means $\tau >> t_{\text{on}} - t_{\text{off}}$ to keep the ripple low. For small ratios $(t_{\text{on}} - t_{\text{off}})/\tau$, the exponential can be represented accurately with the linear term from its Taylor Series expansion,

$$e^x \approx 1 + x, \quad x \text{ small} \qquad (5.10)$$

The time difference $t_{\text{on}} - t_{\text{off}}$ cannot be more than half the period of v_{in}, so the low ripple requirement can be expressed as $RC >> T/2$ if T is the input period. With $\omega = 2\pi/T$, low ripple requires $RC >> \pi/\omega$. Let us study an example related to the previous inductive case.

Example 5.3.1 A 120 V to 10 V 60 Hz transformer supplies a classical rectifier bridge. The load resistance is 13.5 Ω. Low output ripple is desired, so start with $RC = 10(T/2)$ as a possible choice. Find the necessary value of C, then plot the output voltage and input current. What is the average value of $v_{\text{out}}(t)$?

Since $T = 1/60$ s, $RC = 10(T/2) = 1/12$ s. With $R = 13.5$ Ω, C must be about 6200 μF. The product $\omega RC = 31.6$ with this choice. The turn-off angle is the arctangent of $1/(\omega RC)$, equal to 0.0317 rad or 1.81°. The turn-off time is 84.1 μs (notice how close the angle is to the traditional assumption that $\theta_{\text{off}} = 0$). The voltage at time t_{off} is $0.999V_0$. To find t_{on}, we will need to iterate equation (5.9). The turn-on should occur before the next peak of $|v_{\text{in}}|$. As an initial guess for iteration, let us try an angle of 160°, corresponding to $t_{\text{on}} = 7.4$ ms. The iterative scheme for finding t_{on} uses this initial guess as a starting value, then updates an improved guess. One form of (5.9) that works well is given by

$$\frac{\cos^{-1}[\cos(\theta_{\text{off}})\exp(-[t_{\text{on(previous)}} - t_{\text{off}}]/\tau)]}{\omega} = t_{\text{on(next)}} \qquad (5.11)$$

provided we pay attention to the proper quadrant for the arccosine function (the angle should stay between 90° and 180°). After about five repetitions, the result converges to $t_{\text{on}} = 7.25$ ms, or $\omega t_{\text{on}} = 156.6°$. The output voltage waveform is shown in Figure 5.7.

Now that the voltage is known, we can compute the average, based on the waveform period of 1/120 s:

$$\langle v_{\text{out}} \rangle = 120 \int_0^{1/120} v_{\text{out}}(t)\, dt$$

$$\qquad (5.12)$$

$$= 120\left[\int_0^{t_{\text{off}}} v_{\text{in}}(t)\, dt + \int_{t_{\text{off}}}^{t_{\text{on}}} V_0 \cos(\theta_{\text{off}})\exp[-(t - t_{\text{off}})/\tau]\, dt + \int_{t_{\text{on}}}^{1/120} -v_{\text{in}}(t)\, dt \right]$$

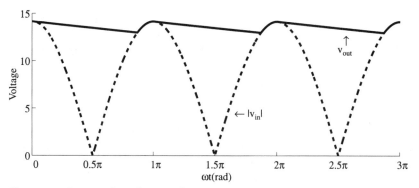

Figure 5.7   Output voltage for Example 5.3.1.

This gives $\langle v_{out} \rangle = 13.6$ V. The average resistor current is $\langle v_{out} \rangle / 13.5 \ \Omega = 1.01$ A. The ripple voltage is $V_0 - v_{out}(t_{on}) = 1.17 \ \text{V}_{\text{peak-to-peak}}$ or 8.6% of the average value.

The diodes are on for an interval of about 25° out of each 180° half-period. Input current flows only during these short intervals, and is $\omega C V_0 \sin(\omega t) + V_0/R \cos(\omega t)$ when 1,1 and 2,2 are on. This gives

$$\text{Diodes 1,1 and 2,2 on:} \quad i_{in}(t) \ (\text{A}) = 33.06 \sin(\omega t) + 1.05 \cos(\omega t) \quad (5.13)$$

The total input current is plotted in Figure 5.8. Notice that although the average output current is only about 1 A, the peak input current exceeds 14 A. The current appears as short spikes of high magnitude—impulses. This result corresponds to the KVL problem: The practice of interconnecting unequal voltage sources, which in effect is present in this circuit, yields impulse currents.

The input current drawn by a classical rectifier can be extreme. The example showed 14 A input spikes for a 1 A output load. Typically, the ratio between peak input current and

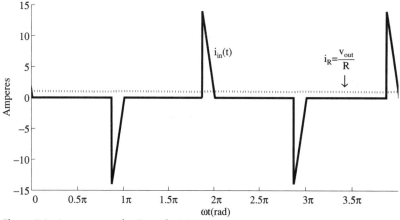

Figure 5.8   Input current for Example 5.3.1.

average output current will be high. The higher the capacitance, the briefer the total on-time, and the higher the current spike will need to be to transfer the necessary energy. To obtain ripple below 1% in the example, the peak input current would be almost 50 A. The power factor associated with such a waveform is poor. For the example, the output power is 13.7 W. Equation (5.13) can be integrated to determine the input RMS current, which yields a value of 3.09 A. The volt-amp product is $S = (3.09 \text{ A})(10 \text{ V}) = 30.9 \text{ VA}$. The power factor is $P/S = 0.442$—compared to the desired value of 1.

The detailed analysis in Example 5.3.1 is direct, although it is rather tedious. Fortunately, there are several reasonable assumptions that can be made about this circuit. The results will provide a simplified design framework. We saw in the example that if the ripple is small, the RC time constant will be much greater than half of the input waveform period. Useful simplifications include:

1. The turn-off time is given in terms of the tangent of a small angle. For small angles, $\tan \theta \approx \theta$. Thus $\omega t_{off} \approx 1/(\omega RC)$.

2. The output voltage at the moment of turn-off is given by the cosine of a small angle. Since $\cos \theta \approx 1$ for small angles, we can assume this voltage to be V_0.

3. The exponential decay will be nearly linear, since the time constant is much slower than the sine wave. Thus after the turn-off time, the voltage will be just about $V_0(1 - t/RC)$.

4. The turn-on time is approximately the point at which $1 - t_{on}/(RC) = |\cos(\omega t_{on})|$.

5. Since the total time of exponential decay never exceeds half the period (for the full-wave rectifier case), the voltage at turn-on will not be less than $V_0[1 - T/(2RC)]$. The peak-to-peak ripple ΔV_{out} will be no more than $V_0 T/(2RC)$. If we substitute the frequency f for the inverse period $1/T$, the ripple is seen to be no more than $V_0/(2fRC)$.

6. The average output will be approximately midway between the maximum and minimum output voltage, or $\langle v_{out} \rangle \approx V_0[1 - 1/(4fRC)]$.

7. When the diodes turn on, the input current is a high spike. Almost all of this flows to the capacitor rather than the resistor, so given a turn-on time t_{on}, the peak input current is approximately $C(dv/dt) = \omega C V_0 \sin(\omega t_{on})$.

All these simplifications require $RC \gg 1/(2f)$. Notice that if this requirement is not met, the ripple will be a large fraction of V_0, and the rectifier will not be especially useful.

The ripple approximation (number 5 above) yields a simple design equation. Given a desired output current $I_{out} = V_0/R$, and a desired ripple voltage ΔV_{out}, we have

$$\Delta V_{out} = \frac{I_{out}}{2fC} \quad \text{or} \quad C = \frac{I_{out}}{2f\Delta V_{out}} \tag{5.14}$$

The capacitor would be selected based on the highest allowed load current, and the actual average output voltage will be the peak value minus half the ripple. Notice that if a half-wave rectifier substitutes for the full bridge, the basic operation of the circuit will not change, except that the maximum decay time will be T instead of $T/2$. The factors of 2 in equation (5.14) will not be present.

Example 5.3.2 Design a classical rectifier to supply $12 \text{ V} \pm 3\%$ to a 24 W load, based on a 120 V, 60 Hz input source.

A transformer will be needed to provide the proper step-down ratio for this design. For completeness, let us include the typical 1 V diode on-state forward drop. The load should draw 24 W/12 V = 2 A, and therefore is modelled with a 6 Ω resistor. The circuit is shown in Figure 5.9. When a given diode pair is on, the output will be two forward drops less than the input waveform, or $|v_{in}| - 2$ V. We need a peak output voltage close to 12 V. Therefore, the peak value of v_{in} should be just about 14 V. The RMS value of v_{in} is almost exactly 10 V for this peak level. Let us choose a standard 120 V to 10 V transformer for the circuit on this basis. The actual peak output voltage will be 12.14 V with this choice.

To meet the ±3% ripple requirement, we do not want the output to exceed 12.36 V, or fall below 11.64 V. For the transformer selected here, the maximum output level is automatically satisfied. To avoid going below the minimum allowed output of 11.64 V, the output ripple should not exceed (12.14 − 11.64) = 0.50 V. The capacitor should be

$$C = \frac{I_{out}}{2f\Delta V_{out}} = \frac{2 \text{ A}}{2(60 \text{ Hz})(0.5 \text{ V})} = 0.0333 \text{ F} \tag{5.15}$$

In practice, the approximate methods overestimate the ripple slightly, so a standard 33000 μF capacitor will probably meet the requirements. The waveforms that result from these choices are shown in Figure 5.10.

To test the assumptions used for simplified rectifier design, compare the approximate results of Example 5.3.2 with more precise calculations. The circuit to be analyzed is a bridge rectifier with 10 V_{RMS} input, 2 V total diode voltage drop, a 6 Ω load resistor, and 33000 μF filter capacitor. The only change from the prior analysis is that the effective full-wave voltage is $|v_{in}| - 2$ instead of just $|v_{in}|$. Results are summarized in Table 5.2.

Important operating values such as the average voltage are well approximated when all the assumptions are used. The error in $\langle v_{out} \rangle$ is only 0.25%. The linear ripple assumption gives close results, and is also notable in that the result is a conservative approximation: The linear ripple assumption always overestimates the actual ripple. The approximations give a useful basis for designing these kinds of circuits.

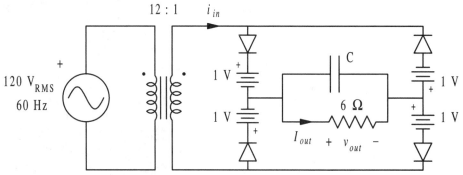

Figure 5.9 Circuit to solve Example 5.3.2.

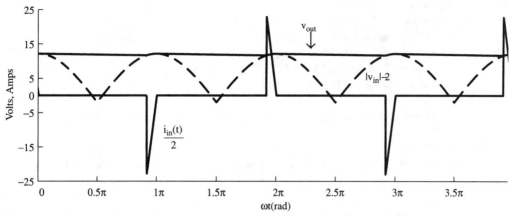

Figure 5.10 Output voltage and input current for Example 5.3.2.

5.4 THE CLASSICAL RECTIFIER—REGULATION

The regulation properties of the classical rectifier circuit are mixed. Since the output voltage is determined by the peak ac level V_0, the circuit has no inherent line regulation. Any change in the ac input line will alter the output on a one-to-one basis. This is a fundamental limitation of the circuit: There are no controllable switches to permit adjustment of the operation. In practice, the regulation can actually be worse than this 100% effect. Since the output is determined primarily by the peak value of the line voltage, any glitches or noise on the line that occur close to the peaks can alter the rectifier output. This can be a significant problem when distorted or low-quality line power is used.

The load regulation is less troublesome. A change in the load resistance R will alter the time constant of the exponential decay. In the examples, a nominal load resistance was used in the analysis. This might represent the rated load of a power supply. Any load resistor higher than this rated value will slow the decay and decrease the ripple value. For loads lighter than the rated value, the average output is nearly the peak output less half the ripple. The ripple is a linear function of the load current, and it is easy to see that the average output will decrease from V_0 as load current is added.

TABLE 5.2
Approximate and Exact Calculations for Rectifier Circuit of Example 5.3.1

Parameter	Simplification Number	Approximate Value	Exact Value
Peak output	None	12.14 V	12.14 V
Peak-to-peak ripple	5	0.505 V	0.460 V
Minimum output	5	11.64 V	11.68 V
Average output	6	11.89 V	11.92 V
Turn-off time	1	35.5 μs	35.5 μs
Turn-on time	4	7.60 ms	7.66 ms
Peak input current	7	48.0 A	46.4 A

Figure 5.11 shows the approximate regulation behavior for loads from 0 to 100% of rated, given the design from Example 5.3.2 as the rated value. For a given change in $\langle v_{out} \rangle$, the ripple changes twice as much. At rated load, the average output has fallen by 0.25 V from the open-circuit value, and the ripple is 0.5 V peak-to-peak. For a $\pm 0.5\%$ ripple design, the load regulation would have been 0.5%. The conclusion is that the regulation value is tied to the ripple value. The output behavior of a classical rectifier varies less with load than with time—the ripple is larger than the regulation.

It is important to keep in mind some of the key characteristics of classical rectifiers:

- Classical diode-capacitor rectifiers have outputs directly linked to the line amplitude, and so do not provide line regulation. Their load regulation properties are good, although tight regulation values such as 0.1% require low ripple and extreme capacitance values.
- Classical rectifiers have poor power factor (the Example 5.3.2 circuit has $pf = 0.3$), and draw input current in brief, high spikes.
- Low-ripple designs require very large capacitors. In the examples, capacitor values of many thousands of microfarads were needed to keep the ripple below 5 or 10%. Ripple much less than 1% is almost out of the question unless the load is tiny.
- The output average value is determined by the choice of magnetic transformer, and is not adjustable. The transformer provides isolation between input and output, which is an advantage.
- The circuits are simple and relatively easy to analyze and design.

The magnetic transformer and capacitor in such a circuit are chosen based on the ac input frequency. At 50 Hz or 60 Hz, the parts can be bulky and heavy. The simplicity, inherent isolation, and good load regulation properties tend to offset the disadvantages at low power levels.

5.5 INDUCTIVE FILTERING

The classical rectifier circuit improves markedly when an output inductor is restored to the bridge circuit, consistent with the source conversion concept. The inductor gives the output

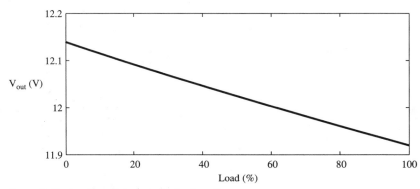

Figure 5.11 Load regulation of classical rectifier.

a tendency to behave as a current source. The high input current spikes can be avoided by adding the proper amount of inductance. If a large inductance is used, a nearly constant current flow will be maintained, and each diode will be on fully half of each cycle. Some results for $L \rightarrow \infty$ are summarized in Figure 5.12. The output of the bridge will be a complete full-wave rectified signal. Since the average voltage across the inductor is zero, the load voltage is the average of $|V_0 \cos(\omega t)|$, or $2V_0/\pi$. If we include the 1 V diode drops, the output will be $2V_0/\pi - 2$. The load resistance and the filter capacitance do not affect the output under these conditions, meaning that the load regulation is nearly perfect.

The filtering properties are probably best explored with an equivalent source approach. If the inductor is large enough, it will always maintain some current flow. The voltage from the bridge will be $|v_{in}|$, less two voltage drops. With this information, the Fourier Series of the full-wave signal can be used to solve for the current and output voltage. The series was found earlier to be

$$|V_0 \cos(\omega_{in}t)| = \frac{2V_0}{\pi} + \frac{4V_0}{\pi} \sum_{n=1}^{\infty} \frac{\cos(2n\omega_{in}t - n\pi)}{1 - 4n^2} \tag{5.16}$$

(diode drops can subtracted if desired).

The dc component plus first harmonic gave an accurate ripple estimate in our previous work. Presumably, the first harmonic

$$v_1(t) = \frac{4V_0}{3\pi} \cos(2\omega_{in}t) \tag{5.17}$$

will provide a good estimate of the ripple voltage and current. We can focus on the equivalent circuit for this specific harmonic to evaluate the results. An equivalent source circuit and first harmonic circuit are given in Figure 5.13.

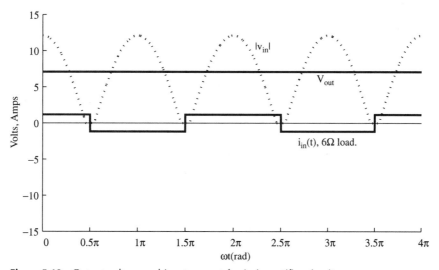

Figure 5.12 Output voltage and input current for L_∞ in rectifier circuit.

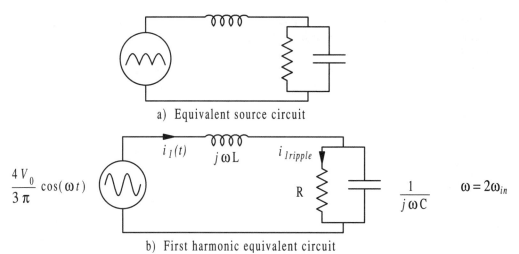

a) Equivalent source circuit

b) First harmonic equivalent circuit

Figure 5.13   First harmonic equivalent circuit for ripple evaluation.

The first-harmonic ripple current phasor can be computed from the impedances at the ripple frequency $\omega = 2\omega_{\text{in}}$. The equivalent impedance is

$$Z_{\text{equiv}} = j\omega L + R\|\frac{1}{j\omega C} = \frac{R}{1 + \omega^2 R^2 C^2} + j\left(\omega L - \frac{\omega R^2 C}{1 + \omega^2 R^2 C^2}\right) \tag{5.18}$$

Since the RC time constant is chosen to be much slower than the period $2\pi/\omega$, the product ωRC will be large compared to 1. For manual computation, the expression simplifies to

$$Z_{\text{equiv}} \approx \frac{1}{\omega^2 R C^2} + j\left(\omega L - \frac{1}{\omega C}\right) \tag{5.19}$$

and the inductor current phasor can be computed as

$$\tilde{I}_1 = \frac{2\sqrt{2}\,V_0}{3\pi} \div Z_{\text{equiv}} \tag{5.20}$$

The output ripple requires the resistor current, i_{ripple}, for which the first harmonic phasor can be obtained based on a current divider as

$$\tilde{I}_{1(\text{ripple})} = \tilde{I}_1 \frac{1/(j\omega C)}{R + 1/(j\omega C)} \tag{5.21}$$

When (5.18) and (5.20) are combined, the ripple is found to be

$$\tilde{I}_{1(\text{ripple})} = \frac{2\sqrt{2}\,V_0}{3\pi(R + j\omega L - \omega^2 RLC)} \tag{5.22}$$

The expression has some interesting properties:

- If the inductor-capacitor pair is chosen to be resonant at ω, such that $\omega^2 = 1/(LC)$, the ripple is maximized for a given inductor.

- When the capacitance value is small, the ripple actually increases over that with no capacitance. In fact, the ripple continues to increase with added capacitance until $\omega^2 LC > 2$. This means the resonant frequency $1/\sqrt{(LC)}$ should be substantially less than the ripple frequency.

- It would be helpful to be able to raise the frequency ω because a higher ω will make $|I|$ smaller, all other things being equal.

An alternative method allows accurate estimation of inductor requirements under certain conditions. Assume once again that the bridge output can be modelled accurately as an equivalent source $|V_0 \cos(\omega_{in}t)|$. If the capacitor is effective, the load sees an approximately constant dc voltage, given by the average value $2V_0/\pi$. The inductor voltage can be approximated with the expression $|V_0 \cos(\omega_{in}t)| - 2V_0/\pi = L\,di/dt$. The remainder of the treatment follows the technique in Example 3.7.3: when $v_L > 0$, the inductor current is rising. Therefore the change in inductor current is given by

$$\Delta i_L = \frac{1}{L}\int_{V_L>0} \left|V_0 \cos(\omega_{in}t)\right| - \frac{2V_0}{\pi}\, dt \tag{5.23}$$

Given a requirement on Δi_L, an inductor value can be chosen based on this expression.

Example 5.5.1   Let us add an inductor to the 12 V classical rectifier designed in Example 5.3.2, and evaluate the ripple result. The output power is intended to be 24 W, and the voltage should be 12 V $\pm$ 3%. In the earlier example, 33000 μF was needed for capacitive filtering. Determine the inductance necessary to bring this value down to a more reasonable 1000 μF.

The addition of significant inductance will make the output voltage equal to the average of the full-wave signal, rather than give it a value close to the peak. The magnetic transformer should provide a turns ratio of 120/15.6 to accomplish the intended result given that 2 V of total diode drop is included. With a 15.6 V secondary, the average of the full-wave signal will be

$$(15.6 \text{ V})\sqrt{2}\frac{2}{\pi} - 2 \text{ V} = 12.04 \text{ V} \tag{5.24}$$

The ripple requirement of 12 V $\pm$ 3% requires that $11.64 < V_{out} < 12.36$ V.

The load current will vary between 1.94 A and 2.06 A. This implies a first-harmonic current ripple of 0.06 A_{peak} and 0.042 A_{RMS}. The inductance can be found from equation (5.22) as

$$0.042 \left|6 + j240\pi L - 240^2 \pi^2 6 \times 10^{-3}L\right| = \frac{2\sqrt{2}\,V_0}{3\pi}$$

$$\left|0.255 + j31.99L - 144.7L\right| = 6.621 \tag{5.25}$$

$$(0.255 - 144.7L)^2 + (31.99L)^2 = 6.621^2$$

The value of L should be at least 46.3 mH to satisfy the relationship. The L, R, and C values have been tested in a SPICE simulation, and the results are shown in Figure 5.14. The figure shows the full-wave signal across the R-L-C combination, the inductor current scaled with the resistance, and the output voltage waveform. If the first harmonic were the only one present, we should expect a peak-to-peak output voltage variation of 6%, or 0.72 V. The actual variation is 0.715 V. The ripple waveform appears to be nearly sinusoidal. This demonstrates that the first harmonic of the full-wave signal provides an excellent estimate of the ripple, and can be used in the design process.

5.6 CHARGE PUMPS

The diode-capacitor rectifier circuit injects a brief large current (a definite amount of charge) into a capacitor, then lets the charge decay slowly. The charging current is high, and suggests a KVL problem. However, if the switches can tolerate the short spikes, the diode-capacitor charging process is useful. This charge pumping action can be applied more generally. Think about a general example: A switch matrix can be used to rearrange the connections in a capacitor grid, and thus allows charge to be transferred among various locations in a circuit.

A typical charge pump application is shown in Figure 5.15. The capacitors are connected in parallel for charging, then reconnected in series for discharging. When n capacitors of equal value are used, the output voltage in the series configuration is n times the voltage in the parallel configuration. The circuit functions as a dc–dc boost converter, with boost ratio $V_{out}/V_{in} = n$. Since charge is transferred immediately in the parallel arrangement, the switch duty ratios do not alter the output value.

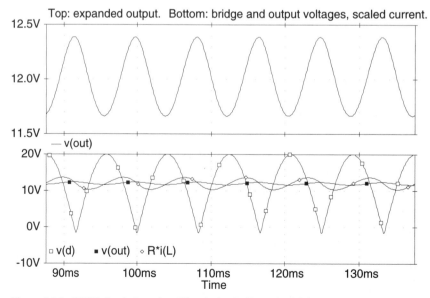

Figure 5.14 SPICE simulation of rectifier design in Example 5.5.1.

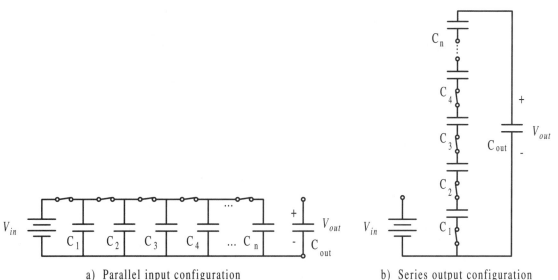

a) Parallel input configuration   b) Series output configuration

Figure 5.15   Parallel input to series output charge pump.

Charge pumps can generate substantial noise levels because of the spike current behavior. However, their simplicity makes them reasonable at low power levels. Figure 5.16
shows a charge-pump voltage doubler circuit. This topology is common, and is an important
adjunct to more complete power converters. Operation of the circuit is as follows:

- Initially, both capacitors are uncharged, and the lower switch 2 is turned on. Capacitor C_1
  accepts charge through the diode D_1 and switch 2 until the voltage $v_{C_1} = V_{in}$. Capacitor
  C_2 accepts charge through diodes D_1 and D_2 until $v_{C_2} = V_{in}$.

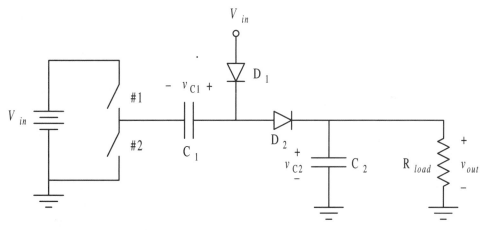

Figure 5.16   Charge-pump voltage doubler.

- Switch 2 is opened, then switch 1 is closed. Once switch 1 is closed, the voltage at the anode of D_2 will be $V_{in} + v_{C_1} = 2V_{in}$. Diode D_2 will turn on and transfer charge to the output capacitor C_2. If the load is light enough to avoid dragging down the output, voltage v_{C_2} will reach $2V_{in}$.

This circuit, in effect, places a capacitor voltage in series with the input voltage to provide a boost function. Each switch is on only long enough to permit the flow of charge to the capacitors.

One application of the charge-pump doubler can be illustrated with a conventional dc–dc buck converter, such as the one in Figure 5.17. To make this circuit work, the MOSFET must be switched on and off. A power N-channel MOSFET typically requires 5 to 10 V at its gate terminal, relative to the source terminal, to switch on. In the buck converter, the FET drain terminal is connected directly to the input. Since the source and drain voltages will be almost the same when the MOSFET is on, a voltage 5 to 10 V higher than the input supply rail is necessary to turn the switch on. A charge pump is one alternative for providing a small amount of energy at this higher voltage.

The doubler in Figure 5.16 can be used to illustrate some of the limitations of charge pumps. After switch 2 turns on, the charge on C_1 builds up to at most $Q_1 = C_1 v_{C_1} = C_1 V_{in}$. This is the maximum charge available for delivery to a load. If the switching frequency of the pump is f_s, charge is transferred at a maximum rate of $f_s C_1 V_{in}$. This product has units of current. If the charge pump is to be effective, the intended load current must be no more than the maximum rate of charge transfer. In fact, it should be much lower, because the charge cannot be removed completely unless v_{C_1} is forced to reach 0 V—in which case no boost action will occur. An example might be of help at this point.

Example 5.6.1 A doubler like the one in Figure 5.16 is to boost a 12 V supply up for a sensor application. The sensor requires 50 mW at a nominal voltage of 24 V, and can be modelled as a resistor. Switching waveforms at 100 kHz are available to drive the active switches. What is the minimum possible capacitance value C_1 to accomplish this? What should C_2 be to provide a high-quality output? If the actual capacitance C_1 is ten times the minimum value, what approximate voltage will be imposed on the sensor?

The sensor draws $(0.05 \text{ W})/(24 \text{ V}) = 2.08$ mA. The effective resistance is 11.5 kΩ. Since $V_{in} = 12$ V and $f_s = 100$ kHz, the minimum capacitance will be

$$f_s C_1 V_{in} > 2.08 \text{ mA}, \qquad C_1 > 1740 \text{ pF} \qquad (5.26)$$

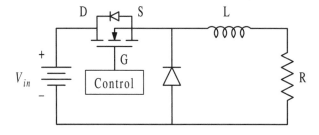

$V_{GS} \approx 10$ V for on, 0 V for off (on requires $V_{GD} \approx 10$ V)

Figure 5.17 Buck converter with gate control.

When the diodes are off (as they will be most of the time), the capacitor C_2 must support the output voltage, just as in the classical rectifier. If the pump output is allowed to drop by no more than 1% over the 10 μs period, the output RC time constant can be obtained from a linear approximation to the exponential, as

$$e^{-t/\tau} \approx 1 - \frac{t}{\tau}, \qquad \frac{10\ \mu s}{\tau} < 0.01, \qquad \tau > 1\ \text{ms} \qquad (5.27)$$

With $RC > 1$ ms, and $R = 11.5$ kΩ, we should choose $C_2 > 0.087\ \mu$F. Let us use standard values of 1800 pF for C_1 and 0.1 μF for C_2 and evaluate the result. There is a problem in this case: since C_1 is so much smaller than C_2, very little charge will transfer to C_2 each cycle. It will take a long time to pump up v_{C2}. If losses in switches and other parts are considered, there is a chance that the circuit will not work at all. Instead, let us choose $C_1 = 10C_{1(\text{min})} = 18,000$ pF, in which case the charge Q_1 is more than a tenth of the charge on C_2, and the output voltage should build up in just a few cycles. A simulated result from SPICE is given in Figure 5.18. It confirms that most of the output voltage buildup occurs within ten cycles. The output reaches a steady state value after about 300 μs. The final value imposed on the sensor—about 22 V—is two diode forward drops below $2V_{\text{in}}$.

A cascade charge-pump arrangement known as a Cockcroft–Walton multiplier, given in Figure 5.19, is often used to create very high dc voltages. The input is an ac waveform, and the circuit takes energy out only near the positive peaks of v_{in}. In the circuit, each capacitor along the upper line develops a potential equal to the peak-to-peak input, but offset by the peak of the preceding stage. Thus C_1 develops 340 V peak if the input is a 170 V peak sine wave, and each succeeding capacitor adds an additional 340 V. In the figure, the final output is approximately 5 times the peak-to-peak input, or about 1500 V in this case.

Figure 5.18 Output voltage vs. time for 18000 pF to 0.1 μF doubler.

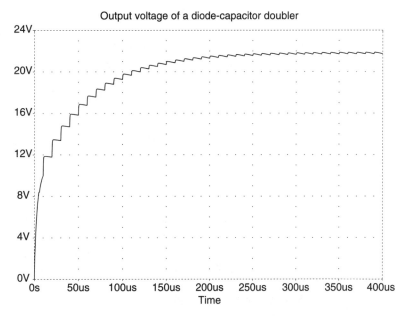

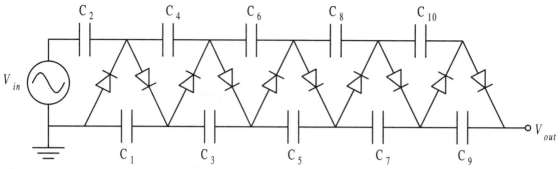

Figure 5.19 Cockcroft–Walton charge pump for high-voltage output.

All the capacitors have equal value. As in the doubler, the capacitor charge and input frequency determine the maximum load power that can be supplied.

Cockcroft–Walton circuits with many stages are used to produce voltages above 100 kV. Typical applications include the excitation voltages for electrostatic precipitators, corona voltages for copiers, and similar low-current high-voltage equipment. They have been used, for example, to provide electrostatic energy for household electronic air cleaners.

5.7 AC–DC SWITCHING POWER CONVERTERS

5.7.1 Introduction

The rectifier and diode-capacitor circuits examined so far do not permit any sort of control. The automatic switch action of a diode precludes even modest adjustments in a converter. Control action cannot be used to improve the line regulation, for instance. Early in the history of rectifiers, this limitation had to be overcome with clever circuit designs that could add time delay to diode action. Today, devices such as the SCR and GTO permit control of rectifier circuits. Since most practical ac sources are intended to provide a fixed phasor voltage, the voltage-to-current converter of Figure 5.2 should be useful in rectifier analysis.

5.7.2 Controlled Bridge and Midpoint Rectifiers

The 2×2 bridge rectifier matrix can be controlled by replacing the diodes with SCRs, as shown in Figure 5.20. Only the turn-on of an SCR can be controlled. Once on, each device acts as a diode, and will remain on until an alternative path is created for the current source. By the trial method, only two configurations are allowed with a current source load: the case with switches 1,1 and 2,2 on, where $v_{out} = v_{in}$, and the case with 1,2 and 2,1 on, where $v_{out} = -v_{in}$. With an SCR, the turn-on time can be delayed. This is equivalent to delaying each switching function by an angle $\alpha = \omega t_{delay}$.

Several possible output voltages are shown in Figure 5.21 given duty ratios of 50%. Since dc output is of interest, and because the output current comes from a dc source, the average voltage $\langle v_{out} \rangle$ needs to be determined. Its value will be

Figure 5.20   SCR single-phase bridge rectifier.

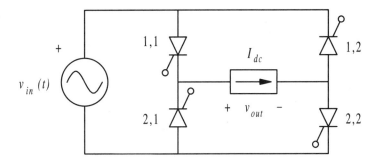

$$\langle v_{\text{out}} \rangle = \frac{1}{\pi} \int_{\alpha - \pi/2}^{\alpha + \pi/2} V_0 \cos \theta \, d\theta = \frac{2V_0}{\pi} \cos \alpha \qquad (5.28)$$

The phase delay angle allows control over the dc output just as duty ratio control permits adjustment of the output in a dc–dc converter.

One reasonable simplification of the circuit of Figure 5.20 is to share a common reference point between input and output. This circuit, illustrated in Figure 5.22, represents the full bridge of Figure 5.20 with $q_{2,2} = 1$ and $q_{1,2} = 0$. The waveforms exhibit half-wave symmetry, with alternate half-cycles removed from the waveforms of Figure 5.21, since only $v_{\text{out}} = v_{\text{in}} > 0$ and $v_{\text{out}} = 0$ are possible output voltages.

The bridge action can be restored by recognizing that the common-reference circuit extends to multiple inputs. In the bridge, both $v_{\text{out}} = v_{\text{in}}$ and $v_{\text{out}} = -v_{\text{in}}$ are possible values. This can be duplicated with a common reference by providing separate v_{in} and $-v_{\text{in}}$ inputs. This idea is illustrated in Figure 5.23. If the switch duty ratios are both 50%, the switch ma-

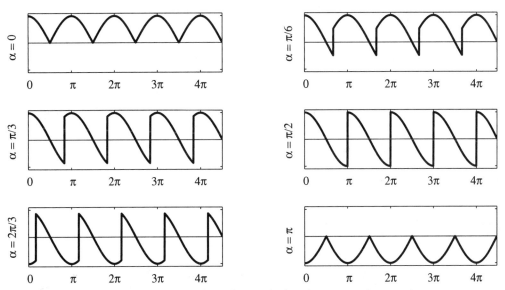

Figure 5.21   Possible output voltage waveforms for SCR bridge, shown as v_{out}/V_0 vs. $\omega t/\pi$.

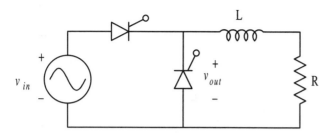

Figure 5.22 Half-wave controlled rectifier.

trix output voltage v_d will be exactly the same as in Figure 5.21, and the average value $\langle v_d \rangle = (2V_0/\pi)\cos \alpha$. Figure 5.24 shows an example for $\alpha = 60°$, with the switching functions shown. Notice that the delay angle α can be represented as a phase shift between the peaks of v_a and the centers of the pulses in $q_a(t)$.

Figure 5.23 is the most basic example of a midpoint rectifier.

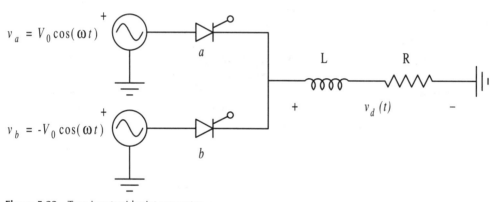

Figure 5.23 Two-input midpoint converter.

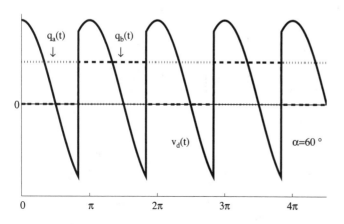

Figure 5.24 Midpoint converter output for $\alpha = 60°$.

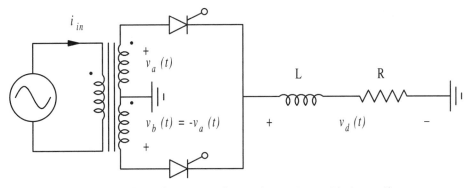

Figure 5.25 Center-tapped transformer to implement the two-input midpoint rectifier.

> **Definition:** A midpoint rectifier *is any rectifier with at least two ac voltage inputs. The inputs and the single output all share a common reference point, normally termed a* neutral.

In the case of two inputs, the midpoint rectifier is commonly implemented with a center-tapped transformer winding, as illustrated in Figure 5.25. This center-tap circuit will produce the same waveforms for both $v_d(t)$ and $i_{in}(t)$ as the bridge rectifier in Figure 5.20. There is one subtle difference: In the bridge, each valid configuration has two switches in series in the closed loop. The center-tap circuit has only one switch in each configuration, so the effect of the switch forward voltage drop is cut in half in the center-tap version. This is an advantage in low-voltage rectifiers.

The midpoint rectifier concept extends directly to any number of sources. As a practical matter, polyphase sources would normally be used, since these are the most convenient form for generation and distribution of bulk electrical energy.

> **Definition:** A balanced polyphase voltage source *consists of* m *sinusoidal sources, each having identical frequency and amplitude, and each sharing a common neutral. The voltages are separated in phase by multiples of the radian angle* $2\pi/$m. *Symbols like* mϕ, *3*ϕ, *and so on indicate such a source. If the voltages differ only slightly from the correct amplitude and phase, the source is termed an* unbalanced polyphase source.

The two-source circuits of Figures 5.23 and 5.25 represent a "two-phase" input with two voltages 180° apart[1]. Three-phase sources are the most common energy form. Six-phase sources can be formed by using three center-tapped transformers with a three-phase input, and represent the three-phase analogue of a bridge. Some high-power rectifiers use transformers to create 12-phase, 24-phase, or even more.

---

[1]It should be pointed out that the term *two-phase* as used in power electronics is distinct from the two-phase source sometimes used in power systems. In the power system version, one input is cos(ωt) while the second is sin(ωt); such an arrangement is more generally termed a *quadrature source*, and is rare in rectifier applications.

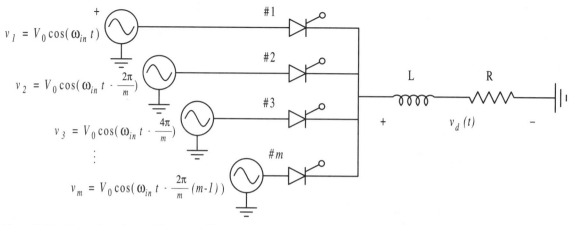

Figure 5.26   General m-phase midpoint rectifier.

The general m-phase midpoint rectifier is shown in Figure 5.26. There is a natural way to operate this circuit: Connect each of the m sources for time T/m, so that the switch duty ratios are all $1/m$. Since the switching frequency must match the input ac frequency to ensure successful energy transfer, and since the natural duty ratio operates the switches symmetrically, only phase remains as an adjustable parameter. To use it, shift each switching function by the delay angle α relative to its associated source. The output voltage $v_d(t)$ is shown in Figure 5.27 for an angular time scale $\theta = \omega_{in}t$. The period of this waveform is T/m, or $2\pi/m$ in the angular time scale. The average value is

$$\langle v_d \rangle = \frac{m}{2\pi} \int_{-\pi/m+\alpha}^{\pi/m+\alpha} V_0 \cos\theta \, d\theta = \frac{mV_0}{\pi} \sin\frac{\pi}{m} \cos\alpha, \qquad m \geq 2 \qquad (5.29)$$

In the limit as m becomes large, L'Hôpital's rule can be used to find the limiting value

$$\lim_{m\to\infty} \frac{mV_0}{\pi} \sin\frac{\pi}{m} \cos\alpha = V_0 \cos\alpha \qquad (5.30)$$

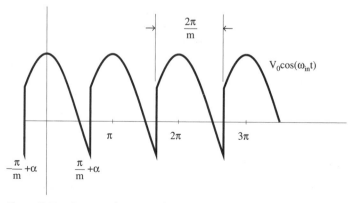

Figure 5.27   Output voltage $v_d(t)$ for m-phase midpoint converter.

The ith source sees a current $i_i(t) = q_i(t)I_{out}$, which will be a pulse train. By KVL and KCL, the midpoint rectifier with an inductive or current-source load requires that one and only one switch is on at any time,

$$\sum_{i=1}^{m} q_i(t) = 1 \tag{5.31}$$

The conversion issue in any converter is whether power is transferred between the input and output. If the output current is approximately a fixed dc value, the power is determined by the dc component of the voltage, and

$$P_{out} = \frac{mV_0 I_{out}}{\pi} \sin \frac{\pi}{m} \cos \alpha, \qquad m \geq 2 \tag{5.32}$$

Notice that this can be positive or negative. Energy can be extracted from the current source by imposing a negative average value on it. The output power would be expected to be shared evenly by the input sources, since each sees identical conditions and waveforms.

Example 5.7.1 (Superconducting energy storage.) A transformer set creates a balanced six-phase source at 480 V_{RMS} 60 Hz for a midpoint rectifier. This rectifier feeds a large superconducting coil, configured to have $L = 1.5$ H. The leads, connections, and switches create resistance of about 0.01 Ω. The maximum rated current of the coil is 4000 A. The coil is used in a prototype system to store magnetic energy for electric utility applications. How long a time is required to bring the coil energy from 0 to 100%, and how will this be accomplished? In steady state, what value of phase delay should be used to maintain full coil energy? How is energy removed from the coil? The circuit is shown in Figure 5.28.

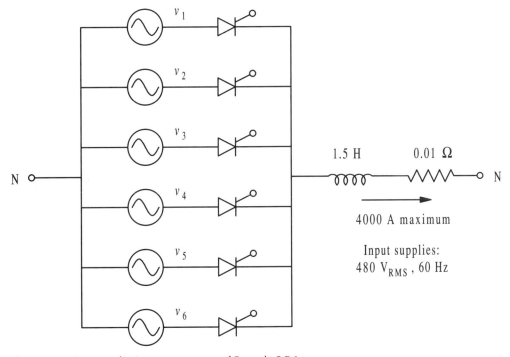

Figure 5.28 Superconducting storage system of Example 5.7.1.

The L/R time constant of 150 s is certainly long enough to make the load look like a dc current source. Of course, it is really an inductor; the time constant is so long that the coil will respond only to the average value of v_d. To increase the energy, the phase angle α will be set low to make $\langle v_d \rangle$ high and thus bring the coil current up. To decrease the stored energy, α will be set close to 180° to give $\langle v_d \rangle$ a large negative value. To keep the energy steady, the value $\langle v_d \rangle$ should be set just high enough to account for the voltage drop in the resistance.

The energy will rise most rapidly when $\langle v_d \rangle$ is maximized, that is, $\alpha = 0$. Since the RMS value is 480 V, the peak value $V_0 = 679$ V. With $\alpha = 0$,

$$\langle v_d \rangle = \frac{6 \times 679 \text{ V}}{\pi} \sin \frac{\pi}{6} \cos 0 = 648 \text{ V} \tag{5.33}$$

and the di/dt value for the inductor will be 432 A/s. It will take 9.26 s to bring the current up to the 4000 A rating limit. At that point, 12 MJ will be stored in the coil (equal to 3.33 kW · hr). When 4000 A is flowing in the coil, the voltage drop across the resistance will be 40 V, and the energy will decay in a few minutes because this 40 V will decrease the coil current. To maintain the energy, $\langle v_d \rangle$ should be set to 40 V so that the net inductor voltage is held at zero. This requires

$$\frac{6 \times 679 \text{ V}}{\pi} \sin \frac{\pi}{6} \cos \alpha = 40 \text{ V}, \qquad \alpha = \cos^{-1}(40/648) = 86.5° \tag{5.34}$$

Notice that the power during this "maintenance voltage" is $40 \times 4000 = 160$ kW. Since the coil only stores 3.33 kW · hr, this coil and circuit are probably appropriate for energy storage applications with time intervals under a minute.

The Fourier Series of output $v_d(t)$ will help with filter design. The voltage can be represented as

$$v_d(t) = \sum_{i=1}^{m} q_i(t) V_0 \cos \left[\omega_{in} t - \frac{(i-1)2\pi}{m} \right] \tag{5.35}$$

Each individual switching function has the Fourier Series

$$q_i(t) = \frac{1}{m} + \frac{2}{\pi} \sum_{n=1}^{\infty} \frac{\sin(n\pi/m)}{n} \cos \left[n\omega_{in} t - n\alpha - n\frac{(i-1)2\pi}{m} \right] \tag{5.36}$$

Trigonometric identities can be used to rewrite (5.35) in terms of (5.36) to provide the Fourier Series for the actual output waveform

$$v_d(t) = \frac{mV_0}{\pi} \sin \frac{\pi}{m} \left\{ 1 + \sum_{n=1}^{\infty} \cos(n\pi) \left[\frac{\cos(nm\omega_{in}t)}{nm+1} - \frac{\cos(nm\omega_{in}t)}{nm-1} \right] \right\} \tag{5.37}$$

for the case $\alpha = 0$. Only frequencies at multiples of $m\omega_{in}$ appear in the output, as expected. The input currents have frequencies at multiples of ω_{in}.

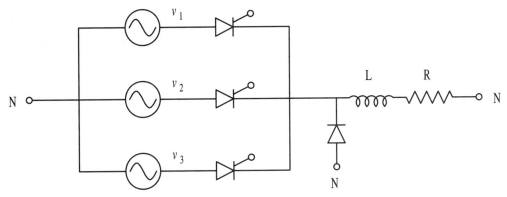

Figure 5.29 Midpoint rectifier with output freewheeling diode.

In some applications, such as certain types of dc drives, a reverse diode is connected in parallel with the load, as in Figure 5.29. The diode is known as a *freewheeling diode* because the load energy in an inductive load or a motor will not change while this diode is on. The waveform $v_d(t)$ will never be negative in this topology, and each SCR will turn off at angle $\omega t = \pi/m$. The waveforms in Figure 5.30 reflect this behavior. These waveforms also appear in a simpler situation: If the load on a midpoint rectifier is purely resistive, the output voltage will never fall below zero with SCRs, because the devices will prevent negative current flow.

For the waveforms of Figure 5.30, the average value will be

$$\langle v_d \rangle = \frac{m}{2\pi} \int_{-\pi/m+\alpha}^{\min(\alpha+\pi/m,\pi/2)} V_0 \cos \theta \, d\theta = \frac{mV_0}{2\pi} \sin \frac{\pi}{m} (1 + \cos \alpha), \qquad 0 \leq \alpha \leq \pi \quad (5.38)$$

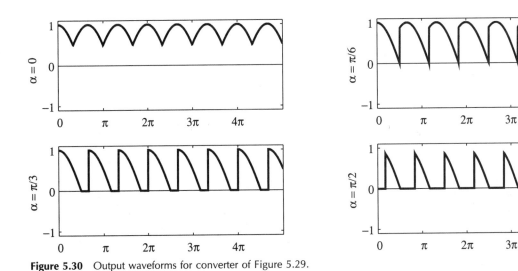

Figure 5.30 Output waveforms for converter of Figure 5.29.

If the converter is to be successful in converting ac to dc, the dc components of voltage and current will be of interest. Power will be transferred only through the dc component of the voltage if the load current is constant. If the load is a pure resistor, the current will not be constant, and the average power will be

$$P_{\text{out}} = \langle v_d(t) i_{\text{out}}(t) \rangle = \left\langle \frac{v_d^2(t)}{R_{\text{load}}} \right\rangle = \frac{v^2_{d(\text{RMS})}}{R_{\text{load}}} \tag{5.39}$$

The RMS value of a waveform such as those in Figure 5.30 is given by

$$
\begin{aligned}
v_{d(\text{RMS})} &= \sqrt{\frac{m}{2\pi} \int_{-\pi/m + \alpha}^{\min(\alpha - \pi/m, \pi/2)} V_0^2 \cos^2 \theta \, d\theta} \\
&= V_0 \sqrt{\frac{1}{2} + \frac{m \cos(2\alpha) \sin(2\pi/m)}{4\pi}}, \qquad \alpha + \frac{\pi}{m} \le \frac{\pi}{2} \\
&= V_0 \sqrt{\frac{1}{4} + \frac{m}{8} - \frac{m\alpha}{4\pi} - \frac{m \sin(2\alpha - 2\pi/m)}{8\pi}}, \qquad \alpha + \frac{\pi}{m} > \frac{\pi}{2}
\end{aligned} \tag{5.40}
$$

Some of these relationships are plotted in Figure 5.31.

5.7.3 The Complementary Midpoint Rectifier

The midpoint rectifier functions with unidirectional output current. Strictly speaking, it functions as an inverter when the average output voltage is negative, since this will reverse the energy flow. The same basic circuit can be used with the opposite current flow. This *complementary midpoint rectifier* is shown in Figure 5.32. A key point in understanding this cir-

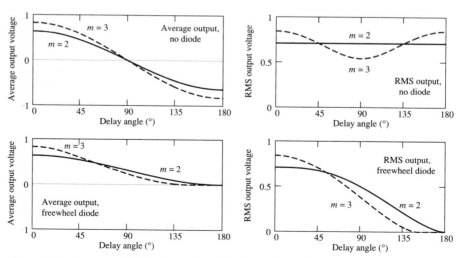

Figure 5.31   Average and RMS voltages for midpoint rectifiers with $m = 2$ and $m = 3$.

Figure 5.32 Complementary midpoint rectifier.

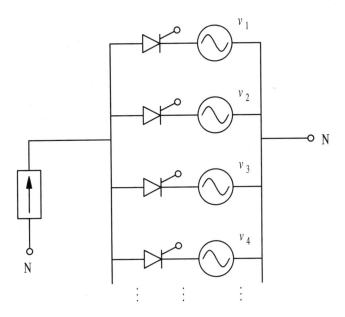

cuit is to recognize that it is essentially identical to the midpoint rectifier of Figure 5.26. The voltage $v_d(t)$ will be identical, since the voltages and switch positions are unchanged. A phase delay angle will still be an appropriate way to provide adjustment of the operation. The only thing that changes is the direction of the current flow. Now the power will be positive into the current source when $\langle v_d \rangle$ is negative, so the circuit can still function as a rectifier. With the change in current direction, it is unclear whether the SCR can be used for switching. Let us examine this question.

Example 5.7.2   A complementary midpoint rectifier connects a 5 A current source to a three-phase source. Each individual source has a peak voltage of 200 V relative to the neutral. Over what range of delay angles α can SCRs be used successfully in this converter? The circuit is given in Figure 5.33.

Figure 5.33 Complementary midpoint rectifier for Example 5.7.2.

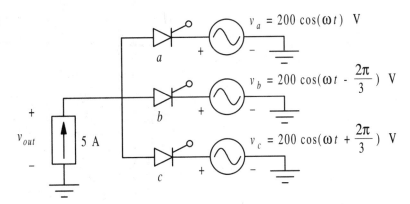

A three-phase waveform set is shown in Figure 5.34 for reference. Consider $\alpha = 0°$ first. In this case, the voltage takes on its highest value at any given time. Each switch carries 5 A when on. Are the intended switching functions consistent with the action of an SCR? To test this with the trial method, assume that switch a is on and that the others are off. Switch a carries positive current. Switch b must block the voltage $v_a - v_b$. This difference is positive during the time we intend switch a to be on, so the configuration will only be valid if switch b is not provided with a gate pulse. At angular time $\pi/3$, a pulse will be applied to switch b to try to turn it on. However, at the same time the voltage $v_a - v_b$ becomes negative, and the device is reverse biased. Recall that the SCR is off until a pulse appears, and then acts like a diode. The reverse bias condition is inconsistent with SCR turn-on, and the device will remain off in spite of the gate pulse.

In fact, the SCR cannot do the job whenever $0° \leq \alpha \leq 180°$, because over that range the voltage $v_a - v_b$ is always negative at the moment switch action tries to turn switch b on. For larger delays, the opposite occurs. For example, when $\alpha = 210°$, v_a will be larger than v_b at the moment we hope to turn on switch b. The device is forward biased when the gate pulse is applied, and it turns on immediately afterward, since the gate pulse changes it to a diode. The valid angle range for SCRs in the complementary midpoint converter is $180° \leq \alpha < 360°$. For the range $180° \leq \alpha < 270°$, the average voltage $\langle v_d \rangle$ is negative, and the circuit is a rectifier. For the range $270° \leq \alpha \leq 360°$, the output voltage is positive and the device is an inverter.

5.7.4 The Multiinput Bridge Rectifier

The most important application of the complementary midpoint rectifier is in generalizing the bridge to multiple ac sources. Figure 5.35 shows a connection of midpoint and complementary midpoint rectifiers to share the same current source. On the left side, one and only one switch can be on at any time, and the average voltage is

$$v_{\text{left}} = \frac{mV_0}{\pi} \sin\left(\frac{\pi}{m}\right) \cos \alpha_{\text{left}} \tag{5.41}$$

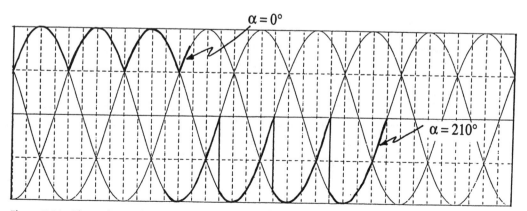

Figure 5.34 Three-phase voltages to help evaluate SCR action in the complementary midpoint rectifier.

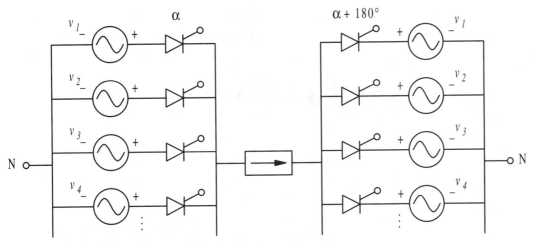

Figure 5.35   Fully controlled bridge rectifier.

On the right side, again one and only one switch can be on, and

$$v_{\text{right}} = \frac{mV_0}{\pi} \sin\left(\frac{\pi}{m}\right)\cos \alpha_{\text{right}} \tag{5.42}$$

What about the choices of α_{left} and α_{right}? If they are chosen to be equal, the average voltage across the load $\langle v_{\text{left}} - v_{\text{right}}\rangle = \langle v_{\text{left}}\rangle - \langle v_{\text{right}}\rangle$ will be zero. The clearest choice would maximize the average output voltage by setting α_{left} and α_{right} 180° apart. This is indeed the action of a fully controlled bridge rectifier.

> **Definition:** A fully controlled bridge rectifier *consists of a midpoint rectifier and a complementary midpoint rectifier with a single dc load in between. The converters are operated together. The phase control angle of the midpoint rectifier is α, while the angle of the complementary midpoint rectifier is $\alpha + 180°$. Often such a circuit is termed a* full-bridge controlled rectifier.

The bridge in Figure 5.35 and the SCR rectifier of Figure 5.20 are identical for the case $m = 2$. The case $m = 3$ yields the three-phase bridge of Example 5.7.3 below. The fully controlled three-phase bridge is one of the most common industrial rectifier circuits.

Alternatives to the fully controlled bridge include the *half-controlled bridge*, in which the complementary midpoint rectifier portion is constructed with diodes. In this case, only the left side phase control angle α can be adjusted. The right side angle is fixed at 180°. The *uncontrolled bridge* is a diode version of the complete circuit, with the left side held at $\alpha = 0°$ while the right side is kept at 180°. The uncontrolled bridge is a generalization of the single-phase bridge.

Example 5.7.3   As illustrated in Figure 5.36, six SCRs, operated as a fully controlled bridge, provide power to a permanent magnet dc motor. The input is a balanced 3ϕ source of

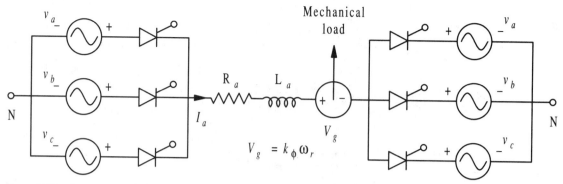

Figure 5.36 dc motor supplied from full-bridge SCR rectifier.

380 $V_{\text{line-to-line}}$, 50 Hz. The figure shows the internal circuit model to be used for the motor. It has series inductance of 25 mH and series resistance of 1 Ω, representing the inductance and resistance of its armature winding. The internal generated voltage represents the electro-mechanical action, and has a voltage proportional to shaft speed. The constant of proportionality is 2.3 V · s/rad, with the speed measured in rad/s. The shaft load represents a constant torque of 25 N · m for any positive speed. Find the shaft speed as a function of the phase control angle α.

This problem is a comprehensive application of the bridge rectifier, and we must consider a number of new dimensions to attack it. First, polyphase sources are traditionally specified in terms of the RMS potential between adjacent lines. Trigonometric identities can be used to show that is the voltage between any two lines is $V_{\ell\ell}$, the voltage between a single line and the common neutral $V_{\ell n}$ is $V_{\ell\ell}/\sqrt{3}$. (To see this, simplify the line difference $v_{ab} = v_a - v_b = V_0 \cos(\omega t) - V_0 \cos(\omega t - 2\pi/3)$ with the trigonometric identities in Appendix A.) Thus the RMS voltage of each source is 220 V. The peak value V_0 is 310 V.

The mechanical power is represented by the power drawn from V_g. Just as force acting through a distance is work (and energy), so a torque acting through an angle is work. Mechanical power is time-rate-of-change of energy. For a fixed force and a moving object, $P = fv$, where v is velocity. In the rotating case with constant torque, $P = T\omega_r$, where ω_r is angular velocity. But the power is given electrically by $V_g I_a$, so

$$V_g I_a = k_\phi \omega_r I_a = T_\ell \omega_r, \qquad k_\phi I_a = T_\ell \tag{5.43}$$

Since the mechanical load is a constant 25 N · m, the motor should draw a continuous (25 N · m)/(2.3 N · m/A) = 10.9 A to properly supply it. The L/R time constant of the load is 25 ms. Since the waveforms ought to be similar to those of a six-phase midpoint rectifier, we expect frequencies at dc and at multiples of 300 Hz to be imposed on the motor. The time constant should be slow enough to do a fair job of filtering at 300 Hz and above, so the dc component of the current should dominate.

If the current is nearly constant, the voltage across R_a is $I_a R_a$, or 10.9 V. Since there is no dc voltage carried by the inductor, the average output of the converter should equal 10.9 V + V_g. The voltage relationships are

$$\langle v_a \rangle = \frac{3}{\pi} 310 \sin\frac{\pi}{3} \cos\alpha - \frac{3}{\pi} 310 \sin\frac{\pi}{3} \cos(\alpha + 180°) = 10.9 \text{ V} + V_g \tag{5.44}$$

This simplifies to give

$$V_g(\text{V}) = 513 \cos \alpha - 10.9 \tag{5.45}$$

Since the speed is $\omega_r = V_g/k_\phi$, the result in rad/s (and also converted to RPM) is

$$\omega_r(\text{rad/s}) = 223 \cos \alpha - 4.73, \qquad \text{speed (RPM)} = 2130 \cos \alpha - 45.1 \tag{5.46}$$

The speed is a linear function of $\cos \alpha$. In this example, the speed can be set between 0 and almost 2100 RPM by adjusting α.

It is important to consider the input current for the bridge. In the midpoint case, the input current was formed as a positive pulse train with duty ratio $1/m$. For the bridge, there are two connections to each ac source, and current flows at intervals half a period apart. During one half-period, positive current flows through the left converter. During the second, negative current will flow through the right converter. The current is symmetrical, as indicated in Figure 5.37. The current pulse centers are shifted by the angle α from the input voltage peaks, so the power factor at the input will be affected by the control angle α. The actual waveform is predetermined (in the sense that its shape is well defined), and the THD will be a fixed number for a given choice of m.

The RMS value of the input current can be found by integration, and is given by

$$i_{\text{in(RMS)}} = I_{\text{out}}\sqrt{\frac{2}{m}}, \qquad m \geq 2 \tag{5.47}$$

Only the fundamental Fourier component of i_{in} contributes to power flow. This component will be

$$i_{\text{in}(1)} = \frac{4I_{\text{out}}}{\pi} \sin \frac{\pi}{m} \cos(\omega_{\text{in}}t - \alpha) \tag{5.48}$$

with RMS value $(2\sqrt{2}I_{\text{out}}/\pi)\sin(\pi/m)$. The total harmonic distortion in the current imposed on v_a is

$$\text{THD} = \sqrt{\frac{f^2_{\text{RMS}} - f^2_{\text{1(RMS)}}}{f^2_{\text{1(RMS)}}}} = \sqrt{\frac{2/m - (8/\pi^2) \sin^2 (\pi/m)}{(8/\pi^2) \sin^2 (\pi/m)}} \tag{5.49}$$

$i_a(t), \alpha = 0°$

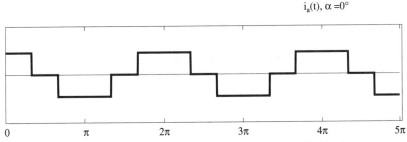

0   π 2π 3π 4π 5π

Figure 5.37 Input current seen by v_a in a three-phase fully controlled bridge.

For $m = 2$, this reduces to exactly the THD of the square wave function, 48.3%. The input power factor $pf = P/S$ will be the power from an individual source divided by the RMS volt-amp product. Since there are m sources, each carries $1/m$ of the output power. The result is

$$P = \frac{2V_0I_{out}}{\pi} \sin \frac{\pi}{m} \cos \alpha, \qquad S = \frac{V_0I_{out}}{\sqrt{m}}$$

$$pf = \frac{2\sqrt{m}}{\pi} \sin \frac{\pi}{m} \cos \alpha \qquad (5.50)$$

For $m = 2$, the power factor is $0.90 \cos \alpha$. For $m = 3$, it is $0.955 \cos \alpha$. While these power factors are good, the high THD is a matter of concern in power systems.

5.8 EFFECTS OF LINE INDUCTANCE

In practical applications, it is not always reasonable to treat the input sources to a rectifier as ideal voltages. The inductance of connections is often substantial, and line inductance is an important factor in most industrial systems. For classical rectifiers, the line inductance helps to avoid the extreme input currents expected from the rectifier analysis. The input waveform in Figure 5.10, for example, suggests infinite di/dt values coincident with the diode turn-on instants. A few microhenries of inductance on the input lines will tend to smooth the current a little, although the harmonic content will still be high and the power factor will still be low.

Other rectifiers are affected by inductance in more subtle ways. Consider that an SCR turns off only once its current reaches zero. Series inductance represents a potential KCL issue, and the inductive energy alters the switch turn-off behavior. As a general example, the operation of a two-phase fully controlled bridge will be examined, with balanced line inductance. The circuit is shown in Figure 5.38. The line inductors $L = L_{in}$ are assumed to be identical, which will be accurate if the circuit has been organized symmetrically.

In the figure, each SCR receives a turn-on command at the appropriate time, based on phase delay α. For discussion purposes, let us assign a reasonable value such as 30° to α. At time $t = 0$, switches 1,1 and 2,2 are conducting; the others are reverse biased. At angular time $\omega t = 100°$, switches 1,2 and 2,1 are forward biased, but no gate signal has been sent, so the devices remain off while 1,1 and 2,2 continue to conduct. At 120°, gate pulses are sent to 1,2 and 2,1. They immediately take on the function of a diode, and turn on because of the forward

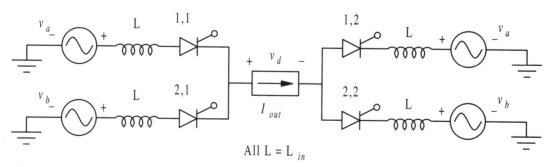

Figure 5.38 Two-phase bridge with symmetrical line inductance.

bias. However, the line inductances mean that switches 1,1 and 2,2 continue to carry current for a time—all the devices are on! This does not violate KVL because of the inductors, but it *does* alter the output voltage v_d. Specifically, a loop is formed, shown in Figure 5.39.

Let us designate the voltage at the left side of the dc current source as v_{left}. This can be written

$$v_{\text{left}} = v_a - L_{\text{in}}\frac{di_{1,1}}{dt}$$

$$v_{\text{left}} = v_b - L_{\text{in}}\frac{di_{2,1}}{dt} \tag{5.51}$$

Adding these expressions,

$$2v_{\text{left}} = v_a + v_b - 2L_{\text{in}}\frac{d(i_{1,1} + i_{2,1})}{dt} \tag{5.52}$$

But $i_{1,1} + i_{2,1} = I_{\text{out}}$ by KCL, and I_{out} is constant. Therefore $v_{\text{left}} = (v_a + v_b)/2$. For the $m = 2$ case, $v_{\text{left}} = 0$ since $v_a = -v_b$.

The currents in the two inductors must add to I_{out}, but they change in opposite ways. Since v_b is positive when $\omega t = 120°$, the current in switch 2,1 will increase with $di/dt = v_b/L_{\text{in}}$. The current in switch 1,1 will decrease with $di/dt = v_a/L_{\text{in}}$. The current change will continue until the current in 1,1 reaches zero, at which time switch 1,1 will finally turn off and v_{left} will increase to v_b. This process is an example of *commutation*. The time required for the process depends on L_{in} and on the load current. The speed of the switching device is not relevant—at least when its speed is faster than the circuit commutation process.

> **Definition:** *The term* commutation *refers to the general process of changing the current flow path from one circuit configuration to another. Diodes and SCRs exhibit what is termed* line commutation, *in which ac sources and line parameters such as inductance define the way in which current switches among paths. Line commutation typically means that a natural "make before break" action occurs in which a new current path is established before the initial path is cut off. Transistors and GTOs exhibit forced commutation, in which the switch can be turned off actively even before another path has been established.*

Figure 5.39 Just after $\omega t = 120°$, all four switches are on, and this circuit forms at the left side.

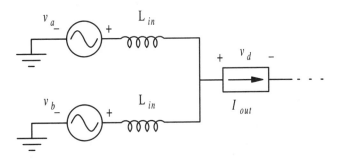

An example of a process application might help illustrate line inductance commutation effects.

Example 5.8.1   A welder for an automotive assembly line generates up to 180 A of dc current at voltages up to about 35 V. The welding load is highly inductive to maintain the current flow. It is controlled with a bridge of four SCRs. The unit is connected to a transformer that provides two 20 V supplies from a 230 V 60 Hz source. The line inductance is 20 μH. Find the phase control angle to give output voltage of 35 V, assuming the inductance has little effect on the average voltage. If the current is the full 180 A, how long will commutation require under this scenario? Plot the output voltage, and comment on whether the commutation process alters the output voltage average value.

Figure 5.40 shows the circuit that the above problem statement describes. The output average voltage is the sum of the left and right contributions, and will be

$$\langle v_{\text{out}} \rangle = \frac{4V_0}{\pi} \sin \frac{\pi}{2} \cos \alpha \tag{5.53}$$

Since the supplies are each 20 V_{RMS}, the peaks will be $V_0 = 28.3$ V, and the output average is $\langle v_{\text{out}} \rangle = 36.0 \cos \alpha$. To make this value 35 V, as requested in the problem statement, the delay angle should be set to $\alpha = 13.6°$. When it is time to turn on switches 1,2 and 2,1, the angular time will be $\omega t = 193.6°$. At that moment, the voltages are $v_a = -8.48$ V and $v_b = 8.48$ V. When the gate of switch 1,2 fires, the switch current will begin to increase at the rate 4.24×10^5 A/s. The full level of 180 A will be reached in about 425 μs. Thus the commutation process takes about 425 μs (from $\omega t = 193.6°$ to $\omega t = 202.8°$), and the output voltage will be zero during the process. This is just over 5% of the 1/120 Hz period of the out-

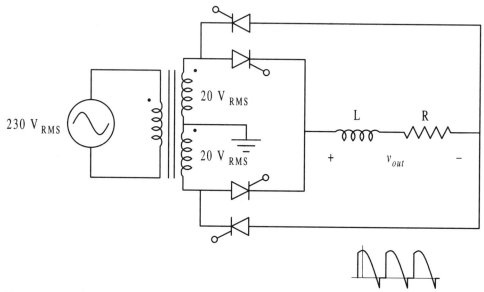

Figure 5.40   Welder rectifier for Example 5.8.1.

put voltage, so the effect on $\langle v_{out} \rangle$ could be as much as 5%. This can be tested, since the necessary angles are known:

$$\langle v_{out} \rangle = \frac{1}{\pi} \int_0^{103.6°} 36 \cos \theta \, d\theta + \frac{1}{\pi} \int_{112.8°}^{180°} -36 \cos \theta \, d\theta \tag{5.54}$$

$$= 34.1 \ V$$

About a 1 V decrease is seen in the output average value under these conditions.

Line inductance is seen to give rise to a *commutation interval*. In the angular time scale, the width of this interval is often given the symbol u. The interval is longest when α is close to zero, since the switch voltages are low near the switching points in this case. High phase delay angles give much shorter values of u. The commutation effect is particularly important in rectifiers with large output currents, such as those used for welding, smelting, and high-voltage dc power transmission. The inductance effect is not necessarily bad, although it does make the output average value load-dependent.

5.9 RECAP

Rectifiers are most commonly built as ac voltage to dc current converters. When a single ac source is involved, a bridge of four diodes or four SCRs can be used to realize a rectifier function. In low-power applications, the inductor required for controlled rectification can be a costly item. Because of this, many low-power converters dispense with the inductor and use direct capacitive filtering instead.

A diode-capacitor bridge or half-wave circuit yields a straightforward power supply design procedure. If the capacitor is large, the rectifier output will be close to the ac peak value V_0. The ripple will be small if the RC time constant of the load-capacitor pair is far slower than the ac frequency. The exponential can be approximated with a linear function when this is true,

$$e^x \approx 1 + x, \quad x \text{ small} \tag{5.55}$$

In the diode-capacitor rectifier, we saw that

- The diodes are all off most of the time. They turn off at a time given by $\tan(\omega t_{off}) = 1/(\omega RC)$.
- The peak output voltage is V_0. The voltage at time t_{off} is very close to this peak value.
- While the diodes are off, the RC pair experiences an exponential decay. However, this decay is nearly linear if $RC \gg T$. Thus the voltage fall closely follows the line $V_0[1 - t/(RC)]$.
- The diodes turn on again when the ac voltage catches up with the exponential decay. This happens approximately when $1 - t_{on}/(RC) = |\cos(\omega t_{on})|$.
- A worst-case estimate for the relative ripple $\Delta V_{out}/V_0$ is given by $1/(2f_{in}RC)$ for a full bridge, and $1/(f_{in}RC)$ for a half-wave rectifier.

- Current is drawn in large, high spikes. These have substantial RMS values, and yield poor power factor.

- Low-ripple designs often require capacitance values of 10,000 μF and above. Supplies designed as diode-capacitor rectifiers tend to be bulky because of the large capacitors. They are heavy because magnetic transformers are normally used to provide the correct value of voltage.

In general, the addition of inductance to a rectifier output improves its performance. However, the value should be large enough to ensure that current will always flow, so that the inductor provides the desired current source function. If the inductor is sufficient, the ripple waveform is well predicted by the first harmonic of the full-wave voltage applied to the load and filter by the rectifier. The relative ripple becomes

$$\frac{\Delta V_{\text{out(peak-to-peak)}}}{V_{\text{out}}} = \frac{4}{3|1 - \omega^2 LC + j\omega L/R|} \tag{5.56}$$

This is maximized when the LC pair is resonant, with $\omega = 1/\sqrt{(LC)}$, so resonance should be avoided in design. Normally the LC pair is chosen to be resonant well below ω so that ripple is reduced.

Diode-capacitor rectifiers suggest the basic form for charge pump circuits, in which brief switch connections of capacitors are used to transfer energy between different voltage levels. The generic charge-pump alternates between parallel and series connections. In parallel connections, low voltages are imposed on a group of capacitors. They can be reconnected in series to provide a step-up ratio equal to the number of storage devices. A doubler and the high-voltage Cockcroft–Walton multiplier circuit were introduced. Triplers and many other configurations have been described in the literature.

More complete controlled rectifiers use SCRs and dc current sources (in the form of L-R series combinations) to achieve consistent operation. *Phase control* is an appropriate way to adjust such a circuit, since both the frequency and duty ratio are determined by other factors. In the common-neutral version, called a *midpoint rectifier*, the average output voltage is given by

$$\langle v_{\text{out}} \rangle = \frac{mV_0}{\pi} \sin \frac{\pi}{m} \cos \alpha \tag{5.57}$$

where V_0 is the peak line-to-neutral voltage, α is the phase-delay control angle, and m is the number of sources (at least two). Often this involves a balanced polyphase source. When a *freewheeling diode* is provided, the average output is instead

$$\langle v_{\text{out}} \rangle = \frac{mV_0}{2\pi} \sin \frac{\pi}{m} \cos(1 + \alpha) \tag{5.58}$$

The midpoint rectifier (without freewheeling diode) can accommodate energy flow in both directions if the dc source can maintain flow with $\langle v_{\text{out}} \rangle < 0$. In this sense, the circuit is also an inverter, although it is unusual to apply it in this manner. This suggests a *complementary midpoint rectifier*, in which the switching devices are turned around to permit reverse current flow. For this converter, the output voltages are identical to the ones in the midpoint rectifier. Successful operation with SCRs, however, limits the phase angles to $180° \leq \alpha \leq 360°$.

When a midpoint rectifier and a complementary midpoint rectifier are combined, the result is a *fully controlled bridge*—an extension and generalization of the basic four-switch bridge at the beginning of the chapter. The bridge will be controlled with phase delays α and $\alpha + 180°$ to maximize the output voltage. In effect, the average output is doubled since the left and right converters create a difference potential across the load.

Line inductance alters the *commutation process*. If inductance is present in the supply lines, the inputs will not be ideal voltage sources. Even infinitely fast switches will require time to switch over to a second current path when inductance is present. The effect is larger with higher load currents and with lower values of the phase control angle.

PROBLEMS

1. A classical rectifier is to be designed to provide 5 V ± 1% for a logic circuit, with output power of up to 12 W. The input is a conventional 120 V, 60 Hz source. Choose a transformer ratio, diode configuration, and capacitor value to meet the requirements.

2. A classical rectifier uses a standard 120 V / 6.3 V 60 Hz transformer. What output dc voltage will result if a diode bridge is used with a large capacitor?

3. Design a classical rectifier for 24 V ± 1% output, at a power level of 4 W or less. Choose the appropriate transformer and capacitor if the circuit is to operate from 230 V at either 50 Hz or 60 Hz. What will be the peak input current if the ac source is ideal?

4. A classical rectifier is to supply 12 V ± 1% at current levels up to 1 A. Design a circuit to meet this need, given an input of 120 V, 60 Hz. What input current rating will be needed for the transformer?

5. Design a classical rectifier to produce 5 V ± 1% at power levels up to 100 W. The source is 120 V, 60 Hz.

6. An unusual rectifier application substitutes SCRs for diodes in a classical rectifier arrangement.

   a. Can this technique provide control over the output voltage? If so, find the average output as a function of the SCR phase delay angle.

   b. Determine a ripple estimate for this circuit approach.

7. A classical bridge rectifier has 15 V output. The input source can be either 50 Hz or 60 Hz. The filter capacitor value is 130 000 μF.

   a. What is the maximum ripple level for loads up to 5 A?

   b. It is desired to use an inductor to augment the output filter so a smaller capacitor can be used. A capacitor value of 20 000 μF is desired. What inductor value will be needed to meet the same ripple performance as the capacitor alone? You may assume that the turns ratio can be altered to account for any change in the dc output.

8. A voltage doubler is to be designed to supply 8 V or more at up to 50 mA for the control circuits of a dc–dc converter. The converter is supplied from a 6 V battery. Choose capacitor values to meet this need. Will the control circuit continue to work if the battery voltage falls to 5 V?

9. Use SPICE or a similar circuit analysis package to simulate a Cockcroft–Walton multiplier with two stages (i.e., a circuit like that in Figure 5.19, but only up through C_3, C_4, and D_4). The input is a 120 V_{RMS} 60 Hz sine wave. Use $C = 0.01$ μF. What maximum load can be supplied without dropping the steady-state output potential by more than 10%?

10. Figure 5.41 shows a "two-stage doubler" multistage circuit. Each switch is on just long enough for the completion of the charge transfer process, but no two switches are on simultaneously. What is the output voltage if the load is light?

11. An SCR bridge takes power from a 120 V_{RMS} 60 Hz source, and provides it to a load at a nominal 72 V dc.

   a. What phase delay angle α will be used to meet this requirement?

   b. Plot the output voltage of the bridge, one of the switching functions, and the input current waveform for your value of α from part (a), given a large output inductor and a 10 A load.

   c. If the output power is to be 1 kW, what value of inductance will be needed to keep the output ripple below 1% peak-to-peak?

12. A full-bridge three-phase SCR rectifier supplies an *L-R* load from a 208 $V_{line-to-line}$ 60 Hz source. The load draws 12 kW when the SCR phase delay angle is 75°. Plot the load power as a function of delay angle for $0° \leq \alpha \leq 75°$.

13. The transformer in a large welder provides single-phase ac output at 30 V, 50 Hz. An SCR bridge is used to control the welder. A large inductor helps to maintain a welding arc.

   a. What phase delay angle should be used to set an initial voltage of 25 V_{dc}?

   b. The phase angle drops back once the arc has formed. What phase delay angle is needed for 8 V_{dc} output?

14. A midpoint converter supplies 60 V at up to 20 A for a chemical coating process. The input is a three-phase source, and each supply is at 120 V_{RMS}, 60 Hz.

   a. What value of inductance would be expected to keep the output current ripple below 1 A peak-to-peak?

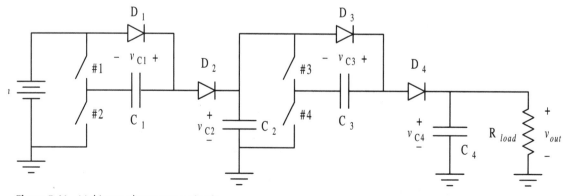

Figure 5.41   Multistage charge-pump circuit.

b. What phase delay angle will be used?

c. What power factor is imposed on the input sources?

15. Consider the full-bridge rectifier in Figure 5.35, with $m = 3$. If the input line-to-line RMS voltage is V_{in}, what is the maximum voltage that must be blocked by any given switch? What switch voltage rating should be used for the case $V_{in} = 230$ V?

16. A full-bridge rectifier with three-phase input supplies a dc motor. The motor has series armature inductance of 0.5 mH and resistance of 0.033 Ω. Rated current is 140 A, and rated output power is 30 kW at 2500 RPM. At full rated speed, the internal voltage $V_g = 225$ V.

   a. The converter should be able to drive this motor at full load current and full speed. Choose an appropriate three-phase voltage (line-to-line potential) from the following possible standard values: 208 V, 230 V, 480 V (all at 60 Hz). What did you consider in making this choice?

   b. What value of α would be appropriate to drive the motor at full load torque and full load speed?

   c. The text example points out that torque and current are proportional. Therefore, a motor's current rating implies a certain torque rating. Consider this motor, driving a load which requires a torque $T_{load} = 10 + 0.7\omega$ in units of newton-meters. What value of α will produce a speed of 1000 RPM with this load? What is the maximum speed with this load if the torque rating is not to be exceeded? At what value of α does this occur?

17. A six-phase midpoint rectifier operates from sources with line-to-neutral voltage of 190 V_{RMS} at 50 Hz. The load is a series R-L combination that draws 40 kW at 200 V. What inductor will be needed to keep the output ripple below $\pm 5\%$? With this inductor, what is the output power of this converter as a function of α?

18. A two-phase full bridge converter supplies a 10 Ω load from sources of 220 V_{RMS} at 50 Hz. No inductor is present, and a freewheeling diode prevents the output voltage from becoming negative.

   a. Find the resistor power as a function of α in this circuit.

   b. What is the power factor measured at the input sources when $\alpha = 45°$?

19. It is suggested to use a full SCR bridge to convert 120 V at 60 Hz for a 12 V dc load. Draw a circuit that can perform this function. Comment on this application from the following perspectives:

   a. What power factor will be seen at the ac input if this circuit is designed for low ripple?

   b. What will happen if a noise spike causes α to change (either up or down) by about 5° for a time? Is the circuit useful?

20. It is desired to use a superconducting energy storage system to store about 1000 MW · hr of energy (the energy produced by a large power plant in 1 hr), equal to 3.6 TJ. You have been asked the assess the power conversion requirements for this device. Example 5.7.1 might be of help. It will be almost impossible to build even a superconducting system with series resistances below 0.01 Ω, and R will probably be much higher. This is necessary to account for normal conductors used in connections, switches, circuit breakers, and other components.

   a. If the superconducting inductor is to store this energy for up to 24 hr, what is the minimum inductance that should be considered in the design? What currents will be involved?

b. It is intended to store the energy over about 10 hr at night, then release it over about 2 hr each afternoon. What voltage will be necessary to change the energy level over these time intervals? What voltage is needed to maintain the energy storage in steady state?

c. For noise purposes, the equivalent of a 12-phase SCR bridge is probably the minimum that would be considered. What line-to-neutral voltage will be needed to support the application, and to provide enough extra that α need not be set to 0° or 180°? With this choice, what is the power factor during the nighttime energy storage process?

d. Line inductance ought to play a role in this system. If the line inductance is 1 mH, how fast will commutation occur? How much will the average voltage be affected at the beginning of the storage process?

e. Can this system function with a freewheeling diode?

21. A six-pulse SCR bridge for a three-phase source draws power from a 500 V, 50 Hz line (the voltage is specified as the RMS line-to-line potential, as is standard in three-phase practice). The rectifier supplies a large dc motor. The motor has series inductance $L_a = 0.4$ mH and series resistance $R_a = 5$ mΩ. The internal voltage $V_g = k\omega$, where k is a field flux constant equal to 3 V $\cdot$ s/rad, and ω is the shaft speed in rad/s. The shaft torque is $T_e = ki_a$, where i_a is the motor current. It is desired to operate this motor with a 250 A current limit to provide a controlled acceleration at start-up. Determine the SCR phase delay angle as a function of motor speed to enforce a 250 A current.

22. An SCR bridge is used for a capacitor charging application for a high-energy pulse system. The bridge takes power from a 50 kV, 60 Hz input, and supplies a capacitor bank with total capacitance of 100 μF. A small series inductance is used to help smooth the operation. The rectifier acts as follows: the delay angle is set so the output matches the capacitor voltage, then the average voltage is raised slowly. Since $i_C = C\, dv/dt$, the charging current can be controlled to set the rate of charge.

a. What is the maximum value of stored energy possible in this circuit?

b. It is desired to charge the capacitor bank from zero in one minute. Find the phase delay angle α, as a function of time, that will support this performance while minimizing the charging current.

c. With a charging time of one minute, what is the maximum power drawn from the ac source during the charging process?

23. Two SCRs are used in a midpoint converter arrangement. The source RMS voltage is 230 V, and the frequency is 50 Hz. The load is a 100 mH inductor in series with 2 Ω. The input lines have series inductance of 50 μH.

a. For a phase delay of 45°, plot the voltage waveform measured at the output of the midpoint converter.

b. What is the output power for 45° delay?

24. A rectifier station for a high-voltage dc transmission system uses a 24-pulse arrangement. This can be represented as an SCR midpoint converter. The input 60 Hz sources each have RMS value 500 kV. The nominal load is 1000 MW, and the intent is to supply a dc line at 600 kV.

a. At nominal load, what phase delay angle will be used, assuming the load acts resistively?

b. Interference with nearby communication lines can be a problem. At nominal load, what value of series inductance should be added to keep the ripple on the dc output line below 100 $V_{\text{peak-to-peak}}$?

REFERENCES

J. D. Cockroft, E. J. Walton, "Further developments in the method of obtaining high velocity positive ions," *Proc. Royal Society A*, vol. 136, p. 619, 1932.

D. J. Comer, *Modern Electronic Circuit Design*. Chapter 15. Reading, MA: Addison-Wesley, 1976.

S. B. Dewan, G. R. Slemon, and A. Straughen, *Power Semiconductor Drives*. New York: Wiley, 1984.

G. K. Dubey, S. R. Doradla, A. Joshi, and R. M. K. Sinha, *Thyristorised Power Controllers*. New York: Wiley, 1986.

R. G. Hoft, *Semiconductor Power Electronics*. Chapter 7. New York: Van Nostrand Reinhold, 1986.

A. W. Kelley, W. F. Yadusky, "Rectifier design for minimum line-current harmonics and maximum power factor," *IEEE Trans. Power Electronics*, vol. 7, no. 2, pp. 332–341, April 1992.

F. G. Spreadbury, *Electronic Rectification*. London: Constable, 1962.

C. K. Tse, S. C. Wong, and M. H. L. Chow, "On lossless switched-capacitor power converters," *IEEE Trans. Power Electronics*, vol. 10, no. 3, pp. 286–291, May 1995.

R. J. Widlar, "IC provides on-card regulation for logic circuits," Application Note 42 in *Linear Applications Databook*. Santa Clara: National Semiconductor, 1986.

CHAPTER 6

Figure 6.1 Industrial inverters for ac motor control.

INVERTERS

6.1 INTRODUCTION

Under the right circumstances, a controlled rectifier manages bidirectional energy flow. In practice, inverters and rectifiers are distinguished more by the nature of their dc sources than by the energy flow direction. *Current-sourced inverters*, in which the dc source tends to maintain a fixed current, usually look very much like the controlled bridge rectifiers of Chapter 5. The waveforms and operating properties follow the rectifier treatment, and FCBB switches such as the SCR or GTO are appropriate. The *voltage-sourced inverter* converts energy from a battery or other fixed dc voltage into ac form. This arrangement functions much differently than the current-sourced case and requires different switch types. As a result, the term *inverter* used alone generally refers to the voltage-sourced case. However, the distinction is still not firm. It is entirely possible to build rectifiers in which the dc terminals act as a voltage source. This type of rectifier is beginning to find application at the line end of many dc power supplies.

In this chapter, crucial inverter issues are discussed first, and the basic circuits are formed from switch matrices. Next, the alternatives for inversion from dc voltage sources are described. Two control methods are discussed. The first uses an inverter to construct a square wave with a particular pulse width. This approach is applied in low-cost inverters, and also at power levels above about 100 kW. The second method modulates the duty ratio of a switch matrix (pulse width modulation, PWM) to reconstruct low-frequency ac waveforms from fast switching. The PWM approach is the basis for most designs up to at least 100 kW, including the ac motor drives in Figure 6.1. The approach is simple to implement; we will begin to see how basic control functions are implemented for power electronics.

A PWM circuit for high-quality rectification is discussed midway through the chapter. This is an extremely important application, given recent standards for power quality and given the need for high power factor. The current-sourced inverter approach is reexamined briefly later in the chapter.

The end of the chapter discusses major applications, with emphasis on PWM circuits for ac motor control. Opportunities are almost unlimited for efficient, high performance motor controls, and this application is one of the most important future growth areas for power electronics.

6.2 INVERTER CONSIDERATIONS

A true inverter takes power from a fixed dc source, and applies it to an ac load such as a utility grid, an ac motor, a loudspeaker, or a conventional product normally powered from an ac line. It is useful to distinguish two types of ac loads.

Definition: *An* active ac load *has the characteristics of an ideal source. It produces a specific waveform as a function of time. The time information in the waveform is available for control, and affects the energy transfer process.*

Definition: *A* passive ac load *has characteristics more like those of an impedance than those of a source. While the waveform might be a good sinusoidal signal, there is no absolute timing information involved. Time information is not available for control, and in general does not alter the energy transfer properties.*

The utility grid is the most familiar active ac load. The utility waveform is controlled very precisely at a central location. A converter connected to the grid cannot alter the timing of the sinusoid. It is possible, for example, to use phase delay control for loads of this type. Many ac motors have active characteristics over short time periods. Active ac loads can be controlled just like the rectifier circuits in Chapter 5, and most systems with these loads use phase delay control as the adjustment tool.

Most devices intended as loads, such as motors under normal circumstances, fluorescent lamps, backup power units, and even radio transmitters, require sine waves but do not have timing requirements. In a series *R-L* load, for instance, any attempt to change the voltage waveform by altering its phase will shift the current waveform to follow the change. A definite phase angle between the voltage and current is always present, and cannot be altered to manipulate energy flow. In this case, phase delay control will not work directly, and other control approaches must be considered.

Real ac loads often include magnetic transformers. The discussion of the magnetic transformer in Chapter 4 pointed out they function only with ac signals. If dc voltage is imposed on a transformer, it can cause the flux to increase until the device no longer functions. This is a key consideration in inverters: Any dc component is unwanted, and in fact can cause considerable trouble. A practical inverter circuit should not produce any dc output component.

Figure 6.2 shows a 2 × 2 switch matrix to transfer energy from a dc voltage source into an ac current source. This circuit is referred to as a *full-bridge inverter*. An ac current source is probably unfamiliar; series inductance is used to provide approximate current-source characteristics. The switches must carry bidirectional current. The voltage source is unidirectional, so only one blocking direction is required. Inverter applications usually use some type of transistor, with a reverse-parallel diode to give it bidirectional current capability. Today, the power MOSFET is used up to 1–10 kW, while IGBTs are used at power levels up to 100 kW. For even higher levels, GTOs with reverse-parallel diodes can be substituted.

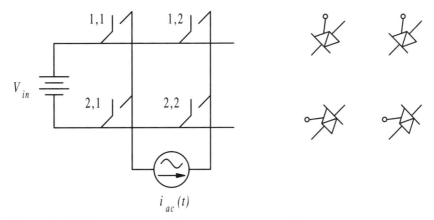

Figure 6.2   Switch matrix for dc voltage to ac current conversion.

In many applications, the complete 2×2 matrix is unnecessary. A "midpoint" version of the inverter can be built if a common neutral connection can be made available. Polyphase dc sources are absurd, but a "two-phase" version can be built in many cases. This is shown in Figure 6.3. The common neutral point shown there means that only two switches are required. The circuit of Figure 6.3 is known as a *half-bridge* inverter. It is used in switching audio amplifiers, small backup power units, and a variety of single-phase loads. The transistors are generic. Any device with a FCFB function can be used with diodes to implement the inverter.

Inverters for ac motors and other industrial applications need three-phase outputs. A matrix for this case, called a *hex-bridge* since it uses six devices, is shown in Figure 6.4. Some manufacturers of MOSFETs and IGBTs sell fully packaged hex-bridges, connected and specified for three-phase inverter applications at levels up to about 10 kW. At even higher power levels, the switching devices are often packaged in half-bridge units. Three half-bridges can be used together to realize a complete hex-bridge.

Figure 6.3   Half-bridge inverter showing the common neutral.

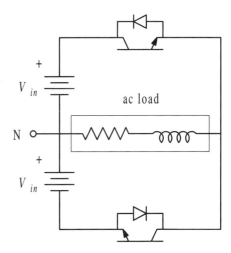

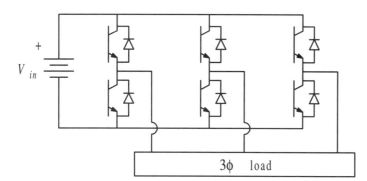

Figure 6.4 Hex-bridge for dc voltage to 3ϕ ac current conversion.

Consider a half-bridge with an active ac load, such as the specific 50 Hz inverter shown in Figure 6.5. For successful energy transfer, the output voltage waveform must contain a 50 Hz Fourier component. As a practical matter, it should not produce any dc component. KVL and KCL requirements force the switches to work in alternation. The output voltage is

$$v_{\text{out}}(t) = q_1(t)V_{\text{in}} + q_2(-V_{\text{in}}), \qquad q_1 + q_2 = 1 \tag{6.1}$$

which simplifies to

$$v_{\text{out}}(t) = (2q_1 - 1)V_{\text{in}} \tag{6.2}$$

The dc component is the average $2D_1 - 1$. The most straightforward way to meet the requirements is to use the switches to create a symmetrical 50 Hz square wave. The duty ratios of one-half will make the dc component zero, while the frequency matching condition for power flow is satisfied by the choice $f_{\text{switch}} = 50$ Hz. The output Fourier Series is

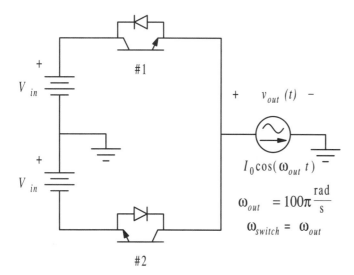

Figure 6.5 Half-bridge with active 50 Hz load.

$$v_{\text{out}}(t) = \frac{4V_{\text{in}}}{\pi} \sum_{n=1}^{\infty} \frac{\sin(n\pi/2)}{n} \cos(\omega_{\text{switch}}t - n\phi) \tag{6.3}$$

The wanted component is the fundamental, since $\omega_{\text{switch}} = \omega_{\text{out}}$, and its amplitude is $4V_{\text{in}}/\pi$. The phase ϕ is the only adjustable parameter, based on using the ac load as an absolute phase reference.

The discussion of the half-bridge circuit is extremely constraining. In essence, only a square wave can do the job. The situation is worse if the load is passive, as in Figure 6.6. In this case, ϕ no longer represents an adjustable parameter: There is no absolute phase reference to be used as the basis for adjustment. Phase must always be measured from some reference signal, and none is present with a passive load. The implication is that no adjustment is possible. Square-wave operation is adequate for certain backup systems, but it does not give the flexibility needed in the majority of inverter applications.

6.3 VOLTAGE-SOURCED INVERTER CONTROL

The half-bridge circuit is sometimes called a *two-level inverter* because the voltages $+V_{\text{in}}$ and $-V_{\text{in}}$ are the only possible outputs. A third switch can be added, or the original full bridge can be used, to permit output of 0. Once three output levels are possible, operating flexibility can be restored. Figure 6.7 shows the two circuits, with waveform examples. The waveforms are exactly the same as those within forward converters—because the front end of a forward converter is an inverter. In the three-level circuit, adjustment is restored by adding another degree of freedom. The full-bridge inverter will generate a square wave when $q_{1,1} = q_{2,2}, q_{2,1} = q_{1,2}$, and the duty ratios are 50%. The waveforms in Figure 6.7 can be produced by adjusting the phase shift between $q_{1,1}$ and $q_{2,2}$. This *relative phase control* provides output adjustment for passive loads.

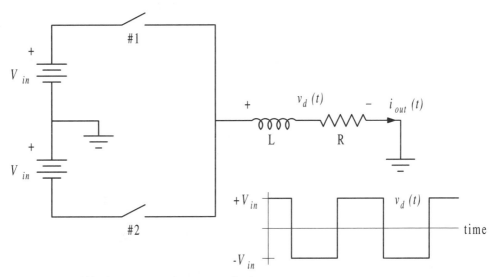

Figure 6.6 Half-bridge inverter with passive ac load.

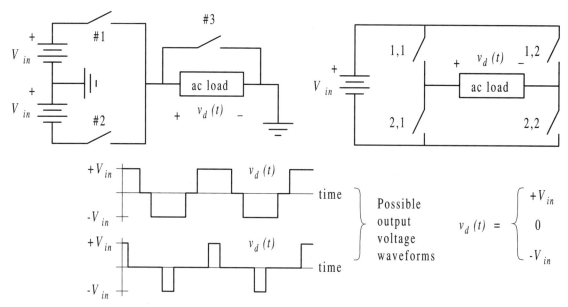

Figure 6.7   Three-level inverter approaches.

Example 6.3.1   A full-bridge inverter supplies a series *L-R* load, as illustrated in Figure 6.8. Define the *displacement angle* δ as the phase shift between $q_{1,1}$ and $q_{2,2}$, and assume the circuit meets other basic inverter requirements. Express the output in terms of switching functions, then determine the amplitude of the fundamental output frequency as a function of δ.

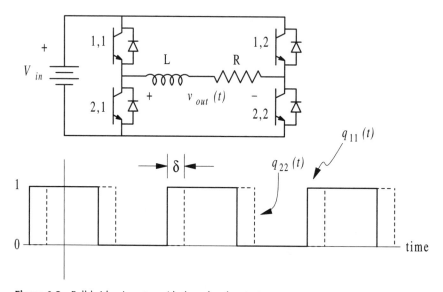

Figure 6.8   Full-bridge inverter with three-level output.

The KVL and KCL conditions require $q_{1,1} + q_{2,1} = 1$ and $q_{1,2} + q_{2,2} = 1$. The output voltage will be nonzero only if either $q_{1,1}$ and $q_{2,2}$ are on together or if $q_{1,2}$ and $q_{2,1}$ are on together. The output can be written

$$v_{\text{out}}(t) = q_{1,1}q_{2,2}V_{\text{in}} - q_{1,2}q_{2,1}V_{\text{in}} = (q_{1,1} + q_{2,2} - 1)V_{\text{in}} \tag{6.4}$$

If the duty ratios are maintained at 50%, there will be no dc component. Let us use $q_{1,1}$ as the phase reference. The function $q_{2,2}$ is delayed from it by δ. The Fourier Series (with $D = \frac{1}{2}$) tells us:

$$V_{\text{in}}(q_{1,1} + q_{2,2} - 1) = \frac{2V_{\text{in}}}{\pi} \sum_{n=1}^{\infty} \frac{\sin(n\pi/2)}{n}[\cos(n\omega_{\text{out}}t) + \cos(n\omega_{\text{out}}t - n\delta)] \tag{6.5}$$

The fundamental should be the wanted component. From trigonometric identities, its value is

$$v_{\text{out(wanted)}} = \frac{4V_{\text{in}}}{\pi} \cos\frac{\delta}{2}\cos\left(\omega_{\text{out}}t - \frac{\delta}{2}\right) \tag{6.6}$$

This is a maximum when $\delta = 0$ (when the output is a full square wave). When $\delta = 90°$, the switching functions overlap only half the time, and the amplitude is reduced to $1/\sqrt{2}$ times the maximum value. When $\delta = 180°$, there is no output.

The term *voltage-sourced inverter* (VSI) often refers to inverters that use relative phase control. Important characteristics of the VSI include:

- The output wanted component amplitude can be adjusted from 0 to $4/\pi = 1.27$ times V_{in}. The extra 27% of possible amplitude above the input value is helpful.

- The switches operate at a low frequency—the output frequency—which is the minimum rate that meets frequency matching requirements.

- The duty ratios are maintained at 50%, so the dc component is always cancelled. In fact, all even harmonics are cancelled when the duty ratios are held at 50%.

- The output is composed of pulses like those of a square wave. Usually, low-frequency harmonics such as the third and the fifth are relatively large. This is a challenge for filtering out unwanted components. It also means that the load is likely to alter filter behavior.

The ability to filter out unwanted components is a key concern. The unwanted frequencies begin at $3\omega_{\text{out}}$, and a low-pass filter should do the job. Let us first examine a series L-R load, and then consider a resonant filter.

Example 6.3.2   A VSI provides energy to a resistive load with an inductor as a filter from a 144 V lead-acid battery set. The load has $R = 2\ \Omega$, while the inductance is 3 mH. The output frequency is 60 Hz. The circuit is shown in Figure 6.9. What is the highest possible wanted current component, and what voltage will it produce at the resistor? When the control angle δ is set to $30°$, what is the Fourier Series of the output current? What is the load power when $\delta = 30°$, taking only the fundamental through fifth harmonics?

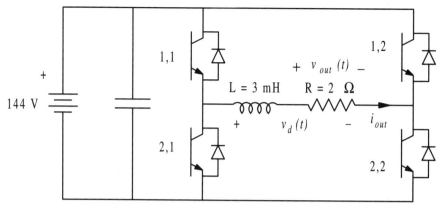

Figure 6.9 VSI circuit for Example 6.3.2.

In this circuit, the voltage waveform at the output of the switch matrix, shown as v_d on the figure, is well-defined by the switch action. It can be taken as an equivalent source for analysis of the problem. The highest possible amplitude will occur when $\delta = 0°$. The wanted voltage component then has amplitude $4V_{in}/\pi = 183.3$ V and RMS value $2\sqrt{2}V_{in}/\pi = 129.6$ V. The current magnitude associated with the wanted component can be found from the 60 Hz equivalent circuit, and the output voltage is just that current times R:

$$\left|I_{out(wanted)}\right| = \frac{129.6}{R + j(120\pi)L} = 56.43 \text{ A}$$

$$V_{out(wanted)} = IR = 112.8 \text{ V} \tag{6.7}$$

This voltage can be taken as the phase reference if desired. The maximum wanted output is 113 V_{RMS}. If δ is set to 30°, the output amplitude is reduced to a factor of $\cos(15°) = 0.966$. Each individual current component is determined by the impedance $R + jn\omega_{out}L$. For the nth component, equation (6.5) can be used to write each Fourier component as a phasor voltage

$$\tilde{V}_n = \frac{2\sqrt{2}\,V_{in}}{\pi} \frac{\sin(n\pi/2)}{n} \cos\left(\frac{n\delta}{2}\right)\angle(-n\delta/2) \tag{6.8}$$

The associated current will be

$$\tilde{I}_n = \frac{\tilde{V}_n}{R + jn\omega_{out}L} \tag{6.9}$$

Here is a short table of the first few values for $\delta = 30°$:

Frequency	Phasor V_d	Phasor Current
60	$125.2\angle-15°$	$54.5\angle-44.5°$
180	$30.6\angle135°$	$7.76\angle75.5°$
300	$6.71\angle-75°$	$1.12\angle-145.5°$

Based on the first, third, and fifth harmonics, the power with $\delta = 30°$ is $(54.5)^2R + (7.76)^2R + (1.12)^2R = 6.06$ kW. We can rewrite to get the actual time functions for the current,

$$i_{out}(t) = 77.1\cos(120\pi t - 44.5°) + 11.0\cos(360\pi t + 75.5°)$$
$$+ 1.58\cos(600\pi t - 145.5°) + \cdots \tag{6.10}$$

The complete Fourier Series for the current will be

$$i_{out}(t) = \frac{4V_{in}}{\pi}\sum_{n=1}^{\infty}\frac{\sin(n\pi/2)\cos(n\delta/2)}{n\sqrt{R^2 + n^2\omega_{out}^2L^2}}\cos\left[n\omega_{out}t - \frac{n\delta}{2} - \tan^{-1}\left(\frac{n\omega_{out}L}{R}\right)\right] \tag{6.11}$$

The resistor voltage is R times this current series.

The waveform quality of the VSI with just an R-L load is not good, since the harmonics are substantial. In Example 6.3.2, if the filter attenuated the harmonics more strongly, it would be hard to get close to 120 V_{RMS} output. Wanted component amplitude must be traded off against unwanted components, and the single-pole low-pass filter cannot attenuate low harmonics very much. Instead, it might be helpful to use a series L-C-R resonant set. The wanted component will not be attenuated at all in this case, but harmonics will be lower. There is no tradeoff between the wanted and unwanted components.

Example 6.3.3   Add a series capacitor to the VSI of Example 6.3.2, then repeat the analysis.

Since $L = 3$ mH, and since we want resonance so 120π rad/s $= 1/\sqrt{(LC)}$, the capacitor should be 2345 μF. The wanted component amplitude will be $(4V_{in}/\pi)\cos(\delta/2)$. The highest possible value for the current amplitude is $183.3/R = 91.6$ A, or 64.8 A_{RMS}. The current phasor relationships at higher harmonics should account for the added capacitor. The impedance magnitudes are:

Frequency	Impedance of L-C-R Combination
60	2 Ω
180	3.62 Ω
300	5.79 Ω

The third and fifth harmonics do not change much, but higher harmonics are attenuated quickly. In Figure 6.10, current waveforms for the L-R and L-C-R filters are compared with $L = 3$ mH. The resonant filter offers a result closer to the intended sinusoid.

Resonant filters are helpful when a specific target frequency is involved. The best examples are fixed-frequency backup systems, in which a known line frequency such as 50 Hz, 60 Hz, or 400 Hz is to be generated. ac link converters for dc–dc applications sometimes use resonant filters either for extreme power levels or when the switching frequency is substantially higher than in other applications. We will revisit the issues in Chapter 8.

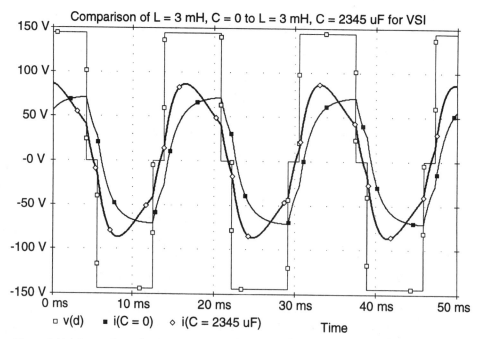

Figure 6.10 Comparison of VSI output current with *L-R* filter and with *L-C-R* filter.

6.4 PULSE-WIDTH MODULATION

6.4.1 Introduction

Some of the most important present applications such as ac motor drives require adjustable frequency. This is also true of backup systems designed to adapt to changing needs and input sources, as well as to more generic inverter applications. In these cases, it is not possible to use a resonant filter. What are the alternatives? The VSI approach varies a relative phase. Frequency adjustment does not seem too promising, since frequency matching is required for energy flow. The duty ratios were held at 50% to avoid any dc output component. The possibilities seem limited, and something different must be identified.

Consider the buck converter and output waveform illustrated in Figure 6.11. The output is $\langle v_{\text{out}} \rangle = D_1 V_{\text{in}}$, but there is no particular value of D_1 required. As in the figure, the duty ratio can be adjusted as needed to provide the desired output. Indeed, D_1 can be adjusted to follow a sinusoid. If the switching is very fast, this approach should be able to track a slow cosine wave as closely as desired. The pulse width is being adjusted as a function of time, so this is a *pulse-width modulation* (PWM) process. The adjustment function is called a *modulating function*, $M(t)$. The basic buck converter only provides positive output, and a bipolar converter must be used to avoid the unacceptable dc offset.

The complete 2×2 matrix buck converter shown in Figure 6.12 allows both positive and negative outputs. When it switches with appropriate symmetry, the dc offset is avoided.

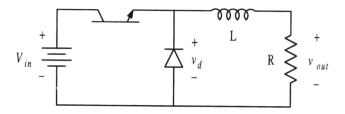

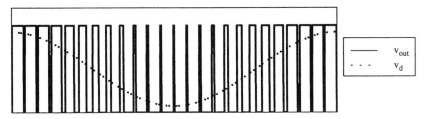

Figure 6.11 Buck converter with slow variation of duty ratio.

In this converter, if switches 1,1 and 2,2 operate together as a pair (as do 1,2 and 2,1), the matrix output $v_d(t)$ will be

$$v_d(t) = (2q_{1,1} - 1)V_{\text{in}} \tag{6.12}$$

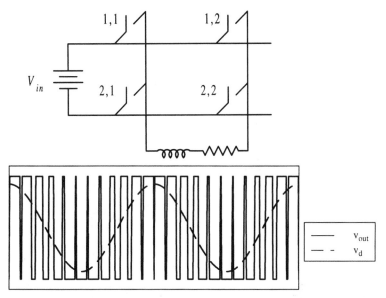

Figure 6.12 Complete 2×2 matrix buck converter, with sample PWM waveform.

In the PWM process the objective is to operate the converter so that the modulating function determines the output. The wanted output is no longer the dc average value, but rather is a unique wanted component. This should be a *moving average* to denote the variation of the average output with time. To distinguish this from other waveform components, the overbar symbol will be used to indicate a moving average. The moving average can be defined by the integral

$$\bar{v}(t) = \frac{1}{T}\int_{t-T}^{t} f(s)\, ds \tag{6.13}$$

For simplicity, the inverter output should be

$$v_d(t) = M(t)V_{\text{in}} \tag{6.14}$$

This should be consistent with the actual converter performance if the duty ratio $d_{1,1}$ is chosen as

$$d_{1,1} = \frac{1}{2} + \frac{M(t)}{2} \tag{6.15}$$

This is a natural choice: the duty ratios must be kept between 0 and 1, so equation (6.15) allows for any modulating function with amplitude up to 1.0. If $M(t)$ is a symmetrical function such as $\cos(\omega_{\text{out}}t)$, the 50% offset in equation (6.15) avoids any dc offset in the circuit output.

The process so far is incomplete. Will a modulating function such as $\cos(\omega_{\text{out}}t)$ really produce the wanted component $M(t)V_{\text{in}}$? What other frequency components are produced? When the Fourier Series of $q_{1,1}(t)$ is substituted into equation (6.12), the result is

$$v_{\text{out}}(t) = (2d_{1,1} - 1)V_{\text{in}} + \frac{4V_{\text{in}}}{\pi}\sum_{n=1}^{\infty}\frac{\sin(n\pi d_{1,1})}{n}\cos(n\omega_{\text{switch}}t)$$
$$\tag{6.16}$$
$$= M(t)V_{\text{in}} + \frac{4V_{\text{in}}}{\pi}\sum_{n=1}^{\infty}\frac{\sin[n\pi/2 + M(t)/2]}{n}\cos(n\omega_{\text{switch}}t)$$

The term $M(t)V_{\text{in}}$ is present; however, this is no longer a Fourier Series, since the term $\sin[n\pi/2 + M(t)/2]$ appears. Consider the particular case in which $M(t) = k\cos(\omega_{\text{out}})$, $0 \le k \le 1$. Equation (6.16) becomes

$$v_{\text{out}}(t) = kV_{\text{in}}\cos(\omega_{\text{out}}t) + \frac{4V_{\text{in}}}{\pi}\sum_{n=1}^{\infty}\frac{\sin[n\pi/2 + (k/2)\cos(\omega_{\text{out}}t)]}{n}\cos(n\omega_{\text{switch}}t) \tag{6.17}$$

The term $\sin[n\pi/2 + (k/2)\cos(\omega_{\text{out}}t)]$ can be analyzed by taking advantage of Bessel functions. The expression for $\sin[x\cos(\omega t)]$ is given in Appendix A. According to that expression, all multiples of modulating function frequencies appear when $\sin(\cos(t))$ is expanded. When the expression is multiplied by $\cos(\omega_{switch}t)$, the result includes sums and differences at frequencies $n\omega_{switch} \pm m\omega_{out}$, for all positive integers m. Fortunately only components with both n and m close to 1 will be large. This establishes the basic Fourier properties of PWM. Since PWM is a standard tool from communication theory, we can draw on that literature to identify the following prior results.

- If the frequencies are chosen so that $\omega_{switch} \gg \omega_{out}$, then the low-frequency component of $v_{out}(t)$ is determined by the modulating function, and will be $M(t)V_{in}$.
- Fourier components near the switching frequency will have as high an amplitude as those near the wanted frequency.
- Low-pass filtering can recover the modulating function, since it represents the low-frequency component.

The PWM process attempts to approximate a low-frequency waveform by adjusting the duty ratio of a fast square wave. Figure 6.13 shows a sample waveform with a 60 Hz cosine modulating function based on 360 Hz switching. The time-varying adjustment of the duty ratio is apparent, and it should not be surprising that a 60 Hz component is present.

The converter need not be a full bridge inverter for the PWM process to work. A half-bridge, as shown in Figure 6.14, is often used in practice. Switching keeps the efficiency high while modulation generates the desired output. There are few requirements on $M(t)$. It must contain only frequency components much lower than the switching frequency, and of course it must maintain the duty ratios between 0 and 100%. Otherwise, any waveform will work, including a single-frequency sine wave, a voice or music signal, a triangle waveform, or any other desired signal. For inverter applications, the modulating function allows the wanted component amplitude, frequency, phase, and even detailed shape to be adjusted, independent of the actual switching process. A low-pass filter will recover the information.

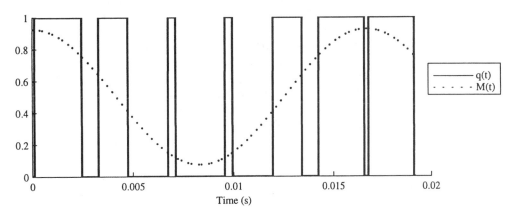

Figure 6.13 Sample PWM waveform; 360 Hz switching, 60 Hz modulation.

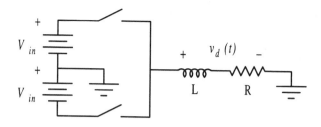

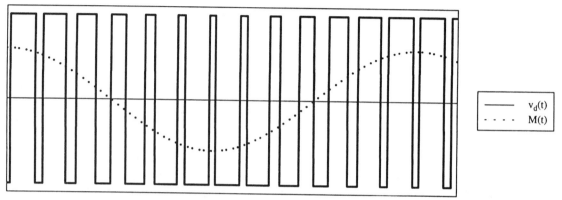

Figure 6.14   Half-bridge PWM inverter with sample output.

6.4.2 Creating PWM Waveforms

The idea of sinusoidal adjustment of duty ratio serves as an excellent example of the way many power converters are controlled. One particularly elegant process is illustrated in Figure 6.15. An oscillator generates a triangular *carrier waveform* at the switching frequency. The triangle function has period T, so let us represent it as tri(t,T). A modulating function $M(t)$ is generated separately, then $M(t)$ and tri(t,T) are both applied to a comparator. The comparator provides a high output if $M(t) >$ tri(t,T), and a low output when $M(t) <$ tri(t,T). The output can be interpreted directly as a switching function! Since the triangle waveform has a voltage linearly dependent on time, the comparator has an output pulse width linearly dependent on the level of $M(t)$.[1]

---

[1]For analysis on a computer, it is often very useful to obtain an expression for the triangle oscillator output. Many programs and languages include a remainder or *modulo* function, where mod(t,T) is the remainder after time t is divided by the period T. With the modulo function, the triangle can be written

$$\text{tri}(t,T) = \left| 2 - \frac{4}{T} \text{mod}(t,T) \right| - 1 \tag{6.18}$$

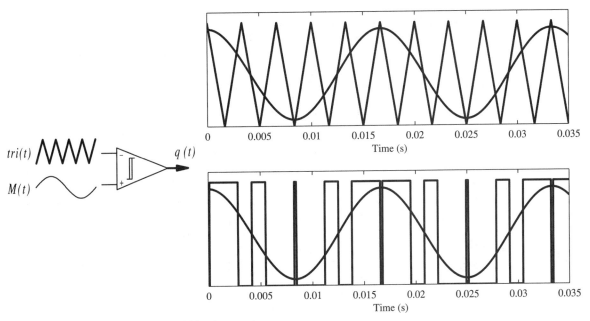

Figure 6.15   Triangle comparison PWM implementation.

Example 6.4.1   A 300 Hz triangle waveform is generated for a PWM process like the one in Figure 6.15. Plot the resulting switching function for $M(t) = 0$, $M(t) = 0.2 \cos(120\pi t)$, and $M(t) = 1.0 \cos(120\pi t)$.

For this example, a very slow switching frequency is used to simplify the analysis. In practical systems, switching frequencies range from about 3 kHz up to 100 kHz or more. The triangle, the three modulating functions, and the switching functions are shown in Figure 6.16. For $M(t) = 0$, the switching function is a fixed 50% duty pulse train. If this is used to control a half-bridge inverter, the output will be a 300 Hz square wave with average value of 0. For $M(t) = 0.2 \cos(120\pi t)$, the switching function shows only a little variation around a 50% mean duty ratio. The half-bridge converter output has a 60 Hz component with amplitude $0.2V_{in}$, in addition to components near 300 Hz. For $M(t) = \cos(120\pi t)$, the duty ratio shows large variations centered around 50%.

In the triangle comparison process, the actual duty ratio $D(t)$ for sinusoidal modulation is given by $\frac{1}{2} + (k/2)\cos(\omega_{out}t)$ where k, the amplitude of $M(t)$, is the *PWM gain* or the *depth of modulation* of the PWM signal. A modulation depth of 100% corresponds to the highest possible variation, and the highest possible output amplitude. A depth of 0 produces only the switching frequency. The triangle is not the only valid carrier waveform. Sawtooth signals are common, and it is even possible to use a sine wave.

Let us consider some of the implementation issues associated with the PWM process.

Example 6.4.2   A 2×2 switch matrix uses PWM to converter energy from a 200 V dc source into an *L-R* load. The output is intended to be 120 V$_{RMS}$ at 60 Hz. The switching frequency is 720 Hz. What types of switches should be used? Determine the appropriate value

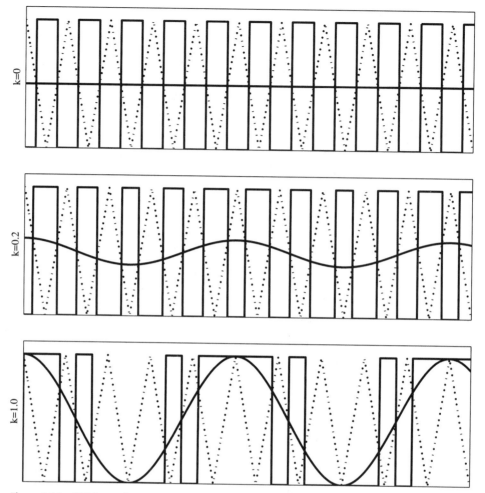

Figure 6.16   PWM waveform construction for Example 6.4.1.

of the modulating function $M(t)$ and determine the $d_{1,1}(t)$ function that results. What is the depth of modulation? Plot the output voltage waveform of the converter. Sketch the output current given an L/R time constant of 2 ms. The circuit is shown in Figure 6.17.

Figure 6.18 shows the switching function construction. A triangle wave at 720 Hz is compared to the modulating waveform to produce the switching function $q_{1,1}(t)$. To keep KVL and KCL satisfied yet provide simple operation, we can set $q_{22}(t) = q_{11}(t)$ and $q_{12}(t) = q_{21}(t) = 1 - q_{11}(t)$, just as in the 2 × 2 buck converter in Figure 6.12. The maximum available peak value is V_{in}, or 200 V. Since the objective is 120 V_{RMS}, or 170 V peak, the depth of modulation should be (170 V)/(200 V) = 0.85. The output will be $(2q_{11} - 1)V_{in}$ and is shown in Figure 6.19. Each switch must carry ac current and needs to block dc voltage. A device with BCFB capability (such as a transistor with reverse parallel diode) is needed. The L/R time constant is somewhat smaller than the switching period 1/720 s. The output current will rise and fall exponentially when $+V_{in}$ or $-V_{in}$ are applied across the

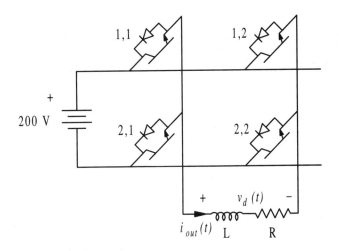

Figure 6.17 Matrix inverter for Example 6.4.2.

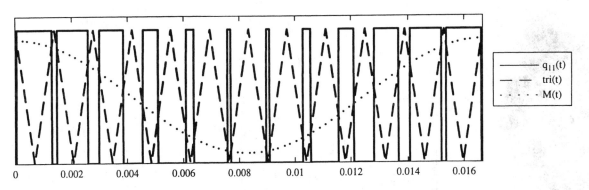

Figure 6.18 PWM waveforms for 60 Hz inverter with 720 Hz switching.

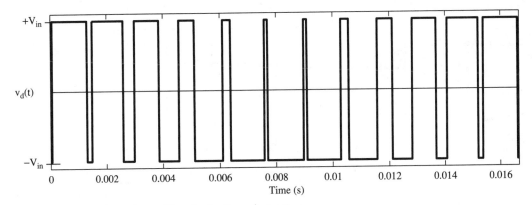

Figure 6.19 Output waveform of inverter for Example 6.4.2.

L-R load. For the moment, let us assume that the exponential is approximately linear during each switch 1,1 on-time. The linear approximation would be very close for more practical fast-switching inverters. As the duty ratio rises and falls, the inductor current on average ramps up and down to follow it. It is not difficult to sketch an approximate current waveform, since the circuit and inductor action are simple. In Figure 6.20*a*, a sketch based on linear approximations provides a rough idea of the current waveform. Figure 6.20*b* shows the results of Mathcad computation as a basis to compare the linear approximation to the actual waveform.

Consider an inverter intended for an ac motor control application. In a modern system, the switching frequency is likely to be at least 6 kHz. The output can range from 0 up to a few hundred hertz, so that the ratio f_{out}/f_{in} is commonly 30 or more. Figure 6.21 shows the output waveform for a half-bridge inverter switching at 6 kHz and modulated at 100 Hz with 80% depth of modulation. Fourier components of this waveform can be computed by integration. The waveform is pieced together from "chunks" of $+V_{in}$ and $-V_{in}$. Figure 6.22 shows the spectrum, c_n vs. f_n, for the frequency range 0 to 50 kHz for this PWM waveform. The spectrum is discrete, meaning that it has nonzero values only at exact multiples of the 100 Hz fundamental, because the switching frequency is an integer multiple of the modulating frequency. The fundamental shows amplitude of $0.8V_{in}$, reflecting the modulating func-

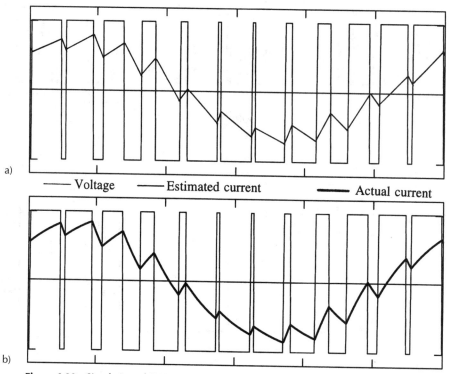

Figure 6.20 Simulation of PWM inverter output.

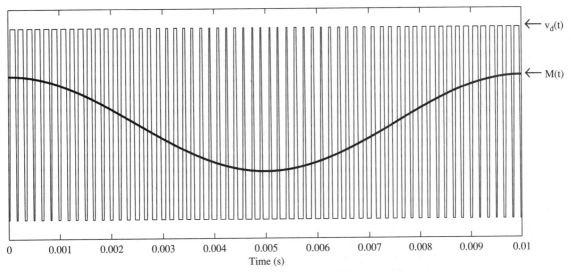

Figure 6.21   Output waveform for inverter with 6 kHz switching and 100 Hz output.

tion. Large components appear close to the 6 kHz switching frequency. Additional components appear near various multiples of 6 kHz.

A practical converter must allow the wanted component to pass while attenuating
unwanted components. With PWM, this is a matter of low-pass filtering. With 100 Hz
output and 6 kHz switching, for instance, a low-pass filter with a corner frequency of 300 Hz
will attenuate the fundamental by only about 5% while reducing the switching-frequency
harmonics by a factor of 20 or more. The output current will be only a little bit distorted.

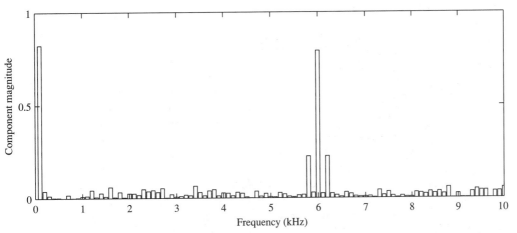

Figure 6.22   Fourier spectrum c_n vs. n for 6 kHz PWM waveform.

6.4.3 Drawbacks of PWM

Pulse-width modulation is a simple, flexible process. It is very common in inverters that need to provide adjustable performance and in small battery powered audio amplifiers. However, it has a few limitations that preclude certain specialized applications.

- The highest output amplitude is equal to the input. This reflects the "buck-type" circuit arrangement. While this is unsurprising, consider that a VSI with a square wave output has a wanted component of $4V_{in}/\pi$, fully 27% larger than possible with standard PWM.
- Numerous switching operations occur each period. In real switches, a small energy loss is incurred every time a switch turns on or off. In a PWM converter with frequency ratio f_{switch}/f_{out}, the energy lost in switching increases by this same ratio, relative to the loss in a VSI inverter. For very high-power converters, the extra switch loss is a serious problem.
- Distortion. Although a low-pass filter should be adequate for recovery of the modulating function, the inverter output has large harmonics extending to very high frequencies. Even small residual components at megahertz frequencies can interfere with communications equipment, sensors, analog signal processing, and sometimes digital logic.

PWM distortion is fairly simple to analyze. In the converters considered so far, the output is always either $+V_{in}$ or $-V_{in}$. The RMS value of any such waveform is V_{in}, regardless of the switch action. The fundamental is $M(t)V_{in}$. The peak value of $M(t)$ is the modulation depth, k. Let us write the RMS value of the fundamental as $c_{1(RMS)} = M_{RMS} = (k/\sqrt{2})V_{in}$. The THD will be

$$\text{THD} = \sqrt{\frac{f^2_{RMS} - f^2_{1(RMS)}}{f^2_{1(RMS)}}} = \sqrt{\frac{V^2_{in} - k^2 V^2_{in}/2}{k^2 V^2_{in}/2}} = \sqrt{\frac{2 - k^2}{k^2}} \tag{6.19}$$

The lowest possible THD value occurs when $k = 1$, with THD of 100%.

6.4.4 Multilevel PWM

In the conventional PWM process discussed so far, the output follows a two-level pattern, switching between $+V_{in}$ and $-V_{in}$. The most important advantage is that any desired modulating function can be used. This allows convenient adjustment of magnitude and frequency, so that almost any ac application can be served effectively. In two-level PWM, the THD is at least 100%, and increases without limit as the depth of modulation decreases. This creates a special problem when zero output is desired: $M(t) = 0$ will command a 50% duty square wave at the switching frequency to appear at the output. This large signal should be attenuated as much as possible by the low-pass filter, but will always expose the load to ripple at the switching frequency.

It is also possible to build a *multilevel* PWM inverter. The simplest example takes advantage of a full-bridge with positive, negative, and zero outputs. The basic process is only a little more complicated than the standard triangle-comparator method. As illustrated in Figure 6.23, when the modulating function is positive, the triangle comparison method is used

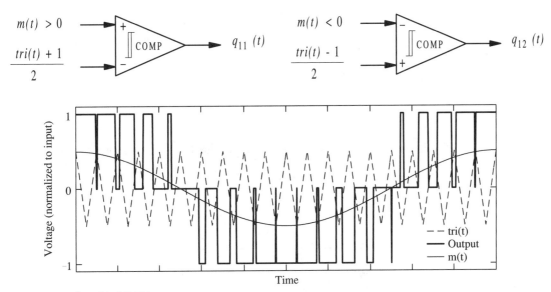

Figure 6.23   Three-level PWM process.

to switch between $+V_{in}$ and zero. When the function is negative, the comparison switches between $-V_{in}$ and zero. The three-level waveform has an important advantage when $M(t) = 0$: There is no switching and no output when zero modulation is commanded. The THD is reduced by using multiple levels.

Figure 6.24 compares Fourier spectra of two-level and three-level PWM processes given $f_{switch}/f_{out} = 25$. In Figure 6.24*a*, the processes are compared for $k = 0.85$. The two spectra are very similar, and in practice there is relatively little difference in the filter requirements at this modulation depth. Figure 6.24*b* compares the processes for $k = 0.10$. In the case of two-level PWM, the wanted component is reduced proportionally, while the unwanted components change little. For three-level PWM, the unwanted components are smaller. A system intended to operate at low modulation can benefit from the three-level approach, although systems for high modulation depths do not show much difference. Some researchers advocate five or more PWM levels for high-power applications with wide variations in k. Most commercial products, however, use two- or three-level PWM for output waveforms.

6.4.5 Inverter Input Current under PWM

The input waveform for PWM is relatively easy to determine. Consider a sinusoidal modulating function. If the switching process is fast and the filter is chosen properly, the output current will be very nearly a pure sinusoid. The output can be treated as an equivalent ac current source $i_{ac}(t)$, given the wanted output voltage $kV_{in} \cos(\omega_{out}t)$. For the full bridge inverter, the input current is $i_{ac}(t)$ when switches 1,1 and 2,2 are on together and $-i_{ac}(t)$ when switches 1,2 and 2,1 are on together. For two-level PWM, the switches operate in diagonal

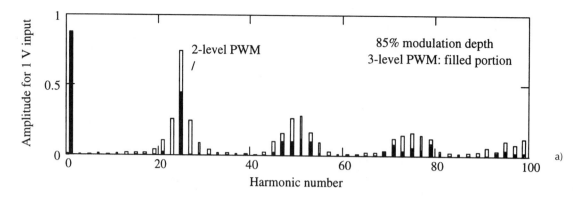

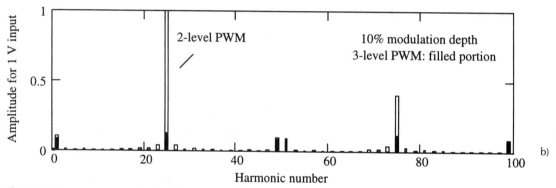

Figure 6.24 Spectra of two-level and three-level PWM.

pairs, and

$$i_{in}(t) = (2q_{1,1} - 1)i_{ac}(t) \tag{6.20}$$

Given $i_{ac}(t) = I_0 \cos(\omega_{out}t - \phi)$, the Fourier Series for $q_{1,1}(t)$ can be used to write

$$i_{in}(t) = (2d_{1,1} - 1)I_0 \cos(\omega_{out}t - \phi) + \frac{4I_0}{\pi} \sum_{n=1}^{\infty} \frac{\sin(n\pi d_{1,1})}{n} \cos(n\omega_{switch}t)\cos(\omega_{out}t - \phi)$$

$$\tag{6.21}$$

$$= M(t)I_0 \cos(\omega_{out}t - \phi) + \frac{4I_0}{\pi} \sum_{n=1}^{\infty} \left(\quad\right)$$

The wanted component is the dc term, since the input source is dc. This component appears in the first term when the appropriate trigonometric identities are applied, given $M(t) = k \cos(\omega_{out}t)$:

$$i_{in}(t) = \frac{kI_0}{2} [\cos \phi + \cos(2\omega_{out}t - \phi)] + \frac{4I_0}{\pi} \sum_{n=1}^{\infty} \left(\quad\right) \tag{6.22}$$

Thus the dc component is $kI_0 \cos \phi$, and the input power is $kV_{in}I_0 \cos \phi$. The current contains an unwanted component at $2\omega_{out}$ as well as those centered around multiples of the switching frequency.

Figure 6.25 shows a typical input current waveform for a two-level PWM inverter, along with its Fourier spectrum. Once again, the unwanted components can be removed with a low-pass filter. The component at $2\omega_{out}$ is the most difficult to filter in this fashion.

6.5 PULSE-WIDTH MODULATED RECTIFIERS

Just as it is possible to use a typical rectifier with reverse energy flow as a current-source inverter, so it is possible to reverse a PWM inverter to function as a rectifier. The concept is shown in Figure 6.26, in which a 2×2 matrix connects an ac current source to a dc voltage source. In this circuit, the voltage at the input node is determined by the PWM process, and ought to be $M(t)V_{out}$ plus unwanted components at multiples of the switching frequency. The current is a pure sinusoid.

Now, take the input source to be an ac voltage source in series with an inductance. The PWM process can be used to ensure a clean sinusoidal input current. This process defines a *power-factor corrected rectifier*, or PFC rectifier. The input, when filtered, has little distortion, while the output is a fixed dc voltage, provided that the switching frequency is much higher than the input (modulating) frequency.

One typical implementation of a PFC rectifier is shown in Figure 6.27, in which a diode bridge allows most of the active switches to be eliminated. This circuit, known as a

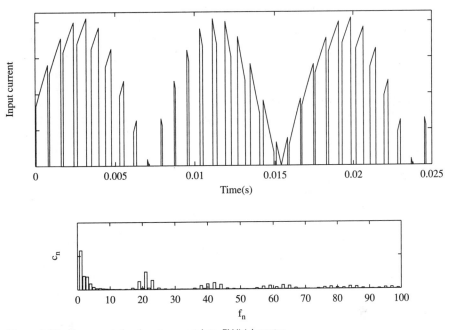

Figure 6.25 Representative input current into PWM inverter.

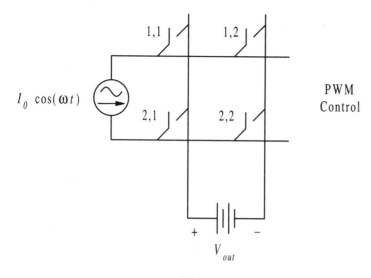

Figure 6.26 Pulse-width modulated rectifier.

PFC boost converter, is growing in popularity for use in the input sections of power converters. The boost converter is appropriate since it can step up the input, no matter how small, into a voltage source with a value well beyond the peak input. Other converter topologies have been tested for power factor correction, but the boost converter's operation is especially easy to follow. The input voltage $v_{in}(t)$ is given by

$$v_{in}(t) = q_2 V_{out} \tag{6.23}$$

The converter provides the desired operation if a modulating function $M(t)$ is chosen to make $d_2(t) = |k\cos(\omega_{in}t)|$. Since switch 2 is uncontrolled, the concept works if switching is fast and produces $d_1(t) = 1 - |k\cos(\omega_{in}t)|$. In this case, the modulating function can be obtained directly from the incoming ac line. Generation of $M(t)$ is convenient and simple.

The only real problem with a boost PFC arrangement is the current applied to the dc source. From the discussion above, this current will contain the wanted dc component, a large component at $2\omega_{in}$, and components near multiples of the switching frequency. The $2\omega_{in}$ component is the most difficult one to filter. Actually, the capacitor size approaches that needed for a classical rectifier if the output ripple is to be kept low. To avoid possible problems, PFC converters are often followed by dc–dc converters such as forward converters.

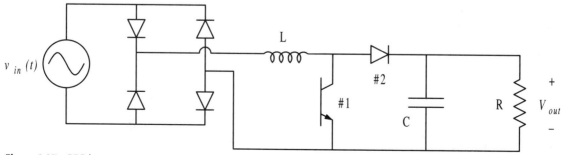

Figure 6.27 PFC boost converter.

The extra adjustment capability allows the PFC process to work successfully even if ripple is large. Figure 6.28 shows some of the waveforms in a typical PFC boost converter.

6.6 CURRENT-SOURCED INVERTERS

The basic current-sourced inverter (CSI) is exactly the same as a controlled rectifier, as discussed earlier. A sample system is shown in Figure 6.29. In this system, the ac voltage sources

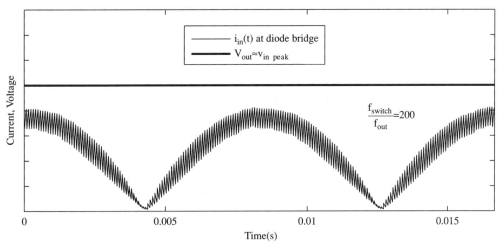

Figure 6.28 Input current and output voltage waveforms for typical boost PFC converter.

Figure 6.29 Current-sourced inverter with active ac voltage load.

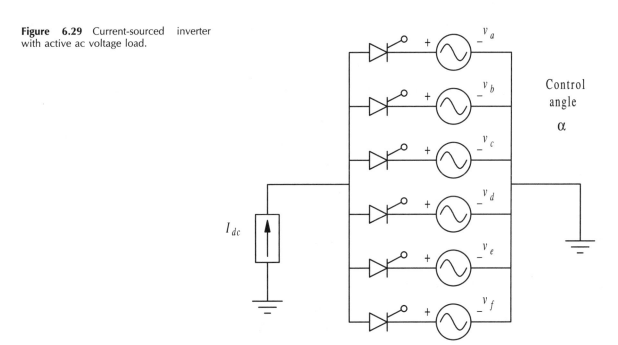

provide timing information that can be used for adjustment. For passive loads, a circuit such as that in Figure 6.30 would be appropriate. This CSI is sometimes used for motor control, especially in the case of large motors. As shown, the switches must be FCBB types, and are often implemented with a GTO or with a series combination of a transistor and diode. It is difficult to use an SCR in the general CSI, because the output voltage waveform is only approximately a sinusoid, and it is hard to set the timing of SCR turn-off.

The CSI can be controlled very much along the lines of voltage-sourced inverters. For example, the switch matrix permits selection of $+I_{in}$, $-I_{in}$, or 0 into the output port. Switching can create a current waveform analogous to the voltage waveforms of Figure 6.5. This current will provide a wanted Fourier component with adjustable amplitude. The output voltage will depend on the impedance properties of the actual load.

Pulse-width modulation can also be used with a CSI. As in the voltage case, this will produce a large fundamental component with substantial harmonics near multiples of the switching frequency. The analysis follows exactly our earlier work for the voltage-sourced case.

There is another, and perhaps simpler, way to operate a CSI. Consider the circuit in Figure 6.31. In this case, a specific voltage cosine wave is desired at the output. Switch action controls the injection of $\pm I_{in}$ into the output. If the output voltage is too low, $+I_{in}$ is injected. If it is too high, $-I_{in}$ is injected. A clock or similar circuit checks the output voltage only at specific times or values, so that the switching frequency does not increase without bound. This simple "voltage following" control method provides an excellent output waveform with relatively simple processing. It is common in high-power ac motor control circuits and is termed *current hysteresis* control.

6.7 A SHORT INTRODUCTION TO CONVERTERS FOR AC DRIVES

Circuits for the operation and control of ac motors are extremely important inverter applications. ac motors are generally more rugged, more efficient, and lower in cost than dc motors. Until late in the 1960s, motor-generator conversion sets were almost the only way to obtain flexible control of ac motors. dc motors were the best alternative in most applications requiring wide-range speed control. Inverters, and particularly PWM inverters, have changed

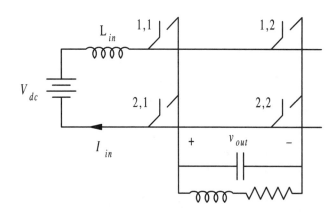

Figure 6.30 Current-sourced inverter with general ac load and voltage-type interface.

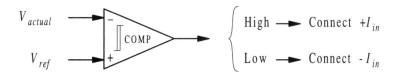

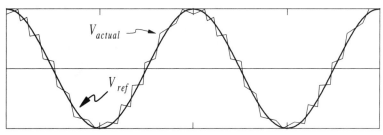

Figure 6.31 Voltage following process for CSI control.

the picture substantially. Within the past few years, it has become possible to match and often exceed the control capabilities of dc motors with simpler ac machines. The combined cost of an inverter and ac motor is now competitive with the cost of a dc machine with its controller. This development is rapidly making dc machines obsolete in many applications.

Frequency variation is one important aspect of ac motor control. Conventional ac machines operate by forming a rotating magnetic field that gives rise to force and motion. The speed of the field is determined by the excitation frequency. Adjustable frequency is a necessity if speed control is to be accomplished. This requirement tends to favor PWM inverters, since low-distortion sinusoids of arbitrary frequency can be created with PWM.

The second major aspect of motor control also is related to the magnetic fields in a machine. The main rotating field is created with electrical energy applied to a coil of wire. By Faraday's Law, the time rate of change of magnetic flux ϕ linking a coil (with flux linkage $\lambda = N\phi$, where N is the number of turns) is equal to the coil voltage. Any material or device has a maximum magnetic flux, beyond which there is no enhancement of the energy conversion process. This *saturation flux* represents the highest reasonable value for a machine. The main rotating field's flux linkage is normally set at this maximum value to provide the highest possible force. Since voltage $v = d\lambda/dt$, for a sinusoidal voltage $V_0 \cos(\omega t)$, we have

$$\int V_0 \cos(\omega t)\, dt = \lambda, \qquad \frac{V_0}{\omega} \sin(\omega t) = \lambda \qquad (6.24)$$

To make λ as high as possible, we should require

$$\frac{V_0}{\omega} = \lambda_{\max} \qquad (6.25)$$

As the frequency changes, the ratio V_0/ω should be held constant.

In summary, an inverter for an ac motor should have at least two characteristics:

1. It should allow frequency adjustment to reflect the entire desired range of speeds. For example, if a standard 60 Hz, 3600 RPM motor is to be used over the range 100 RPM–6000 RPM, the frequency must be adjustable over the range $1\frac{2}{3}$ Hz to 100 Hz.

2. It should maintain maximum flux linkage at all times. The ratio V/f is held constant at the maximum value to accomplish this.

An inverter controlled to operate with these constraints is called a *constant volts per hertz* inverter.

Many modern *adjustable speed drives* for ac motor control operate by meeting these basic requirements. Speed control determines the desired frequency, while the ratio V/f is held at its rated value. This combination is adequate for control of fans, pumps, compressors, conveyers, and a host of other mechanical systems. It is also helpful for more demanding applications such as elevators, electric vehicles, and many industrial processes. Precision applications, such as machine tools and machining, benefit from more sophisticated closed-loop *vector controls* built around a motor.

PWM provides significant benefits for ac motor control. PWM easily accommodates wide frequency ranges while maintaining constant volts per hertz operation. Any reasonable frequency can be selected. It is also possible to add extra features such as adjustable acceleration during transients, shutdown for overspeed or underspeed, precise speed control, or other aspects. Of course, the square-wave approach with a VSI or a current-sourced inverter can also be used to provide a fixed ratio of V/f with variable frequency, but no method other than PWM provides easy filtering of unwanted components over the full operating range of an ac motor controller.

Figure 6.32 illustrates the action of constant volts per hertz operation of an ac motor. In Figure 6.32a, the output shaft torque as a function of speed is shown for a conventional squirrel-cage polyphase induction motor. The rated operating speed is just under 3600 RPM. This motor is intended to operate in the steep near-linear region close to the torque zero crossing. In Figure 6.32b, the effects of holding V/f constant while manipulating the frequency are apparent. The inverter allows the torque-speed characteristic to be moved anywhere to the left or right along the speed axis. Maximum torque can be obtained at any desired speed in this arrangement.

The basic ac motor control methods are not limited to squirrel-cage machines. Permanent-magnet ac machines and other synchronous ac motors and generators can be controlled in the same manner. The possibilities are broad enough that the term *brushless dc motor* is common to describe a combination of inverter and permanent-magnet synchronous motor. The motor itself is an ac machine, but the inverter's dc input and the control flexibility make the combination a good substitute for a true dc machine.

6.8 INVERTER DESIGN EXAMPLES

Let us consider a few inverter design problems as a way to further illustrate the issues and possibilities in these systems.

Example 6.8.1 A utility company intends to operate an experimental photovoltaic power unit. The solar cells will cover more than one hectare (10,000 m²), and can produce peak

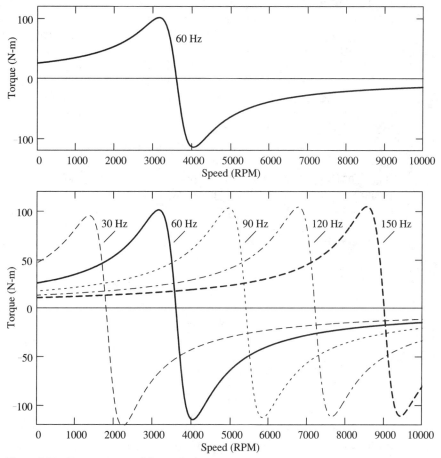

Figure 6.32   Torque vs. speed for squirrel-cage ac motor.

output power of 1.5 MW. Individual cells provide about 0.5 V at up to 3 A, and cover an area of 0.01 m². The cells are arranged in 10 m² panels, each connected in series to produce a nominal 500 V bus. The 10 m² panel sections ultimately connect in parallel to supply three inverters, one for each phase of the utility three-phase power. The ac side is a standard 480 $V_{\text{line-to-line}}$ 60 Hz supply. Propose an inverter design for this application. How will it be controlled? When maximum power is available, what will the operating conditions of the inverter be? Under heavy cloud cover, the cells maintain about the same voltage, but the output current can drop by a factor of ten. What are the operating conditions of the inverter when heavy clouds obscure the entire hectare?

Each of the three inverters must handle up to 500 kW. Since the idea is to supply energy into the power grid, efficiency should be as high as possible. These factors suggest a VSI as opposed to a PWM inverter. The power level is high for PWM, and the waveform advantages of PWM are outweighed by the extra losses associated with fast switching in this application. Given a VSI circuit, let us consider just one of the three phases. On the ac side, we have the 480 $V_{\text{line-to-line}}$ value, which means a peak voltage between any two phases of $480\sqrt{2}$ V = 679 V. Each individual phase voltage will have an RMS value of $480/\sqrt{3}$ V = 277 V, for a peak value of 392 V. When the voltage-sourced inverter provides a square wave, it produces a peak wanted component of $4V_{\text{in}}/\pi$, which gives 637 V with the 500 V connection of the panels. For proper control, the VSI output will need to be adjusted so that its wanted component is close to the ac supply voltage. The component can be adjusted to any peak value below 637 V, so it makes sense to connect to the ac side line-to-neutral voltage (392 V peak) rather than to the line-to-line voltage. The circuit for one phase is shown in Figure 6.33. In this application there are two degrees of freedom: the relative delay angle δ controls the wanted component amplitude, and a switching function phase term ϕ controls the component phase. relative to the ac line.

The VSI output can be set for a specific waveform v_{VSI}, so the equivalent source method is appropriate. Only the 60 Hz current component transfers energy to the ac supply, so let us consider the 60 Hz equivalent circuit. The angle ϕ can be defined conveniently as the phase shift between the center of the positive VSI output pulse and the peak of v_{ac}, as indicated in Figure 6.33. The wanted component, from equation (6.6), becomes

$$v_{\text{VSI(wanted)}} = \frac{4V_{\text{in}}}{\pi}\cos\left(\frac{\delta}{2}\right)\cos(\omega t - \phi) \tag{6.26}$$

With $V_{\text{in}} = 500$ V, the associated phasor is

$$\tilde{V}_{\text{VSI(wanted)}} = 450\cos\left(\frac{\delta}{2}\right)\angle{-\phi} \tag{6.27}$$

while the ac source phasor is $277\angle 0°$. There is no very definite way to choose the inductor, but some guidance is available. The inductor should be large for good filtering. On the other hand, at 500 kW flow, the expected RMS current will be more than 1800 A. The impedance of the inductor should be low enough to permit this current flow. If $|Z| = 0.05\ \Omega$, the inductor drop will be about 90 V at full current—probably well within the voltage capability, since the phasor in equation (6.27) has a much higher maximum value than the ac line. On this basis, let us test the value $L = (0.05\ \Omega)/(\omega\ \text{rad/s}) = 133\ \mu\text{H}$. The current is

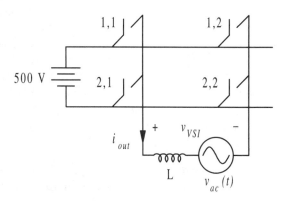

Figure 6.33 Voltage-sourced inverter for Example 6.8.1.

$$\tilde{I}_{\text{out}} = \frac{\tilde{V}_{\text{VSI(wanted)}} - \tilde{V}_{\text{ac}}}{j\omega L} = \frac{450 \cos\left(\frac{\delta}{2}\right)\angle(-\phi) - 277\angle 0°}{j0.05} \tag{6.28}$$

The output average power is $\langle v_{\text{ac}} i_{\text{out}} \rangle$, or

$$P_{\text{out}} = |\tilde{V}_{\text{ac}}| |\tilde{I}_{\text{out}}| \cos(\angle V - \angle I)$$

$$= |\tilde{V}_{\text{ac}}| \left| \frac{\tilde{V}_{\text{VSI(wanted)}} - \tilde{V}_{\text{ac}}}{j\omega L} \right| \cos(\angle V - \angle I) \tag{6.29}$$

The current and power results can be manipulated to find the outcome in terms of δ and ϕ. After some complex algebra, the output power and reactive power P_{out} and Q_{out} associated with the wanted component can be written

$$P_{\text{out}} \text{ (W)} = \frac{450 \cos(\delta/2)277}{0.05 \ \Omega} \sin(-\phi)$$

$$Q_{\text{out}} \text{ (var)} = \frac{450 \cos(\delta/2)277}{0.05 \ \Omega} \cos(-\phi) - \frac{(277 \text{ V})^2}{0.05 \ \Omega}$$

$$P_{\text{out}} \text{ (W)} = 2.49 \times 10^6 \cos\left(\frac{\delta}{2}\right)\sin(-\phi) \tag{6.30}$$

$$Q_{\text{out}} \text{ (var)} = 2.49 \times 10^6 \cos\left(\frac{\delta}{2}\right)\cos(-\phi) - 1.54 \times 10^6$$

This result is interesting because the power and reactive power can be adjusted independently. If we choose to make $Q_{\text{out}} = 0$ for the wanted component when $P_{\text{out}} = 500$ kW, for example, the expressions reduce to

$$2.49 \times 10^6 \cos\left(\frac{\delta}{2}\right)\sin(-\phi) = 500\,000 \text{ W} \tag{6.31}$$

$$\cos\left(\frac{\delta}{2}\right)\cos(-\phi) = \frac{1.54 \times 10^6}{2.49 \times 10^6}$$

These can be divided to provide an expression for $\tan \phi$. Once ϕ is known, the value of δ can be determined easily. The values $\phi = -18.05°$ and $\delta = 99.31°$ provide the desired maximum power and reactive power. In the case of heavy cloud cover, the maximum available power reduces to 50 kW. If we again choose to make $Q_{\text{out}} = 0$, the same procedure gives $\phi = -1.866°$ and $\delta = 103.97°$. Figure 6.34 shows the actual voltage waveforms and the output current waveform with $P_{\text{out}} = 500$ kW, $Q_{\text{out(wanted)}} = 0$, $\phi = -18.05°$, and $\delta = 99.31°$.

Example 6.8.2 A VSI is to be used for a backup power application. This unit, shown in Figure 6.35, draws power from a set of lead-acid batteries with nominal output of 144 V and supplies up to 2 kW into a load. A resonant filter is to be used to provide an output sinusoid at 120 V_{RMS}, 60 Hz, with THD < 3% at full load. If the actual battery voltage will vary be-

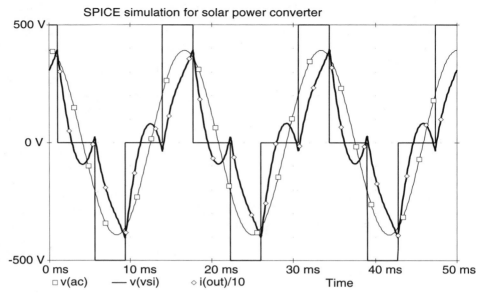

Figure 6.34 Output voltage and current waveforms for Example 6.8.1.

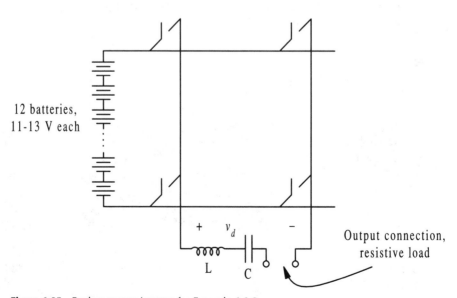

Figure 6.35 Backup power inverter for Example 6.8.2.

tween 132 V and 156 V, what range of δ values will be used for this application? Select a series *L-C* pair to meet the requirements.

The inverter will produce a wanted component with peak value $(4V_{in}/\pi)\cos(\delta/2)$. The desired output of 120 V_{RMS} requires 170 V peak. The operating values for δ should be:

Battery Bus Voltage	δ for 170 V peak
132	0°
144	44.0°
156	62.3°

The bus value of 132 V produces an output about 1 V below the desired level, but this is within the range of variation of a standard utility supply (±5% or more). Thus the relative phase angle δ could be between 0° and 62.3°.

The full-load output power of 2 kW requires $V_{RMS}^2/R = 2000$ W, and corresponds to a load resistance of 7.2 Ω. If the THD value is to be below 0.03, the harmonics need to be very low. The largest unwanted component, the third harmonic, will have to be less than 0.03 (120 V) = 3.6 V_{RMS} even if all other harmonics are ignored. Equation (6.10), computed for a purely *R-L* load, provides the attenuation for each harmonic. This is approximately

$$\frac{R}{n\sqrt{R^2 + n^2\omega^2 L^2}} < 0.03 \tag{6.32}$$

When $n = 3$, $R = 7.2$ Ω, and $\omega = 120\pi$ rad/s are substituted in, the result suggests that L should be at least 70 mH. Let us use $L = 100$ mH to attenuate enough to account for higher harmonics. Then the resonance condition $1/\sqrt{(LC)} = 120\pi$ rad/s requires $C = 70$ μF. For this choice, the voltage waveforms $v_d(t)$ and $v_{out}(t)$ are shown in Figure 6.36. The THD can

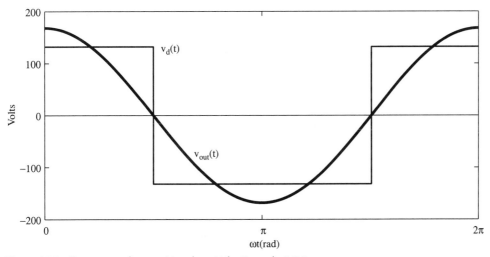

Figure 6.36   Output waveforms $v_d(t)$ and $v_{out}(t)$ for Example 6.8.2.

be computed from a Fourier series or transform, or it can be approximated from the first several harmonics. As a good approximation, the first 31 harmonics were used to estimate the THD. The value with $V_{in} = 132$ V and $\delta = 0°$ is 2.56%. The figure confirms that there is minimal distortion in the output waveform.

Example 6.8.3 A PWM inverter is to supply a small ac motor. The motor has an input inductance of about 40 mH, and draws 500 W at full load. The input dc voltage is 325 V, and the motor requires a nominal 230 V at 50 Hz. What modulation depth will be used when 50 Hz is applied to the motor? Can you suggest a switching frequency? What modulation depth will be used if the motor is adjusted to run at 80% of rated speed?

If the switching frequency is much faster than the modulating function, the motor will draw a sinusoidal current as if a voltage $M(t)V_{in}$ is applied. The nominal voltage of 230 V_{RMS} implies a peak of 325 V, so the modulation depth should be 100% at the 50 Hz condition. The motor can be modelled roughly as a series LR combination. If a pure cosine wave at 230 V_{RMS} and 50 Hz is applied to the motor when it is fully loaded, the current will be $(230 \text{ V})/Z = (230 \text{ V})/(R + j100\pi L)$, and the power will be 500 W $= |I|^2 R$. With PWM, we should get the same result if switching is fast. The value of R can be found:

$$|\tilde{I}|^2 R = 500 \text{ W}, \qquad |\tilde{I}| = \frac{230 \text{ V}}{\sqrt{R^2 + \omega^2 L^2}}$$

$$\frac{(230 \text{ V})^2 R}{R^2 + \omega^2 L^2} = 500 \text{ W}, \qquad 230^2 R = 500 R^2 + \omega^2 L^2 \tag{6.33}$$

With $L = 40$ mH and $\omega = 100\pi$ rad/s, the quadratic in R has a solution $R = 104.3$ Ω (the second solution is unrealistic). The L/R time constant is therefore 0.384 ms. If the switching is chosen to be much faster than this time constant, the LR pair will effectively filter out the switching frequency while passing the 50 Hz component. Let us select a switching period of 40 μs, for a switching frequency of 25 kHz. This is a factor of almost ten faster than the L/R time constant, and a factor of 500 faster than the modulating frequency. The waveforms that result are shown in Figure 6.37.

To run the motor at 80% of rated speed, the frequency should drop to 40 Hz, or 80% of nominal. Since this is an ac motor application, the ratio V/f should be held constant. Therefore, the voltage should also drop to 80%, or 184 V. The modulation depth will be 80% to accomplish this.

6.9 RECAP

Inverters in general are circuits for dc to ac conversion. In practice, the term *inverter* most commonly refers to a circuit for energy transfer from a dc voltage source to an ac current source. The adjustment possibilities for inverters depend on whether the load is *active*, meaning that the load is an ac source with its own inherent timing properties, or *passive*, meaning that the load acts as an impedance. For active loads, the phase shift between switching

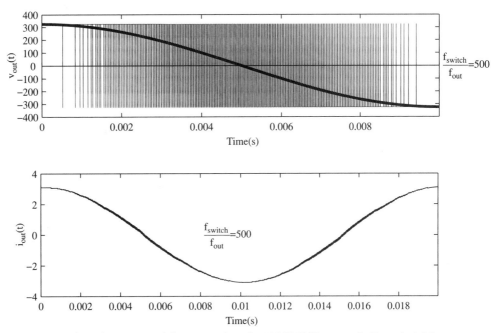

Figure 6.37 The voltage $v_d(t)$ and the current $i_{out}(t)$ for 230 V 50 Hz output in Example 6.8.3.

functions and the ac source can be adjusted to control energy flow. For passive loads, the relative phase shifts within the switch matrix can be adjusted.

Inverters can be formed as full-bridge 2×2 switch matrices. When the dc source permits it, an input center tap allows a half-bridge configuration to be used. Half-bridges are used in many backup power applications. A set of three half-bridges produces a complete three-phase inverter called a hex-bridge. In any of these converters, it is important to avoid any dc output component, since dc can damage transformers and other types of ac loads. The simplest way to do this in either the full-bridge or half-bridge circuits is to operate the switches with duty ratio of $\frac{1}{2}$ at all times.

The term *voltage-sourced inverter* (VSI) most often refers to a quasi-square-wave inverter. A VSI uses possible outputs of $+V_{in}$, $-V_{in}$, and 0 to form portions of a square wave. A *displacement angle*, symbol δ, determines the angular time interval in each half-cycle during which the output is zero. The nth Fourier component at the output of a VSI is

$$v_{out(n)} = \frac{4V_{in}}{\pi} \cos\left(n\frac{\delta}{2}\right) \frac{\sin(n\pi/2)}{n} \cos\left(n\omega_{out}t - n\frac{\delta}{2}\right) \tag{6.34}$$

where $n = 1$ normally defines the wanted component. This type of inverter is often used for fixed-frequency applications such as standby power. When the output frequency is known, it is possible in principle to provide a resonant filter to enhance the quality of the output waveform. In examples, we saw THD values below a few percent with resonant filtering.

For flexibility and conceptual simplicity, *pulse-width modulation* (PWM) is an excellent inverter control alternative. In this method, the switch duty ratios are adjusted slowly so that the moving average of the output follows a desired time waveform. If the switching frequency is much higher than the modulating frequency, the output will track the modulating signal. Unwanted components are shifted to the switching frequency and its multiples, and can be attenuated with a low-pass filter. The PWM process requires only an oscillator such as a triangle or a ramp and a comparator that provides a discrete-valued output when the modulating signal exceeds the triangle. The PWM process is simple enough that it is the basis for many types of inverter and dc–dc converter control. Many commercial power converter control ICs implement the key components of a PWM circuit.

The *depth of modulation* or *modulation index* reflects the ratio of the modulating function to the maximum possible value. A modulation depth of 100% implies that the peak value of the wanted component matches the input voltage value. As the depth of modulation decreases, the variation of pulse widths in a waveform reduces accordingly.

The PWM inverter permits easy control of output frequency and amplitude. However, it is not without limitations. For a given output power level, a PWM inverter is less efficient than a VSI since the much more frequent switch action is associated with energy loss for each switching. Distortion is high in the unfiltered waveforms, and harmonics extend to extreme frequencies. Multilevel PWM avoids part of the distortion problem by switching voltages as closely matched to the output as possible, rather than by switching back and forth between $+V_{in}$ and $-V_{in}$. The input current from the dc voltage source looks like a set of pulses within a full-wave rectified envelope. This means that the input current can be a significant filtering problem.

The operation of a PWM inverter suggests utility for the reverse process. A full-bridge PWM inverter resembles a complete 2×2 matrix buck converter. When input and output roles are reversed, we obtain a type of boost converter that takes energy from the ac line and converts it for a dc load. The unusual aspect of this process is that the current from the ac source is a sinusoid with low distortion. This is an important advantage for converters connected to the utility grid. It is possible to provide excellent power factor and low distortion in the current drawn from the ac source when a PWM rectifier is used. The most common form is a boost converter that draws a clean rectified sine wave from a diode bridge rectifier.

Current-sourced inverters (CSI) resemble controlled rectifiers, especially with an active load. In the active load case, a CSI can be built like a complementary midpoint rectifier and operated with phase delay control. When a passive load is involved, parallel capacitance or a more sophisticated source interface must be used to ensure that the load acts like an ac voltage source.

Ac motor drive circuits are an inverter application of growing importance. The advantages of ac motors make them highly sought for the full range of industrial applications. Their speed control is more complicated than that of a dc motor. The input frequency must be adjusted to alter the rotation speed of the main magnetic field. When this is done, it is crucial to keep the same magnetic field strength, which is proportional to the volts per hertz ratio *V/f*. An inverter, such as a PWM inverter, with the capability for broad variation in output frequency provides an excellent basis for ac motor control.

1. A simple inverter imposes a square wave on a load. Filtering is accomplished by including a series 25 mH inductor at the output. The load resembles a resistor. Given a 50 Hz circuit application, plot the amplitude of the wanted component of the output voltage as a function of the load resistance in this case. Comment on load regulation.

2. Find the total harmonic distortion from a VSI as a function of the relative phase angle δ.

3. A VSI connects a large battery storage system to an ac utility grid. One phase of the circuit is shown in Figure 6.38. The inverter bridge has relative phase angle δ, and is set so its wanted component lags the ac voltage by an angle ϕ. For $\delta = 45°$ and $\phi = -30°$, what is the average power per phase flowing to the ac source? What if $\phi = +30°$?

4. For the VSI circuit in Figure 6.38, the relative phase delay angle δ is set to 60°. If the wanted component has a value $V_0 \cos(\omega t + \pi/8)$,

   a. Find the value of V_0 and the average power into the ac source.
   b. Find the RMS current in the ac source. What power factor is seen by the ac source?

5. For the VSI circuit in Figure 6.38, the relative phase angle δ is set to 15°. The wanted component is set to be in phase with the ac supply voltage. Find and plot the current into the ac source as a function of time. What is the THD of the current waveform?

6. In a VSI, the choice of relative phase delay angle δ is sometimes open. However, equation (6.11) suggests that certain values of δ will cancel out some of the harmonic effects. In a VSI with a series LR load, what values of δ will cancel out the third harmonic? The fifth harmonic? The seventh harmonic?

7. The voltage-source inverter in Figure 6.38 uses relative phase control. Under a particular circumstance, the relative phase delay angle is 36°. For phase angle ϕ between the ac source and the wanted component,

   a. Find the power flow to the ac source as a function of ϕ.
   b. Estimate the RMS current by computing the first, third, and fifth harmonics. What is the RMS current as a function of ϕ?

Figure 6.38   VSI to interface between a large battery and a 2000 V_{RMS} 60 Hz source.

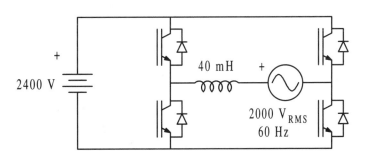

c. In the inverter's full three-phase configuration, the triplen harmonics (multiples of three) in the current cancel out. Comment on the distortion in the line current for this inverter.

8. A VSI operates with angle $\delta = 90°$ and input of 72 V. It supplies a series L-R load with $L = 20$ mH and $R = 1 \, \Omega$. The switching frequency is 50 Hz.

a. Draw the circuit and plot the voltage and current at the L-R combination.

b. Plot the current waveform drawn from the dc supply.

9. A backup power system can take in a battery voltage between 132 V and 300 V, or it can take in rectified, filtered three-phase power from a generator set. The generator output voltage is 208 $V_{\text{line-to-line}}$ at 60 Hz, and is converted through a conventional diode midpoint rectifier with large inductance to filter the output, followed by a large capacitance to provide a voltage-source interface. The backup system's inverter operates as a VSI to produce 120 V_{RMS}, 60 Hz, single phase. Draw a possible circuit for this application. What values of δ will be used for each of the various input sources? What is the highest value of current that any of the inverter switches will carry when the load draws 5000 W with a resonant filter?

10. Remote power equipment is another important application of inverters. Many villages in East Central Africa do not have electrical energy supplies. However, electric water pumps are an important part of the sanitation system. Consider a village with a solar-powered well pump. The pump fills a tank so that water is available at night or on overcast days. A typical system has about 3 m^2 of solar panels, with peak output power of about 320 W—enough to supply a 1/3 HP motor. If the panels are configured with the cells in series, the nominal dc voltage is about 150 V. A VSI is to be designed for this application to power a small single-phase motor.

a. Suggest a motor voltage (based on standard 120 V, 208 V, 230 V, 380 V, or 460 V machines) that will be consistent with the VSI output. For this motor, what angle δ will be used? Plot the voltage waveform for this choice.

b. If the motor is represented by a series L-R connection, and it shows a power factor of 0.8 to an ideal incoming ac line, plot the motor current when 60 Hz output is used.

c. In an emergency, if part of the panel fails or if there are long periods of heavy overcast, it is desirable to run the pump at reduced speed and power so that water can be obtained. Assume the pump draws a power that is linear with respect to motor speed. At 50% power, what frequency and value of δ should be used for the inverter, if the bus voltage is still about 150 V? What if the bus voltage is reduced to 100 V when a panel is removed for service?

11. A PWM inverter operates with 50% depth of modulation from an input voltage of 72 V. It supplies a series L-R load with $L = 1$ mH and $R = 1 \, \Omega$. The modulating frequency is 50 Hz, while the switching frequency is 500 Hz.

a. Draw the circuit. Based on triangle PWM, plot the voltage and current at the L-R combination.

b. Plot the current waveform drawn from the dc supply.

12. A PWM inverter has modulation depth k. The output is variable. The inverter operates from a 300 V dc bus and supplies a load with 90 μH in series with 0.8 Ω. The switching frequency is 18 kHz.

    a. What is the load power as a function of k for 60 Hz output?

    b. The modulation frequency varies between 0 and 400 Hz. Is load power affected by the modulation frequency? What is the output power for $k = 0.7$ at 10 Hz? What is the output power with $k = 0.7$ at 400 Hz?

13. A PWM inverter operates from a dc bus at 625 V $\pm$ 5%. It supplies a motor rated at 50 kW, 460 $V_{\text{line-to-line}}$, 60 Hz, 1800 RPM. It is desired to operate the motor at 1400 RPM instead (speed is a linear function of frequency). What is the modulation index under these conditions with nominal bus voltage? What are the modulation indexes at the high and low bus values?

14. A three-phase line source provides a voltage of 380 $V_{\text{line-to-line}}$ at 50 Hz. This source is rectified with diodes in a full bridge configuration. The rectifier output is a high capacitance. This dc bus serves as the input for a PWM inverter. The inverter switching frequency is 12 kHz.

    a. Estimate the dc bus voltage. How big should the capacitor be to keep the voltage ripple below 5% peak to peak, if the nominal power level is 15 kW?

    b. The inverter serves a motor rated at 380 $V_{\text{line-to-line}}$, 15 kW at 50 Hz. What is the modulation index when the inverter provides rated output for the motor?

    c. The inverter serves a motor rated at 230 $V_{\text{line-to-line}}$, 15 kW (20 HP) at 60 Hz. What is the modulation index with the motor operating at rated conditions?

15. In three-level PWM, the modulating process operates on each portion of the modulating function separately. The output peak value is still given by the depth of modulation, but the RMS value of the square wave is not V_{in}. Compute the THD for three-level PWM as a function of the depth of modulation. (*Hint*: Each pulse-width is well-defined based on the modulating function; the square of a pulse function is easy to compute.)

16. In a PWM inverter, the output ripple around the desired modulating function is nearly a triangle superimposed on the sinusoidal modulation. With a simple L-R load, it is possible to pick out the portion of the waveform with the worst-case ripple as if we were analyzing ripple for a buck converter. Given a PWM inverter with L-R load such that $L/R = 100$ μs, and given $V_{\text{in}} = 240$ V, $M(t) = 170 \cos(120\pi t)$, $f_{\text{switch}} = 50$ kHz, estimate the maximum peak-to-peak ripple around the desired modulating function, as measured at the output.

17. A PWM inverter is to provide power for a 380 V, 50 Hz three-phase motor (the voltage is an RMS line-to-line value). Suggest an input dc bus voltage to support this application. Draw a candidate inverter circuit. What is the maximum current and voltage that the switches must carry or block if the motor draws 100 A_{RMS} at full load and rated speed?

18. A boost converter draws power from a diode bridge and steps it up with a PWM process for a

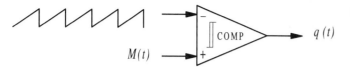

$M(t)$

$q(t)$

Figure 6.39   Sawtooth PWM process.

dc–dc converter. If the input is taken from either a 120 V 60 Hz line or a 220 V 50 Hz line, suggest a specific value of dc voltage that can be used for both applications.

19. Explain the benefits of PWM rectification. You should be aware that many power supply designers now use such rectifiers at the input side to meet certain international power quality requirements.

20. Pulse-width modulation can be performed based on a sawtooth waveform, as illustrated in Figure 6.39. Consider a system with 170 V input dc voltage that is to provide 120 V_{RMS} output at 60 Hz. For 720 Hz switching, plot the inverter output voltage waveforms for both triangle and sawtooth oscillators. Discuss how the results compare.

21. Some workers advocate five-level PWM, in which extra voltages $\pm V_{in}/2$ are used. Here is one way to implement it: When the wanted waveform is greater than $V_{in}/2$, switch between $V_{in}/2$ and V_{in}; when the wanted waveform is between 0 and $V_{in}/2$, switch between 0 and $V_{in}/2$, and so on. The carrier triangles will need to be in the desired range as the comparator process evolves. A battery system provides $V_{in} = 360$ V. It is desired to use five-level PWM and a switching frequency of 500 Hz to create a 230 V, 50 Hz output. The load is 20 mH in series with 5 Ω. Plot the load voltage and current. Comment on distortion. Is it significantly different than for the three-level case?

22. A boost PWM rectifier draws 1500 W from a 120 V, 60 Hz source for a battery charging application. The switching frequency is 25 kHz. The power factor has been corrected as close to $pf = 1$ as possible.

  a. Ignoring ripple, plot the current that flows in the inductor.

  b. Determine an output capacitance sufficient to make $V_{out} = 250$ V $\pm$ 2%.

REFERENCES

B. D. Bedford, R. G. Hoft, *Principles of Inverter Circuits*. New York: Wiley, 1964.

B. K. Bose, *Power Electronics and Ac Drives*. Englewood Cliffs, NJ: Prentice Hall, 1986.

R. Chauprade, "Inverters for uninterruptible power supplies," *IEEE Trans. Industry Applications*, vol. IA-13, no. 4, pp. 281–295, July 1977.

D. M. Divan, T. A. Lipo, T. G. Habetler, "PWM techniques for voltage source inverters," presented at the *IEEE Applied Power Electronics Conf.*, 1990.

G. K. Dubey, *Power Semiconductor Controlled Drives*. Englewood Cliffs, NJ: Prentice Hall, 1989.

M. Ehsani, "A tutorial on pulse width modulation techniques in inverters and choppers," presented at the *IEEE Applied Power Electronics Conf.*, 1981.

D. C. Griffith, *Uninterruptible Power Supplies*. New York: Marcel Dekker, 1993.

M. H. Kheraluwala, D. M. Divan, "Delta modulation strategies for resonant link inverters," *IEEE Trans. Power Electronics*, vol. 5, no. 2, pp. 220–228, April 1990.

H. W. Van der Broek, H. C. Skudelny, G. V. Stanke, "Analysis and realization of a pulsewidth modulator based on voltage space vectors," *IEEE Trans. Industry Applications*, vol. 24, no. 1, pp. 142–150, January 1988.

P. Vas, *Vector Control of Ac Machines*. Oxford: Clarendon Press, 1990.

P. Wood, *Switching Power Converters*. New York: Van Nostrand, 1981.

CHAPTER 7

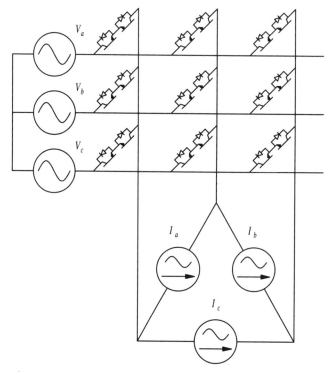

Figure 7.1 Matrix converter for general ac to ac conversion function.

AC TO AC CONVERSION

7.1 INTRODUCTION

Ac to ac conversion is the most general energy transfer possibility in power electronics. The ac to ac matrix shown in Figure 7.1 can transfer energy between arbitrary sources. It can be used for dc–dc conversion (both the input and the output are single-phase ac sources with zero frequency), for rectification, and for inversion. Many practical converter systems have an ac–ac function. For example, a commercial inverter rectifies an ac utility source, then performs a dc–ac conversion on the result.

Given the broad functionality, it is natural to ask why ac–ac converters are not used generally for switching power conversion. There are two key reasons.

- ac–ac conversion requires switches that can carry or block without regard to polarity. In particular, the switches must be able to switch on or off while current is flowing in either direction. This is nontrivial for a semiconductor device, and switches for the purpose are usually fabricated as combinations of more familiar unilateral devices.
- With true bilateral switches, the switching action is determined solely by design. There are no passive diodes to automatically provide KCL paths, for example. A designer must pay close attention to the various circuit configurations to ensure that KVL and KCL problems are not created as switches interact. It is not possible to produce switching functions that are perfectly synchronized or exactly complementary.

The first of these issues tends to make direct ac–ac solutions expensive, since several devices are required for each switch. The second has fundamental implications for reliability.

It is important to be mindful of the drawbacks, but ac–ac conversion certainly has a significant and growing place in power electronics. Bilateral semiconductor switches are used in fields as diverse as telephone switching, analog multiplexing, and low-level ac conversion. High-power ac–ac methods are used for some motor drives and for utility applications. The utility area in particular has strong potential for growth over the next decade or so.

In this chapter, the major techniques of ac–ac conversion are examined. Both phase control methods and PWM apply to this function. General ac–ac matrix converters are examined. The cycloconverter, an ac–ac method based on controlled rectifier combinations, is studied. It can be modelled with a particular form of phase control. Most of the important ac–ac applications, like commercial inverters, use a dc link arrangement. A few dc link systems are considered here.

One common ac–ac application—the *ac regulator*—uses its load to create the switch control action. This is important for controlling power levels in a fixed-frequency system, but it does not support transfer between sources at different frequencies. Even though this limits the function, ac regulators are common.

7.2 FREQUENCY MATCHING CONDITIONS

Nonzero average power flow is contributed only when both the voltage and current have Fourier components with matching frequency. Consider the simple, but general, direct ac–ac converter in Figure 7.2. The input source is a voltage with frequency f_{in}. The output load is represented by a current source at frequency f_{out}, consistent with the idea of source conversion. The switch matrix between the two provides a multiplication function, $V_{out} = V_{in}Q$, as in equation (2.2). Remember that Q is the switch state matrix, and represents all the switching functions in the converter.

Average power flow requires that the output voltage has a frequency component at f_{out} and that the input current has a component at f_{in}. In this converter, let us assign a switching frequency f_{switch}, and write the output voltage as

$$v_{out}(t) = [q_{1,1}q_{2,2} - q_{1,2}q_{2,1}]v_{in}(t) \tag{7.1}$$

To illustrate the matching issues, let us take $q_{1,1} = q_{2,2}$ and $q_{1,2} = 1 - q_{1,1}$. This simplifies the expressions without loss of generality for the purposes of evaluating the requirements for nonzero power flow. Then

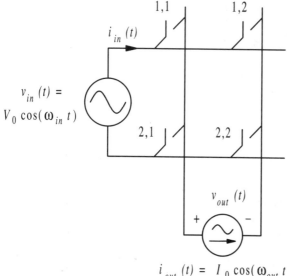

Figure 7.2   Direct ac voltage to ac current converter.

$$v_{out}(t) = (2q_{1,1} - 1)v_{in}(t) \tag{7.2}$$

Substituting the Fourier series forms of the input voltage and switching function, a complete expression for the output voltage can be obtained,

$$v_{out}(t) = \left\{(2D - 1) + \frac{4}{\pi} \sum_{n=1}^{\infty} \frac{\sin(n\pi D)}{n} \cos[n\omega_{switch}(t - t_0)]\right\} V_0 \cos(\omega_{in}t)$$

$$\tag{7.3}$$

$$= (2D - 1)V_0 \cos(\omega_{in}t) + \frac{4V_0}{\pi} \sum_{n=1}^{\infty} \frac{\sin(n\pi D)}{n} \cos[n\omega_{switch}(t - t_0)]\cos(\omega_{in}t)$$

In this expression, the term $(2D - 1)V_0 \cos(\omega_{in}t)$ will not meet frequency requirements except in the special case $f_{in} = f_{out}$. The summation will be rewritten based on the trigonometric identity $2 \cos A \cos B = \cos(A + B) + \cos(A - B)$ to show the frequency components,

$$v_{out}(t) = (2D - 1)V_0 \cos(\omega_{in}t) + \frac{2V_0}{\pi} \sum_{n=1}^{\infty} \frac{\sin(n\pi D)}{n}$$

$$\times [\cos(n\omega_{switch}t + \omega_{in}t - n\phi_0) + \cos(n\omega_{switch}t - \omega_{in}t - n\phi_0)]$$

$$\tag{7.4}$$

with $\phi_0 = n\omega_{switch}t_0$. Average power can flow only when there is a voltage Fourier component that matches f_{out}. This requires

$$\omega_{out} = n\omega_{switch} + \omega_{in} \quad \text{or} \quad \omega_{out} = n\omega_{switch} - \omega_{in} \tag{7.5}$$

Since power conversion is the objective, we want the component to be as large as possible. This normally occurs for the $n = 1$ term. The final result is the frequency matching condition.

Definition: *The* frequency matching condition *in an ac–ac converter requires that the switching functions contain a Fourier component at frequency f_{switch} such that $f_{out} = f_{switch} \pm f_{in}$. Nonzero average power flows only if the condition is met.*

The frequency matching condition can be rewritten with simple algebra as $f_{switch} = f_{in} \pm f_{out}$. For this discussion, there is no need to distinguish between positive and negative frequency. After all, $\cos \theta = \cos(-\theta)$, so negative and positive frequency have the same effect in a power converter.

7.3 DIRECT-SWITCHING FREQUENCY CONVERTERS

Given the frequency matching condition, the most direct way to perform ac–ac conversion is to connect a source and load through a switch matrix, then operate so that $f_{switch} = f_{in} \pm f_{out}$. Thus, a converter for 50 Hz input and 60 Hz output would operate by switching at either 10 Hz or 110 Hz. This approach is clear enough, and does indeed provide a valid conversion function. In this section, some examples of this approach will be developed.

7.3.1 Slow-Switching Frequency Converters: The Choice $f_{in} - f_{out}$

The switching frequency choice $f_{switch} = f_{in} - f_{out}$ is the lowest frequency that directly provides a desired ac–ac conversion. It has been termed *slow-switching frequency conversion* (SSFC) in some of the literature. Slow-switching conversion can be useful at very high power levels, when the switching devices incur significant losses each time they operate. Let us consider two application examples.

Example 7.3.1 An electrical machine tool designed for use in North America is needed for a European application. The project manager proposes that a 50 Hz to 60 Hz direct-switching converter offers a simple alternative to provide the necessary input power. If slow switching is used, losses caused by switching will be low. Develop a possible conversion circuit for this application, and evaluate its operation. Discuss the wanted and unwanted Fourier component behavior.

The problem statement provides only general information, so let us consider the issues. Focus on single-phase power (or one phase out of several). The converter might draw power from a 230 V, 50 Hz voltage source and convert it to 120 V, 60 Hz for the tool. The usual KVL and KCL restrictions require the output to resemble a current source if the input is a true voltage source. A bridge converter that might be a reasonable choice is shown in Figure 7.3. The converter's output voltage, v_{out}, is the 50 Hz input chopped with the 10 Hz switch action.

As for a VSI, let us operate the switches in diagonal pairs and use a duty ratio of 50% to begin the evaluation. The output waveform, based on a switch phase delay of 0, is shown in Figure 7.4. It is certainly not obvious that this choice achieves 60 Hz output, so Fourier analysis needs to be applied. The expression in equation (7.4) gives the necessary result. Since the duty ratio is 50%, the lead term in (7.4) is zero. The series leads to Table 7.1. Nonzero components appear at 0 Hz, 20 Hz, 40 Hz, 60 Hz, ... This is consistent with the 20 Hz fundamental frequency apparent by inspection of the waveform. Odd multiples of 10 Hz have been cancelled because of the symmetry of the 50% duty ratio. The conversion does

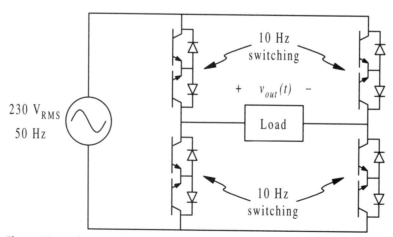

Figure 7.3 Bridge converter for 50 Hz input and 60 Hz output.

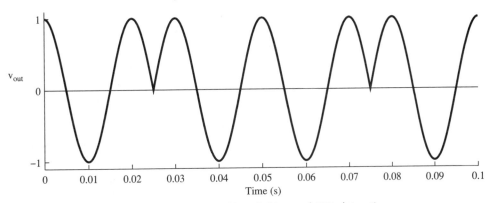

Figure 7.4 Output voltage for 50 Hz input, 10 Hz switching, and 50% duty ratio.

indeed provide a 60 Hz output. The amplitude is 188 V at 60 Hz (133 V_{RMS}), but the waveform has some undesirable properties.

- A dc component appears at the output. This will cause trouble if a transformer is used to set the necessary amplitude.
- There are unwanted frequency components, called *subharmonics*, below 60 Hz.

In this case, a phase delay can help the situation. For example, if the switching function associated with Figure 7.4 is delayed by 1/4 of the 50 Hz cycle, the dc component cancels. In general, however, a resonant filter or similar method will be needed to provide an interface to the load. The dc component must be avoided, and of course the low-frequency unwanted components cannot be removed with a low-pass filter.

Slow-switching methods are not used often for applications requiring high-quality waveforms when the input and output frequencies are close together because of the subharmonic terms. The situation is considerably better when low frequencies are needed. For example, certain types of ac motor controllers use a converter to connect frequencies as low as a few hertz to the rotor circuit. Let us examine this type of application to consider the application of the slow-switching approach.

TABLE 7.1
Coefficients and Frequencies for 50 Hz to 60 Hz Conversion with 10 Hz Switching

n	Frequency, $nf_{switch} + f_{in}$ Term of (7.4)	Amplitude, $nf_{switch} + f_{in}$ Term of (7.4)	Frequency, $nf_{switch} - f_{in}$ Term of (7.4)	Amplitude, $nf_{switch} - f_{in}$ Term of (7.4)
1	60 Hz	$2V_0/\pi$	40 Hz	$2V_0/\pi$
2	70	0	30	0
3	80	$2V_0/(3\pi)$	20	$2V_0/(3\pi)$
4	90	0	10	0
5	100	$2V_0/(5\pi)$	0	$2V_0/(5\pi)$
6	110	0	10	0
7	120	$2V_0/(7\pi)$	20	$2V_0/(7\pi)$

Example 7.3.2   ac motors in applications above a few hundred horsepower often are built with slip rings to allow electrical connections to the rotor circuits. Consider a three-phase induction motor rated at 1000 HP (750 kW), 60 Hz, 4160 V. The rotor circuit rating is about 75 kW. An ac–ac converter is under consideration to connect the rotor circuit (typical frequency: 5 Hz) to the 60 Hz power line for control and energy recovery purposes. The rotor circuit is also a three-phase arrangement. Draw an appropriate circuit. Evaluate the waveform and harmonic performance imposed on one of the rotor phases. There is no common neutral, because the rotor connection is a delta.

A candidate circuit (showing a rotor circuit model) appears in Figure 7.5. The circuit requires nine switches, each fully bilateral, and each with ratings on the order of 6 kV and 20 A. Each column of the matrix can have one and only one switch on at a time because of the KVL and KCL restrictions. Given the 5 Hz output objective, the switching frequency should be $60 - 5 = 55$ Hz. The devices would be expected to have a duty ratio of 1/3. Consider one of the rotor phase voltages, shown as v_{ab} in the figure. This voltage can be one of the six choices v_{AB}, v_{AC}, v_{BC}, v_{BA}, v_{CA}, v_{CB}, where $v_{BA} = -v_{AB}$ and so on. The logical sequence of switch action, based on switching serially among these six choices, would follow the order 1,1 and 2,2 on; 1,1 and 3,2 on; 2,1 and 3,2 on; 2,1 and 1,2 on; 3,1 and 1,2 on; 3,1 and 2,2 on. This produces the waveform shown in Figure 7.6.

Inspection shows that this piecewise sinusoidal waveform attempts to track a 5 Hz signal. In some ways, the waveform resembles a PWM voltage, and in fact the distortion is rel-

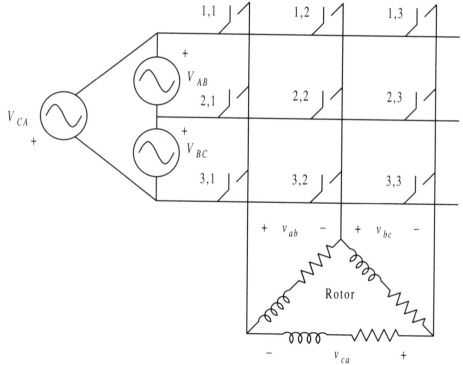

Figure 7.5   Possible converter for 60 Hz to 5 Hz rotor interface.

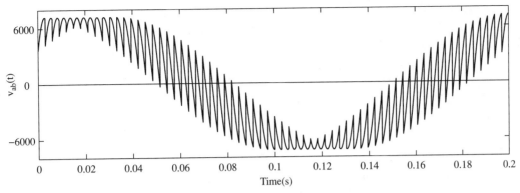

Figure 7.6   Output voltage $v_{ab}(t)$ of rotor converter circuit.

atively easy to filter in this case. Once again we might use results as in equation (7.4) to evaluate the possible Fourier component frequencies. This polyphase case will have a more complicated expression than equation (7.4) for the output components, but the frequencies will still be determined by $nf_{switch} \pm f_{in}$, as shown in Table 7.2. The symmetry cancels out all even harmonics, and also those that are multiples of 3. In this example, the waveform harmonics are low enough that a simple low-pass filter will do an excellent job of reducing distortion, and the rotor currents will be nearly sinusoidal.

The second example shows that slow-switching methods are useful for low output frequencies. In practice, the method requires output frequencies below about 25% of the input. For instance, a 400 Hz to 60 Hz converter might use 340 Hz switching to perform the necessary conversion. As we shall see, however, the cycloconverter offers similar performance with less stringent device requirements. The slow-switching method requires true bilateral switches.

7.3.2 The Choice $f_{switch} = f_{in} + f_{out}$

When the sum frequency is used in an ac–ac application, the result is sometimes called an *unrestricted frequency converter* (UFC). This is because the limitation $f_{out} \ll f_{in}$ imposed

TABLE 7.2
Coefficients and Frequencies for 60 Hz to 5 Hz Conversion with 55 Hz Switching

n	Frequency, $nf_{switch} + f_{in}$	Amplitude, from Fourier Analysis	Frequency, $nf_{switch} - f_{in}$	Amplitude, from Fourier Analysis
1	115 Hz	0	5 Hz	$3V_0/\pi$
3	225	0	105	0
5	335	$3V_0/(5\pi)$	215	0
7	445	0	325	$3V_0/(7\pi)$
9	555	0	435	0
11	665	$3V_0/(11\pi)$	545	0
13	775	0	655	$3V_0/(13\pi)$

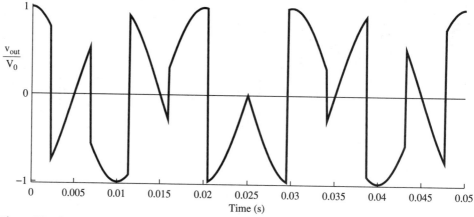

Figure 7.7 Output voltage for 50 Hz to 60 Hz converter with 110 Hz switching.

by the slow-switching case is removed by switching at the higher rate. The slow-switching method also has a problem when $f_{out} = f_{in}$ is needed, as it might be in a variable-frequency motor drive. The slow-switching method suggests $f_{switch} = 0$ in this case. This is a valid choice (no switch action), but it eliminates possibilities for control. In the unrestricted case, the choice $f_{out} = f_{in}$ leads to $f_{switch} = 2f_{in}$, and no special problems occur.

The UFC typically has better Fourier component properties than its slow-switching counterpart. Consider once again the 50 Hz to 60 Hz and 60 Hz to 5 Hz examples.

Example 7.3.3 Evaluate a single-phase UFC as an alternative for 50 Hz to 60 Hz conversion. Compare the results to those of Example 7.3.1. The circuit is exactly the same, and the only change is to set the switching frequency to 110 Hz.

The output voltage waveform for this case is given in Figure 7.7. It is a bit easier to pick out the 60 Hz symmetry in this waveform than in that of Figure 7.4. Table 7.3 shows the possible frequencies, along with the amplitudes from equation (7.4). In this case, the lowest component occurs at 60 Hz, and the next component does not appear until 160 Hz. Subharmonics have been avoided. In the slow-switching case, we found that the switching function phase had a significant effect, and that a dc offset could be created with the wrong

TABLE 7.3
Coefficients and Frequencies for 50 Hz to 60 Hz Conversion with 110 Hz Switching

n	Frequency, $nf_{switch} + f_{in}$ Term	Amplitude, $nf_{switch} + f_{in}$ Term	Frequency, $nf_{switch} - f_{in}$ Term	Amplitude, $nf_{switch} - f_{in}$ Term
1	160 Hz	$2V_0/\pi$	60 Hz	$2V_0/\pi$
2	270	0	170	0
3	380	$2V_0/(3\pi)$	280	$2V_0/(3\pi)$
4	490	0	390	0
5	600	$2V_0/(5\pi)$	500	$2V_0/(5\pi)$
6	710	0	610	0
7	820	$2V_0/(7\pi)$	720	$2V_0/(7\pi)$

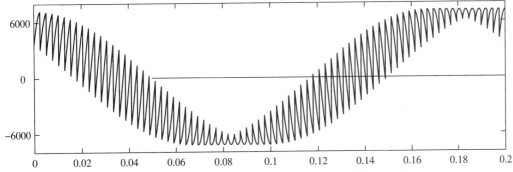

Figure 7.8 Output voltage waveform for 60 Hz to 5 Hz converter given 65 Hz switching.

choice. In the UFC case, there is no dc offset, and the specific timing of the switching functions has no effect on the component amplitudes. From the standpoint of filter design, the UFC is a better choice than the SSFC for this conversion application.

Example 7.3.4 Consider again the 60 Hz to 5 Hz converter for the ac machine rotor. Use a switching frequency of 65 Hz, and compare the results with those of Example 7.3.2.

The new output waveform is presented in Figure 7.8. By inspection, it does not seem much different from the result in Figure 7.6. The component frequencies are rather different, however, as shown in Table 7.4. Each unwanted component is shifted higher, with some benefits for filtering. One subtle point in the waveform is worthy of note: In the SSFC waveform, the output was delayed somewhat from the zero time axis because of the specific choice of switching function phases. In the UFC waveform here, the same switching function phases generate an output lead instead of an output lag. It is true in general that the UFC and SSFC methods affect phase in opposite directions. The two switching methods are not much different when the output frequency is low, but the UFC method always offers some benefit in the filter design.

The UFC should be able to support conversion between arbitrary frequencies. One such application might create a single-phase 60 Hz output from a three-phase 60 Hz input for an ac motor drive. Figure 7.9 shows a midpoint-type converter for this application. The output waveform is given in Figure 7.10a. Figure 7.10b shows a sample output waveform for a

TABLE 7.4
Coefficients and Frequencies for 60 Hz to 5 Hz Conversion with 65 Hz Switching

n	Frequency, $nf_{switch} + f_{in}$	Amplitude, from Fourier Analysis	Frequency, $nf_{switch} - f_{in}$	Amplitude, from Fourier Analysis
1	125 Hz	0	5 Hz	$3V_0/\pi$
3	255	0	135	0
5	385	$3V_0/(5\pi)$	265	0
7	515	0	395	$3V_0/(7\pi)$
9	645	0	525	0
11	775	$3V_0/(11\pi)$	655	0
13	905	0	785	$3V_0/(13\pi)$

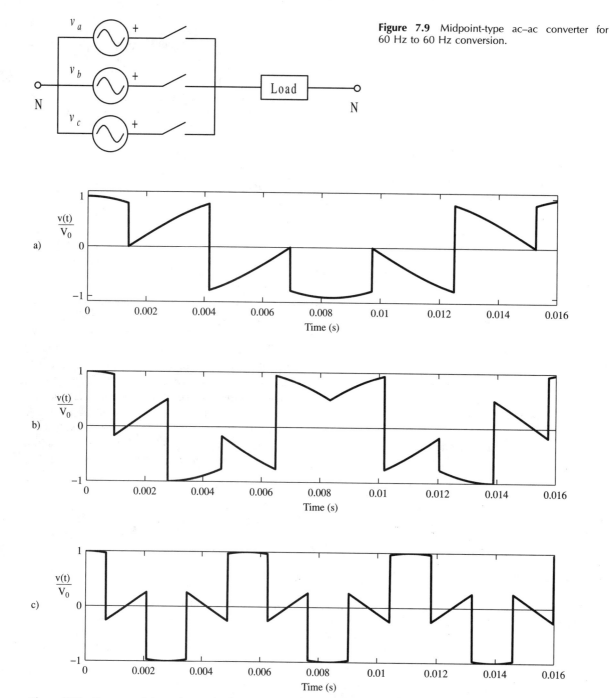

Figure 7.9 Midpoint-type ac–ac converter for 60 Hz to 60 Hz conversion.

Figure 7.10 Output waveforms for 3ϕ 60 Hz to 1ϕ converter. Plot (a) uses 120 Hz switching for 60 Hz to 60 Hz. Plot (b) uses 180 Hz switching for 60 Hz to 120 Hz conversion. Plot (c) uses 240 Hz switching.

60 Hz to 120 Hz converter, while Figure 7.10c shows the output from a 60 Hz to 180 Hz converter.

7.3.3 Unifying the Direct Switching Methods: Linear Phase Modulation

For purposes of implementation, it is convenient to express the direct ac–ac conversion methods in a somewhat different form. Consider that the circuit designer of an ac–ac converter requires timing information to produce the necessary switching frequency given an input frequency. One reasonable approach is to use the input waveform as a reference signal, then add or subtract a bias to produce the appropriate switch action.

The Fourier Series of a switching function illustrates the process:

$$q(t) = D + \frac{2}{\pi} \sum_{n=1}^{\infty} \frac{\sin(n\pi D)}{n} \cos\{n[\omega_{in}t + M(t)]\} \tag{7.6}$$

The slow-switching method uses the control function $M(t) = -\omega_{out}t$, while the UFC approach uses $M(t) = +\omega_{out}t$. Notice that the function $M(t)$ appears as a phase. Equation (7.6) suggests that the switching function is derived from $v_{in}(t)$, with a time-varying adjustment of phase. This process is *phase modulation*. The functions $M(t) = \pm\omega_{out}t$ both vary linearly with time, so that the direct switching converters use *linear phase modulation*.

The concept of linear phase modulation is useful for implementation. If the input sine wave is used to create a square wave, then this square wave is successively delayed by a phase angle, the rate of delay allows direct adjustment of the output. When the linear phase modulation form is substituted into equation (7.4), the result is

$$v_{out}(t) = (2D - 1)V_0 \cos(\omega_{in}t) + \frac{2V_0}{\pi} \sum_{n=1}^{\infty} \frac{\sin(n\pi D)}{n}$$
$$\times \{\cos[(n + 1)\omega_{in}t \pm nM(t)] + \cos[(n - 1)\omega_{in}t \pm nM(t)]\} \tag{7.7}$$

The wanted component appears in the last term for $n = 1$, and has the value

$$v_{out(wanted)} = \frac{2V_0}{\pi} \sin(\pi D)\cos[M(t)] \tag{7.8}$$

This is formally equivalent to substituting a time function $M(t)$ for the phase delay angle α in a phase-controlled rectifier. In principle, we can substitute a function for $M(t)$ to produce a desired output.

Linear phase modulation is not the only alternative. There are a few important nonlinear modulation approaches used for ac–ac conversion, as described in the section that follows.

7.4 THE CYCLOCONVERTER

The term *cycloconversion* is sometimes used to describe the general ac–ac conversion process. Often, the term refers more specifically to an ac–ac converter that attempts to follow a low-frequency wanted component, such as the 60 Hz to 5 Hz converters of the preceding section. In practice, low-frequency conversions are often accomplished with a specialized form of nonlinear phase modulation, adapted so that rectifier-like circuits can support the operation. This arrangement is often implied when the term *cycloconverter* is used.

The wanted component in equation (7.8) suggests that ac–ac converters are similar in some control sense to rectifiers. If the phase angle α in a rectifier is adjusted slowly, a desired output result can be generated. The basic idea is much like PWM, except that phase is being modulated in place of duty ratio. The SCR midpoint rectifier in Figure 7.11 shows some of the issues and limitations associated with the phase modulation idea in its simplest form. In the figure, the load current must be positive. ac–ac conversion only works if an offset current can be accepted.

A subtle limitation of the SCR rectifier in the figure is that the value of α is restricted. If $0° \leq \alpha \leq 180°$, the rectifier follows the phase delay control pattern, and a variable angle α can be used. However, if $180° < \alpha < 360°$, the devices will not switch at the turn-on command point because the voltage bias is inconsistent with SCR behavior. Since the angle $\alpha = \pm\omega_{out}t$ has no limit, simple linear phase modulation is not feasible to make a rectifier provide an ac output.

Waveforms for 90° delay and for 270° delay are shown in Figure 7.12 to demonstrate the issue. In the 90° case, the output voltage makes a transition from low to high at the switching point. Each SCR will be blocking a positive forward voltage when the gate signal is applied. By the trial method, the SCR becomes a diode when its gate is pulsed. Each device then must turn on immediately if it is blocking forward voltage when the gate signal arrives.

In the 270° case, the output voltage makes a transition from high to low at the intended switching time. Each SCR is therefore reverse-biased when the gate signal is applied. Reverse bias is consistent with an off device, so the SCRs in a midpoint arrangement of Figure 7.11 will not operate to give a waveform like that of Figure 7.12b.

The problem of limits on α can be solved by recognizing that the converter output wanted component is determined by $\cos\alpha$. What if, instead of $\alpha = \pm\omega_{out}t$, the value is set to $\alpha = \cos^{-1}[\cos(\pm\omega_{out}t)]$? The converter output will still be proportional to $\cos(\pm\omega_{out}t)$, but the in-

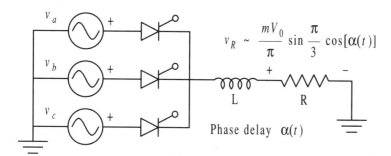

$$v_R \sim \frac{mV_0}{\pi} \sin\frac{\pi}{3} \cos[\alpha(t)]$$

Figure 7.11 Midpoint rectifier with phase modulation for ac output.

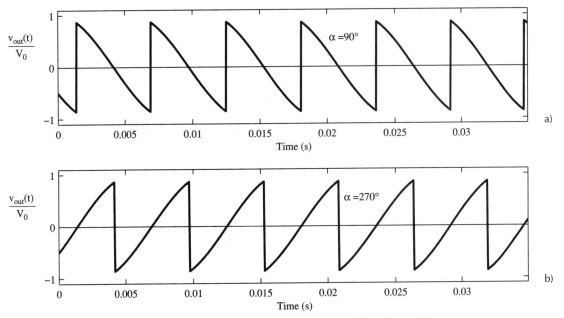

Figure 7.12 Hypothetical output voltages for midpoint converter with $\alpha = 90°$ and $\alpha = 270°$.

verse cosine limits the range of $\alpha(t)$. By convention, the inverse cosine can be taken to give a result between 0° and 180°—exactly what is needed to make a rectifier operate properly.

If a control circuit can set $\alpha = \cos^{-1}[\cos(\pm\omega_{out}t)]$, it can also insert an amplitude adjustment k such that $\alpha = \cos^{-1}[k\cos(\pm\omega_{out}t)]$. As long as $0 \le k \le 1$, the value of α will be reasonable for phase control. Figure 7.13 shows this function for $k = \frac{1}{2}$ and for $k = 1$. The

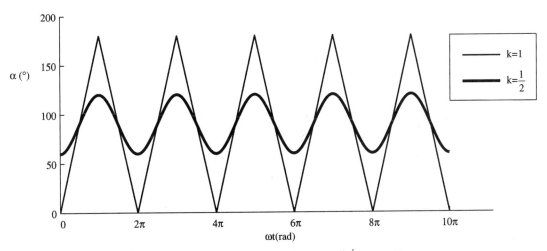

Figure 7.13 Phase control functions $\alpha = \cos^{-1}[\cos(\omega_{out}t)]$ and $\alpha = \cos^{-1}[\frac{1}{2}\cos(\omega_{out}t)]$.

SSFC and UFC converter schemes from the previous section did not provide any clear way to adjust the amplitude of the output wanted component. The value k remedies this limitation.

The choices of α from Figure 7.13 give rise to what is known as a *positive converter*, an ac–ac converter in which all transitions move from a low voltage to a high voltage. An analogous choice maintains $180° \le \alpha \le 360°$ (or $-180° \le \alpha \le 0°$). This control creates a *negative converter*. The current-sourced inverter in Figure 7.14 is an example of a converter that benefits from negative converter control. In this case, the reverse direction of each SCR ensures that each will have positive forward bias at the turn-on point if $-180° \le \alpha \le 0°$.

The final step in creating a cycloconverter is to combine the configurations: Use a positive converter to supply the load when $i_{\text{load}} > 0$, and use a negative converter to accept load current when $i_{\text{load}} < 0$. This combination circuit, shown in Figure 7.15, allows unidirectional devices to perform complete ac–ac conversion. A representative output voltage waveform is shown in the figure as well. This approach is not new; it was proposed in the 1930s shortly after grid-controlled mercury arc tubes became available.

Cycloconverters are most likely to be found in low frequency high-power applications, such as controllers for ac machine rotors and low-speed motor drives. They tend to be more common than either slow-switching or UFC arrangements, since they support the SCR as a device for ac–ac conversion.

Example 7.4.1   A cycloconverter uses the function $\alpha(t) = \cos^{-1}[\cos(40\pi t)]$ to convert a 60 Hz three-phase input to single-phase 20 Hz output. Draw a candidate circuit. Plot the functions $\pm\alpha(t)$ for the positive and negative converters. What would the output voltage waveforms be for each of these converters acting alone? If the load is an ideal current source

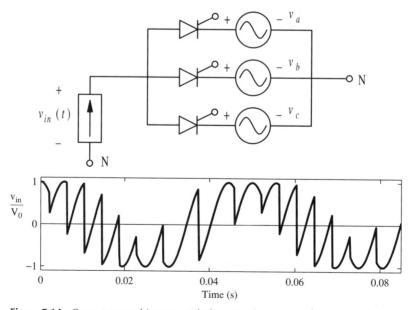

Figure 7.14   Current-sourced inverter with the control $\alpha = -\cos^{-1}[k\cos(\omega_{\text{out}}t)]$.

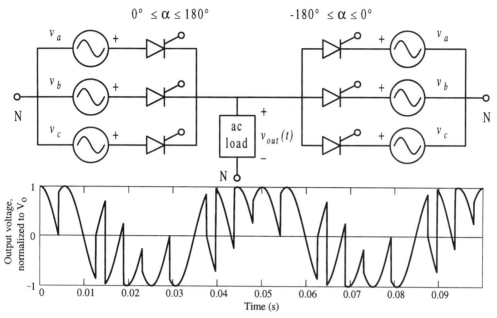

$0° \le \alpha \le 180°$

$-180° \le \alpha \le 0°$

Figure 7.15 A cycloconverter: positive and negative converters combine to support ac–ac conversion with SCRs.

$I_0 \cos(40\pi t)$ (with phase reference $\angle v_a(t) = 0°$), determine the actual output voltage.

The circuit is given in Figure 7.16. The control waveforms are triangular, with a frequency of 20 Hz, as shown in Figure 7.17. The positive and negative converters alone would provide outputs consisting of successively delayed pieces of the inputs. Here, let us assume that switching takes place based on the instantaneous value of $\alpha(t)$. Figure 7.18 shows the positive and negative converter outputs. If the output current is positive, the SCRs in the negative converter will be off, and the positive converter will determine the waveforms. Similarly, the negative converter determines the waveform when i_{out} is negative. The actual output voltage waveform is given in Figure 7.19.

Figure 7.16 Cycloconverter for Example 7.4.1.

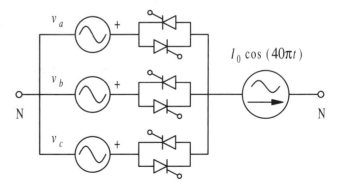

$I_0 \cos(40\pi t)$

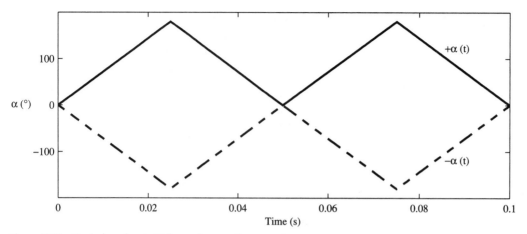

Figure 7.17 Control angles $\pm\alpha(t)$ for cycloconverter example.

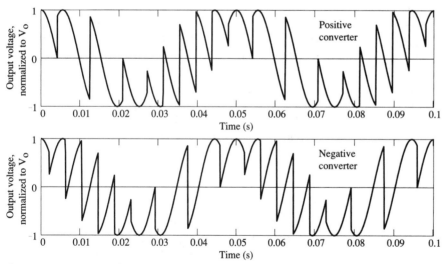

Figure 7.18 Positive and negative converter voltages for Example 7.4.1.

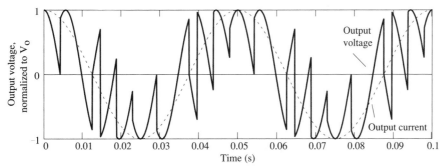

Figure 7.19   Cycloconverter output, 60 Hz to 20 Hz conversion.

7.5 OTHER NONLINEAR PHASE MODULATION METHODS

The cycloconverter is the primary example of nonlinear phase modulation applied to converters. There are other possibilities as well. The control angle $\alpha(t)$ in Figure 7.13 appears nearly sinusoidal for $k = \frac{1}{2}$, with an average value centered at 90°. What about the choice $\alpha(t) = 90° - k60° \cos(\omega_{out}t)$? (The 60° value is chosen to be a good match to the inverse cosine when $k = \frac{1}{2}$.) This is a little easier to generate than the inverse cosine function, yet still meets the operating constraints of a positive converter. This function is compared to the waveform for $\alpha(t) = \cos^{-1}[\frac{1}{2}\cos(\omega_{out}t)]$ in Figure 7.20. They are nearly identical (and in fact remain similar for k below about 0.8). Almost any periodic function with a restricted range can be used for nonlinear phase modulation.

If cosines and similar periodic functions work for cycloconverter control, a square wave might be a reasonable possibility as well. In Figure 7.21, an output waveform is given for a

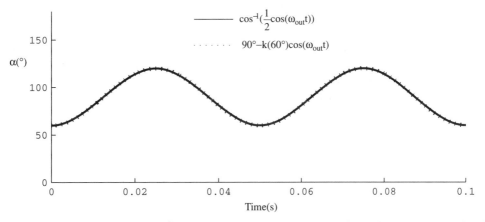

Figure 7.20   Comparison of $\cos^{-1}[k\cos(\omega_{out}t)]$ to $90° - k60° \cos(\omega_{out}t)$ for cycloconverter control with $k = \frac{1}{2}$.

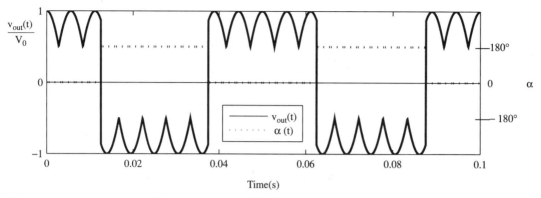

Figure 7.21   ac–ac converter with α(t) a square wave from 0° to 180°.

converter that uses a three-phase input and a square wave between 0° and 180° for $\alpha(t)$. This waveform is the same as would be obtained if an unfiltered rectifier is cascaded with an inverter. The output shifts back and forth between full positive output and full negative output. Once again, SCRs can support this process if both positive and negative converters are provided.

   With the square wave, the output amplitude can be altered by reducing the square-wave amplitude. In a rectifier-inverter cascade, this is equivalent to controlling the output amplitude by adjusting the rectifier phase, and creating the basic output waveform by switching back and forth between positive and negative output connections. A circuit and waveforms for 30° to 150° variation are given in Figure 7.22. This process provides a useful model for *dc link* ac–ac converters, in which a rectifier-inverter cascade converts to an intermediate dc transfer source. dc link converters will be considered in a subsequent section.

7.6 PWM AC–AC CONVERSION

Phase modulation methods beg the question of PWM. If phase can be adjusted to produce a given ac output, why not use pulse-width modulation to accomplish the same thing? Since almost any duty ratio (between 0 and 1) can be set, a value suitable for ac–ac conversion should be feasible. In practice, we need to keep in mind that converter operation is given by the vector product $\mathbf{V_{out}} = \mathbf{Q V_{in}}$. For both dc–dc and dc–ac converters, the wanted component of the output was determined by the duty ratio and the input voltage or current. In an ac–ac application, the input voltages will be functions of time such as $v_{in}(t) = V_0 \cos(\omega_{in}t)$, so it is important to consider the product $d(t)v_{in}(t)$. The frequency matching condition requires a component $f_{in} \pm f_{out}$ in the switching functions. With PWM control, switching is rapid, and the wanted component is imposed on the duty ratio.

   Consider a case in which the duty ratio is the time-varying function $d(t) = \frac{1}{2} + \frac{1}{2} \cos(\omega_{in}t + \omega_{out}t)$. The product $d(t)v_{in}(t)$ represents some of the important components of the output,

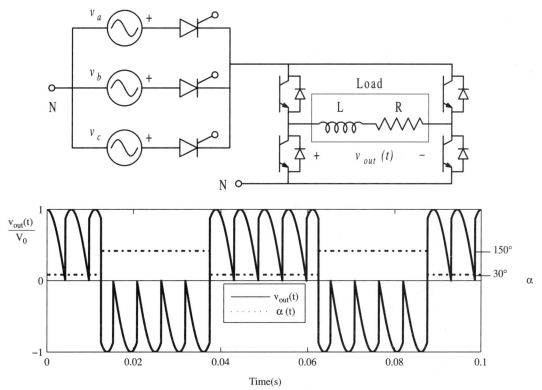

Figure 7.22   Rectifier-inverter cascade with 30° phase delay.

$$v_{\text{out}}(t) = d(t)v_{\text{in}}(t) = \frac{V_0}{2}\cos(\omega_{\text{in}}t) + \frac{V_0}{4}\cos(2\omega_{\text{in}}t + \omega_{\text{out}}t) + \frac{V_0}{4}\cos(\omega_{\text{out}}t) + \cdots \quad (7.9)$$

with the third term as the actual wanted component. The first term, $\frac{1}{2}V_0\cos(\omega_{\text{in}}t)$, can be avoided by using a symmetrical bridge arrangement. Figure 7.23 shows a bridge converter and the associated output waveform based on 60 Hz to 20 Hz conversion at 100% depth of modulation. The switching frequency is relatively low, at 720 Hz, to help make the complete waveform clear. In this converter, there is a large unwanted component at $2f_{\text{in}} + f_{\text{out}} = 140$ Hz. The other unwanted frequencies are shifted up by multiples of the switching frequency, and are relatively easy to filter out.

When PWM is used for ac–ac conversion, it shares the standard disadvantages of PWM: high distortion and extra switching losses. The advantages are less compelling for the ac–ac case, since at least one large low-frequency unwanted component will be present.

7.7 DC LINK CONVERTERS

Cascaded rectifier-inverter combinations are the most widely used type of ac–ac conversion. Indeed, this arrangement is so common that it is rarely studied as an ac–ac design problem.

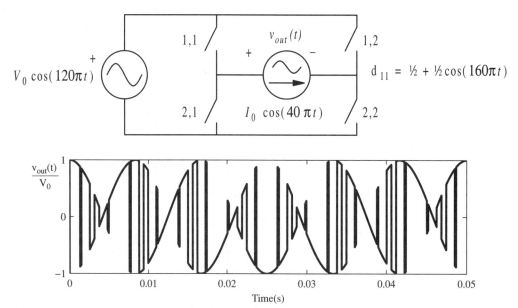

Figure 7.23 PWM ac–ac converter and output waveform.

In general, the rectifier and inverter are designed separately, and the methods of Chapters 5 and 6 determine the configuration and operation of the converters. Most applications use simple diode rectifiers, since PWM inverters offer excellent control capability.

Rectifier-inverter cascades are just one example of link converters. The link can be in the form of a capacitive dc constant-voltage bus, an inductive constant-current bus, or even a resonant LC circuit that might act more like an ac link. Formally, these are indirect converters with an internal transfer source, along the lines of indirect dc–dc converters.

dc link converters have specific limitations that might not be present in more complete ac–ac implementations. For example, the input rectifier normally prevents energy flow back to the ac source. Consider the following motor drive example, in which a separate converter must be provided for regeneration energy.

Example 7.7.1 A dc link converter for a 100 HP (75 kW) ac motor drive consists of a six-pulse SCR rectifier set, an LC interface with large capacitance, and a PWM inverter for three-phase output. While the overall behavior is similar to phase modulation with a square wave for the phase angle α, the actual converter has separate control of the rectifier and inverter. In this case, the rectifier acts to maintain a tight 600 V dc bus level on the capacitor, so that the inverter can operate from a regulated voltage source. The application is a package retrieval system for a large warehouse. As the system moves about, direction and speed change rapidly. Comment on the issues of regeneration, and propose an approach to actively brake the equipment when necessary.

Figure 7.24 shows the circuit arrangement described in the problem. The rectifier senses the dc link voltage, then adjusts its phase control angles to maintain the voltage at exactly 600 V dc. When the inverter is either accelerating or driving the motor, energy flows from the dc link into the load, and the filtered current i_{bus} shown in the figure is positive. In this case, the system can be understood along the same lines as examples in Chapter 6.

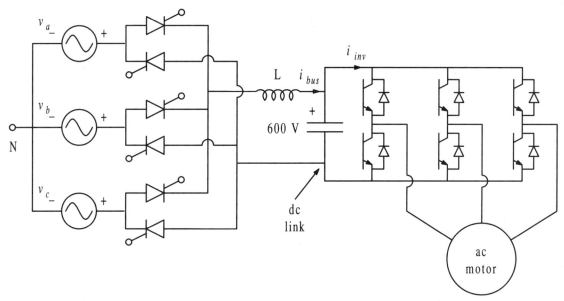

Figure 7.24   dc link motor drive for factory application.

If the motor is actively decelerated, or if it is absorbing energy as it might be in bringing a heavy pallet down from a high shelf, the energy flow in the inverter ought to reverse. The inverter itself will not alter its behavior. After all, the devices can handle bidirectional current, and they have no "knowledge" of the power flow direction. Energy flowing back from the inverter toward the source should make the current i_{inv} negative. The SCRs will not support a reversal of the current i_{bus}, so the current i_{inv} must flow into the capacitor. This will increase the bus voltage above 600 V. The SCRs will not be able to bring the voltage down by phase delay, because all the devices are turned off as soon as i_{bus} tries to reverse. How big would the capacitor need to be? Well, imagine the motor trying to provide enough negative force to keep a 200 kg load from falling from a height of 5 m. The potential energy in the load is the weight times the height, or (200 kg)(9.807 m/s²)(5 m) = 9807 J. If the capacitor voltage is allowed to change from 600 V to no more than 660 V (a 10% increase), we would have

$$\frac{1}{2}C(660 \text{ V})^2 - \frac{1}{2}C(600 \text{ V})^2 = 9807 \text{ J} \tag{7.10}$$

This requires capacitance of more than 250,000 μF. Capacitors of this value and voltage are extremely expensive and bulky.

A common alternative is to connect a voltage-sensitive switch or even a buck converter to the dc bus. If the bus voltage increases by more than an allowed amount, a connection is made to a separate load resistance. The excess energy is dissipated rather than recovered. This provides the required braking action. The voltage-sensitive switch approach is called a resistive *crowbar*, since it forces action to drop the bus voltage if the voltage rises too much. The crowbar arrangement is shown in Figure 7.25. A more complicated alternative is to add a second set of SCRs, directed for current-sourced inversion rather than rectification. If the

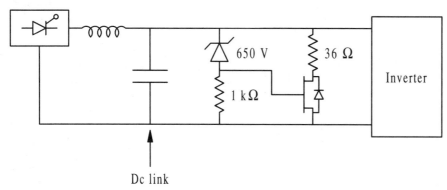

Figure 7.25   dc link converter with link crowbar for resistive braking.

bus voltage rises too much, this second converter begins to operate. It can send energy back to the ac source, so that regenerative energy recovery is possible.

dc link systems are by far the most important approach to ac–ac conversion for industrial purposes. The cascade arrangement, however, means that they are rarely studied as ac–ac applications. The most common circuits actually intended for ac–ac applications are considerably simpler, but do not show the flexibility of the dc link approach. These common circuits will be considered in the next section.

7.8 AC REGULATORS

In many common ac applications, the conversion issue is not frequency, waveform, or even distortion. Instead, the focus is on control of actual energy flow. While energy flow control is an appropriate power electronics application, the circuits used for this purpose differ conceptually from most switching converters. Switching circuits that directly manipulate power flow in an ac system without formally performing a conversion are called *ac regulators*, since they add a control capability to a given power circuit.

The SCR circuit of Figure 7.26 is a simple example. If the gate phase control angle is delayed, the time allotted for input-output connection can be adjusted. Adjustment of the in-

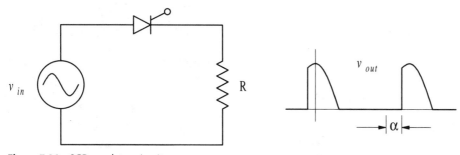

Figure 7.26   SCR regulator circuit.

terconnect time alters the energy flow. This circuit also has a rectification function. The source current will have a dc component that will be unacceptable in most cases. A better example is shown in Figure 7.27, in which two SCRs work together to manipulate an ac waveform. As long as both devices have the same gate delay, the waveform will be symmetrical, and a dc component will be avoided.

The triac is a switching device equivalent to reverse parallel SCRs sharing a common gate. The triac lacks full gate control, so it is not adequate for true ac–ac conversion. However, it works as an ac regulator component. ac regulators typically supply resistive loads, such as incandescent lamps or heaters, or inductive loads such as motor windings. Triac circuits are the basis for consumer products such as lamp dimmers, variable-speed power tools, and small adjustable-speed appliances. In most cases, an RC circuit is used with a potentiometer to create a variable time delay for the gate.

In practical ac regulators, the switching device (e.g., triac) operates each half-cycle of the input. This guarantees that switching is taking place at twice the input frequency—exactly the rate required for conversion between two sources with identical frequency. Switch action is not fully controlled. Neither an SCR nor a triac will shut off until its current reaches zero. The timing of the current zero depends on the load. For an inductive load, the current zero will be delayed by an amount proportional to the L/R time constant. Unless the load characteristics are known precisely, it is not possible to determine the action of an ac regulator.

Consider an ac regulator with a resistive load, shown in Figure 7.28. Since the load is well defined, the waveform can be determined given the delay angle α. Notice in the figure that α is usually measured relative to the zero-crossing for this circuit. Since the switching function duty ratio is load-dependent, the time of switch turn-on is more logical as a reference point than the pulse center. The average output power is the square of the output RMS voltage divided by R. The RMS voltage is

$$v_{\text{out(RMS)}} = \sqrt{\frac{1}{\pi}\int_{\alpha}^{\pi} V_0^2 \sin^2 \theta \, d\theta} = V_0 \sqrt{\frac{1}{2} - \frac{\alpha}{2\pi} + \frac{\sin 2\alpha}{4\pi}} \tag{7.11}$$

with $0 \le \alpha \le \pi$. When $\alpha = 0$, the expression reduces to $V_0/\sqrt{2}$. A plot of the RMS output as a function of α is given in Figure 7.29.

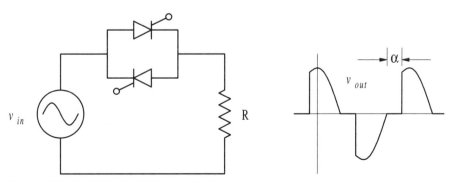

Figure 7.27   Dual SCR circuit for ac regulation.

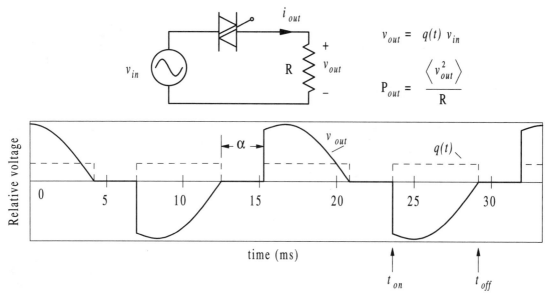

$$v_{out} = q(t)\, v_{in}$$

$$P_{out} = \frac{\langle v_{out}^2 \rangle}{R}$$

Figure 7.28   ac regulator with resistive load.

With inductive loads, the regulator action is harder to predict. The switch turn-off time, which itself is load dependent, is needed to construct the output waveform. Let us consider a general *LR* load, as shown in Figure 7.30. The turn-on time still occurs at angle α, but the turn-off is at an angle $\beta > \pi$. The differential equation is

$$V_0 \sin(\omega t) = L\frac{di_{out}}{dt} + i_{out}R, \qquad i_{out}(\omega t = \alpha) = 0 \tag{7.12}$$

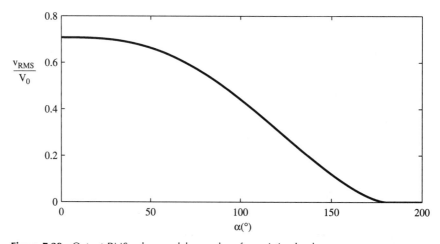

Figure 7.29   Output RMS value vs. delay angle α for resistive load.

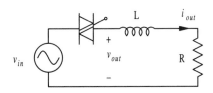

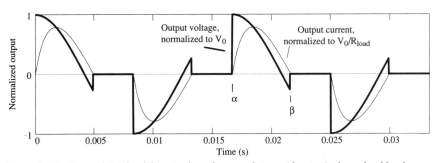

Figure 7.30 General *L-R* load for triac-based ac regulator, with a typical graph of load current and output voltage.

This must be solved for the time $\omega t_1 = \beta$ when the current returns to zero. The solution of equation (7.12), at angle β, must equal zero,

$$\frac{V_0}{R}\; \frac{\xi e^{\alpha/\xi} \cos \alpha - \xi e^{\beta/\xi} \cos \beta - e^{\alpha/\xi} \sin \alpha + e^{\beta/\xi} \sin \beta}{(1 + \xi^2)e^{\beta/\xi}} = 0 \tag{7.13}$$

with $\xi = \omega L/R$. There is no convenient analytical solution, but we can substitute actual circuit parameters to get an answer. Notice that the value ξ is essentially a ratio of time constant to period. A "resistive load" has $\xi \ll 1$, while a load with "large" inductance has $\xi \gg 1$. Figure 7.31 shows current waveform examples for a few values of α and ξ.

The average output power with the *L-R* load will be $i^2_{\text{out(RMS)}}R$. This is given by the square of the RMS integral,

$$P_{\text{out}} = R\frac{1}{\pi}\int_{\alpha}^{\beta} i^2_{\text{out}}(t)\, dt \tag{7.14}$$

Figure 7.32 shows the output power (for $V_0 = R = 1$) with various values of α and ξ.

ac regulators are useful in noncritical applications such as lighting controllers. They can also be used with feedback control to adjust output power if necessary. When control is applied, it is possible to maintain constant power to an *L-R* load even when the input voltage and load impedance are subject to variation. The main drawback of ac regulators is the harmonic spectrum imposed on the incoming line. Dimming circuits today usually have line filters to avoid imposing excessive harmonics on the source.

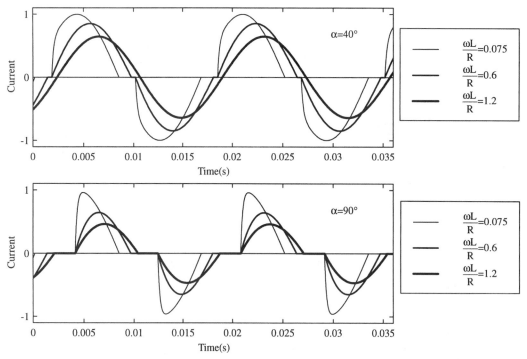

Figure 7.31   ac regulator load current waveforms for various α, ξ.

Example 7.8.1   An ac regulator provides power for a baseboard electric heater. The heater provides up to 5 kW output with 240 V input. What is the output power as a function of delay angle for this application? For a power output of 100 W, plot the current waveform and compute the input power factor. Comment on power quality.

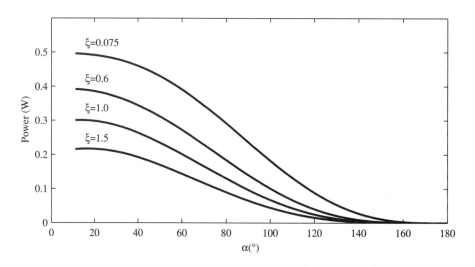

Figure 7.32   Average output power from ac regulator for various α, ξ, given $V_0 = R = 1$.

In this circuit, a resistive load should be a reasonable model. Since the load draws 5000 W at 240 V_{RMS}, the resistance value is 11.52 Ω. The variation of power with α is shown in Figure 7.29, and is given by

$$P_{out} = (5000 \text{ W})\left[1 - \frac{\alpha}{\pi} + \frac{\sin(2\alpha)}{2\pi} \right] \tag{7.15}$$

so $P_{out} = 2500$ W when $\alpha = 90°$, and so on. For output of 100 W, the value of α is 153.6°. The current is shown in Figure 7.33. Since $i_{RMS}^2 R = 100$ W with this phase delay, the RMS current for 100 W is 2.95 A. The power factor at the input is P/S, where $S = V_{RMS}I_{RMS}$. Here, the input voltage is 240 V, so $S = 707$ VA. The power factor is $(100 \text{ W})/(707 \text{ VA}) = 0.14$, a poor value. Harmonic distortion is also not good. The current THD can be computed as 188.6%. When the output power is low, this circuit is a significant power quality problem.

Example 7.8.2 Typical lamp and heater filaments have strongly temperature-dependent resistances. When these are connected directly to an ac line, a large in-rush current flows until the filament heats up. In some systems, the in-rush is hard to tolerate. A ceramic heater for a high-temperature metals processing system has cold resistance of 4 Ω and hot resistance of 15 Ω. This heater is to be used with a 120 V 60 Hz source, but it heats slowly enough that a 15 A circuit breaker always trips when the heater is connected. Suggest an ac regulator approach that might resolve the problem.

If a conventional ac regulator is used, the phase delay will allow adjustment of the RMS current to avoid overloading the circuit. When cold resistance of only 4 Ω, a setting of $\alpha = 0°$ permits 30 A to flow, and there is a problem. Instead, α might be set initially to a high value (an angle of 114° will limit the RMS current to 15 A with a 4 Ω load). As the filament heats up, the phase delay can decrease, until it ultimately reaches $\alpha = 0°$. This is not difficult to implement: the phase delay can be set initially to 180°, and an RC decay can be used to drop α slowly during the warm-up process.

The load dependency of the switch action in ac regulators complicates design problems. When the loads are not resistors, the energy flows determine the turn-off action in a

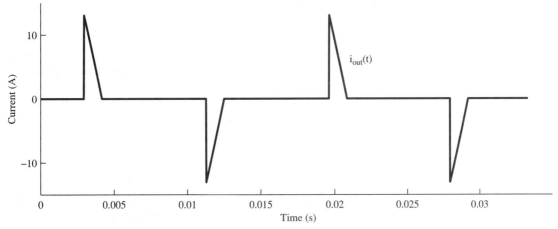

Figure 7.33 Current waveform given 100 W output for Example 7.8.1.

tightly coupled manner that is hard to address mathematically. It will be difficult to design filters or other auxiliary components for this regulator, since the load might not be predictable. ac regulators are helpful for adjusting power flow, but may not be an adequate substitute for a more complete switching converter if good quality and performance are necessary.

7.9 INTEGRAL CYCLE CONTROL

A switching technique similar to ac regulation, called *integral cycle control* (see Problem 2.15), is sometimes used for power control at high levels. It is also typical for longer-term switching such as the power cycling control of many electric ranges and microwave ovens. In integral cycle control, a specific number of periods of the input waveform are delivered to the output. All switching occurs at zero crossings. This allows SCRs and triacs to function effectively for control with minimal loss. With integral cycle waveforms, only discrete choices of output frequencies are possible. For example, a one-cycle-on one-cycle-off sequence provides an effective halving of the frequency. One-cycle-on two-cycle-off sequences given one-third of the input frequency. Two-cycle-on one-cycle-off provides two-thirds of the input frequency, and so on. Typical waveforms are shown in Figure 7.34. With this method, severe subharmonics can be generated if frequency conversion is the objective. The output frequency in general must be lower than the input, and subharmonics will be generated in virtually every possible integral cycle control sequence. Integral cycle control is not used extensively for frequency conversion partly because of the subharmonic problem. Like ac regulation, it can be appropriate when power adjustment for a load is necessary.

Since the waveform is applied in single-cycle pieces, the power behavior of integral cycle circuits is not hard to predict. If the load inductance is not high, each cycle provides a definite output energy. The average power is the switch duty ratio times the power when the source is applied at all times. If half of all cycles are delivered to the output, then half as much power as the maximum will be delivered. The relationship

$$P_{\text{out}} = DP_{\text{max}} \tag{7.16}$$

works for almost any load if the switching period is long enough. A microwave oven, for example, is a complicated load. But if the switching period is 10 s or so, as it is in a typical product, the circuit is in the sinusoidal steady state almost all the time.

At first, it might seem that switching periods of several seconds would avoid problems on the incoming power line. However, large power transients occurring every few seconds can cause objectionable dimming of incandescent lights and other subtle problems because of voltage drops. Integral cycle control is not widely favored as a conversion technique because of the power quality problems it creates.

7.10 RECAP

ac–ac conversion in the context of a direct matrix switching approach is unusual, both because of the problems of bilateral switches and because of switching reliability concerns. The two most direct operating strategies—the slow-switching method and the universal method—can be represented as linear forms of phase modulation. Switching occurs at either

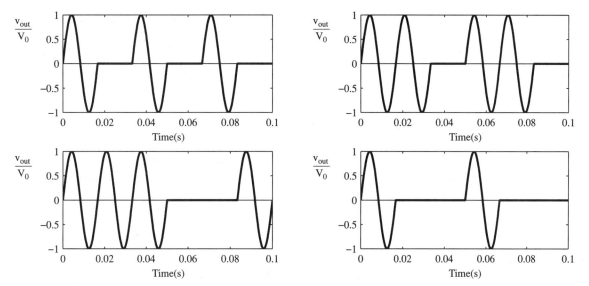

Figure 7.34 Some typical integral cycle control output voltages.

the sum or the difference of the input and output wanted frequencies under these approaches. The universal method, in which $f_{\text{switch}} = f_{\text{in}} + f_{\text{out}}$ offers a better match for filtering than the slow-switching approach. Slow switching can produce subharmonics: unwanted components at frequencies below that of the wanted component. Subharmonics can be extremely difficult to filter, and are usually avoided by ensuring that $f_{\text{out}} \ll f_{\text{in}}$. In general, when phase modulation is employed, the output wanted component for an m-phase input source is given by

$$v_{\text{out(wanted)}} = \frac{mV_0}{\pi} \sin\frac{\pi}{m} \cos(\alpha(t)) \qquad (7.17)$$

Most practical ac–ac converters use some form of nonlinear phase modulation. The basic concept is to adjust the phase delay angle of a rectifier-like circuit to produce a desired result. The phase angle control choice

$$\alpha(t) = \cos^{-1}[k \cos(\pm \omega_{\text{out}} t)] \qquad (7.18)$$

is especially useful, since the range of angles given by the arccosine function can be arranged to fall within the limits of conventional SCR rectifiers and inverters. The cycloconverter is a circuit that takes advantage of this nonlinear method to implement low-frequency ac–ac converters with unilateral devices.

Many motor drives and other industrial converters are cascaded rectifier-inverter circuits. In effect, these are ac–ac converters in which the phase angle modulation is a square wave. Usually, extra internal filtering is added, and the cascade arrangement is called a *dc link* converter to emphasize the action of an internal transfer source. The inverter in these applications is often controlled by PWM. It is also possible to perform ac–ac conversion directly with PWM, but this brings back the problem of bilateral devices.

The most common ac–ac applications are not really ac–ac converters in the strictest sense. These *ac regulators* take advantage of the load circuits to apply phase delay control.

Essentially, power is alternately connected and disconnected to the load, with minimal regard for the waveform. These converters often have no obvious wanted component at the output, because a resistor can absorb energy at any frequency. The input power factor can be low as a result. For a resistive load, the RMS output voltage of a single-phase ac regulator is

$$v_{out(RMS)} = V_0 \sqrt{\frac{1}{2} - \frac{\alpha}{2\pi} + \frac{\sin 2\alpha}{4\pi}} \tag{7.19}$$

Typical ac regulators have poor power factor at light loads, and most have high harmonic distortion. Integral cycle controllers perform a similar function, although the strict zero-crossing switch action makes them easier to analyze.

PROBLEMS

1. List two possible switching frequencies for each of the following conversions:

   a. 50 Hz to 400 Hz   **b.** 0 Hz to 400 Hz   **c.** 400 Hz to 0 Hz

   d. 60 Hz to 12 Hz   **e.** 180 Hz to 180 Hz   **f.** 25 Hz to 18 Hz

2. Consider a bridge-type converter with an ac voltage source input and an ac current source output. Assume the bridge operates symmetrically, so that the output voltage is either $+v_{in}(t)$ or $-v_{in}(t)$, each with 50% duty. Plot the output voltage waveform and the input current waveform for the following choices, based on SSFC:

   a. 30 Hz in, 60 Hz out   **b.** 60 Hz in, 20 Hz out

   c. 60 Hz in, 15 Hz out   **d.** 60 Hz in, 75 Hz out

3. Repeat Problem 2 for the UFC method, and also consider a 60 Hz to 60 Hz conversion.

4. Consider a circuit with 60 Hz input and 80 Hz switching. What are the most likely wanted component frequencies? What are their amplitudes relative to the input amplitude, if the input is a single-phase source and the circuit is a bridge?

5. A bridge arrangement is used with an ac voltage source input and a current source output to provide 50 Hz to 20 Hz conversion. The output voltage can be $\pm v_{in}$ or it can be 0.

   a. Plot the output voltage given that the switches function in diagonal pairs. What is the RMS value of the output? Does the phase of any given switching function relative to v_{in} make a difference?

   b. It is proposed that output control be performed with *relative phase control*, along the same lines as the voltage-sourced inverter in Section 6.3. If the diagonal pair of switching functions $q_{1,1}(t)$ and $q_{2,2}(t)$ are displaced such that $q_{2,2}$ is delayed from $q_{1,1}$ by $\frac{1}{4}$ of a switching period, plot the output voltage and compute the new RMS value. Has the wanted component amplitude changed?

6. A midpoint arrangement converts a three-phase voltage source to an ac current source. If the output is a 40 Hz ideal current source and the input is a 60 Hz voltage set, plot the output voltage for either the SSFC or UFC method. Plot the input current from phase a for the same choice of switching functions. What is the amplitude of the wanted component of v_{out} and of $i_{in(a)}$? Does switching function phase have any effect?

7. A special-purpose triangle oscillator is required for a communications application. The oscillator maintains a symmetrical current of 10 A peak at a frequency of 600 Hz. It is desired to use a switching ac–ac converter to transfer power into this oscillator from a 120 V_{RMS} 60 Hz voltage source.

    a. Suggest an appropriate circuit and switch action for the proposed application.

    b. How much power does your proposed converter transfer to the oscillator?

8. A hex-bridge arrangement uses fully bilateral switches to convert 60 Hz input into 400 Hz output. The input is a three-phase 60 Hz source at 230 $V_{line-to-line}$.

    a. Recommend a switching frequency and an operating method.

    b. What is the amplitude of the 400 Hz output component based on your suggested method?

9. A full-bridge arrangement converts a three-phase 50 Hz input into 50 Hz for a backup application. The circuit is shown in Figure 7.35.

    a. Based on the UFC approach, find and plot the output voltage.

    b. What is the amplitude of the wanted output component?

    c. Find and plot the total current in the source v_a.

10. A set of interface circuits is used to create a three-phase 400 Hz ac current source at the output of an aircraft generator. This current set is to be converted to 120 V_{RMS} 60 Hz for the on-board appliances. The SSFC method is to be explored, since $f_{out} \ll f_{in}$.

    a. Keep in mind the KCL issues associated with an input current source. The three currents of a 3ϕ set sum to zero, so the input sources are connected together at a neutral node. Begin with a switch matrix, and suggest a switching circuit for this application.

    b. What switching frequency will be used? What is the output current for the proposed circuit?

    c. What should the current amplitudes be to support transfer of 10 kW to the output source?

11. The negative converter input voltage waveform in Figure 7.14 is based on a desired 20 Hz output. Examine the waveform, and determine the switching frequency. How does this compare to the UFC and SSFC methods?

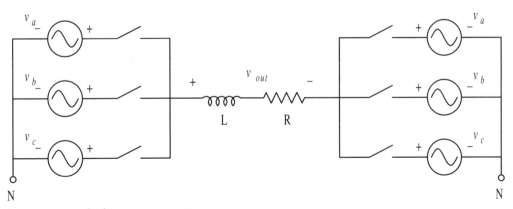

Figure 7.35 Hex-bridge ac–ac converter.

12. Compare the positive converter modulating function $\cos^{-1}[k\cos(\omega_{out})]$ to the function $90° - k60°\cos(\omega_{out})$ for $k = 0.2, 0.4, 0.6, 0.8, 1.0$. Plot the functions. If possible, compare converter output waveforms for a few of the values as well. Does the cosine function appear to be a useful alternative?

13. A cycloconverter, shown in Figure 7.36, operates as follows: the functions $\pm\alpha(t)$ are constructed for $k = 1$ (i.e., they are triangle waveforms). Each SCR receives a gate pulse based on the value of $\alpha(t)$ at the moment it is to be gated. For example, the gate pulse delay will be zero only if $\alpha(t) = 0°$ at the instant of the voltage zero crossing. The gate pulse delay will be 45° only if $\alpha(t) = 45°$ at the moment when $v_{in}(t) = 0.7071V_0$, and so on. For 60 Hz input and 10 Hz output, construct the output voltage waveform based on this switch action. What is the switching frequency? If the output wanted component really is

$$v_{out(wanted)} = \frac{2V_0}{\pi}\cos(\alpha(t)) \qquad (7.20)$$

what is the TUD of the output waveform?

14. A PWM ac–ac converter is used to provide power from a 50 Hz source to a 400 Hz load. The switching frequency is 10 kHz, and the output filter attenuates the high-frequency components very effectively. If the converter is a single-phase bridge with $v_{out}(t) = \pm v_{in}(t)$, what is the THD of the filtered output waveform?

15. Two alternative dc link converters are to be compared. In the first, the rectifier is a six-pulse diode bridge, and the inverter uses PWM to control the output amplitude and frequency. In the second (identical to Figure 7.24 except that the voltage need not be 600 V), the rectifier is a six-pulse SCR bridge, and the inverter PWM control alters only the output frequency but does not adjust the amplitude. Each converter is adjusted so that with $\alpha = 0°$ and 100% depth of modulation, the desired maximum output is obtained at an output frequency of 60 Hz. The converters are used in an ac motor application with a constant V/f ratio at the output.

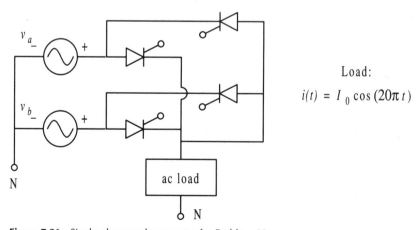

Load:

$$i(t) = I_0\cos(20\pi t)$$

Figure 7.36   Single-phase cycloconverter for Problem 13.

a. For each of these two, determine the phase delay angle (always zero for the diode case, of course), and the PWM modulating function $m(t)$ for 60 Hz, 30 Hz, and 10 Hz output.

b. Assume that the load power varies as the square of the output frequency, and that the dc link filter provides a constant current load for the rectifier and a constant voltage source for the inverter. Plot the current in the phase *a* voltage source for each of the two converters for 60 Hz and 30 Hz outputs. Comment. You may choose specific voltage and power levels to assist in preparing a plot if desired.

16. An ac regulator as shown in Figure 7.27 is to control the heating resistor for a clothes dryer. The resistance is 5 Ω when the heater is cold and 10 Ω when it is hot. It is desired to provide output power levels of 1 kW and 5 kW for various feature settings. The dryer is intended to function on a 230 V 30 A circuit.

a. For the 1 kW output power level, what should be the phase delay angle setting at cold start? What is the phase setting once the resistance is hot?

b. For the 5 kW setting, what should be the phase delay setting at cold start? What is the value once the resistor is hot?

c. Consider a "rapid start" mode, in which the resistor is provided with a full 30 A_{RMS} for 2 s to heat it quickly from a cold start. What phase delay angle will be used for this mode?

17. A microwave oven uses a simple control scheme to adjust its power flow: During each 15 s interval, a triac is used to connect the ac line to the microwave generator a time fraction of 10% to 100%. Assume that the oven draws 1500 W from the wall plug, and that the source is 120 V_{RMS} at 60 Hz.

a. If the power factor is close to 1 with a setting of 100% (it often is not for this type of appliance), what current is drawn from the wall plug for the 100% setting?

b. What RMS current is drawn from the wall plug at the 10% power setting? You may assume that all switch action occurs in an integral-cycle fashion. Are current subharmonics drawn for the 10% setting? If so, what is the amplitude of the largest subharmonic?

c. Find and plot the input power factor as a function of the load setting.

REFERENCES

T. H. Barton, *Rectifiers, Cycloconverters, and Ac Controllers*. Oxford: Clarendon Press, 1994.

L. Gyugyi, B. R. Pelly, *Static Power Frequency Changers*. New York: Wiley, 1976.

W. McMurray, *The Theory and Design of Cycloconverters*. Cambridge, MA: MIT Press, 1972.

B. R. Pelly, *Thyristor Phase-Controlled Converters and Cycloconverters*. New York: Wiley, 1971.

J. E. Quaicoe, S. B. Dewan, "A clamped ac–ac frequency converter for induction heating," *IEEE Trans. Industry Applications*, vol. IA-22, no. 6, pp. 1018–1026, November 1986.

C. Rombaut, G. Seguier, R. Bausiere, *Power Electronic Converters, Volume 2: Ac–Ac Conversion*. New York: McGraw-Hill, 1987.

W. Shepherd, P. Zand, *Energy Flow and Power Factor in Non-Sinusoidal Circuits*. Cambridge: Cambridge Univ. Press, 1976.

P. D. Ziogas, Y.-G. Kang, V. R. Stefanović, "Rectifier-inverter frequency changers with suppressed dc link components," *IEEE Trans. Industry Applications*, vol. IA-22, no. 6, pp. 1026–1036, November 1986.

CHAPTER 8

Figure 8.1 Many miniature power supplies use resonance to reduce size.

INTRODUCTION TO RESONANCE IN CONVERTERS

8.1 INTRODUCTION

In this chapter, we will consider a few ways in which resonance is used to advantage in power electronics. Resonant circuits commonly perform one of two major functions in a converter. The first, which we have examined mostly for inverters, is concerned with the interface problem. A resonant circuit used as a filter can separate the wanted component even when unwanted frequencies are low or closely spaced.

The second function of resonance relates more directly to switching action. Any switch generates a certain amount of loss when it operates. In general, the losses caused by switch operation depend on the power handling level at the moment of switching. Any semiconductor that tries to switch on a large current just after blocking a large voltage will incur significant energy loss during the process. A semiconductor that switches to alter an energy-related value such as a capacitor voltage or an inductor current will also incur high loss if proper attention is not given to current paths or voltage loops.

The integral cycle controller discussed at the end of Chapter 7 represents the opposite of high-power switching. In an integral cycle controller, the switches act only at zero crossings of the incoming waveform, and very little loss occurs during switching. In general, if the switch voltage is near zero just before turn-on is commanded, the current change comes about with minimal loss. Similarly, if the switch current can be brought close to zero just prior to turn-off, little loss will occur during the turn-off transition. However, few types of converters provide natural opportunities to switch under low voltage or low current conditions. When properly designed resonant combinations are added to dc–dc converters, low-voltage or low-current instants suitable for switching can be created.

Both major resonance functions are explored in this chapter. After a review of resonance and its mathematics, issues of resonant filters are examined. Design examples are provided, related to both input and output filters. The components needed for resonance at line frequency tend to be large, so resonant filters are most often applied when power levels are very high or at frequencies well above conventional line values.

Resonant switch action, often termed *soft switching* in the literature, is introduced in the second half of the chapter. Virtually any type of converter can make use of resonant switch action. New ideas for resonant methods and applications are published each month. Over the past five years or so, the steady improvement in power electronics owes much to innovative uses of resonance. In particular, resonance can reduce losses and energy storage requirements so that only very small inductors or capacitors are needed. Most modular dc–dc converters produced for high-performance equipment or military applications, like the ones in Figure 8.1, use resonant methods to improve efficiency while reducing size.

8.2 REVIEW OF RESONANCE

8.2.1 Characteristic Equations

A circuit or system with oscillatory behavior has its own resonant frequency. At this frequency, energy is exchanged among the elements of the system rather than lost or gained externally. A pendulum, for example, exchanges kinetic and potential energy as it swings back and forth. A crystal oscillator used for a digital clock exchanges energy between a piezoelectric crystal and a circuit. The energy exchange properties of resonance are very useful in power electronics. For a load that requires a sinusoidal voltage, a resonant filter can support the time-varying power flow without loss, while switches process only the average power. In resonant switching, switch action is timed to coincide with voltage or current zero crossings as energy moves back and forth.

Resonant applications in power converters often use a series *RLC* circuit combination. A filter of the series type uses the arrangement as a voltage divider to separate wanted and unwanted components. A resonant switch arrangement of the series type takes advantage of current zeroes.

The series *RLC* circuit shown in Figure 8.2 has a current determined by the differential-integral equation

$$v_{in}(t) = L\frac{di}{dt} + iR + \frac{1}{C}\int i\, dt \tag{8.1}$$

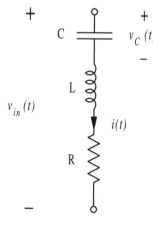

Figure 8.2   Series *RLC* circuit.

Since the current is governed by the relationship $i = C(dv_C/dt)$, this can be rewritten as a second-order differential equation

$$v_{in}(t) = LC\frac{d^2v_C(t)}{dt^2} + RC\frac{dv_C(t)}{dt} + v_c(t) \tag{8.2}$$

Alternatively, an expression for $i(t)$ can be obtained by differentiating equation (8.1) with respect to time:

$$\frac{dv_{in}(t)}{dt} = L\frac{d^2i}{dt^2} + R\frac{di}{dt} + \frac{1}{C}i \tag{8.3}$$

Notice that if the input $v_{in}(t)$ is zero, equations (8.2) and (8.3) are identical.

To solve equation (8.2), we begin with the *homogeneous solution*—the solution when the input is zero. Let us try a solution of the form $v_C(t) = c_1 e^{st}$. The constant c_1 allows matching of the value of the capacitor voltage at time $t = 0$. With this substitution, equation (8.2) becomes

$$LCs^2 c_1 e^{st} + RCsc_1 e^{st} + c_1 e^{st} = 0 \tag{8.4}$$

Dividing by $LCc_1 e^{st}$, it becomes clear that s represents a solution of the *characteristic equation*,

$$s^2 + \frac{R}{L}s + \frac{1}{LC} = 0 \tag{8.5}$$

The solutions are

$$s = \frac{-R/L \pm \sqrt{R^2/L^2 - 4/(LC)}}{2} \tag{8.6}$$

Let the ratio $R/(2L)$ be denoted as a *damping parameter*, ξ. Define the radian resonant frequency $\omega_r = 1/\sqrt{L/C}$. Then equation (8.6) becomes

$$s = -\xi \pm \sqrt{\xi^2 - \omega_r^2} \tag{8.7}$$

The solutions for s will be real and distinct if $\xi^2 - \omega_r^2 > 0$, a complex pair if $\xi^2 - \omega_r^2 < 0$, and will be real repeated roots if $\xi^2 = \omega_r^2$. In the special case of $\xi = 0$, the solutions for s are purely imaginary, and the exponential solution $c_1 e^{st}$ becomes sinusoidal.

It is convenient to define the *characteristic impedance*, $Z_c = \sqrt{L/C}$. The *quality factor*, Q, is the ratio of this impedance to the resistance, $Q = Z_c/R$. With this definition, it is possible to show that the characteristic equation can be written

$$s^2 + \frac{\omega_r}{Q}s + \omega_r^2 = 0 \tag{8.8}$$

and that the damping factor ξ is $\omega_r/(2Q)$. The two solutions can be presented in the form

$$s = \xi(-1 \pm \sqrt{1 - 4Q^2}) \tag{8.9}$$

and the case $Q = \frac{1}{2}$ represents repeated real roots.

The basic behavior changes substantially as Q is altered. When $Q < \frac{1}{2}$, the solutions of the homogeneous equation show exponential decay. A pendulum with this behavior will swing from its starting point to its lowest point and stop without travelling farther. For $Q > \frac{1}{2}$, the solutions for s form complex conjugates. Since $e^{\xi+j\theta} = e^{\xi}(\sin\theta + j\cos\theta)$, the behavior is that of a damped sinusoid. A pendulum with this behavior will swing with exponentially decreasing amplitude.

8.2.2 Step Function Excitation

A nonzero input voltage for equation (8.2) provides a *forcing function* for the solution, and a particular solution must be found to correspond to the specific form. For many simple functions, it is relatively easy to pick out the necessary form for the extra terms in the solution. In power electronics, a step or a square wave are common forcing functions. A circuit for discussion is shown in Figure 8.3.

Consider the effect of a step input, in which $v_{in}(t)$ is zero for $t < 0$, then set to V_{in} at time $t = 0$. Initially, both $v_C = 0$ and $i_L = 0$. The differential equation is

$$V_{in} = \frac{1}{\omega_r^2} \frac{d^2 v_C}{dt^2} + \frac{1}{\omega_r Q} \frac{dv_C}{dt} + v_C, \quad t \geq 0 \tag{8.10}$$

$$v_C(0) = 0, \quad i_L(0) = 0$$

A solution of the form

$$v_C(t) = c_1 e^{s_1 t} + c_2 e^{s_2 t} + c_3 \tag{8.11}$$

should satisfy the equation, with values of s corresponding to equation (8.7), when $Q \neq \frac{1}{2}$. In the repeated root case with $Q = \frac{1}{2}$, a form

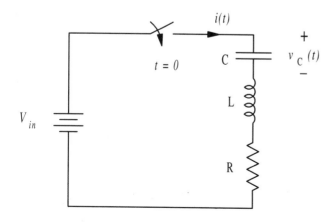

Figure 8.3 Series *RLC* combination with step input.

$$v_C(t) = (c_1 + c_2 t)e^{st} + c_3 \tag{8.12}$$

will work. To match the dc input, it is required that $c_3 = V_{in}$. The unknown constants c_1 and c_2 can be found by matching the initial conditions $v_C(0) = 0$ and $i_L = C(dv_C/dt) = 0$.

For the real, distinct case with $Q < \frac{1}{2}$, the initial conditions require

$$v_C(0) = c_1 e^0 + c_2 e^0 + c_3 = 0, \qquad c_1 + c_2 + c_3 = 0, \qquad c_3 = V_{in},$$

$$\left.\frac{dv_C}{dt}\right|_0 = s_1 c_1 e^0 + s_2 c_2 e^0 = 0, \qquad s_1 c_1 + s_2 c_2 = 0 \tag{8.13}$$

These can be solved simultaneously to give

$$c_1 + c_2 = -V_{in} \tag{8.14}$$

$$c_1 = \frac{s_2 V_{in}}{s_1 - s_2}, \qquad c_2 = \frac{-s_1 V_{in}}{s_1 - s_2}$$

The final result for the capacitor voltage becomes

$$v_C(t) = V_{in}\left(\frac{s_2}{s_1 - s_2} e^{s_1 t} - \frac{s_1}{s_1 - s_2} e^{s_2 t} + 1\right) \tag{8.15}$$

The inductor current $i_L(t) = C(dv_C/dt)$ is

$$i_L(t) = CV_{in}\left(\frac{s_1 s_2}{s_1 - s_2} e^{s_1 t} - \frac{s_1 s_2}{s_1 - s_2} e^{s_2 t}\right) \tag{8.16}$$

In the case of $Q = \frac{1}{2}$, the initial condition $v_C(0) = 0$ requires $c_1 + c_3 = 0$. The initial condition on $i_L(t)$ requires

$$i_L(0) = \left.\frac{dv_C}{dt}\right|_0 = c_2 e^0 + c_1 s e^0 = 0, \qquad c_2 = -c_1 s \tag{8.17}$$

Combining the two requirements, the solution becomes

$$v_C(t) = V_{in}(1 - e^{-\omega_r t} - \omega_r t e^{-\omega_r t}) \tag{8.18}$$

since s must match ω_r. The current is

$$i_L(t) = CV_{in}\omega_r^2 t e^{-\omega_r t} \tag{8.19}$$

For the case of complex roots with $Q > \frac{1}{2}$, the solution of equation (8.11) can be written more conveniently in the form

$$v_C(t) = e^{-\xi t}[c_1 \sin(\omega t) + c_2 \cos(\omega t)] + c_3 \tag{8.20}$$

where $\xi = \omega_r/(2Q)$ as before and $\omega = \sqrt{\omega_r^2 - \xi^2}$. Once again $c_3 = V_{in}$ to match the dc conditions. The initial condition on voltage requires $c_2 + c_3 = 0$. The initial condition on current requires $-\xi c_2 + \omega c_1 = 0$. Thus $c_2 = -V_{in}$ and $c_1 = -\xi V_{in}/\omega$. The final result is

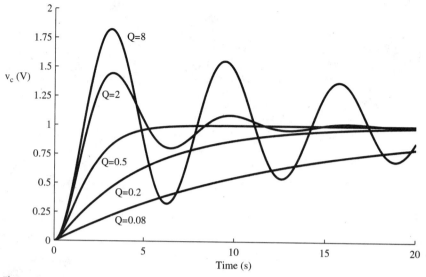

Figure 8.4 Response of $v_C(t)$ in an *RLC* series circuit to a 1 V step.

$$v_C(t) = V_{in} - V_{in}e^{-\xi t}\left[\frac{\xi}{\omega}\sin(\omega t) + \cos(\omega t)\right] \tag{8.21}$$

and the current is

$$i_L(t) = \xi C V_{in}e^{-\xi t}\left[\frac{\xi}{\omega}\sin(\omega t) + \cos(\omega t)\right] - \omega C V_{in}e^{-\xi t}\left[\frac{\xi}{\omega}\cos(\omega t) - \sin(\omega t)\right] \tag{8.22}$$

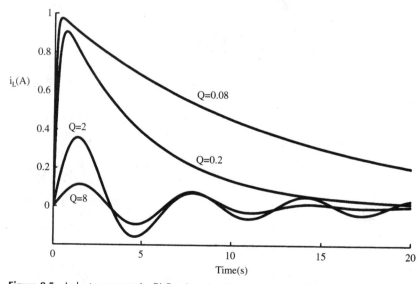

Figure 8.5 Inductor current in *RLC* series circuit exposed to a 1 V step.

The voltage and current solutions, respectively, for various cases are summarized in Figures 8.4 and 8.5. The resonant frequency ω_r has been set to one radian per second. Other values are $V_{in} = 1$ V and $R = 1$ Ω. Low values of Q lead to exponential decay, while larger values lead to increasingly underdamped oscillations. Figure 8.6 summarizes the energy behavior in the circuit for four values of Q. When $Q > \frac{1}{2}$, the inductor shows wide energy swings. The capacitor ultimately stores the energy $\frac{1}{2} CV_{in}^2$. Comparison of Figures 8.6a and 8.6b shows the energy exchange between the two storage elements.

8.2.3 Phasor Analysis of Series-Resonant Filters

When the forcing function is a periodic waveform, the current that flows can be found by applying the Fourier Series of the input term by term, just as in Chapter 2. In periodic steady-state, the RLC combination forms a divider network at each frequency. The nth voltage com-

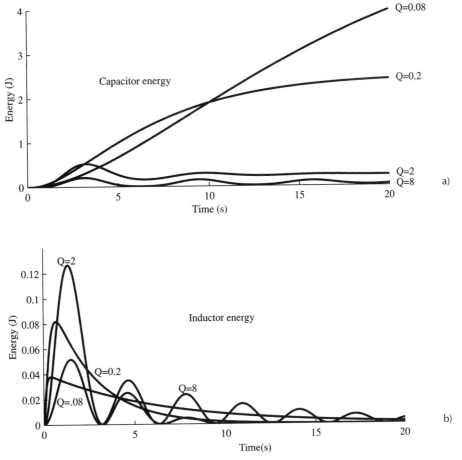

Figure 8.6   Capacitor and inductor energy in RLC circuit.

ponent $v_{in(n)}(t)$, has phasor value $\tilde{V}_{in(n)}$. Analysis of the divider gives the nth resistor voltage component phasor value

$$\tilde{V}_{out(n)} = \tilde{V}_{in(n)} \frac{R}{jn\omega L - j/(n\omega C) + R} = \tilde{V}_{in(n)} \frac{R/Z_c}{jn\omega/\omega_r + j/(n\omega/\omega_r) + R/Z_c} \quad (8.23)$$

where again Z_c is the characteristic impedance $\sqrt{L/C}$. It is of interest how the circuit will react when excited at a particular frequency. Figure 8.7 shows a log-log plot of the ratio of output magnitude to input magnitude as a function of frequency for various values of Q. The x axis in the figure is scaled in terms of the ratio of the input frequency to the resonant frequency. It is not surprising that the response peaks at the point where the input frequency matches the resonant frequency. The higher the value of Q, the sharper the peaking of the response at ω_r. High values of Q, such as $Q = 100$, can make the action very selective: the output–input ratio is close to one near ω_r, but becomes small quickly as the frequency moves away from ω_r. An RLC circuit with high Q makes a good filter for this reason. The value of Q determines a resonant filter's *selectivity*.

High values of Q have interesting effects in addition to good filter action. Consider the capacitor voltage phasor,

$$\tilde{V}_{C(n)} = \frac{\tilde{V}_{in(n)}}{1 - n^2\omega^2 LC + jn\omega RC} = \frac{\tilde{V}_{in(n)}}{1 - n^2(\omega^2/\omega_r^2) + jn\omega/(\omega_r Q)} \quad (8.24)$$

When the input frequency $n\omega$ exactly matches the resonant frequency, the capacitor voltage magnitude is Q times the input magnitude. For a 120 V_{RMS} input, a Q of 100 means that 12 000 V_{RMS} is imposed on the capacitor! The inductor similarly sees a voltage gain factor of Q at the resonant frequency. The large inductor and capacitor voltages are 180° out of phase, so the net voltage across the series combination cancels to zero at resonance.

Filter design must trade off selectivity against component cost and reliability. High selectivity requires extreme capacitor and inductor voltage ratings and high values of Z_c. High

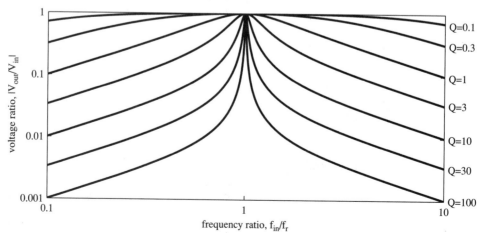

Figure 8.7   Ratio of output to input vs. frequency at various values of Q.

characteristic impedance in turn requires high inductance values. It is difficult and expensive to achieve very selective resonant filters in power applications. In practice, the values of Q are not usually very high ($Q = 10$ is high for a typical power filter). This helps to keep device ratings reasonable. Selectivity is limited by a related issue: An individual component, such as an inductor, introduces its own series resistance in wires and connections. The quality factor for a single energy storage element is defined as the ratio $Q = X/R$, where X is the reactance. Individual inductors for power applications rarely show $Q > 100$. At line frequencies, Q values of 30 or less are not unusual. Capacitors tend to show higher Q values if $C < 10$ μF. At line frequencies, capacitors of 1000 μF or more can exhibit $Q < 10$.

Component quality factor limits the overall Q value that can be obtained in the complete filter. Consider, for example, a resonant filter intended for a 60 Hz, 120 V, 1 kW application. This power level corresponds to a resistive load of 14.4 Ω. A resonant filter with $f_r = 60$ Hz and $Q = 10$ requires $\sqrt{L/C} = 10(R_{\text{load}}) = 144$ Ω. The expressions

$$\frac{1}{\sqrt{LC}} = 120\pi\,\frac{\text{rad}}{\text{s}}, \qquad \sqrt{\frac{L}{C}} = 144\ \Omega \tag{8.25}$$

can be solved simultaneously for L and C, suggesting that $L = 382$ mH and $C = 18.4$ μF will work. This inductor value is quite high, at least for this power level. If the inductor and capacitor have $Q = 30$, each part will introduce an additional series resistance equal to 1/30 of its reactance. For $L = 382$ mH, this means the inductor has series resistance of 4.8 Ω. For $C = 18.4$ μF, this value of Q means an additional 4.8 Ω exists in the circuit. The complete circuit, shown in Figure 8.8, has substantial losses in the storage elements. The overall value of Q is reduced, since the total circuit resistance is now 24 Ω. With $Z_c = 144$ Ω, the filter Q becomes 6 instead of 10, and filter performance is not up to expectations.

Line-frequency inverters, such as those for backup power, are important applications for resonant filters. In applications intended for very high power levels, the good selectivity of a resonant filter is often worth the component cost. An example illustrates the utility of such a filter for an inverter.

Example 8.2.1 A voltage-sourced inverter produces a square wave at 60 Hz. The waveform is processed through a series-resonant filter. Use the relative magnitude of the third harmonic to compare filter performance at various values of Q.

Figure 8.8 Series resonant filter with device quality factor $Q = 30$.

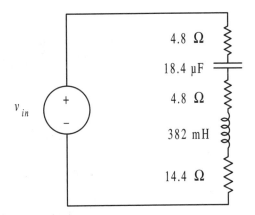

The Fourier Series of a 60 Hz square wave with peak value V_0 is

$$v(t) = \frac{4V_0}{\pi} \sum_{n=1}^{\infty} \frac{\sin(n\pi/2)}{n} \cos(n120\pi t) \tag{8.26}$$

With no filter, the third harmonic amplitude is 1/3 that of the fundamental. When the RLC filter with resonant frequency $1/\sqrt{LC} = 120\pi$ rad/s is added, the resistor voltage third harmonic amplitude V_3, taken from equation (8.23) at radian frequency 360π, becomes

$$V_3 = \frac{4V_0}{3\pi} \left| \frac{R}{j360\pi L - j/(360\pi C) + R} \right|$$

$$= \frac{4V_0}{3\pi} \left| \frac{1}{j3Q - jQ/3 + 1} \right| = \frac{4V_0}{3\pi} \frac{1}{\sqrt{1 + (64/9)Q^2}} \tag{8.27}$$

with V_0 and V_3 in volts. If Q is large in this expression, the third harmonic amplitude is about $1/(8Q)$ times the fundamental amplitude. The third harmonic can be attenuated below 3% of the fundamental if $1/(8Q) < 0.03$, or $Q > 4.2$. Higher harmonics are attenuated even more.

Series resonance can be extremely effective in converters with high-frequency links. At 60 kHz, for instance, a characteristic impedance of 144 Ω or more can be achieved with

$$L > 382 \ \mu\text{H}, \qquad C = 0.0184 \ \mu\text{F} \tag{8.28}$$

These values are much more reasonable than the corresponding values at 60 Hz.

8.3 PARALLEL RESONANCE

Parallel-resonant combinations of inductors and capacitors can be used in a manner dual to the series combination, blocking a specific frequency f_r while allowing other frequencies to pass. Two parallel-resonant combinations are shown in Figure 8.9. In Figure 8.9a, all three elements are in parallel. This arrangement is useful if the input is a current source: the circuit is a current divider, and the LC pair controls the current flow into the resistor. The combination in Figure 8.9b uses the LC pair to block flow at the resonant frequency. Blocking behavior is sometimes termed *antiresonance*, since it is the opposite of the filter characteristic in the series resonant case.

Consider first the parallel RLC combination in Figure 8.9a. Treat the circuit as a current divider. Then the magnitude of the ratio of resistor current to input current at a source frequency ω_s is

$$\left| \frac{\tilde{I}_{\text{out}}}{\tilde{I}_{\text{in}}} \right| = \left| \frac{Z_L Z_C}{Z_L Z_C + RZ_C + RZ_L} \right| = \left| \frac{L/C}{L/C + \dfrac{R}{j\omega_s C} + j\omega_s LR} \right| \tag{8.29}$$

In the parallel case, there is a *characteristic admittance*, $Y_c = \sqrt{C/L}$. The resonant frequency ω_r is unchanged. It is appropriate to define the quality factor as $Q = Y_c/G$, where $G = 1/R$ is the load conductance. Then equation (8.29) can be reduced to

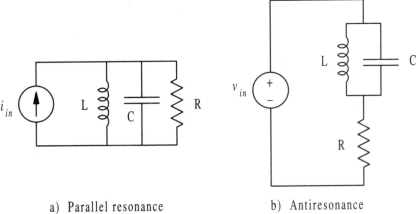

a) Parallel resonance   b) Antiresonance

Figure 8.9 Parallel resonant circuits.

$$\left|\frac{\tilde{I}_{\text{out}}}{\tilde{I}_{\text{in}}}\right| = \left|\frac{\omega_s/\omega_r}{j(\omega_s/\omega_r)^2 Q - jQ + \omega_s/\omega_r}\right| \tag{8.30}$$

The magnitude of the current ratio as a function of frequency is plotted for various Q values in Figure 8.10. Equation (8.30) gives exactly the ratio that was found for voltage in the series resonant case, so Figure 8.10 shows the same behavior as Figure 8.7.

The parallel RLC arrangement is appropriate for filtering in a current-sourced inverter. To obtain good selectivity, the characteristic admittance should be high. This requires a large capacitance—the dual of the series resonant case.

Example 8.3.1 A current-sourced inverter produces a 50 Hz square-wave of current with peak value $I_0 = 20$ A. The resistive load is intended to develop 230 V_{RMS}. Design a parallel-resonant filter with $Q = 10$ for this situation. What is the resistor value? How much is

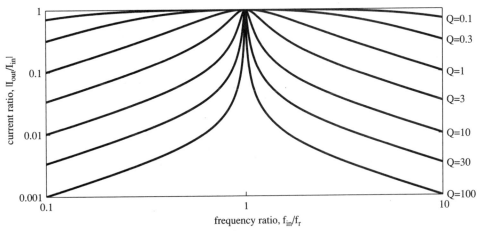

Figure 8.10 The ratio $|\tilde{I}_{\text{out}}/\tilde{I}_{\text{in}}|$ vs. input frequency ratio for various Q in parallel resonance.

the third harmonic attenuated at the load? What is the magnitude of 50 Hz current flow in the inductor?

A square wave with 20 A peak will provide a 50 Hz fundamental with amplitude $4I_0/\pi = 25.5$ A and RMS value 18.0 A. An ideal parallel-resonant filter with $f_r = 50$ Hz will pass this current to the resistor unattenuated. To develop 230 V_{RMS}, the resistance should be 12.8 Ω. The conductance is 0.0783 S. The load will consume 4.14 kW. A parallel-resonant filter with $Q = 10$ will require $Y_c = 0.783$ S and $\omega_r = 100\pi$ rad/s. Solving $Y_c = \sqrt{C/L}$ and $\omega_r = 1/\sqrt{LC}$ simultaneously (and neglecting the quality factor of the individual elements), $L = 4.07$ mH and $C = 2491$ μF.

From equation (8.30), the third harmonic at 150 Hz is attenuated by a factor of about 27 by the filter. Since a square wave has a third harmonic 1/3 as large as the fundamental, the output 150 Hz component will be only 1/81 as large as the output 50 Hz component. What about the inductor current at 50 Hz? The inductor is exposed to a voltage of 230 V_{RMS} at 50 Hz, since it is in parallel with the resistor. The inductor's reactance is (4.07 mH)(100π rad/s) = 1.28 Ω. Its RMS current is (230 V)/(1.28 Ω) = 180 A, the same as Q times the resistor current.

Analogous to the series resonant case, at the resonant frequency each storage element in a parallel-resonant filter is exposed to a current larger than the input current component by a factor of Q. The inductor and capacitor in a filter with $Q = 100$ carry one hundred times the load current. This behavior requires extreme ratings for L and C parts, just as in the series resonant case. In practice, the value of Q must be low enough to avoid high stress and excessive loss in the energy storage components.

The action of an antiresonant circuit, like the one in Figure 8.9b, contrasts with the preceding resonant systems. The antiresonant combination provides an output–input voltage ratio of

$$\left|\frac{\tilde{V}_{out}}{\tilde{V}_{in}}\right| = \left|\frac{R}{R + \dfrac{L/C}{j\omega_s L + 1/(j\omega_s C)}}\right| \tag{8.31}$$

With $Q = Y_c/G$, equation (8.31) can be written in terms of the ratio ω_s/ω_r as

$$\left|\frac{\tilde{V}_{out}}{\tilde{V}_{in}}\right| = \left|\frac{Q(\omega_s^2/\omega_r^2 - 1)}{Q(\omega_s^2/\omega_r^2 - 1) - j\omega_s/\omega_r}\right| \tag{8.32}$$

This ratio is plotted for several values of Q in Figure 8.11. Notice that the vertical scale is linear instead of logarithmic. The resonant frequency is completely blocked with this ideal inductor–capacitor pair. When Q is high, nonresonant frequencies are passed without any attenuation.

The performance of a real antiresonant filter is limited by the quality factors of the individual L and C elements, since any stray resistance forms an RLC loop. As might be expected, the presence of resistance with the LC pair allows part of the resonant frequency component to pass through to the output. For example, if the inductor and capacitor both exhibit component quality factors $Q = 100$, then a filter combination with $Y_c/G = 10$ passes 1/6 of the resonant frequency component. Figure 8.12 shows the attenuation characteristics of a parallel resonant blocking filter, assuming that L and C both show $Q = 100$ at the source frequency.

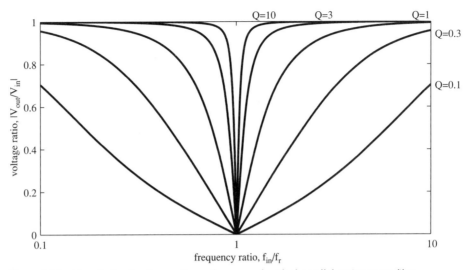

Figure 8.11 Magnitude of voltage ratio vs. frequency for ideal parallel antiresonant filter.

The parallel blocking combination has possibilities in power converters. An excellent example is a PWM inverter with variable modulating frequency and fixed switching frequency. A blocking filter, designed to resonate at the switching frequency, is effective in attenuating components near the switching frequency without altering the wanted component.

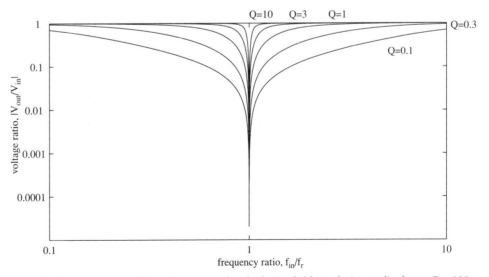

Figure 8.12 Antiresonant filter performance when both L and C have device quality factor $Q = 100$.

8.4 SOFT SWITCHING TECHNIQUES—INTRODUCTION

8.4.1 Soft Switching Principles

Every time a switch is thrown in a power converter, a certain amount of energy is lost. During each switching cycle, the transition from zero current to on-state current and back again, and the transition from off-state voltage to zero voltage and back must be made. Since a real switch requires time to make these transitions, there is a nonzero power loss as the currents and voltages change. The effects of the loss are very important in fast-switching or high-power circuits. As much as 50% of the losses in a modern dc–dc converter can be caused by the switching transitions.

The loss situation is a fundamental concern for a PWM inverter. In this converter, the switching frequency is high for convenience in matching a desired waveform. The losses scale with frequency, since each extra transition adds more loss. A PWM inverter switching at 60 kHz can have 1000 times the loss of a VSI switching at 60 Hz for similar devices.

One way to avoid this *switching loss* is to make sure that the voltage or current does not need to change as the switch operates. A common example is the diode. A diode will switch off only when its forward current reaches zero. If the current can be held near zero during the transition, turn-off loss will be very low (keeping the current low is not trivial, since most diodes require a negative *recovery current* to remove internal charge). A switching transition that takes place with low current represents *zero-current switching* (ZCS). In the low-voltage alternative, the voltage drop across the switch is held close to zero during the turn-on transition. The current increases while the voltage is low, so the loss during the transition is low as well. This action is called *zero-voltage switching* (ZVS).

Switching devices themselves cannot provide ZCS or ZVS action in general. In a dc–dc converter, for instance, each switch must block full voltage during its entire off state, and must carry full current during its on state. Instead, ZCS or ZVS action in dc converters requires additional parts that manipulate switch voltages and currents while supporting the general needs of the power converter. Resonant combinations can be used for these purposes.

Converters with low switching losses, based on ZCS or ZVS, require two things:

1. If no ac source is present, resonant combinations are added to generate zero crossings of switch voltages or currents. The zero crossings are opportunities for low-loss switch transitions.
2. The switching device action must be timed for the correct behavior. For ZCS, the switch must turn off only when the current crosses zero. For ZVS, the switch must turn on only when the voltage crosses zero.

The behavior of ideal diodes and thyristors supports natural zero-current switching, since turn-off corresponds to a current zero crossing. An ideal diode also acts naturally as a zero-voltage switch: When the forward voltage rises through a zero crossing, the device turns on. MOSFETs and other power semiconductors can support ZCS or ZVS only if external gate controls operate them at the right instant.

In a few ac applications, ZCS or ZVS behavior can be obtained without resonant elements. ac regulators, for example, use the zero crossings of the source, combined with triac or SCR behavior, to give zero-current switching. Integral-cycle controllers attempt to provide both ZVS and ZCS action to minimize loss. Most such converters operate naturally at line frequency.

When switching frequencies are high, as in dc–dc converters and PWM inverters, resonance can be used to create the conditions for ZCS or ZVS. The approach is called *soft switching*, since the switching devices alter a resonant network rather than forcing an abrupt change in current or voltage.

8.4.2 Basic Configurations

A basic soft switching circuit appears in Figure 8.13. In this case, resonant action of the RLC set allows SCRs to implement an inverter if the switching frequency is below the resonant frequency. Circuit action is as follows: Starting from rest, the top SCR is triggered. This applies a step dc voltage to the RLC set. If $Q > \frac{1}{2}$, the current will be underdamped, and will oscillate. When the current swings back to zero, the SCR will turn off. After that point, the lower SCR can be triggered for the negative half-cycle.

Figure 8.14 is a SPICE simulation for the choices $Q = 2$ and $\omega_r = 4000$ rad/s. Figure 8.14*a* shows the resistor voltage and the capacitor voltage. Figure 8.14*b* shows the voltage across the upper SCR and the SCR current. The specific values are $L = 5$ mH, $C = 12.5$ μF, $R = 10$ Ω, and $V_{\text{in}} = 200$ V. The SCRs are triggered 1 ms apart. When the top SCR first switches on, the capacitor charges and current flows in the inductor. Because of resonance, the current returns to zero at about $t = 0.8$ ms, at which point the SCR switches off. The circuit reaches periodic steady state after about 5 ms.

The simulation brings out a number of limitations of resonant switching. For example, in this circuit the output frequency must meet the constraint that $\omega_{\text{out}} < \omega_r$. Each SCR trigger control must wait until the current in the other has returned to zero and the other device has turned off. The capacitor develops a peak voltage about twice that across the resistor, consistent with $Q = 2$. This voltage affects the switch device off-state. The two SCRs must block the full capacitor voltage at times. In this circuit configuration, the switch action represents ZCS. There is some residual switching loss, however, because the voltage is high at the points of both turn-on and turn-off.

Even with its limitations, the circuit of Figure 8.13 has two advantages: It provides ZCS action in an inverter, and it supports voltage-sourced inversion using SCRs. The effi-

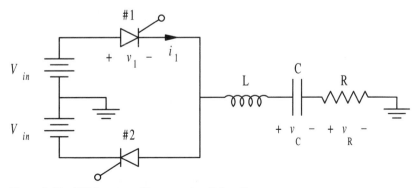

Figure 8.13   SCR inverter with resonant switch action.

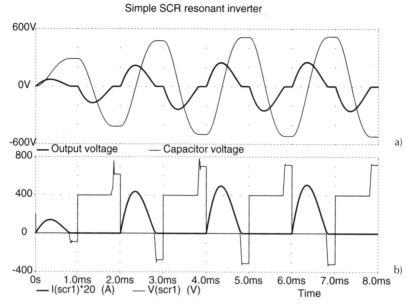

Figure 8.14 Simulation of *SCR* inverter for $Q = 2$ and $\omega_r = 4000$ rad/s.

ciency is limited mainly by the on-state voltage drop of the switches. A particularly interesting feature of the SCR circuit—and of many soft-switching configurations—is that there is little advantage to high values of Q. In Figure 8.15, the SPICE simulation is repeated for $Q = 0.80$. The triggering interval has been increased to 1.2 ms since the current zero-cross-

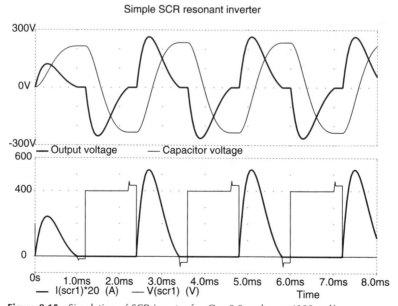

Figure 8.15 Simulation of SCR inverter for $Q = 0.8$ and $\omega_r = 4000$ rad/s.

ing occurs later at this low value of Q. The basic operation is unchanged. Because Q is low, however, the capacitor develops less voltage than the output resistor. The switches are exposed to off-state voltages only slightly higher than the input. The switching transitions are less lossy because the off-state voltages are significantly lower, at the expense of distortion in the output waveform.

The SCR inverter represents a *series resonant switch* configuration. The natural ZCS properties of SCRs are advantageous in this particular circuit.

8.4.3 Parallel Capacitor as a Soft Switching Element for the dc–dc Case

In Figure 8.16, a capacitor has been added to a buck converter to form a type of soft-switching dc–dc converter. The figure also shows some of the configurations of the circuit. When the transistor is on, the capacitor is shorted, and the transistor carries full inductor current. At time t_{off}, the transistor is shut off. The capacitor provides a path for the inductor current, and slows the change in voltage across the transistor as it turns off. This result is illustrated in Figure 8.17. In essence, the capacitor slows the rate of rise of the switch voltage during turn-on, and can be chosen for approximate ZVS action. The capacitor voltage climbs until the diode is forward biased, then the diode turns on.

The single capacitor is useful only during turn-off. When it is time for the transistor to turn on, the capacitor has voltage V_{in} across it. The stored energy dissipates in the transistor, causing extra energy loss $\frac{1}{2}CV_{in}^2$ during the turn-on transition. The problem is that the capacitor has not created a resonant combination, and no zero crossings of voltage or current occur. In the next section, we consider how a resonant combination like that in the SCR inverter can be used for soft switching in the dc–dc case.

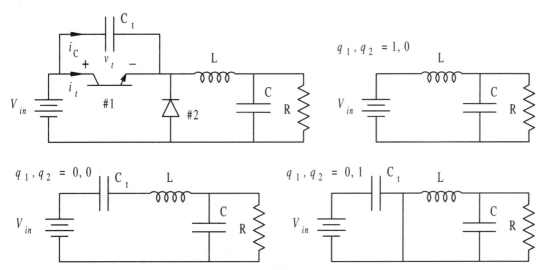

Figure 8.16 Buck converter with capacitor for soft turn-off.

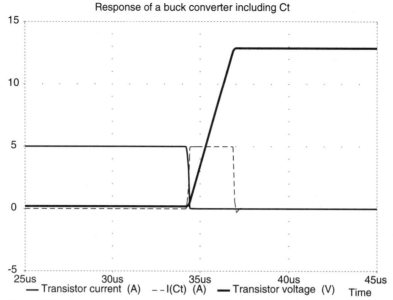

Figure 8.17 Transistor voltage and current in soft switching circuit.

8.5 SOFT SWITCHING IN DC–DC CONVERTERS

8.5.1 Description of Quasi-Resonance

The circuit of Figure 8.18 provides an interesting arrangement for dc–dc conversion. In this case, both an inductor and a capacitor have been added to alter the transistor action. A similar *LC* pair is added at the diode. The combinations offer several possibilities for resonant action. The actual behavior will depend on the relative values of the parts. Consider one possibility:

- When the transistor turns off, the input voltage excites the pair L_t and C_t. The input inductor current begins to oscillate with the capacitor voltage. The capacitor voltage initially holds the voltage low as the transistor turns off.

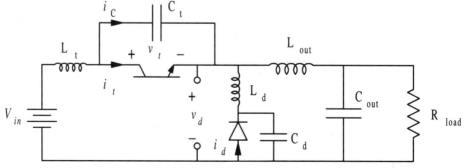

Figure 8.18 Resonant components added to dc–dc converter.

- The capacitor voltage swings well above V_{in}, and the diode turns on.
- The capacitor voltage swings back down. There might be an opportunity for zero-voltage turn-on of the transistor when the capacitor voltage swings back to zero.

Other combinations provide an opportunity for zero-current switching at transistor turn-off.

In any of these soft-switching cases (including the SCR inverter example in the preceding section), switch action at a zero crossing cuts off the ringing resonant waveforms. In the literature, this technique is often termed *quasi-resonance*. Converters based on circuits such as Figure 8.18 have been called both resonant converters and *quasi-resonant converters*.

Figure 8.19 shows some of the possible waveforms in the circuit of Figure 8.18. In Figure 8.19a, the parts are chosen so that C_t and L_d are very small and have minimal effect on circuit action. The inductor L_t and capacitor C_d form a series LC combination, and the transistor can take advantage of current zero crossings for ZCS. In Figure 8.19b, the values of C_t and L_d are significant while L_t and C_d are small. This supports ZVS action for the transistor. It is possible in principle to use all four parts to support ZVS and ZCS action together. This technique, called *multiresonance*, is not common because it tends to constrain converter control.

In any soft-switching converter, switch action must be coordinated to match the zero crossings. The duty ratio no longer serves as the primary control parameter. In Figure 8.19a, for example, the on-time of the transistor should be approximately half the period of resonance, or about π/ω_r, to match the objective of zero-current switching. The transistor off-time becomes the control parameter. Since the on-time is fixed, adjustment of the off-time represents adjustment of the switching frequency. This is true of most resonant converters: The output magnitude is adjusted by altering f_{switch}.

8.5.2 ZCS Transistor Action

Although the control approach must be different from that in the hard-switched dc–dc converters of Chapter 4, soft-switching dc–dc converters can be analyzed conveniently with the appropriate mathematical tools. To see this, consider a buck converter as in Figure 8.18 with C_t and L_d negligible. The circuit is shown in Figure 8.20. The diode internal to the power MOSFET is shown explicitly, since it will affect the ZCS operation. If the output inductor is large, it can be treated as a current source. The main analytical challenge in this circuit is that additional configurations are possible. With the extra L and C, it is possible to have both switches off or on together without causing a KCL or KVL violation.

The sequence of configurations of this buck converter is shown in Figure 8.21, beginning from condition (0,1) with the transistor off and the diode on. When transistor turn-on occurs, the diode will remain on until its current reaches zero. Thus both switches are on until the current in L_t climbs to match I_{out}. In configuration (1,1), the current in L_t rises linearly since $v_L = V_{\text{in}}$. Once the diode turns off, L_t and C_d are free to ring resonantly. The resonant LC pair begins from the initial condition $v_d = 0$ and $i_t = I_{\text{out}}$. If the inductor and capacitor have been chosen correctly, the inductor current swings up, then down until it is negative. The diode within the MOSFET will turn on. At this point, to enforce soft-switch-

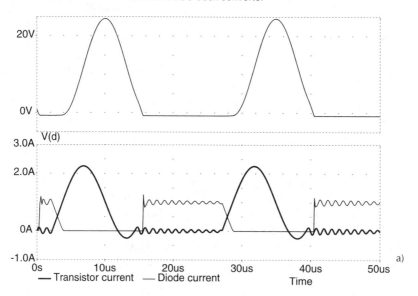

Resonant ZCS buck converter

20V

0V

V(d)

3.0A

2.0A

0A

-1.0A

0s   10us   20us   30us   40us   50us

— Transistor current — Diode current

Time

a)

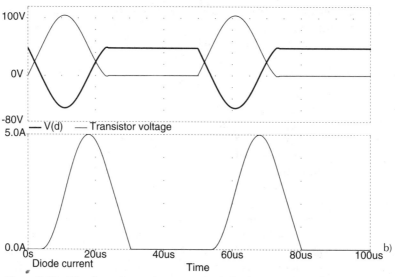

Resonant ZVS buck converter

100V

0V

-80V

— V(d)   — Transistor voltage

5.0A

0.0A

0s   20us   40us   60us   80us   100us

Diode current

Time

b)

Figure 8.19 Waveforms in quasi-resonant dc–dc buck converter. *a.* Quasi-resonant action with L_t and C_d. *b.* Quasi-resonant action with L_d and C_t.

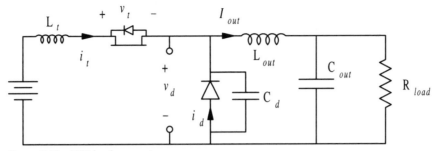

Figure 8.20 ZCS buck converter.

ing action, the gate signal is shut off. When the inductor current attempts to swing positive again, the FET's diode will shut off. Since the gate signal has been removed, configuration (0,0) occurs. Current I_{out} flows from C_d, and the voltage v_d falls until the main diode becomes forward biased and turns on. This completes the cycle. The load voltage is $V_{out} = \langle v_d \rangle$, but now v_d is a quasi-resonant waveform. An example might better illustrate the details.

Example 8.5.1   The resonant buck converter of Figure 8.20 has $L_t = 20~\mu H$ and $C_d = 0.2~\mu F$. The input voltage is 12 V. Will this combination of L and C provide resonant converter action? If so, determine the duration of the gate signal and the switching period needed to provide 5 V output into a 5 W load. What is the highest possible output voltage that preserves resonant action given a 1 A load?

For the 5 W load, let us model the load as a 1 A current source. Start from configuration (0,1). The transistor turns on, causing the current in L_t to ramp at $V_{in}/L_t = 6 \times 10^5$ A/s.

Figure 8.21 Sequence of configurations in ZCS buck converter.

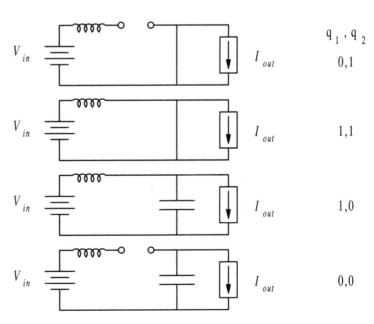

The diode current will drop to zero 1.67 μs after transistor turn-on, then the circuit enters configuration (1,0). The circuit equations in this configuration give

$$V_{\text{in}} - L_t \frac{di_t}{dt} - v_d = 0$$

$$i_{Cd} = C_d \frac{dv_d}{dt} = i_t - I_{\text{out}}$$

(8.33)

Differentiating the KVL expression with respect to time and using the KCL equation to substitute for dv_d/dt, we see that

$$L_t C_d \frac{d^2 i_t}{dt^2} + i_t - I_{\text{out}} = 0$$

(8.34)

The solution has the form $i_t(t) = c_1 \sin(\omega_r t) + c_2 \cos(\omega_r t) + c_3$, where ω_r is the resonant frequency. If the time when the diode turns off is used as the reference time $t = 0$, the initial conditions for (8.34) are $i_t(0) = I_{\text{out}}$ and $di_t/dt(0) = V_{\text{in}}/L_t$. The solution is an undamped sinusoid,

$$i_t(t) = \frac{V_{\text{in}}}{Z_c} \sin(\omega_r t) + I_{\text{out}}$$

(8.35)

where $Z_c = \sqrt{L_t/C_d}$. Notice that we need to ensure $V_{\text{in}}/Z_c > I_{\text{out}}$ to guarantee that the current will return to zero. This is an important detail: The characteristic impedance must be low enough to provide large variations in the inductor current when the resonant circuit is operating. In this example, the inductor and capacitor give $Z_c = 10\ \Omega$. The ratio $V_{\text{in}}/Z_c = 1.2$ A, and this condition for resonant converter action is met. However, if the load increases above 1.2 A, this *LC* combination will no longer support soft switching.

Resonant converter action is generated as follows:

- Keep the FET gate signal high until the current $i_t(t)$ reverses polarity. When the current reverses, the internal diode of the MOSFET will carry i_t.
- Turn the gate signal off while the current is negative (this will make the internal diode control the turn-off process).
- The MOSFET will turn off at the current zero crossing from negative to positive when its internal diode current reaches zero.

This action is illustrated in Figure 8.22. The on-time of the FET is the time from $t = 0$ until the rising zero-crossing of the current occurs. This is just slightly less than one cycle of the resonant waveform of Figure 8.22. This point of FET turn-off t_{off} is the solution of the set of equations

$$0 = \frac{V_{\text{in}}}{Z_c} \sin(\omega_r t_{\text{off}}) + I_{\text{out}}$$

$$\frac{di_t}{dt}(t_{\text{off}}) > 0, \qquad 0 < t_{\text{off}} < \frac{2\pi}{\omega_r}$$

(8.36)

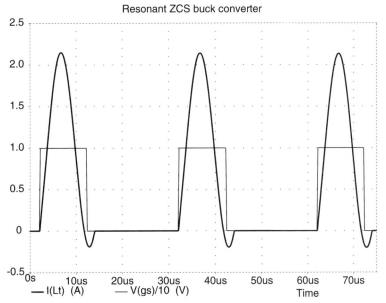

Figure 8.22 Inductor current and possible gate voltage signals.

In this example, $\omega_r = 5 \times 10^5$ rad/s and $2\pi/\omega_r = 12.6$ μs. The angle $\omega_r t_{\text{off}}$ can be found to be 303.6°, corresponding to $t_{\text{off}} = 10.6$ μs. The MOSFET gate signal should be held high until the falling zero crossing occurs (at 236.4° or $t = 8.25$ μs), then turned off before t_{off} is reached. A gate signal duration of 10 μs will work well in this converter.

To find the output voltage, the voltage v_d is needed. The configurations of interest are (1,0) and (0,0), since $v_d = 0$ otherwise. In configuration (1,0), the solution in equation (8.35) can be used in (8.33) to give

$$v_d(t) = V_{\text{in}} - V_{\text{in}} \cos(\omega_r t) \tag{8.37}$$

This is valid from time $t = 0$ to time t_{off}. At time t_{off}, configuration (0,0) becomes valid, and the entire load current I_{out} flows out of the capacitor. The capacitor voltage drops linearly according to $-I_{\text{out}} = C_d(dv_d/dt)$. When the voltage reaches zero, the diode turns on and the converter returns to configuration (0,1). Over a switching period T, the average diode voltage will be

$$\langle v_d \rangle = \frac{1}{T}\left[\int_0^{t_{\text{off}}} V_{\text{in}} - V_{\text{in}} \cos(\omega_r t)\, dt + \int_{t_{\text{off}}}^{t_{\text{diode on}}} v_d(t)\, dt \right] \tag{8.38}$$

The second integral is the area of a triangle. The triangle peak value is $v_d(t_{\text{off}}) = V_{\text{in}} - V_{\text{in}} \cos(\omega_r t_{\text{off}}) = 5.37$ V. Since the slope of v_d during the second integral is $-I_{\text{out}}/C_d$, the time until v_d reaches zero (the base of the triangle) is $[v_d(t_{off})]/(I_{out}/C_d) = 1.07$ μs. The triangle area is $\frac{1}{2}C_d v_d^2(t_{off})/I_{out} = 2.88 \times 10^{-6}$ V·s. The first integral has a value of 1.47×10^{-4} V·s. Thus the average output is

$$\langle v_d \rangle = \frac{1.47 \times 10^{-4} + 2.88 \times 10^{-6}}{T} = \frac{1.50 \times 10^{-4}}{T} \tag{8.39}$$

A switching period of 30 μs will produce an average output of 5 V. If the gate signal is held high for 10 μs, the converter operates with a gate duty ratio of $\frac{1}{3}$. The relevant waveforms are shown in Figure 8.23.

How high an output can be produced, assuming resonant action and a 1 A load? As long as the circuit steps through all configurations (0,1), (1,1), (1,0), (0,0), in sequence, the above analysis is accurate. Equation (8.39) will be valid as long as there is enough time for the circuit to operate in each configuration. In this example, the circuit (1,1) exists for 1.67 μs. Configuration (1,0) exists for 10.6 μs, and configuration (0,0) lasts about 1.07 μs. Provided the switching period T is longer than the sum, 13.4 μs, the circuit will have enough time to operate in each of these three configurations, with time left over for configuration (0,1). The time spent in (0,1) is a free parameter: the switching period can be altered to manipulate this time. The constraint $T > 13.4$ μs corresponds to $f_{switch} < 74.6$ kHz. Equation (8.39) can be rewritten in terms of the switching frequency to give

$$V_{out} = 1.50 \times 10^{-4} f_{switch}, \quad f_{switch} < 74.6 \text{ kHz} \tag{8.40}$$

The maximum possible output occurs at the highest allowed switching frequency, which gives $V_{out} = 11.2$ V. In effect, there is a maximum duty ratio limit in the ZCS converter.

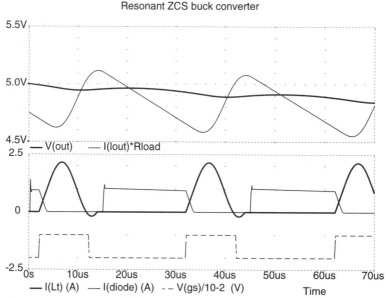

Figure 8.23 Current and voltage waveforms for 12 V to 5 V resonant converter.

It is possible to produce a fairly general analysis of the ZCS buck converter, based on the approach in the preceding example. Assuming that the output filter inductor has been chosen large (for low ripple), the analysis in Mathematica could be written as provided in Table 8.1.

A particularly interesting application of resonant switching uses stray capacitances and inductances inherent in switching power semiconductor device packages. For example, a typical power MOSFET has inductance of about 10 nH because of the packaging geometry. If a diode capacitance of 100 pF is provided (or is present within the diode), the value of Z_c will be 10 Ω—low enough to support ZCS action in many low-power converters without a separate inductor. The capacitance C_t usually cannot be neglected in this context, since it has a value of several hundred picofarads for typical MOSFETs. However, C_t has no effect during configuration (1,0), and it does not alter the output voltage very much. The basic ZCS action is maintained even with this capacitor present. High-performance switching power supplies sometimes make use of the stray elements in resonant converter design. The technique improves efficiency compared to square-wave switching conversion, and complicates only the control rather than the conversion circuit.

8.5.3 ZVS Transistor Action

The dual of the ZCS case uses transistor capacitance along with diode inductance to produce zero-voltage switch action. In a ZVS buck converter such as that shown in Figure 8.24, transistor turn-off generates a series LC resonant loop. The capacitor voltage rises, then swings back to zero if the element values are chosen correctly. The voltage zero-crossing provides an opportunity for low-loss turn-on of the switch. In the ZCS case discussed earlier, resonant action determined the transistor on-time, and the off time was adjusted to alter the con-

TABLE 8.1
Mathematica Analysis of ZCS Buck Converter

(* Analysis of the ZCS resonant buck converter. P. T. Krein, 1996. *)
(* The resonant component values are Lt and Cd. The output is Vout/Rload. It is assumed that the user has chosen the inductor and capacitor to meet the condition that the Vin/zc > Vout/Rload. *)
(* The zero time point is taken as the moment of diode turn-off. *)

wr = 1/Sqrt[Lt Cd]; zc = Sqrt[Lt/Cd]; iout = Vout/Rload
it[t_] = Vin/zc Sin[wr t] + iout
vd[t_] = Vin − Vin Cos[wr t]

(* Time toff assumes that the inverse sine function maps to the interval −Pi to +Pi. Note that a solution requires Vin/zc > iout. *)

toff = (ArcSin[−iout zc/Vin] + 2 Pi)/wr
vdoff = vd[toff]

(* Things are a bit circular here because formally the average diode voltage (which is Vout) depends on Vout, but the dependence is not strong, and a couple of manual iterations will generally provide a good value. A different symbol for the average output will be used to prevent any trouble. *)

vave = fswitch (Integrate[vd[t],{t,0,toff}] + 1/2 Cd vdoff ∧ 2/iout)
tmin = toff + vdoff Cd/iout + iout Lt/Vin
fswmax = 1/tmin

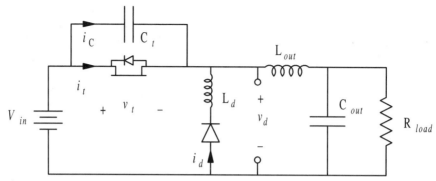

Figure 8.24 Buck converter with ZVS capability.

verter output. In the ZVS case, the opposite is true. Resonance determines the off-time, and the on-time must be adjusted to change the average output. As before, the switching frequency is the basis for control.

Following the earlier analysis of the ZCS case, consider four configurations of the ZVS buck converter shown in Figure 8.25. Let us start with the (1,0) arrangement, and assume that resonant transients have died out and that the output inductor is large. The transistor imposes voltage V_{in} on the diode. The voltage across C_t is zero, and the current in L_d is also zero. Then the transistor is turned off, and the converter enters the (0,0) configuration. In this case, the output current flows in C_t since the diode is off. The voltage buildup in C_t causes v_d to fall linearly until v_d reaches zero and the diode becomes forward biased. At that time, the circuit enters configuration (0,1). The pair C_t and L_d rings resonantly. The sinusoidal variation causes the transistor voltage to swing back to zero, when the internal FET diode turns on so that the configuration becomes (1,1). This imposes voltage V_{in} across L_d,

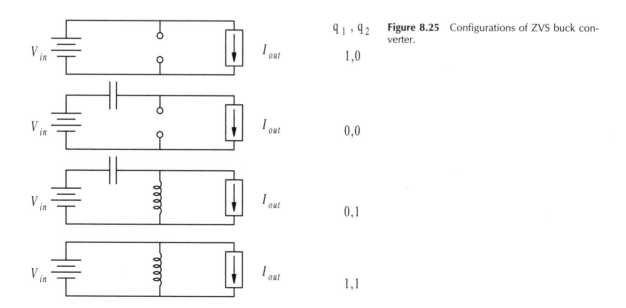

q_1, q_2

Figure 8.25 Configurations of ZVS converter.

1,0

0,0

0,1

1,1

and the inductor current ramps to zero linearly. The circuit enters configuration (1,0), and the cycle can be repeated. Typical waveforms showing this sequence are given in Figure 8.26.

The most interesting of these configurations is (0,1), during which the inductor and capacitor are in resonance. At the beginning of this configuration, the diode has just turned on, and current $i_d = 0$. The voltage on the capacitor is equal to V_{in} at the moment the diode turns on. Let us define time $t = 0$ to coincide with diode turn-on. Circuit laws require

$$V_{in} - v_t - L_d \frac{di_d}{dt} = 0$$

$$C_t \frac{dv_t}{dt} = I_{out} + i_d$$

(8.41)

When the voltage expression is differentiated and the current expression is used to substitute for dv_t/dt, the result is

$$L_d C_t \frac{d^2 i_d}{dt^2} + i_d + I_{out} = 0$$

(8.42)

With $i_d(0) = 0$ and $v_t(0) = V_{in}$, the solution to equation (8.42) is

$$i_d(t) = I_{out}[\cos(\omega_r t) - 1]$$

$$v_t(t) = V_{in} + I_{out} Z_c \sin(\omega_r t)$$

(8.43)

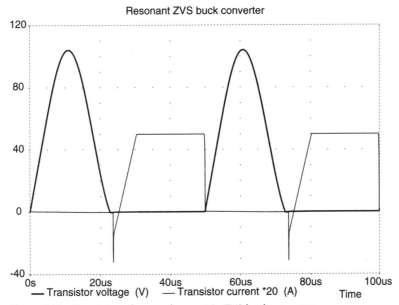

Figure 8.26 Transistor voltage and current in ZVS buck converter.

where Z_c is the square root of the ratio L_d/C_t. Notice that the inductor current has been defined positive downward for convenience in computing the output voltage. The solution for $i_d(t)$ is consistent with the diode direction, and gives $i_d(t) < 0$.

For ZVS action, the characteristic impedance needs to be large enough to ensure $I_{out}Z_c > V_{in}$ so that a voltage zero crossing will occur in equation (8.43). When the diode turns on, the voltage v_t swings up to a peak value of $V_{in} + I_{out}Z_c$, then swings down toward zero. At the moment when $v_t = 0$, the transistor can be switched on for ZVS action, and the converter will enter configuration (1,1). A MOSFET supports ZVS action quite well: Its reverse diode will turn on as the voltage attempts to change sign. A gate signal applied after the zero crossing will maintain the on condition as long as necessary.

In configuration (1,1), the inductor L_d carries voltage V_{in}. The inductor current rises linearly from its negative value until it reaches zero. At that point, the diode turns off, and configuration (1,0) is restored. Resonant action ceases, and configuration (1,0) will persist as long as the transistor gate signal is held high. In this converter, the transistor off-time is approximately one period of resonance, while the on-time is determined primarily by gate action. There is a minimum off-time—the time required to cycle through all the configurations.

The ZVS buck converter illustrates one important characteristic of resonant conversion: To ensure that v_t swings back to zero, we require $I_{out}Z_c > V_{in}$. This means that the peak off-state voltage across the transistor, given by $V_{in} + I_{out}Z_c$, will be at least double the input voltage. The transistor ratings must reflect this very substantial extra voltage. In the preceding ZCS case, the on-state current swings to a high value. This extra current will increase losses, since any semiconductor switch has a nonzero forward voltage drop when on. Resonant converters save on switching loss at the expense of higher ratings and higher on-state or off-state losses.

Example 8.5.2   A ZVS buck converter like the one in Figure 8.24 is to be designed for 48 V to 12 V conversion at a power level of 30 W. The component values are $L_d = 1\ \mu H$ and $C_t = 2000$ pF. The output inductance is large enough to be modelled as a 2.5 A current source. Find the gate signal duration, the transistor off-time, and the total period associated with a 12 V output. What device ratings will be needed in this converter? What is the lowest possible output voltage into a 2.5 A load?

For these parameters, the characteristic impedance is $Z_c = 22.4\ \Omega$, and the resonant radian frequency is 2.24×10^7 rad/s. The period of a resonant cycle will be 0.281 μs. Consider the converter initially in configuration (1,0). The MOSFET gate signal is dropped, and the converter enters configuration (0,0). The load current of 2.5 A will flow in C_t. The capacitor voltage increases linearly from zero at a rate of $(2.5\ A)/C_t = 1.25 \times 10^9$ V/s. In 38.4 ns, the capacitor voltage will reach 48 V and the diode will become forward biased. The converter enters configuration (0,1). Let us define this time point as $t = 0$. During configuration (0,1), the inductor current in L_d and capacitor voltage in C_t can be determined as in equation (8.43), and

$$i_d(t) = (2.5\ A)[\cos(\omega_r t) - 1]$$
$$v_t(t) = 48\ V + (55.9\ V)\sin(\omega_r t) \tag{8.44}$$

The capacitor voltage starts at 48 V when $t = 0$, and rises to a peak of 104 V in about 70 ns. It then falls sinusoidally, returning to 48 V at $t = 140$ ns. The capacitor voltage reaches zero when $55.9 \sin(\omega_r t) = -48$ V, which occurs at $\omega_r t = 239°$ and $t = 187$ ns. Let us designate this point as $t = t_{on}$. The internal diode of the FET will turn on, and the MOSFET gate should be set high soon after to maintain configuration $(1,1)$.

At the moment when configuration $(1,1)$ is entered, the inductor current is $2.5(\cos 239° - 1) = -3.78$ A. In this arrangement, the inductor sustains a voltage $v_L = V_{in}$. The current rises linearly at a rate of $(48 \text{ V})/(1 \ \mu\text{H}) = 4.8 \times 10^7$ A/s. Since $(3.78 \text{ A})/(4.8 \times 10^7 \text{ A/s}) = 78.8$ ns, after another 78.8 ns, the current will reach zero. This occurs at time $t = 187$ ns $+ 78.8$ ns $= 265$ ns. At that moment, the diode current is zero and it turns off. The configuration has returned to $(1,0)$.

During configurations $(1,1)$ and $(1,0)$, the voltage v_d in Figure 8.24 is simply V_{in}. During $(0,0)$, the voltage decreases linearly from 48 V to 0 V in 38.4 ns. During $(0,1)$, the voltage is $V_{in} - v_t$, or $-55.9 \sin(\omega_r t)$. The average value is

$$\langle v_d \rangle = \frac{1}{T}\left(\int_{-38.4 \text{ ns}}^{t=0} V_{in} - v_t(t) \, dt + \int_{t=0}^{t_{on}} -55.9 \sin(\omega_r t) \, dt + \int_{t_{on}}^{T-38.4 \text{ ns}} V_{in} \, dt \right) \quad (8.45)$$

where T is the total switching period. The first integral is a triangular area with height 48 V and width 38.4 ns, and contributes 0.922 μV·s to the total. The second integral yields -4.76 μV·s, while the last is $48(T - 38.4 \text{ ns} - 187 \text{ ns}) = 48T - 10.8 \times 10^{-6}$ V·s. The output voltage is

$$\langle v_d \rangle = \frac{1}{T}(0.922 \times 10^{-6} - 4.76 \times 10^{-6} + 48T - 10.8 \times 10^{-6}) \text{ V},$$

$$\quad (8.46)$$

$$V_{out} = 48 - f_{switch}(14.6 \times 10^{-6}) \text{ V}$$

This is a linear function of the switching frequency, just as in the ZCS case. To produce 12 V, a switching frequency of 2.46 MHz is needed. This gives a total period of 407 ns. The MOSFET gate should be set high when $t = 187$ ns, and set low 38.4 ns before the end of the period. Thus the gate is high for about 180 ns. The transistor will be off for a total of 187 ns $+ 38.4$ ns $= 225$ ns. Voltage and current waveforms for this converter, based on 12 V output, are given in Figure 8.27.

To ensure resonant operation, the transistor must be on long enough to cycle through all four configurations. The total time spent in $(0,0)$, $(0,1)$, and $(1,1)$ is 225 ns $+ 79$ ns $= 304$ ns. Therefore

$$T > 304 \text{ ns}, \qquad f_{switch} < 3.29 \text{ MHz} \quad (8.47)$$

The lowest possible period will be associated with the lowest possible output. When $f_{switch} = 3.29$ MHz, the output voltage is $48 - 3.29(14.6) = 0$ V. The waveforms are such that the requirements for a minimum on-time and minimum period do not limit the output at the low end. Higher voltages are obtained simply by keeping the transistor on longer.

The ZVS buck converter's characteristic impedance gives it load properties opposite to those of the ZCS converter. In the zero-current-switching case, there is a maximum load current above which the transistor current will not swing back to zero. For the ZVS circuit, there is a minimum load to support zero-voltage action. In the converter of the above example, a load current above 2.15 A is needed to ensure that $I_{out}Z_c > 48$ V.

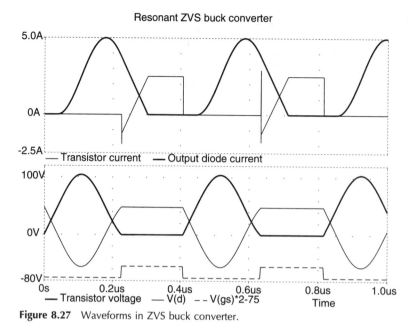

Figure 8.27   Waveforms in ZVS buck converter.

Resonant switch action is certainly not confined to buck-type converters. Any dc–dc converter can make use of the techniques, although of course the various equations will be different. Resonant switching, for example, is very helpful in flyback converters. Recall that leakage inductance can cause trouble in flyback circuits. Instead, the leakage inductor can be used intentionally as part of a resonant switching circuit. One possible arrangement is that in Figure 8.28. Here the capacitor C_t can be selected to produce ZVS behavior in the MOS-FET. Flyback converters can be built with significant efficiency improvements when resonant methods are applied.

8.6 RESONANCE USED FOR CONTROL—FORWARD CONVERTERS

Recall from Chapter 4 that forward converters can be modelled as ac link circuits: An inverter-rectifier cascade. Many resonant switching techniques take advantage of ac links for

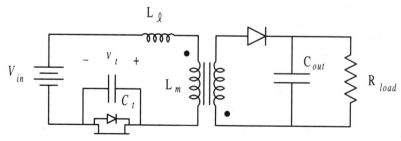

Figure 8.28   ZVS arrangement for flyback converter.

operation. Figure 8.29 illustrates a series-resonant forward converter, in a half-bridge configuration. The LC pair in the transformer primary provides important advantages: The capacitor prevents the flow of any dc component, and oscillations due to the LC pair support low-loss switching, just as in the SCR circuit of Figure 8.13. When a transistor turns on initially, the inductor L_{in} ensures that the current stays low while the voltage across the switch drops to zero. Transistor turn-off is controlled to coincide with a current zero-crossing.

There are two alternatives available in operating this circuit:

- Operation can be constrained to be synchronized with resonant behavior. In this case, each switch must have both the duty ratio and frequency adjusted together to provide the correct average output while matching the zero crossings.
- We can take advantage of the strong function of the ratio between the excitation frequency and the resonant frequency in a series resonant filter.

The second alternative is interesting. As the switching frequency is raised above the resonant point, the wanted component is attenuated quickly. A relatively small frequency range provides easy adjustment of the output voltage, at the cost of some loss in the switches since ZCS and ZVS action are no longer supported. Figure 8.30 shows some typical waveforms for the converter of Figure 8.29.

8.7 RECAP

Resonant component combinations are often used in power electronics. Such combinations serve either as filters or to ensure that switch action takes place at low voltage or current to minimize loss. The series combination of inductance and capacitance is one of the most common resonant configurations for both major applications. As a filter, the series RLC combination provides a bandpass characteristic with a bandwidth inversely proportional to the quality factor. As a resonant switch support circuit, a series LC combination can be designed to introduce current and voltage zero-crossings in dc circuits and inverters.

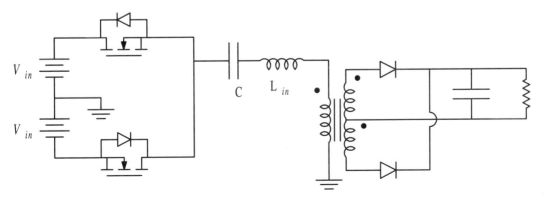

Figure 8.29 Forward converter with series resonant input switching.

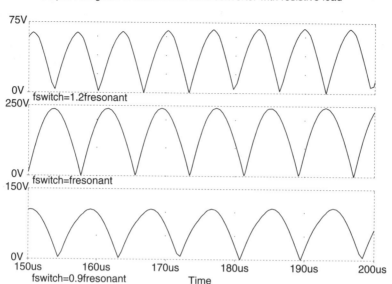

Figure 8.30 Transformer primary voltage waveforms at off-resonant frequencies for LC forward converter, $L = 250\ \mu H$, $C = 0.025\ \mu F$, $Q = 10$.

The series RLC circuit has voltage and current solutions that can be written based on exponentials e^{st}, with

$$s = \xi(-1 \pm \sqrt{1 - 4Q^2}) \tag{8.48}$$

where ξ is the *damping parameter* $R/(2L)$ and Q is the *quality factor*. The quality factor for a series RLC set is the ratio of *characteristic impedance* to resistance,

$$Q = \frac{Z_c}{R}, \qquad Z_c = \sqrt{\frac{L}{C}} \tag{8.49}$$

For a circuit with $Q > \frac{1}{2}$, a step input will cause the current and voltage to ring at a frequency ω related to the resonant frequency ω_r, where

$$\omega = \sqrt{\omega_r^2 - \xi^2}, \qquad \omega_r = \frac{1}{\sqrt{LC}} \tag{8.50}$$

When a series RLC arrangement is used as a filter, the phasor voltage at a given frequency $n\omega$ across the resistor can be found from that across the full combination as

$$\tilde{V}_{R(n)} = \tilde{V}_{RLC(n)} \frac{1}{1 + jQ(n\omega/\omega_r - \omega_r/(n\omega))} \tag{8.51}$$

If the value of Q is high, the resistor voltage matches the full voltage only at the resonant frequency, and is small at all other frequencies. High values of Q have important drawbacks, however. At the resonant frequency, both the capacitor and the inductor experience voltage peaks given by Q times the full input voltage peak. If $Q = 10$, for instance, an input voltage of $200 \cos(\omega_r t)$ will expose both the capacitor and inductor to $2 \text{ kV}_{\text{peak}}$.

Parallel LC combinations can be used to block flow at the resonant frequency or to provide selective filtering of current sources. When an equivalent current source is applied to a parallel RLC combination, the resistor current can be made sinusoidal at the resonant frequency, while other harmonics do not flow to the resistor. In the parallel case, the quality factor is defined based on the *characteristic admittance Y_c* as

$$Q = \frac{Y_c}{G}, \qquad Y_c = \sqrt{\frac{C}{L}} \tag{8.52}$$

where $G = 1/R$ is the load conductance. The phasor resistor current can be written in terms of the input source current magnitude and frequency ω_{in} as

$$\tilde{I}_R = \tilde{I}_{\text{in}} \frac{1}{1 + jQ \ (\omega_{\text{in}}/\omega_r - \omega_r/\omega_{\text{in}})} \tag{8.53}$$

This current ratio is identical to the voltage attenuation characteristic of the series RLC combination. Dual to the series case, the capacitor and inductor currents at the resonant frequency in the parallel circuit are amplified by a factor of Q above the input current. This is commonly a limiting factor in making high Q parallel filters.

The parallel LC combination has a net series impedance that becomes infinite at resonance. This behavior allows specific frequencies to be blocked. In PWM inverters, for example, parallel LC arrangements can serve to block the flow of switching frequency to the load. This in a circuit that has a resistor in series with a parallel LC pair, this *antiresonant* behavior gives a ratio of resistor voltage to input voltage of

$$\tilde{V}_R = \tilde{V}_{\text{in}} \frac{Q(\omega_{\text{in}}/\omega_r - \omega_r/\omega_{\text{in}})}{Q(\omega_{\text{in}}/\omega_r - \omega_r/\omega_{\text{in}}) - j} \tag{8.54}$$

This produces a bandstop filter with bandwidth inversely proportional to Q.

The design of both series and parallel resonant filters must consider that individual components are imperfect. A given single component has its own quality factor, defined as the ratio of reactance to resistance. Real inductors and capacitors have lead resistance and other losses that limit their individual Q values. Typically, a filter's Q will not be higher than the factors of its individual parts.

The practice of using resonant elements to reduce switching loss is termed *soft switching*. In this broad application, LC combinations are arranged to produce sinusoidal currents or voltages within a converter. The waveform zero crossings provide opportunities to turn a device on or off with very low loss. This technique is important at high power levels and at high frequencies, where significant losses can be incurred in a real semiconductor as it switches. Soft switching has been used to support the use of SCRs in dc circuits and inverters in which they would not normally be able to function. It is common in very high frequency dc–dc conversion. In some circumstances, it is used in high-power rectifiers and inverters to reduce power loss.

One of the simplest soft-switch combinations is an SCR in a series *RLC* string supplied from a dc voltage source. When the SCR turns on, the current begins to swing sinusoidally. Provided $Q > \frac{1}{2}$, the current will return to zero after a time. At the current zero, the SCR will turn off. The switch on-time duration is determined by resonant action, but the gate signal can be selected to set almost any desired duration of off-time. Since an SCR turns off only at current zeroes, the turn-off loss can be very low.

Converters that make use of soft switching have different KVL and KCL restrictions compared to more conventional hard switching circuits. The extra circuit elements provide current paths and create voltage loops such that there might be intervals in which all switches are on or off together. These extra configurations make the analysis more complicated, since the input–output relationships are no longer simple functions of duty ratios.

In MOSFET dc–dc converters, the combination of transistor series inductance with diode parallel capacitance supports zero-current turn-off switching similar to that in the SCR circuit. The combination of transistor parallel capacitance with diode series inductance supports zero-voltage turn-on switching. While the SCR is only appropriate at low switching frequencies, the most important difference between transistor-based and SCR-based resonant circuits is that the gate signal for the MOSFET must be timed correctly to synchronize with the resonant action.

Zero-current switching requires an upper limit on current flow. To maintain resonance, the characteristic impedance of the *LC* pair must be low enough to permit large swings in current. In a typical converter, the ratio V/Z_c must be higher than *I*, where *V* and *I* represent, respectively, the conventional off-state voltage and on-state current in the switching device. Zero-current switch action works best with low characteristic impedance and low on-state current.

Zero-voltage switching requires a lower limit on current instead. In this case, the voltage must swing widely to generate zero crossings. In general, the product IZ_c must exceed *V*, where again *I* and *V* are the on-state current and off-state voltage. Zero-voltage switch action works best with high characteristic impedance and high load current.

PROBLEMS

1. Design an *LC* series filter to take an input square wave of 320 V peak at 50 Hz and provide a sinusoidal output into a 10 Ω load. The third harmonic amplitude in the output should be no more than 1% as large as the fundamental amplitude. What is the value of *Q* for this filter?

2. A voltage-sourced inverter uses relative phase control, with a control angle δ, to produce a quasi-square wave output. The general waveform is given in Figure 8.31. In this application, $T = 1/60$ s and $V_0 = 240$ V. The load is a 20 Ω resistor.

   a. For δ = 45°, design a series *LC* filter to pass the fundamental while attenuating the third harmonic to no more than 3% of the fundamental.

   b. Is it possible to find *L* and *C* values that can ensure a third harmonic below 3% of the fundamental for any value of δ between 0° and 180°? If so, give the values.

3. A backup power supply is shown in Figure 8.32. It generates a 60 Hz square wave from a 12 V battery.

   a. Given a 300 W load, design a series *LC* filter to pass the fundamental while bringing the third

Figure 8.31   Quasi-square wave output from a VSI.

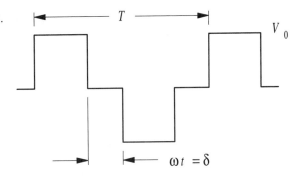

harmonic down below 5%. What are the peak voltages on the inductor and capacitor? Assume ideal elements.

b. Consider the situation in part (a), except that the available parts have device quality factors $Q = 30$. Design the filter under these conditions. What power is lost in the inductor and capacitor?

4.   Components with low quality introduce significant resistive drops in a circuit. A simple inverter produces a 60 Hz square wave with a peak value of 125 V. Filter components with component quality factor $Q = 20$ are available. The inverter is intended to support loads between 5 W and 500 W. It is desired to have no more than 10% amplitude drop over the full load range (that is, the load regulation should be better than 10%).

a. Find values of L and C for a series filter that will meet the regulation requirements while providing a useful filter function.

b. What is the highest level of attenuation that can be achieved for the unwanted components without violating the regulation requirement?

5.   An inverter produces 400 Hz for an aircraft application. The desired voltage is 120 $V_{RMS} \pm 5\%$. What square wave amplitude should be used? For a load range of 1000 W to 5000 W, design a series LC filter.

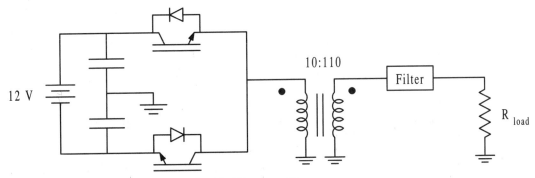

Figure 8.32   Battery backup power supply with LC resonant filter.

    a. Assume ideal components, and try to achieve third harmonic distortion below 2%.

    b. Given an inductor and capacitor with $Q = 20$, design an LC filter that will achieve less than 2% third harmonic distortion while meeting the output specification of $\pm 5\%$ over the entire load range.

  6. A PWM inverter switches at 2 kHz, and produces 50 Hz output at 380 V_{RMS} for a highway lighting application. The nominal load is 20 kW. Design a series LC filter to pass the fundamental while attenuating unwanted components by at least a factor of 100. Use values of $Q = 25$ at 2 kHz for the parts.

  7. An antiresonant filter is to be designed for a PWM inverter. The inverter switches at 2 kHz, and produces 50 Hz at 380 V_{RMS}. The nominal load is 20 kW. Available inductors and capacitors have $Q = 25$ at 2 kHz. Find L and C to attenuate switching frequency unwanted components by at least a factor of 100 while causing less than 3% drop in the fundamental.

  8. A three-phase PWM inverter is designed for a motor drive application. The output frequency can vary between 0 and 500 Hz. The input dc voltage is 300 V. The switching frequency is 12.5 kHz. The motor draws up to 50 A_{RMS} per phase.

    a. Design a parallel LC antiresonant filter to block the switching frequency while attenuating the wanted component by no more than 2%. Assume ideal L and C.

    b. The largest unwanted components in this inverter occur at $f_{switch} \pm f_{out}$. For the design of part (a), how much are the components at 12.0 kHz and 13.0 kHz attenuated when the output is 500 Hz?

  9. In a high-power rectifier application, antiresonant filters can help block large components. Consider a 24-pulse rectifier with average output of 4800 V when the phase control angle is 0°. The output power is 12 MW. Design a parallel antiresonant filter to block the largest unwanted component. The input frequency is 60 Hz.

    a. For inductors and capacitors with $Q = 100$, find parts that can attenuate the ripple to less than 1% with minimum power loss.

    b. What is the lowest value of device quality factor that can be tolerated in this application to keep the power loss below 1%?

  10. The current-sourced inverter in Figure 8.33 is sized to create a square wave current of 500 A peak at 60 Hz into a 0.5 Ω load. Notice the parallel LC filter shown in the figure.

    a. Choose L and C values to make the current in the resistor sinusoidal. The desired third harmonic distortion is less than 0.5%. Assume ideal parts.

    b. What is the value of V_{in} for proper power balance with the LC filter in place?

  11. A PWM inverter switches at 25 kHz, and produces an output wanted component of $325 \cos(100\pi t)$ volts from a 500 V dc bus. A parallel LC antiresonant filter is used to block unwanted components. The load is resistive, and draws 15 kW.

    a. Find ideal L and C values that attenuate the unwanted components at 25 kHz $\pm f_{out}$ by at least a factor of 100. Choose values that minimize the required voltage and current ratings.

Figure 8.33   Current sourced inverter.

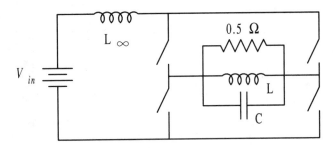

b. The load changes, and now is partly inductive. It still draws 15 kW, but includes series inductance of 5 mH. Will the filter design from part (a) still provide the same level of attenuation?

12.   A half-bridge PWM circuit has been built as an efficient audio amplifier. However, there is too much switching noise at the output. The switching frequency is 100 kHz. The output load is approximated as an 8 Ω resistor. The input dc voltage is 30 V. It is desired to use an LC parallel combination to block unwanted components while passing audio information at frequencies up to 20 kHz.

a. Design an LC filter to block $f_{switch} \pm f_{out}$ while passing the audio output. Is it possible to attenuate these unwanted terms by a factor of 100 without diminishing the audio signal by more than 10% (about 1 dB)? Assume ideal elements.

b. Examine the design more broadly. What range of inductor and capacitor Q will allow significant attenuation of switching noise without excessive drop of the wanted signal? Find and plot the highest signal-to-switching noise ratio as a function of device Q.

13.   Figure 8.34 shows a general quasi-resonant ZCS buck converter. Consider a design to convert 20 V to 12 V for a sensor circuit. Other values are $L_t = 2 \ \mu H$ and $C_d = 0.01 \ \mu F$.

a. Find the resonant frequency associated with the transistor inductance and diode capacitance. Also find the characteristic impedance Z_c for this pair.

b. What range of values for R_{load} will support resonant action in this converter?

c. Let $R_{load} = 30 \ \Omega$. What switching frequency is needed to produce 12 V output?

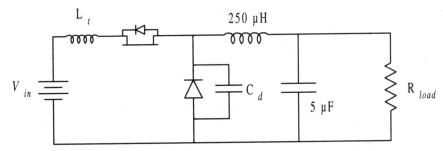

Figure 8.34   Quasi-resonant ZCS buck converter.

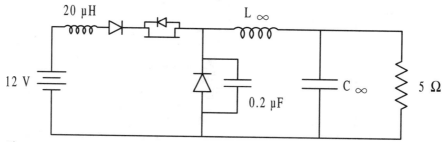

Figure 8.35   Quasi-resonant ZCS converter with blocking diode at the input.

14. A quasi-resonant buck converter like that in Figure 8.34 has $L_t = 4\ \mu H$, $C_d = 0.02\ \mu F$, and $V_{in} = 48$ V. The load is $R_{load} = 50\ \Omega$. What is V_{out} when the switching frequency is 200 kHz?

15. In a high-frequency dc–dc converter, the device stray inductance and capacitance cannot be neglected. Such a converter (like the one in Figure 8.34) might have $L_t = 50$ nH and $C_d = 80$ pF for realistic parts.

   a. With a 12 V input, what is the maximum load current that supports ZCS action with these values?

   b. Given a 12 V input, and assuming the load draws the maximum current with ZCS action, what is the switching frequency for 5 V output?

16. A blocking diode is used in a ZCS quasi-resonant converter to prevent current backflow in the MOSFET. The circuit is shown in Figure 8.35. Follow along Example 8.5.1. In this circuit, what is the sequence of configurations? How much time is spent in each configuration? What is the output voltage as a function of the switching frequency?

17. It is desired to prepare a ZCS buck converter for 48 V to 12 V conversion at power levels up to 200 W. The intended switching frequency is 400 kHz. Choose an L_t and C_d combination to meet these requirements. Find the output voltage as a function of the switching frequency for your choice.

18. A ZCS boost converter is desired, along the lines of Figure 8.36. The input current is 2 A, and the input voltage is 12 V.

   a. What is the sequence of configurations if ZCS action takes place? Which configuration is free to be extended for control?

   b. Will the component values shown in the figure support ZCS action of the transistor? If so, how much time is spent in each configuration?

   c. What is f_{switch} for 48 V output?

19. A general quasi-resonant ZVS buck converter is shown in Figure 8.37. A converter suitable for 48 V to 12 V conversion with output power of 200 W and a switching frequency of 400 kHz is desired. Find values of C_t and L_d to meet these requirements. What is the gate control signal duty ratio?

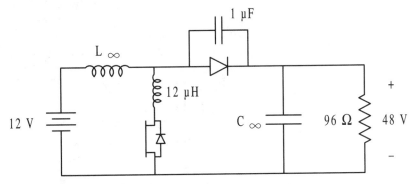

Figure 8.36   A proposed ZCS boost converter.

20.   A buck converter intended for ZVS action has $C_t = 0.1\ \mu F$ and $L_d = 0.1\ \mu H$. The input is 12 V, and the output is intended to be 5 V at 100 W. Find the operating frequency and gate signal duty ratio to produce the desired action. Will ZVS action continue if the load drops to 10 W?

21.   A ZVS buck converter has $V_{in} = 300$ V, $C_t = 2\ \mu F$. and $L_d = 4\ \mu H$. What is the minimum load to support ZVS action if the intended output voltage is 100 V? Find the output voltage as a function of the switching frequency.

22.   A ZVS circuit has been proposed for 5 V to 2 V conversion, at currents between 5 A and 25 A. Find values of C_t and L_d that support this conversion. With your choice, what is the peak voltage that must be blocked by the transistor and diode? A switching frequency above 200 kHz is recommended.

23.   Figure 8.38 shows a flyback converter, with its leakage inductance, intended to operate with transistor ZVS action.

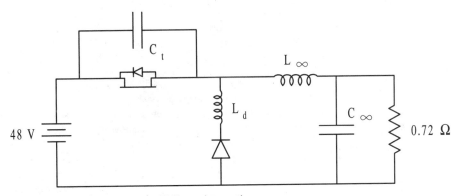

Figure 8.37   Buck converter for ZVS transistor action.

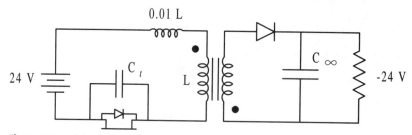

Figure 8.38 Flyback converter with leakage inductor and C_t for soft switching.

    a. Assuming ZVS action occurs, describe the sequence of configurations and the circuit action in each configuration. Establish rules for Z_c values that will support soft switching.

    b. Choose a value for C_t that will support ZVS action. What is the average output voltage as a function of the switching frequency?

24. An inverter for a naval application is to supply up to 2 MW at 2 kV$_{RMS}$ and 400 Hz. The power and current levels are such that an SCR series-resonant inverter is being considered. You have been asked to present a brief analysis and a base design.

    a. Draw the circuit. The load can be represented with a resistor.

    b. Select L and C values to support resonant action. SCRs with peak blocking capability of 6 kV are available. The output waveform should have as little distortion as possible, but the SCR rating limits must not be exceeded and resonant action must be supported. Determine the voltage and current ratings needed for the inductor and capacitor.

    c. By simulation or other means, plot the output voltage waveform as a function of time.

25. In a certain boost converter, none of the values L_t, C_t, L_d, or C_d can be neglected. The converter has 5 V input, and is intended to produce 15 V output at loads ranging between 0.1 A and 0.5 A. The values are $L_t = 15$ μH, $C_t = 3500$ pF, $L_d = 1$ μH, $C_d = 5000$ pF. Do these values support either ZCS or ZVS action? If so, what switching frequency should be used to produce 15 V at the low and high load levels? If not, can ZCS or ZVS action be supported through the addition of extra inductance in series with L_t? What value?

REFERENCES

A. K. S. Bhat, "A resonant converter suitable for 650 V bus operation," *IEEE Trans. Power Electronics*, vol. 6, no. 4, pp. 739–748, October 1991.

R. Farrington, M. M. Jovanović, and F. C. Lee, "A new family of isolated converters that uses the magnetizing inductance of the transformer to achieve zero-voltage switching," *IEEE Trans. Power Electronics*, vol. 8, no. 4, pp. 535–545, October 1993.

S. Freeland, R. D. Middlebrook, "A unified analysis of converters with resonant switches," in *IEEE Power Electronics Specialists Conf. Rec.*, 1987.

M. K. Kazimierczuk, "Steady-state analysis and design of a buck zero-current-switching resonant dc/dc converter," *IEEE Trans. Power Electronics*, vol. 3, no. 3, pp. 286–296, July 1988.

M. K. Kazimierczuk, D. Czarkowski, *Resonant Power Converters*. New York: Wiley, 1995.

A. I. Pressman, *Switching Power Supply Design*. New York: McGraw-Hill, 1991.

W. A. Tabisz, F. C. Lee, "Zero-voltage-switching multiresonant technique—a novel approach to improve performance of high-frequency quasi-resonant converters," *IEEE Trans. Power Electronics*, vol. 4 no. 4, October 1989.

R. Watson, F. C. Lee, G. C. Hua, "Utilization of an active-clamp circuit to achieve soft switching in flyback converters," *IEEE Trans. Power Electronics*, vol. 11, no. 1, pp. 162–169, January 1996.

CHAPTER 9

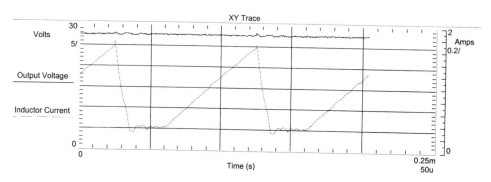

Experimental traces for boost converter in discontinuous mode.

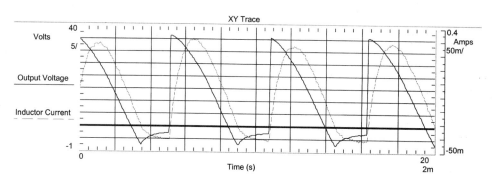

Experimental traces for three-pulse rectifier in discontinuous mode.

Figure 9.1 Internal voltage and current waveforms in several types of converters.

DISCONTINUOUS MODES

9.1 INTRODUCTION

In the converters examined so far, KVL and KCL play perhaps a dominant role in determining switch action. Almost all these converters have switch action that is carefully predetermined and orchestrated with controls to perform a desired function. For nonresonant converters, the duty ratios and phases provided the ability to make adjustments. In the resonant converters, the additional restriction of creating a close match to circuit resonance requirements helped to determine the switch operation.

An important exception was found in Section 7.8. There, the ac regulator was discussed. This circuit takes advantage of load properties to alter power flow. It does not really do switching conversion in a general sense, since there is no way to convert frequency with this circuit.

In this chapter, we will consider power electronic circuits, with primary focus on the dc–dc case, in which the load is important to the switch action. KVL and KCL determine switch action when sources are to be interconnected; here we consider cases in which source interconnection is not always a factor. Converters of this type are said to operate in *discontinuous mode* because voltages and currents reach zero during some time intervals, and new switch combinations not normally allowed by KVL and KCL become valid. Rectifier and dc–dc converter waveforms typical of this mode are shown in Figure 9.1. Discontinuous mode operation involves design tradeoffs different from those in other converters. This can be an advantage in some special cases, and has been used for innovative dc–dc converters.

In many practical converters, discontinuous mode occurs when loads are light. The new switch combinations alter the output behavior; often the output voltage jumps up significantly at light load.

9.2 DC–DC CONVERTERS ACTING IN DISCONTINUOUS MODE

9.2.1 The Nature of Discontinuous Mode

Consider the buck-boost dc–dc converter shown in Figure 9.2. The analysis in Chapter 4 treats the inductor as a current source. However, the switching action of the converter does not require any particular value of current. Indeed, any positive inductor current will cause this circuit to act in much the same way as an ideal transfer current source.

Discontinuous mode behavior occurs when the inductor is too small to maintain the current flow. When the diode turns on, energy is drawn from the inductor to supply the load. If the inductor is small, the energy is quickly exhausted, and the inductor current reaches zero. In Figure 9.2, an inductor current of zero will cause the diode to turn off prior to transistor switching. There will be a time interval in which neither switch is on. The switching functions q_1 and q_2 will no longer cover a full period, and $q_1 + q_2 < 1$. This simple change has a substantial effect on the converter output. With both switches off, the output shows an exponential decay. The relationships used in Chapter 4 to describe the converter may not be valid under these conditions.

Consider two examples, given below. In the first, it is shown that if the inductor is large enough to maintain current flow, the relationships from Chapter 4 will apply even though the inductor is not infinite. In the second, smaller values of inductance are explored.

Example 9.2.1   The buck-boost converter in Figure 9.2 is to be considered with various inductor values. Assume that the output capacitor is large. The input voltage is 12 V and the output is -12 V. The load current $I_{\text{load}} = 20$ A is nearly constant since the capacitor is large. Determine the converter's operation with 100 kHz switching for (a) an infinite inductor, (b) a 10 μH inductor, and (c) a 0.75 μH inductor. Comment.

If the inductor is very large, it can be represented in periodic steady state as a current source. The switches manipulate the transfer voltage v_T, the input current i_{in}, and the diode current i_d. The inductor current must have a path, so the switches must work in alternation with $q_1 + q_2 = 1$. As in Chapter 4,

$$v_T = q_1 V_{\text{in}} + q_2 V_{\text{out}}, \qquad \langle v_T \rangle = D_1 V_{\text{in}} + D_2 V_{\text{out}}$$

$$i_{\text{in}} = q_1 I_L, \qquad \langle i_{\text{in}} \rangle = D_1 I_L \tag{9.1}$$

$$i_d = q_2 I_L, \qquad \langle i_d \rangle = D_2 I_L$$

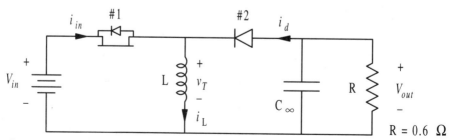

Figure 9.2   Buck-boost converter for Example 9.2.1.

For the choice of voltage shown, the duty ratios must be 50%, since $\langle v_T \rangle = 0$. The load current is 20 A, so the value $\langle i_d \rangle = 20$, and the inductor current will be a constant 40 A. For the 10 μH inductor, some ripple current is expected. The average voltage across the inductor must still be zero, so the relationship

$$0 = D_1 V_{in} + D_2 V_{out} \tag{9.2}$$

is still valid. Consider the currents. The inductor current is no longer constant, and

$$i_{in} = q_1 i_L(t), \qquad \langle i_{in} \rangle = \langle q_1 i_L \rangle$$
$$i_d = q_2 i_L(t), \qquad \langle i_d \rangle = \langle q_2 i_L \rangle \tag{9.3}$$

The value $\langle q_1 i_L \rangle$ is needed, and it is not obvious that it will be $D_1 \langle i_L \rangle$. Take the case when the transistor is on. Then the inductor voltage is 12 V, and the current will change linearly according to $v_L = L(di/dt)$ during the 5 μs interval $D_1 T$. The peak-to-peak current ripple is therefore 6 A—a small fraction of 40 A. Since the current is always well above zero, the switches must act to provide a current path, and $q_1 + q_2 = 1$, just as before. The duty ratios will still be 50%. In Figure 9.3, the inductor current, input current, and diode current are shown. The diode current must still have an average value of 20 A to be consistent with the load current, but of course the actual waveform is not a square wave. The actual inductor current varies between 37 A and 43 A, although the average is still 40 A. Notice that the triangular waveform gives the result

$$\langle i_d \rangle = \langle q_1 i_L \rangle = D_1 \langle i_L \rangle \tag{9.4}$$

Even though the waveforms exhibit current ripple, the converter with $L = 10$ μH still has exactly the same average behavior as the converter with $L \to \infty$.

What about $L = 0.75$ μH? With this choice, and with $D_1 = 0.5$, the configuration with the transistor on means $v_L = L\, di/dt$, or 12 V $= (0.75 \times 10^{-6}$ H$)\, di/dt$. The current rises by 80 A. When the diode is on, the inductor current falls by 80 A. The average diode current must be 20 A, and the current waveforms therefore will be those of Figure 9.4. In spite of the extreme variation, the figure shows that the switches must maintain a current path for the inductor to satisfy KVL and KCL. Thus $q_1 + q_2 = 1$, and

$$0 = D_1 V_{in} + D_2 V_{out}, \qquad D_1 + D_2 = 1, \qquad \frac{D_1}{1 - D_1} V_{in} = -V_{out} \tag{9.5}$$

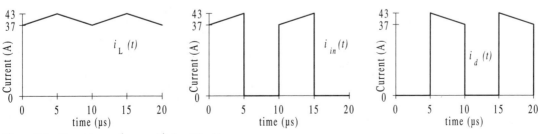

Figure 9.3 Current waveforms with $L = 10$ μH.

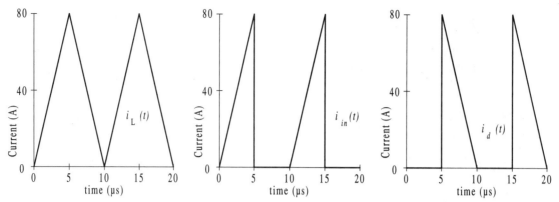

Figure 9.4   Current waveforms with $L = 0.75\ \mu H$.

The output is still -12 V, the average inductor current is still 40 A, and in an average sense, the converter performs in essentially the same manner as when $L \to \infty$.

The example demonstrates that the basic behavior of a buck-boost converter is maintained as long as the inductor ensures some current flow. This is true of any dc–dc converter: If the inductor in a given converter is sufficiently large, the switch matrix must provide a path for its current at all times. The switches must operate just as they would for an infinite inductor—or an ideal dc current source. If the inductor is not sufficient, converter operation changes substantially. The inductor current will ramp down all the way to zero during part of a period. The following example explores the results.

Example 9.2.2   Consider again the buck-boost converter of the previous example. The transistor duty ratio is set to 50%, while the load is a large capacitor in parallel with a 0.6 Ω resistor. The switching frequency is 100 kHz. If the inductor is 0.5 μH, how will the circuit operate?

In this case, the inductor current varies in the extreme. To decipher the action, let us start with no inductor current, then turn the transistor on for 5 μs. The inductor voltage is 12 V during this configuration, so the current changes by (12 V) $\Delta t/L = 120$ A. After 5 μs, the current has ramped to 120 A. The energy stored in the inductor after 5 μs is $\frac{1}{2}Li^2 = 3.6$ mJ, and the average input current is 30 A. Once the transistor turns off, the diode can conduct and the inductor energy transfers to the load.

Let us guess that the output voltage is still -12 V. Then when the transistor is off, the inductor current would ramp down to 0 A in 5 μs. The current i_L would be a triangle waveform with peak value of 120 A. The average inductor current would be 60 A. But this is inconsistent: Earlier an average inductor current of 40 A was found. A higher current requires a higher output voltage. If the load current is forced to be higher than 20 A, the current waveforms given in Figure 9.5 will appear. In this case, the inductor current is steady at zero during the final portion of each cycle.

How can the output voltage be determined in this new case? The average input current can be found by integrating $i_{in}(t)$ from Figure 9.5, and $\langle i_{in} \rangle$ has a value of 30 A. The input power is therefore 360 W, and the output power must match this. If $P_{out} = 360$ W $= V_{out}^2/R$, then the output with a 0.6 Ω resistor will be $-\sqrt{360 \times 0.6} = -14.7$ V. The average power in the inductor is zero, so the average inductor voltage must be zero. This relationship still requires

$$0 = D_1 V_{in} + D_2 V_{out} \tag{9.6}$$

Since now there are times when $i_L = 0$, during these times the diode carries no current, and both switches are off. In this case, $D_1 + D_2 < 1$. All energy is removed from the inductor in a time less than $(1 - D_1)T$. In this example, the time D_2T is the time required to remove 3.6 mJ from the inductor. With the diode on, $v_L = -14.7$ V $= L\ \Delta i/\Delta t$. Since the current must ramp down to zero, the change $\Delta i = -120$ A, so $\Delta t = 120 \times 0.5 \times 10^{-6}/14.7 = 4.08\ \mu s$. Thus $D_1 = 0.5$ and $D_2 = 0.408$; the ratio $V_{out}/V_{in} = -0.5/0.408 = -1.22$, and the output magnitude is higher than would be expected just from D_1.

In Example 9.2.2, the inductor current is said to be *discontinuous* (although strictly speaking it is not discontinuous in the mathematical sense). The converter action has changed from the large inductor case, and the new action defines the discontinuous mode.

> ***Definition:*** *A dc–dc converter is defined as operating in* discontinuous mode *if its inductors are too small to maintain a current flow at all times. In discontinuous mode, there are times when extra switches will turn off because no current path is needed for KCL.*

One key property of discontinuous mode is that the converter's output becomes load dependent. In the first buck-boost example above, a load change from 0.6 Ω to 0.5 Ω will have no effect on the output voltage or basic converter operation. In the discontinuous ex-

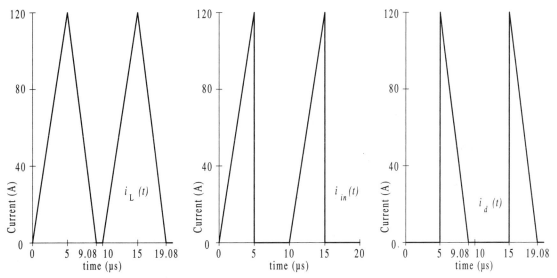

Figure 9.5 Current waveforms with $L = 0.5\ \mu H$.

ample, a load change from 0.6 Ω to 0.5 Ω will alter the output voltage from -14.7 V to -13.4 V (this computation is a useful exercise).

9.2.2 Discontinuous Mode Relationships for dc–dc Converters

Basic relationships and waveforms for discontinuous mode are given in Figure 9.6 for the three most common dc–dc converters. In each case, $D_1 + D_2 < 1$. The value of D_2 becomes an additional unknown: The diode turns off as soon as the energy in the inductor is exhausted. The new information that makes it possible to find this added unknown is the fact that all the inductor's energy is removed during each cycle. Since the switches are lossless, the power balance $P_{in} = P_{out}$ provides an additional equation.

One caveat is very important to remember. If all the switches in a converter are off, "stub elements" appear: unterminated energy storage devices that are expected to be open circuits. In reality, there are small capacitances and inductances throughout a circuit. The converter will begin to act as a resonant circuit, just as was done by design in Chapter 8. The results in this section neglect the effects of this stray element resonance. The implicit assumption here is that stray storage components store negligible energy, and have little effect on the conversion process.

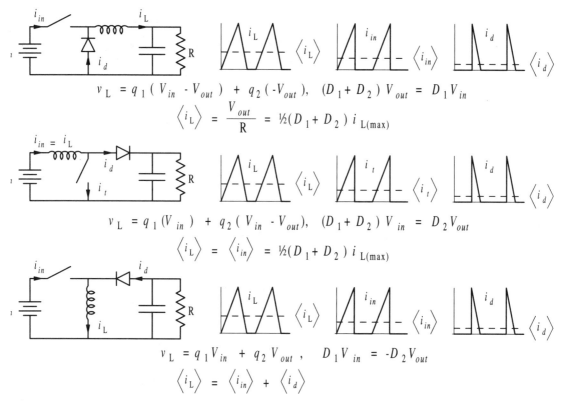

$$v_L = q_1(V_{in} - V_{out}) + q_2(-V_{out}), \quad (D_1 + D_2)V_{out} = D_1 V_{in}$$

$$\langle i_L \rangle = \frac{V_{out}}{R} = \tfrac{1}{2}(D_1 + D_2)\, i_{L(max)}$$

$$v_L = q_1(V_{in}) + q_2(V_{in} - V_{out}), \quad (D_1 + D_2)V_{in} = D_2 V_{out}$$

$$\langle i_L \rangle = \langle i_{in} \rangle = \tfrac{1}{2}(D_1 + D_2)\, i_{L(max)}$$

$$v_L = q_1 V_{in} + q_2 V_{out}, \quad D_1 V_{in} = -D_2 V_{out}$$

$$\langle i_L \rangle = \langle i_{in} \rangle + \langle i_d \rangle$$

Figure 9.6 Discontinuous mode relationships for buck, boost, and buck-boost converters.

For the individual converters, relationships can be found between input and output. When its active switch (the transistor) is on, the buck converter ramps the inductor current up from zero to a peak value i_{peak}, to give the current relationship

$$V_{in} - V_{out} = L\frac{di}{dt} = L\frac{i_{peak}}{D_1 T}, \qquad i_{peak} = \frac{D_1 T}{L}(V_{in} - V_{out}) \tag{9.7}$$

The input current $i_{in} = q_1 i_L$ is a sequence of triangles just like the input current of the buck-boost converter. The average current can be computed as

$$\langle i_{in} \rangle = \frac{D_1}{2} i_{peak} \tag{9.8}$$

Combining equations (9.7) and (9.8), the average input current becomes

$$\langle i_{in} \rangle = (V_{in} - V_{out})\frac{D_1^2 T}{2L} \tag{9.9}$$

Since the input voltage is constant, the average input power $P_{in} = V_{in}\langle i_{in} \rangle$. Substituting equation (9.9), the input power is

$$P_{in} = (V^2_{in} - V_{in}V_{out})\frac{D_1^2 T}{2L} \tag{9.10}$$

The output power, $P_{out} = V_{out}^2/R$, must match P_{in}. This provides an additional relationship for the converter,

$$\frac{V^2_{out}}{R} = D_1^2 \frac{T}{2L}(V^2_{in} - V_{in}V_{out}) \tag{9.11}$$

The output voltage for low inductance is the solution of the quadratic equation in V_{out} represented by (9.11). Of course V_{out} must be positive because switch 2 is normally a diode, so the result is

$$\text{buck:} \quad V_{out} = \frac{-D_1^2 V_{in} RT}{4L} + D_1 V_{in}\sqrt{\frac{RT}{2L} + \frac{R^2 T^2 D_1^2}{16L^2}} \tag{9.12}$$

For the boost converter, the transistor ramps the inductor current from 0 to a peak value. Here is a summary of the peak current, average input current, and input power relationships:

$$V_{in} = L\frac{di}{dt} = L\frac{i_{peak}}{D_1 T}, \qquad i_{peak} = \frac{V_{in}D_1 T}{L}$$

$$\langle i_{in} \rangle = (D_1 + D_2)\frac{i_{peak}}{2}, \qquad \langle i_{in} \rangle = (D_1 + D_2)\frac{V_{in}D_1 T}{2L} \tag{9.13}$$

$$P_{in} = V_{in}\langle i_{in} \rangle, \qquad P_{in} = (D_1 + D_2)\frac{V^2_{in}D_1 T}{2L} = \frac{V^2_{out}}{R}$$

In these expressions, both V_{out} and D_2 are unknown, and $D_1 + D_2 < 1$. A second equation for D_2 can be obtained by considering the diode current. This is $q_2 i_L$, a sawtooth wave with a peak value of i_{peak} and time width of $D_2 T$. The average diode current matches the load current V_{out}/R, so

$$\langle i_{out} \rangle = \frac{V_{out}}{R} = D_2 \frac{i_{peak}}{2}, \qquad D_2 = \frac{2 V_{out} L}{R V_{in} D_1 T} \tag{9.14}$$

The expression for $P_{in} = P_{out} = V_{out}^2/R$ can be used to eliminate D_2, giving the quadratic expression

$$V_{out}^2 = \left(D_1 + \frac{2 V_{out} L}{R V_{in} D_1 T} \right) \frac{V_{in}^2 D_1 T R}{2L} = \frac{V_{in}^2 D_1^2 T R}{2L} + V_{out} V_{in} \tag{9.15}$$

The final result for V_{out} with small L is

$$\text{boost:} \quad V_{out} = \frac{V_{in}}{2} + \frac{V_{in}}{2} \sqrt{1 + \frac{2 D_1^2 R T}{L}} \tag{9.16}$$

For the buck-boost converter, the transistor causes the inductor current to ramp up to a current i_{peak}. Following the expressions from the preceding cases, the relationships are

$$V_{in} = L \frac{di}{dt} = L \frac{i_{peak}}{D_1 T}, \qquad i_{peak} = \frac{V_{in} D_1 T}{L}$$

$$\langle i_{in} \rangle = D_1 \frac{i_{peak}}{2}, \qquad \langle i_{in} \rangle = \frac{V_{in} D_1^2 T}{2L} \tag{9.17}$$

$$P_{in} = V_{in} \langle i_{in} \rangle, \qquad P_{in} = \frac{V_{in}^2 D_1^2 T}{2L} = \frac{V_{out}^2}{R}$$

Since V_{out} is known to be negative, the output for small inductance will be

$$\text{buck-boost:} \quad V_{out} = -D_1 V_{in} \sqrt{\frac{RT}{2L}} \tag{9.18}$$

In each of these cases, a "small inductor" is defined as one in which the inductor current is zero at least part of each cycle, such that when the transistor turns on, its current always ramps up from zero. Notice that the parameter RT/L is present in all the final expressions. The time constant ratio T/τ, $\tau = L/R$, plays a role in determining whether an inductor is small in the sense used here.

Discontinuous mode is avoided by some power electronics engineers, but embraced by others. Some of its characteristics show why there is disagreement:

Discontinuous Mode Advantages:

- Very fast response. The converter starts each cycle with no stored inductor energy, and makes its full current transition immediately.
- The inductor values are low. Given the cost and expense of inductors, the reduced value of L implies lower cost.
- Discontinuous mode tends to provide a higher output voltage. If boosting is needed, the extra boost can be useful.

Discontinuous Mode Disadvantages:

- Load regulation problems. The load resistor R appears in all the expressions. At light loads, the output voltage tends to increase. This can be handled with feedback (Chapter 15), but negates the naturally good load regulation behavior of dc–dc converters.
- High current variation in an inductor will increase losses in its magnetic parts, and can lead to magnetic saturation.
- The tendency of the output voltage to increase at low power levels can add reliability problems at the load.

In converters that operate at very low power levels, discontinuous mode usually must be planned into the operating strategy.

If a designer wants to avoid discontinuous mode completely, the usual solution is a ballast load.

> **Definition:** A ballast load *is a resistive load, permanently connected across the output terminals of a converter, designed to keep the ratio* RT/L *low enough to avoid discontinuous mode.*

The ballast load resistor in a typical converter serves as a guaranteed minimum load, and allows the designer to choose an inductor to avoid discontinuous operation. Typical ballast loads use a small fraction of rated converter output power, so that full-load efficiency is not sacrificed too much.

9.2.3 Critical Inductance

There is an operating boundary in inductance between normal and discontinuous operating modes. If L is large enough, a KCL path is necessary at all times. If L is too small, its current reaches zero. The term *critical inductance* defines the transition value.

> **Definition:** The critical inductance, L_{crit}, *is defined as the lowest value of inductance for which* $i_L > 0$ *at all times.*

From the preceding examples, the critical inductance is just sufficient to avoid discontinuous mode. If $L > L_{crit}$, the inductor current will be greater than zero when the transistor turns on, and the converter relationships are the average expressions from Chapter 4. If $L < L_{crit}$ the relationships in equations (9.12), (9.16), and (9.18) will apply instead.

The critical inductance is usually straightforward to calculate. When $L = L_{crit}$, the inductor waveform is a triangle with a minimum current $i_L = 0^+$, so both the square root relationships and the conventional duty ratio relationships are valid simultaneously. The inductor current will ramp from 0 to a peak value exactly equal to twice its average. For a buck converter, the critical inductor can be found by setting $V_{out} = D_1 V_{in}$ in equation (9.11). This gives

$$\text{buck:} \quad \frac{D_1^2 V_{in}^2}{R} = D_1^2 \frac{T}{2L_{crit}}(V_{in}^2 - D_1 V_{in}^2), \qquad L_{crit} = \frac{RT}{2}(1 - D_1) \qquad (9.19)$$

For a boost converter, the value $V_{out} = V_{in}/(1 - D_1)$ can be substituted into equation (9.15), to give

$$\text{boost:} \quad \frac{V_{in}^2}{(1 - D_1)^2} = \frac{V_{in}^2 D_1^2 TR}{2L_{crit}} + \frac{V_{in}^2}{1 - D_1}, \qquad L_{crit} = \frac{D_1(1 - D_1)^2 RT}{2} \qquad (9.20)$$

For a buck-boost converter, the value $V_{out} = V_{in} D_1/(1 - D_1)$ can be substituted into (9.18) to give

$$\text{buck-boost:} \quad \frac{V_{in}^2 D_1^2}{(1 - D_1)^2} = \frac{V_{in}^2 D_1^2 RT}{2L_{crit}}, \qquad L_{crit} = \frac{(1 - D_1)^2 RT}{2} \qquad (9.21)$$

These expressions show that critical inductance is affected by load and by switching time. Faster switching means smaller inductors for a given converter performance. In many problems it is almost as easy to derive the critical inductance directly rather than trying to apply equations (9.19)–(9.21).

Example 9.2.3 A dc–dc converter provides 48 V to 5 V conversion at a nominal power level of 100 W. If the switching frequency is 100 kHz, what is the critical inductance for this converter?

The converter is most likely a buck circuit, as in Figure 9.7. If $L > L_{crit}$, the duty ratio $D_1 = 5/48$, and $D_2 = 43/48$. The average inductor current should be 20 A to provide the 100 W output. If $L = L_{crit}$, the inductor current ramps from 0 to twice the average value. The peak current is therefore 40 A. With the transistor, the inductor voltage is $48 - 5 = 43$ V, and

$$43 \text{ V} = L\frac{di}{dt} = L\frac{\Delta i}{\Delta t}, \qquad 43 \text{ V} = L_{crit}\frac{40}{D_1 T}$$

$$(9.22)$$

$$L_{crit} = \frac{43}{40} \times \frac{5}{48} \times 10^{-5} = 1.12 \ \mu\text{H}$$

Example 9.2.4 Consider a buck-boost converter with 24 V input and -12 V output, shown in Figure 9.8. The load is 60 W. Find the critical inductance as a function of switching frequency.

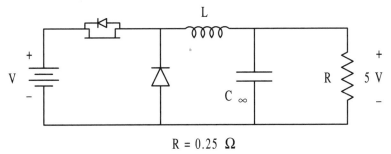

Figure 9.7 Buck converter for Example 9.2.3.

In this converter, the average output current must be 5 A to supply the 60 W load. The input current must be 2.5 A by conservation of energy. The average inductor current is the sum, 7.5 A, provided $L \geq L_{\text{crit}}$. If the inductor is exactly equal to its critical value, the inductor current will be a triangular waveform between 0 and 15 A. The duty ratio D_1 should be 1/3 to provide 24 V to 12 V conversion. With the transistor on,

$$V_{\text{in}} = L\frac{di}{dt} = L\frac{\Delta i}{\Delta t}, \qquad 24\text{ V} = L_{\text{crit}}\frac{15}{D_1 T}, \qquad L_{\text{crit}} = \frac{8}{15}T \tag{9.23}$$

The frequency relationships are given in Table 9.1.

Example 9.2.5 A boost converter, illustrated in Figure 9.9, provides 48 V to 200 V conversion. The inductor value is 15 μH, and the switches operate at 50 kHz. What is the minimum load power that must be maintained so that $L \geq L_{\text{crit}}$?

Since $L \geq L_{\text{crit}}$ by design in this problem, the duty ratio $D_2 = 48/200 = 0.24$, and $D_1 = 0.76$. When the active switch is on, the inductor sees the 48 V input. When $L = L_{\text{crit}}$, the inductor voltage ramps from 0 to a peak value equal to twice the average inductor current. Then

$$48\text{ V} = L\frac{di}{dt} = L\frac{\Delta i}{\Delta t}, \qquad 48\text{ V} = L_{\text{crit}}\frac{2\langle i_{\text{in}}\rangle}{D_1 T} \tag{9.24}$$

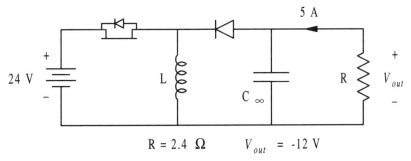

Figure 9.8 Buck-boost converter for Example 9.2.4.

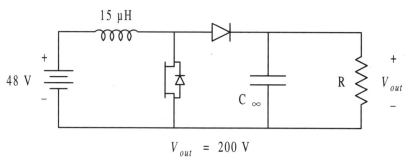

15 µH

48 V

C_{∞}

R V_{out}

V_{out} = 200 V

Figure 9.9 Boost converter for Example 9.2.5.

The lowest average current that can be maintained with $L \geq L_{crit}$ is that for which $L_{crit} = 15\ \mu H$, so

$$48\ V \times 0.76 \times 20\ \mu s = 15\ \mu H \times 2\langle i_{in} \rangle, \qquad \langle i_{in} \rangle = 24.32\ A \qquad (9.25)$$

The input power (and hence the load power) must be at least $(48\ V) \times (24.32\ A) = 1167\ W$ to ensure $L \geq L_{crit}$.

The examples so far address converters operating under a specific set of conditions. In an actual application, a converter will encounter input voltage variation, load variation, or other variable conditions. The definition of critical inductance is not quite applicable, and needs to be extended.

> **Definition:** *The* critical inductance for a converter, L_{crit}, *is the smallest inductance for which* $i_L > 0$ *at all times, and* under all allowed operating conditions *of the converter.*

The implications should be clear. If, for a converter, $L \geq L_{crit}$, then the converter will never operate in discontinuous mode. To avoid this mode, it is sufficient to ensure that $L \geq L_{crit}$. When $L < L_{crit}$, there will be some conditions under which discontinuous mode operation is observed. As the notation implies, it is rare to distinguish between critical inductance and critical inductance for a converter. In general, L_{crit} will refer to the latter.

TABLE 9.1
Critical Inductance for Various f_{switch} from
Example 9.2.4

Switching Frequency	Critical Inductance
1 kHz	533. μH
10 kHz	53.3 μH
50 kHz	10.7 μH
100 kHz	5.33 μH
200 kHz	2.67 μH

Example 9.2.6 A buck converter is designed for nominal 48 V input and 5 V output. It switches at 100 kHz. In reality the input can be anywhere between 30 V and 60 V. The load power ranges between 10 W and 200 W. What is L_{crit} for this converter? Conversely, what is the maximum inductance that will ensure discontinuous mode under *all* allowed conditions? The circuit is the same as Figure 9.7, except for the line and load variations.

The input voltage and output load current are no longer known, but it is still true that $V_{out} = D_1 V_{in}$ if $L \geq L_{crit}$. With the active switch on, $D_1 = 5/V_{in}$, and

$$V_{in} - V_{out} = L\frac{di}{dt}, \qquad V_{in} - 5 = L\frac{\Delta i}{(5/V_{in})T} \tag{9.26}$$

Under one particular set of conditions, the inductor $L = L_{crit}$ will produce the familiar triangle wave with a minimum at zero and a maximum at $2\langle i_{out}\rangle$. Under these conditions,

$$V_{in} - 5 = L_{crit}\frac{2\langle i_{out}\rangle}{(5/V_{in})T}, \qquad \langle i_{out}\rangle = \frac{P_{out}}{5} \tag{9.27}$$

The critical inductance can be found from (9.27) as

$$L_{crit} = \left(5T - \frac{25T}{V_{in}}\right)\frac{5}{2P_{out}} = \frac{1.25 \times 10^{-4}}{P_{out}} - \frac{6.25 \times 10^{-4}}{P_{out}V_{in}} \text{ H} \tag{9.28}$$

The converter never operates in discontinuous mode if we choose

$$L \geq L_{crit} = \frac{1.25 \times 10^{-4}}{P_{out}} - \frac{6.25 \times 10^{-4}}{V_{in}P_{out}} \text{ H} \tag{9.29}$$

In this case, the worst-case combination is for the minimum output power and maximum line voltage. Let $P_{out} = 10$ W, and $V_{in} = 60$ V. Then $L \geq 11.46$ μH. The critical inductance for this converter is 11.46 μH.

We might also choose to ensure discontinuous mode under all circumstances. In this case, the choice would be

$$L < \frac{1.25 \times 10^{-4}}{P_{out}} - \frac{6.25 \times 10^{-4}}{V_{in}P_{out}} \tag{9.30}$$

The worst-case combination now is at the maximum output power and minimum input voltage. Then $L < 0.521$ μH. An inductor below 0.521 μH will ensure discontinuous mode operation under all allowed conditions. For inductors between 0.521 μH and 11.46 μH, some but not all combinations of line and load will give discontinuous mode.

Critical inductance is very useful for design in a variety of ways. For example, the worst-case current ripple when $L = L_{crit}$ is $\pm 100\%$. An inductor equal to $100L_{crit}$ will produce $\pm 1\%$ current ripple under worst-case conditions. It is fairly easy to specify inductor requirements once L_{crit} is known.

9.2.4 Critical Capacitance

Just as a large inductor stores enough energy to require attention to KCL throughout a switching period, so a large capacitor stores enough energy to enforce KVL restrictions. As was shown in Chapter 5, there are practical rectifiers that attempt to violate KVL with large capacitors present. The large current spikes that result from this practice are undesirable. The concept of critical capacitance, exactly analogous to critical inductance, provides a way to examine the KVL issues in a converter.

> **Definition:** Critical capacitance, C_{crit}, is defined as a capacitance large enough to maintain $v_C > 0$ at all times in a given power electronic circuit. The critical capacitance for a converter is that value of capacitance that maintains $v_C > 0$ under all allowed operating conditions.

Critical capacitance is encountered much less in practice than L_{crit} for two reasons. First, capacitors often serve as primary interfaces for voltage sources. Since a user usually expects voltage source behavior at the output of a power electronic circuit, the voltage ripple should be low. This requires $C \gg C_{crit}$. Second, capacitors are not often used as transfer sources. The exception is the boost-buck converter.

When subcritical capacitors are used, KVL requirements do not always need to be enforced. Typically, this means that switches normally acting in alternation will be on simultaneously once the capacitor voltage reaches zero. In general, the behavior is the dual of discontinuous mode.

Example 9.2.7   Figure 9.10 shows a boost-buck converter intended for 24 V to −12 V conversion. The switching frequency is 200 kHz, and the output power is a resistive load that draws 120 W under these conditions. What is the critical capacitance? If the duty ratio is kept constant at 1/3, what is the output voltage if $C = C_{crit}/2$?

If $C = C_{crit}$, the converter should follow conventional average behavior. Since the output power is 120 W, the output current source has a value of 10 A, while the input source

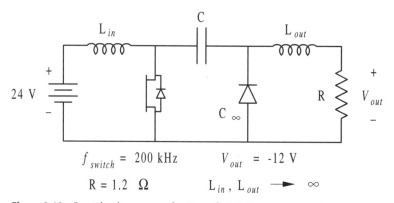

Figure 9.10   Boost-buck converter for Example 9.2.6.

is 5 A. The load resistance is 1.2 Ω. The transfer voltage on average should be given by V_{in}/D_2, or 36 V (the sum of the input and output). If $C = C_{crit}$, the capacitor voltage will vary with a triangular waveform, and a peak value equal to double the average. The peak will be 72 V. When the transistor is on, the diode is off. The 10 A output current will flow in the capacitor. The relationships are

$$i_C = C\frac{dv}{dt} = C\frac{\Delta v}{\Delta t}, \qquad 10 = C_{crit}\frac{72}{D_1 T}, \qquad C_{crit} = 0.2315 \ \mu F \qquad (9.31)$$

Now, set $C = 0.1157 \ \mu F$, and examine the behavior. Since $C < C_{crit}$, the voltage v_C will ramp down to zero while the transistor is on. If v_C ramps down any farther, the diode will become forward biased, and both switches will be on simultaneously. The duty ratios give $D_1 + D_2 > 1$. During the time when the diode alone is on, the capacitor current is the input current P_{in}/V_{in}, and the capacitor voltage ramps from 0 to $2(V_{in} + V_{out})$. This time interval is $(1 - D_1)T$, or 3.333 μs. Thus

$$\frac{P_{in}}{V_{in}} = C\frac{\Delta v}{\Delta t} = C\frac{2(V_{in} + V_{out})}{3.33 \ \mu s}, \qquad P_{out} = P_{in} = \frac{V_{out}^2}{R}$$

$$\frac{V_{out}^2}{V_{in}R} = 0.06944(V_{in} + V_{out}), \qquad V_{out}^2 = 48 + 2V_{out} \qquad (9.32)$$

$$V_{out} = 8 \ V$$

The input and output power have dropped off to $53\frac{1}{3}$ W, and the input current is 2.22 A. Unlike the inductor-based converters, the boost-buck converter output falls if the capacitor is too small.

9.3 RECTIFIERS AND OTHER CONVERTERS IN DISCONTINUOUS MODE

9.3.1 Rectifiers

The concepts of critical inductance and capacitance are usually considered in the context of dc–dc converters. However, the basic ideas are relevant whenever dc source characteristics are necessary. Consider the midpoint rectifier in Figure 9.11. This rectifier will act according to ideal phase control relationships if the inductor is big enough to maintain $i_L > 0$. There is a problem, however. The value L_{crit} is proportional to the switching period. All other things being equal, L_{crit} for a 60 Hz circuit is 1000 times that for a 60 kHz circuit.

Many practical rectifiers operate with subcritical inductance simply because the critical value of L can be prohibitive. As in the dc–dc case, an inductance $L < L_{crit}$ will create time intervals during which KCL paths are not needed, and all switches are off. Once the switches are off, the output voltage is determined by the load rather than by the switch action. The output voltage simply reflects the open-circuit behavior of the load during all-off intervals. When SCRs are used for rectification, subcritical inductance means that each device will turn off prior to the turn-on signal for the subsequent switch. When diodes are used, the trial method provides the necessary tool. In any case, small inductance ensures that the

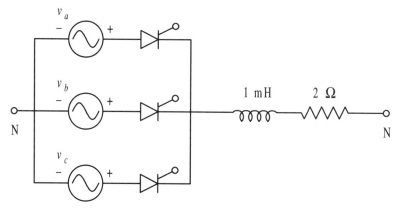

Figure 9.11 Midpoint rectifier with subcritical inductor.

current always starts from $i_L = 0$. The waveform will have periodic steady-state action, but each cycle can be examined entirely independent of other cycles.

Consider the midpoint rectifier, acting with subcritical inductance, with a series L-R load and SCR switches. For an m-phase input based on $V_0 \cos(\omega t)$, the turn-on time t_{on} corresponds to an angle $\omega t_{on} = -\pi/m + \alpha$. The current flow is governed by the differential equation

$$V_0 \cos(\omega t) = L\frac{di}{dt} + Ri, \qquad i(t_{on}) = 0 \tag{9.33}$$

As a general example, let $V_0 = 170$ V, $\omega = 120\pi$ rad/s, $R = 10\ \Omega$, and $m = 3$ phases. Figure 9.12 shows voltage and current waveforms for inductor values of 3 mH, 6 mH, 9 mH, 12 mH, 15 mH, and 18 mH, with $\alpha = 60°$. The figure shows that the critical inductance for this case is between 15 mH and 18 mH. The actual value is 16.34 mH.

It is possible to estimate the critical inductance for a rectifier by using the lowest unwanted component of voltage and current. The unwanted voltage fundamental is the largest contributor to current. If the current ripple is $\pm 100\%$ for a certain value of L, that value should approximate the critical inductance. Let us test this for the circuit used to generate Figure 9.12.

Example 9.3.1 A three-phase midpoint rectifier has inputs at 60 Hz and 170 V_{peak}. The load is a series LR combination, with $R = 10\ \Omega$. Use the lowest unwanted component to estimate the critical inductance when $\alpha = 60°$.

The Fourier components can be found through the use of equation (2.7) or with computer code in Appendix C. The coefficient c_1 is of interest in this example. For the 60° delay waveform, $c_1 = 93.00$ V. The average current should be $\langle v_{out}\rangle/R$:

$$\langle v_{out}\rangle = \frac{3V_0}{\pi} \sin\frac{\pi}{3} \cos\alpha = 70.30\text{ V}, \qquad \langle i_L\rangle = \frac{\langle v_{out}\rangle}{R} = 7.030\text{ A} \tag{9.34}$$

If the inductor is the critical value, the current should vary approximately between 0 and 14.06 A. The peak of the fundamental therefore should be about 7.03 A. This requires

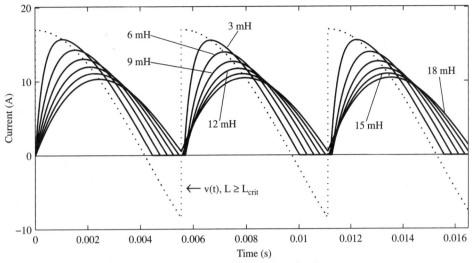

Figure 9.12 Voltage and current waveforms for subcritical *R-L* load.

$$\frac{c_1}{|R + j\omega L_{\text{crit}}|} \approx 7.03 \text{ A}, \qquad \frac{93}{7.03} \approx \sqrt{R^2 + \omega^2 L^2_{\text{crit}}}$$

$$\omega L_{\text{crit}} \approx 8.66 \ \Omega, \qquad \omega = 360\pi, \qquad L_{\text{crit}} \approx 7.66 \text{ mH}$$

(9.35)

This estimate of L_{crit} is about half the correct value—mostly because the highly distorted waveform makes it hard to neglect the higher frequencies. However, the estimate is useful to determine how critical inductance varies with load or phase angle. A plot of the current waveform based on the fundamental plus the dc component is given in Figure 9.13 for an inductance of 8 mH. The actual current is shown for comparison.

Figure 9.13 Current waveform for *L* = 8 mH compared to first-harmonic estimate.

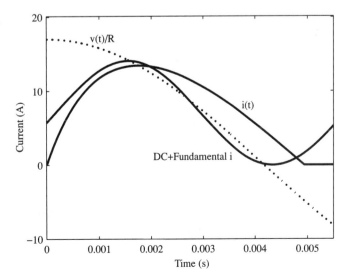

An exact analysis of the critical inductance can be performed by solving the appropriate differential equation. Modern mathematical computer tools provide powerful tools to find solutions. Here is Mathematica code to analyze the series LR configuration and determine L_{crit}:

```
(* Code to find L_crit for an m-phase midpoint rectifier with series L-R load.*)
(* P. T. Krein, 1996 *)

(* First solve the differential equation, starting from zero current. *)
iall = DSolve[{v0 Cos[w t] == L i'[t] + R i[t],i[(-Pi/m + alpha)/w] == 0},i[t],t]

(* Now put in some circuit values. *)
f=60; w=2 Pi f; m=3; v0=170; R=10; tstart=(-Pi/m + alpha)/w /.alpha → Pi/3
iall2[t_,L_] = Simplify[iall[[1]][[1,2]] /. alpha → Pi/3]

(* The actual current can't be negative, so clamp it at zero. *)
iall0[t_,L_]:=  If[N[iall2[t,L]] > 0, iall2[t,L], 0]

(* Plot for a few values of inductor, and get the average as well. *)
Plot[{iall0[t,0.01],iall0[t,0.015]},{t,tstart,tstart + 1/(f m)}]
iave = NIntegrate[iall0[t,0.01],{t,tstart,tstart + 1/(f m)}]

(* The critical inductance is the value that just barely allows the
current to return to zero during one period. Find this value. *)
Lcrit = FindRoot[iall2[tstart + 1/(f m),L] == 0,{L,0.01}]
```

The critical inductance value drops as the load current average increases. This is a matter of slowing the L/R time constant. In rectifiers, a problem occurs as the phase delay angle increases. As the delay approaches $90°$, the critical inductance value goes to infinity. A practical rectifier would include both capacitive and inductive filtering, as discussed in Section 5.5. When both L and C components are used, higher frequencies are usually attenuated much more than in the LR case, so that the first harmonic component gives a better approximation of the ripple.

When a rectifier operates in discontinuous mode, the output tends to rise, since the time while the switches are off is a time when the input voltage is less than the load voltage. The average output value is

$$\langle v_{out} \rangle = \frac{m}{2\pi} \int_{\theta_{on}}^{\theta_{off}} V_0 \cos(\theta)\, d\theta, \qquad \theta_{on} = -\frac{\pi}{m} + \alpha \qquad (9.36)$$

where θ_{off} is determined by the turn-off point, and depends on L/R. Since the turn-off time cannot be expressed in closed form, the average is normally computed numerically, as in Example 9.3.1.

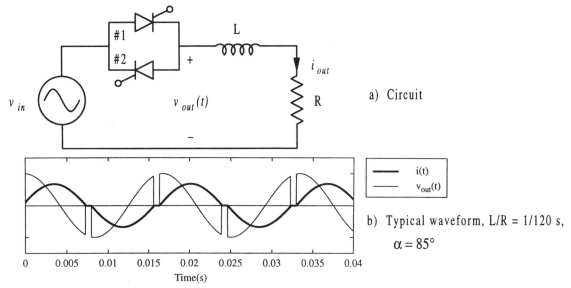

a) Circuit

b) Typical waveform, L/R = 1/120 s,
    $\alpha = 85°$

Figure 9.14 ac regulator circuit based on SCRs.

9.3.2 ac Regulators Revisited

Ac regulators deliberately take advantage of low inductances or capacitances to allow control of the power in an ac circuit. A circuit like the one in Figure 9.14 can control energy flow only by turning both switches off. When SCRs are used, it is not possible to force both switches off; they will turn off only when the current through them drops to zero. Inductance has strong influence on the behavior since it tends to maintain current flow and keep an SCR on longer. Figure 9.14 also shows current and voltage waveforms given an L/R time constant of 1/120 s (the input frequency is 60 Hz) and phase delay angle of 90°.

Imagine an ac regulator application with significant inductance. The inductance delays SCR turn off by a substantial angle. In a single-phase regulator there are two SCRs. If one is carrying forward current, the other cannot be turned on until the current returns to zero. An inductor might create a delay in one SCR current sufficiently long that current flow lasts beyond a subsequent gate trigger. In this case, the other SCR will not turn on at all. If a small phase delay angle is applied in an effort to reduce energy flow, one SCR might carry all the current, while the second device is prevented from turning on because its gate signal arrives only when reverse current would have to flow. This creates a *commutation failure*, and circuit action is delayed to the next half-cycle. Figure 9.15 compares current waveforms at delay angles of 0°, 45°, and 90° for $L/R = 1/120$ s. For this time constant value, correct ac regulation action does not take place unless the delay angle is greater than about 72°, and the delay of 90° is the only one that gives a true ac regulator output. Commutation fails below $\alpha = 72°$ for $L/R = 1/120$ s.

When commutation fails, the results can be a very significant problem. The currents shown in Figure 9.15 for $\alpha = 0°$ and $\alpha = 45°$ are effectively rectifier currents, since only

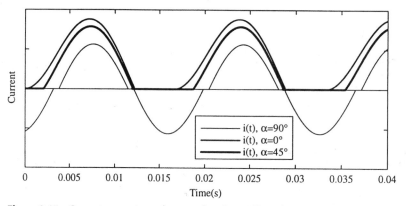

Figure 9.15   Currents at various phase angles given $L/R = 1/120$.

one of the SCRs can operate. A dc component of current will flow. The dc flow is unacceptable in most electric power systems. To avoid this dc flow problem, a practical ac regulator must have very low inductance levels. A time constant of just $1/1200$ s in a 60 Hz circuit, for instance, requires at least $18°$ of delay to avoid commutation failure. An overhead light with a 500 W rating on a 120 V circuit has a nominal resistance of about $28 \ \Omega$. A time constant of $1/1200$ s corresponds to 24 mH. If the inductance can be limited to perhaps $20 \ \mu H$ (the inductance of the wiring, for instance), then the time constant will be less than $1 \ \mu s$, and commutation failure will not be an issue. Figure 9.16 compares the output current for a delay setting of $\alpha = 30°$ given time constants of $4 \ \mu s$, $800 \ \mu s$, and 2.5 ms. The choice of $\alpha = 30°$ leads to commutation failure for the highest time constant.

9.4 RECAP

It is not an absolute requirement that converters use large energy storage elements. If small storage elements are used, a converter can exhibit extra configurations, such as "all off" or

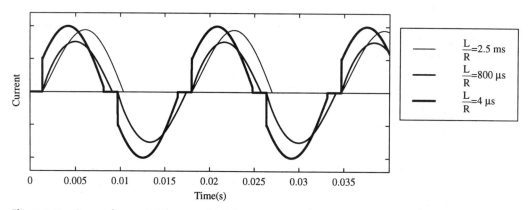

Figure 9.16   Current for $\alpha = 30°$ for several L/R time constants.

"all on" switch behavior. The concepts of critical inductance and critical capacitance provide a convenient way to determine whether storage components are small in the sense of power electronics. The critical inductance for a converter, L_{crit}, is defined as the smallest inductance for which $i_L > 0$ under all allowed conditions in a power electronic circuit. The value of L_{crit} is extremely useful. It defines the boundary between normal and discontinuous modes, and can be used as the basis for inductor ripple design. Current ripple is inversely proportional to inductance, with ripple of $\pm 100\%$ at L_{crit}, $\pm 10\%$ at $10L_{crit}$, and so on. Critical capacitance for a converter is the value C_{crit} such that $v_C > 0$ under all conditions. Critical capacitance is used much less often than critical inductance, since voltage ripple is undesirable in most converters.

Converters can be built with $L < L_{crit}$. This design choice provides very fast dynamic response traded off against load regulation. In a dc–dc converter or rectifier, all the switches connected to an inductor will be off during some portion of a cycle if $L < L_{crit}$. The length of this off time depends on the ratio of the L/R time constant to the switching period. This makes the behavior load-dependent once $L < L_{crit}$. Approximate relationships between output and input can be derived for discontinuous mode.

If a designer wishes to avoid discontinuous mode, two techniques are commonly used:

1. Specify a minimum load for a converter. If $P_{out} > 10$ W can be guaranteed, for instance, a value of L_{crit} can be calculated and installed in the circuit. As $P_{out} \to 0$, the critical inductance approaches infinity.

2. Provide a *ballast load*, which is a small resistor installed inside a converter across the load terminals. This in effect guarantees a definite minimum load, yet the user can operate the converter with an open circuit output without trouble.

The value of L_{crit} can be decreased by raising the switching frequency.

In rectifiers, the switching frequency is usually low, and the value L_{crit} tends to be relatively high. Many industrial rectifiers operate with subcritical inductance because of the low frequency. When $L < L_{crit}$, most converters exhibit output voltages higher than when $L \geq L_{crit}$. In commercial rectifier systems, this behavior is compensated through the use of larger phase delay angles and automatic control.

ac regulators actually require subcritical inductance for proper operation. If the current does not return to zero during a half-cycle, commutation failure can occur. Commutation failure defines a situation in which the terminal conditions on a switch are inconsistent with attempts to control the switch. If an SCR is reverse biased when a gate pulse arrives, the device will not turn on, and a commutation failure has occurred. Commutation failure in ac regulators can produce dc average output values. This can be a serious problem, and ac regulators are normally intended for resistive loads to avoid trouble.

PROBLEMS

1. A boost dc–dc converter switches at 100 kHz. The input is $+12$ V and the duty ratio of the transistor is 75%. The input inductor is 25 μH, and the output capacitor is 100 μF. If the load draws 50 W, the output voltage is found to be 48 V. What is the output voltage for a 230Ω load?

2. A buck dc–dc converter uses a 10 μH inductor and a 1000 μF output capacitor. It switches at 160 kHz. The load is a 5 Ω resistor. Compute and plot the output voltage as a function of the transistor duty ratio for $0 \leq D \leq 1$.

3. A flyback converter uses a coupled inductor with an input–output turns ratio of $10:1$. Find an expression similar to equation (9.18) for output voltage as a function of the load, switching period, and other parameters, if L is small.

4. It is desired to assure continuous mode operation in a buck converter. The converter switches at 400 kHz, and supplies output power ranging from 0 W to 100 W. The input is 120 V and the output is 48 V. Given a ballast load that draws 1 W, find the smallest inductor value that assures continuous mode for all output power levels.

5. A buck converter takes a $+12$ V input and provides a $+9$ V output. The switching frequency is 200 kHz, and the load is 25 W.

   a. What is the critical inductance for this converter if a large output capacitor is used?

   b. If $L = L_{crit}/2$, what duty ratio is needed for 9 V output? Plot the inductor current as a function of time.

6. A buck converter allows inputs ranging from 11 V to 18 V, and provides a $+9$ V output. The switching frequency is 200 kHz, and the load can range from 5 W to 40 W.

   a. What is the critical inductance for this converter if a large output capacitor is used?

   b. What is the maximum value of inductance for which discontinuous mode operation is assured under all conditions?

7. Consider the converter of Problem 6. The inductor is set to the value found in part 6(b), so that the converter is always in discontinuous mode. Find and plot the duty ratio as a function of input voltage (to provide the $+9$ V output) for loads of:

   a. 5 W   b. 25 W   c. 40 W

8. A flyback converter uses a $20:1$ turns ratio to convert 150 V to 5 V output. The desired load range is 0 W to 200 W. The target switching frequency is 120 kHz.

   a. Select a ballast load, then choose the smallest inductor such that $L > L_{crit}$. The ballast load should not drop the full-load efficiency by more than 1%.

   b. For your ballast load, what is L_{crit} if the input voltage can vary by $\pm 20\%$?

9. A boost-buck converter switches at 200 kHz. The input range is 4 V to 8 V, and the output is -9 V. The load is in the 1 W to 5 W range. What is the critical capacitance for this converter?

10. A boost-buck converter has been designed for $+12$ V to -12 V conversion, with 60 W output. The switching frequency is 100 kHz. If the transfer capacitor is half the critical value, what is the duty ratio?

11. Find a general expression for V_{out} as a function of V_{in}, duty ratio, and circuit parameters for a boost-buck converter with $C < C_{crit}$.

12. A boost-buck converter switches at 100 kHz. The input ranges from 10 V to 20 V. The output is -12 V. The load is ballasted so that its effective range is 1 W to 50 W.

  a. Find the input critical inductance for this converter.

  b. Find the critical capacitance for this converter.

  c. Assume that the input inductor is equal to its L_{crit} value, and that the transfer capacitor matches C_{crit}. The output inductor is large. In this case, does the converter still follow the average relationships of the same circuit with infinite values for L and C? What duty ratio range will be used to meet the operating requirements?

13. Discontinuous mode provides a boosting effect, in which the output is higher than would be expected for a given duty ratio. Is it possible to design a buck converter with $V_{out} > V_{in}$ based on discontinuous mode? If so, show a circuit that could provide a 15 V output from a 12 V input given a 3 W load.

14. A small battery charger is capable of charging any battery between 1.2 V and 24 V. To protect against excessive current, the charger places a 10 Ω resistor in series with the battery. The charger input is a constant +12 V supply. The unit controls its duty ratio to charge the attached battery at the rate of 0.1 A.

  a. Draw a switching converter capable of performing this function. Be sure to show the inductors and capacitors needed for interfaces.

  b. Find the critical inductance for the converter given $f_{switch} = 50$ kHz.

15. A boost converter allows inputs of 15 V to 40 V, and gives output of 120 V at 12 kW for an industrial process. The inductor is 10 μH.

  a. What value of switching frequency is needed to keep $L \geq L_{crit}$ for this converter? At this frequency, what is the inductor's maximum peak-to-peak ripple current?

  b. The design engineer proposes that subcritical inductance be used to provide fast response because of the high power level. What switching frequency is needed to ensure discontinuous mode operation for this converter? What is the inductor's maximum peak-to-peak ripple current at this frequency?

16. A buck converter for an aircraft system is to allow an input voltage between 22 V and 40 V, and provides tightly regulated 12 V output. The circuit should use an output capacitor large enough to guarantee $\pm1\%$ output ripple at most. The switching frequency is 50 kHz, and the load can range between 10 W and 500 W. Power levels below 200 W are unusual, and inductors are large and costly. It has been suggested that the inductor be chosen so that $L \geq L_{crit}$ if $P_{out} > 200$ W.

  a. Draw a circuit that can meet these requirements. What is the range of duty ratios when $P_{out} > 200$ W?

  b. What value of inductance will ensure $L \geq L_{crit}$ for $P_{out} > 200$ W?

  c. Is it possible to maintain $V_{out} = 12$ V at the minimum 10 W load? If so, what duty ratio range will be needed at $P_{out} = 10$ W? If not, suggest a way to correct the trouble without changing the value of L.

17. Given a six-phase midpoint rectifier (equivalent to a three-phase bridge) with a series L-R load for which $L \ll L_{crit}$, what is the average output voltage as a function of the delay angle α? (*Hint:* Consider the action with a purely resistive load.)

18. A lamp dimmer is to be designed for a domestic application, rated to work for loads between 0 and 1200 W from 120 V_{RMS} 60 Hz input. A 1 W ballast load would be reasonable if necessary. The inductance will depend on the wiring layout and some other details, but is not expected to exceed 50 μH. The triac gate pulse width is 100 μs (or 2.16° on a 60 Hz time scale). Commutation failure will not occur if the triac current goes to zero before or during the gate pulse. The phase delay angle is measured from the voltage zero crossing to the beginning of the pulse. Is there any load value or situation in which commutation failure can occur in this application? It is desired to allow all values $0° < \alpha < 180°$. If commutation failure can occur, will the ballast load help the situation? If not, what is the minimum value of α that ensures no failure? (*Hint*: The gate pulse width and action essentially ensures $\alpha \geq 2.16°$ for the purposes of evaluating commutation failure.)

19. A buck-boost converter provides -15 V output from $+15$ V input. The load is a resistor $R = 15 \, \Omega$. The output capacitor is 100 μF. It is desired to compare the dynamic performance under several conditions, given a switching frequency of 100 kHz. The duty ratio is held fixed at 50%.

   a. Find L_{crit} for this converter.

   b. Find and plot the inductor current and the capacitor for the first ten switching periods, starting from $V_{out} = 0$ and $i_L = 0$, for $L = 10L_{crit}$. After this time, how close is the output to the desired -15 V?

   c. Repeat part (b) given $L = L_{crit}$. Obviously, the inductor current changes much more quickly in this case, but does the capacitor prevent faster output response?

   d. What should the duty ratio be if $L = L_{crit}/4$? Set the duty ratio to this value, then examine the start-up behavior as in part (b). Has the response been speeded up?

20. A six-pulse rectifier has been designed for a dc motor drive application. The line inputs are 60 Hz, 480 V_{RMS}. The motor draws power in the range of 500 W to 20 kW. Over a phase angle range from $0° \leq \alpha \leq 75°$, what is the critical inductance for this converter?

21. In multi-pulse rectifiers, it is possible to have $L_{crit} = 0$. For example, a three-pulse rectifier with $\alpha = 0°$ enforces $i_{out} > 0$ even into a purely resistive load. Find the maximum value of α for which $L_{crit} = 0$ as a function of the number of pulses m in a polyphase bridge rectifier.

22. A 24-pulse rectifier bridge uses phase control to supply a load. The input frequency is 50 Hz. The maximum output is 4800 V_{dc}, and the maximum load current is 6000 A_{dc}. The load is a fixed impedance, with a certain resistance and an inductor $L = 100 \, \mu$H.

   a. What is the highest value of delay angle α for which $L > L_{crit}$?

   b. It is desired to control the output dc current down to 20 A_{dc} for this load. What phase angle will be used?

24. A small appliance motor has $L = 80$ mH and $R = 24 \, \Omega$. The motor operates for 120 V_{RMS} through an ac regulator. What is the lowest delay angle that avoids commutation failure?

25. A theater lighting application uses a 230 V_{RMS} 50 Hz supply. The wiring is very lengthy, and measurements show series inductance of 1 mH. The minimum desired phase angle is 5°. Determine the range of loads for which commutation failure will not occur. the desired power level at low phase delay is 20 kW. Will the system work without commutation failure?

26. Commutation failure in ac regulators is often avoided by lengthening the SCR pulse duration. A single-phase ac regulator for a heating application has input of 280 V_{RMS} at 60 Hz. The load is resistive, with an approximate value of 12 Ω. The SCR pulses are 100 μs in duration to try to avoid commutation failure. What is the highest value of series inductance that can be tolerated without commutation failure in this application for delay angles down to 0°?

REFERENCES

S. Ćuk, R. D. Middlebrook, "A general unified approach to modelling switching dc-to-dc converters in discontinuous conduction mode," in *IEEE Power Electronics Specialists Conf. Rec.*, 1977, pp. 36–57.

N. Femia, V. Tucci, "On the modeling of PWM converters for large signal analysis in discontinuous conduction mode," *IEEE Trans. Power Electronics*, vol. 9, no. 5, pp. 487–496, September 1994.

D. Maksimović, S. Ćuk, "A unified analysis of PWM converters with discontinuous modes," *IEEE Trans. Power Electronics*. vol. 6, no. 3, pp. 476–490, July 1991.

N. Mohan, T. M. Undeland, W. P. Robbins, *Power Electronics*, 2nd ed. New York: Wiley, 1995.

V. Vorperian, "Simplified analysis of PWM converters using model of PWM switch, Part II: Discontinuous conduction mode," *IEEE Trans. Aerospace Electronic Systems*, vol. 26, no. 3, pp. 497–505, May 1990.

PART III

REAL COMPONENTS
AND THEIR EFFECTS

CHAPTER 10

Figure 10.1   Batteries, motors, and other real sources and loads.

REAL SOURCES AND LOADS

10.1 INTRODUCTION

The principle of source conversion demands that a power converter have source-like behavior at any port. The basic ideas of using inductance to create constant current loops and capacitance to create constant voltage nodes have guided much of the design work so far. Real devices do not behave as ideal sources. In this chapter, we will explore characteristics of a few real-world sources and loads like those pictured in Figure 10.1. Some of the details of choosing inductors and capacitors for source interfaces will be examined. Parallel capacitance and series inductance will be used to enhance dc sources. Resonant circuits will be used to enhance the characteristics of ac sources. Basic impedance arguments about source properties will guide the development.

The chapter begins with an overview of real loads. The emphasis in power electronics is usually on the steady-state behavior of a load or on the behavior of very fast dynamic load changes such as those in digital circuits. These two extremes lead to relatively simple load models, intended to help a designer determine the basic issues of power converter operation. Both dc and ac loads are introduced.

Almost all loads tend to exhibit current source behavior on short time scales. An important reason for this is that wiring and connections introduce inductance. The self-inductance of an isolated wire is reviewed, and some of the implications for design are examined.

The concepts of critical inductance and capacitance, introduced in Chapter 9, can be used to characterize the relative quality of a dc source. For example, if a series inductance much larger than L_{crit} is used, very good current-source behavior will be exhibited. In the case of ac sources, the source impedance provides a helpful indication of quality. Source interfaces, analogous to series inductance for dc current or parallel capacitance for dc voltage, are developed for ac sources.

10.2 REAL LOADS

With few exceptions, the actual electrical loads supplied from a switching converter (or any other electrical source) are not well modelled simply as a resistance. Wires have inductance

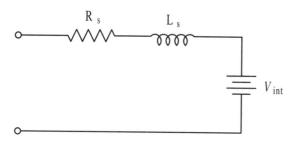

Figure 10.2   Simple series *R-L-V* model for static load.

and capacitance. Loads frequently contain electrical sources, interface filters, or other features that complicate their circuit model. Many loads are not linear, and as such are difficult to model in the form of a circuit diagram. Even tungsten lamp filaments and resistive heaters typically show strong temperature dependence, and act very differently during their initial warm-up than in steady state.

Many load models start with the simple *R-L-V* series combination in Figure 10.2. In this model, the internal action of a battery under charge, a dc motor, an electrochemical industrial process, or some other load is associated with an ideal source. The resistance and inductance model the properties of connecting wires and internal parts. The model serves as a useful representation for almost any load that does not vary rapidly. A second type of load, shown in Figure 10.3, adds internal switch action. This is a useful representation of a digital circuit, a radar system, or certain other loads that draw energy in impulse fashion.

A more detailed example is the battery model given in Figure 10.4. The voltage V_{int} represents the internal electrochemical potential, which has a known temperature dependence and a dependence on the stored energy level. The resistor R_{int} represents the internal battery structure and the electrolyte resistance. The resistor R_{dis} is an internal self-discharge process that causes the battery to lose energy even when disconnected. The effective capacitance C_{eff} models the effects of internal construction as well as the tendency of the battery to provide its own filtering function. The external connections are shown as the series pair R_s and L_s. A model like this one offers the ability to do design: Given these parameters, a buck converter or phase-controlled rectifier could be set up to give a desired current level. Key parameters such as the total charge injected into V_{int} can be monitored and controlled.

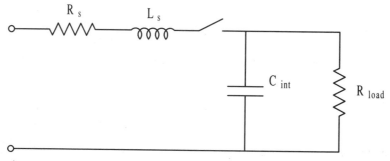

Figure 10.3   Load with internal switch draws short bursts of current.

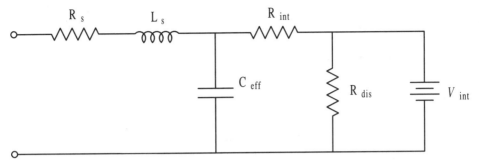

Figure 10.4 Circuit model for a battery on charge.

Most dc loads fall into one of two broad categories:

- Quasi-steady loads. These do not change their properties or power requirements much, at least on the time scales of switch action. Simple circuit models generally give good results for design purposes. Converters that drive such loads usually provide excellent performance if the load voltage or current can be sensed and used for converter control.
- Transient loads. These vary on time scales similar to a switching period. Large digital circuits, for example, switch hundreds or thousands of capacitors simultaneously with each clock pulse. Each clock pulse draws a brief current spike. Energy storage interfaces are required to provide good results with such loads, since the switches cannot react fast enough.

Quasi-steady loads will be the main focus of discussion here. Transient loads move energy so quickly that their analysis and design are almost like those of charge pumps. When a dc–dc converter serves a transient load, the typical practice is to provide enough output capacitance to account for essentially all of the charge that the load will need during its fast current spike.

Let us consider two typical power electronics applications and evaluate their basic load action.

Example 10.2.1   A dc motor is driven aggressively to give rapid mechanical response. The motor has rated torque of 10 N · m and moment of inertia $J = 0.04$ kg · m^2. The circuit model, from Figure 10.2, has an internal voltage $V_{int} = 1.1\omega$, where ω is the shaft speed in rad/s. The series R-L pair has a time constant of 12 ms. At time t_0, the motor is running at 100 rad/s. The dc–dc converter that controls power to this motor imposes a high input voltage to quickly drive the current up and bring the motor output torque up to five times the rated value. How much does the internal voltage change in 1 ms?

The R-L time constant suggests that a dc–dc converter switching at 1 kHz or more will see the machine as a current source. The load would be considered quasi-steady if the internal voltage changes very little during one switch period. The action of the mechanical portion of this system is governed by Newton's Second Law for a rotational system,

$$\sum T = J\frac{d\omega}{dt} \tag{10.1}$$

where T is torque. To get the fastest possible performance, the converter should impose the highest possible torque. At time t_0, the voltage is changed to bring the torque up to its limit of 50 N · m. This change requires some time because of the L-R time constant. However, in principle it is possible to use a high voltage to make the torque change very rapidly. When the torque is 50 N · m, Newton's Second Law gives $50 = 0.04(d\omega/dt)$, and the speed change is 1250 rad/s². In one millisecond, the speed changes by 1.25 rad/s, or 1.25% of the running speed. The internal voltage changes little, from 110 V to 111.4 V. Therefore, a circuit with a resistor, inductor, and internal dc voltage source $V_{int} = 1.1\omega$ will model the actual behavior well over short time intervals, such as those of a switching period. The load is quasi-steady even though the mechanical system is being driven hard.

Example 10.2.2 A certain clocked CMOS digital circuit switches a total capacitance of 1000 pF at the output of a 5 V dc supply. The connections from the supply to the chip have about 25 nH of inductance. The capacitance is initially uncharged. How quickly does the load current change when capacitor switching occurs? Discuss the effect on the power supply.

The capacitors on the digital chip are initially uncharged, so the current drain prior to clock switching is low. If the dc supply uses a buck configuration, its output inductor current would be low as well. The inductor's current does not change quickly, so the circuit can be represented with the circuit shown in Figure 10.5. The left switch in the figure represents the internal clock action of the digital circuit, while the right switch acts in complement to it to remove the charge on C_{LOAD}. When clock action occurs, the time-rate-of-change of current is $v_L/L = di/dt = 5$ V/(25 nH) $= 2 \times 10^8$ A/s. Current rates of change are often given in A/μs; 200 A/μs is considered a very high value.

The power supply effect is determined entirely by its interface capacitor, C_{out}, since no switch action occurs inside the power converter during the brief current surge into the digital circuit. If $C_{out} >> C_{load}$, the charge transferred from the filter capacitor to the load will be small compared to the total stored charge. For example, if $C_{out} = 10$ μF, its value is

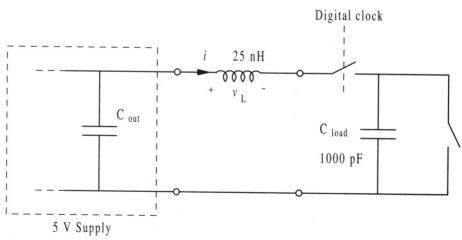

Figure 10.5 Circuit model for Example 10.2.2.

Figure 10.6 Synchronous motor quasi-steady model.

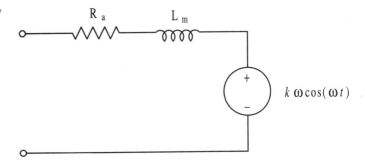

$k\,\omega\cos(\omega t)$

a factor of 10^4 larger than the load capacitance. Each time the digital clock operates, the load will draw a fraction 10^{-4} of the total charge in C_{out}. The voltage will drop by 0.5 mV on average each time the digital clock operates. If the digital clock is less than one hundred times faster than the converter switching frequency, the total effect on the output voltage will be less than 1%.

ac loads have many similarities to the dc models. In Figure 10.6, a simplified model for a synchronous motor is given. The internal voltage is a speed-dependent ac source, but the connections and internal structure still follow the basic *R-L* behavior. More detailed models can be found in the literature. In Figure 10.7, the circuit model of an induction motor operating under steady conditions is shown. This model has a speed-dependent resistor (which can be negative under some circumstances) rather than an internal voltage source.

The incandescent lamp model in Figure 10.8a shows connection inductance in series with a strongly temperature-dependent resistor. The resistance rises when the filament is first heated. The low cold resistance means that the current is initially very high when the lamp is switched on. A loudspeaker is often modelled as a simple *R-L* series connection, as in Figure 10.8b. This model has limited accuracy, since a loudspeaker is a type of electric machine with its own internal voltages, but it is useful for understanding power flows into a speaker.

All the loads examined so far have a common feature: series inductance. This is a property of virtually any load that might be served with power electronics. The inductance has

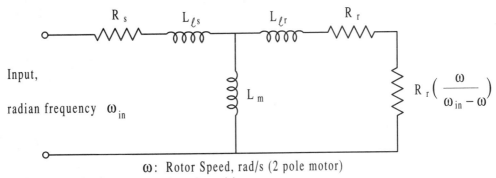

ω: Rotor Speed, rad/s (2 pole motor)

Figure 10.7   Induction motor quasi-steady model.

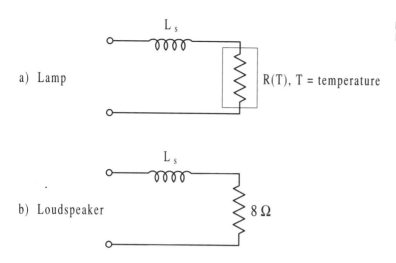

a) Lamp

b) Loudspeaker

Figure 10.8 *R-L* models: incandescent lamp and loudspeaker.

an interesting implication: Although voltage sources are familiar, real loads act like current sources on time scales shorter than their series *L/R* time constants.

10.3 WIRE INDUCTANCE

Self-inductance of a wire arises because the magnetic field created by any moving charge stores energy and interacts with other charges. The inductance can be divided into internal inductance, by which magnetic flux inside the wire interacts with the current, and external inductance due to flux created outside the wire. A complete analysis can be found in an electromagnetics text. The self-inductance of a long cylindrical wire is a classical fields analysis example,[1] with the result

$$L_{\text{wire}} = \left(\frac{\mu_{\text{wire}}}{8\pi} + \frac{\mu}{2\pi} \ln \frac{D}{R} \right) l \tag{10.2}$$

where *l* is the length of the wire, μ_{wire} is the magnetic permeability of the conducting material, μ is the permeability of the surrounding air or insulation, *R* is the wire radius, and *D* is the distance to the center of the return wire. The first term is the internal self-inductance, while the second is the external self-inductance. Copper and aluminum conductors exhibit $\mu_{\text{wire}} \approx \mu_0$, the permeability of free space, with $\mu_0 = 4\pi \times 10^{-7}$ H/m. Air and most insulators also have $\mu = \mu_0$. The internal self-inductance for most conductors is therefore 5×10^{-8} H/m, while the external value depends on spacing. Table 10.1 gives examples of spacing values.

The logarithmic behavior means that self-inductance is not a strong function of spacing. Often, the value is given in nH/cm, and values of about 5 nH/cm would be expected for a typical configuration. This can be reduced by up to a factor of 2.5 through the use of tightly

[1]C. T. A. Johnk, *Engineering Electromagnetic Fields and Waves.* New York: Wiley, 1975, pp. 316–329.

TABLE 10.1
Wire Inductance

Wire Configuration	Ratio *D/R*	Self-Inductance
Isolated wire	100	1000 nH/m
Typical wire runs	10	500 nH/m
Close spacing	5	350 nH/m
Tightly twisted, thick insulation	3	250 nH/m
Tightly twisted, thin insulation	2.1	200 nH/m

twisted wire (although equation (10.2) is only approximate if *D/R* is not large). A very open configuration is unlikely to show inductance of more than 10 nH/cm. A complete circuit has both the wire and a return path. When a pair of wires is used, the total self-inductance will be twice that of each individual conductor.

The internal self-inductance of 0.5 nH/cm is difficult to alter. One wiring method that reduces this value follows the twisted pair concept. With *Litzendraht wire*, or just *litz wire*, an individual conductor is formed from many strands of small mutually insulated wires. These wires are carefully braided to ensure that each is along the outer edge as much as possible. This way, the internal flux interacts with the minimum possible current and self-inductance is reduced. Flat rectangular conductors or hollow tubes are also used when internal inductance must be reduced. Litz wire, twisted pair, and flat bus bar alternatives are illustrated in Figure 10.9.

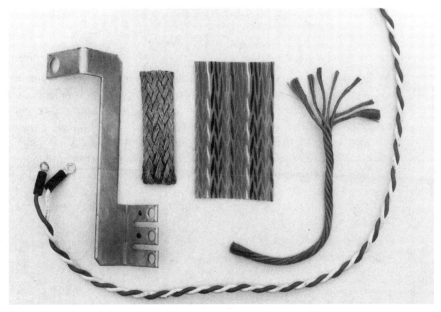

Figure 10.9 Special types of wires designed to reduce internal self-inductance.

> **Empirical Rule:** *The inductance per centimeter length of a single wire in a typical geometry is about 5 nH/cm. A wire pair yields about 10 nH/cm if not twisted, while a tightly twisted pair yields about 4 nH/cm total self-inductance.*

Wire inductance is an important issue in converter design as well as in sources. Consider the following example.

Example 10.3.1   A battery is connected to a buck converter for a dc motor control application. The battery is large. Two wires, each 20 cm long and with 1 mm radius, connect the converter to the battery. The wires are about 10 cm apart. The buck converter's output current is 50 A, and the active switch operates in 200 ns. What is the input wire inductance, and how will it affect converter operation?

The total input wire length is 40 cm. The ratio of D/R is 100, so the total series inductance is about 0.40 m $\times$ 1000 nH/m = 400 nH. (A rule of thumb of 5 nH/cm would give 200 nH, which is probably a low estimate in this case because of the wide spacing of the wires.) Since the inductance is in series with a switch, it represents a *KCL* problem. Switch action must turn the inductor current off, but the voltage $L(di/dt)$ will oppose this action. In this example, the switch is not ideal, but instead requires about 200 ns to act. If the current reduces during turn-off at a rate of 50 A/200 ns, the value of di/dt will be -2.5×10^8 A/s. With a 400 nH inductor, a voltage $v_L = -100$ V will occur during turn-off. The switch will need to be able to block $V_{\text{battery}} + 100$ V if it is to function properly in this application. If the battery is 12 V, a switch rated for at least 120 V is required.

In Example 10.3.1, a connection length of only 20 cm gives enough inductance to dominate the selection of switch voltage rating. A converter can easily induce voltages of 100 V or more along lengths of even the most conductive wire.

10.4   CRITICAL VALUES AND EXAMPLES

In Chapter 9, the concepts of critical inductance and capacitance were examined. The concept of critical component values gives a convenient way to examine source performance. To illustrate the concepts, consider a buck converter like that shown in Figure 10.10. If the load really behaves as a current source, the output voltage on average will be $D_1 V_{\text{in}}$. Recall that the critical inductance is the lowest value that ensures $i_L(t) > 0$ under all conditions. Any larger value ensures current source behavior and maintains the usual average relations.

For an interface circuit, the critical inductance and capacitance have special meaning:

- The *critical inductance* is the minimum value of series inductance that supports a current-source model of a given load.

- The *critical capacitance* is the minimum value of parallel capacitance that supports a voltage-source model of a given load.

If near-ideal source behavior is desired, $L \gg L_{\text{crit}}$ or $C \gg C_{\text{crit}}$ are necessary conditions.

Recall from Chapter 9 that the critical inductance value depends on the converter load, and that $L_{\text{crit}} \rightarrow \infty$ as the average inductor current approaches zero. Many commercial power

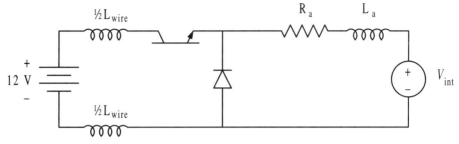

Figure 10.10   Source wires create an input inductance for a buck converter.

converters exhibit higher output ripple at low load power than at rated load power. The *ballast load* idea in Chapter 9 is often used when a power supply must function right down to the no-load condition.

Example 10.4.1   A buck converter is to be designed for 15 V input and 5 V output. The load can be anywhere in the range 0 to 100 W. The output ripple should be no more than ±1%. A switching frequency of 100 kHz is reasonable. Suggest a ballast load, inductor, and capacitor to meet these requirements while avoiding the control issues of discontinuous mode. The full-load efficiency should be as high as possible.

The arrangement is shown in Figure 10.11. To avoid discontinuous mode, choose $L \geq L_{crit}$. However, the converter is to keep working down to no load. It is impossible to have $L \geq L_{crit}$ in a true no-load situation. Instead, a small ballast load can be provided to provide an effective minimum load "invisible" to the user. If the ballast load is limited to $\frac{1}{2}$ W, for example, the efficiency at 100 W output will drop by only half a percentage point, and the resistor will be cheap and small. Let us choose $R_{ballast} = 50 \ \Omega$ based on this idea. A higher value can be used, although it will lead to higher inductance requirements.

For 15 V to 5 V conversion and $L \geq L_{crit}$, the duty ratio $D_1 = \frac{1}{3}$. The average inductor current equals the load current, $P_{out}/5$. For the value $L = L_{crit}$, the inductor ripple current will be double the average value, so $\Delta i = 2P_{out}/5$. When the MOSFET is on, the inductor voltage is $V_{in} - V_{out} = 10$ V, so

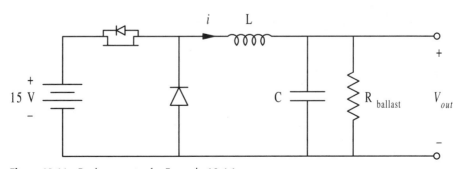

Figure 10.11   Buck converter for Example 10.4.1.

$$10 \text{ V} = L_{\text{crit}} \frac{2P_{\text{out}}/5}{D_1 T} = 1.2 \times 10^5 L_{\text{crit}} P_{\text{out}}$$

$$L_{\text{crit}} = \frac{8.33 \times 10^{-5}}{P_{\text{out}}}$$

(10.3)

To ensure $L \geq L_{\text{crit}}$, examine the worst case—the lowest power setting. This is $\frac{1}{2}$ W for the ballast. Then $L = 167 \ \mu$H. The actual inductor current ripple can be computed with the MOSFET on to be

$$10 \text{ V} = 167 \times 10^{-6} \frac{\Delta i}{D_1 T}, \qquad \Delta i = 0.2 \text{A}$$

(10.4)

This current change is independent of load. Notice that at $P_{\text{out}} = \frac{1}{2}$ W, the average output current is 0.1 A, and ripple of 0.2 A is consistent with the choice $L \geq L_{\text{crit}}$. The capacitor current is the inductor current less the average load current if output ripple is low. Therefore, i_C is a triangle waveform with peak-to-peak value of 0.2 A, as shown in Figure 10.12. As in Example 3.7.3, the variation in capacitor voltage is $1/C$ times the integral of the positive current half-cycle. This integral is the triangular area $\frac{1}{2}bh = \frac{1}{2}(5 \ \mu\text{s})(0.1 \text{ A}) = 2.5 \times 10^{-7}$. To keep the output variation below $\pm1\%$, we need Δv 0.1 V, so $2.5 \times 10^{-7}/C \leq 0.1$, and $C \geq 2.5 \ \mu$F. The choices $R_{\text{ballast}} = 50 \ \Omega$, $L = 167 \ \mu$H, and $C = 2.5 \ \mu$F are not the only possibilities, but they will meet all the requirements of this converter. It is interesting to notice that both the output voltage and output ripple are independent of load for this design.

The relative quality of a given dc source is determined by the ratio of the actual storage element to the critical value. This can be very helpful for design. Consider the critical capacitance for a converter, as in the following example.

Example 10.4.2 A boost converter provides 24 V output from a 6 V source. The input inductor is large, and the output power is to be 60 W. Given a switching frequency of 40 kHz, what is the critical capacitance for this converter? What is the output voltage ripple if $C = C_{\text{crit}}$? What about $C = 100C_{\text{crit}}$?

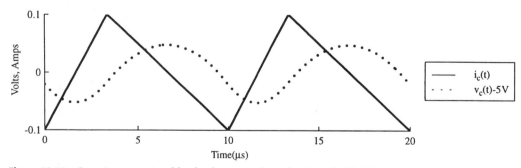

Figure 10.12 Capacitor current and load voltage waveforms for Example 10.4.1.

The critical capacitance by definition is the minimum value for which $v_C > 0$ under all allowed conditions. In the boost converter, a nonzero output voltage will ensure that the switches work in alternation, with $D_1 + D_2 = 1$. With 24 V output and 6 V input, the duty ratio $D_2 = 1/4$, while $D_1 = 3/4$. If the inductor is large and the load really is drawing 60 W, the inductor current must be $P_{in}/V_{in} = 10$ A. The average output current must be $D_2 I_L = 2.5$ A. To find C_{crit}, let us assume that the load current matches this average value. Then, with the diode on, the net capacitor current will be 7.5 A. The capacitor voltage should ramp from 0 to 48 V while the diode is on if $C = C_{crit}$. The circuit and a few major waveforms are given in Figure 10.13.

The capacitance relationships are

$$i_C = C\frac{dv}{dt}, \qquad 7.5 \text{ A} = C_{crit}\frac{48}{D_2 T} = C_{crit}\frac{48}{0.25 \times 25 \ \mu s}, \qquad C_{crit} = 0.977 \ \mu F \quad (10.5)$$

With $C = C_{crit} = 0.977 \ \mu F$, the output will be 24 V $\pm$ 100%. If $C = 97.7 \ \mu F$, the currents and basic relationships are unchanged, and

$$7.5 \text{ A} = C\frac{\Delta v}{\Delta t} = 97.7 \ \mu F\frac{\Delta v}{0.25 \times 25 \ \mu s}, \qquad \Delta v = 0.48 \text{ V} \quad (10.6)$$

to give $V_{out} = 24$ V $\pm$ 1%. While it is extremely unlikely that a designer would use a capacitor on the order of the critical capacitance, the value of C_{crit} can be used to design for a given ripple level. Ripple of $\pm 0.2\%$, for example, requires $C = 500 C_{crit} = 489 \ \mu F$ for this boost converter.

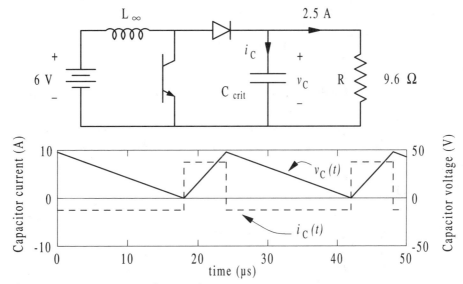

Figure 10.13 Boost converter for Example 10.4.2.

The inductor and capacitor values can be computed based on source requirements as well as on a converter design. The question of how closely a load matches ideal characteristics can be addressed directly. Consider the following example.

Example 10.4.3   An inductor is connected in series with a resistor to create a load with near-ideal current source behavior. A current source should be able to handle almost any voltage without changing. The source is intended for an application with nominal current of 10 A. If $R = 10 \ \Omega$, what value of inductor will be needed if a 50 Hz voltage input level of 10 kV peak is to change the current by less than $\pm 1\%$?

A current source of 10 A that maintains this level even when subjected to 10 kV ought to be ideal for many applications. The circuit is just a series $L\text{-}R$ combination, and an ac voltage of $10\,000 \cos(100\pi t)$ should not force a current of more than 0.1 A peak to flow. An applied dc voltage of just 100 V will produce the 10 A constant flow described in the problem statement. Since only the ac peak current is of interest, the value $i_{peak} = v_{peak}/|Z|$ will give the necessary result. Then

$$i_{peak} = \frac{v_{peak}}{\sqrt{R^2 + \omega^2 L^2}} \leq 0.1 \ \text{A} \tag{10.7}$$

so $10^{10} \leq R^2 + (100\pi)^2 L^2$ and $L \geq 318$ H. This is truly a large inductor!

10.5 REAL SOURCES AND INTERFACES FOR THEM

10.5.1 Impedance Behavior of Sources

Real sources in general are similar to real loads. Circuit models associated with batteries, synchronous machines, and some of the other devices commonly used as sources have been introduced so far. As in the load case, almost any real source, including the voltage at a wall plug, has series inductance, and tends to act as a current source on short time scales. The impedance properties of sources are helpful in evaluating quality and designing interfaces. In this section, the impedance behavior of both ac and dc sources is considered, then interfaces are designed on this basis.

An ideal voltage source maintains a specific $v(t)$ value regardless of the current it carries. The familiar ideal properties can be restated in terms of impedance.

> **Definition:** *An* ideal voltage source *provides a defined* v(t) *value at* any *current. If a current is imposed on such a source, that current has no effect on the voltage value. Since current flow does produce any corresponding voltage drop, a voltage source displays zero impedance to current flow.*

> **Definition:** *An* ideal current source *maintains a defined current* i(t) *at any* voltage. *A high voltage imposed on the source will not change the current. This means that a current source displays infinite impedance to external voltage: An arbitrary voltage cannot change the current.*

TABLE 10.2
Ideal and Real Voltage Sources Compared

dc Voltage—Ideal	dc Voltage—Real
Fixed voltage, all currents	Series resistance causes terminal voltage to decrease as load increases
Power = VI, no extra loss	I^2R loss in resistor
$Z = 0$, all frequencies	Inductor causes Z to increase with frequency

ac Voltage—Ideal	ac Voltage—Real
Definite $v(t)$, all currents	Series resistance causes voltage to decrease with load power.
No loss	I^2R loss in resistor
dc current has no effect	dc current causes problems with magnetic components
Waveform is defined absolutely	Phase might be undefined, and will depend on switch action or load
$Z = 0$; short circuit except that power can flow at the source frequency	Impedance increases with frequency

These source impedance properties manifest themselves in many ways. A battery or good quality dc power supply acts as a short circuit for ac waveforms or audio signals. Current sources block signal flows.

The energy storage elements used so far for dc loads are consistent with an impedance interpretation. An inductor can be designed to show high impedance, while still allowing the flow of dc power, consistent with a current source. A large capacitor shows low impedance to externally imposed currents. The impedance concept is general, and provides convenient extension to ac sources. Source impedance offers a way to evaluate the "idealness" of a given source, particularly an ac source. Table 10.2 compares ideal and real source properties for voltages, based on the models introduced in Section 10.2.

Real sources show increasing impedance as frequency rises. This is consistent with current source behavior on short time scales. Although most sources show this behavior, few actual power sources can provide current source properties over long time intervals. For both dc and ac sources, addition of series inductance makes the behavior more "current-like" since an inductor increases the impedance.

10.5.2 dc Source Interfaces

A battery with current-source interface is shown in Figure 10.14. If ac voltage is imposed on the terminals, the inductor can be chosen to minimize the ac current flow. The series re-

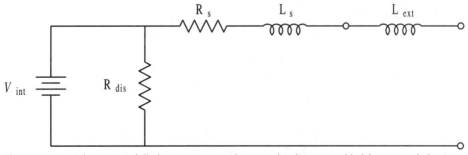

Figure 10.14 A battery modelled as a source, with external inductance added for current behavior.

sistance cannot be avoided, and even an infinite inductor will not counteract the decrease in dc output voltage with increasing load. However, series inductance reduces the resistive loss very substantially.

Example 10.5.1 A battery with internal voltage 13.2 V and series resistance of 0.1 Ω and $L_s = 1$ μH is exposed to a time-varying load. The load (basically a current-sourced inverter) draws 400 W from the internal voltage, and exhibits a 60 Hz ripple voltage of 6 V peak. What is the power loss in the resistor? A series inductor, $L = 10$ mH, is added between the battery and this load. What is the resistor power with the inductor in place?

The load draws average current of $P_{in}/V_{in} = 30.3$ A from the battery. The internal voltage appears as a short circuit to the 60 Hz external ripple. The circuit can be decomposed by superposition into a dc and an ac equivalent combination, as in Figure 10.15. The resistor power is the sum of powers in the two circuits. In the dc circuit, the resistor dissipates $I^2R = 91.8$ W. In the ac circuit, the resistor sees an RMS current of

$$I_{RMS} = \frac{V_{RMS}}{|Z|} = \frac{6/\sqrt{2}}{\sqrt{0.1^2 + (377 \times 10^{-6})^2}} = 42.4 \text{ A} \qquad (10.8)$$

The ac power is $I_{RMS}^2 R = 180$ W. The total loss 180 W + 91.8 W = 272 W.

When the 10 mH inductance is added, the dc equivalent circuit is unchanged. The ac equivalent shows an added series impedance of $j3.77$ Ω. In the dc circuit, the resistor still dissipates 91.8 W. In the ac case, the RMS current is

$$I_{RMS} = \frac{V_{RMS}}{|Z|} = \frac{6/\sqrt{2}}{\sqrt{0.1^2 + 3.77^2}} = 1.12 \text{ A} \qquad (10.9)$$

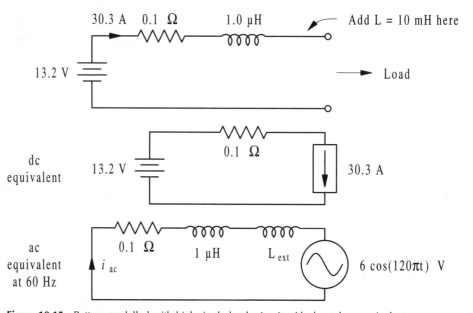

Figure 10.15 Battery modelled with high-ripple load: circuit with dc and ac equivalents.

The ac I^2R loss is 0.127 W, and the total resistor loss is 91.9 W. By making the current source behavior more nearly ideal with an inductor, 180 W of loss was avoided.

To make a source more "voltage-like," parallel capacitance can be added. It is possible to choose capacitors to meet power loss requirements as well as ripple requirements, as the next example shows.

Example 10.5.2 A battery with internal voltage of 13.2 V and with 0.1 Ω in series with 1 μH at the terminals supplies a buck converter with output voltage of 5 V at 300 W. The output inductor L is much greater than L_{crit}. The switching frequency is 50 kHz, and the small battery series inductance does not alter the converter behavior substantially. What is the power loss in the resistor? A parallel capacitor $C = 100$ μF is added at the converter input terminals as a source interface. With this in place, what is the resistor loss?

This is the same battery as in the preceding example. With 5 V 300 W output, the buck converter's inductor must carry 60 A in this application. A circuit diagram is given in Figure 10.16. With no source interface capacitor, the battery is exposed to a switching function waveform $q_1 I_L$. This current has a peak value of 60 A, and causes a 6 V drop inside the battery. Only 7.2 V is available at the battery terminals with this drop, so the duty ratio D_1 is (5 V)/(7.2 V) = 0.694. The RMS value of a switching function is the square root of its duty ratio, so the RMS current in the resistor is (60 A) $\sqrt{0.694}$ = 50 A. The resistor loss is I^2R = 250 W.

With 100 μF in place, Figure 10.17 shows the equivalent source method applied to this problem. The battery current ripple should be lower. For low ripple, the battery current should match the output average value. The duty ratio changes, because now the resistor carries much less than 60 A, and the voltage drop is lower. For very low ripple,

$$P_{\text{in}} = V_{\text{in}} I_{\text{in}} = P_{\text{out}} + P_{\text{loss}} = 300 \text{ W} + I^2_{\text{in}} R$$

$$P_{\text{out}} + P_{\text{loss}} - P_{\text{in}} = 300 + 0.1 I^2_{\text{in}} - 13.2 I_{\text{in}} = 0$$

(10.10)

The solution of this quadratic equation is $I_{\text{in}} = 29.2$ A. With only 2.92 V of internal battery drop, now 10.28 V is available for the converter. The duty ratio is 0.486. The capacitor must absorb 29.2 A while the diode is on. This will change its voltage:

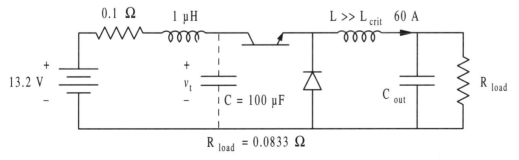

Figure 10.16 Battery and buck converter for Example 10.5.2.

$$i_C = C\frac{dv}{dt}, \qquad 29.2 \text{ A} = 100 \times 10^{-6}\frac{\Delta v}{D_2 T}, \qquad \Delta v = 3.00 \text{ V} \qquad (10.11)$$

The equivalent source method allows the resistor loss to be considered as the sum of loss in two separate circuits. One is a dc circuit with $I = 29.2$ A, while the other is an ac circuit at the 50 kHz fundamental, with a triangle waveform of 2.92 $V_{\text{peak-to-peak}}$. The dc circuit generates loss of 85.3 W in the resistor. From equation (3.9), the triangle waveform has a fundamental amplitude equal to $8/\pi^2$ times its peak. This produces 50 kHz current of

$$I_{\text{RMS}} = \frac{\dfrac{2.92}{2\sqrt{2}}\dfrac{8}{\pi^2}}{\sqrt{0.1^2 + (2\pi\,50000)^2(10^{-6})^2}} = 2.6 \text{ A} \qquad (10.12)$$

Higher harmonics are attenuated. The resistor's ac loss is 0.68 W, for total loss of 86.0 W. The addition of a 100 μF capacitor has made this system far more efficient. Additional capacitance would have little effect, since the internal resistor loss due to ripple is a small fraction of the unavoidable dc loss due to average power flow.

The 100 μF capacitor in the example has impedance of 0.03 Ω at 50 kHz. Therefore it shows a lower impedance to the switching ripple than the battery's 0.1 Ω resistor. In general, a parallel capacitance selected to provide an interface function for a real voltage source will improve circuit operation if its impedance is lower than the real source's internal impedance.

Source interfaces are designed with knowledge of the expected unwanted frequencies. A voltage source interface should act as a short circuit for unwanted components. When switching frequencies are high, the capacitors required for good performance are moderate in size. Good performance means that the internal parts in the source are not exposed to ac waveforms. The energy savings can be significant.

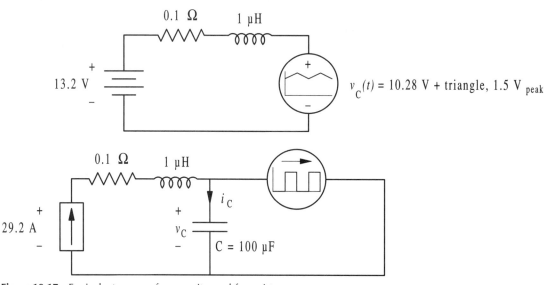

Figure 10.17 Equivalent sources for capacitor and for resistor.

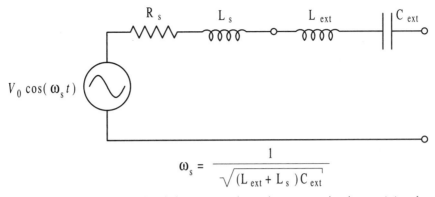

$$\omega_s = \frac{1}{\sqrt{(L_{ext} + L_s)C_{ext}}}$$

Figure 10.18 Series *LC* interface helps give a real ac voltage source the characteristics of an ideal current source.

10.5.3 Interfaces for ac Sources

For ac sources, pure capacitance or inductance will not do the job of providing appropriate impedance without interfering with energy flow. Chapter 8 already addressed the basic issues of resonant filters, and those might be more effective. Here sources are considered specifically. A sinusoidal ac current source provides average power only at its own frequency. It has infinite impedance to dc or any ac frequency, except that power can flow at its own frequency. A series interface circuit should block all frequencies except the source frequency f_s. A series *LC* pair can meet the need. The pair should meet the same requirements as any resonant filter:

1. The resonant frequency must match f_s, so $2\pi f_s = 1/\sqrt{LC}$.
2. The characteristic impedance $\sqrt{(L/C)}$ should be much greater than the series impedance of the real source, and ideally should be high enough to prevent significant power flow.

A series *LC* pair added to a real ac source will give it current source characteristics, as in Figure 10.18.

For a voltage source, a circuit is needed that provides low parallel impedance to all frequencies, except that the voltage source frequency f_s should be allowed to provide power without hindrance. In this case, a parallel-resonant *LC* pair will do the job. Figure 10.19

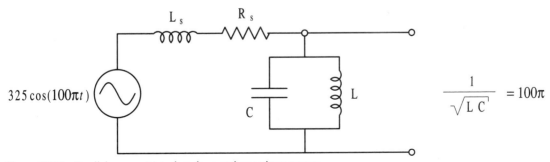

Figure 10.19 Parallel resonant interface for a real ac voltage source.

shows an example based on a 230 V 50 Hz source. The resonant frequency must match f_s. The ratio $(\sqrt{(L/C)})$ should be low compared to the real source's own series impedance.

Example 10.5.3 A real ac voltage wall source, nominally 230 V_{RMS} at 50 Hz, displays series resistance of 0.12 Ω and series inductance of 4 mH. It supplies a rectifier load for which the power is 150 W and the RMS current is 5 A. The current is drawn in brief spikes near the voltage peaks. Design a source interface circuit for this situation. Compare the loss in the 0.12 Ω resistor before and after the interface is in place.

At 50 Hz, the real source's impedance is 0.12 Ω + $j1.26$ Ω, with magnitude 1.26 Ω. The rectifier draws 5 A_{RMS} and apparent power $V_{RMS}I_{RMS}$ = 1150 VA. Without an interface, the I^2R loss in the resistance is 3 W. This is not large, but it is 2% of the output power. The rectifier's internal loss is probably also about 2%. The power factor P/S = (150 W)/ (1150 VA) = 0.130. A parallel LC pair is needed, with characteristic impedance on the order of 0.12 Ω. Given resonance at 50 Hz and $\sqrt{(L/C)}$ = 0.12 Ω, the values could be C = 27 000 μF and L = 375 μH. The rectifier power of 150 W suggests that the 50 Hz current component has an ideal RMS value of (150 W)/(230 V) = 0.652 A. This current continues to flow in the source after the interface is added.

To see the effect of the interface, take the third harmonic of the current, which has a value of several amps. The third harmonic equivalent circuit, shown in Figure 10.20, treats the ideal part of the source as a short circuit, and treats the third harmonic of the rectifier current as if it is an ideal source. The circuit is a current divider. If the rectifier draws a third harmonic current of about 4 A_{RMS}, almost all of this will flow in the interface. The input source series impedance supplies only 60 mA of third harmonic current with the interface in place. The resistor power is approximately the sum $(0.652 \text{ A})^2 R + (0.06 \text{ A})^2 R$ = 0.051 W. This reduction reflects a change in power factor seen at the source from 0.130 to 0.996. The total system loss has been cut in half, since only rectifier internal losses remain.

ac source interfaces share the problems of resonant filters: Since the frequencies are often 50 or 60 Hz, the parts need to be large. Also, the Q value should be high, meaning that large voltages or currents are present within the resonant pair. The 27 000 μF capacitor in Example 10.5.3 has such a low impedance that it carries almost 2000 A at 50 Hz! The proposed design might be impractical given the extremes in resonant circuits. In general, lower Q values must be used.

One important advantage in designing ac source interfaces is that the unwanted frequencies are often well known. For example, the rectifier in Example 10.5.3 exposes the input source only to odd multiples of 50 Hz. Few power electronic loads other than cy-

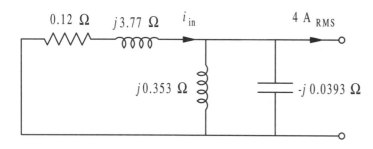

Figure 10.20 Third harmonic equivalent circuit for Example 10.5.3.

cloconverters or integral cycle controllers expose a source to significant subharmonics. If there are no subharmonics, the source interface only needs to deal with unwanted frequencies above f_s.

In the series LC case, the pair shows inductive reactance above the resonant frequency. In the parallel case, the pair shows capacitive reactance above f_r. A simple approximate source interface would use series inductance for current or parallel capacitance for voltage. The parts can be chosen so as not to alter the behavior of the circuit at f_s too much. At least high switching harmonics will be filtered out with this method.

Example 10.5.4   Revisiting the 230 V 50 Hz source supplying a rectifier, from Example 10.5.3, a 27 000 μF capacitor is out of the question for this application. The part is very large (its volume exceeds a liter), and the extreme resonant currents are unrealistic. Consider instead a simple parallel capacitor as a source interface. The power factor target is 0.8, and an estimate of the value of C to do the job is needed.

From the previous example, the current that contributes to 50 Hz power flow has an RMS value of 0.652 A. With no interface, the total RMS current is 5 A. Input power of 150 W and power factor of 0.8 implies $S = P/pf = 187.5$ VA, and an RMS current of 0.815 A. Ignoring the harmonics for a moment, this current is consistent with a capacitive current of $[(0.815^2 - 0.652^2)]^{1/2}$ A $= 0.489$ A. For 230 V input, the capacitor for this current is 6.8 μF. Unfortunately, this capacitor has too much impedance to be very effective below about 2 kHz. It also generates a resonance with the source's series inductance near the nineteenth harmonic. This resonance can amplify the nineteenth harmonic flow in the input source, and might actually increase the loss. A larger capacitor is necessary to avoid this trouble, but then the power factor will be lower.

The examples illustrate the dilemmas encountered when trying to create useful ac source interfaces. The essence of the trouble is the same issue encountered in Chapter 8: Resonant filters for power line frequencies are fundamentally difficult to implement. The most common alternative is to focus more closely on the unwanted components and direct the filter action toward avoiding them.

Consider an ac voltage source that supplies a PWM rectifier, as in Figure 6.26. The unwanted components occur close to the switching frequency, which is likely to be 20–200 kHz. A parallel capacitance with low impedance at 20 kHz will provide a useful source interface. In the previous examples, a capacitor of only a few microfarads would provide good source qualities for 20–200 kHz switching. Figure 10.21 shows both the PWM rectifier application and the PWM inverter circuit with the appropriate interface components in place. The frequency separation inherent in the PWM process is an important advantage for interface design, since it can easily reduce the required component values by a factor of 1000. In addition, high-frequency PWM can be filtered without resort to resonant combinations.

For power line frequency applications, a designer can focus on specific unwanted components. In rectifiers, the line currents are square waves containing odd harmonics of the source frequency. A 60 Hz rectifier can use circuits tuned to eliminate 180 Hz, 300 Hz, 420 Hz, and higher components in place of a 60 Hz resonant set. The necessary configuration, given in Figure 10.22, uses *tuned traps* to shunt large unwanted components to ground. The tuned trap concept offers crucial advantages over a direct resonant interface: The resonant frequencies are higher, so the L and C values are lower. The value of Q does not need to be

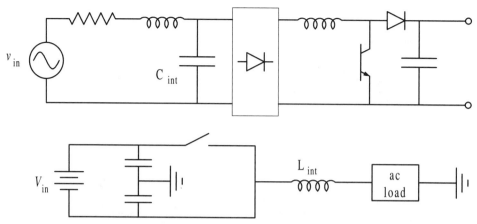

Figure 10.21   PWM rectifier and inverter with source interfaces.

very high, since each trap is intended primarily to minimize impedance only at one frequency. Tuned traps are relatively common in high-power applications.

Example 10.5.5   A 3ϕ source is rated at 208 $V_{\text{RMS}(l-l)}$ and 60 Hz. Each of the three individual phases has series resistance of 0.08 Ω and series inductance of 1.6 mH. This source supplies a six-pulse rectifier. The rectifier line current has a peak value of 20 A, and there is no phase shift between a given phase voltage and the corresponding current. A tuned trap filter arrangement is to be used to reduce two harmonics. It is desired that the displacement factor (the power factor for the 60 Hz component) be at least 96%. It is also desired to minimize distortion. Suggest a circuit configuration to meet the requirements, and evaluate its performance.

Figure 10.23 shows the circuit described by this problem. The symmetry of the current waveform means that no third harmonic is present, so the tuned traps should remove the fifth and seventh harmonics. Since the line-to-line voltage is 208 V, the individual sources

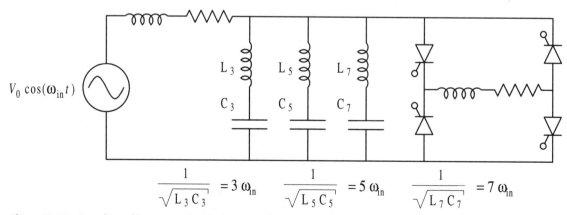

$$\frac{1}{\sqrt{L_3 C_3}} = 3\,\omega_{\text{in}} \qquad \frac{1}{\sqrt{L_5 C_5}} = 5\,\omega_{\text{in}} \qquad \frac{1}{\sqrt{L_7 C_7}} = 7\,\omega_{\text{in}}$$

Figure 10.22   Tuned trap filter to provide ideal voltage characteristics for a 60 Hz source.

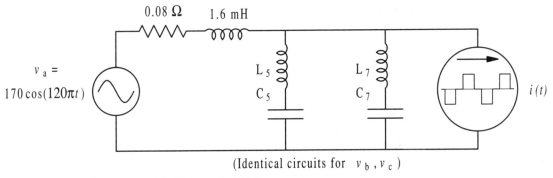

Figure 10.23   Interfaces associated with Example 10.5.5.

have an RMS value of 120 V. The six-pulse rectifier current is shown as an equivalent source. The design problem involves tuning the traps to eliminate fifth and seventh harmonics, meeting 60 Hz requirements, and minimizing distortion. Define the line radian frequency $\omega_{in} = 120\pi$ rad/s. The fifth harmonic trap must be resonant at 300 Hz, so $1/\sqrt{(L_5C_5)} = 5\omega_{in}$. Similarly, $1/(L_7C_7) = 7\omega_{in}$. The 60 Hz equivalent circuit is needed to test the displacement factor. It is shown for phase A in Figure 10.24a. From Chapter 5, equation (5.48), the 60 Hz component of the output current is found to be $(4I_{out}/\pi)\sin(\pi/3)\cos(\omega_{in}t)$, or $22\cos(120\pi t)$. Loop and node equations can be written for the source current to allow computation of the power factor:

$$120\angle 0° - j\omega_{in}L_s\tilde{I}_{in} = \tilde{V}_s,$$

$$\tilde{I}_{in} = j\omega_{in}\left(\frac{25}{24}C_5 + \frac{49}{48}C_7\right)\tilde{V}_s + \frac{22}{\sqrt{2}}\angle 0 \tag{10.13}$$

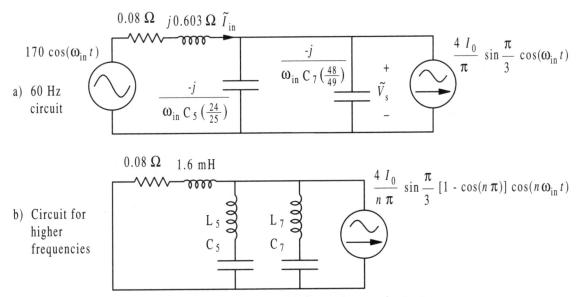

Figure 10.24   Equivalent circuits for 60 Hz and for higher frequencies for Example 10.5.5.

It is required that

$$\cos(\angle \tilde{V}_s - \angle \tilde{I}_{in}) \geq 0.96 \qquad (10.14)$$

Solving these relationships simultaneously (with a computer program or similar math tool) gives a value of about 71.6 μF for the equivalent capacitance $25C_5/24 + 49C_7/48$. The distortion is measured by THD. In this case, the appropriate quantity is the THD of the current I_{in}. With no filter, the output current gives a THD value of about 31% from equation (5.49). With the filter, the THD can be found from term-by-term Fourier analysis. The equivalent circuits above 60 Hz are current dividers, as in Figure 10.24b.

A little experimentation shows that the lowest distortion occurs when C_7 accounts for almost all of the 71.6 μF. The following circuit elements can be used for the end result (C_7 has been decreased a little to make C_5 slightly bigger for a reasonable value of L_5):

$$C_5 = 10 \ \mu F \quad L_5 = 28.1 \ mH$$

$$C_7 = 59 \ \mu F \quad L_7 = 2.43 \ mH$$

These values reduce the THD from the unfiltered 31% to a value of 9.22%. The input current waveform is shown in Figure 10.25. The distortion is still present, but the waveform looks more sinusoidal. The voltages across the individual circuit elements are important for ratings. In the case of the fifth harmonic trap, the current component at 300 Hz generates about 165 V_{RMS} across C_5 and L_5. This is not nearly as extreme as the expected behavior from Example 10.5.3. Tuned traps are a relatively practical way to reduce distortion.

The difficulties associated with resonant filters mean that PWM techniques have a significant filtering advantage. Because of this, the power levels addressed with PWM inverters and rectifiers continue to climb as devices improve.

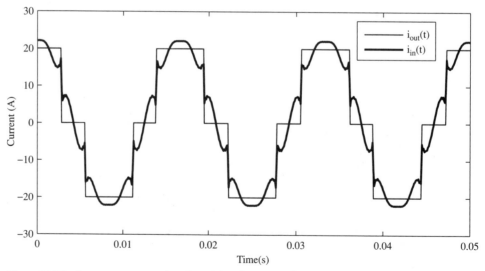

Figure 10.25 Source current as a function of time for a six-pulse rectifier with fifth and seventh harmonic traps.

10.6 RECAP

Real sources and loads have internal resistance, series inductance, and sometimes parallel capacitance. These extra circuit elements contribute to power loss, and also tend to give most power circuits the characteristics of current sources when frequencies are high. dc loads can be treated with simple steady-state models if their behavior does not change much over the time scale of a switching period. Examples of such loads include dc motor drives and electronic loads with large input capacitors. dc loads with fast variations include many digital circuits. In these cases, the load dynamics can be much faster than the switching frequency, and the output filter elements of a converter determine the behavior.

ac loads are fundamentally similar to dc loads, although the impedances of internal L and C values do not become trivial at the wanted frequency. Some common loads, such as resistive heating elements, have unusual nonlinear behavior. Loads as simple as incandescent lamps require temperature-dependent models for proper analysis of their behavior.

The self-inductance of a single wire was discussed. The following is a helpful rule of thumb.

Empirical Rule: *The self-inductance of a single wire in a typical geometry is about 5 nH/cm. A wire pair gives about 10 nH/cm if not twisted, while a tightly twisted pair yields about 4 nH/cm.*

Alternative wire geometries, such as flat strips or multistrand Litz wire, can be used to bring the inductance down somewhat. These special arrangements are helpful in high-frequency applications. They reduce the internal self-inductance. This decreases the skin effect and helps provide a better current distribution as frequencies increase.

The critical inductance and capacitance values introduced in Chapter 9 provide a way to characterize a dc source. A real source can be modelled as ideal if the series inductance or parallel capacitance is higher than the critical value, and an ideal model becomes very accurate when $L \gg L_{\mathrm{crit}}$ and $C \gg C_{\mathrm{crit}}$. Values of critical inductance and capacitance help establish specifications such as minimum load for a given dc–dc converter or rectifier.

Any power electronic system with real sources and loads must address the interface problem. If a real source is connected to a converter without regard for unwanted components, it is likely that substantial losses will occur in the real source's resistances. A properly designed interface minimizes the exposure of a real source or load to unwanted harmonics. The basic issue in addressing the source interface problem is to match the impedance properties of ideal sources: Impedance is high for a current source and near zero for a voltage source.

dc sources require series inductance to give them current-source characteristics, or parallel capacitance to give them voltage-source characteristics. ac sources in principle require resonant circuits to provide the proper impedance properties. For an ac voltage source, the simplest interface is a parallel LC circuit. In this case, the source frequency is matched to the LC resonance. At this frequency, the net impedance is infinite, so the power transfer process is unaffected by the interface. At all other frequencies, the LC pair exhibits low impedance, and shunts off the unwanted component currents. For an ac current source, a series LC pair functions in similar fashion: The wanted frequency passes through unaltered, while other frequencies see a very high impedance.

Unfortunately, resonant interfaces bring problems of their own. In particular, very high Q values are desired. This results in extreme voltages across the filter elements. Two alternatives are available:

- Use PWM to give a wide separation between wanted and unwanted components. This eliminates the need for resonant pairs.
- Use tuned traps to eliminate specific unwanted frequencies, rather than trying to eliminate all possible frequencies. This allows the use of much lower Q values, and makes resonant interface circuits practical.

It is important to resolve interface problems in real systems to obtain high efficiencies and good performance.

PROBLEMS

1. Tests on a lead-acid battery show the following characteristics with full charge: The open-circuit terminal voltage is 13.4 V, the terminal voltage is 13.3 V with a 10 A load, and 13.5 V during 10 A charging. The manufacturer reports that the battery loses the equivalent of about 1200 C of charge each week when it is in storage. Based on Figure 10.4, but ignoring the L and C values, determine a circuit model for this battery.

2. A stereo amplifier covers the full audio range from 20 Hz to 20 kHz. It can provide a total of up to 200 W into an 8 Ω speaker. The amplifier is about 50% efficient. It is supplied from a forward converter that produces ± 60 V rails from rectified ac line voltage. The forward converter switches at 100 kHz. Can the amplifier be treated as quasi-steady, or is it a transient load? What is the highest value of di/dt that you would expect at the forward converter output?

3. A boost converter is wired to a remote load. The converter has 48 V input and 120 V output. The load draws 4 A. For reasons of space, the converter's capacitor is located at the load instead of adjacent to the output switch. The switching devices are rated for 200 V and 5 A. Estimate the highest wire inductance allowed in the connection to the load without causing excessive voltages on the switches. Use a switching time of 100 ns. What length of wire does this represent?

4. An advanced microprocessor runs at 2.5 V. The effective total capacitance inside the device is 80 000 pF. The digital clock runs at 50 MHz. The power supply uses a buck converter to step 5 V down to the necessary 2.5 V level.

   a. If the buck converter switches at 200 kHz, what should its output capacitance be to ensure no more than 50 mV voltage drop while the buck converter's diode is on?

   b. What is the average current drawn by the microprocessor?

   c. The connection between the converter and microprocessor has enough length to produce a significant inductance. If the processor supply current can swing as fast as 100 A/μs, what is the highest value of inductance that can be tolerated without allowing the processor supply voltage to drop below 2.0 V even momentarily? What connection length does this suggest?

5. Your company purchases 200 W dc–dc converters from a well-known manufacturer. Your experience is that they work well, and are highly reliable. Lately, however, failures have been reported on the line. On investigation, you find that a new technician has been reconnecting the power supply input stage temporarily so a power meter can be used for measurements. The measurements are made, then the standard connection is restored and the supplies are placed on the line. The

meter is on a shelf 50 cm above the bench. Explain a possible cause in terms of wire inductance. Use sample computations to bolster your arguments.

6. A buck converter supplies a 480 V dc motor from a 500 V source. The switching frequency is 10 kHz. The motor has armature inductance $L_a = 1$ mH and resistance of 0.05 Ω. At full speed, the motor's internal voltage is 480 V. What is the minimum motor output power that will ensure $L_a \geq L_{crit}$ at full speed? At 10% speed?

7. Figure 10.26 illustrates a PWM inverter supplying one phase of an ac induction motor. The motor is intended for a pure 60 Hz source at 120 V_{RMS}. The inverter output voltage is approximately $170 \cos(120\pi t) + 170 \cos(24\,000\pi t)$. Compare the circuit power with this PWM waveform to the intended 60 Hz power.

8. Your employer intends to obtain an uninterruptible power unit capable of supplying up to 10 kW at 230 V and 50 Hz from a 12 V battery bank. The switches in the inverter bridge are rated at 600 V and 100 A. It is desired to store the batteries in a separate room, and you have some concern about connection inductance.

   a. Estimate the longest allowable connection length if the switches require 200 ns to act.

   b. Suggest a way to use a capacitor to permit longer connection lengths.

9. Estimate the voltage induced across a one meter length of copper wire when a resonant power converter draws current $i(t) = 20 \cos(400\,000\pi t)$ through the wire. What if a steel wire with $\mu = 1000\mu_0$ is substituted?

10. A given load draws between 40 W and 200 W from a 48 V to 12 V dc–dc converter. The switching frequency is 100 kHz. Suggest an appropriate circuit. What is the critical inductance?

11. A 150 V battery bank supplies a backup system for a major communications company. The battery bank has series resistance of about 0.04 Ω and total series inductance of 8 μH. The backup system is a PWM inverter. It draws average power of 3 kW. There is an unwanted 120 Hz current component of 10 A peak, and also a component at 25 kHz with a peak value of 20 A. The arrangement is shown in Figure 10.27.

   a. What is the loss in the battery resistance as shown?

   b. Suggest an interface element to improve the situation. With your interface in place, what is the loss in the battery resistance?

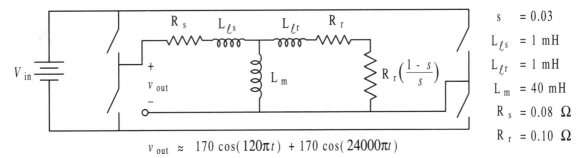

$$v_{out} \approx 170 \cos(120\pi t) + 170 \cos(24000\pi t)$$

Figure 10.26 Induction motor powered from PWM inverter for Problem 7.

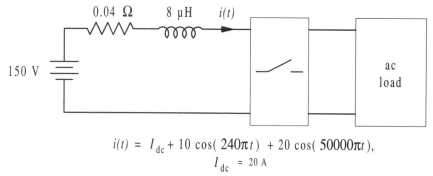

$$i(t) = I_{dc} + 10 \cos(240\pi t) + 20 \cos(50000\pi t),$$
$$I_{dc} = 20 \text{ A}$$

Figure 10.27   Battery bank supplying a PWM inverter.

12. A rectifier draws 12 A_{RMS} from a standard 120 V_{RMS} 60 Hz wall plug. The power factor is 10%. The wall source has series resistance of 0.2 Ω and series inductance of 50 μH.

   a. What is the loss within the wall source as this system is described? What is the loss as a percentage of the rectifier load?

   b. With a properly designed interface, what current will the rectifier draw from the wall source? What will the loss in the wall source become with the interface?

13. A 60 Hz source with series resistance of 0.2 Ω and series inductance of 50 μH is to be provided with a resonant interface to make it nearly ideal. Draw the necessary circuit, and determine L and C values if the interface should have an impedance of less than 10 Ω at 120 Hz.

14. A synchronous motor is supplied with 400 Hz from a 60 Hz to 400 Hz converter switching at 460 Hz. This converter produces a large unwanted component at 520 Hz. It is hard to provide a high enough value of Q to pass the 400 Hz output while providing effective blocking for the 520 Hz component. One alternative to consider is a parallel-resonant tuned trap. The combination, including parameters of the machine, is shown in Figure 10.28. Both the 400 Hz and 520 Hz voltage components have an amplitude of 200 V. Without the interface in place, what is the ratio of the 520 Hz current component to the 400 Hz component amplitudes? What is this ratio with the interface in place?

15. A buck converter has a 200 μH output inductor. The switching frequency is 25 kHz. The input can be between 20 and 30 V, while the output is to be held to +12 V $\pm$ 2%. The input source has series resistance of 0.2 Ω. Find the minimum output load power which will allow these specifications to be met. What is the loss in the source resistance at this load level?

16. A buck-boost converter operates with an input battery. It converts +12 V to −12 V at a power level of about 75 W. The switching frequency is 120 kHz. The battery has an internal series resistance of 0.2 Ω and series inductance of 200 nH.

   a. What is the operating value of the duty ratio? What power is lost in the battery resistance?

   b. Propose an interface structure to improve operation and decrease losses. What are the duty ratio and battery resistance loss with your interface in place?

17. A filter is needed for a large uninterruptible supply. The supply generates a quasi-square wave at five levels ($\pm V_{peak}$, $\pm V_{peak}/2$, and 0), which is controlled to eliminate the third and fifth harmonics. The fundamental component of the output voltage has an RMS value of 460 V and a frequency of 60 Hz. The nominal load is 40 kW.

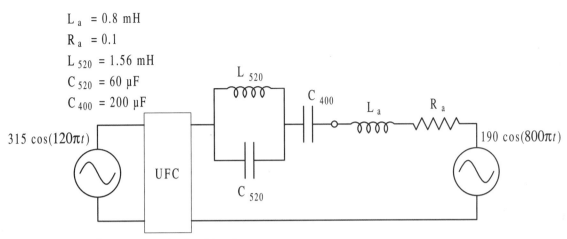

$L_a = 0.8$ mH

$R_a = 0.1$

$L_{520} = 1.56$ mH

$C_{520} = 60$ μF

$C_{400} = 200$ μF

Figure 10.28 Synchronous motor load and interface.

   a. Plot the waveform described by this problem statement.

   b. Design a tuned trap filter that eliminates seventh and eleventh harmonics.

   c. If the load is resistive, what is the current waveform before and after the trap filter is added?

18. A PWM inverter switches at 50 kHz. The output ranges from 0 Hz to 100 Hz. The input is an 80 V_{dc} source. The load is nominally 8 Ω at 200 W. Suggest appropriate interfaces for both the input and output. Design an input interface to make the current ripple less than 20%. If the input source has resistance of 0.2 Ω, how much will the interfaces improve efficiency?

19. A particular large utility system has a typical load of 8000 MW. The power factor of the load on average is 0.85, and the distribution network has overall efficiency of 96%. How much power will be saved if ideal interfaces can correct the load power factor to 1.00?

20. In high-voltage dc systems, a high pulse number is used to minimize filter requirements. Consider a 24-pulse midpoint rectifier, with input frequency 60 Hz. The load is resistive. You are asked to compare series and shunt interface filtering. In one case, an *LC* trap will eliminate the ripple fundamental at 1440 Hz. In the other case, a parallel *LC* combination will block 1440 Hz to the load. The load is 200 MW, at a voltage of 100 kV_{dc}. Compare the two alternatives. Does either one require less stored energy in the interface, or do they match?

REFERENCES

T. R. Bosela, *Introduction to Electrical Power System Technology*. Upper Saddle River, NJ: Prentice Hall, 1997.

D. Brown, E. P. Hamilton III, *Electromechanical Energy Conversion*. New York: Macmillan, 1984.

V. Del Toro, *Basic Electric Machines*. Englewood Cliffs, NJ: Prentice Hall, 1990.

C. T. A. Johnk, *Engineering Electromagnetic Fields and Waves*. New York: Wiley, 1975.

R. D. Shultz, R. A. Smith, *Introduction to Electric Power Engineering*. New York: Harper & Row, 1985.

Specialty Battery Division, *Charging Manual*. Milwaukee: Johnson Controls, 1990.

P. Wood, *Switching Power Converters*. New York: Van Nostrand Reinhold, 1981.

CHAPTER 11

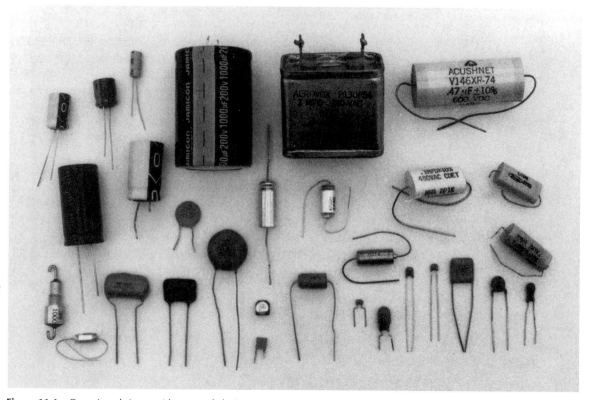

Figure 11.1 Capacitors bring a wide array of choices.

CAPACITORS AND RESISTORS

11.1 INTRODUCTION

The capacitors, inductors, and resistors used in circuit analysis have ideal properties. Capacitors and inductors follow precise derivative relationships. Real components do not follow such simple models, just as in the case of sources. The nonideal behavior of energy storage components is very important in the context of power electronics. After all, we seek large values of inductance and capacitance for energy storage and source interfaces. In a converter, components are exposed to high and rapidly changing voltages and currents. High derivatives and extremes of voltage and current can produce losses and complicated behavior.

Many of the nonideal effects are straightforward: wires have resistance and inductance, coils of wire exhibit capacitance between adjacent turns, insulators have leakage resistance and might exhibit losses. The designer of power electronic circuits must consider many of these effects. Capacitors, for example, have temperature and current ratings because of resistive losses. If losses are ignored, circuit conditions can cause a component failure. If other effects are ignored, there will be unexpected operating results.

In this chapter, the characteristics of real capacitors and resistors will be considered. The capacity of copper wire is discussed. In the case of capacitors, like the devices shown in Figure 11.1 a standard circuit model is developed. The circuit is similar to models used in RF design, but is important in power converters switching at 20 kHz or even less. Inductors will be addressed later, in the larger context of magnetic design in Chapter 12.

11.2 CAPACITORS—TYPES AND EQUIVALENT CIRCUITS

11.2.1 Major Types

Real capacitors come in many forms, but almost all are built from two conducting plates or films, separated by an insulating layer. From field theory, the parallel plate geometry in Figure 11.2 gives $C = \epsilon A/d$ for the capacitance, where ϵ is the electrical permittivity of the insulating layer, A is the plate surface area, and d is the plate spacing. By definition, the charge

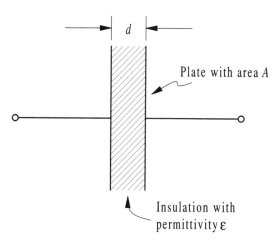

Figure 11.2 The basic parallel-plate capacitor geometry.

Plate with area A

Insulation with permittivity ε

on either plate will be $Q = CV$, where V is the applied voltage. In the parallel plate arrangement, the electric field is given by V/d. The time derivative of $Q = CV$ gives the standard property $i = C(dv/dt)$.

The parallel plate arrangement provides some challenges. Consider that two plates, each with one square meter of area, separated by a 1 μm air space form a capacitor of less than 10 μF. Typical capacitors needed in power converters are on the order of 100 μF or more. However, the permittivity of free space or air is small, with $\epsilon_0 = 8.854$ pF/m. Evidently, very large plates are needed to give significant values of C. One challenge is how to get this large surface area into a small package. Capacitance can be raised by decreasing the spacing. However, if the insulation is too thin, it will not sustain a useful voltage level, but instead will fail because of dielectric breakdown.

Because of the packaging challenge, there has been a long and continuing history of trying new materials and arrangements for capacitors. Virtually every known polymer has been tested as insulation for capacitors. A wide range of ceramic materials has been tried, and many are commonly used. Paper, air, glass, and most other common insulation materials have been attempted as well. Today, there are dozens of different kinds of capacitors, distinguished primarily by method of construction and by the specific insulating materials. These types can be listed in three general classes:

- *Simple dielectric capacitors.* These are the most basic type, formed directly with parallel plates and an insulator. They might be constructed as a sandwich of foil and plastic layers, then rolled or folded to fit in a small package. Or metal might be plated directly on an insulating material. Simple dielectrics come in an enormous variety. A few examples are shown in Figure 11.3.

- *Electrolytic capacitors.* In this case, one plate is formed as a porous metal mass. Aluminum and tantalum are the usual choices for this material. The insulator is formed by oxidizing the metal surface, and the second electrode is a liquid or solid electrolyte. The familiar capacitor cans are usually electrolytic. Figure 11.4 gives examples, and shows a cross section of one.

- *Double-layer capacitors.* The classic example is the membrane of a biological cell, in which the chemical properties of a polar molecule in solution can maintain a substantial charge separation over a very short distance. A few examples, based on activated carbon

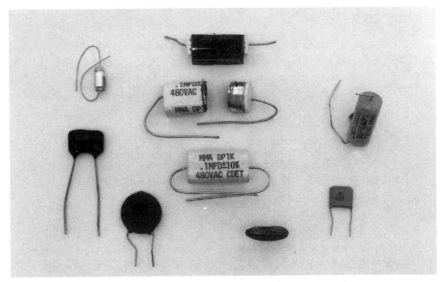

Figure 11.3 Simple dielectric capacitors, showing a typical construction.

or other substrates, are available commercially. These are sometimes called *ultracapacitors* or *supercapacitors* because it is possible to obtain very high capacitance values with this arrangement. The technology is developing, and applications are expected to grow.

All three major classes are used for power conversion. Some comparisons will be made in this chapter. Another type, called a *feedthrough capacitor*, uses a coaxial geometry to provide a capacitor with reduced wire inductance.

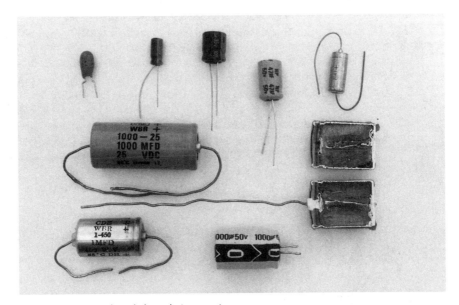

Figure 11.4 Examples of electrolytic capacitors.

11.2.2 Equivalent Circuit

In a real capacitor, the wires and plates have resistance and inductance. The insulation is not perfect, and will have a leakage resistance. These direct properties can be summarized in a circuit model, shown in Figure 11.5. These properties are inherent in the process of making and using a capacitor. They can be minimized, but not avoided altogether. This equivalent has a number of crucial properties:

- The current is not really consistent with $i = C(dv/dt)$.
- Current flows even when dc voltage is applied.
- The combination of L and C creates a resonance. Above the resonant frequency, the device behaves as an inductor!
- There is a nonzero power loss.

The inductance is expected to be in the nanohenry range, since it represents wire inductance and geometric effects. The leakage resistance should be very high, and the time constant of the insulation, $\tau = R_{leak}C$ should be very long.

In many power electronics applications, a capacitor is to be used at a specific frequency. A boost converter, for instance, exposes a capacitor to a large signal at the switching frequency. Given a radian frequency ω, the circuit model can be evaluated as a set of impedances, then simplified. The steps and relationships are given in Figure 11.6. The derivation begins with a capacitor C in parallel with the leakage resistance R_{leak}. Each step retains the wire resistance R_w and impedance of the wire inductance L_w. The parallel combination of leakage and capacitor is transformed into a series equivalent. Since $R_{leak}C$ is high, at frequencies above about 1 Hz it should be true that $\omega^2 R_{leak}{}^2 C^2 \gg 1$. Then the reactance portion simplifies to $-j/(\omega C)$. The series resistance portion simplifies to $1/(\omega^2 R_{leak}C^2)$.

The equivalent circuit has been simplified to a series R-L-C combination, valid for frequencies above 1 Hz. The inductor in the circuit is termed *equivalent series inductance* or *ESL*. The capacitor represents the internal ideal capacitance effect. The resistor, termed *equivalent series resistance* or ESR, has the value

$$ESR = R_w + \frac{1}{\omega^2 R_{leak}C^2} \tag{11.1}$$

The combination, shown in Figure 11.7, is sometimes called the *standard model of a capacitor*. It is widely used by manufacturers as the basis for specifications. The standard model

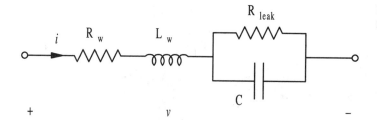

Figure 11.5   General circuit model for a real capacitor.

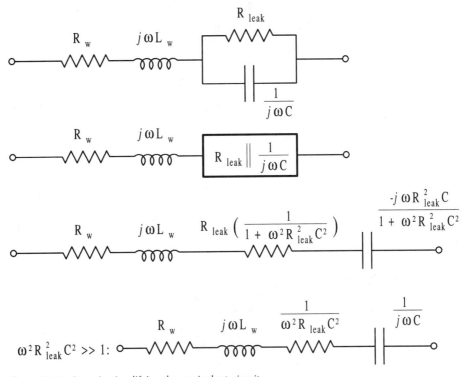

Figure 11.6 Steps in simplifying the equivalent circuit.

neglects dc leakage, and this behavior is usually specified separately. Since the ESR is obtained through a transformation, it is a nonlinear frequency-dependent resistance. The value is typically given at a particular frequency (120 Hz is common) on data sheets.

The *dissipation factor*, *df*, is often used to indicate the quality of a capacitor. It represents the ratio of resistance to reactance. For frequencies well below $1/\sqrt{(ESL)C}$, the Standard Model's reactance is approximately $1/(\omega C)$, and $df = R/X = (ESR)\omega C$. This ratio is also called the *loss tangent*, $\tan \delta$, since it is the complement of the impedance angle $\phi = \tan^{-1}(X/R)$.

If the wire resistance is small, the ESR becomes the second term in (11.1), and the dissipation factor can be written

$$\tan \delta = \omega(ESR)C \approx \frac{\omega C}{\omega^2 R_{leak}C^2} = \frac{1}{\omega R_{leak}C} \tag{11.2}$$

Figure 11.7 The standard model of a capacitor.

$$ESR = R_w + \frac{1}{\omega^2 R_{leak}C^2} \qquad ESL = L_w$$

The leakage resistance can be computed from the resistivity of the insulation, ρ, since the plate geometry is known. The relationships are

$$C = \frac{\epsilon A}{d}, \qquad R_{\text{leak}} = \frac{\rho d}{A} \qquad (11.3)$$

Notice that the product $\omega R_{\text{leak}} C$ simplifies from (11.3) to give $\omega R_{\text{leak}} C = \omega \rho \epsilon$, and $\tan \delta = (\omega \rho \epsilon)^{-1}$. Thus the loss tangent is independent of geometry, and can be considered a material property associated with the insulator. The ESR value depends strongly on the choice of insulation material.

Insulators are not well characterized by constant resistivity. An interesting fact is that for many insulators, the value $\tan \delta$ is roughly constant over a substantial frequency range. Given the loss tangent of a particular material, the ESR value can be determined as

$$\text{ESR} \approx \frac{\tan \delta}{\omega C} \qquad (11.4)$$

with $\tan \delta$ taken as a characteristic value for a given insulation. In power electronics applications, low ESR values are desired to minimize losses and other effects. Low ESR is achieved either through the use of very large capacitors or through the use of high frequencies. The ESR value can be a significant problem. At low voltages (5 V and below), capacitors are often selected to provide a target ESR value rather than to store extra energy.

The following examples will begin to provide some perspective on capacitor properties.

Example 11.2.1   A 100 μF electrolytic capacitor is connected to a circuit through about 5 cm of wire, total. It has additional internal inductance of about 15 nH. The insulation has a roughly constant loss tangent of about 0.2 (20%). Estimate the resonant frequency. What is the ESR for a dc–dc converter application switching at 40 kHz?

The rule of thumb from Chapter 10 is 5 nH/cm for the wire, so the total ESL in this application should be about 40 nH. For 100 μF, this value of ESL and C give a resonant frequency of 80 kHz. The ESR value will be $(\tan \delta)/\omega C$ with $\omega = 40\,000(2\pi)$ rad/s. This gives 0.0080 Ω or 8.0 mΩ, neglecting wire resistance.

Example 11.2.2   A 2 μF simple dielectric capacitor has $C = 2$ μF and ESL of 25 nH. The insulator loss tangent is roughly 1%. What is the resonant frequency? What is the ESR for (a) a full-wave rectifier with 60 Hz input, and (b) dc–dc switching at 150 kHz?

The resonant frequency corresponding to 25 nH and 2 μF is 712 kHz. If this capacitor is applied at the rectifier output, it will see a large 120 Hz component. The ESR will be $(\tan \delta)/(\omega C)$, with $\omega = 240\pi$ rad/s, giving 6.63 Ω. In the dc–dc converter application at 150 kHz, the ESR is computed at $\omega = 300\,000\pi$ rad/s, to give 0.0053 Ω.

11.2.3 Impedance Behavior

When capacitors are used as source interfaces, their key property is an impedance magnitude $1/(\omega C)$ that falls with frequency. On a log-log plot of $|Z|$ vs. frequency, the ideal behavior will follow a straight line with slope -1. Real capacitors show much different behavior: the impedance falls until the resonant frequency is reached, then it rises as the ESL begins to dominate. A plot of the phase $\angle Z$ vs. frequency shows a quick transition from $-90°$ to $+90°$ at the resonant value. The basic impedance behavior does not depend on the capacitor type: even the best available parts follow this characteristic.

Figure 11.8 shows laboratory data for $|Z|$ vs. frequency for a 1000 pF polystyrene capacitor. From the resonant frequency of about 36 MHz, the inferred value of ESL is 20 nH. An ESR value of about 0.15 Ω can be obtained immediately at the resonant frequency, since the impedance is purely resistive there. This capacitor has tan δ below 1% (at least up to a few megahertz), and is considered a nearly ideal part.

When high capacitance values are needed, the device packages can be large. This means that the connection lengths are likely to increase. With both higher C and higher ESL, large capacitors have limited resonant frequencies. Consider an example.

Example 11.2.3 A 2700 μF capacitor rated for 400 V is needed for an inverter source interface application. The device is intended to absorb the unwanted components in a PWM process. The connections to this large part total 20 cm in length. Estimate the resonant frequency, and comment.

This capacitor should have an ESL of at least 100 nH, just because of the wire connections. The resonant frequency is $1/\sqrt{(2700 \ \mu F)(100 \ nH)} = 6.1 \times 10^4$ rad/s, or 9.7 kHz. This part will not function as a capacitor for frequencies above 9.7 kHz. Its use in a PWM system makes sense only if the switching frequency is well below 10 kHz.

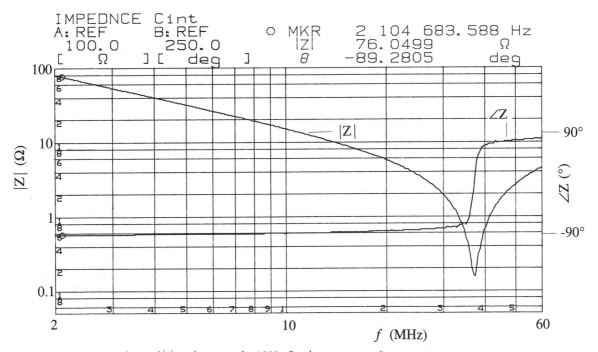

Figure 11.8 Experimental test of $|Z|$ vs. frequency for 1000 pF polystyrene capacitor.

This example illustrates the difficulties in applying large capacitors in high-power switching applications.

The impedances have immediate implications because of ESR. For example, the 100 μF capacitor of Example 11.2.1 has an ESR of about 8 mΩ at 40 kHz. The reactance at this frequency is $1/(\omega C) = 40$ mΩ, so the ESR value is 20% of the reactance, and not necessarily negligible. Now, apply this capacitor to reduce ripple in a 5 V, 500 W application. The load current is 100 A, and the load resistance is only 50 mΩ. The load is not much different from the capacitor's impedance, and in fact the filtering action might not be very effective.

Certain circuit combinations can reduce the impedance troubles. One common technique is to use parallel combinations of capacitors to achieve a desired component behavior.

Example 11.2.4   Two capacitor values are made by a well-known manufacturer. The two have similar size, and each has an ESL value of 20 nH and tan $\delta = 0.2$. One is marked 1000 μF, and the other 100 μF. Compare the single 1000 μF to a parallel combination of ten 100 μF parts.

The individual 1000 μF part has a resonant frequency of 36 kHz, while the 100 μF parts resonate at 113 kHz. At 10 kHz, the ESR value is 3.2 mΩ for the 1000 μF parts and 32 mΩ for the 100 μF parts. Since each of the 100 μF parts is identical, their parallel combination will have a tenth the ESL, a tenth the ESR, and ten times the capacitance. This means that the combination has ESR = 3.2 mΩ, $C = 1000$ μF, and ESL = 2 nH. The combined ESR value offers no special advantage, but the combined ESL value is a big improvement: the parallel combination is resonant at 113 kHz instead of 36 kHz. There is some indication here that parallel capacitor combinations can enhance operating frequency limits at the expense of the larger size of a combination.

The resonant behavior of real capacitors tends to limit their application at very high frequencies. It is fundamentally difficult to create capacitors of more than 1000 μF or so that retain capacitive behavior above 100 kHz. In commercial power supplies, multiple parallel units can be used for energy storage, but of course extra connections and longer wires are necessary to interconnect a parallel set.

11.2.4 Simple Dielectric Types and Materials

The performance of a capacitor is tightly linked to the insulating material used to make it. A high value of capacitance $\epsilon A/d$ requires thin insulation, large area, and high permittivity. An ideal material should be easy to form into thin sheets and should have a high value of ϵ. Thin material implies high electric fields, so dielectric breakdown strength is important. Good insulators can withstand fields well above 10^7 V/m, so a layer only a few microns thick can support several tens of volts.

The most important groups of materials used in simple dielectric capacitors include ceramics and polymers. Some ceramic materials, such as barium titanate, have relatively high ϵ values. It is possible to find materials with $\epsilon > 1000\epsilon_0$ for use in capacitors. Figure 11.9 shows two basic structures for a ceramic capacitor. Figure 11.9a shows a conventional ceramic capacitor, with a single block of ceramic coated with metal to provide connections. In Figure 11.9b, a multilayer ceramic structure is shown. Single-layer structures with capaci-

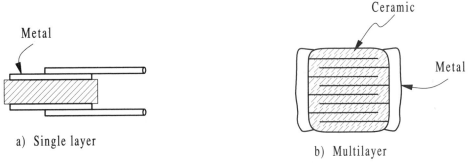

Figure 11.9 Ceramic capacitor construction.

tances of up to about 5 μF are available. In principle, there is no limit to the capacitance of a multilayer structure, but devices designed for 20 μF or more tend to be expensive. As the figure implies, the ceramic layers are difficult to make thin. The high permittivity values help offset the relatively thick insulation layer so that capacitors up to the low microfarad range can be constructed.

Example 11.2.5 A zirconium titanate and barium titanate mixture is used to form a ceramic block 0.1 mm thick by 3 mm square. It has $\epsilon = 3500\epsilon_0$. Estimate the capacitance value when this block is used in a simple dielectric capacitor.

  This gives $\epsilon A/d = 3500(8.854 \times 10^{-12}\,\text{F/m})(9 \times 10^{-6}\,\text{m}^2)/10^{-4}\,\text{m} = 2790\,\text{pF}$. Larger blocks or extra layers will be needed to reach even as high as 0.1 μF.

  Ceramic materials tend to have significant leakage effects, and typical values of tan δ for these materials can range from 1% to about 5%. The materials are affected by moisture and by temperature. Expensive ceramics must be used when tight tolerances or wide temperature ranges are needed.

  Mica is a natural ceramic that was previously used widely for manufacture of capacitors. Small brown silver-mica parts were once one of the most common high-quality types. Mica does not have the high permittivity of the titanates or other electrically active materials, but it is an excellent insulator that forms uniform thin layers. Mica capacitors usually have tan δ well below 1%. They are available in values up to about 10,000 pF.

  Polystyrene, polyethylene, polypropylene, polytetrafluoroethylene (Teflon), and polycarbonate are among the many polymers that are available in capacitors. In general, these materials have permittivities just a few times that of air, with $\epsilon \approx 3\epsilon_0$ for many polymers. Their great advantages are the ease of forming thin layers, and the very low insulator leakage. However, the low ϵ value means that high capacitance requires very large parts. While some higher values can be obtained, it is unusual to find polymer capacitors rated above 10 μF, and 0.1 μF or less is more usual. These materials can have tan δ values below 0.1%, and rarely have tan $\delta > 1\%$. The low leakage makes them very useful in high-frequency converter interface applications, if the low C values are not a problem. Polymer capacitors are often called *film capacitors* since they are constructed from plastic films.

  High-power ac applications, such as utility power factor correction, often use oil-impregnated paper as the dielectric material. This tends to be inexpensive compared to many of the film types, and sacrifices only about one percentage point on tan δ. For 50 Hz and

60 Hz applications, paper–oil combinations can be built to withstand very high currents. The devices tend to be large, and are rarely encountered in high-frequency power conversion applications.

11.2.5 Electrolytics

None of the capacitor arrangements so far has provided convenient values above about 10 μF. There are no simple tradeoffs for ϵ, A, or d that will help enhance C. For example, it ought to be possible to increase the capacitance of a polymer capacitor by making the insulation thinner. Below a certain thickness, however, it becomes almost impossible to manufacture or even to handle the material. This limits the ability to use thin films to obtain high C values.

The electrolytic capacitor avoids the limitations of simple dielectrics by forming the insulation in a direct way. The basic structure is shown in Figure 11.10. A small block of metal is etched to form a porous material, or is sintered from a powder base. The metal is then exposed to an oxidation process to form a thin insulating layer. A liquid or solid electrolyte is used to provide electrical contact across the outside of the oxide layer, so that capacitance is created between the metal and the electrolyte. The structure is sealed in a can or encased in epoxy to form a complete part.

The big advantage of an electrolytic device is the extreme ratio of A/d that can be achieved. The surface area is large, and is limited only by the ability to form a porous material. The oxide layer can be almost as thin as desired: The thickness can be controlled simply by adjusting the time spent during the oxidation step. Since the oxide thickness can be controlled, it is possible to trade off voltage ratings and capacitance values. Electrolytic parts rated at 6 V are common, while it is rare to find simple dielectrics rated at less than 50 V.

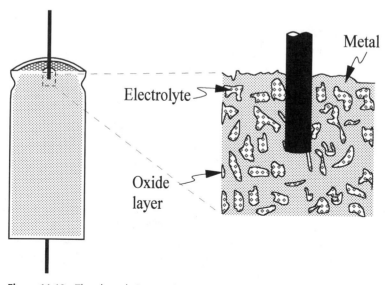

Figure 11.10 The electrolytic capacitor.

The oxide material is crucial to electrolytic technology. It must be an excellent insulator, must be easy to form, and must have perfect physical integrity as a surface coating. Both aluminum and tantalum form surface oxides with appropriate properties. Aluminum is inexpensive, and aluminum electrolytics are the most common type. Ta_2O_5 has a higher permittivity than Al_2O_3, so tantalum capacitors are smaller for a given capacitance. Each metal is anodized with a conventional process until the desired oxide thickness is produced. The trouble with this approach is that it is reversible: If the polarity reverses, the oxide layer becomes thinner, and dielectric breakdown might occur.

The anodizing process is the basis for electrolytic capacitor polarity. Only a unipolar field can be used without damaging the oxide. Oxide damage can be a very quick process. Even if the oxide layer degrades at only one point within the structure, that point will tend to carry more leakage current. The extra current and loss can lead to further local degradation, and the layer fails as a short circuit. This "fail short" behavior represents a highly undesirable property. When a liquid electrolyte is used, for example, the liquid can boil and vaporize if oxide failure occurs. New designs attempt to keep local degradation from growing into global failure.

The electrolyte and thin insulation make electrolytic capacitors much lossier than simple dielectrics of similar value. The tan δ value for a typical electrolytic is at least 5%, and can approach 30% or even more for low voltage parts in high frequency applications. The high leakage means that significant dc current flows, and manufacturers routinely specify dc leakage currents for electrolytics. Because of the high value of tan δ, the impedance behavior of electrolytics is significantly different than for simple dielectrics. In particular, an electrolytic capacitor forms an *RLC* resonant circuit with low *Q*. The sharp resonance seen in Figure 11.8 will not appear with an electrolytic because of the low *Q* value. Figure 11.11

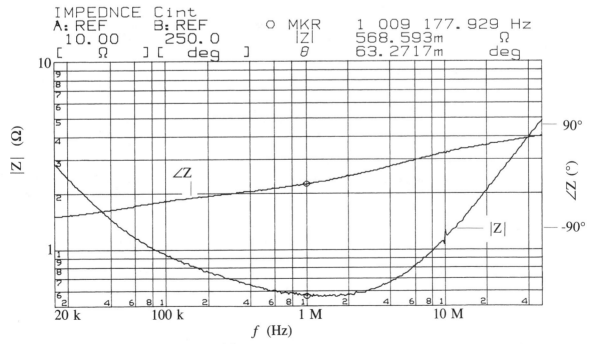

Figure 11.11   Measured electrolytic capacitor |Z| vs. frequency.

shows an experimental measurement of $|Z|$ vs. frequency for a 22 μF tantalum electrolytic capacitor.

11.2.6 Double-Layer Capacitors

Any material surface has a natural tendency to form a Helmholtz double layer—a charge separation region across the surface. Double-layer capacitors make use of this behavior to create energy storage. The approach typically involves very low voltages, since excessive voltage causes electric conduction through the surface. Practical devices use many units in series to create working voltages on the order of 5 V or more.

The double-layer phenomenon can be used to produce microscopic charge separation distances across enormous areas. Separation distances of about 1 nm are realistic. Activated carbon is frequently used as the surface-active material, and porous structures with surface areas on the order of 1000 m^2 have been reported for packages with volumes as low as a few cubic centimeters. The values are high: $\epsilon_0 A/d$ can be several farads for this geometry. The high capacitance of double-layer structures makes them suitable for certain types of energy storage applications. For example, a double-layer device can provide backup energy for power interruptions up to a few seconds long. As double-layer technology develops, a wider range of applications can be expected.

11.3 EFFECTS OF ESR

Below the resonant frequency, a real capacitor will show a resistance (the ESR) in series with a capacitive reactance. The Standard Model circuit then becomes a simple RC series combination. The effects of such a circuit are straightforward to analyze, but lead to some unusual results. The ESR will exhibit a voltage drop called the *ESR drop*. In applications such as a dc–dc boost converter, a capacitor is often exposed to a square wave of current. Ideally, this produces a triangular voltage across the part. With the ESR in place, a small square wave appears in series with the triangle. This creates an abrupt voltage change called an *ESR jump*. In this section, the ESR jump will be studied for a few sample converters. In some cases, it can dominate over ripple during the design process. For example, a converter with 5 ± 0.05 V output at 1000 W cannot meet its specifications if the output capacitor has ESR > 0.0005 Ω.

Example 11.3.1   Consider an electrolytic capacitor with very low ESL, low wire resistance, and tan $\delta = 0.20$, placed at the output of a boost converter with switching frequency of 100 kHz. The input voltage is 10 V, and the load current is 10 A. What value of C will be needed to keep V_{out} in the range 50 V $\pm$ 1%?

Since the ESL is low, the capacitor operates well below resonance. As shown in Figure 11.12, the ideal C value should be replaced with an R-C combination, where $R = $ ESR $= $ (tan δ)/ωC. The largest unwanted component of the output will occur at the switching frequency, 100 kHz. At this frequency, $\omega = 2\pi 10^5$ rad/s, and the ESR is $(3.2 \times 10^{-7}/C)$ Ω. The diode current $i_d = q_2 I_L$. By conservation of energy, if the power in the ESR is small, the input power should be 500 W, and the inductor current will be 50 A. The capacitor current is

$$i_C = i_d - i_{\text{load}} = q_2 I_L - i_{\text{load}} \tag{11.5}$$

Figure 11.12 Boost converter for ESR example.

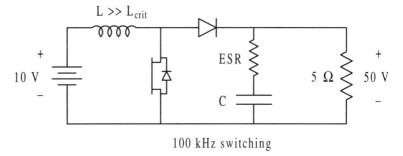

$$L \gg L_{\text{crit}}$$

ESR

10 V 5Ω 50 V

C

100 kHz switching

The capacitor current is 40 A when the diode is on, and -10 A when it is off. The total change in load voltage is the sum of change of v_C and the change in v_{ESR},

$$\Delta v_{\text{load}} = \Delta v_C + \Delta v_{\text{ESR}} = \Delta v_C + \text{ESR}\Delta i_C \tag{11.6}$$

The change in current is the difference, 50 A, between the two configurations. The capacitor voltage change is familiar. With the diode on, the capacitor current is 40 A, so

$$i_C = C\frac{\Delta v_C}{\Delta t}, \qquad \Delta v_C = (40 \text{ A})\frac{D_2 T}{C} \tag{11.7}$$

The duty ratio $D_2 = 0.2$ for this 5 : 1 voltage ratio. The total change is the sum

$$\Delta v_{\text{load}} = \left(40\frac{0.2 \times 10^{-5}}{C} + 50\frac{3.2 \times 10^{-7}}{C}\right) \text{V} = \frac{9.6 \times 10^{-5}}{C} \text{V} \tag{11.8}$$

The total voltage change must be less than 1 V, so

$$\frac{9.6 \times 10^{-5}}{C} \text{V} \leq 1 \text{ V}, \qquad C \geq 96 \ \mu\text{F} \tag{11.9}$$

The ESR jump will be $(50 \text{ A})(3.2 \times 10^{-7}/C) \ \Omega = 0.17$ V. The output voltage with $C = 96 \ \mu\text{F}$ is shown in Figure 11.13. What about conservation of energy? The input current of 50 A ignored the power consumed by the ESR. In fact, the ESR carries current of 40 A during

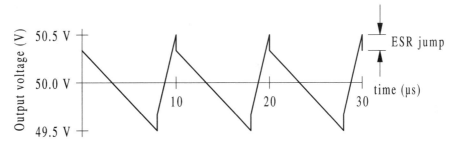

Figure 11.13 Output voltage, showing ESR, with $C = 96 \ \mu\text{F}$ in boost converter.

20% of each cycle and -10 A during 80% of each cycle. The average power is the total energy lost during each cycle divided by the cycle period. This gives

$$P_{ESR} = \left(\frac{1}{T} \int_0^{D_1 T} 10^2 \text{ ESR } dt + \frac{1}{T} \int_{D_1 T}^{T} 40^2 \text{ ESR } dt \right) W = 400 \text{ ESR W} \quad (11.10)$$

With $C = 96$ μF, the ESR is 3.3 mΩ, and $P_{ESR} = 1.33$ W. This is a tiny fraction of 500 W, so the input current will be very close to 50 A.

In large capacitors, the wire and connection resistance tends to place a floor on the ESR value. A 1000 μF part with tan $\delta = 0.1$, for instance, should have ESR $= 0.4$ mΩ at 40 kHz. However, even a small amount of wire is likely to add several milliohms to the ESR [recall that equation (11.2) ignores the wire resistance contribution]. This effect can be extremely important, as a further example shows.

Example 11.3.2   From the previous example, it is proposed to use a 100 μF capacitor as the output component for a 10 V to 50 V boost converter at 500 W. The part has tan $\delta = 0.12$ at the 100 kHz switching frequency. The ESL is 5 nH, and the wire resistance is 5 mΩ. What is the expected value of the ripple?

This capacitor has a resonant frequency of 225 kHz, so the model of an ESR in series with a 100 μF capacitor should be appropriate. If wire resistance is small, then ESR $=$ $(\tan \delta)/\omega C = 1.9$ mΩ. In this case, R_{wire} should not be ignored, since it is bigger than $(\tan \delta)/\omega C$. The total ESR from equation (11.1) is 6.9 mΩ. In this converter, the capacitor current is 10 A when the diode is on. The capacitor current changes by 50 A as the switches act. The change in load voltage is

$$\Delta v_{load} = \Delta v_C + \Delta v_{ESR} = \left(40 \frac{D_2 T}{C} \right) V + (50 \text{ A})(6.9 \text{ m}\Omega) \quad (11.11)$$

for a total change of 1.15 V.

At low voltages, the ESR can become a limiting issue. Look at a typical computer supply application to see the issues.

Example 11.3.3   A flyback converter, shown in Figure 11.14, has duty ratios of 50%. It supplies up to 100 W at 5 V. The output capacitor has wire resistance of 5 mΩ and tan $\delta = 0.10$

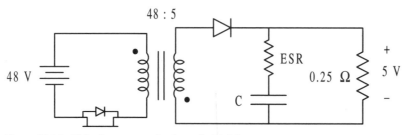

Figure 11.14 Flyback converter for Example 11.3.3.

at the 100 kHz switching frequency. What should the capacitor value be to keep the output in the range of 5 V ± 1%?

The total allowed change in the output is 2% peak-to-peak, or 0.1 V. The output capacitor must supply the full load current of 20 A when the transistor is on. If the load current is 20 A and the diode duty is 50%, the diode will carry the 40 A current expected for an equivalent buck-boost converter when it is on. The ESR will see a current change of 40 A. The effect of the wire resistance is to introduce a voltage jump of (40 A)(5 mΩ) = 0.2 V. The total change in load voltage will be

$$\Delta v_{\text{load}} = \Delta v_C + \Delta v_{\text{ESR}} = \left(20\frac{D_2 T}{C} + 40(0.005) + 40\frac{0.1}{\omega C}\right) \text{V} \qquad (11.12)$$

Even an infinite capacitor value will not meet the requirements because of the voltage jump in the wire resistance. A capacitor with lower wire resistance must be found to allow this converter to meet the specifications.

This example demonstrates the critical nature of ESR. It also suggests that careful management of connection inductance and resistance is very important, especially for low-voltage applications. In almost any interface application, the capacitor is intended to have the lowest possible impedance at the switching frequency. The impedance should be low in comparison with the load if successful filter action is to occur. For the 5 V, 100 W converter, the rated load resistance is only 0.25 Ω. The filter capacitor should have both an ESR and an X_C value much lower than this rated resistance. A more extreme case is a 2 V, 100 W converter, which has a rated load resistance of only 0.04 Ω. In this last case, it will be difficult to obtain a useful filter capacitor.

The ESR jump is an approximation of the actual behavior. After all, the true behavior is frequency dependent. ESR values based only on the switching frequency actually provide good predictive results. An experimental output voltage from a boost converter is shown in Figure 11.15. The ESR jump is clearly evident. Some of the ringing that occurs at the switching times can be attributed to the ESL value.

11.4 WIRE RESISTANCE

In a power converter, in which a great deal of power is handled in a small volume, wire resistance introduces important limits. In this section, rules of thumb are developed for wire current carrying. A metal wire with resistivity ρ develops a resistance $R = \rho\ell/A$, where ℓ is the length and A is the cross-sectional area. Given a current I, the loss in this resistor is $I^2 R = \rho\ell I^2/A$. The volume of the material is $\text{Vol} = \ell A$. The power per unit volume is $I^2 R/\text{Vol} = \rho I^2/A^2$. The current density J is current per unit area, so the loss can be written $I^2 R/\text{Vol} = \rho J^2$. Any given volume of conductor dissipates energy at the rate of ρJ^2. Since resistivity is a material property, a given material can handle only a limited current density without overheating.

The resistivity of conductor-grade copper at room temperature is 1.724×10^{-8} Ω · m. To develop an understanding of the power loss issue, consider a bit of engineering judgment: We might expect that a 1 cm^3 cube of conductor can dissipate 1 W without much trouble.

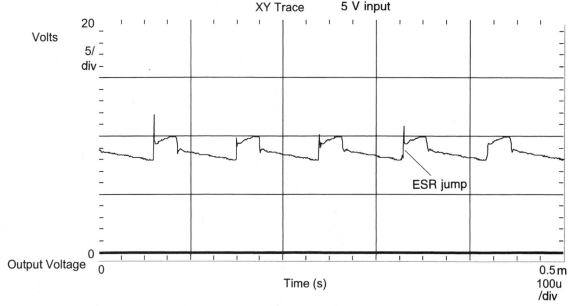

Figure 11.15 Boost converter output, showing the ESR jump.

This gives a power density of 10^6 W/m^3. For copper, this corresponds to current density $J = 7.62 \times 10^6$ A/m^2. Many designers try to keep current densities below this level to avoid excessive power loss and heating. The current density is frequently given in A/cm^2. A design rule results.

Empirical Rule: *Copper conductor can handle current densities in the range of 100 to 1000 A/cm^2 (10^6 to 10^7 A/m^2). The low end of this range is used for tight, enclosed inductor windings with minimal cooling. The high end of the range applies to open wires with good local air flow.*

In North America, copper wire is generally specified according to the *American Wire Gauge* or AWG size. The AWG number is based on a geometric progression between two defined sizes. Two features can be used to help remember it: #18 wire has a diameter very close to 1 mm (actually, it is 1.02 mm), and every three steps down in gauge number approximately doubles the wire area. Thus #15 wire has twice the area of #18 wire, #12 wire has four times the area of #18 wire, #24 wire has one quarter the area of #18 wire, and so on. A wire diameter of 1 mm gives area of 7.85×10^{-3} cm^2. Thus a #18 wire should handle between 0.8 A and 8.0 A. This is consistent with North American commercial wiring practice: a 15 A circuit is normally built with #14 wire, a 20 A circuit uses #12 wire. Open-air line cords sometimes push #18 wire up to about 10 A, but in most applications the level is closer to 5 A. Table 11.1 summarizes wire sizes and capabilities. The table lists even gauge sizes from #4 to #40. Odd and fractional sizes exist, but are unusual except in magnetics and

TABLE 11.1
Standard Wire Sizes and Current Capabilities

AWG Number	Diameter (mm)	Cross-Sectional Area (mm²)	Resistance, mΩ/m at 25°C	Current Capacity, 500 A/cm²	Current Capacity, 100 A/cm²
4	5.189	21.15	0.8314	105.8	21.15
6	4.115	13.30	1.322	66.51	13.30
8	3.264	8.366	2.102	41.83	8.366
10	2.588	5.261	3.343	26.31	5.261
12	2.053	3.309	5.315	16.54	3.309
14	1.628	2.081	8.451	10.40	2.081
16	1.291	1.309	13.44	6.543	1.309
18	1.024	0.8230	21.36	4.115	0.8230
20	0.8118	0.5176	33.96	2.588	0.5176
22	0.6438	0.3255	54.00	1.628	0.3255
24	0.5106	0.2047	85.89	1.024	0.2047
26	0.4049	0.1288	136.5	0.6438	0.1288
28	0.3211	0.08098	217.1	0.4049	0.08098
30	0.2546	0.05093	345.1	0.2546	0.05093
32	0.2019	0.03203	549.3	0.1601	0.03203
34	0.1601	0.02014	873.3	0.1007	0.02014
36	0.127000	0.0126677	1389.	0.06334	0.0126677
38	0.1007	0.007967	2208.	0.03983	0.007967
40	0.07987	0.005010	3510.	0.02505	0.005010

in motor manufacturing. Gauges below #1 use the numbers 0, 00, 000, and 0000. For wires bigger than #0000, a separate size scale applies.

Consider an electrolytic capacitor with two 5 cm leads made of #22 copper (a fairly typical geometry). This wire has series resistance of about 5.4 mΩ. This resistance places a lower limit on the capacitor's ESR. The following examples help emphasize some of the issues.

Example 11.4.1   A dc power supply provides 5 V at 200 W for a computer. The power is connected to the load through 50 cm total wire length. What gauge size should be used to safely handle the current and to avoid a voltage drop of more than 1%? What power is lost in the connection line for this choice of gauge?

The supply provides up to (200 W)/(5 V) = 40 A to the load. If the voltage drop along the connection wire is to be less than 1% of 5 V, a 40 A current should not produce more than 50 mV of drop. The resistance should not exceed 1.25 mΩ. Since the connection length is $\frac{1}{2}$ m, the wire should show no more than 2.5 mΩ/m resistance. From Table 11.1, this requires #8 wire, with a diameter of 3.26 mm. With 40 A flowing, the power lost in the wire is $I^2R = 2$ W. It is easy to imagine that 2 W of loss, dissipated over 50 cm of length, will not cause excessive heating. However, the design sacrifices 1% of the converter's output just to cover loss in the connecting wires.

Example 11.4.2   You were recently hired by XXX Computers as the resident power electronics engineer. In recent weeks, XXX has been having problems with their 5 V, 200 W supply board. It seems that somehow the ripple has increased, and the logic circuits in the computers are not always operating properly. In talking to a line technician, you discover a change in the output capacitor mounting. Figure 11.16 shows the configuration. An older

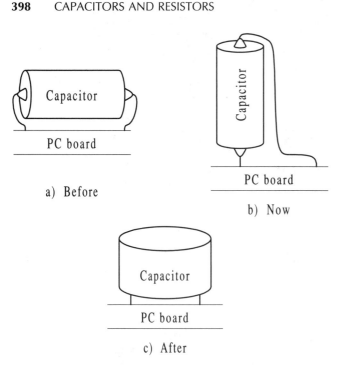

Figure 11.16 Capacitor configurations for Example 11.4.2.

a) Before

b) Now

c) After

"good board" at the technician's bench has the capacitor mount shown in Figure 11.16a. A board fresh from the line shows the mount of Figure 11.16b. The technician reports that the change was made to make assembly easier. With a quick phone call, you obtain some cheaper parts that mount according to Figure 11.16c. They solve the problem, yet cost less. Please explain what was done and why the boards have improved.

Based on the examples in the preceding ESR section, it seems reasonable that the capacitor ESR jump dominates the output ripple of this 5 V, 40 A converter. Furthermore, the wire resistance might be an important portion of the ESR. In Figure 11.16a, the total capacitor lead length is about 3 cm of #24 wire, with resistance of about 2.6 mΩ. In Figure 11.16b, the lead length is close to 5 cm, for resistance of 4.3 mΩ. If the change in capacitor current is 40 A (it could easily be double this if a flyback converter is being used), the extra 1.7 mΩ of resistance translates into an extra 68 mV of output ripple.

The capacitor of Figure 11.16c has more direct radial leads. The total lead length is reduced to only about 1 cm, with a fraction of a milliohm of resistance, such that the capacitor's internal ESR becomes the most important effect. The reduction in lead resistance allows the capacitance to be reduced by perhaps 20% while still reducing the output ripple. The lower capacitance makes the part somewhat cheaper than the original.

Wire resistance plays an important role in limiting the ratings of inductors and capacitors as well as the circuits themselves. For example, a capacitor with #22 leads cannot realistically handle 80 A ripple currents. Such a high current corresponds to almost 25,000 A/cm^2 and contributes 4.1 W of loss per centimeter of length. Similarly, an inductor rated to carry 20 A will require #12 wire or larger.

Example 11.4.3   A 5 V to 2 V converter is to be designed for the Wunderlux Stage H microcomputer. This new product uses the lower voltage because it helps to conserve power. It draws 12 W at 2 V (if it is run on 5 V instead, the power consumption climbs to 50 W). What wire sizes would you recommend?

Most likely a dc–dc buck converter will be used. The output inductor current matches the load current, 6 A. The wire size should be about #16, although there is a voltage drop issue because of the low output level. The connection wires could easily total 10 cm in length. With #16 wire, this would create resistance of about 1.3 mΩ. At 6 A, this is a voltage drop of 7.8 mV. This drop has implications for load regulation, and it might be sensible to use larger wire. With #14 wire, the resistance decreases to about 0.85 mΩ. This produces 5 mV of output variation in addition to the ESR jump.

Example 11.4.4   The owner of a recreational vehicle wishes to find a long extension cord for electrical hookup. The vehicle draws up to 5 kVA at 230 V and 60 Hz. Suggest a wire size for a 30 m cord. (In many cases, the size would need to meet specific code requirements. Here, consider only the physical constraints on the wire.)

The cord will need to carry up to about 22 A to provide 5 kVA at 230 V. Since the cord will probably not be tightly coiled, a current density almost up to 1000 A/cm^2 should be feasible. From Table 11.1, the next smaller even size is #12 AWG. Before suggesting this, consider the load regulation. Wire of this size has resistance of about 5.3 mΩ/m. With 30 m of wire, the total path length through both conductors is 60 m, and the total series resistance will be about 0.32 Ω. With current of 22 A, the voltage drop along the wire will be at least 7.0 V. This is just over 3% of the 230 V nominal value. A regulation value of 3% is on the same order as the utility voltage tolerance, so this drop is probably reasonable. An extension cord made from 30 m of #12 AWG cord should give acceptable results in this context. Larger wire can be selected, but it will be heavier.

Resistance has both temperature and frequency dependence. The temperature dependence for metals is nearly linear. Copper, for example, changes its resistance upward by a factor of 0.39% for every 1°C increase in its temperature. Table 11.2 lists conductivity and temperature coefficient for several metals used for wires or electrical loads. Consider the value for tungsten. The metal is a good conductor, and the temperature coefficient is less than 0.5% per °C. Yet if it is heated to 2000°C, the resistance would be expected to increase by a factor of 9. This is exactly the cold resistance vs. hot resistance issue that was raised in Example 7.8.2.

TABLE 11.2
Resistivity and Temperature Coefficient for Several Metals

Metal	Resistivity, $\mu\Omega \cdot$ m at 20°C	Temperature Coefficient of Resistance, Change per °C
Copper (conductor grade)	0.01724	0.0039
Silver	0.0159	0.0041
Aluminum (conductor grade)	0.0280	0.0043
Nichrome	1.080	0.0001
Tin	0.120	0.0046
Tantalum	0.1245	0.0038
Tungsten	0.0565	0.0045

Frequency dependence is introduced through the *skin effect*, an interaction between current in a wire and its own magnetic field. The *skin depth*, δ, in general provides a measure of the portion of the material that actively carries current. It is given by

$$\delta = \sqrt{\frac{2\rho}{\omega\mu}} \tag{11.13}$$

where ω is the radian frequency and μ is the magnetic permeability. As frequency increases, current flows more and more at the surface of a material. For conductor grade copper, the skin depth is given by

$$\delta_{Cu} = 0.166 \sqrt{\frac{1}{\omega}} \text{ meters} \tag{11.14}$$

A wire 1 mm in diameter will not see much skin effect at frequencies below about 15 kHz. When frequencies reach 100 kHz, however, current flows mainly in the outer 0.2 mm layer of a copper wire. This implies increased resistance, since the surface area carrying current is significantly reduced. Skin effect can be minimized by using thin conductors. The copper plating on a circuit board, for example, will not show strong skin effects. Litz wire is another way to reduce the effect.

11.5 RESISTORS

Elements designed specifically as resistors are intended to dissipate energy rather than store it, and have a fundamentally different function from other parts in a power electronic system. Ohm's Law applies primarily for metals and certain semiconductors. However, the relationship between current and voltage is nonlinear for insulators and most slightly conducting materials.

Resistors are distinguished by construction. They might be *composition* types, in which a material of a desired resistivity is formed into a bulk resistor, *film* types, in which carbon, a metal, or a metal oxide film is deposited on a ceramic substrate, or *wirewound* types, formed from lengths of resistive alloys. Composition resistors have fallen out of favor because they are imprecise, and tend to be sensitive to moisture and other environmental stresses. Most resistors rated for 1 W or less use either carbon films or metal oxide films. Wirewound resistors are common at higher power levels.

The basic resistor construction introduces series inductance and even some capacitance across the device. Figure 11.17 shows possible circuit models for film and wirewound resistors. The capacitance is usually well below 1 pF, so resonance effects are rare in practical resistors and the capacitive portion of the model is normally ignored. The *equivalent series inductance* can be important when a resistor is exposed to high di/dt levels. In the case of wirewound resistors, the construction resembles that of an inductor. Standard wirewound types often exhibit inductances of 10 μH or more. Film types generally follow the 5–10 nH/cm rule of thumb developed earlier for wire self-inductance.

All resistors show temperature dependence. Recall that copper changes resistance by

Figure 11.17 Circuit models for major types of resistors.

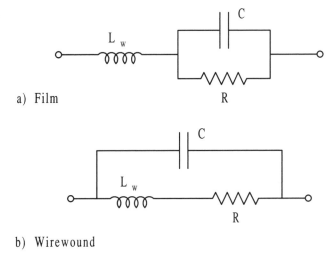

a) Film

b) Wirewound

a factor of 0.0039 per °C. This translates to 3900 parts per million (ppm) per °C. Practical resistors are much less sensitive to temperature. Typical values for temperature coefficient are about ±300 ppm, and precision resistors that vary by less than 10 ppm/°C are commercially available. Wirewound types often use temperature-insensitive alloys to keep thermal coefficients low. Common examples of resistor wire materials include:

- *Nichrome*, an alloy of 80% nickel and 20% chromium with a temperature coefficient of 100 ppm/°C.
- *Constantan*, an alloy of 55% copper and 45% nickel with a temperature coefficient below 20 ppm/°C.
- *Inconel*, an alloy of 76% nickel, 16% chromium, and 8% iron that changes resistance by about 1 ppm/°C.

Nichrome is very common in standard wirewound applications.

The high inductance of wirewound resistors is a serious issue when they must be used in power electronics. While resistors are normally avoided as much as possible, they are needed for sensing output voltages and currents, for current-limiting applications, for power semiconductor control, and for other control-related functions in a power converter. It is possible to make "noninductive" wirewound resistors by providing two separate windings wrapped in opposite directions. This cancels a substantial fraction of the extra wire inductance. The most effective noninductive designs do not eliminate the inductance entirely, but can reduce it to match the self-inductance of an equivalent length of isolated wire (10 nH/cm).

Example 11.5.1   A wirewound resistor is formed from a one meter length of nichrome wire with a size equivalent to #30 AWG. What is the resistance value at 20°C? What resistance would be expected at 100°C?

From Table 11.2, nichrome wire has a resistivity of 1.08 $\mu\Omega \cdot$ m. Size AWG #30 corresponds to a diameter of 0.0254 cm, from Table 11.1. The wire resistance is $\rho\ell/A$, with $\ell =$

1 m. The area is 5.07×10^{-4} cm^2, or 5.07×10^{-8} m^2. The resistance should be 21.3 Ω. This is probably a standard 22 Ω resistor (the bonds, tabs, and other parts will add a fraction of an ohm to the total). The temperature coefficient of nichrome is $+100$ ppm/°C, so the change from 20°C to 100°C should change the resistance by a factor of 8000 ppm, or 0.8%. The resistance at 100°C will be 21.5 Ω.

Example 11.5.2   A standard wirewound resistor with a value of 100 Ω is used for the ballast load of a 12 V dc–dc converter. The resistor is found to have series inductance of 10 μH. The converter load is transient in character, and can vary at a rate of up to 5 A/μs. What effect will the resistor's ESL have?

Since the resistor has significant inductance, a change in current will tend to induce a voltage across it. Ignore any output capacitance for the moment. Since the resistor serves as a ballast, any transient change in load current will affect the resistor prior to the effects of switch action. The induced voltage $v = L(di/dt)$ will be 50 V if the entire 5 A/μs current ramp appears at the ballast—more than 400% of the rated converter output voltage. When the output capacitor is considered, other interesting effects can occur. If $C = 100$ μF, for instance, the resistor's ESL will create a resonance at 5 kHz. This could cause significant ringing as a result of a load transient.

11.6 RECAP

Most capacitors are constructed with two conductive layers separated by an insulator. This construction introduces wire inductance and resistance, and also implies a leakage resistance across the insulating layer. Many manufacturers use the Standard Model of a Capacitor—an ESR in series with an ESL and the capacitor—to describe and specify parts. When wire resistance is neglected, the ESR value is determined by the *dissipation factor*, tan δ, of the insulator,

$$\text{ESR} = \frac{\tan \delta}{\omega C} \tag{11.15}$$

The ESL means that real capacitors exhibit self-resonance, and act as inductors above their resonant frequency.

Capacitors are most commonly built as simple dielectric types, with metal films or plating coating a specific insulation layer, or as electrolytic types, in which an oxide insulator is anodized on a porous metal substrate. Electrolytic capacitors require unipolar voltage so that the insulation layer does not degrade. They have higher values of tan δ than simple dielectric types, partly because a liquid or solid electrolyte must be used to make electrical contact with the outside of the oxide layer. Double-layer types take advantage of the active charge layer formed at a surface. They have very low voltage ratings, but can exhibit capacitances of many farads.

For simple dielectrics, ceramic or polymer materials are commonly used as the insulation layer. Ceramics with high ϵ are used to provide high capacitance per unit volume. These tend to have dissipation factors on the order of 1–5%. Polymers typically have $\epsilon \approx 3\epsilon_0$, but their excellent insulating properties can provide tan δ as low as 0.01%. The polymer types are often called *film capacitors* by their manufacturers.

The ESR value is very important in power converter design. This resistance introduces an *ESR jump* in voltage when a capacitor is exposed to a square wave of current. In low-voltage applications, the ESR jump can be higher than the voltage variation across the ideal internal capacitance, and the capacitor must be selected to provide low ESR rather than high C. This becomes a severe problem at 5 V and below, since even the resistance of wire connections cannot be ignored when voltage ripple requirements are tight and currents are high. In one example, even an infinite capacitance could not meet requirements because of the ESR value.

Wire resistance is a function of material, temperature, and frequency. A rule was developed:

> **Empirical Rule:** *The current carried by a wire should be limited to avoid excessive heating. Typically, a copper wire can carry between 100 A/cm^2 and 1000 A/cm^2 (10^6 to 10^7 A/m^2) without difficulty. The lower value applies to tightly wound coils with little cooling, while the higher value applies to open conductors.*

Materials other than silver can carry less current density than copper, since their resistivity and loss per unit volume is higher. Resistors, lamp filaments, and other wires intended to become hot normally carry much higher current densities.

Resistors exhibit series inductance, which can be substantial for wirewound designs. The capacitive effects in resistors are usually negligible. Wirewound resistors can be provided with a second counter winding to minimize their inductance, although this raises the cost relative to a simple single-winding configuration. Resistors have thermal characteristics that can be approximated well as a linear change of resistance with temperature.

PROBLEMS

1. A capacitor is known to exhibit an impedance of $0.100\angle -74°$ Ω at 150 kHz. It exhibits a resonant frequency of 1.4 MHz. Determine the ESL, ESR, and C values at 150 kHz.

2. A simple dielectric capacitor exhibits a resonant frequency of 5.7 MHz. At this frequency, the measured impedance is $4\angle0°$ mΩ. The label value is 0.33 μF. Assume that tan δ is constant.

   a. What is tan δ for the part, from ESR $= (\tan\delta)/(\omega C)$?

   b. Determine the ESR, ESL, and C values associated with a model of this capacitor at 200 kHz.

3. An engineer for an audio company is asked about large simple dielectric capacitors. A given dielectric has permittivity of $2.3\epsilon_0$, and can support voltages of up to 50 V/μm of thickness. Film thicknesses are available in 5 μm increments. Suggest a combination of plate and film area and thickness that will create a 100 μF capacitor with a voltage rating of at least 200 V.

4. A 10 μF capacitor exhibits a resonant frequency of about 300 kHz with a certain lead configuration. Estimate the resonant frequency if the mounting is changed to reduce total lead length by 1 cm.

5. In most cases, it is difficult to reduce the ESL of a capacitor below a minimum value. The package has some inductance, and lead length cannot be made negligible. If this minimum ESL is 5 nH, plot the highest feasible self-resonant frequency vs. capacitance over the range 0.1 μF to 1000 μF. A logarithmic scale is suggested.

6. A 2.2 μF capacitor has ESR of 35 mΩ at 100 kHz. Assuming that tan δ is constant, what is the tan δ value for this part? What ESR will be measured at frequencies of 50 kHz and 200 kHz?

7. It is desired to produce a "high ripple" capacitor capable of 30 A$_{RMS}$ for use in power supply applications. What lead wire size would you recommend? If the tan δ value is 0.2, what power loss is expected for a 500 μF part at 50 kHz?

8. A buck-boost converter has +12 V input and −400 V output. It has $L \gg L_{crit}$, f_{switch} = 50 kHz, and P_{out} = 1 kW.

   a. Find C necessary to maintain V_{out} to −400 V ± $2\frac{1}{2}$%, if the capacitor has $f_{resonance} \gg$ 50 kHz and tan δ = 0.30.

   b. What is the ESL of your capacitor if $f_{resonance}$ = 100 kHz? What power is lost in the device?

9. A given set of capacitors has tan δ = 0.20 and a labelled value of 1000 μF. The resonant frequency is found to be 100 kHz. A second set of capacitors also has tan δ = 0.20, but the value is 100 μF with a resonant frequency of 250 kHz.

   a. Find the ESL and ESR of each of the two types at 100 kHz.

   b. Ten of the 100 μF units are wired in parallel to form an equivalent 1000 μF device. Find the ESL and ESR of this combination at 100 kHz. What is the resonant frequency?

   c. Ten of the 1000 μF units are wired in series to form an equivalent 100 μF device. Find the ESL and ESR of this combination at 100 kHz. What is the resonant frequency?

10. A buck-boost converter uses a large inductor, and supplies −12 V output from a 12 V input. At full load, the output current magnitude is 20 A. The switching frequency is 100 kHz. What value of output capacitor will be needed if available devices have tan δ = 0.10 at this frequency? Use a reasonable ripple specification.

11. A lawn mower draws 12 A from a standard 120 V, 60 Hz outlet. It is desired to use a line cord 100 m long for the mower. What size wire would you recommend?

12. A flyback converter for a power supply input stage has 400 V input and 20 V output. The turns ratio has been chosen so that the duty ratio is close to 50%. The maximum load is 100 W, and the switching frequency is 100 kHz. What wire sizes would you recommend for the two windings on the flyback inductor?

13. A boost converter has 12 V input and 100 V output for a display application. The nominal load is 100 W. The output capacitor is a 100 μF electrolytic type. It has tan δ = 0.12 at the switching frequency of 40 kHz, and the leads total 5 cm of #24 AWG wire. The converter inductor is large. Compare the wire resistance to the value tan $\delta/(\omega C)$. Can wire resistance be neglected in the ESR? What is the output ripple at nominal load? How much does ESR jump contribute to this?

Figure 11.18 Voltage divider test circuit for capacitor.

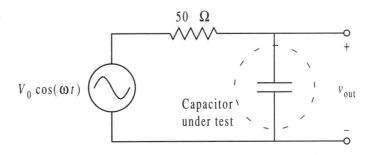

$V_0 \cos(\omega t)$

50 Ω

Capacitor under test

v_{out}

14. A buck converter for a new 64-bit microprocessor provides 2.2 V at 10 W from a 5 V input. The switching frequency is 250 kHz. It is desired to achieve ripple of no more than 50 mV on the 2.2 V supply. An output capacitor will be needed to do the job. If the inductor has been chosen so that $L = L_{crit}$, evaluate the capacitor requirements. What value of C would you recommend if $\tan \delta = 0.03$? What if $\tan \delta = 0.30$? Should wire resistance be considered in the design?

15. A flyback converter has an input voltage of 170 V (from a rectifier), and provides a 5 V output at up to 75 W. The switching frequency is 100 kHz. The turns ratio is set up to allow a 50% duty ratio. A supply of capacitors having a loss tangent of 0.3 is available for this application. What value of capacitance should be used to keep the peak-to-peak output ripple below 1%? How big is the ESR jump at full load?

16. A capacitor, presumably one of high quality, is tested in the divider circuit shown in Figure 11.18. The input is a cosine with 10 V amplitude. Measured data are shown in Table 11.3. Find $\tan \delta$, the ESL value, and the value of capacitance from the table data. (*Hint*: Analyze the impedance divider based on the assumption that the part is an *RC* series combination below resonance and an *RL* series combination above resonance.)

17. The power loss of a certain type of capacitor is to be tested. The capacitor has nominal value of 20 μF and a 10% dissipation factor. The resonant frequency is just over 1 MHz. It is exposed to a low-voltage sine wave of 10 mV$_{p-p}$ for testing. Compute the loss in the ESR as a function of frequency, from 1 Hz to 1 MHz.

TABLE 11.3
Data from RC Divider for Problem 16

Frequency (kHz)	Output Voltage Amplitude (mV)	Output Voltage Phase
10	39.6	−88.9°
20	19.4	−88.5°
50	6.84	−86.6°
100	1.77	−78.0°
133	0.37	13.9°
200	2.53	82.0°
500	10.5	88.0°
1000	22.1	89.0°

18. A capacitor has ESL $= 10$ nH, $\tan \delta = 0.20$, and $C = 200$ μF. Find:

    a. The self-resonant frequency.

    b. The quality factor (based on the complete *RLC* circuit) near the resonant frequency.

    c. The magnitude of the device impedance at 20 kHz, 50 kHz, 100 kHz, 200 kHz, and 500 kHz.

    d. If the part uses #20 AWG wire for its leads, what is the resistance of a total of 3 cm of wire connections? How will this extra resistance affect the results in parts (a) to (c)?

19. Capacitors are not normally designed for significant power dissipation, and a power level of 1 W/cm^3 would be excessive for most capacitors. A more reasonable value is 0.1 W/cm^3. Several electrolytic capacitors are available, and it is of interest to find out their ripple current capability for a full-bridge rectifier application. In other words, we need to determine the highest RMS current that can be permitted while keeping power loss below 0.1 W/cm^3, given a ripple frequency of 120 Hz. Estimate the ripple current limit (in A$_{RMS}$) for the following parts. Assume that each will be equipped with 2 cm of #20 AWG wire, and that the wire loss should be included in the total limit.

    a. 47 μF, cylindrical. Diameter: 10 mm. Length: 12.5 mm. $\tan \delta$: 0.175 at 120 Hz.

    b. 1500 μF, cylindrical. Diameter: 12.5 mm. Length: 20 mm. $\tan \delta$: 0.25 at 120 Hz.

    c. 330 μF, cylindrical. Diameter: 16 mm. Length: 25 mm. $\tan \delta$: 0.175 at 120 Hz.

    d. 10 μF, rectangular. Width, height: 4.5 mm. Length: 6 mm. $\tan \delta$: 0.40 at 120 Hz.

    e. 22 μF, cylindrical. Diameter: 19 mm. Length: 28 mm. $\tan \delta$: 0.11 at 120 Hz.

    f. 18 000 μF, cylindrical. Diameter: 75 mm. Length: 120 mm. $\tan \delta$: 0.16 at 120 Hz.

20. A new type of film capacitor has a value of 10 μF, and has been rated for maximum ripple current of 100 A$_{RMS}$. The connections are made with tabs 5 mm wide. How thick should the tabs be to support this current rating?

21. In heater applications, wire current density is often very high, since extreme power dissipation is desired. A certain application uses nichrome material equivalent in size to #24 AWG.

    a. What is the resistance per meter of this wire?

    b. What length will be needed to dissipate 200 W from a 12 V source?

    c. If the wire heats up to 1000°C, how much does the resistance change from cold to hot conditions?

22. A test load is needed for a 5 V, 1200 W power converter. Suggest a size and length of nichrome wire to meet this need. You may use several wires in parallel, since there will be large bus bars connected to the 5 V source.

23. Some people think speaker wire is troublesome. Consider a loudspeaker, approximately modelled as 8 Ω in series with 350 μH. Conventional lamp cord has inductance on the order of 4 nH/cm per wire. What length of #18 AWG lamp cord can be tolerated without dropping the speaker voltage by more than 0.1 dB for audio frequencies between 20 Hz and 20 kHz? What about #16 AWG lamp cord?

24. Student laboratories often make use of #22 AWG wire for projects. A buck converter is to be built for a project. The input is 24 V and the output is 5 V at up to 40 W. There is no input filter capacitor. What total length of #22 wire can be tolerated in the circuit without dropping the efficiency by more than one percentage point?

25. A dc–dc boost converter accepts 12 V from a car battery, and produces 150 V at up to 200 W for a computer backup arrangement. Initially, the connections are made with #20 AWG wire, with a total length of 20 cm. What is the total power loss in the wire? Recommend a more appropriate wire size. What is the loss with the new size?

REFERENCES

"Aluminum electrolytic capacitors," *Catalog H7*, Rosemont, IL: United Chemi-Con, 1995.

Capacitor Catalog and Engineering Guide. Lincolnwood, IL: Illinois Capacitor, 1996.

Electrical Engineering Pocket Handbook. St. Louis: Electrical Apparatus Service Association, 1988.

H. F. Littlejohn, Jr., *Handbook of Power Resistors*. Mount Vernon, NY: Ward Leonard Electric, 1959.

T. Longland, T. W. Hunt, W. A. Brecknell, *Power Capacitor Handbook*. London: Butterworth, 1984.

J. R. Miller, D. A. Evans, "Design and performance of high-reliability double-layer capacitors," in *Proc. 40th Electronic Components and Technology Conf.*, 1990, pp. 289–297.

National Electrical Safety Code. IEEE/ANSI Standard C3-1993. New York: IEEE, 1993.

W. J. Sarjeant, D. T. Staffiere, "A report on capacitors," *Proc. IEEE Applied Power Electronics Conf.*, 1996, pp. 12–17.

"Tantalum capacitors," *Engineering Bulletin, E-100-001A*. Huntington Beach, CA: Matsuo Electronics, 1985.

D. M. Trotter, "Capacitors," *Scientific American*, vol. 259, no. 1, July 1988, pp. 86–90B.

CHAPTER 12

Figure 12.1 Cores ready to be formed into inductors or transformers.

CONCEPTS OF MAGNETICS FOR POWER ELECTRONICS

12.1 INTRODUCTION

Inductors and transformers are important components of most power electronic systems. These elements are always built with magnetic materials, so the design and application of them requires an understanding of the basic issues in magnetic devices. This is important because limitations of magnetic devices are important considerations in design. A device will have a maximum dc current rating, determined by the nonlinear saturation effect, and might have frequency and voltage limitations as well. In this chapter, we will examine the fundamental properties of magnetic devices, with the objective of trying out a few designs. In contrast to other components in power electronics, an engineer working on a magnetic device will usually design it from certain raw materials like the cores in Figure 12.1, rather than obtain it directly. We will consider only a few types of raw materials, and will focus on magnetic circuit elements. Permanent magnets, specific types of materials, and more detailed applications can be studied by means of the reference list at the end of the chapter.

The chapter begins with a brief review of the relevant versions of Maxwell's equations and magnetic circuits. Hysteresis and the basic physical behavior of ferromagnetic materials are also reviewed. Based on material properties, almost all the energy stored in a typical inductor is present in an air gap. Both saturation and energy requirements are used to establish design rules for inductors. Transformer design as it relates to high-frequency switching converters is discussed.

12.2 MAXWELL'S EQUATIONS WITH MAGNETIC APPROXIMATIONS

Maxwell's equations can be written in many different ways. To establish notation, and to assist in the quasistatic approximations below, let us consider the integral forms shown as equation (12.1), with $\mathbf{H}$ defined as magnetic field intensity, $\mathbf{B}$ as flux density, $\mathbf{E}$ as electric field, $\mathbf{J}$ as current density, and ρ_v as charge density per unit volume.

$$\oint_s \epsilon \mathbf{E} \cdot ds = \int_v \rho_v \, dv \qquad \text{Gauss's Law}$$

$$\oint_s \mathbf{B} \cdot ds = 0 \qquad \text{Gauss's Law (magnetic fields)}$$

$$\oint_l \mathbf{E} \cdot dl = \frac{-d}{dt} \int_s \mathbf{B} \cdot ds \qquad \text{Faraday's Law} \qquad (12.1)$$

$$\oint_l \mathbf{H} \cdot dl = \int_s \mathbf{J} \cdot ds + \frac{\partial}{\partial t} \int_s \epsilon \mathbf{E} \cdot ds \qquad \text{Ampère's Law}$$

$$\oint_s \mathbf{J} \cdot ds = \frac{-d}{dt} \int_v \rho_v \, dv \qquad \text{Conservation of charge}$$

In a magnetic field system, certain simplifications are possible. Consider Ampère's Law. In a power converter, current densities on the order of 10^6 A/m^2 are expected. The electric fields are low, and the frequencies are likely not to exceed a few megahertz. The last term in Ampère's Law will be small: The time rate of change is associated with a radian frequency ω, the value of ϵ_0 is less than 10 pF/m, and only inside the plates of a capacitor do the electric fields exceed about 10^3 V/m. In magnetic elements, the last term will almost surely not be more than of order $(10^7 \text{ rad/s})(10^{-11} \text{ F/m})(10^3 \text{ V/m}) = 0.1$ A/m^2—far below the magnitude of $\mathbf{J}$. When the last term is neglected, Ampere's law can be simplified to

$$\oint_s \mathbf{H} \cdot dl = \int_s \mathbf{J} \cdot ds \qquad (12.2)$$

Faraday's Law, considered in a magnetic system, is the basis for a key circuit relation: Kirchhoff's Voltage Law around a circuit loop. When KVL is written $\Sigma v_{\text{loop}} = 0$, it represents an electric field relation around the loop. At any point along the loop, the local voltage is given by the negative of the local electric field voltage $-\mathbf{E}$ times a differential length dl. Thus KVL with only electric fields present can be written:

$$\oint_l \mathbf{E} \cdot dl = 0 \qquad (12.3)$$

Let us define magnetic flux $\phi = \int \mathbf{B} \cdot ds$. Then $\mathbf{B}$ is flux per unit area, or *flux density*. Faraday's Law can be written in terms of ϕ to give

$$\oint_l \mathbf{E} \cdot dl = \frac{-d\phi}{dt} \qquad (12.4)$$

The left-hand side is a sum of voltages associated with electric fields. If the time-rate-of-change of flux, $-d\phi/dt$, is interpreted as a voltage source $v = d\phi/dt$, then the right-hand side is also a voltage, and the net result is $\Sigma v_{\text{loop}} = 0$. Thus KVL can be interpreted as Faraday's

Law if time changing flux is considered equivalent to voltage. Such an interpretation is completely consistent with the mathematics.

For magnetic systems, the relevant expressions become

$$\oint_s \mathbf{B} \cdot ds = 0 \qquad \text{Gauss's Law (magnetic fields)}$$

$$\oint_l \mathbf{E} \cdot dl = \frac{-d}{dt} \int_s \mathbf{B} \cdot ds \quad \text{Faraday's Law} \qquad (12.5)$$

$$\oint_l \mathbf{H} \cdot dl = \int_s \mathbf{J} \cdot ds \qquad \text{Ampère's Law}$$

These expressions form the basis of magnetic circuit analysis.

12.3 MATERIALS AND PROPERTIES

Three general types of magnetic materials can be identified:

Diamagnetic—the material tends to exclude magnetic fields slightly.

Paramagnetic—the material is slightly magnetized by a magnetic field.

Ferromagnetic—the material contains small regions, known as domains, that are strongly magnetized.

Each general type represents groups of materials with a wide range of special magnetic properties. Ferromagnetic materials in particular can be divided into a range of different classes, such as permanent magnet materials, ferromagnetic metals, and so on. Here, we will be concerned primarily with ferromagnetic ceramics and metals, since these are of special significance in the design of magnetic components. Superconductors form a fourth distinct type of magnetic material, and have the property that they tend to exclude all exterior fields. It is likely that they will come into increased use as active magnetic materials.

In a given material, the magnetic field intensity, $\mathbf{H}$, depends on the interaction between electrons in the material and the local flux density $\mathbf{B}$. A constituent relation can be defined for a given material, such that $\mu\mathbf{H} = \mathbf{B}$, where μ is the magnetic permeability. For vacuum and many nonferromagnetic materials, μ is constant and represents a linear relationship between $\mathbf{H}$ and $\mathbf{B}$. For vacuum, $\mu = \mu_0 = 4\pi \times 10^{-7}$ H/m. The relative permeability μ_r, defined according to $\mu = \mu_0\mu_r$, is used to describe most materials. Diamagnetic materials exhibit μ_r slightly less than one. Paramagnetic materials display μ_r slightly greater than one. Ferromagnetic materials can have μ_r values up to 10^5 or more. The field exclusion, or *Meissner effect* in superconductors is equivalent to $\mu_r \rightarrow 0$.

Units are an issue in magnetics. Many manufacturers still use cgs units to specific magnetic materials and parts, while this chapter uses primarily SI units. The SI and cgs units and conversions can be found in Appendix B.

In practice, diamagnetic and paramagnetic materials are not much different than air in their magnetic properties since they have $\mu_r \approx 1$. Superconductors have not yet been widely used for their magnetic properties, so ferromagnetic materials are of primary interest to the magnetics designer. Ferromagnetic materials are made up of small regions called *domains*. In each such region, the magnetic moments associated with the atoms are aligned. This produces a strong local magnetic field. For any ferromagnetic material, there is a characteristic temperature, known as the *Curie temperature*, T_c, above which the alignment is disrupted. When such a material is heated above T_c, it becomes paramagnetic rather than ferromagnetic. When it is cooled again, the domains will reform, but will be oriented at random relative to each other. The net external magnetic field is then zero. Heating a material above T_c therefore provides a way to demagnetize it. Temperature can be an important issue in magnetic devices, since their Curie temperatures are usually lower than temperatures at which obvious physical damage occurs.

When external fields are applied to a ferromagnetic substance, the domains tend to realign with the imposed fields. This realignment of large groups of atoms produces the high permeability value. Once all domains have realigned, the material is said to be saturated, and the permeability drops to a paramagnetic value not much different from μ_0.

Several metallic elements and alloys act as ferromagnetic materials. Table 12.1 lists the ferromagnetic elements. Many alloys containing these elements are also ferromagnetic. Common alloying elements include aluminum, manganese, zinc, chromium, and several rare-earth metals. There are even a few ferromagnetic alloys and compounds, such as chromium dioxide, that contain no ferromagnetic elements. The fact that permeability changes as a substance saturates makes these materials nonlinear.

12.4 MAGNETIC CIRCUITS

12.4.1 The Circuit Analogy

Consider again Faraday's Law

$$\oint_l \mathbf{E} \cdot dl = \frac{-d\phi}{dt} \tag{12.6}$$

which is KVL when the right-hand side is associated with a voltage source. The terms $\mathbf{E} \cdot dl$ and $d\phi/dt$ are called *electromotive force*, EMF. Similarly, Ampère's Law requires

TABLE 12.1
Ferromagnetic Elements and Their Curie Temperatures

Element	Curie Temperature (°C)
Iron	770
Cobalt	1130
Nickel	358
Gadolinium	16
Dysprosium	−168

$$\oint_l \mathbf{H} \cdot dl = \int_s \mathbf{J} \cdot ds. \tag{12.7}$$

In the case where currents are confined to wires, each carrying I, the right-hand side can be written as NI, where N is the number of wires passing through the contour integral loop. The NI term can be taken as a kind of source equivalent to $\mathbf{H} \cdot dl$. The terms $\mathbf{H} \cdot dl$ and NI are *magnetomotive force*, MMF, and Ampère's Law can be interpreted as requiring the sum of MMFs around a closed loop to be zero, $\Sigma MMF_{loop} = 0$. This is analogous to KVL, written in terms of MMF rather than EMF.

Gauss's Law for magnetic fields provides another relationship. Since a given surface s has a magnetic flux $\phi = \int_s \mathbf{B} \cdot ds$, the law

$$\oint_s \mathbf{B} \cdot ds = 0 \tag{12.8}$$

requires that the sum of all magnetic fluxes ϕ into a closed region of space must be zero. A small region of space can be considered as a *node*, so $\Sigma \phi_{node} = 0$. The flux relationship is analogous to KCL if magnetic flux substitutes for electric current.

12.4.2 Inductance

The analogies can be extended further, and render a view of a magnetic field system as a *magnetic circuit*. Such a circuit can be analyzed in a familiar manner. Let us explore some of the other analogies by means of a general example. Consider a toroid made of ferromagnetic material, with a square cross section. A coil of wire is wrapped around it and a slice has been cut out to form an air gap, as shown in Figure 12.2. A current i flows in the coil. Faraday's Law along the wire gives

$$\oint_l \mathbf{E} \cdot dl = \frac{-d}{dt} \int_s \mathbf{B} \cdot ds \tag{12.9}$$

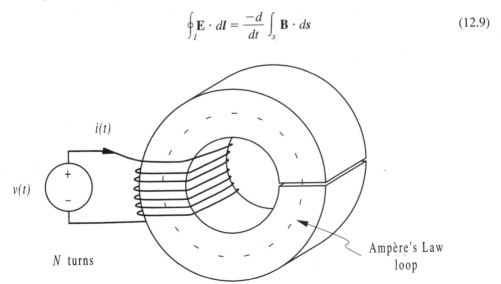

Figure 12.2 Gapped toroidal core.

If the wire has negligible voltage drop, the wire loop yields $-v_{in}$ for the left side of this equation. For the right-hand side, ϕ and **B** are not known. Only one flux value exists, and it crosses the coil N times. The value $\lambda = N\phi$ is called *flux linkage*, and equation (12.9) reduces to $v_{in} = N(d\phi/dt)$, or

$$v_{in} = \frac{d\lambda}{dt} \quad \text{Faraday's Law} \tag{12.10}$$

Now, consider Ampère's Law taken around the core. Define a loop which passes along the center of the core to form a circle with circumference l. The right-hand side of the equation is just Ni, an MMF source, while the left-hand side can be split into an integral within the core material plus an integral within the air gap. Assuming that H is nearly uniform in each material, the integrals give $H_{core}l_{core} + H_{air}l_{air}$ for the total of $\int_l \mathbf{H} \cdot dl$. The full expression becomes

$$H_{core}l_{core} + H_{air}l_{air} = Ni \tag{12.11}$$

but $\mu H = B$, so

$$\frac{B_{core}}{\mu_{core}}l_{core} + \frac{B_{air}}{\mu_{air}}l_{air} = Ni \tag{12.12}$$

If B does not change much across the toroid cross section, then $\phi = \int \mathbf{B} \cdot ds = BA$, where A is the cross-sectional core area. Then

$$\frac{\phi_{core}l_{core}}{\mu_{core}A_{core}} + \frac{\phi_{air}l_{air}}{\mu_{air}A_{air}} = Ni \tag{12.13}$$

As discussed earlier, Ni is analogous to a voltage source and ϕ is analogous to current in the sense that Ampère's Law yields the equivalent of KVL for MMFs and Gauss's Law yields the equivalent of KCL for fluxes. The quantity $l/(\mu A)$ must be analogous to resistance. For a magnetic circuit, this quantity is called *reluctance*, and the usual symbol is a script $\mathscr{R}$. When (12.13) is rewritten in terms of reluctance, the result is

$$\underset{\substack{\text{MMF} \\ \text{drop}}}{\phi_{core}\mathscr{R}_{core}} + \underset{\substack{\text{MMF} \\ \text{drop}}}{\phi_{air}\mathscr{R}_{air}} = \underset{\substack{\text{MMF} \\ \text{source}}}{\text{MMF}_{in}} \tag{12.14}$$

In general, an Ampère's Law loop can be selected, then an equation in MMFs can be written for that loop.

Equation (12.14) includes the reluctances as fixed parameters, independent input Ni, and two unknown fluxes. A second equation is needed for these two unknowns, and Gauss's Law will provide it. An expanded view of the air gap region is shown in Figure 12.3. A very thin surface is shown wrapped around one of the exposed faces of the core material. At the face, the flux makes the transition from the core material to the air. Gauss's Law requires that

$$\int_s \mathbf{B}_{core} \cdot ds + \int_s \mathbf{B}_{air} \cdot ds = \phi_{core} - \phi_{air} = 0 \tag{12.15}$$

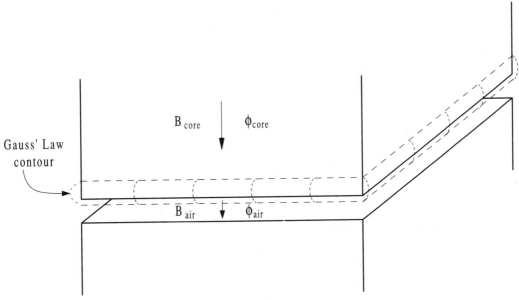

Figure 12.3 Expanded view of toroid air gap region.

The negative sign results because the core flux flows into the surface of integration, while the air flux flows out. Thus $\phi_{core} = \phi_{air}$, and there is only one flux, $\phi_{core} = \phi$. Since flux resembles current in its behavior, it is not surprising that it is conserved as it flows through an interface. Equation (12.14) now reduces to

$$\phi(\mathcal{R}_{core} + \mathcal{R}_{air}) = Ni \qquad (12.16)$$

A total reluctance, $\mathcal{R}_{tot} = \mathcal{R}_{core} + \mathcal{R}_{air}$ can be defined, and now $\phi\mathcal{R}_{tot} = Ni$. The time derivative of this expression gives

$$\mathcal{R}\frac{d\phi}{dt} = N\frac{di}{dt}, \qquad \frac{\mathcal{R}}{N}\frac{d\lambda}{dt} = N\frac{di}{dt} \qquad (12.17)$$

Since the input voltage is equal to $d\lambda/dt$, the voltage can be substituted, and the result is

$$v_{in} = \frac{N^2}{\mathcal{R}_{tot}}\frac{di}{dt} \qquad (12.18)$$

The parameter $N^2/\mathcal{R}_{tot}$ is an inductance L, such that $v_{in} = L(di/dt)$. For a given core with a winding, the magnetic fields produce an inductance proportional to the square of the number of turns and inversely proportional to the total reluctance to the winding's MMF.

This derivation followed the details of Maxwell's equations. Future analysis can make use of the MMF and flux relationships established here. In summary, a number of analogies make the analysis of a magnetic device very similar to that of an electric circuit. Magnetically, the core and coil combination follows the dc circuit shown in Figure 12.4, and the net flux is the input MMF divided by the total reluctance. Table 12.2 summarizes some of the analogies. Notice the term *permeance*, analogous to conductance.

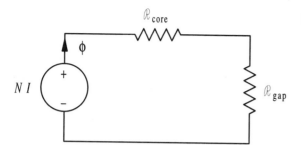

Figure 12.4   Electric circuit analogous to Figure 12.2.

Example 12.4.1   Find the inductance at the coil terminals for the structure shown in Figure 12.5.

First, let us redraw this structure as a magnetic circuit, shown in Figure 12.6. The various legs represent reluctances along the two Ampère's Law loops in Figure 12.5. The winding represents an MMF source. The right-hand side loop in Figure 12.6 has total reluctance $\mathcal{R}_{\text{right}} = \mathcal{R}_3 \parallel (\mathcal{R}_1 + 2\mathcal{R}_2)$. Each reluctance is $l/(\mu A)$. Each loop is a square 5.5 cm across, so in this case $l = 0.055$ m for each reluctance. The areas differ. Following the loops, the reluctances are

$$\mathcal{R}_1 = \frac{0.055 \text{ m}}{1000\mu_0(4 \times 10^{-4} \text{ m}^2)}, \qquad \mathcal{R}_2 = \frac{0.055 \text{ m}}{1000\mu_0(3 \times 10^{-4} \text{ m}^2)},$$

$$\mathcal{R}_3 = \frac{0.055 \text{ m}}{1000\mu_0(2 \times 10^{-4} \text{ m}^2)} \tag{12.19}$$

Since inductance is proportional to $1/\mathcal{R}$, the units of reluctance are inverse henries. The values are $\mathcal{R}_1 = 1.094 \times 10^5 \text{ H}^{-1}$, $\mathcal{R}_2 = 1.459 \times 10^5 \text{ H}^{-1}$, $\mathcal{R}_3 = 2.188 \times 10^5 \text{ H}^{-1}$, and $\mathcal{R}_{\text{right}}$ becomes $1.416 \times 10^5 \text{ H}^{-1}$. The equivalent single loop is shown in Figure 12.7. The total reluctance is $\mathcal{R}_{\text{tot}} = \mathcal{R}_1 + 2\mathcal{R}_2 + \mathcal{R}_{\text{right}} = 5.43 \times 10^5 \text{ H}^{-1}$. The inductance is $L = N^2/\mathcal{R}_{\text{tot}} = 10\ 000/(5.43 \times 10^5 \text{ H}^{-1}) = 18.4$ mH.

Example 12.4.2   Figure 12.8 shows a two-winding core. If the windings are connected in series so that they provide a total MMF $= N_1 i_1 + N_2 i_2$, what is the inductance of the combination?

TABLE 12.2
Magnetic Analogues to Electric Circuits

Electric Circuit	Magnetic Circuit
Electromotive force $-\int \mathbf{E} \cdot d\ell$	Magnetomotive force (MMF) $\int \mathbf{H} \cdot d\ell$
Voltage source $d\lambda/dt$	MMF source Ni
KVL, $\Sigma v_{\text{loop}} = 0$	MMF law, $\Sigma \text{MMF}_{\text{loop}} = 0$
KCL, $\Sigma i_{\text{node}} = 0$	Gauss's Law, $\Sigma \phi_{\text{node}} = 0$
Current	Magnetic flux
Resistance $R = \rho\ell/A$	Reluctance $\mathcal{R} = \ell/(\mu A)$
Conductance $G = 1/R$	Permeance $\mathcal{P} = 1/\mathcal{R}$
Conductivity $\sigma = 1/\rho$	Permeability μ
Conductor $\sigma \to \infty$	Ferromagnetic material $\mu \to \infty$
Insulator $\sigma \to 0$	Diamagnetic material μ small

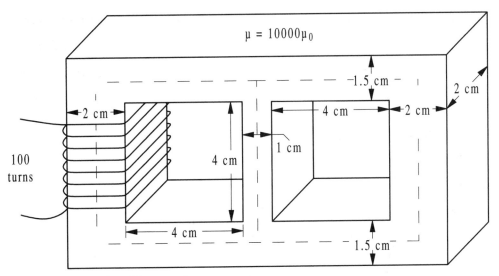

Figure 12.5 Core with coil for example.

The magnetic circuit equivalent of this structure is given in Figure 12.9, with the various reluctance values given there as well. Gauss's Law requires $\phi_1 + \phi_2 - \phi_3 = 0$. The two MMF loops give the simultaneous equations

$$N_1 i - \phi_1 \mathcal{R}_{\text{left}} - \phi_3 \mathcal{R}_{\text{right}} = 0$$
$$N_2 i - \phi_2 \mathcal{R}_{\text{cent}} - \phi_3 \mathcal{R}_{\text{right}} = 0 \qquad (12.20)$$

To find the inductance, the relationships between voltage and current are needed in each loop. The voltages can be determined from the flux linkages $\lambda_1 = N_1 \phi_1$ and $\lambda_2 = N_2 \phi_2$. The fluxes ϕ_1 and ϕ_2 are

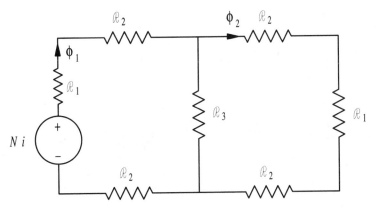

Figure 12.6 Magnetic circuit for core of Example 12.4.1.

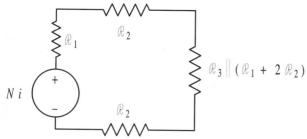

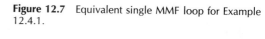

Figure 12.7 Equivalent single MMF loop for Example 12.4.1.

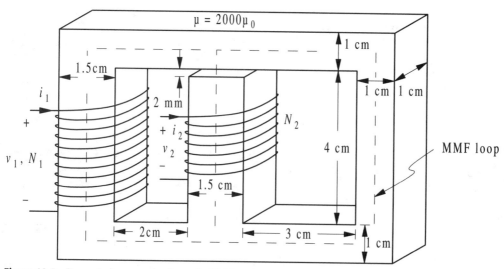

Figure 12.8 Two-winding core for Example 12.4.2.

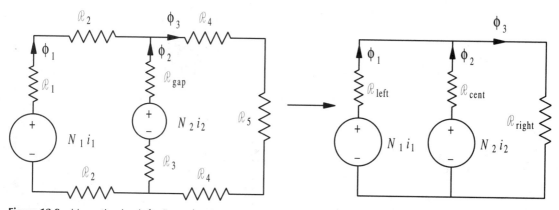

Figure 12.9 Magnetic circuit for Example 12.4.2.

$$\phi_1 = i\,\frac{N_1\mathcal{R}_{\text{cent}} + N_1\mathcal{R}_{\text{right}} - N_2\mathcal{R}_{\text{right}}}{\mathcal{R}_{\text{cent}}\mathcal{R}_{\text{left}} + \mathcal{R}_{\text{cent}}\mathcal{R}_{\text{right}} + \mathcal{R}_{\text{left}}\mathcal{R}_{\text{right}}}$$

$$\phi_2 = i\,\frac{N_2\mathcal{R}_{\text{left}} - N_1\mathcal{R}_{\text{right}} + N_2\mathcal{R}_{\text{right}}}{\mathcal{R}_{\text{cent}}\mathcal{R}_{\text{left}} + \mathcal{R}_{\text{cent}}\mathcal{R}_{\text{right}} + \mathcal{R}_{\text{left}}\mathcal{R}_{\text{right}}} \tag{12.21}$$

The total flux linkage seen by the series combination of the two windings is

$$\lambda_1 + \lambda_2 = i\,\frac{N_1{}^2\mathcal{R}_{\text{cent}} + N_2{}^2\mathcal{R}_{\text{left}} + N_1{}^2\mathcal{R}_{\text{right}} - 2N_1N_2\mathcal{R}_{\text{right}} + N^2{}_2\mathcal{R}_{\text{right}}}{\mathcal{R}_{\text{cent}}\mathcal{R}_{\text{left}} + \mathcal{R}_{\text{cent}}\mathcal{R}_{\text{right}} + \mathcal{R}_{\text{left}}\mathcal{R}_{\text{right}}} \tag{12.22}$$

When the time derivative is taken, the left-hand side of (12.22) becomes the total winding voltage, while the right side defines $L(di/dt)$. Therefore, the inductance is

$$L = \frac{N_1{}^2\mathcal{R}_{\text{cent}} + N_2{}^2\mathcal{R}_{\text{left}} + N_1{}^2\mathcal{R}_{\text{right}} - 2N_1N_2\mathcal{R}_{\text{right}} + N_2{}^2\mathcal{R}_{\text{right}}}{\mathcal{R}_{\text{cent}}\mathcal{R}_{\text{left}} + \mathcal{R}_{\text{cent}}\mathcal{R}_{\text{right}} + \mathcal{R}_{\text{left}}\mathcal{R}_{\text{right}}} \tag{12.23}$$

The reluctance $\mathcal{R}_{\text{left}}$ is the sum of a reluctance along the 5 cm path associated with $\mathcal{R}_1$ in Figure 12.9 plus $2\mathcal{R}_2$, along a path 7 cm long. The Ampère's Law loops are shown as dotted lines in Figure 12.8. The numbers give $\mathcal{R}_{\text{left}} = 4.11 \times 10^5\ \text{H}^{-1}$. The center reluctance $\mathcal{R}_{\text{cent}}$ is the air gap reluctance $1.06 \times 10^7\ \text{H}^{-1}$ plus the center leg core reluctance along a 4.8 cm path, $1.27 \times 10^5\ \text{H}^{-1}$, so $\mathcal{R}_{\text{cent}} = 1.07 \times 10^7\ \text{H}^{-1}$. The total right-hand side reluctance represents a path 13.5 cm long through an area of 1 cm^2, so $\mathcal{R}_{\text{right}} = 5.37 \times 10^5\ \text{H}^{-1}$. Let us take, for instance, $N_1 = 20$ and $N_2 = 10$. Then (12.22) gives the final result for equivalent inductance, $L_{\text{equiv}} = 0.422$ mH. It is interesting to consider the physical issues in this value. For example, the air gap reluctance is much bigger than any other. In effect, the center leg is a high-impedance path to magnetic flux. We might expect that almost all of the flux will flow in the large outer loop, with N_1i_1 as the MMF source. This path should exhibit an inductance $L_1 = N_1{}^2/(\mathcal{R}_{\text{left}} + \mathcal{R}_{\text{right}})$. The value of L_1 is found by computation to be 0.422 mH—the same as the total. Physically, the center leg does very little in the structure of Figure 12.8.

At this point, it is important to point out a few limitations of the magnetic circuit viewpoint (as distinct from the underlying Ampère and Faraday relationships). The first, mentioned earlier, is the nonlinear nature of ferromagnetic permeability. Reluctances are not as linear as resistances, and will change somewhat as the MMF value changes. A second is the large size of practical magnetic cores, relative, say, to a wire in an electric circuit. The resistance relationship $R = l/(\sigma A)$ assumes A small. This is not often true for reluctance. The underlying assumption is that **B** and **H** are uniform throughout the cross section of the core. If the Ampère loop is drawn differently, different values of reluctance might result. While the examples followed a convention of drawing the path through the centerline of the material, manufacturers try to specify an *equivalent magnetic path length*, l_{equiv}, and an *equivalent magnetic area*, A_{equiv}, to provide an accurate relationship for reluctance in an actual core.

A third difference concerns the relative reluctance values. The current-carrying wires of an electric circuit are far more conductive than the surrounding insulation and air. Even resistors are much more conductive than air. Indeed, the ratio of copper conductivity to air conductivity is more than 10^{20}. This extreme value means that the air surrounding a circuit plays no role in the circuit's operation. For a magnetic circuit, the standard flux-conducting ferromagnetic materials have permeabilities no more than 10^5 times that of air. The analog is an electric circuit, stripped of insulation and operated immersed in seawater. In the magnetic circuit, some of the flux flows into and through the surroundings, and is not confined to the magnetic material. The examples neglect additional MMF paths in the air around the cores.

12.4.3 Ideal and Real Transformers

Even with the limitations, it is possible to make excellent predictions of magnetic device behavior through magnetic circuit calculations. In this section, the ideal transformer is considered, this time from a magnetic circuit viewpoint. Given a toroid core with two windings, as shown in Figure 12.10, the MMF expression gives $N_1 i_1 - \phi \mathcal{R} = N_2 i_2$. If μ is sufficiently large, then $\mathcal{R}$ will be small, and $N_1 i_1 = N_2 i_2$. The flux ϕ is given by $v_1 = N_1 (d\phi/dt)$. For the second coil, the flux must be the same, and $v_2 = N_2 (d\phi/dt)$. Dividing these two,

$$\frac{v_1}{v_2} = \frac{N_1}{N_2} \quad \text{provided} \quad \frac{d\phi}{dt} \neq 0 \qquad (12.24)$$

This describes an *ideal transformer*, with the important caveat that the flux must have nonzero time derivative. Input and output powers are equal, and current and voltage are modified by the turns ratio.

In reality, the reluctance of the core is not negligible. Current will flow in the #1 winding even if $i_2 = 0$. Setting $i_2 = 0$ makes the core an inductor $L = N_1^2/\mathcal{R}$. This value is the *magnetizing inductance* L_m. Since the core does not have infinite permeability, some of the flux will flow in the surrounding air. The #1 winding, for example, will give rise to flux $\phi_1 = \phi_{l1} + \phi$, where ϕ_{l1} is the portion that leaks through the air while ϕ is the portion that flows through the core and couples with the #2 winding. These two separate fluxes can be associated with separate inductances, since each gives rise to flux linkage. The *leakage inductance* L_{l1} will be the inductance $N_1^2/\mathcal{R}_{\text{leak}}$ associated with the leakage flux. A similar effect occurs in the #2 winding. Finally, each winding has a resistance that depends on the size

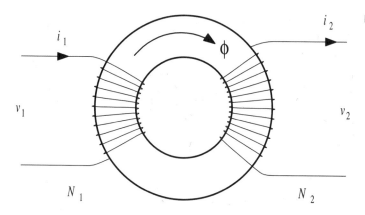

Figure 12.10 Toroid core with two coils.

Figure 12.11 Real transformer.

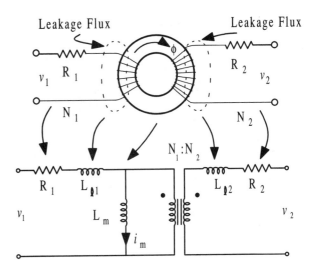

and composition of the wire and also on temperature and frequency. Figure 12.11 shows again the real transformer.

Example 12.4.3 A toroidal core with outside diameter of 15 cm and inside diameter of 10 cm is 2.5 cm thick. It is made of ferrite material with $\mu = 5000\ \mu_0$. The core is wound with 100 turns of #12 AWG wire on the primary, and 500 turns of wire on the secondary. As shown in Figure 12.12, the windings are "separated," meaning that they are confined to opposite sections of core rather than overlapped (coil separation permits higher isolation between input and output, at the expense of higher leakage). Suggest a wire size for the secondary, and compare the two windings. Develop a circuit model for the complete transformer, based on an estimate

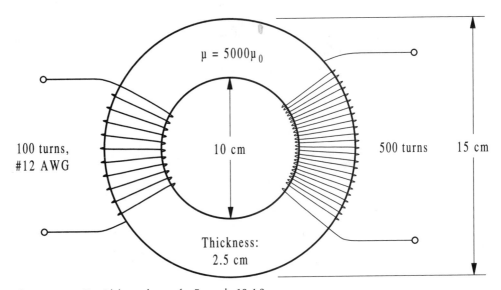

Figure 12.12 Toroidal transformer for Example 12.4.3.

that because $\mu_r = 5000$, about 0.02% of the flux follows a leakage path. If the transformer is used with an input converter waveform that is approximately a 120 V_{RMS} sine wave at 2500 Hz, what will the input current be with no load? Suggest a load rating.

Given low leakage, the secondary voltage should be 500/100 = 5 times the primary voltage. The current at the secondary should be 1/5 of that at the primary. For a given current density, the secondary wire size should be about 1/5 that of the primary. Since the #12 wire on the primary has an area of about 3.3 mm^2 from Table 11.1, the secondary wire should be about 0.66 mm^2, which corresponds to just about #19 wire (area 0.653 mm^2). If smaller wire were used, it would have higher current density than the primary, and would get hotter than the primary. If larger wire is used, the primary will run hotter than the secondary. In either case, mismatch in the current densities will make a portion of the whole transformer reach a thermal limit before the rest of the material. In general, it is true that the material is used most effectively if the current densities in primary and secondary are matched. The total area of the primary winding will be 100(3.309 mm^2) = 331 mm^2, while that of the secondary will be 500(0.653 mm^2) = 326 mm^2. The coils should be matched in size to make $J_1 = J_2$.

The core reluctance is needed to determine L_m. The length is approximately the circumference of the Ampère's Law loop in Figure 12.12, so $l = 0.125\pi$. The magnetic area is 6.25 cm^2. The core reluctance is $\mathcal{R} = l/(\mu A) = 10^5$ H^{-1}. From the primary, the magnetizing inductance is $N_1^2/\mathcal{R} = 100$ mH. Total leakage flux of 0.02% suggests 0.01% for each coil. Since flux is given by $Ni/\mathcal{R}$, the leakage reluctance should be a factor of about 10 000 higher than the core reluctance. Thus the primary leakage inductance is about $N_1^2/\mathcal{R} = 10^4/(10^9$ H$^{-1}) = 10$ μH. The secondary value is $(500)^2/(10^9$ H$^{-1}) = 250$ μH.

The resistances require knowledge of wire length. The shortest possible wire loop would traverse the perimeter of a core leg, and would be 10 cm long. The winding process is not perfect, and some wires might be stacked to allow all to fit. Let us guess a length of 11 cm for a typical primary turn, and 12 cm for a secondary turn (the secondary is longer since it will probably be a few layers thick). Then the primary total length is about 11 m and the secondary length is about 60 m. The primary resistance will be about 60 mΩ. Wire size #19 AWG has a resistance of 26.9 mΩ/m at 25°C, so the secondary resistance will be about 1.6 Ω. Notice that both the resistance and inductance ratios are close to the square of the turns ratio. Ideally, a properly designed transformer will show $L_{l1}/L_{l2} = R_{s1}/R_{s2} = (N_1/N_2)^2$ if the leakage fluxes and current densities are properly matched.

The complete equivalent circuit is shown in Figure 12.13. (Notice that it could have been developed from the secondary side, in which case a magnetizing inductance of 2.5 H

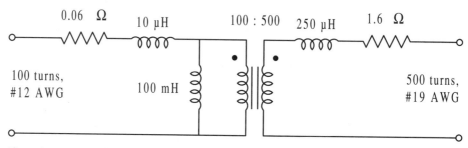

Figure 12.13 Equivalent circuit for transformer of Figure 12.12.

would appear on the secondary instead of on the primary.) If 120 V at 2500 Hz is imposed at the input with no secondary load, the input impedance will be $0.06\ \Omega + j(2\pi2500\text{rad/s})$ (100 mH + 10 μH) = $0.06 + j1571\ \Omega$. The primary will draw a *magnetizing current* of (120 V)/(1571 Ω) = 0.076 A_{RMS}. The rated load is limited to a large degree by the wire capacity. If the current is held to 100 A/cm^2, the primary should be able to handle 3.3 A_{RMS}, supporting a power rating of 400 W. The secondary current would be about 0.66 A, and the losses in the wires would be $(3.3\ A)^2(0.06\ \Omega) + (0.66\ A)^2(1.6\ \Omega) = 1.35$ W, or 0.34% of the 400 W rating.

12.5 THE HYSTERESIS LOOP AND LOSSES

The nonlinear permeability and saturation effects in a ferromagnetic material can be shown in a B vs. H curve, called a *magnetization characteristic*. Let us start with a typical ferromagnetic core. First, heat it above T_c to eliminate all magnetization. The internal structure will consist of small, fully magnetized domains oriented at random, as in Figure 12.14. The net flux measured outside the material is zero. Now, apply MMF to the material through a coil. As the current is slowly raised, domains begin to reorient. The flux rises quickly. At sufficiently high current, most domains have flux aligned with the imposed H, and the rate of increase declines. Finally, virtually all domains are aligned, and further increases in flux are governed by the material's inherent paramagnetism. This behavior is shown in Figure 12.15. The slope of the curve, by definition, is permeability. The slope close to the origin is termed *initial permeability*.

When the MMF is reduced, many domains tend to remain aligned. Hence H must be reduced significantly before the domains begin to return to their initial unaligned arrangement. The magnetization curve is not fully reversible, a phenomenon known as *hysteresis*.

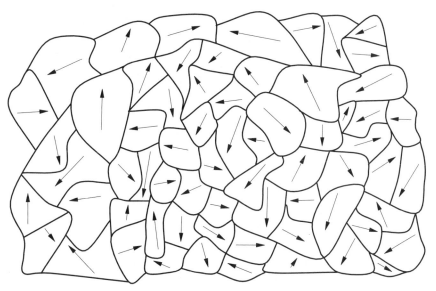

Figure 12.14 Internal structure of demagnetized ferromagnetic object.

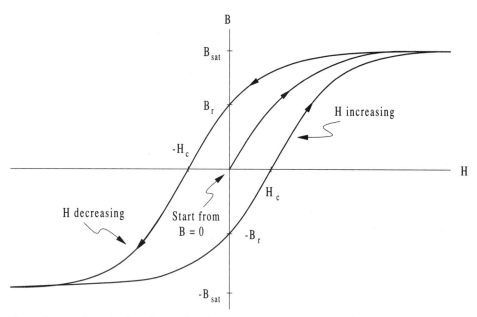

Figure 12.15 Magnetization characteristic for initially unmagnetized sample.

When the applied MMF is reduced to zero, some residual alignment remains, and an external *remanent flux* with density B_r can be detected. The MMF must be reversed and driven somewhat negative before the net flux is again zero. The value of H needed to force the flux to return to zero is the *coercive force*, H_c. Figure 12.16 shows a typical complete magnetization curve, or *hysteresis loop*, measured experimentally for a transformer core.

The B vs. H loop is characteristic of a given material, and is often provided by manufacturers of magnetic materials. The loop can also be expressed as λ vs. Ni, if the geometry and number of turns are known. It is important to recognize that the size of the loop will depend on the range of H. The smaller the variation in imposed field, the narrower the loop. Figure 12.17 shows a sequence of loops for a specific core, measured at various peak values of H.

The irreversibility of domain alignment associated with hysteresis gives rise to an energy loss. It can be shown that the area inside the curve, which is an integral of $B\, dH$, has units of energy per unit volume. The area inside the λ vs. Ni loop is the actual energy loss for a specific core. Any two distinct points along the hysteresis curve have distinct energies, and thus any movement along the curve requires energy input. As in Figure 12.17, the higher the range of variables on the hysteresis curve, the larger the enclosed area of the loop. The effect is nonlinear: Loop size increases rapidly as the maximum flux density rises, to the extent that energy loss increases almost in proportion to B^2 up to saturation.

Power loss occurs as the loop is traversed. When the MMF excitation is sinusoidal,

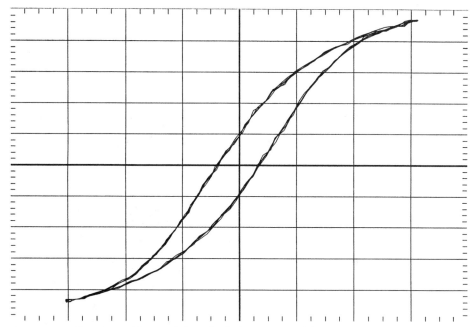

Figure 12.16 Experimental hysteresis loop for ferrite core material.

each cycle of the waveform moves the material around the hysteresis loop. Since the loop area is an energy loss, the loss per unit time is the energy enclosed by the loop times the rate of traversal. Hysteresis loss therefore is approximately proportional to frequency. Magnetic hysteresis is observed primarily in ferromagnetic materials, and does not occur in air or other materials with $\mu = \mu_0$.

In addition to hysteresis loss, ohmic loss occurs in most magnetic materials. After all, the ferromagnetic elements and alloys are metals, and we should expect that the core materials represent additional circuits that interact with the fields through Faraday's Law. The flux within the core material produces an internal voltage $d\lambda/dt$ that drives *eddy current* circulation within the material. The losses because of eddy currents depend on the flux amplitude and frequency, and on the internal resistivity of the core. The amplitude and frequency effects are quadratic since loss is proportional to v^2. The internal resistance must be kept as high as possible, especially if high frequencies are involved.

The eddy current effect is important in determining appropriate core materials and construction methods for power electronics. Steel for magnetic cores, for example, is alloyed with silicon to make it more resistive. In metals more generally, the resistance to eddy current loops is made high by constructing the core from thin insulated plates, or *laminations*, oriented parallel to the flux. Voltage is determined by flux, not flux density; thin laminations have little flux in any individual layer. The geometry decreases $d\lambda/dt$ and increases resistance. An illustration in Figure 12.18 shows the effect of lamination. The lamination struc-

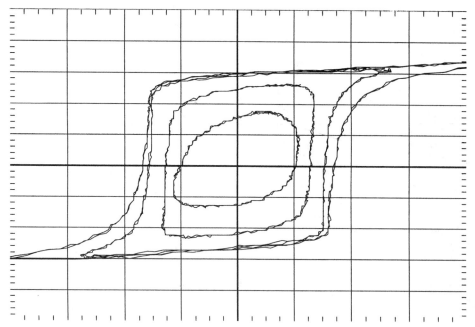

Figure 12.17 Sequence of measured hysteresis loops for a ferrite toroid with $\mu = 5000\mu_0$.

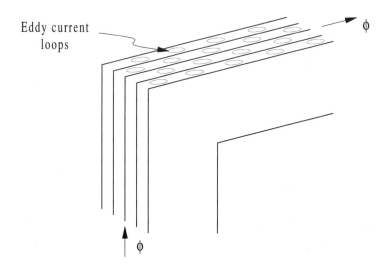

Eddy current loops

ϕ

ϕ

Figure 12.18 Lamination geometry for metallic core material.

ture in small cores is often obtained conveniently by winding up a ferromagnetic ribbon. These *tape-wound* cores are easy to manufacture.

Even with laminations, it is hard to make metal cores that are useful above about 100 kHz. The eddy current problem can also be addressed by using high-resistance magnetic ceramics, or *ferrites*, as core materials. Ferrites provide excellent core materials for frequencies from 10 kHz to about 10 MHz, with different groups of materials for the 10 kHz to 1 MHz range and for the 500 kHz to 10 MHz range. Another alternative is to use powdered metallic core materials embedded in a nonconductive matrix. Powdered iron cores use this approach to combine high values of flux with low losses.

The net result of hysteresis and eddy current loss is that the power loss in a magnetic core is given approximately by

$$P_{\text{loss}} = P_0 f^a B^b \qquad (12.25)$$

where a and b are empirical constants for a given material. Typically, a is somewhat greater than 1, while b approaches 2 or even more. For example, at close to 50 kHz and 0.1 T, the ferrite material known as 3F3[1] exhibits $a \approx 1.4$ and $b \approx 2.4$ (as for any ferrite, the material is useful because the base loss P_0 is low). Manufacturers usually specify loss in watts per unit volume or watts per unit weight for given values of maximum flux density and frequency.

12.6 SATURATION AS A DESIGN CONSTRAINT

12.6.1 Saturation Limits

As the hysteresis loop reaches a maximum value of B, saturation occurs. The permeability drops quickly toward μ_0 as the saturation level, B_{sat}, is reached. Extra MMF will still increase B, but the rate of increase reduces drastically for a ferromagnetic material above B_{sat}. Manufacturers routinely specify a value of B_{sat} at a particular value of H, since the saturation level is somewhat ambiguous. For good magnetic steels, the saturation value is about 2 T. For typical ferrites, $B_{\text{sat}} \approx 0.3$ T. Powdered iron cores and many other magnetic alloys can operate to at least 1 T.

The effects of saturation in general are undesirable and are to be avoided, although there are specialized applications in which saturation provides switch-like action. The main reason for avoiding saturation in an inductor or transformer is that when the saturated permeability approaches μ_0, a core becomes "magnetically transparent"—indistinguishable from the outside air. Most of the flux then leaks through the air, and the core offers no benefit. For a transformer, the relationship between input and output can break down under saturation, since the leakage flux does not couple the windings.

To avoid saturation, the flux density in a given core needs to be known. An inductor with a single coil, for instance, has $\phi = Ni/\mathcal{R}$. The flux density is $B = Ni/(\mathcal{R}A)$. When there are multiple coils, each current contributes to the total flux, so $B = (Ni)_{\text{net}}/(\mathcal{R}A)$. To keep

---

[1]*Ferrite Materials and Components Catalog, 8th ed.* Saugerties, NY: Philips Components, 1992, pp. 1–17.

$B < B_{sat}$, we need to set a limit on the current (or, more strictly, on the MMF value). To avoid saturation in an inductor,

$$Ni < B_{sat}\mathcal{R}A, \qquad Ni < B_{sat}\frac{l}{\mu} \tag{12.26}$$

Thus there is an MMF limit, often called an *ampere-turn limit*, associated with any inductor with a ferromagnetic core. The MMF limit has interesting implications. Consider that the inductor's stored energy is $\frac{1}{2}Li^2 = \frac{1}{2}i^2N^2/\mathcal{R}$. If the MMF is set to the maximum value, Ni_{max}, the energy is $W_{max} = \frac{1}{2}(Ni_{max})^2/\mathcal{R}$. Since Ni_{max} is determined by saturation, a given core has a definite energy limit.

> **Empirical Rule:** *The maximum energy that can be stored with a given magnetic core is*
>
> $$W_{max} = B^2_{sat}\frac{l_{core}A_{core}}{2\mu} \tag{12.27}$$
>
> *This energy is proportional to the core volume and inversely proportional to the permeability. In most practical cases, the easiest way to make the maximum energy a significant amount is to include an air gap in the magnetic circuit, so the air gap volume divided by μ_0 determines the maximum energy, and* $W_{max} = \frac{1}{2}B_{sat}^2 V_{gap}/\mu_0$.

Wound cores (with known N) have a current limit implied by Ni_{max}, and are normally given a current rating for this reason.

For transformers, the net MMF imposed on a core is intended to be zero: The total sum $N_1 i_1 + N_2 i_2 = 0$. In these cases, the flux linkage λ provides an alternative way to determine B. Given that $v = d\lambda/dt$, we can write $\int v\,dt = \lambda = N\phi = NBA$. As long as $\int v\,dt < NB_{sat}A$, the material should not saturate (when the average net MMF is zero). The integral differs from the actual flux by an integration constant proportional to the dc current. In a transformer, dc current is avoided in part because it tends to saturate the core. If for any reason an imbalance occurs such that nonzero dc current is allowed to flow, the total flux becomes $\int v\,dt/NA + Ni_{dc}/(\mathcal{R}A) < B_{sat}$.

The integral $\int v\,dt$ represents a volt-second product. The maximum value of $\int v\,dt/N$ is often called a *maximum volt-second rating*. It suggests that a dc voltage can be applied to a coil only for a short time before saturation occurs. For a sinusoidal transformer voltage $v = V_0\cos(\omega t)$, the volt-second integral $\int V_0\cos(\omega t)\,dt$ suggests that $V_0\sin(\omega t)/(\omega NA) < B_{sat}$. The peak value of flux gives a constraint,

$$\frac{V_0}{\omega NA} < B_{sat} \tag{12.28}$$

This relation is often called *maximum volts per turn* by power transformer designers. In summary:

Empirical rule: *An inductor or transformer has a maximum volt-second rating, given by*

$$\frac{\int v\, dt}{NA_{core}} < B_{sat} \begin{cases} dc: & V_{dc}\Delta t < B_{sat}NA_{core} & \text{max. volt-sec.} \\ ac: & \dfrac{V_{peak}}{N} < B_{sat}\omega A_{core} & \text{max. volts per turn} \end{cases} \qquad (12.29)$$

In a power converter, any attempt to apply excess volt-seconds to a coil will cause its core to saturate. As the core saturates, the permeability and inductance both decrease. A dc–dc converter, for example, tends to enter discontinuous mode if excess flux is attempted. In this mode, the inductor currents rise rapidly. Losses increase, and often the converter becomes inefficient.

Example 12.6.1 A ferrite core, shown in Figure 12.19, has an Ampère's Law loop length $l_{equiv} = 10$ cm and cross-sectional area of 2 cm². Its saturation flux density is 0.3 T, and the permeability is $1250\mu_0$. What inductive energy can be stored in this core? With ten turns of wire, what is the maximum allowed dc current? How many turns would be needed if this core is to be used instead for a transformer with 120 V, 60 Hz input?

The core as described has reluctance

$$\mathcal{R} = \frac{l}{\mu A} = \frac{0.10\ \text{m}}{(1250 \times 4\pi \times 10^{-7}\ \text{H/m})(2 \times 10^{-4}\ \text{m}^2)} = 3.18 \times 10^5\ \text{H}^{-1} \quad (12.30)$$

To avoid saturation, keep $B < 0.3$ T. This is equivalent to

$$\frac{Ni}{\mathcal{R}} < B_{sat}A, \qquad Ni < (0.3\ \text{T})(2 \times 10^{-4}\ \text{m}^2)(3.18 \times 10^5\ \text{H}^{-1}), \qquad Ni < 19.1\ \text{A} \cdot \text{turns} \quad (12.31)$$

The maximum stored inductive energy is the square of the amp-turn limit divided by twice the reluctance. For this core, the energy cannot exceed 0.573 mJ. If ten turns of wire are used, the inductor coil should not carry more than 1.91 A.

Figure 12.19 Ferrite toroid core for Example 12.6.1.

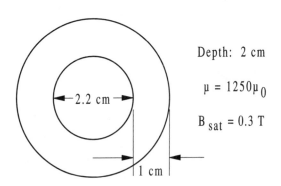

Depth: 2 cm

$\mu = 1250\mu_0$

$B_{sat} = 0.3$ T

2.2 cm

1 cm

For a transformer, the net dc current is low, but the voltage imposed on the primary induces flux in the core. To avoid saturation, the flux linkage λ must be kept below a limiting value. Given 120 V_{RMS} input at 60 Hz,

$$\int 170 \cos(120\pi t)\, dt = \lambda = NBA < NB_{sat}A$$

$$\frac{170 \text{ V}}{120\pi NA} < 0.3 \text{ T}, \qquad 7516 < N \tag{12.32}$$

This huge number of turns is unwieldy for a core of this size. Very tiny wire would be required, and the wire resistance would be high. Notice that if $N = 7516$ turns, the coil will be able to handle no more than 2.54 mA of dc current because of the dc limit on amp-turns. This suggests that the core will be extremely sensitive to any unwanted dc component. In practice, the core is small for a 120 V, 60 Hz transformer application.

12.6.2 General Design Considerations

We can identify three common types of static (stationary) magnetic devices: permanent magnets, transformers, and inductors. Let us examine each in light of our knowledge about magnetic materials.

A permanent magnet (PM) produces flux even with no imposed MMF. Common uses include establishment of constant fields in motors and generators, sensing devices, and a range of industrial applications. In any application, a PM is exposed to MMF sources as it interacts with windings and other sensing mechanisms. This will cause hysteresis loss in the magnet, and might even tend to diminish the flux density over time. Important material properties are high B_R values and high H_C values. To keep both of these values high, the most common permanent magnet materials tend to have square hysteresis loops, as in Figure 12.20.

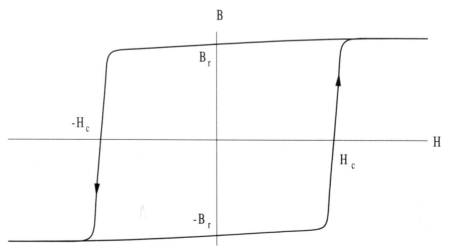

Figure 12.20   Magnetization curve appropriate for permanent magnet.

Newer rare-earth materials, especially samarium-cobalt and neodymium-iron-boron, have revolutionized the application of permanent magnets. These materials have much higher coercive forces than other PM materials, and exhibit remanent flux values beyond 0.5 T. The hysteresis loops are not square, like those of most permanent magnet materials. Instead, the loops resemble the conventional shape in Figure 12.15, except for the scale values. In practice, this loop shape tends to give very low hysteresis loss even with wide swings in the externally imposed MMF. This is because the portion of the loop in the quadrant with positive flux and negative MMF is almost linear. We can expect much wider use of PM devices in the future. The applications in power electronics focus mainly on motors, which are beyond the scope here.

Transformers require high permeabilities to minimize leakage flux and to keep the magnetizing current as low as possible. Losses must also be low. A constant value of μ is not as important, as long as saturation is avoided. Materials used for transformers generally show magnetization curves that are as narrow as possible, reflecting the desire to minimize loss. An example is shown in Figure 12.21.

The issue for inductors is *linearity*. A value of inductance L presumes that $\mathcal{R}$, and hence μ, does not change as imposed voltage, MMF, flux, and frequency change. For ferromagnetic materials, linear behavior is inconsistent with the basic domain action that leads to high permeability. For air and nonferromagnetic materials, the value $\mu = \mu_0$ offers excellent linearity but low values of inductance. These fundamental characteristics create a dilemma: It is very difficult to make inductors that have constant, substantial values of L. Good linearity can be achieved only if air or a similar linear magnetic material is used in place of a ferromagnetic core. The dilemma is resolved by adding an air gap to a core. The ferromagnetic core material helps make μ as high as possible, while an air gap helps keep

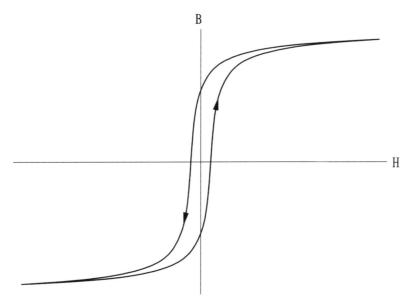

Figure 12.21 Magnetization curve appropriate for a transformer.

μ constant. Thus an air gap in a ferromagnetic core serves not only to store the magnetic energy, but also to help make the inductance behavior linear. Powdered metal cores in effect provide a *distributed air gap* to produce the same results.

The hysteresis loop slopes in Figure 12.17 illustrate the linearity issue. In the figure, the permeability changes by a factor of two or more as the MMF amplitude changes. To see the problem, consider a hypothetical ferromagnetic material in which the permeability varies from a nominal value by $\pm 50\%$. For an inductor, the relationship is

$$L = \frac{N^2}{\mathcal{R}_{\text{total}}} = N^2 \frac{\mu A_{\text{equiv}}}{l_{\text{equiv}}} \tag{12.33}$$

with L proportional to μ. For this hypothetical material, the inductance will vary by $\pm 50\%$ from a nominal value. The air gap solution to this dilemma is examined further in the following example.

Example 12.6.2   A core suggested for an inductor application has a loop length of 10 cm and cross-sectional area of 2 cm^2. The permeability is $1250\mu_0 \pm 50\%$, given $B_{\text{sat}} = 0.3$ T. It is proposed to alter the core by adding a 1 mm air gap, as shown in Figure 12.22. For the gapped core, find the maximum inductive energy storage. Define an *effective permeability* from the total reluctance, such that $\mu_{\text{eff}} = \mathcal{R}_{\text{total}} A/l$. What is the variation of this μ_{eff}? If an inductor is formed by wrapping ten turns around the core, what are the value and tolerances of L?

From Example 12.6.1, the ungapped core has a reluctance of 3.18×10^5 H^{-1} and a limit $Ni < 19.1$ A $\cdot$ turns. The energy storage capability was a fraction of one millijoule. With the air gap added, the total reluctance is

$$\mathcal{R}_{\text{total}} = \mathcal{R}_{\text{core}} + \mathcal{R}_{\text{gap}} = \frac{0.099 \text{ m}}{1250\mu_0(2 \times 10^{-4} \text{ m}^2)} + \frac{0.001 \text{ m}}{\mu_0(2 \times 10^{-4} \text{ m}^2)} = 4.29 \times 10^6 \text{ H}^{-1}$$

$$\tag{12.34}$$

The extra reluctance of the air gap gives an amp-turn limit of 258 A $\cdot$ turns. The maximum energy becomes 7.73 mJ. The increase in energy storage capability is proportional to the reluctance increase, and is more than an order of magnitude.

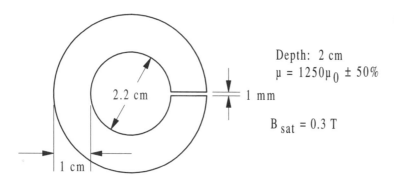

Figure 12.22   Gapped core for Example 12.6.2.

Depth: 2 cm

$\mu = 1250\mu_0 \pm 50\%$

2.2 cm

1 mm

$B_{\text{sat}} = 0.3$ T

1 cm

The *effective permeability* is often specified by manufacturers of inductor cores. For this core, μ_{eff} is defined such that

$$\mathscr{R}_{\text{total}} = 4.29 \times 10^6 \text{ H}^{-1} = \frac{0.1 \text{ m}}{\mu_{\text{eff}}(2 \times 10^{-4} \text{ m}^2)}, \qquad \mu_{\text{eff}} = 92.7\mu_0 \qquad (12.35)$$

Thus the total permeability is $92.7\mu_0$ with the air gap, and the core is functionally equivalent to a core material having a relative permeability of about 93. The variation can be found with a little conventional error analysis, but let us consider it in some detail: The permeability of the ferrite could be as low as $625\mu_0$ or as high as $1875\mu_0$. With the gap, the total reluctance follows equation (12.34), and will be somewhere between 4.19×10^6 H^{-1} and 4.61×10^6 H^{-1}. These values correspond to effective permeabilities of $95.0\mu_0$ and $86.3\mu_0$, respectively. The total width of the variation is less than 10%—a factor of ten lower than without the air gap. In practice, the value might be given as $90 \pm 5\%$, since a symmetric error distribution is easier for computation. The value $90\mu_0$ gives $\mathscr{R}_{\text{total}} = 4.4 \times 10^6$ H^{-1}. A ten-turn coil therefore produces $L = 23$ μH $\pm 5\%$. This is usually considered to be an excellent tolerance level for an inductor.

12.7 DESIGN EXAMPLES

The process of magnetic design involves selection of a core material and properties, selection of a core geometry and air gap, evaluation of limits, choice of wire size, and winding configuration. In this section, the design issues are considered broadly. A host of examples can be found in specific magnetic design texts[2]. Those presented here are representative.

12.7.1 Core Material and Geometry

The selection of a material follows from the previous discussion of losses. In a line-frequency application, eddy current loss can usually be managed adequately with a laminated structure. Steel or other metal alloys are cheap, easy to manufacture, and have very high values of B_{sat} and μ. These materials are common in line-frequency devices. For frequencies between 1 kHz and 100 kHz, less lossy materials such as powdered iron or ferrites would be preferred. Much above 100 kHz, the inherent high resistivity of ferrites makes them almost the only reasonable choice. One trouble with ferrites is their inherently low saturation flux density. A ferrite core might be a factor of three or four larger than an equivalent metal laminated core to accommodate the same flux level. All of the materials used for transformers and inductors are called *soft* ferromagnetic materials, meaning that the coercive force needed to cancel the remanent flux is very low.

Core geometry is chosen for practicality, cost, coil winding convenience, and air gap considerations. One point not considered earlier is that a core must provide space for windings. The *window area*, the area of the opening available for coils, is an important geomet-

---

[2]A classic design manual is W. T. McLyman, *Transformer and Inductor Design Handbook*. (New York: Marcel Dekker, 1978). The focus is on applications in power electronics.

ric factor beyond those for Faraday's and Ampère's Laws. The window area A_{window} must provide sufficient space for each winding, for wire insulation, for air spaces (since the packing of the wire is not complete), and for any extra winding hardware. The noncopper elements in the window define a *fill factor*, a_{fill}, the ratio of actual copper to total window area. In a typical core, a fill factor below 50% would not be unusual. In a transformer, multiple windings must be handled. The current rating of a winding must be consistent with current density limits for copper as well as saturation amp-turn limits. Figure 12.23 shows two core configurations with the window area emphasized.

A toroid is an excellent choice for a core because the shape tends to minimize leakage flux, it is easy to mold, and the window is directly accessible. There is one very important drawback: toroids are hard to wind. In many instances, toroids are wound by hand, with the attendant problems of high cost and uncertain repeatability. Core winders mechanically similar to sewing machines can shuttle a wire around a toroid. Such machines are temperamental, and the shuttle mechanism limits the fill factor. Toroids are good alternatives when the number of turns is low and expensive core materials are involved. They are common in power converters, especially when frequencies are above 100 kHz.

Most nontoroidal core geometries are designed for convenient winding. Several arrangements are shown in Figure 12.24. Each of these types forms the core in separate sections that can be assembled around a prewound bobbin. The core designations rely on shape. Arrangements such as E-E cores, U-I cores, E-I cores and others are common both for metal laminated materials and for ferrites. Pot-cores are a rotated version of an E-E core configuration. They have certain advantages for magnetic shielding, and are used in high-performance converters. Square cores are similar to pot cores.

Once the basic geometry has been selected, the general size of the core can be determined. For an inductor, the air-gap volume is a primary consideration. Given the desired energy level, it is easy enough to determine a minimum gap volume. Manufacturers usually provide preselected gap lengths, since ferrites are difficult to machine for alternative values. For a transformer, the core must provide sufficient window area to hold the necessary number of turns. The window also must hold enough copper to carry the desired current.

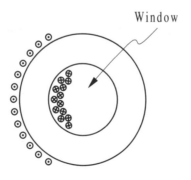

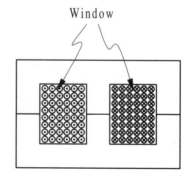

a) Toroid, fill factor ~ 10%   b) E - E core, fill factor ~ 50%

Figure 12.23 Core geometries, showing the window area and fill factor.

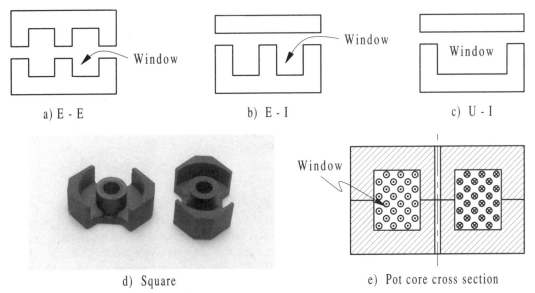

a) E - E   b) E - I   c) U - I

d) Square   e) Pot core cross section

Figure 12.24 Alternate core geometries.

Example 12.7.1 A typical geometry for an E-I core is shown in Figure 12.25. Notice that the two outer core legs have half the cross-sectional area of the center, since the flux should divide between them. The core is available in a variety of dimensions, governed by the parameter d. It is desired to use such a core to produce a 500 VA isolation transformer with 220 V, 50 Hz input and output. If the wire size is chosen to keep $J \leq 200$ A/cm^2, what size core will be needed for this application?

Figure 12.25 E-I laminated core for Example 12.7.1.

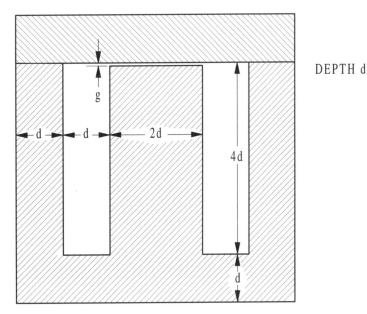

DEPTH d

Since this transformer is intended for line frequency use, a laminated metal structure will be chosen. The window area in the Figure is $4d^2$. For 500 W at 220 V, the wire should be able to carry 2.27 A. From Table 11.1, this implies a wire size of about #16 AWG, which has a cross-sectional area of 1.309 mm². There are two windings of N turns each, for a total of $2N$. With a fill factor of 0.5, the window area should be double the copper area, or $2 \times 2N \times 1.309$ mm². Thus $4d^2 > 5.236N$ mm². With d given in meters, $N < 7.64 \times 10^5 d^2$.

The magnetic area based on the center leg is $2d^2$, and the core should be chosen large enough to avoid saturation. For metal laminations, we might expect $B_{sat} = 1.5$ T. The sinusoidal input of 220 V_{RMS} has a peak value of 311 V. From (12.28),

$$\frac{311 \text{ V}}{100\pi N 2d^2} < 1.5 \text{ T}, \qquad N > \frac{0.330}{d^2} \qquad (12.36)$$

It can be shown by direct computation that if d is less than about 2.5 cm, it is not possible to meet both constraints on N simultaneously. In this example, the solution d equal to one inch (many magnetic material manufacturers use U.S. units for their products) is very close to the requirements. Substituting $d = 0.0254$ m, saturation can be avoided with 512 turns. This number of turns will fit if the fill factor rises to 0.52. If a lower fill factor is vital, a larger core must be used. This is the smallest value of d, and therefore the smallest core of this type of material, that meets the requirements.

The windings will be wound around the center post rather than the two legs for two reasons. First, the transformer will be smaller if the windings are located in the center. Second, if both windings are on the center post, leakage flux effects are much lower than if the windings are on separate legs. The airgap would be set to $g = 0$.

Example 12.7.2   An E-E core with air gap is to be chosen for an inductor for a dc–dc converter. The geometry is exactly the same as in Figure 12.25, except that an air gap is provided by grinding down the center post. It is suggested that the air gap g be kept less than about $d/10$ to help avoid leakage flux in the air surrounding the gap. The inductor is to carry 5 A on average, and will be used in a 10 V to 5 V buck converter running at 100 kHz. Current ripple should not exceed 1% peak-to-peak.

Given the high frequency, a ferrite core is a logical choice for the material. It is likely to exhibit $B_{sat} = 0.3$ T. Furthermore, a value $\mu = 2000\mu_0$ might be typical for a material in this frequency range. First, let us determine the necessary energy storage and air-gap volume. The circuit is given in Figure 12.26. When the transistor is on, the inductor voltage is +5 V, and its current should change by no more than 50 mA. The duty ratio is 5/10, and the transistor is on for 5 μs each cycle. Thus

$$5 \text{ V} = L\frac{di}{dt}, \qquad 5 \text{ V} = L\frac{0.05 \text{ A}}{5 \text{ }\mu s}, \qquad L = 500 \text{ }\mu H \qquad (12.37)$$

The inductor should store $\frac{1}{2}Li^2 = 6.25$ mJ. If all this energy is stored in the air gap, (12.27) suggests that the gap volume, at a minimum, should be given by

$$6.25 \text{ mJ} = (0.3 \text{ T})^2\frac{V_{gap}}{2\mu_0}, \qquad V_{gap} = 1.75 \times 10^{-7} \text{ m}^3 \qquad (12.38)$$

Figure 12.26 Buck converter and magnetic circuit for Example 12.7.2.

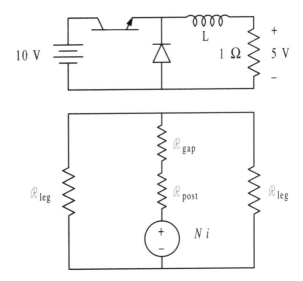

For the maximum gap of $d/10$, the gap volume is $0.2d^3$. To provide the required volume, $d > 9.56$ mm. Let us choose $d = 10$ mm, and examine the results. With $d = 10$ mm and $g = 1$ mm, the reluctance can be determined with the simple magnetic circuit in Figure 12.26. The length of a large loop around both outer legs will be $20d$, the length through the center post is $4.9d$, the outside legs have magnetic area of d^2, and the core and gap have area $2d^2$. The leg reluctances will be determined by a loop half the total length, so $l = 10d$, and $\mathcal{R}_{leg} = 10d/(\mu d^2) = 10/(2000\mu_0 d)$. With $d = 10$ mm, $\mathcal{R}_{leg} = 3.98 \times 10^5$ H^{-1}, while the center post reluctance will be $\mathcal{R}_{post} = 4.9d/(2000\mu_0 2d^2) = 9.75 \times 10^4$ H^{-1}. The gap reluctance of 3.98×10^6 H^{-1} is a factor of ten higher than the others. The total reluctance is the parallel combination of the outer legs in series with the gap and center post reluctance, so $\mathcal{R}_{total} = 4.28 \times 10^6$ H^{-1}. To meet the requirement that $L > 500$ μH, the number of turns must provide $N^2/\mathcal{R}_{total} > 500$ μH. To avoid saturation, the ampere-turns must not exceed the limit from (12.26), so that $Ni < 0.3(4.28 \times 10^6)2d^2$. The inductance requirement means $N > 43$ turns. If saturation is to be avoided at 5 A of current, the number of turns should not exceed 51. Thus it appears that 43 turns of wire on this core will avoid saturation and provide adequate energy storage with a 1 mm gap.

What about wire size? Given the 5 A current value, wire on the order of #16 AWG or #14 AWG will be needed, based on Table 11.1. The window area $4d^2$ is 400 mm^2. With #14 wire, 43 turns will require 90 mm^2. A fill factor of (90 mm^2)/(400 mm^2) = 0.225 is involved. This is a good practical value. Notice that if d is made much smaller, there might not be enough room for the wire.

Notice the tightly coupled nature of magnetics design problems. The geometry must provide adequate space for wire and space for an air gap if the device is an inductor. The size and choice of material must be properly selected to avoid saturation. The losses should be low. In the preceding examples, the size of a core was an important determining factor in device design.

12.7.2 Design Checks and Capacity

The examples in Section 12.7.1 led to complete devices, based on the underlying size requirements. Manufacturers, in general, provide a relatively limited set of core types and geometries. Consider how a few types of cores might be used and what their limitations reflect. Cores with very high permeability, for example, store little energy, and therefore are not appropriate for inductors. Cores with gaps or powdered cores with distributed gaps are more appropriate when inductance is needed.

Example 12.7.3   A certain toroid (Ferroxcube 846T250), shown in Figure 12.27, has an outside diameter of 22 mm, an inside diameter of 13.7 mm, and thickness of 6.4 mm. The manufacturer reports that the magnetic path length is 54.2 mm, while the magnetic area is 25.9 mm^2. The permeability is $2700\mu_0$, and $B_{sat} = 0.32$ T. Find the inductance for a one-turn winding. What is the dc current rating of such an inductor? Discuss the likely applications for this core. How much power or energy can it handle in a likely application?

It appears that this core is a ferrite, by virtue of its saturation characteristics. The permeability is high, and there is no air gap. The inductance of a single turn is exactly the inverse of the reluctance, $L = 1/\mathcal{R}$. For this core, the length, area, and permeability are all given by the manufacturer, so

$$\mathcal{R} = \frac{l}{\mu A} = \frac{0.0542 \text{ m}}{2700\mu_0(25.9 \times 10^{-6} \text{ m}^2)} = 6.17 \times 10^5 \text{ H}^{-1} \tag{12.39}$$

and the single-turn inductance is 1.62 μH. Many manufacturers use the symbol A_L for the *specific inductance*, the inductance for a given number of turns. One common unit is nanohenries for a single turn. More formally, the units would be nH/turn^2 to reflect the $N^2/\mathcal{R}$ inductance relation. This toroid has $A_L = 1620$ nH/turn^2. The current rating is given by $Ni/(\mathcal{R}A) < B_{sat}$. In this case, $Ni < 5.1$ A·turns. For a single turn, the dc current rating is 5.1 A. For 10 turns, the rating drops to 0.51 A.

What about the volt-second rating? The integral $\int v \, dt/(NA)$ should not exceed the saturation value. For one turn, this requires a volt-second product of less than 8.29 μV·s. While this seems small, it could be appropriate for a switching converter. As an inductor, this core is very limited. The energy storage capacity, limited by saturation, is $\frac{1}{2}Li^2_{max} = \frac{1}{2}(1.62\ \mu$H) $(5.1$ A$)^2$. This amounts to only 21.2 μJ. While high inductance values are possible with a substantial number of turns, 100 turns will allow only 51 mA—probably not enough for a power application.

As a transformer, the core is well suited for high-frequency applications. Consider a square wave at 48 V peak and 100 kHz—the sort of waveform that might be typical for a

Outside diameter: 22 mm
Inside diameter: 13.7 mm
Depth: 6.4 mm

Figure 12.27   Toroid core for Example 12.7.3.

forward converter in a telephone application. The waveform applies 48 V to the core for 5 μs, or 240 μV·s. If a primary winding of 29 turns is used, the core's capacity will be at least 29×8.29 μV·s = 240 μV·s (actually, the capacity effectively doubles because the flux can swing between $\pm B_{sat}$, not just 0 to B_{sat}). This number of turns would avoid saturation. For a transformer, at least two windings are needed. The primary winding should not use more than half of the available window area so that there will be room for other windings. The window radius is half the inside diameter, or 6.85 mm. The total window area is 147 mm^2. With a fill factor of 0.5, this means that 36.8 mm^2 are available for primary winding copper. With 29 turns, each turn can have area of 1.27 mm^2 without causing trouble. This corresponds to about #16 wire, and provides current capacity of about 5 A$_{RMS}$. Since the voltage is 48 V$_{RMS}$, the power capacity of this core as a transformer is 240 W—a substantial value for a small core.

Example 12.7.4 A *pot core* is essentially a rotated E-E core arrangement. Pot cores are popular for many types of power converters because they are easy to assemble and mount. The rotated geometry means that the windings are surrounded by high-permeability material. This helps to confine the magnetic fields. An air gap can be provided by grinding down the center post; manufacturers typically offer various gap values. The pot core shown in Figure 12.28 is one of several standard sizes, ranging from less than 10 mm diameter up to more than 60 mm. The specific pot core for this example has been prepared with an air gap of about 0.23 mm. The manufacturer rates the core to have inductance of 400 mH with 1000 turns. The ferrite material used to form the core has $\mu = 2000\mu_0$, and saturates somewhat above 0.3 T. Additional geometric and magnetic information is provided in Table 12.3. The core is wound with 20 turns of wire. Determine the inductance value and the dc current rating of the inductor. Find the equivalent permeability. What is the inductance tolerance if the ferrite permeability is $2000\mu_0$ +100% −20%? Recommend a wire size.

If the manufacturer's data are consistent, a single turn on this core will correspond to 400 mH/1000^2 = 400 nH. Twenty turns would have 400 times this inductance, so $L = 160$ μH. Let us confirm this result. If the core has length 37.6 mm and area 94.8 mm^2, its reluctance should be $l_{core}/(2000\mu_0 A_{core}) = 1.58 \times 10^5$ H^{-1}. The gap reluctance should be $l_{gap}/(\mu_0 A_{gap})$. With a gap length of 0.23 mm, this gives 2.39×10^6 H^{-1}. The total reluctance is 2.55×10^6 H^{-1}. The inverse of this value is the specific inductance, $A_L = 392$ nH/turn^2.

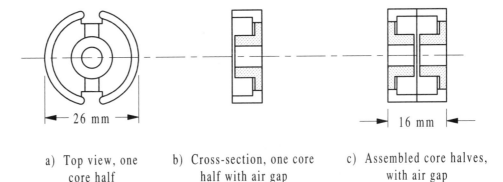

a) Top view, one   b) Cross-section, one core   c) Assembled core halves,
   core half   half with air gap   with air gap

Figure 12.28 Pot core, standard size 2616 (26 mm diameter, 16 mm thickness).

TABLE 12.3
Information for Pot Core of Example 12.7.4

Characteristic	Value
Magnetic path length l_{core}	37.6 mm
Magnetic core area A_{core}	94.8 mm^2
Effective air gap area A_{gap}	76.5 mm^2
Core volume	3.53 cm^3
Specific inductance, A	400 nH/turn^2
Curie temperature, T_{curie}	200°C minimum
Core loss, 100 kHz	100 mW/cm^3 at RMS flux of 0.1 T
Core loss, mathematical fit	$3.2 \times 10^{-8} B^{2.6} f^{1.8}$ W/cm^3 for a given frequency and RMS flux
Window area	57.4 mm^3, or 40.6 mm^3 with bobbin in place
Mean length of wire turn	53 mm

This is only 2% different from the 400 nH/turn^2 value reported by the manufacturer. The equivalent permeability is the value that produces $\mathcal{R} = 2.55 \times 10^6$ H^{-1} when the core geometry values are used. Thus $l_{core}/(\mu_e A_{core}) = 2.55 \times 10^6$ H^{-1}, and the value should be $\mu_e = 1.56 \times 10^{-4}$ H/m. This is $124\mu_0$ (the manufacturer's value is $\mu_e = 125\mu_0$).

The inductance tolerance can be evaluated based on the core reluctance. If the core permeability can vary between $1600\mu_0$ and $4000\mu_0$, the core reluctance will fall between 7.89×10^4 and 1.97×10^5. The total reluctance will fall between 2.47×10^6 and 2.59×10^6 H^{-1}. The specific inductance falls between 386 and 405 nH/turn^2. With twenty turns, the inductance is between 154 μH and 162 μH. The designer will probably label the part as 160 μH $\pm$ 5%, especially since error in the air gap spacing has not been included in the analysis.

To avoid saturation, the flux density should be kept below 0.3 T. This requires $Ni < (0.3 \text{ T})A_{core}\mathcal{R}_{total}$, or $Ni < (0.3 \text{ T})(94.8 \times 10^{-6} \text{ m}^2)(2.55 \times 10^6 \text{ H}^{-1})$; $Ni < 72.5$ A·turns. With 20 turns, the dc current should not exceed 3.63 A. From Table 11.1, this current requires wire size on the order of #16 AWG. There is little reason not to fill the window with as much wire as possible to keep losses low. A plastic bobbin will be used in most cases to make winding easy. This will leave 40.6 mm^2 for wire. A fill factor of 0.7 allows just over 28 mm^2 for copper (the fill factor is higher than 0.5 because the bobbin area was already accounted for). Twenty turns of #16 AWG wire requires 26 mm^2, and should fit. The maximum current density will be about 275 A/cm^2.

The examples demonstrate that, given a core arrangement, magnetic circuit analysis can be combined with saturation limits to identify the core properties. Cores with high permeabilities and no air gaps are suitable for transformers. Typically, transformers must be large for low frequencies. The core size needed for a transformer falls approximately linearly as the frequency increases.

Cores with air gaps are usually required for inductor designs. The gap serves as the primary energy storage element. Pure air-core inductors have extremely low permeabilities. When a ferromagnetic core is used with an air gap, effective permeabilities on the order of $100\mu_0$ are possible, and practical values of inductance can be obtained. Even though core materials exhibit wide variations in their μ values, the addition of an air gap allows inductor designs with tolerances of $\pm$5% or better.

12.7.3 Losses

Wire size has been an important aspect of the inductor design examples so far. From Chapter 11, we see that a given wire can handle only a limited current density to avoid excessive power loss. The window of a given core must hold enough copper to avoid excessive heating of the wire. Hysteresis and eddy current effects are also important. The issue of wire loss influences key details in magnetic core geometry. For an inductor design, saturation limits the amp-turn value. The current density limit in the wire also represents an amp-turn limit. A rule results from this equivalence:

> **Empirical Rule:** *An inductor core has an amp-turn limit imposed by saturation and an amp-turn limit imposed by wire size. The window area is designed so that these two limits are equal.*

Consider what happens if the rule is violated. If the core window is too small, the wire could overheat before the saturation amp-turn limit is reached. If the window is too large, then saturation will limit the core's capability, and the copper might be underutilized.

Hysteresis and eddy current effects complicate the analysis of magnetic device losses. These losses are caused by time-varying flux. An inductor is an interesting case: If an inductor carries a constant dc current below its saturation limit, the core flux will be constant, and the hysteresis and eddy current losses will be zero. This is not realistic in practice. Any converter has current ripple, and the flux in an inductor core will vary at the switching frequency. The variation will be proportional to the current ripple. Just like wire, magnetic material can handle only a limited loss per unit volume. Losses of 1 W/cm^3 are usually considered high for ferrites, and values in the range of 0.2 W/cm^3 are more common.

The losses in any magnetic device divide between copper and core loss. Copper loss is an I^2R value, representing the RMS current squared times the total wire resistance. If the wire volume is known instead, then from Section 11.4 the loss is (ρJ^2)Vol, where J is the RMS current density and Vol is the volume of copper. Manufacturers often assist in the computation of copper loss by reporting an average wire length per turn. Without such a number, it can be difficult to estimate length or volume. Given the average length per turn, resistance becomes straightforward:

$$R_{\text{wire}} = (\text{resistance per meter}) \times (\text{length per turn}) \times N \qquad (12.40)$$

The hysteresis and eddy current losses require more information about the magnetic core material. For many common ferrites, graphs and tables that show loss as a function of flux density variation and frequency can be obtained from the manufacturer. In many cases, flux density variation is represented as the RMS value of B, after the dc component has been removed.

The inductor of Example 12.7.4 will serve to illustrate loss computations. The manufacturer's loss data has been fitted to a power law expression, given in Table 12.3. A candidate dc–dc converter design will be used as the basis for a sample analysis.

Example 12.7.5   The pot-core inductor of Example 12.7.4 is wound with #16 AWG wire. It is used in a dc–dc converter with dc current of 3 A and peak-to-peak current ripple of 20%. Estimate the copper and magnetic losses with a switching frequency of 100 kHz.

Table 12.3 reports a wire length of 53 mm/turn. With 20 turns, the total length should be about 1.06 m. Given #16 AWG wire, with resistance of 13.4 mΩ/m, the total wire resistance will be about 14.2 mΩ. The copper loss reflects the total RMS current. Since the current is 3 A $\pm$ 10%, the RMS value is just slightly over 3 A (for triangular ripple, the RMS value is 3.005 A). The copper loss is $I^2R = 0.128$ W.

The magnetic loss is governed by the frequency and the RMS value of the ac portion of the flux density. The flux density is given by $B = Ni/(\mathscr{R}A)$. Since $\mathscr{R} = 2.55 \times 10^6$ H^{-1} and $A = 94.8$ mm^2, the flux density is 0.0827 times the current. With current varying between 2.7 A and 3.3 A, B varies between 0.223 and 0.273 T. Subtracting the dc value of 0.248 T, the ac portion of the flux density is a triangle waveform of about 0.05 T peak-to-peak. This waveform has an RMS value of 0.014 T. The core loss can be estimated as

$$P = 3.2 \times 10^{-8}B^{2.6}f^{1.8} \text{ W/cm}^3$$
$$= 3.2 \times 10^{-8}\ 0.014^{2.6}\ 100\ 000^{1.8} \text{ W/cm}^3 \qquad (12.41)$$
$$= 0.48 \text{ mW/cm}^3$$

Since the core volume is 3.53 cm^3, the core loss should be about 1.7 mW. This is a tiny fraction of the copper loss. The total loss in the inductor should be on the order of 0.13 W—low given the volume of the part. Indeed, this particular material can operate at more than 500 kHz before the core losses become a significant issue.

Core losses increase rapidly as flux variation increases. In Example 12.7.5, the inductor had total flux variation of $\pm$10%. If another converter design had exhibited flux variation of $\pm$100%, the core losses would be higher by a factor of $10^{2.6} \approx 400$, and the core loss would be several times the copper loss. In most cases, it is desirable to operate a core with the smallest possible variation of flux density to limit the losses.

Example 12.7.6   A transformer core has saturation flux density of 1.5 T. It is used for a 60 Hz application. The core area is 1000 mm^2, and the window area is 3800 mm^2. The magnetic path length is 400 mm, and the core volume is 500 cm^3. The mean length per turn is 200 mm. The core material has power loss given by

$$P = 7 \times 10^{-6}B^2f^2 \text{ W/cm}^3 \qquad (12.42)$$

Choose wire sizes to provide 120 V to 20 V transformation at the highest possible power. Estimate the copper and core losses.

To avoid saturation, equation (12.28) requires that the peak volts per turn not exceed $\omega B_{sat}A_{core}$. The 120 V winding requires no more than 0.565 V/turn. With 170 V peak, at least 300 turns will be necessary. For the 20 V winding, 50 turns will be necessary. To maximize power, the largest possible wire should be used. With a fill factor of 0.5 and two windings, an area of (3800 mm^2)/4 = 950 mm^2 is available for the copper in each winding. On the 120 V side, the wire area can be up to (950 mm^2)/300 = 3.17 mm^2. The nearest even size in Table 11.1 is #14 AWG, with an area of 2.08 mm^2. On the 20 V side, the wire area can be $(950 \text{ mm}^2)/50 = 19 \text{ mm}^2$. This corresponds approximately to #6 wire, with an area of 13.3 mm^2. Given the high number of turns, the current density must not be too high. If $J \le 200 \text{ A/cm}^2$, then the 120 V winding can support up to 4.16 A. The transformer rating will be about 500 VA.

Copper loss requires currents and wire resistances. With mean wire length of 200 mm, the primary wire length should be about 60 m, while the secondary wire should be 10 m long. Since #14 wire has about 8.45 mΩ/m resistance, the primary winding will have $R_1 = 0.507 \ \Omega$. The secondary's #6 wire has resistance of 1.32 mΩ/m, so $R_2 = 0.0132 \ \Omega$. With primary current of 4.16 A, the copper loss on the primary side will be 8.77 W. The secondary current should be 25 A, based on the turns ratio. This current produces loss of 8.25 W. Notice that the two windings have losses that match.

The magnetic loss reflects the total variation in flux. In this case, the flux varies from -1.5 T to $+1.5$ T as the voltage swings between negative and positive peaks. The RMS flux density is $1.5/\sqrt{2} = 1.06$ T. Substituting into equation (12.42), the power loss per cubic centimeter will be

$$P = 7 \times 10^{-6} \ 1.06^2 \ 60^2 \ \text{W/cm}^3$$
$$= 0.028 \ \text{W/cm}^3 \tag{12.43}$$

With total volume of 500 cm^3, the core loss should be about 14.2 W. At full load, this transformer can handle 500 W, and exhibits total losses of 31.2 W. This suggests an efficiency of 94%—a fairly typical value for a transformer of this power rating.

Core losses estimated along the lines of this section are only approximate, and are strongly dependent on the properties of the specific core material. It is very important to recognize the strong effects of frequency and flux variation on core loss. In a transformer, flux variation is typically large to take full advantage of saturation limits. In an inductor, flux variation might be relatively low; high frequencies tend to offset the benefits of low flux variation. Losses and saturation limits make magnetic design a significant challenge. The need for smaller, more efficient cores continues to drive the development of new materials and geometries.

12.8 RECAP

We have seen that:

- Magnetic devices can provide an inductance function $L \ di/dt = v$.

- Magnetic devices can provide a transformer function.

- Only minor assumptions allow the relevant equations of Maxwell for a magnetic device to emulate those for an electric circuit.

- Inductors and transformers built with ferromagnetic materials exhibit:
  - hysteresis and associated loss
  - eddy currents and associated loss

- winding loss
- leakage flux and associated inductance
- Design considerations for magnetic devices include:
  - core size and permeability
  - effects of air gaps on reluctance and linearity
  - core loss limits
  - saturation values of flux
  - core window size (for windings)
  - current density allowed in windings
  - maximum values of ac voltage, dc volt-seconds, and dc current
  - operating frequency

The analysis began with *magnetic circuits*, in which Ampère's Law provides an MMF relation analogous to KVL. Gauss's Law guarantees that the net flux into an isolated node is zero, analogous to KCL. Faraday's Law provides the link between magnetic circuits and electric circuits. A winding with amp-turn product Ni serves as an MMF source for a magnetic circuit. The constituent relation $\mu\mathbf{H} = \mathbf{B}$ provides the link between MMF and magnetic flux. A *reluctance*, analogous to resistance, was defined for magnetic elements. The flux ϕ in a magnetic material is given by $Ni/\mathcal{R}$. Faraday's Law showed that a magnetic device creates an inductance. For a single winding with N turns, the inductance is $L = N^2/\mathcal{R}$, where $\mathcal{R}$ is the total reluctance of the device. Table 12.4 summarizes some of the magnetic circuit relations.

Ferromagnetic materials, with high permeability, are of special interest for engineering applications. They are the only room-temperature materials for which μ is significantly different from $\mu_0 = 4\pi \times 10^{-7}$ H/m, the value for air or vacuum. Most ferromagnetic materials are alloys or compounds containing iron, nickel, cobalt, or rare-earth elements. *Ferrites* are ceramic compounds with the basic chemical formula XFe_2O_4. They are ferromagnetic, and have much lower conductivity than metallic materials. Ferrites are common in high-frequency magnetic applications.

A magnetic core with two windings serves as a transformer. A material with high μ provides a near-ideal transformer, except that a magnetic device requires a time-varying flux to operate. Magnetic devices do not directly support dc transformation. In a realistic trans-

TABLE 12.4
Summary of Magnetic Circuit Relationships

Expression	Significance
$\Sigma\phi_{node} = 0$	A form of Gauss's Law. Analogous to KCL for magnetics.
$\Sigma MMF_{loop} = Ni$	A form of Ampère's Law. Analogous to KVL for magnetics.
$v = d\lambda/dt$	A form of Faraday's Law. Relates flux linkage $\lambda = N\phi$ to voltage.
$\mu\mathbf{H} = \mathbf{B}$	Definition of permeability. For vacuum, $\mu = \mu_0 = 4\pi \times 10^{-7}$ H/m.
$\mathcal{R} = \ell/(\mu A)$	Definition of reluctance. Length ℓ is defined with an Ampère's law loop. Area A is defined as an area enclosed by a coil.
$L = N^2/\mathcal{R}$	Result for inductance. *Specific inductance $A_L = 1/\mathcal{R}$.*

former, each winding has an individual leakage flux associated with a leakage inductance, in addition to flux that couples the turns for transformation. The windings have resistance, and time-varying flux contributes to core losses.

For a ferromagnetic material, the basis of high μ values is *magnetic domains*, in which large groups of electrons tend to align together. A small change in MMF brings a rapid change in flux density. However, the nature of domains makes flux density change nonlinear. Usually μ is not well characterized by a single constant value in the case of a ferromagnetic material. A *magnetization curve*, rather than a value of μ, is often needed for a good description of the behavior of ferromagnetic materials. At temperatures above the *Curie temperature*, T_c, the domains are thermally disrupted, and the value of μ for a material drops close to μ_0. Above a limiting flux density value, B_{sat}, all domains are aligned and μ also drops from its high value down to μ_0. If an MMF is applied and then removed, some of the alignment remains, and a *remanent flux B_r* persists in the material. A reverse value of MMF, called the *coercive force H_c*, must be applied to return the flux density to zero.

Remanent flux and alignment dynamics mean that a magnetization curve shows irreversibility. This means that energy is lost when MMF and flux vary in a magnetic material. The losses include hysteresis loss, which represents the actual effect of magnetic nonlinearity, and eddy current loss, in which the core functions as a winding with its own internal I^2R loss. In metallic materials, *laminations* are used to reduce eddy current loss. They increase internal resistance and decrease flux within a magnetic core. Resistive materials such as ferrites offer an alternative. The power loss generally takes the form

$$P_{loss} = P_0 f^a B^b \tag{12.44}$$

Losses increase approximately with the square of the flux amplitude and with a power of the frequency between 1 and 2.

Magnetic saturation is one of the most important limiting factors for device design. In an inductor, saturation limits the dc current level or the applied volt-second product. This in turn limits the energy storage capacity of a given core. In a transformer, saturation limits the voltage that can be supported by each turn of a winding. For inductors, the need for energy storage usually requires an air gap. This raises the dc current limit and increases the energy capacity of a ferromagnetic core. For a transformer, the core must be large enough to avoid saturation under the highest allowed voltage. Saturation design rules are summarized in Table 12.5.

Air gaps are important for inductors not just because of the higher energy storage capability they bring, but also because they improve the linearity of a given magnetic device.

TABLE 12.5
Summary of Saturation Design Rules

Rule	Interpretation
$Ni < B_{sat}\mathcal{R}A$	Amp-turn limit for an inductor.
$W_{max} = \frac{1}{2}B_{sat}{}^2 \ell_{core}A_{core}/\mu$	Maximum energy that can be stored in a given core.
$W_{max} = \frac{1}{2}B^2_{sat}V_{gap}/\mu_0$	Maximum energy is determined by air gap volume if the core has high μ.
$V_0/N < \omega B_{sat}A$	Maximum volts per turn (for a transformer) at frequency ω.
$\int v\,dt < NB_{sat}A$	Maximum volt-seconds for an inductor or transformer.

Ferromagnetic materials can easily exhibit ±50% variation in permeability. The addition of an air gap reduces the variation by producing an *effective permeability* μ_e intermediate between the precise μ_0 of air and the imprecise high value of a core material. In powdered metal core materials, the air gap is distributed throughout the volume by embedding the ferromagnetic material in a non-magnetic matrix.

For design purposes, a core size and geometry must be selected. Saturation constraints are evaluated to determine how they limit winding designs and ratings. Wire size is selected to support the desired level of current and to fit into the core's *window area*. Metal core materials are common for line-frequency applications. At higher frequencies, ferrites or powdered metal cores are a common choice. Cores might be in the form of toroids or various types of sectioned core structures. Sectioned structures tend to be easier to wind than toroids, and are common unless a low number of turns is desired.

All the wire wound on a given core must fit into its *window*, the opening for wire. The window also must hold insulation and any structure on which the wire is mounted. In practice, only about 50% of the window area can actually carry active conductor. This fraction is called the *fill factor*. In a two-winding transformer, this means that each winding can fill not more than 25% of the total window area. For toroid cores, windings are often designed to form a single layer of copper material around the inside of the core. This keeps the device small and minimizes leakage. For other core shapes, windings often use the largest wire that will fit conveniently into the window. This minimizes losses, and maximizes the power rating for a transformer.

Losses can be divided in *copper loss*, the I^2R loss in the wire wound on a magnetic device, and *core loss*, the combination of hysteresis and eddy current loss. When several windings are used, they are usually sized so that copper losses match among them. This practice provides the optimum use of material in a magnetic device.

PROBLEMS

1. Find the inductance of the magnetic circuit shown in Figure 12.29. What is the maximum allowed dc current that avoids saturation?

2. A toroid has thickness of 4 mm, outside diameter of 16 mm, and inside diameter of 8 mm. It has $\mu = 25 \mu_0$. What is the inductance of a ten-turn winding on this toroid? What is the maximum current to keep $B < 0.5$ T?

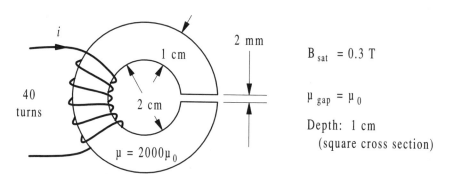

Figure 12.29   Inductor for Problem 1.

3. A manufacturer rates a certain ferrite core to have a per-unit inductance $A_L = 400$ nH/turn^2. The core has an effective magnetic area of 1.5 cm^2.

    a. What is the amp-turn limit to avoid saturation?

    b. What is the highest inductance value possible with this core if the current rating is to be at least 1 A?

    c. If powdered iron could be substituted to provide the same A_L value, what would the amp-turn limit and maximum inductance for 1 A rating become?

4. A large powdered iron toroid has outside diameter of 100 mm, inside diameter of 50 mm, and thickness 40 mm. The permeability is $\mu = 50\mu_0$. What is the inductance of a twenty-turn coil on this core? What current rating would be allowed?

5. A powdered iron toroid core has $\mu = 25\mu_0$. The outside diameter is 30 mm, the inside diameter is 15 mm, and the thickness is 10 mm.

    a. What energy can be stored in an inductor built on this core?

    b. In a certain dc–dc converter application, the maximum volt · second value is 50 μV·s. Choose a number of turns to allow this core to meet this limit. What are the inductance and current rating?

    c. What is the highest current rating that can be achieved in an inductor built on this core?

6. A gapped ferrite core has effective permeability of $200\mu_0$. It has a toroidal shape. The outside diameter is 30 mm, while the inside diameter is 15 mm. The core is 10 mm thick. What energy can be stored in an inductor built on this core? What is the highest possible current rating?

7. A 10 mH inductor is needed for a drive application. The inductor must be able to handle 75 A$_{dc}$. Powdered iron cores rated at either $\mu = 25\mu_0$ or $\mu = 75\ \mu_0$ are available. What volume of magnetic material will be needed in each case to provide the necessary energy storage and inductance value? Assuming a toroidal shape, propose core dimensions and provide a design based on each of the two materials. Which type would you recommend?

8. A certain ferrite core is known to have loss of 4 W in a dc–dc converter application with a switching frequency of 100 kHz. If the switching frequency in this same converter doubles, about how much will the power loss change?

9. The standard pot core in Figure 12.28 is to be used to build a 250 μH inductor. The total air gap is 1 mm, and the Figure is an accurate scale drawing. How many turns will be needed? What is the dc current rating of this inductor?

10. A steel laminated core is to be used for an inductor. The purpose of the inductor is to help filter out switching frequency ripple at the output. Except for the air gap, the core and dimensions are identical to those in Figure 12.8. A 100 μH inductor is needed, at currents up to 10 A. Choose an air gap and number of turns to meet the need. (*Hint*: The winding must go on the center leg to insure that the air gap is in the flux path.)

11. Ferrite pot cores are usually numbered based on both geometry and inductance. For example, a 1408-250 pot core has an outside diameter of 14 mm, a thickness of 8 mm (for two sections

mounted together), and inductance of 250 nH for one turn of wire. The center post area covers about 15% of the total cross section. It is desired to build a 100 μH inductor with a current rating of 5 A. Which of the following cores can be used to build such an inductor? How many turns should be used in the case of cores that can meet the requirements?

   a. 1408-250   **b.** 2010-200

   c. 3018-400   **d.** 3622-400

12. A small battery charger unit is capable of charging any battery from 1.2 V to 24 V. To protect against excessive current, the charger puts a 10 Ω resistor in series with the battery. The charger input is a constant +12 V supply. The unit is set to charge the attached battery at the rate of 0.1 A.

   a. Draw a switching converter that can perform this function. Be sure to include the inductors and capacitors necessary in a good design.

   b. Find L_{crit} for this converter if $f_{switch} = 40$ kHz.

   c. Suggest a core arrangement to implement L_{crit}.

13. A boost converter allows inputs of 15 to 40 V_{dc}, and gives output of 120 V at 12 kW. The inductor is 10 μH.

   a. What value of f_{switch} is needed to keep $L \geq L_{crit}$ under all input voltages?

   b. An inductor core has high μ_r, $A = 0.001$ m^2, and an air gap with spacing g. Find a number of turns N and a value of g which gives $L \geq 10$ μH and $B \leq 1$ T.

14. A ferrite core with $\mu = 5000\mu_0$ uses an E-E geometry. Each leg of the two E sections is 6 mm wide and 15 mm long. The side member of the E is 36 mm long. The thickness is 6 mm. This core will be used for a transformer in a forward converter application. At a switching frequency of 50 kHz, what is the maximum power that this transformer can handle?

15. A transformer is to be built for a power application. The frequency is 60 Hz. The device is to step an input of 12 kV down to 240 V, at power levels up to 75 kW. A steel laminated core with $B_{sat} = 1.8$ T is proposed. The permeability is $10^5 \mu_0$.

   a. Considering the general E-I geometry in Figure 12.25, how big will d need to be to allow this transformer to function?

   b. What wire sizes and numbers of turns will be used?

   c. Estimate the wire resistance and the copper losses at full load.

16. A 500 μH inductor is needed for a dc–dc converter application. The input is 24 V, and the output is 5 V at loads from 1 W to 200 W. Propose an inductor design. Specify a possible geometry, material, number of turns, and wire size.

17. A certain ferrite material has power loss given by $P_0 f^{1.4} B^{2.4}$, where B represents the RMS magnitude of the flux density ripple. This material is used for an inductor for a boost dc–dc converter running at 150 kHz. The flux ripple is $\pm 5\%$ of B_{sat}. At 200 kHz, is the power loss higher or lower than at 150 kHz? What is the ratio of lost power at these two switching frequencies?

18. Magnetic energy storage has been proposed for electric automobiles. It is inconvenient to use superconducting materials in such circumstances. You have been asked to examine conventional magnetics as an alternative. A powdered iron material with $\mu = 25\mu_0$ is available.

  a. How much material would be needed to store 500 kJ (the typical kinetic energy recovered when a car slows down from 100 km/hr)?

  b. If such a core is built, can a winding be designed to handle 300 V_{dc} for 20 s (this might represent an acceleration)?

  c. For copper wire and the candidate core, what energy storage time constant might be expected?

19. In Chapter 5, it was found that a 1.18 H inductor is needed for useful filtering in a bridge rectifier with 60 Hz input and 9 V output at 6 W. Given a steel laminated E-I core like the general one in Figure 12.25, except that a small gap can be added in the center leg, design a 1.18 H inductor that can handle 0.67 A. What is the smallest possible core, given that the air gap length should not exceed $0.5d$? Estimate the copper losses.

20. A certain powdered iron core is wound into a 50 μH inductor rated at 20 A. It shows a total loss of 1 W when the flux swings between $\pm B_{sat}$ at 200 kHz. The loss expression is $P_{loss} = P_0 f^{1.5} B^2$, where B is the RMS value of the flux density variation. What is the quality factor $Q = X/R$ of this inductor when it is applied in a 50 kHz circuit with less than 10% flux density ripple (compared to B_{sat})?

REFERENCES

K. H. Carpenter, S. Warren, "A wide bandwidth, dynamic hysteresis model for magnetization in soft ferrites," *IEEE Trans. Magnetics*, vol. 28, no. 5, pp. 2037–2041, September 1992.

"Iron powder cores," *Catalog 4 for EMI and Power Filters.* Anaheim: Micrometals, 1993.

D. C. Jiles, D. L. Atherton, "Ferromagnetic hysteresis," *IEEE Trans. Magnetics*, vol. 19, no. 5, pp. 2183–2185, September 1983.

Col. W. T. McLyman, *Transformer and Inductor Design Handbook, 2nd ed.* New York: Marcel Dekker, 1993.

MIT EE Staff, *Magnetic Circuits and Transformers.* New York: Wiley, 1943.

N. Mohan, T. Undeland, W. P. Robbins, *Power Electronics, 2nd ed.* New York: Wiley, 1995.

Philips Components, *Ferrite Materials and Components Catalog, 8th ed.* Saugerties, NY: Philips Components, 1992.

CHAPTER 13

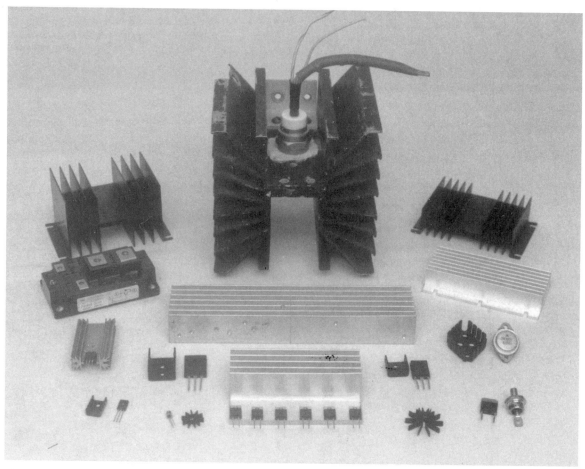

Figure 13.1 A variety of power semiconductor devices, packages, and heat sinks.

POWER SEMICONDUCTORS IN CONVERTERS

13.1 INTRODUCTION

So far, the semiconductor devices of a power converter have been considered primarily for their major function as a switch. Circuit polarities gave rise to the restricted switch concept, and it has been possible to associate the needs of a specific power electronic circuit with a particular type of device. Some of the capabilities of modern semiconductor switching devices and the major types have been considered. The SCR—a switch restricted in its control as well as in polarity—was considered operationally in the context of rectifiers.

In this chapter, power semiconductors will be considered at the component application level. What aspects of each device are relevant to the operation of a power electronic circuit? The issues of the switching process are considered, and also static effects such as forward voltage drops. Device speed is an important issue in any switching network, and some lumped-parameter circuits will be introduced that provide an approximate view of the switch speed effects in several types of devices. Both forward voltage and speed effects contribute to device losses. The loss processes and values will be studied. Simplified methods for thermal analysis of semiconductor devices are introduced. These give an idea of whether a given device is suited to a particular converter.

For design purposes, most manufacturers provide extensive information about power devices like the ones depicted in Figure 13.1. Commercial data sheets are used in examples. The data needed to determine appropriate applications and to estimate device losses will be considered. What matters is the operation of a device in a real system. At the end of the chapter, a sample dc–dc converter design illustrates some of the specification sheet issues.

13.2 SWITCHING DEVICE STATES

Any switching device must be either on or off in operation. In fact, there are three modes of operation:

- The *on* state, in which current is carried with little voltage. Current ratings are important when selecting a device to meet a set of on-state requirements. While $V_{\text{switch}} = 0$ ideally, in reality a small *residual voltage* will be present. The diode forward drop is the most familiar example, but all devices exhibit some amount of residual voltage.

- The *off* state, in which current is close to zero and a high voltage is being blocked. All devices have voltage ratings to give some guidance for off-state design. Although the current $I_{switch} = 0$ in a perfect device, real semiconductors exhibit a small *residual current* in the off state. This current, also known as a *leakage current*, is very low in modern silicon devices, and often can be neglected.

- The *commutation* state, in which the device makes the transition from on to off or the reverse. In more general terms, *commutation* refers to the transfer of current from one conductor to another. Speed ratings are particularly important for design aspects of commutation.

Each of the three operating modes is associated with energy loss. The on and off modes are termed *static* states since the device remains in the state for a substantial fraction of each converter switching period, and the device action does not change much during these static intervals. Devices in static states can be represented with relatively simple circuits. In contrast, commutation is a *dynamic* state, and requires models with energy storage elements and controlled sources to track the observed transitions. As the device types are considered, models for both static and dynamic states will be discussed.

For the on state, manufacturers often provide several kinds of current ratings:

1. Continuous current rating. This should be a safe handling level under the stated conditions. The continuous rating is important when the duty ratio is not known in advance, or when a worst-case design is to be performed.

2. Average current rating. In a device with fixed residual voltage, the on-state loss is the residual voltage times the average current. Therefore, average current is often a good predictor of power loss. This rating is important for rectifiers as well as in dc–dc converters.

3. RMS current rating. Any real device exhibits some resistance during its on state. The RMS current governs the loss in this resistance. This value can be important in inverter design, and is sometimes considered in rectifiers and dc–dc converters as well.

4. Peak current rating. A given piece of silicon has definite limits on the maximum current level. Above this level, high currents might produce excessive local heating or physical damage to metal contacts or wires. In some types of devices, high currents tend to initiate internal instability, which can lead to failure.

5. Current-time values. A device should be able to handle the rated peak current briefly and infrequently, and the continuous current indefinitely. In devices intended for wide duty ratio and frequency ranges, details of the amount of current that can be handled for a particular time are helpful. Many power MOSFET manufacturers provide current-time curves for designers.

For the off state, the information is highly device-dependent. Diodes and MOSFETs, for example, have well-defined voltage ratings. An attempt to exceed the rated off-state voltage produces avalanche current. If avalanche current is not limited externally, the device fails rapidly (often more quickly than a fuse). The time involved is so brief that the peak off-state voltage rating is usually a single number, without any time factor. The off-state residual current commonly is specified at the full rated off-state potential.

SCRs include several bipolar junctions, and sometimes can tolerate very brief high off-state reverse voltages. They also have off-state capabilities that depend on the voltage polarity. In the reverse direction, an SCR behaves much like a diode, and avalanche currents can flow if the reverse rating is exceeded. In the forward direction, the SCR will actually switch *on* if the maximum voltage is exceeded, as will be discussed later. An SCR data sheet normally specifies maximum forward and reverse off-state voltages separately, and might also indicate a momentary peak rating.

Bipolar transistors have slightly more complicated off-state ratings. The off-state voltage that a given device can block depends on the base signal. Manufacturers often provide two separate off-state voltage rating values:

1. The maximum off-state collector-emitter voltage when there is no base current, V_{CEO}. This is the voltage rating when the base is an open circuit.
2. The maximum off-state collector-emitter voltage with a specified base input, V_{CEX}. Often, the specification calls for a base-emitter short circuit, or possibly even a negative base bias voltage. The condition helps prevent any collector leakage current from being amplified, so $V_{CEX} > V_{CEO}$ for a device.

The dynamic behavior of a device during the commutation state can be complicated, and depends strongly on the external circuit. Basic switch operation is depicted in Figure 13.2. The figure shows current vs. voltage for an FCFB switching device such as a BJT. The on state shows high current and low residual voltage, while the off state shows high voltage and small residual current. The commutation process must drive the device from one static point to the other. The pathway followed is a function of both the device and the external circuit. The transition requires a definite amount of time, which is not reflected on this *I-V* plot. The paths followed between the on and off states define the *switching trajectory*. There is a *turn-on trajectory* that follows the changing voltages and currents as the transition is made from the off state to the on state, and a *turn-off trajectory* defining the opposite change.

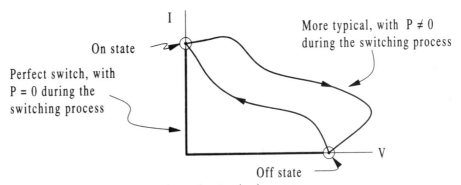

Figure 13.2 Switch current vs. voltage, showing the three states.

Manufacturers typically provide two types of data with respect to the commutation state. The first involves the time behavior of a device. Switching times such as rise and fall times (based on the current), internal time delays, and dynamic device characteristics called *recovery times* are often provided. These times provide estimates of how long the device will remain in commutation. The second type of data is more directly tied to the switching trajectory. For fully controlled devices such as transistors, manufacturers specify a *safe operating area* (SOA). This is a region of the *I-V* plane that the switching trajectory must fall inside. If the converter application produces a switching trajectory that moves outside the safe operating area, the device might be damaged or even destroyed. Figure 13.3 shows typical switching trajectories for a dc–dc converter, along with the safe operating area for a MOSFET chosen properly for this converter.

13.3 STATIC MODELS

The on and off properties of real switching devices have direct effects on the operation of power converters. We have considered these before in some contexts, such as low-voltage rectifiers. It is often convenient to use the restricted switch idea to provide a circuit model for power semiconductors. The on-state and off-state properties can be captured and included in analysis and design. A real diode, for instance, can be represented as a voltage drop in series with an ideal diode. The concept is called *static modelling*, and is well known in electronics.

Consider the diode forward characteristic and possible models shown in Figure 13.4. The first model is a simple forward-drop. If the current level is known, the voltage occurring at that current would be used for the model. The second model uses a resistor to track the slope of the curve as the current increases. The third model suggests that a piecewise linear static model can be constructed to follow the actual curve as closely as desired. In power

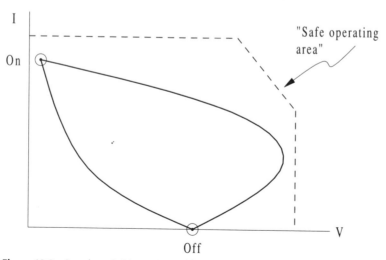

Figure 13.3 Sample switching trajectories, with an appropriate device safe operating area.

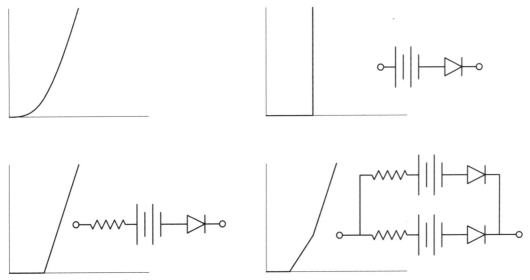

Figure 13.4 Typical rectifier diode static forward characteristic.

electronics, only the forward-drop and forward-drop with resistor models are common. More complicated models introduce additional switches into a circuit, complicating the analysis with little extra benefit.

Reverse current can be modelled with a leakage resistor, with an appropriate ideal diode to confine its effect to the off state. In power conversion, this is not usually necessary. A typical power diode rated at 10 A and 200 V has residual current of only 10 μA at 200 V in the off state. Since the residual current is a factor of one million below the rated level, it is unlikely to have a measurable effect on a converter circuit. The low residual current means that power lost during the off state is usually negligible compared to the total power loss in the switching device. This assertion is tested in some of the examples that follow.

Example 13.3.1 The measured forward characteristic of an MUR3040PT silicon diode is shown in Figure 13.5. The device is rated at 30 A and 400 V. The manufacturer's data show leakage current of 10 μA at 400 V in the off state. From the curve and the data provided, develop a static model for this device. What is the static resistance for 100 A peak?

The leakage resistance for this device should be about 400 V/10 μA, or 40 MΩ. The dotted line drawn on Figure 13.5 shows a visual fit to the characteristic curve. The dotted line intersects the $I = 0$ axis at 0.94 V, and has a slope of 100 A/1.5 V = 67 S, corresponding to 0.015 Ω. A static model based on these data is given in Figure 13.6.

Example 13.3.2 Use the MUR3040PT with an ideal transistor in a buck dc–dc converter for 12 V to 5 V conversion into a 100 W load. What should the transistor duty ratio be? What are the power losses in the diode, ignoring the effects of commutation?

A buck converter for these requirements is shown with the diode's static model in Figure 13.7. According to the procedures of Chapter 4, the switches manipulate the input current and the voltage v_d into the L-R load. With the transistor on, the configuration is shown in Figure 13.8a. When the transistor turns off, the inductor current imposes a positive bias

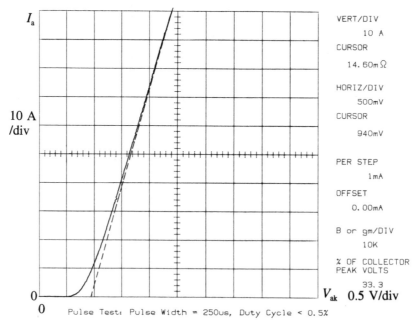

I_a

10 A
/div

0

0   Pulse Test: Pulse Width = 250us, Duty Cycle < 0.5%

V_{ak} 0.5 V/div

Figure 13.5 Forward current vs. forward voltage for MUR3040PT fast diode.

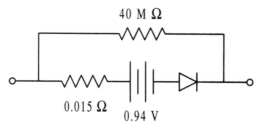

40 M Ω

0.015 Ω   0.94 V

Figure 13.6 Static model for MUR3040PT diode.

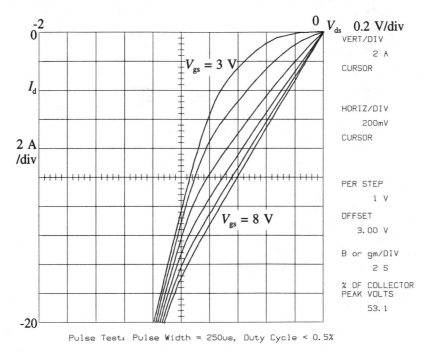

Figure 13.12 MOSFET reverse characteristic for MTP33N10.

Figure 13.13 Static model for MOSFET. The leakage resistance R_{leak} is normally ignored.

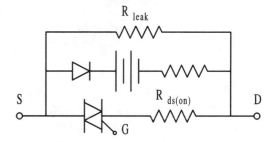

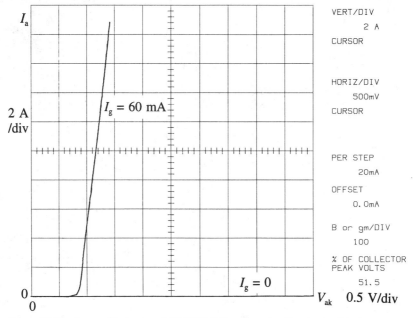

Figure 13.14   Forward behavior of 2N6508 SCR with and without gate signal.

13.4 SWITCH ENERGY LOSSES AND EXAMPLES

13.4.1 General Analysis of Losses

The nonideal properties of each of the three switch states mean that losses will be generated. During a switching period T, the three states will be associated with four different times: the on time $D_{on}T$, the off time $D_{off}T$, and the turn-on and turn-off times for commutation. The energy lost in the switching device, W_{loss}, can be divided into the loss during the static states W_{static} and the commutation loss or *switching loss* W_{switch}. These can be obtained by integration:

$$W_{static} = \int_{on\ time} i_{on}(t)v_{residual}(t)\,dt + \int_{off\ time} i_{residual}(t)v_{off}(t)\,dt, \qquad (13.5)$$

$$W_{switch} = \int_{turn-on} i(t)v(t)\,dt + \int_{turn-off} i(t)v(t)\,dt \qquad (13.6)$$

The static loss usually follows directly from the converter operation. For example, the diode in a typical dc–dc converter will conduct a fixed current I_{on} when on, with a well-defined residual voltage V_{resid} corresponding to that current. The energy lost per period will be $(D_{on}T)(V_{resid}I_{on})$. The off-state loss will be a small leakage current times the off-state voltage, or $(D_{off}T)(V_{off}I_{resid})$. The switching loss requires integration of the current and voltage as they evolve along a switching trajectory.

The average power lost over a period T is the total energy lost during a period divided by T. Commutation loss occurs each time the switch operates, so the power lost in commutation is proportional to switching frequency. Consider a dc–dc converter in which the on-state current and off-state voltage are constant. The average power loss is

$$P_{loss} = \frac{D_{on}TV_{resid}I_{on} + D_{off}TV_{off}I_{resid} + W_{switch}}{T} \tag{13.7}$$

The off-time $D_{off}T$ should be $(1 - D_{on})T$, and (13.7) reduces to

$$P_{loss} = DV_{resid}I_{on} + (1 - D)V_{off}I_{resid} + f_{switch}W_{switch} \tag{13.8}$$

The result shows that the power lost is the sum of average powers in the on and off states, plus a commutation term proportional to the switching frequency.

Example 13.4.1 The 12 V to 5 V, 100 W buck converter of Example 13.3.2 is to be re-examined with different switches. The transistor is a MOSFET with $R_{ds(on)} = 0.25\ \Omega$, and leakage current of 10 μA with 100 V reverse voltage. The diode is modelled as a fixed forward drop 1 V and leakage current of 1 μA with 100 V in reverse. The devices are very fast. Draw the circuit with static models. Find the on-state and off-state losses and compare them. What is the converter efficiency?

Since the switches are very fast, the time spent in commutation is negligible, and only the on-state and off-state models and power losses are meaningful. The circuit with static models is shown in Figure 13.15. Current will never flow through the MOSFET's reverse diode, so this component has not been shown in the figure. The inductor current must be 20 A to provide 100 W of output. It can be shown that the leakage resistances do not have a measurable effect on the duty ratios. The average input current will be D_1I_L, and the voltage v_d will be given by

$$\begin{aligned} v_d &= q_1(V_{in} - I_LR_{ds(on)}) + q_2(-V_{diode}), \\ \langle v_d \rangle &= V_{out} = D_1(V_{in} - I_LR_{ds(on)} + V_{diode}) - V_{diode} \end{aligned} \tag{13.9}$$

The duty ratio D_1 is found to be 0.75, while the diode is on 25% of each period. The transistor on-state loss should be $D_1I_L{}^2R_{ds(on)} = 75$ W. The off-state voltage across the transistor is $V_{in} + V_{diode} = 13$ V, so leakage current of about 1.3 μA is expected. The off-state loss should be $(1 - D_1)I_{leak}V_{off} = 0.25(1.3\ \mu A)13\ V = 4.23\ \mu$W. The diode on-state loss should be $0.25(20\ A)(1\ V) = 5$ W. The diode off-state voltage is only 7 V, representing the input

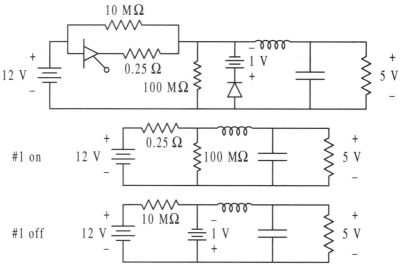

Figure 13.15 Buck converter for Example 13.4.1 with static models.

voltage less the drop across the FET. The expected leakage current is about 70 nA, to give off-state loss of 0.75(70 nA)(7 V) = 0.368 μW. The total losses in the on states are 80 W, while the total in the off states is only 4.60 μW. The off-state loss is more than a factor of ten million smaller than the on-state loss, and certainly should be ignored. The converter input power should be 180 W to account for the losses. The efficiency is (100 W)/(180 W) = 55.6%.

The FET on-state drop of 5 V is very substantial compared to other voltages in this converter. The poor efficiency, combined with this large drop, suggest that a designer should be careful to consider the effects of $R_{ds(on)}$ when choosing a MOSFET for a given application.

13.4.2 Losses during Commutation

Switching loss is difficult to obtain from a data sheet. The actual voltages and currents in the devices determine this loss, and these depend in large part on the external circuit. Let us consider a set of simplified waveforms for a dc–dc converter, shown in Figure 13.16. In the figure, the switch voltages and currents make a linear transition between the two static states. The current shows a well-defined rise time, and it is easy to integrate to find the switching energy loss. If the time curves of Figure 13.16 are plotted as switching trajectories, the result, shown in Figure 13.17, is a straight line connecting the two static states. This behavior is called *linear commutation* by virtue of the switching trajectory shape, and represents idealized switching action.

The switching energy losses during linear commutation can be integrated as follows: During turn-on, the current rises with slope $I_{on}/t_{turn-on}$, and the voltage falls with slope $-V_{off}/t_{turn-on}$. The turn-off case is similar. When the linear expressions are written as the product of slope and time, the integrals give

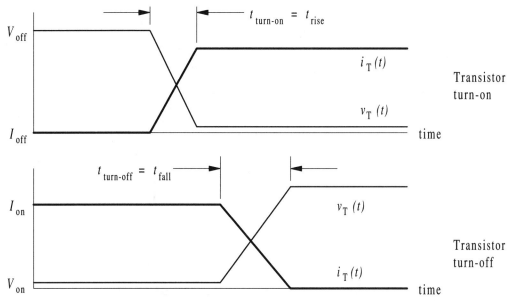

Figure 13.16 Switch voltages and currents in a dc–dc converter, showing hypothetical linear transitions between static states.

$$W_{\text{switch}} = \int_0^{t_{\text{turn-on}}} \frac{I_{\text{on}}}{t_{\text{turn-on}}} t \left(V_{\text{off}} - \frac{V_{\text{off}}}{t_{\text{turn-on}}} t \right) dt + \int_0^{t_{\text{turn-off}}} \frac{V_{\text{off}}}{t_{\text{turn-off}}} t \left(I_{\text{on}} - \frac{I_{\text{on}}}{t_{\text{turn-off}}} t \right) dt \quad (13.10)$$

The result is

$$W_{\text{switch}} = \frac{V_{\text{off}} I_{\text{on}} t_{\text{turn-on}}}{6} + \frac{V_{\text{off}} I_{\text{on}} t_{\text{turn-off}}}{6} \quad (13.11)$$

Let us define a total switching time $t_{\text{switch}} = t_{\text{turn-on}} + t_{\text{turn-off}}$. Then $W_{\text{switch}} = V_{\text{off}} I_{\text{on}} t_{\text{switch}}/6$ for linear commutation.

Figure 13.17 Switching trajectory for linear transitions. This behavior is called *linear commutation*.

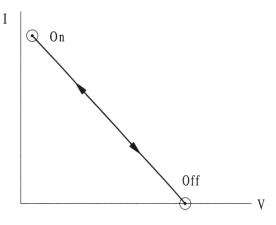

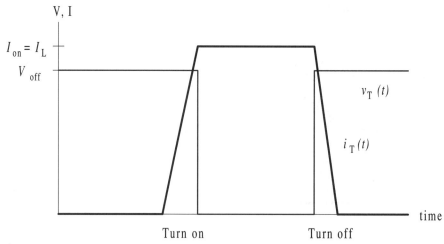

Figure 13.18 Transistor commutation in a buck converter with ideal diode.

Consider another case, in which a transistor in a buck dc–dc converter operates in conjunction with an ideal diode. The sequence should be as follows: Before transistor turn-on is attempted, the diode is carrying the inductor current, and the transistor voltage is $V_{in} = V_{off}$. The transistor is commanded to turn on, but the current requires time to rise to $I_L = I_{on}$. During the rise time, the diode must remain on to satisfy KCL, so the transistor voltage remains V_{in} until the current reaches its full value. At that point, the diode turns off, and the transistor voltage drops immediately to zero. At the point of turn-off, the current begins to fall in the transistor. The inductor will generate negative voltage since the current is falling and $v_L = L(di/dt)$. This causes the diode to turn on immediately, and quickly imposes the full V_{in} across the transistor. This type of inductor-dominated commutation means that the transistor voltage will always have a high value equal to V_{off} while the current is changing. The time plot for this case is shown in Figure 13.18, and the switching trajectory is shown in Figure 13.19. From the plots, it is not surprising that this is sometimes called *rectangular commutation*. The switching loss with this type of commutation is

Figure 13.19 Rectangular commutation illustration.

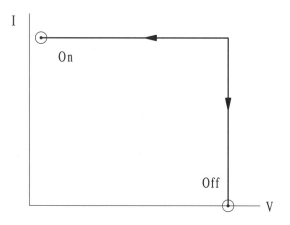

$$W_{\text{switch}} = \int_0^{t_{\text{turn-on}}} V_{\text{off}} \frac{I_{\text{on}}}{t_{\text{turn-on}}} t \, dt + \int_0^{t_{\text{turn-off}}} V_{\text{off}} \left(I_{\text{on}} - \frac{I_{\text{on}}}{t_{\text{turn-off}}} \right) t \, dt$$

$$= \frac{V_{\text{off}} I_{\text{on}} t_{\text{switch}}}{2}$$

(13.12)

Rectangular commutation is a reasonable model of converter behavior when inductive loads are switched, provided some other device (the ideal diode in this case) prevents the inductor voltage from swinging to an extreme negative value. The behavior is also termed *clamped inductive commutation*. If the diode is not ideal, its limited speed can result in large inductive voltage swings during transistor turn-off. A turn-off trajectory plotted for a real buck converter is shown in Figure 13.20, along with the associated transistor voltage and current waveforms. During turn-off, the inductor reacts quickly to the falling current with a negative voltage that significantly increases the voltage across the switch. This high voltage holds until the diode is able to respond. This is often referred to as *unclamped inductive commutation*. In the unclamped case, the transistor turn-off losses are commonly 50% to 100% higher than in the clamped case. If the diode turn-on is slow and the inductor is large, even higher losses can be generated. Unclamped transistor turn-off can cause the switching trajectory to swing outside the safe operating area, and is one of the most common causes of device failures in dc–dc converters.

For turn-on, the situation changes significantly: If the diode is not ideal, an attempt to turn the transistor on will cause the inductor current to rise. A positive voltage is induced, which tends to keep the transistor voltage low until the current transition is complete. This can make transistor turn-on even less lossy than under linear commutation. Turn-on behavior for a real buck converter (the same one as in Figure 13.20) is shown in Figure 13.21. The trajectory nearly follows the current and voltage axes. This suggests another case: A no-loss case in which current and voltage transitions occur separately, such that voltage is low while current is changing, or current is low while voltage is changing. This is one principle of res-

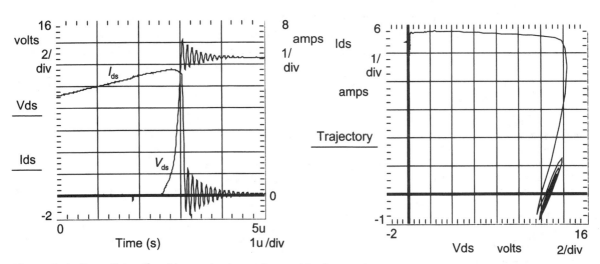

Figure 13.20   Turn-off time waveforms and trajectory for a real buck converter.

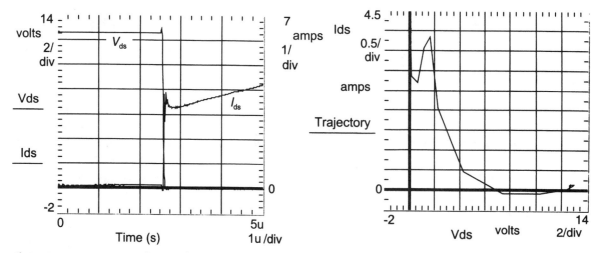

Figure 13.21 Turn-on waveforms and switching trajectory for a real buck converter.

onant switch action as presented in Chapter 8. In unclamped inductive commutation, the turn-on loss is often low, while the turn-off loss approaches $W_{\text{turn-off}} = V_{\text{off}}I_{\text{on}}t_{\text{turn-off}}$. If turn-on and turn-off times are similar, the net switching loss for such a converter is approximately

$$W_{\text{switch}} = \frac{V_{\text{off}}I_{\text{on}}t_{\text{switch}}}{2} \tag{13.13}$$

The commutation behaviors examined thus far suggest a general form:

$$W_{\text{switch}} = \frac{V_{\text{off}}I_{\text{on}}t_{\text{switch}}}{a} \tag{13.14}$$

where a is called the *commutation parameter*. Experience shows that the linear case $a = 6$ is overly optimistic, and represents a best-case estimate for many practical power converters. The value $a = 2$ is more typical, and some converters show poorer behavior because of inductive commutation. Table 13.1 lists some values. In making estimates of switching loss when limited information is available, the value $a = 2$ provides a useful first approximation.

TABLE 13.1
Commutation Parameter Values

Type of Commutation	Commutation Energy Loss	Comments
Linear	$V_{\text{off}}I_{\text{on}}t_{\text{switch}}/6$	A "best-case" loss value.
Rectangular (clamped inductive)	$V_{\text{off}}I_{\text{on}}t_{\text{switch}}/2$	Inductive load with one ideal switch. This provides a good estimate of commutation loss when information is limited.
Inductive	$V_{\text{off}}I_{\text{on}}t_{\text{switch}}/1.5$	When neither switch is ideal, inductance can cause substantial voltage overshoot.

13.4.3 Examples

Recall that the average switching power loss is given by $f_{\text{switch}}W_{\text{switch}}$. Given the general form in equation (13.14), the switching loss for a hypothetical dc–dc converter is

$$P_{\text{switch}} = \frac{V_{\text{off}}I_{\text{on}}t_{\text{switch}}f_{\text{switch}}}{a} = \frac{V_{\text{off}}I_{\text{on}}}{a}\frac{t_{\text{switch}}}{T} \tag{13.15}$$

The product $V_{\text{off}}I_{\text{on}}$ is the *power handling* value for the switch, and therefore switching loss is proportional to the power handling value times the ratio of the total switching time to the converter switching period. Low switching loss requires either low power handling or a switching time much faster than the switching period. For example, the MOSFET and diode in a dc–dc converter switching at 100 kHz should commutate in no more than about 100 ns to make switching loss values reasonable. Let us revisit the preceding buck converter example, this time considering the additional effects of switching loss.

Example 13.4.2   Consider again a buck converter for 12 V input and 5 V, 100 W output. The converter switches at 100 kHz. A MOSFET with $R_{\text{ds(on)}} = 0.05\ \Omega$ will be used in place of the previous 0.25 Ω device, since the voltage drop was excessive in Example 13.4.1. The diode exhibits residual drop of 1.0 V, and each device requires 100 ns to switch on or off. Estimate the total power loss and efficiency of this converter.

Based on equation (13.9), the duty ratio D_1 is now 50%. The transistor on-state loss is $D_1I_L^2R_{\text{ds(on)}} = (0.5)(20\ \text{A})^2(0.05\ \Omega) = 10$ W. The off-state loss is still negligible. The diode on-state loss is $(0.5)(20\ \text{A})(1\ \text{V}) = 10$ W. The on-state losses therefore total 20 W. Since the details of commutation are not known, the switching loss will need to be estimated. A value such as $a = 2$ might be selected as a representative value. The transistor has on-state current of 20 A and off-state voltage of 13 V. The diode also carries 20 A when on, and blocks about 11 V when off. Each device has $t_{\text{switch}} = t_{\text{turn-on}} + t_{\text{turn-off}} = 200$ ns. The transistor switching loss is approximately (13 V)(20 A)(200 ns)(100 kHz)/2 = 2.6 W. The diode loss is about (11 V)(20 A)(200 ns)(100 kHz)/2 = 2.2 W. The total converter losses are about 20 + 2.6 + 2.2 = 25 W, to give efficiency of (100 W)/(125 W) = 80%. The switching loss represents 20% of the total loss, and certainly should not be neglected when estimating efficiency.

In circuits other than dc–dc converters, losses are analyzed in very much the same fashion, with the only complication being the waveforms. Let us consider several cases to practice some of the issues.

Example 13.4.3   Evaluate the losses in a three-phase midpoint rectifier. The input source is a 440 $V_{\text{line-to-line}}$ source at 60 Hz, and the output has a large enough inductance to be modelled as a 50 A current source. The load is rated to draw 9 kW. The switching devices are SCRs, and may be modelled in the on state as constant 1.5 V drops. In the off state, the worst-case leakage current is 5 mA. The devices switch in approximately 20 μs.

The circuit is shown in Figure 13.22. Since the input voltage is 440 $V_{\text{line-to-line}}$, each individual phase of the midpoint converter sees the phase voltage $440/\sqrt{3} = 254$ V_{RMS}, and a peak voltage of $254\sqrt{2} = 359$ V. For the 50 A load to draw 9 kW, the average output voltage should be 180 V. Since each device drops a constant 1.5 V, the output average value

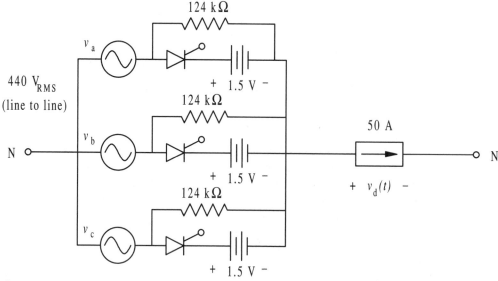

Figure 13.22   Midpoint rectifier for Example 13.4.3.

will be

$$\langle v_d \rangle = \frac{3V_{\text{peak}}}{\pi} \sin \frac{\pi}{3} \cos \alpha - 1.5 \tag{13.16}$$

and α should be chosen based on

$$181.5 \text{ V} = \frac{3 \times 359}{\pi} \sin \frac{\pi}{3} \cos \alpha, \qquad 181.5 \text{ V} = 297 \cos \alpha \tag{13.17}$$

or $\alpha = 52.3°$. Since each SCR is on 1/3 of a period, the on-state loss in each is (50 A)(1.5 V)/3 = 25 W. In the off state, the loss can be estimated by using the worst-case leakage of 5 mA as if it were a constant value, and then computing loss from the actual off-state voltage. Figure 13.23 shows the voltage $v_d(t)$, along with the voltage $v_a - v_d$, the voltage across the top SCR. The leakage current has the same polarity as the switch voltage, since this voltage generates it. This means the absolute value of the switch voltage is used for loss computation. The off-state loss is

$$P_{\text{off}} = \frac{1}{T} \int_{\text{off state}} v_{\text{switch}} i_{\text{switch}} \, dt = D_{\text{off}} 0.005 \langle |v_a - v_d| \rangle \tag{13.18}$$

The average of the absolute value of the switch voltage can be computed as 259 V, so that the off-state loss for each device is 1.3 W. This is probably a substantial overestimate, since the maximum leakage of 5 mA would normally occur only at maximum rated off-state voltage. The off-state loss is a small fraction of the on-state loss.

Commutation loss requires knowledge of the waveforms during switching. In this circuit, the current will change between 0 and 50 A during switching, while the voltage will probably hold close to the value just before or just after switching. From Figure 13.23, the voltage across switch *a* just before turn-on is 491 V, while that just after turn-off is −491 V. A commutation estimate would be based on the absolute value of this voltage (treated as the "off-state" voltage just at the time of switching), since commutation is a lossy process whatever the voltage or current polarities. Using a commutation parameter of 2 for estimation purposes, the energy loss should be

$$Q_{switch} = \frac{(491 \text{ V})(50 \text{ A})t_{switch}}{2} = (1.23 \times 10^4 \text{ W})(40 \text{ } \mu s) = 0.491 \text{ J} \qquad (13.19)$$

Based on the 60 Hz switching frequency, the switching loss for each device should be about 29.5 W. In this converter, the total loss per switch should be about 25 W + 1.3 W + 29.5 W = 55.8 W. The loss in all devices therefore should be about 167 W, which implies an efficiency of (9000 W)/(9167 W) = 98.2%. This efficiency is fairly typical of SCR-based controlled rectifiers.

Example 13.4.4   A full-bridge inverter uses PWM to convert a 330 V dc voltage source to a 230 V_{RMS} 50 Hz output. The switching frequency is 5 kHz. The four devices are IGBTs. The manufacturer reports that the on-state voltage drop is typically 2.0 V, the leakage current is typically 50 μA, and an inductive commutation causes loss of about 4 mJ under conditions similar to those in this circuit. Estimate the losses in this converter, based on a 20 kW output load with 0.8 power factor. What is the efficiency?

The inverter for this example is shown in Figure 13.24. The peak value of a 230 V sine wave is 325 V. The 2 V drops across the devices mean that only 330 V − 2(2 V) = 326 V is available for the PWM output. The modulation index will be close to 1.0 for this particular inverter. The *L-R* output load (which also performs the filtering functions) atten-

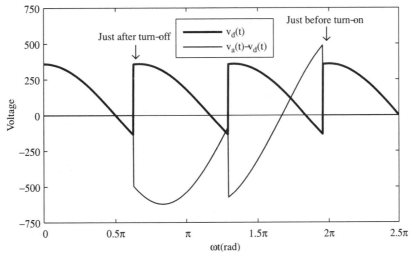

Figure 13.23 The converter voltage waveform $v_d(t)$ and the voltage across switch *a* for a three-phase midpoint rectifier with $\alpha = 52.3°$.

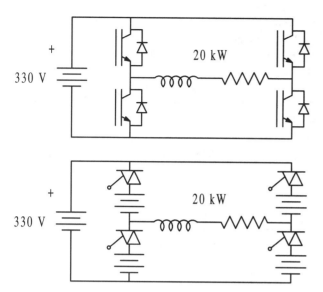

Figure 13.24 PWM inverter for Example 13.4.4.

uates the switching frequency while passing the 50 Hz modulated signal with minimal attenuation. The output 50 Hz voltage component should be 230 V_{RMS}, while the current component will be $(230 \text{ V})/(R + j\omega L) = 107 \angle -35.9° \text{ V}$. This current phasor can be written as $152 \cos(100\pi t - 35.9°)$ A.

The on-state loss in each 200 μs switching period for a given device is the 2 V switch drop (directed so as to oppose current flow) times the on current, times the duty ratio during that interval. The duty ratio is determined by the modulating function. Substituting the duty ratio from equation (6.15) for the modulating function $1.0 \cos(100\pi t)$, the loss in each 200 μs interval is

$$P_{\text{on state}} = d_{\text{on}}V_{\text{on}}i_{\text{on}} = \left[\frac{1}{2} + \frac{\cos(100\pi t)}{2}\right] \times 2 \text{ V} \times |152 \cos(100\pi t - 35.9°)| \text{ A} \quad (13.20)$$

where the absolute value reminds us that the switch voltage drop and the current always match in polarity. The concern is the average over a full 50 Hz period,

$$P_{\text{on state}} \text{ (W)} = 50 \int_0^{1/50} [1 + \cos(100\pi t)]|152 \cos(100\pi t - 35.9°)| \, dt \quad (13.21)$$

This integral is the same as the average of a full-wave rectified current, and the power loss value is $(152 \text{ A}) \times 2/\pi$, times the overall average duty ratio of 0.5, times the 2 V switch drop. This gives 96.8 W. In the off state, each switch must block 330 V. Given the 50 μA leakage, along with the average duty ratio of 50% over a 1/50 s period, the off-state loss should not exceed $330 \text{ V} \times 50 \text{ }\mu\text{A} \times \frac{1}{2} = 8.25 \text{ mW}$, a factor of ten thousand less than the on-state loss.

Since the manufacturer provides direct information about commutation loss under similar conditions, this information allows estimation of switching loss. Each switch turns on or off a total of 10 000 times each second. If each action causes 4 mJ of loss, the average loss will be (4 mJ)(10 000 Hz) = 40 J/s or 40 W. In this converter, each transistor loses 96.8 W + 40 W = 137 W. The off-state contribution is negligible. The total loss in all four devices is 548 W, and the converter's efficiency is (20 000 W)/(20 548 W) = 97.3%.

In the above examples, the on-state static behavior alters the converter action slightly—usually in the sense that a small voltage drop must be compensated. The commutation waveforms were not analyzed in terms of their effect on converter operation (often the effect is small), but loss during commutation is certainly not insignificant. Consider that if the switching frequency in the last example is raised to 100 kHz, the commutation loss in each device would reach 800 W.

Off-state losses are rarely more than 1% of the total losses in a given device, and off-state static models are ignored in most cases for the purposes of loss estimation. Commutation, on the other hand, must be taken into account. Since the actual switching trajectories depend on the external circuit, commutation losses can be a major problem if the details of devices and circuit components are not taken into account.

13.5 SIMPLE HEAT TRANSFER MODELS FOR POWER SEMICONDUCTORS

One very important open question at this point is loss management. Is 136 W of loss in a single IGBT realistic? What will this do to the device? Will a heat sink be needed, and if so what size? In this section, some of the fundamentals in thermal design of power converters are discussed. A full treatment is subject matter for a separate text, but it is not possible to talk about power converters and their design without at least a basic treatment of the thermal issues.

Most engineers would expect a small transistor to be destroyed if it must dissipate several hundred watts. The very rough value in Chapter 11 (1 W/cm^3 as a reasonable dissipation value) applies here, and a device that must dissipate more than 100 W ought to be physically large. Ineffective heat transfer should cause parts to get hot, and at some point this will lead to failure.

The basic issues can be addressed in two points:

1. When power is lost in a part, it represents thermal energy injected into the part. This energy must be removed. In almost any power supply or power converter other than those on a spacecraft, lost energy must be emitted into the surrounding air.

2. In a power semiconductor, power loss occurs inside the material near the various P-N junctions. Any semiconductor has a maximum *junction temperature*, $T_{j(max)}$. Above this value, the part might be damaged or destroyed, as dopant atoms begin to diffuse, materials begin to soften or melt, or other undesired effects occur.

The problem in most power converter designs is to remove waste heat to the atmosphere effectively enough to keep all temperatures below $T_{j(max)}$.

There are three mechanisms by which heat can be transferred among objects. These are conduction, convection, and radiation. In conduction heat transfer, heat flows between two solid objects in direct contact. The rate of heat flow is governed by Fourier's Law of

Heat Conduction, in which heat flow q is determined by a temperature gradient ∇T, such that

$$q = k\nabla T \tag{13.22}$$

Here the units of q are W/m^2, temperature is in kelvins, and the *thermal conductivity k* has units of W/(m·K). Thermal conductivity is a material property, and takes on a characteristic value for copper, aluminum, epoxy, or the material being used. In an electronic circuit, thermal conduction models the heat transfer to heat sinks, mounting hardware, or other physical objects attached to parts. It is important to recognize that the rate of heat removal by thermal conduction depends on the size of an object and on a temperature difference between objects or within a given material.

In convective heat transfer, heat flows from one place to another by means of a fluid flow. A common example is forced-air cooling with a fan, in which heat is carried away from an object by the motion of the air. Heat convection is governed by the relation

$$q = h(T_{object} - T_{ambient}) \tag{13.23}$$

where h is a *heat transfer coefficient*, and $T_{ambient}$ is the temperature in the fluid far away from the object being cooled. Convection is a critical process in almost any power converter, since heat must ultimately reach the surrounding air (or water in marine applications). The value of h is a system property rather than a material property. It depends on flow rates, surface geometry, and other parameters as well as on the properties of the object and the fluid. As in the case of conductive transfer, convective transfer is proportional to a temperature difference.

In radiative heat transfer, heat flows by electromagnetic radiation between bodies at distinct temperatures. Spacecraft applications require radiation as the final step in rejecting waste heat to the environment. In most terrestrial applications, heat flow by radiation is relatively unimportant. It can be significant for very small or very hot devices. The expression for heat flow in this case is

$$q = \epsilon\sigma T^4 \tag{13.24}$$

where T is absolute temperature. The net flow between two bodies is the difference between the result of equation (13.24) for each. The value σ is the *Stephan–Boltzman constant* $\sigma = 5.670 \times 10^{-8}$ W/(m^2·K^4), and ϵ is a parameter called the *emissivity* of the hot object. The emissivity is 1 for an ideal black body—the most efficient emitter of radiation. It can be less than 0.05 for white or highly reflective objects. Even though radiation is not a major component of heat transfer in most cases, it is beneficial to use dark objects to maximize this type of heat flow. Heat sinks and semiconductor parts are black when possible for this reason.

Since conduction and convection are the major contributors to heat flow in most applications, and because heat flow is linearly dependent on a temperature difference in these cases, an electrical analogue of heat flow,

$$P_{heat} = G_{thermal}\Delta T \tag{13.25}$$

can be defined. Here P_{heat} is the total flow in watts (the geometry has been factored in), ΔT is the temperature difference, and G_{thermal} is a *thermal conductance* that expresses the linear relationship. Notice that G is a function of parameters such as h, and that it also takes geometry into account. The electrical analogue is as follows: heat flows in a manner analogous to current along a path. The flow is driven by a potential difference ΔT, and the path has a certain conductance. Equation (13.25) is used to develop *thermal resistance*, R_θ or R_{th}, given by

$$P_{\text{heat}} = \frac{\Delta T}{R_\theta} \tag{13.26}$$

This last expression attempts to capture the basic physics of heat flow: The path along which flow occurs resists heat transfer. If this resistance is high, a high temperature difference is needed to drive a given flow, and a given power loss will heat up a part. If the resistance is low, heat can be removed easily, and parts will not get much hotter than the surroundings even in the presence of loss. Units of R_θ are those of temperature difference divided by power. In the SI system, the units are K/W. Most manufacturers report thermal resistances in °C/W. These units are the same, since only the temperature difference is relevant.

The simple heat flow expression in (13.26) is often used to determine temperature differences rather than heat flows. In any converter system, there is a definite amount of heat loss in any individual device. This loss must flow to the ambient environment. If somehow the flow P_{heat} is less than the loss, the device temperature will rise until ΔT is high enough to force $P_{\text{heat}} = P_{\text{loss}}$. A conceptual application is illustrated in Figure 13.25. Heat is generated at the semiconductor junction, and must flow to the ambient environment. The flow

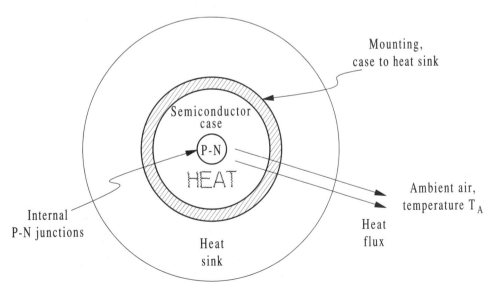

Figure 13.25 A power semiconductor and heat flow.

must pass through the case of the part, the interface between the case and the heat sink, and finally to the outside air. Each step adds resistance to the flow.

The important thermal resistance values are shown in Table 13.2. Recall that a given semiconductor has a well-defined maximum junction temperature. Ultimately, heat flows between the semiconductor at temperature T_{junction} and the outside environment at temperature T_{ambient}. Given a certain power loss P_{loss}, the flow is represented by

$$P_{\text{loss}} = \frac{T_{\text{junction}} - T_{\text{ambient}}}{R_{\theta(\text{ja})}} \tag{13.27}$$

where $R_{\theta(\text{ja})}$ is the total thermal resistance between the P-N junction and the ambient environment. To avoid trouble, it is important to maintain $T_{\text{junction}} \le T_{\text{j(max)}}$. This requires

$$P_{\text{loss}}R_{\theta(\text{ja})} + T_{\text{ambient}} \le T_{\text{max}} \tag{13.28}$$

Some of the implications of (13.28) include these rules:

- To avoid overheating, either the power loss or the thermal resistance must be low.
- It becomes increasingly difficult to keep a part cool as the surrounding temperature increases.

It is much harder to meet $T_{\text{junction}} \le T_{\text{j(max)}}$ on a sunny desert day under the hood of an automobile than in an air-conditioned office. The expression shows just how much harder. Let us consider some examples.

Example 13.5.1   A power MOSFET in a dc–dc converter application must sustain power loss of 30 W. The device is manufactured in a TO-220 package. It has typical specifications for a MOSFET in this sort of package: $R_{\theta(\text{jc})} = 1.2$ K/W, $T_{\text{j(max)}} = 150°C$. The manufacturer reports that without a heat sink, the value $R_{\theta(\text{ja})} \approx 45$ K/W. Can the power loss value be handled without a heat sink? If not, assume $R_{\theta(\text{cs})} = 1.0$ K/W, and suggest a specification for $R_{\theta(\text{sa})}$ to meet the requirements.

TABLE 13.2
Thermal Resistances for a Power Semiconductor

Thermal Resistance	Symbol	Comments
Junction to case	$R_{\theta(\text{jc})}$	A property of a given part. This models the resistance to heat flow from the inside of the part to its surface. It is specified by the part manufacturer.
Case to sink	$R_{\theta(\text{cs})}$	The part will not be in perfect contact with the heat sink. This models the thermal resistance of the mounting arrangement and any insulators. Heat sink manufacturers usually provide advice on good mounting methods.
Sink to ambient	$R_{\theta(\text{sa})}$	This value is determined by the convection performance of the heat sink, and is specified by the manufacturer of the sink. It depends strongly on air flow rates.
Junction to ambient	$R_{\theta(\text{ja})}$	The sum $R_{\theta(\text{jc})} + R_{\theta(\text{cs})} + R_{\theta(\text{sa})}$, representing the overall heat flow resistance. Semiconductor manufacturers sometimes provide this value for a part without heat sink.

The heat flow model tells us

$$P_{\text{loss}} = \frac{T_{\text{junction}} - T_{\text{ambient}}}{R_{\theta(\text{ja})}}, \qquad 30 R_{\theta(\text{ja})} = T_{\text{junction}} - T_{\text{ambient}} \qquad (13.29)$$

If no heat sink is used, the thermal resistance from junction to ambient of 45 K/W requires a 45 K temperature rise above ambient to drive 1 W of heat flow. With 30 W of loss, a temperature difference of 1350 K is required. This is impossible with $T_{\text{junction}} \leq 150°C$, so a heat sink is required.

For heat sink specification, let us start with an ambient temperature of 25°C. The maximum allowed temperature difference is 125 K, so

$$(30 \text{ W}) R_{\theta(\text{ja})} \leq 125 \text{ K}, \qquad R_{\theta(\text{ja})} \leq 4.17 \text{ K/W} \qquad (13.30)$$

The total resistance $R_{\theta(\text{ja})}$ is the sum of all thermal resistances in the flow path, and since the junction to case and case to sink values are given, the heat-sink value $R_{\theta(\text{sa})} \leq 1.97$ K/W to meet the requirements. Notice that a better sink is needed if the ambient temperature is higher. For example, if $T_{\text{ambient}} = 75°C$ (a value that might be encountered in an industrial plant or on a military vehicle), the heat sink must provide enough flow to keep the temperature difference below 75 K. This requires $R_{\theta(\text{ja})} \leq 2.5$ K/W, and $R_{\theta(\text{sa})} \leq 0.3$ K/W. This last value is very aggressive, and will most likely require a large heat sink and an external cooling fan.

Example 13.5.2 A power semiconductor in a TO-220 package has $R_{\theta(\text{jc})} = 1.2$ K/W. Without a heat sink, it exhibits $R_{\theta(\text{ja})} = 45$ K/W. What is the maximum power that can be dissipated by this part without a heat sink? If an "infinite heat sink" (a device with $R_{\theta(\text{ca})} = 0$) is used, how much power can be dissipated? Keep the junction temperature below 150°C, and assume a 50°C ambient.

With no heat sink, this part has $R_{\theta(\text{ja})} = 45$ K/W. With an infinite heat sink, the part shows $R_{\theta(\text{ja})} = 1.2$ K/W. With $T_{\text{junction}} = 150°C$ and $T_{\text{ambient}} = 50°C$, the allowed temperature rise at the junction is 100 K over the ambient. In the respective cases,

$$P_{\text{loss(no sink)}} \leq \frac{100 \text{ K}}{45 \text{ K/W}}, \qquad P_{\text{loss(sink)}} \leq \frac{100 \text{ K}}{1.2 \text{ K/W}} \qquad (13.31)$$

Without a heat sink, the part can safely dissipate 2.2 W. With an infinite sink, it can dissipate 83 W. This provides a helpful rule of thumb for this popular package: A semiconductor in a TO-220 package cannot be expected to dissipate more than about 2 W without a heat sink. Even with a sink, it is hard to dissipate more than about 50 to 75 W in such a package.

The thermal resistance heat flow model helps to elucidate some of the basic thermal specifications for a power converter, but gives an incomplete picture of the thermal design problem. For example, heat flow is a relatively slow process, and a given power requires time to make the temperature of an object change. Semiconductor manufacturers try to capture some of this behavior through *thermal capacitance*, C_θ, that models the time constants involved in a heat transfer problem. Typical time constants are measured in seconds or even minutes. In power electronics, the slow dynamics of heat transfer usually make the average power more germane than the instantaneous power in computing the temperatures. However, commutation can generate high power for very short time intervals. When switching loss is high, for example, extreme short-term power levels will heat a device up quickly, possibly above the allowed limits.

Dynamic heating effects can be important when switching frequencies are low, such as in rectifiers, or when short-term power levels are high with very small semiconductor packages. The instantaneous power during commutation can be as high as $V_{off}I_{on}$—the power handling value for the semiconductor. In an SCR, this can easily be several tens of kilowatts, even if the average loss is only a dozen watts or so. Figure 13.26 shows simulated time traces of switch power loss and junction temperature for a TO-220 package in a dc–dc converter application. The temperature trace, like an inductor current trace, is a triangular signal filtered from the power spikes during commutation.

Heat transfer has many interesting implications in power converters. Many power supplies, for example, are specified with a temperature derating. As the above analysis shows, the allowed power loss level goes down as ambient temperature goes up. A simple way to address this problem is to require the load power (and therefore the loss) to decrease linearly with temperature. Another issue is heat sink size. Low values of thermal resistance require large heat sinks—consistent with the notion that heat loss on the order of 1 W/cm^3 is reasonable. It is difficult to stay below a certain volume when air is the cooling medium. Output power per unit volume tends to be one figure of merit for a power converter, but many manufacturers specify values without considering the heat sink requirements. If a converter requires forced-air cooling, for example, the power needed to operate the fan represents an

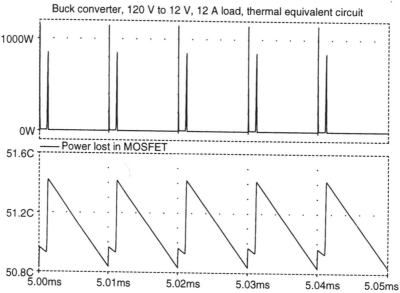

Figure 13.26   Power and temperature vs. time for a MOSFET in a TO-220 package; dc–dc converter application.

additional loss, and the fan and heat sink add to the overall size, weight, and cost. Thermal management is an important overall factor in practical converter design.

At very high power levels, liquid cooling for semiconductor devices can be beneficial. Water has excellent properties for this purpose, and is used if its electrical conductivity is not a problem. The liquid cooling process is very effective in removing heat from a small transistor with liquid and carrying it to a larger radiator where it can be dissipated to the atmosphere. The drawback of liquid cooling is in system design. Normally closed-loop plumbing is needed, and pumps and fans must move the materials. Extra hardware and weight are required. Liquid cooling tends to be considered for large, stationary systems for these reasons. Small power supplies need to be simple, and direct air cooling is almost always simpler than liquid cooling. In any case, it is crucial to minimize the loss in semiconductors, magnetic materials, and other power converter components to minimize the heat transfer design requirements.

13.6 THE P-N JUNCTION AS A POWER DEVICE

Semiconductor devices are formed from junctions of dissimilar material. The junctions can be formed between P and N-type doped materials, or between a metal and a semiconductor, as in Schottky diodes. The characteristics of P-N junctions are especially important in gaining an understanding of device operation. In this section, an introduction to the behavior of P-N junctions under conditions in a power converter is presented.

Figure 13.27 shows blocks of P and N material as they might appear in a diode. The dopant atoms in each region provide free holes or electrons, corresponding to the P and N materials. With no external connection, charges diffuse throughout the material. Near the junction, diffusing charges combine and are neutralized. This creates a *depletion region* in the vicinity of the junction, in which net free charge is low. Notice that the depletion region represents a low-charge region between regions of free charge. Therefore, it is associated with a capacitance, and also a voltage since charge separation over some distance defines an electric field and voltage.

Statically, the energy relationships within the depletion region give rise to the familiar exponential current-voltage behavior of a diode, with anode current

$$I_a = I_s(e^{qV/kT} - 1) \tag{13.32}$$

Figure 13.27 Simplified P-N junction showing charge segregation.

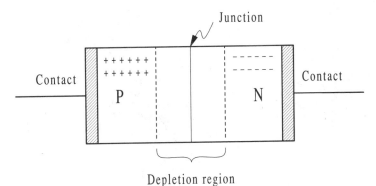

where q_e is the charge on an electron, $q_e = 1.602 \times 10^{-19}$ C, k is Boltzman's constant, and T is the temperature in kelvins. The static relationship shows that the *saturation current, I_s,* can be treated as the off-state residual current, and suggests that very little forward voltage is generated even with large forward values of anode current.

Dynamically, the picture is more complicated. When reverse bias is imposed on the junction, the charges are driven apart by the external electric field. The depletion region widens, and supports off-state voltage. The blocking voltage of a P-N junction is related to the size of the depletion region, as well as to the intrinsic electric field strength of silicon. When forward bias is applied and the junction is carrying current, the depletion region narrows. Charge is driven into the depletion region, where it recombines with opposite charge to generate current flow. Each bias arrangement creates spatially separated layers of charge formed around the junction. In each case, a corresponding capacitance can be used to help model the behavior.

In the case of forward bias, the tight layers of charge near the junction give rise to a *diffusion capacitance.* Charges build up in the P and N regions, so there is a substantial charge present on this capacitance. To turn a P-N junction off and restore its blocking capability, the charge must be removed through a *reverse recovery process* so the depletion region can grow and carry a significant off-state voltage. This *reverse recovery charge* flows out of the junction region when turn-off is attempted. Until the charge has been removed, the junction will not support reverse voltage. Reverse recovery usually dominates the dynamic behavior of power junction diodes, and many manufacturers specify certain aspects of the behavior. Reverse recovery charge, sometimes denoted Q_{rr}, leads to *reverse recovery current i_{rr}* and to *reverse recovery time t_{rr}.* A sample of junction turn-off behavior is illustrated in Figure 13.28 for a typical power diode. The reverse recovery charge is the integral of current over the interval when negative current flows, and reverse recovery current is often defined as the peak reverse current. This current is unrelated to off-state residual current, but instead represents the process of removing stored charge in the diffusion capacitance and restoring a full-size depletion region.

Reverse recovery is only approximately represented as a capacitive discharge, although this serves as a useful model and allows a helpful circuit representation of the behavior. The charge is eliminated from the depletion region during turn-off primarily through recombi-

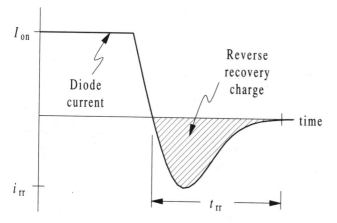

Figure 13.28   Diode turn-off behavior, showing reverse recovery.

nation. Most power diodes are doped with gold or platinum to form extra recombination sites that can speed the charge neutralization process. Without such treatment, P-N junction diodes are not especially fast as power switches. Standard rectifier-grade diodes with no such doping have t_{rr} values on the order of a few microseconds even for on-state ratings as low as one ampere. With extra recombination sites, speed improves markedly. A typical *fast recovery* diode can have a t_{rr} value on the order of 50 ns at current ratings of 10 A or more. The standard types are useful at line frequencies, but fast recovery types are necessary in inverters and dc–dc converters.

In the reverse-biased configuration, the wide charge separation around the junction forms a *depletion capacitance*. Its value is much less than that of the diffusion capacitance, since the charge separation distance is larger, but it still has a measurable effect on junction dynamics. In order to change the bias from reverse to forward, for example, the depletion capacitance must charge accordingly. Charges are delivered to the depletion region by drift and diffusion, leading to a time constant for turn-on. As turn-on is attempted, the junction itself will not carry much net current until the charging process has been completed. The time involved is termed the *forward recovery time* t_{fr}. It is usually much briefer than t_{rr}, and is not provided in basic specification data in most cases. The P-N junction forward recovery process can be illustrated by imposing a current source on a diode, with results shown in Figure 13.29. The forward voltage shows a large initial overshoot as the depletion region becomes charged. Once the junction's internal charge structure is in place, the voltage drops to its residual level, and net current begins to flow.

Diode reverse recovery can be tailored for dynamic characteristics other than speed alone. For example, *snap recovery* diodes have high values of i_{rr} in an attempt to minimize turn-off time. *Soft recovery* diodes exhibit relatively low i_{rr} values. Snap recovery types tend to produce strong harmonics because of the impulsive current behavior. They can be inconvenient in a converter system because of their need for high short-term currents. Many designers prefer soft recovery devices unless the very fastest speeds are necessary.

Figure 13.29 Turn-on behavior of a diode, showing forward recovery characteristics.

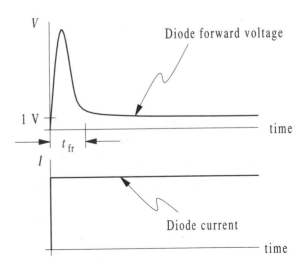

Forward recovery times and reverse recovery times are determined in part by the external circuit. The turn-on behavior in Figure 13.29 is forced with a current source, and represents nearly the fastest value of t_{fr} possible for the particular device. Even though a diode is an uncontrolled device, the external circuit can have influence on its dynamics. In more complicated structures, such as bipolar transistors, several junctions work together to influence the device dynamics.

13.7 P-N JUNCTION DIODES AND ALTERNATIVES

A P-N junction represents an excellent diode, as we have seen. A few of the issues involved in preparing a bipolar diode include the device size (which is governed by the current density limit to avoid excessive heating), the doping geometry and structure (which influence the voltage rating and the recovery times), and connections and packaging. Many power diodes are constructed as P-i-N structures, with a lightly doped *intrinsic layer* formed between the P and N regions. This layer helps enhance reverse blocking capability, and can also speed up the device by reducing the diffusion capacitance. Further discussion is beyond the scope here, but it should be mentioned that the many hundreds of different types of power diodes make slightly different tradeoffs between blocking voltage, speed, forward current rating, and packaging convenience.

The Schottky barrier diode is an important alternative to a P-N junction bipolar diode. In a Schottky diode, a block of semiconductor (usually N type) contacts a metal conductor. The junction between the dissimilar materials represents a potential barrier to the flow of charge. The barrier voltage is related to the relative *work function* between the semiconductor and metal, representing the energy difference between conduction-band electrons in the two regions. Schottky diodes show exponential current-voltage relations, but with much higher values of saturation current than bipolar diodes. While a typical silicon bipolar diode might show residual current on the order of 10 μA, a similar Schottky diode might exhibit residual current of 10 mA or more. In many power converter applications, the extra leakage current is not significant, so Schottky diodes can be considered.

The high saturation current and low potential barrier of Schottky diodes provide both benefits and drawbacks for their application. A major benefit is low residual voltage, which can be as low as 0.5 V for many power devices, The lack of a well-defined depletion region tends to minimize internal charge storage, so Schottky diodes are very fast and often do not show the recovery characteristics exhibited by junction devices. The primary drawback of a Schottky diode is its limited off-state voltage capability. The relative work function represents a rather weak barrier against reverse flow. Until recently, Schottky diodes with reverse ratings above 50 V were rare.

Schottky diodes are common today in low-voltage dc–dc converters. In general, they can reduce losses when their limited blocking voltages are not a problem. Over the next few decades, new materials are likely to make Schottky diodes increasingly practical for power electronics applications. Gallium arsenide Schottky diodes, for instance, support blocking voltages of 200 V or more, and retain a substantial speed advantage over bipolar diodes. Silicon carbide shows promise for this type of diode as well.

A more sophisticated alternative to a diode can be created with the *synchronous rectifier* technique. In this approach, a device acting as a near-ideal switch is controlled to turn

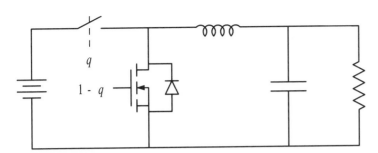

Figure 13.30   Synchronous rectification.

on under forward bias and turn off under reverse bias just like a diode. The power MOSFET is a candidate for this technique. In Figure 13.11, it was shown that the static behavior of a MOSFET is bidirectional, except that reverse operation is limited by the internal diode. In a MOSFET circuit such as that in Figure 13.30, reverse current will flow through the transistor rather than through the diode if the transistor gate signal is high and forward drop $I_d R_{ds(on)}$ is lower than the diode forward bias voltage V_{diode}. The full-bridge rectifier in Figure 13.31 can be made more efficient by setting the transistor gates high when the diodes are forward biased. Synchronous rectification is useful when voltage drops significantly lower than those of Schottky devices are necessary. Converters operating below 3 V often make use of the technique.

Example 13.7.1   A designer wishes to compare a bipolar diode, a Schottky diode, and a synchronous rectifier for a flyback converter design. The design calls for $V_{out} = 3.3$ V and $P_{out} = 25$ W. The input source is nominally 170 V_{dc}. What device current and voltage ratings will be needed, assuming that the inductor turns ratio is chosen for a nominal duty ratio of 50%? Compare a bipolar diode with a static model of 0.9 V in series with 0.015 Ω, a Schottky diode with 0.4 V in series with 0.025 Ω, and a power MOSFET used for synchronous rectification. What value of $R_{ds(on)}$ will be needed to provide an advantage to the synchronous rectifier?

For 50% duty ratio, the turns ratio will be 170 : 3.3 to properly match the voltages. The circuit is given in Figure 13.32. The current $I_{out} = (25$ W$)/(3.3$ V$) = 7.58$ A. For an equivalent buck-boost converter, the secondary inductor current would be $I_{out}/D_2 = 15.2$ A. With the

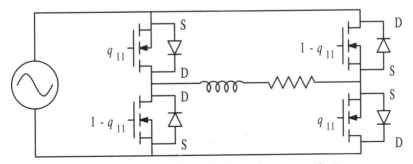

Figure 13.31   Full-bridge rectifier enhanced with synchronous rectification.

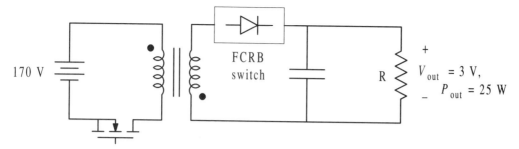

Figure 13.32   Flyback converter for Example 13.7.1.

transistor on, 170 V is applied across the inductor primary while -3.3 V appears on the secondary. The diode blocks the output less the secondary voltage, for a total of 6.6 V, in its off state. Diode device ratings of about 10 V and 10 A (average) should be adequate in this converter.

Now consider the implications of the static models. When the diode is on, it carries 15.2 A. The bipolar diode will show a residual drop $V_{\text{resid}} = (0.9 \text{ V}) + (0.015 \ \Omega)(15.2 \text{ A}) = 1.13$ V. The Schottky diode exhibits $V_{\text{resid}} = (0.4 \text{ V}) + (0.025 \ \Omega)(15.2 \text{ A}) = 0.78$ V. A MOSFET used as a synchronous rectifier will have a lower drop if $(15.2 \text{ A})R_{\text{ds(on)}} < 0.78$ V. This requires $R_{\text{ds(on)}} < 0.051 \ \Omega$. Ignoring commutation loss, the bipolar diode will dissipate $(1.13 \text{ V})(15.2 \text{ A})D_2 = 8.54$ W. The Scottky diode loss will be $(0.78 \text{ V})(15.2 \text{ A})D_2 = 5.90$ W. If a MOSFET with $R_{\text{ds(on)}} = 0.04 \ \Omega$ is chosen, its loss will be $(15.2 \text{ A})^2(0.04 \ \Omega)D_2 = 4.59$ W. When switching losses are taken into account, the Schottky diode will show a relative improvement because of its fast action and low stored charge. Total losses will be lowest with the synchronous rectifier if $R_{\text{ds(on)}}$ is low enough. For output voltages less than 3.3 V, synchronous rectifiers become almost necessary.

13.8 THE THYRISTOR FAMILY

The term *thyristor* is used generically here to represent a multijunction device with *latching* switch behavior. The thyristor function was first realized with mercury tubes. Today, the term *thyristor* refers to the semiconductor approach, while tubes of this type are called *thyratrons*. The most basic thyristor is the four-layer P-N-P-N junction device. When a gate terminal is provided, this structure functions as an SCR. While it might look like a simple series combination of P-N diodes (and indeed this is a good on-state model for the device), the extra junction in the center provides forward blocking capability under the right conditions.

The action of the four-layer structure can be understood with the well-known two-transistor model of the SCR. The development is shown in Figure 13.33. The middle layers can be split and associated with separate PNP and NPN transistors, connected base to collector as shown in the figure. With no gate current applied, the NPN transistor should have no collector current. The two devices are back-to-back, and can block both forward and reverse flow. When a gate current I_G is injected, the NPN transistor will carry a collector current given by $I_{C(\text{NPN})} = \beta_{\text{NPN}}I_G$. This collector current provides a base current for the PNP transistor, and $I_{C(\text{PNP})} = \beta_{\text{PNP}}\beta_{\text{NPN}}I_G$. If the device is externally biased for positive anode-cath-

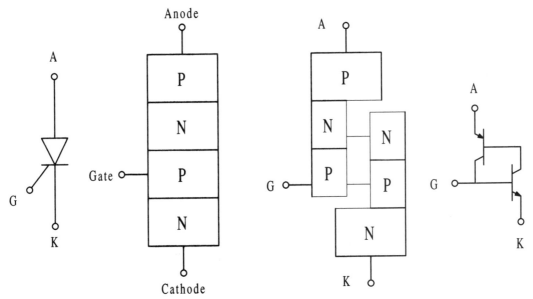

Figure 13.33 Two-transistor model of the SCR.

ode voltage (to provide a source for the collector flows) and if the product $\beta_{PNP}\beta_{NPN} > 1$, the collector current in the PNP transistor can take the place of the gate current. Currents will continue to flow, and the device will *latch* into its on state. The self-sustaining gate current process is termed *regenerative gate current*. Only a brief current pulse is needed to trigger the action.

What are the conditions required for successful latching? Given the three junctions within the structure, it might be expected that significant time will be required to set up the various depletion regions and charge structures needed for turn on. Furthermore, there must be an external source to support the flow of current. The conditions for turn-on of an SCR include:

- The external bias must support forward current flow large enough to provide the necessary gate regeneration. In practical devices, there is a minimum *holding current* I_H that must be present to maintain the on state.

- To turn the device on, the external gate pulse must have enough duration to allow complete formation of the internal charge structure. Since there are three internal junctions, the time involved is at least several microseconds for all but the very smallest devices.

Since gate current is driven into the base of an NPN transistor, the current must overcome one base-emitter drop to guarantee flow.

The requirement $\beta_{NPN}\beta_{PNP} > 1$ is modest. This is just as well, since the power current in the SCR must flow through the base of at least one of the two transistors. In practice, it is a challenge to avoid SCR behavior in unintended PNPN structures formed into integrated circuits or other semiconductor devices. Practical power SCRs require only a few tens of milliamps to initiate the latching process (the value must be kept high enough to avoid noise sensitivity), and are easy to use in controlled rectifier applications.

Consider one aspect of the model: If the SCR is on, the gate represents an internal access into the circuit. A negative gate current signal might turn the device off, for example, if it exceeds the value of $I_{C(PNP)}$ such that the net base current into the NPN device drops to zero. There are practical limits. Consider that if each device has $\beta = 1$, for instance, the necessary gate current value will be half as big as the total anode current. A device rated for 50 A of forward flow could be turned on with a 100 mA pulse, but a 25 A negative pulse would be needed to turn it off. In practice, the negative current capability is useful if the NPN gain is high while the PNP gain is low. For example, if $\beta_{NPN} = 8$ and $\beta_{PNP} = 0.2$, the condition $\beta_{NPN}\beta_{PNP} > 1$ is satisfied with a current $I_{C(PNP)}$ equal to about 1/4 of the total anode current. A negative gate current pulse of amplitude $I_A/4$ should turn the device off. This behavior is the basis of the *gate turn-off thyristor* or GTO. In a GTO, the two transistors have different gains to support the turn-off process. GTOs have a *turn-off gain* that represents the ratio of anode current to gate current for a successful turn-off process. Devices with gains up to about 5 are available.

Commercial SCRs vary tremendously in forward rating and capability, from 10 mA devices intended for precision triggering applications up to 10,000 A devices for electrochemical rectifiers. Switching time varies with current rating. A device with a 15 A forward current rating requires about one microsecond to switch on. Turn-off is slower, since it involves a reverse recovery process in the junctions. It can easily take about twenty times the turn-on interval to shut a device off completely. Full-wafer devices require far longer intervals for switching, and turn-off times can approach 1 ms for the largest devices.

The regenerative behavior of thyristors supports simple control processes. For example, a small *pilot thyristor* can be used to switch current into the gate of a large device, yielding a composite SCR that can be controlled with less than a milliamp. Large devices often use a pilot thyristor integrated directly with the main structure. Light pulses can be used to generate carriers for sensitive-gate SCRs, yielding the *light-fired SCR*. Light-fired devices are especially useful in high-voltage applications, in which series strings of switches are used to reach the desired blocking voltage.

The self-sustaining turn-on behavior of the SCR makes it vulnerable to certain external effects. One such effect is *voltage breakover*. When the off-state voltage is positive and high, the leakage current can become high enough to initiate turn-on. This method of switching is used intentionally in the *four-layer diode*, a gateless SCR that turns on when the forward voltage reaches a predetermined threshold. In conventional SCRs, the ratings are normally selected to avoid voltage breakover. In protection applications, breakover can be simulated in the gate circuit. Figure 13.34 shows a Zener diode arrangement in which the SCR remains off when $V_{ak} < V_z$. If V_z is reached, the SCR will latch on to provide a low-

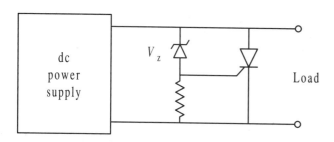

Figure 13.34 SCR with Zener diode gate control for crowbar application.

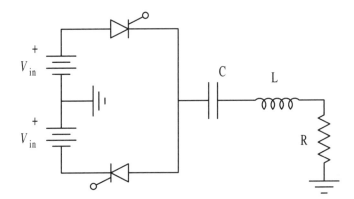

Figure 13.35 Resonant SCR inverter for Example 13.8.1.

impedance path and divert current from the anode voltage supply. This simple breakover *crowbar* SCR circuit is a common protection feature for power supply outputs.

A less desirable turn-on mode can be created if the voltage across a thyristor rises rapidly. A high value of dv/dt will induce current in the junction capacitances, and can generate enough current to cause switching. A device with 10 pF coupling to the gate will be exposed to 1 mA if the forward voltage rises at 100 V/μs. Typical ratings for 20 A SCRs fall in the range of 10 to 100 V/μs. Given the typical voltage levels and speeds in practical rectifier and inverter circuits, device dv/dt limits are often troublesome.

Example 13.8.1 Two SCRs are to be used in a resonant inverter application, shown in Figure 13.35 (the same circuit as in Figure 8.13). If the resonant combination has $Q = 2$ and each SCR turns on in 2 μs, what is the highest value of input voltage that can be supported if the device dv/dt ratings are 50 V/μs?

Recall the action of the resonant inverter from Chapter 8. When an SCR turns on, it causes the series RLC circuit to ring at the resonant frequency. A half-wave sinusoidal current flows, then the SCR turns off at the current zero crossing. After a short delay, the other SCR turns on to provide the negative half-wave of current. Since $Q = 2$, the capacitor develops a peak voltage double that of the output-wanted component, or $8/\pi$ times the input voltage.

Consider the top SCR in the circuit in this scenario. When it turns off, the current is zero and both the inductor and resistor voltages are zero. The device blocks the input voltage less the peak capacitor voltage, so $V_{off} \approx V_{in} - 8V_{in}/\pi = -1.55V_{in}$. This negative voltage is consistent with the SCR off condition. When the bottom SCR turns on, it imposes a voltage $-V_{in}$ on the RLC combination, and the top SCR must block $+2V_{in}$. The top SCR voltage waveform, given in Figure 13.36, shows that the blocking voltage on the upper SCR changes abruptly from $-1.55V_{in}$ to $+2V_{in}$ when the lower switch turns on. If the turn-on process takes 2 μs, the upper SCR is exposed to a rising voltage transient of $3.55V_{in}/(2\ \mu s)$. To keep $dv/dt < 50$ V/μs, the voltage V_{in} should be kept below 28.2 V. This low value can be a limiting factor in applying such a circuit.

Example 13.8.2 Consider an SCR in a midpoint rectifier. If the input line-to-line voltage is 480 V, what are the limiting factors given a dv/dt limit of 5 V/μs for the devices?

In a midpoint rectifier built with SCRs, the output voltage has a positive transition

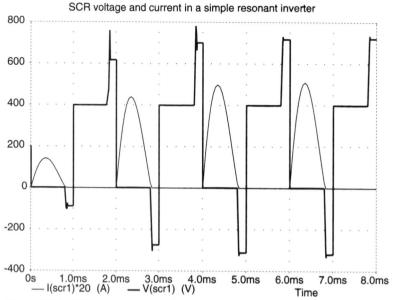

Figure 13.36 SCR inverter voltages for $Q = 2$ and $\omega_r = 4000$ rad/s.

when a switch is turned on (this is necessary to be consistent with SCR behavior). Since each SCR blocks an input line-to-neutral voltage less the output voltage, switch action always causes a decrease in blocking voltage. Unlike the inverter of the previous example, SCR turn-on does not expose devices to positive dv/dt. However, each SCR must block the input line-to-line voltage when off. The worst case dv/dt value is given by the highest derivative of the line potential, ωV_0. In this example, $V_0 = 480\sqrt{2} = 679$ V. To ensure that $\omega V_0 < 5$ V/μs, the radian frequency must be low enough, and $\omega < 7.37 \times 10^3$ rad/s. This corresponds to 1.17 kHz. These devices should not cause much trouble for line frequencies up to this value. They should be suitable for conventional 50 Hz, 60 Hz, and 400 Hz sources.

13.9 BIPOLAR POWER TRANSISTORS

Bipolar-junction power transistors (BJTs) have become unusual in recent years, supplanted at low voltages by MOSFETS and at high voltages by IGBTs. Insight into their operation is useful, however, since they form the basis for thyristors as well as IGBTs. Although BJTs are not often used in new power electronics designs, many existing inverters use them, and they are common in power analog applications such as audio amplifiers.

Power BJTs are normally of the NPN type. These devices allow higher current densities than PNP types because of mobility differences between electron-based and hole-based current flows. The base is narrow—on the order of the depletion region size in the reverse-biased collector-base junction—so current in the base-emitter junction alters the charge distribution in the collector-base region. The device is turned on by establishing an electron gradient within the base: Forward-bias of the base-emitter junction strongly alters the distribution of base-region minority carriers (electrons in an NPN device). The gradient drives a diffusion current of electrons into the collector. Some of the electrons recombine with holes in the base region; these result in base current flow. The base current modulates flow into the collector, giving rise to the familiar current gain behavior $I_C = \beta I_B$.

The base region in power transistors tends to be wider than in small-signal transistors. In addition, the objective of providing low forward voltage drop across the transistor in the on state limits the bias voltage at the collector-base junction. Both of these effects tend to increase recombination action within the base region. As a result, the gain values can be extremely low in power BJTs. A 30 A device rated for 400 V blocking, for example, has $\beta \approx$ 5 to provide a residual voltage of about 1 V. For "hard" turn-on, excess charge is needed in the base, and $\beta I_B > I_C$ is necessary.

The dynamics of the turn-on process are governed in part by the base-emitter junction. To turn the transistor on, the B-E junction must become forward biased to allow delivery of charge into the base region. Thus there is a diode forward recovery process, followed by time needed to set up the charge gradient between collector and emitter. Fast turn-on is usually accomplished by injecting a high current ($i_b \approx I_C$) into the base until collector flow is initiated, then dropping the base current to $I_B = I_C/\beta_f$. The symbol β_f in this expression indicates a *forced beta* chosen so β_f is slightly lower than the actual device gain. The lower the value of β_f, the lower the residual voltage. The initial forward recovery time can be associated with a *turn-on delay time*, $t_{d(on)}$, during which no collector current flows. When current begins to flow, it builds up according to a current *rise time*, t_r. For a 30 A transistor, turn-on delays on the order of 200 ns are typical. Rise-time values are often in the range of 500 ns.

The turn-off process is significantly slower than turn-on. The base-emitter junction must undergo a reverse recovery process. Once the junction has recovered, the charge within the base region can be removed. The removal process tends to be slow because recombination is the primary mechanism. In circumstances where β_f is kept low, the collector-base junction becomes slightly forward biased. The BJT is saturated, and there is excess charge in the collector region. The more deeply saturated the device, the more charge must be removed at turn-off. The reverse recovery process yields a *turn-off delay time*, $t_{d(off)}$, while the charge removal interval is termed *storage time*, t_{store}. When the storage time is complete, the actual current turn-off process begins, defining the *fall time*, t_{fall}. Turn-off delay is usually lumped with storage time, since it is hard to make this distinction in a real circuit. For the 30 A device, storage times on the order of 2 μs are typical, although much longer times are observed in heavy saturation. The fall time is similar to the rise time.

Figure 13.37 shows a set of base and collector current traces based on a design for very fast switching. A high initial pulse of base current is used to minimize the turn-on delay. For rapid turn-off, a negative bias is applied to the base. This draws negative current to remove stored charge and speed the recovery process. The circuits are discussed further in Chapter 14.

In contrast to diodes and thyristors, transistors of all types provide an intermediate *active state*—usually an undesired behavior in power electronics. In a BJT, if the base current is insufficient for full turn-on, collector current will flow with high collector-emitter voltage. The resulting high power dissipation can lead to rapid destruction. This is especially problematic in switching circuits exposed to external faults. For instance, if the collector current rises unexpectedly because of a problem at converter load terminals, the base current will not be sufficient to ensure low residual voltage. External faults are often destructive to tran-

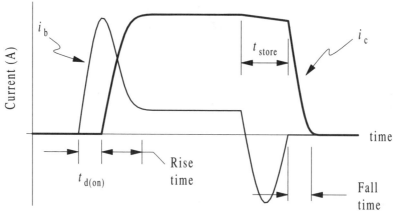

Figure 13.37 Base and collector currents for fast BJT switching.

sistors partly for this reason.

Bipolar transistors have an additional failure mode based on *current crowding*, often termed *second breakdown*. In a BJT, the thermal coefficient of resistance in the doped regions is negative. At high collector current levels, a local increase in current causes local heating. Local heating drops the local resistance, causing current to crowd into small channels. This can become a runaway process that leads to local thermal failure or to high local electric fields. Second breakdown limits the safe operating area. It is necessary to avoid simultaneous high voltage and current on a BJT.

The low gain of power BJTs means that high base currents are needed for switch operation. A device rated at 100 A can require 20 A into the base for operation, and rapid switching might benefit from 100 A pulses. In effect, the high base current will require its own separate power converter. One common way to avoid this trouble is a Darlington connection, shown in Figure 13.38. The main transistor has low β because of the high current, but the input transistor usually has a much higher β. The combination readily achieves current gain of several hundred or more. The main drawback of the Darlington arrangement is low speed, especially at turn-off. The input device cannot draw negative base current from the main device, and charge removal is slow. The input device adds its own dynamics to the switching process as well. Power Darlington BJTs are often integrated as a single device. In this case, internal resistances or even a direct connection to the main device base terminal are provided to help improve speed.

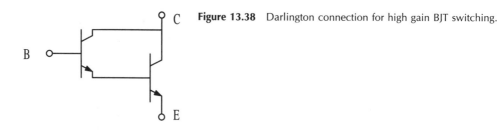

Figure 13.38 Darlington connection for high gain BJT switching.

13.10 FIELD-EFFECT TRANSISTORS

Figure 13.39 shows a simplified cross section of a lateral metal-oxide-semiconductor field-effect transistor (MOSFET) of the N-channel type. In this version of the MOSFET, the bulk P material is lightly doped and highly resistive. The N-type wells at the source and drain terminals are highly doped and exhibit high conductivity. If a voltage is applied between gate and source, the electric field draws charge into the region under the gate electrode near the source terminal. The doping level in the P-type substrate is low enough that the charge effectively converts it locally to N-type material, with excess electrons. As the gate-source voltage increases, this *inversion effect* extends across the gate region. At a specific *threshold voltage* V_{th}, a complete N-type *channel* forms between source and drain. This conductive channel provides a current path. The electric field *enhances* flow between drain and source, and this MOSFET therefore is an *enhancement mode* device.

As the electric field between gate and source increases further, the channel becomes larger, and the channel resistance drops. The device can be considered a *voltage-controlled resistor*, since the gate-source voltage V_{gs} alters the resistance between drain and source. When current is allowed into the drain, it creates a voltage drop that opposes the gate-source field effect. The drain current cannot be raised arbitrarily, since eventually the voltage drop will close the channel and turn off the device. At a given value of V_{gs} the opposing voltages lead to a *saturation current*, and the drain current will not exceed a maximum level. Notice that a saturated FET operates in its active region, in which the drain current is a function of gate-source voltage. This contrasts with a BJT, which is said to be saturated when there are excess base carriers and $I_C/I_B < \beta$. In the active region, the MOSFET exhibits a *transconductance* g_m, defined as

$$g_m = \frac{\partial I_{DS}}{\partial V_{GS}} \tag{13.33}$$

at a specific value of drain-source voltage. Typical values for power MOSFETs range from about one to ten siemens (inverse ohms).

The current-voltage behavior is reviewed in Figure 13.40. When V_{gs} is below the threshold, no current flows between drain and source. With $V_{gs} > V_{th}$, the current-voltage behavior is linear nearly up to the saturation current level. For the device characteristics shown in the figure, the saturation current is very high and the device resistance reaches a minimum

Figure 13.39 Basic structure of lateral MOSFET.

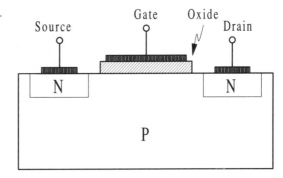

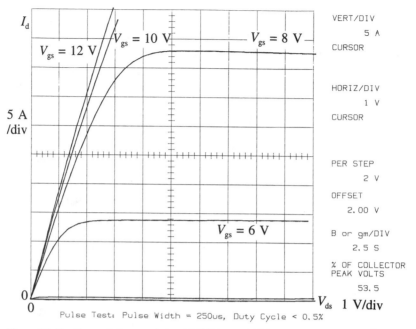

Figure 13.40 Forward characteristic of 20N20-type MOSFET (20 A, 200 V, N-channel).

level when V_{gs} is about 10 V. In most conventional power MOSFETs, the gate requires about 8 to 12 V for low-resistance turn-on. Some manufacturers provide *logic-level* MOSFETs with more sensitive gate structures. These might require only 4 V for turn-on.

For a P-channel device, highly doped P-type material is located at the drain and source terminals, and the bulk material is lightly doped N-type semiconductor. The gate-source electric field must draw holes into the bulk material to invert it to P-type and form a channel. The gate-source voltage must be negative to provide the appropriate field polarity. In the channel, holes flow from source to drain. The on-state current therefore is negative into the drain terminal.

The dynamic behavior of a MOSFET is simple compared to other switching devices. Although the electric field must be established and the channel must be created, there are no P-N junctions directly involved in the process. No storage time is encountered, and current rise and fall times can be very rapid. The dynamic behavior can be understood at a basic level by recognizing the capacitance of the gate-source interface. For a power device, this capacitance is relatively large; 2000 pF is not unusual for a device intended to carry 20 A and block 200 V. Since the gate-source voltage levels are consistent with those in conventional analog circuits, it is plausible to use op-amps or similar analog ICs to drive MOS-

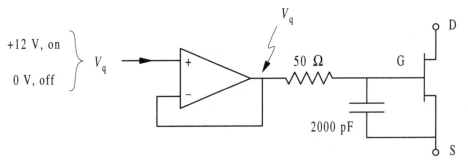

Figure 13.41 Op-amp used to operate the gate of a power MOSFET.

FETs. A typical op-amp with output resistance of 50 Ω forms an RC circuit, as illustrated in Figure 13.41. The circuit time-constant is 100 ns. In this configuration, the gate drive circuit becomes the limiting factor for switching speed: nothing can happen until sufficient voltage is applied to form the channel. The gate drive shown generates turn-on delay and turn-off delay times. The rise and fall times for this device are each about 100 ns, so the gate drive can be considered an important factor. Gate drive issues and a more complete discussion of the capacitive effects in power MOSFETs are addressed in Chapter 14.

In power devices, the large packages and connections introduce inductance in a MOS-FET. Figure 13.42 shows a circuit-based dynamic model of a MOSFET, with package inductances for a TO-220 size included. This simple model captures the basic static and dynamic behavior of the device. The model shows that the gate-source voltage should be about double the threshold value to ensure hard turn-on. Unlike a BJT base, there is little penalty

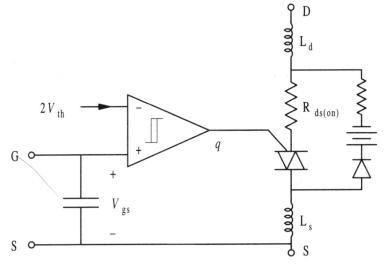

Figure 13.42 Circuit model for MOSFET static and dynamic behavior.

for overdriving the gate input to a MOSFET, provided the breakdown strength of the oxide layer does not come into play.

The internal reverse diode of a power MOSFET has been mentioned many times as a fundamental characteristic of these devices. To see why, consider the lateral N-channel geometry in Figure 13.39. The N-P-N regions combine to form a *parasitic bipolar transistor* in parallel with the FET. Although there is no circuit connection to the base, capacitive currents caused by high dv/dt values or residual currents can turn on the parasitic BJT. In practice, this parasitic effect is an important limiting factor on the dv/dt specifications of a power MOSFET. Even a small base current can produce enough collector current to destroy a device in its off state. To avoid the possibility of turning on the parasitic device, an internal short is created between the bulk material (the *body*) and the source. This short circuit creates a P-N pair from source to drain, termed the *reverse body diode*. Since the N-P-N combination is fundamental to the construction of a MOSFET, the presence of the diode cannot be avoided without compromising reliability. The diode is an asset in bidirectional applications, including inverters. Modern devices often are fabricated so that the reverse body diode has similar dynamic performance and power handling capability to the MOSFET itself. In a P-channel device, the effects are the same, except that the reverse body diode anode is connected to the drain, and the cathode N-type material is shorted to the source terminal.

The failure modes of MOSFETs tend to be simpler than those of other power devices. The channel configuration is simple, and turn-off involves closing the channel rather than waiting for internal recombination. Some of the important failure modes are associated with the parasitic bipolar transistor rather than with the MOSFET itself. The MOSFET acts as a *majority carrier device*, since the carriers are the free charges in the channel. Silicon exhibits increasing resistivity with temperature for majority carriers, and the thermal runaway behavior exhibited in BJTs is avoided in MOSFETs. The temperature behavior makes it easy to use multiple MOSFETs in parallel: Any device carrying excess current will heat up and become more resistive, diverting current into parallel paths. Real devices are formed as parallel combinations of many thousands of individual MOSFET cells to take advantage of the thermal behavior. Excessive loss still produces thermal failure in MOSFETs, but there is no unstable runaway effect if the parasitic BJT does not act. In the off state, the blocking voltage capability is based on the ratings of the reverse body diode. This voltage is determined in part by the distance from source to drain. High blocking voltage capability implies high channel resistance because of the geometry, so there is a tradeoff between low $R_{ds(on)}$ values and device voltage capability.

Example 13.10.1   The forward characteristic curves of a MOSFET are shown in Figure 13.43. Determine the values of V_{th}, $R_{ds(on)}$, and g_m for this device. The transconductance at 10 V of forward drop is desired.

The characteristic curves show that a positive gate-source voltage is needed to turn the device on. Evidently this is an N-channel MOSFET. The threshold voltage cannot be determined precisely, but is about 4 V for this set of curves. To turn the device on, use $V_{gs} \approx 2V_{th}$. In this case, $V_{gs} = 9$ V is suitable. The curve shows a slope of $(25\ \mathrm{A})/(1.5\ \mathrm{V})$ when $V_{gs} = 9$. This corresponds to $R_{ds(on)} = 0.06\ \Omega$. With 4 V forward drop, the value of g_m can be estimated from the $V_{gs} = 5$ V and $V_{gs} = 6$ V curves. The value $\partial I_{DS}/\partial V_{GS}$ for this choice gives $g_m = 10$ S.

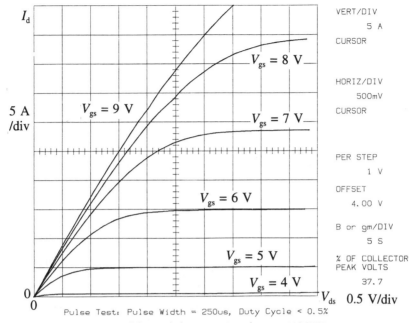

Figure 13.43   Experimental forward characteristics of power MOSFET.

13.11 INSULATED-GATE BIPOLAR TRANSISTORS

One drawback of the power MOSFET is that only a fraction of the material serves as the current-carrying channel. The current density in a given MOSFET is only about one-fifth that in a BJT of similar ratings. This tends to make MOSFETs relatively expensive when high voltage ratings and low on-state resistances are desired. An alternative is to use a Darlington combination of an FET and a BJT, as shown in Figure 13.44. This arrangement has two important advantages: It makes better use of the semiconductor material than a MOSFET alone (and can carry a larger current for a given size), and it uses the convenient gate-source control behavior of the MOSFET to avoid high base currents.

This Darlington combination as shown has its own drawbacks. The device is likely to be slow, particularly during turn-off since negative base bias cannot be applied. An additional junction will be needed to provide both the NPN arrangement for the BJT and the separate NPN arrangement for the MOSFET. It seems likely that undesired parasitic elements will be present. The insulated-gate bipolar transistor (IGBT) is based on the Darlington FET-BJT combination, although the integration is more complete to avoid some of the drawbacks of the Darlington structure. A vertical device structure is shown in Figure 13.45.

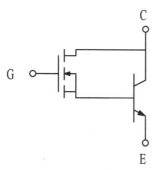

C

G

E

Figure 13.44 FET-BJT Darlington connection for power switching.

If an electric field is provided between the gate and the emitter terminals, a channel can be created through the upper P region. This channel will not carry bulk current, but serves to provide base current into the internal N-region. The PNP combination produces a wide-base bipolar transistor that carries bulk current in the device.

Notice that the IGBT has a PNPN structure—that of a thyristor. It is possible to cause latching behavior if gate current flows in this thyristor. The emitter terminal forms a short circuit across the top PN junction to help avoid turn-on of the parasitic SCR. A more complete circuit model is provided in Figure 13.46. The resistance R_B is residual internal resistance after the emitter is considered. If it is low enough, the thyristor is unlikely to act. The avoidance of thyristor latchup is a critical issue in IGBT design. Many of the details in the semiconductor processing involve extra structures or doping schemes to reduce the possibility of thyristor action. The extra junction does provide advantage as well: It provides in reverse–blocking behavior. Unlike a MOSFET, an IGBT has inherent reverse blocking capability. This is an advantage in rectifier applications. Manufacturers often add an internal reverse diode in IGBTs intended for inverters.

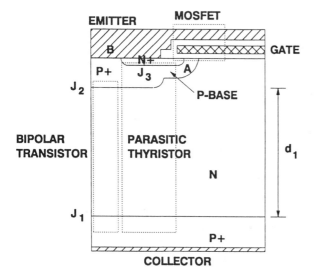

Figure 13.45 IGBT cross section. (From B. J. Baliga, *Power Semiconductor Devices*. Boston: PWS Publishing, 1996, p. 428. Reprinted by permission.)

Figure 13.46   Device-based model for IGBT.

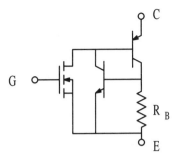

The dynamic behavior of the IGBT is similar to that of a power BJT. The MOSFET does not add a speed penalty. For typical devices, when a gate-emitter voltage of about 10 V is applied, the combined delay time and rise time is about 500 ns—close to the values observed in power BJTs when a high base current is injected at turn-on. Turn-off is slower, and has two components. A substantial fraction of the current flows in the MOSFET channel, since the effective gain of the internal PNP transistor is low. To turn the IGBT off, the gate-emitter voltage is set to zero. Once the internal gate-source capacitance has discharged, the channel current falls quickly. However, the current flow in the transistor falls more slowly, since the carriers in the base region are eliminated by recombination. This results in a current profile called *current tailing*. A typical set of current waveforms is given in Figure 13.47. The current tail usually occurs while high off-state voltage is imposed on the device. It is a major source of switching loss. In some devices, turn-off times on the order of 20 μs are observed. These types of IGBTs are common in backup power applications, in which inverter switching frequencies do not exceed a few kilohertz. For PWM applications, faster turn-off is needed. Processing methods can reduce the turn-off time to about 500 ns, although higher forward drop is generated as a consequence of the methods.

For the purposes of static models, a useful representation for the IGBT is a diode in series with a power MOSFET. This captures the forward voltage behavior of the device, the reverse blocking capability, and the gate-source control function. A typical static model is provided in Figure 13.48.

Figure 13.47   IGBT turn-off current waveforms.

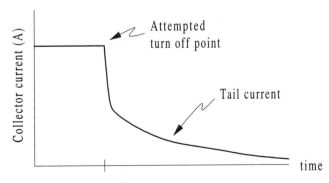

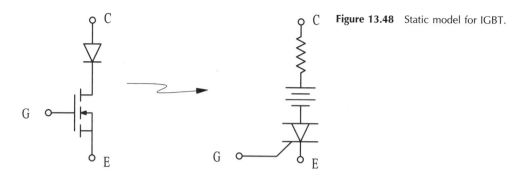

Figure 13.48   Static model for IGBT.

Example 13.11.1   The processes for making IGBTs produce a trade-off between forward voltage drop and turn-off speed. Many manufacturers provide devices in two or three different speed ranges for this reason. Two such devices are to be compared for a dc drive application, shown in Figure 13.49. The drive input voltage is 500 V, and the output current is 100 A. The slow IGBT has a forward drop of 1.5 V at 100 A. Its current tail generates loss of 12 mJ per switching at 500 V and 100 A. The fast IGBT has a forward drop of 2.7 V at 100 A, but current tailing consumes only 4 mJ per switching. Over what switching frequency range will the slow device give lower total loss?

Each device has on-state loss given by $DV_{\text{resid}}I_{\text{on}}$. Assuming the turn-on loss is low, turn-off losses will be the switching frequency multiplied by the energy lost during each turn-off action. For each device, total loss is

$$\text{Slow:}\quad P_{\text{loss}} = D(1.5\text{ V})(100\text{ A}) + f_{\text{switch}}(12\text{ mJ})$$

$$\text{Fast:}\quad P_{\text{loss}} = D(2.7\text{ V})(100\text{ A}) + f_{\text{switch}}(4\text{ mJ})$$

(13.34)

The losses will match when $f_{\text{switch}} = D(15\text{ kHz})$. For a duty ratio of 50%, the losses will match at 7500 Hz. Depending on the detailed drive requirements for low-speed torque, it can be argued that the slow IGBT is the better choice for switching frequencies below about 5 kHz, while the fast IGBT will be needed above about 12 kHz. In this drive, a MOSFET is probably not a very good alternative. MOSFETs rated at 500 V tend to have high $R_{\text{ds(on)}}$ values, and it will be difficult to provide forward drops below 3 V with a MOSFET in this application.

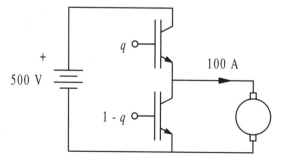

Figure 13.49   dc drive application for IG-BTs.

13.12 SNUBBERS

13.12.1 Introduction

With all types of power semiconductors, switching loss is an important consideration. Static-state losses can be addressed directly with circuit models, but switching losses usually need to be considered separately. The switching trajectory represents voltage and current evolution. In this section, auxiliary circuits intended to alter the switching trajectory and reduce losses are considered. Two extreme cases have been addressed in previous discussion. When a semiconductor switch acts alone, the inductance in a converter can impose high transient voltages, especially during turn-off. Excessive voltage ratings might be required to avoid failures. Alternatively, resonant methods, introduced in Chapter 8, avoid changes in voltages or currents during switching.

Auxiliary circuits intended to manipulate the switching trajectory are termed *snubbers* since their primary function is often to suppress overshoot in voltage or current.

> ***Definition:*** *A snubber is a circuit connected around a power semiconductor device for the purpose of altering its switching trajectory. Snubbers usually have the objective of reducing power loss in the semiconductor device.*

Snubber circuits act to prevent fast change of voltage or current during switching, so that the commutation process can become more nearly linear. A simple example is shown in Figure 13.50. In this case, a parallel capacitor prevents the switch voltage from rising rapidly during turn-off. The effect on the trajectory is to avoid the voltage overshoot caused by the inductor. The circuit is too simple, however: At turn-on, the charge stored in the capacitor is dissipated through the switch, and the turn-on trajectory develops a high current overshoot. Trajectories with and without the capacitor are compared in Figure 13.51.

13.12.2 Lossy Turn-Off Snubbers

The turn-on overshoot can be avoided by making the snubber unidirectional, as in Figure 13.52, to form a *turn-off snubber*. The resistor has been added to make the capacitor dis-

Figure 13.50 Inductive switching with parallel snubber capacitor.

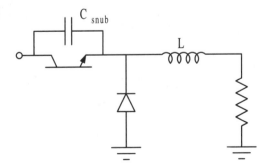

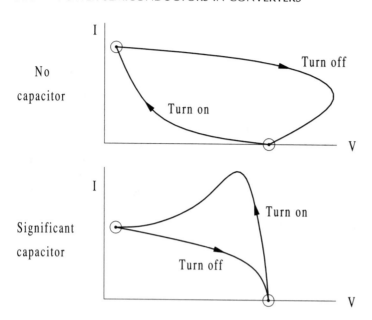

Figure 13.51 Trajectory comparison with and without parallel capacitor.

charge process gradual and to ensure that the discharge energy is dissipated outside the semi-conductor. The circuit of Figure 13.52 represents a *lossy snubber*, since energy is dissipated in the resistor. It trades off resistor loss and semiconductor loss. If the capacitor is chosen correctly, the total loss will be less than with no snubber in place.

The capacitor and resistor selection process must address two requirements: The capacitor must be sufficient to avoid voltage overshoot during the current fall, and the RC time constant must allow the stored energy to dissipate completely during the switch on-time. A useful way to approach the design is to assume that the current falls linearly during the switch fall time, t_f. The external inductor acts to maintain constant total flow. During turn-off, the currents will be approximately

$$i_{\text{switch}} = I_L\left(1 - \frac{t}{t_f}\right), \qquad i_C = I_L - i_{\text{switch}} = I_L\frac{t}{t_f} \tag{13.35}$$

Figure 13.52 Turn-off snubber circuit.

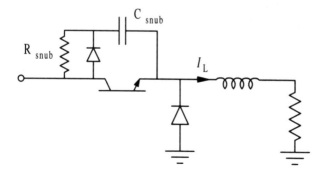

The capacitor voltage is found by integrating the current, and will be a function of t^2 during the interval $0 < t < t_f$ if the capacitor is large enough. If C is small, the capacitor voltage will reach the full off-state value, and the other switch will turn on to carry the remaining current. For sufficient C, the energy lost during turn-off is the integral of $v_C i_{switch}$, or

$$W_{switch} = \int_0^{t_f} v_C i_{switch}\, dt = \int_0^{t_f} \left(\frac{I_L}{C}\frac{t^2}{2t_f}\right) I_L \left(1 - \frac{t}{t_f}\right) dt$$
$$= \frac{I_L^2}{24C} t_f^2 \tag{13.36}$$

In addition, an energy equal to $\frac{1}{2}CV^2_{off}$ will be lost in the resistor as the charge dissipates. Compared to an inductive commutation, with loss $(V_{off}I_L t_{f}\!f_{switch})/2$, the loss with the snubber is

$$P_{switch} = \left(\frac{I_L^2}{24C}t_f^2 + \frac{1}{2}CV^2_{off}\right) f_{switch} \tag{13.37}$$

The snubber will decrease the switching loss compared to the inductive case if

$$\frac{I_L^2}{12C}t_f^2 + CV^2_{off} < I_L V_{off} t_f \tag{13.38}$$

Figure 13.53 shows the effect of capacitor choice on switching loss. Without the capacitor, inductive commutation is assumed. The capacitor adds its discharge loss, but decreases the switching loss. There is an optimum choice that minimizes total loss. It can be found with a partial derivative of the left-hand side of equation (13.38) to be

$$C_{opt} = \frac{I_L t_f}{\sqrt{12}\, V_{off}} \tag{13.39}$$

Figure 13.53 Switching power loss as a function of snubber capacitor value.

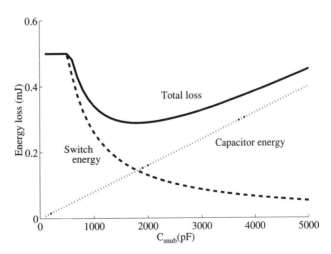

This is only approximate since the fall current has been assumed to be linear. The "optimum" snubber produces a commutation parameter equal to $\sqrt{12}$, or about 3.5. This is substantially better than inductive commutation. Figure 13.54 shows the hypothetical switching trajectory when the capacitor is near its optimum value.

The resistor should dissipate the capacitor's stored energy during the turn-on interval so that the capacitor will be ready to snub the next turn-off. If the RC time constant is less than half the on interval, more than 98% of the capacitor energy will be dissipated prior to turn-off. This means

$$RC < \frac{DT}{2}, \qquad R < \frac{DT}{2C} \tag{13.40}$$

The power rating needed for the resistor is given by the discharge energy times the switching frequency,

$$P_r = \frac{1}{2}CV^2_{\text{off}}f_{\text{switch}} \tag{13.41}$$

An example can help provide ideas about typical values.

Example 13.12.1   A buck-boost converter produces 5 V output from a 3 V battery input. The switching frequency is 300 kHz. The load is 25 W. Assume that the MOSFET turn-on loss is low. The turn-off process requires 100 ns. The device has $R_{\text{ds(on)}} = 0.02\ \Omega$. A small heat sink provides $R_{\theta(\text{ja})} = 25$ K/W for the MOSFET. The diode is a Schottky device, with low switching loss, and can be modelled as a fixed forward drop of 0.5 V. Estimate the power loss in this converter without a snubber. Design a turn-off snubber, and estimate the loss with this circuit in place. For an ambient temperature of 25°C, what junction temperature exists in the MOSFET? The polarity reversal of the buck-boost converter can be addressed in this application by reversing the battery voltage. The circuit, with snubber shown, is given in Figure 13.55.

Assuming the inductor and capacitor are large and the battery is ideal, the circuit can be analyzed with the static model given in Figure 13.56. The switches manipulate the in-

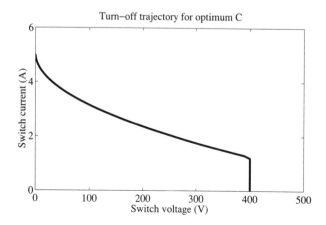

Turn–off trajectory for optimum C

Figure 13.54 Switching trajectory to minimize total switching power loss.

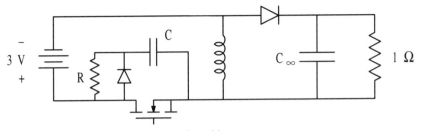

Figure 13.55 Buck-boost converter with snubber.

ductor voltage and diode current, and

$$v_t = q_1[-3 \text{ V} + (0.02 \text{ }\Omega)I_L] + q_2(5.5 \text{ V})$$

$$i_d = q_2 I_L \tag{13.42}$$

The average values are 0 and 5 A, respectively. Thus

$$I_L = \frac{5 \text{ A}}{D_2} \quad \text{and} \quad 0 = D_1\left[-3 \text{ V} + (0.02 \text{ }\Omega)\frac{5 \text{ A}}{D_2}\right] + D_2(5.5 \text{ V}) \tag{13.43}$$

The solution is $D_1 = 0.671$, $D_2 = 0.329$, $I_L = 15.2$ A. The on-state loss in the switches is $D_1I_L^2R_{ds(on)} + D_2I_L(0.5 \text{ V}) = 8.1$ W. The switch off-state voltages are about 8 V, and the on-state currents are 15.2 A. If the commutation parameter is estimated as 2, the switching loss without a snubber is $I_{on}V_{off}t_f f_{switch}/2 = 1.8$ W. In this circuit, the optimum snubber capacitor can be estimated as

$$C = \frac{I_L t_f}{\sqrt{12} \, V_{off}} = 0.055 \text{ }\mu\text{F} \tag{13.44}$$

The on-time is 2.24 μs, so the snubber resistor should be no more than 20.4 Ω. The resistor will need to dissipate 0.53 W. A reasonable design choice is to use a 1 W, 12 Ω resistor with a 0.05 μF capacitor. The switching loss should drop to about 1.1 W with this snubber in

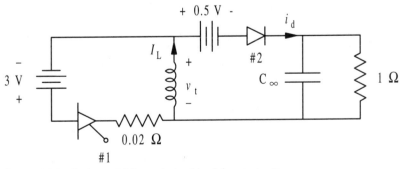

Figure 13.56 Static model for analysis of buck-boost circuit.

place. Without the snubber, the total circuit loss is almost 10 W, and the efficiency is 71%. With the snubber, the loss is 9.2 W, and efficiency is 73%. The efficiency improvement is not dramatic. However, the thermal issue is more significant. Without the snubber, the total loss in the FET is $D_1I_L^2(0.02\ \Omega) + 1.8\ \text{W} = 4.9\ \text{W}$. The junction temperature with 25 K/W from junction to ambient will be

$$T_j = 25°C + (25\ \text{K/W})(4.9\ \text{W}) = 148°C \tag{13.45}$$

This offers virtually no headroom under the typical 150°C maximum temperature for the MOSFET junction. With the snubber in place, the loss in the MOSFET is 3.1 W in the on state and only 0.53 W because of switching, for a total of 3.6 W. The junction temperature will be

$$T_j = 25°C + (25\ \text{K/W})(3.6\ \text{W}) = 115°C \tag{13.46}$$

still high but within the capability of the device.

The above analysis is only approximate. The best value of capacitance will be somewhat different compared to the result in equation (13.39). It is also important to recognize that the optimum capacitor depends on the converter load: it must be selected for a specific value of inductor current. In practice, the capacitor is often chosen higher than the optimum value. This reduces dissipation in the switch at the cost of higher total loss. The turn-off snubber in Figure 13.52 therefore supports a tradeoff between reliability (low switch dissipation) and efficiency. In the example, the small change in efficiency but large change in operating temperature motivates the use of a larger than optimum capacitor. In the circuit of Example 13.12.1, $C = 0.1\ \mu\text{F}$ and $R = 10\ \Omega$ would have further reduced junction temperature without significant penalty in total loss.

13.12.3 Turn-On Snubbers

The capacitive snubber of Figure 13.52 reduces the losses caused by the turn-off transition. Equally important, it avoids voltage overshoot and makes it easier to keep the semiconductor within its safe operating area during switching. The capacitor is needed because it stores energy as the switch enters its off state, when voltage appears across the switch. For turn-on snubbing, the dual of this behavior is needed: series inductance can store energy as the switch turns on. An appropriate inductor circuit should be able to alter the switching trajectory during the turn-on transient.

Figure 13.57 shows an inductor combined with a discharge path through a diode, connected as a turn-on snubber in a buck converter. The inductor limits the rate of rise of switch current during switching. After the switch turns off, the stored energy in the inductor dissipates in the resistor. Snubber action can be estimated by assuming that the switch voltage falls linearly during the turn-on transient, over a time interval t_{fv}, the "voltage fall time." (This time interval is determined mainly by circuit factors, but it is common practice to estimate it as equal to the current rise time t_r.) During the turn-on transient, the main diode in the converter must remain on, since its current is positive. The total voltage $V_{off} = V_{switch} + v_L$ must be blocked by the series combination of inductor and switch. Thus $v_L = V_{off} -$

Figure 13.57   Lossy turn-on snubber.

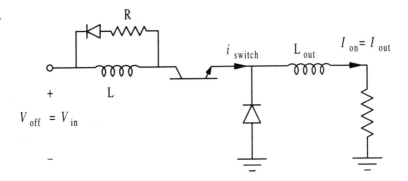

V_switch, and the linear approximation gives

$$v_L = V_\text{off}\frac{t}{t_\text{fv}} \tag{13.47}$$

The current is found by integration. During turn-on, the inductor current matches the switch current. The energy loss in the switch is

$$W_\text{switch} = \int_0^{t_\text{fv}} V_\text{switch}i_\text{switch}\,dt = \int_0^{t_\text{fv}} V_\text{off}\left(1 - \frac{t}{t_\text{fv}}\right)\left(\frac{V_\text{off}}{L}\frac{t^2}{2t_\text{fv}}\right)dt$$
$$= \frac{V^2_\text{off}t^2_\text{fv}}{24L} \tag{13.48}$$

As in the turn-off snubber, the switch loss can be made arbitrarily low by increasing the inductance. The total energy loss, however, includes the stored energy in the inductor, which must dissipate in the resistor while the switch is off. The total switching power loss, including snubber loss, is

$$P_\text{switch} = \left(\frac{V^2_\text{off}t^2_\text{fv}}{24L} + \frac{1}{2}LI^2_\text{on}\right)f_\text{switch} \tag{13.49}$$

As before, there is an optimum inductor, given by

$$L_\text{opt} = \frac{V_\text{off}t_\text{fv}}{\sqrt{12}\,I_\text{on}} \tag{13.50}$$

that minimizes total turn-on loss. The resistor must provide an L/R time constant fast enough to remove almost all the energy during the off interval. The value

$$\frac{L}{R} < \frac{(1 - D)T}{2}, \qquad R > \frac{2L}{(1 - D)T} \tag{13.51}$$

will ensure that at least 98% of the energy is removed.

The turn-on snubber analysis tends to be more approximate than for the turn-off snubber because the linear fall representation of voltage is simplistic. However, the analysis can establish a range for reasonable inductor values. Consider the following example.

Example 13.12.2   The IGBT in a PWM inverter application must block 325 V when off, and carries sinusoidal current of up to 200 A peak. The inverter operates a motor drive, and switches at 12 kHz. The IGBT current rise time is 500 ns. The device is provided with an internal reverse diode. Suggest a turn-on snubber for this application. Estimate the loss caused by the turn-on transient before and after the addition of this snubber.

Consider one half-bridge leg of the circuit (which is most likely a hex bridge for a three-phase motor, as shown in Figure 13.58). Since the load is a motor, the load current will not change much over the 500 ns switching interval. Without the snubber, turn-on at best requires a transition from 325 V to near zero, and from zero current to the full motor load current. The load current is a sinusoid with peak value of 200 A. If the current rises before the voltage falls, the energy loss at each turn-on point is

$$W_{switch} = \frac{V_{off}I_{on}t_{switch}}{2} = \frac{|200\cos(\omega_{out}t)\,A|(325\,V)(500\,ns)}{2} \tag{13.52}$$

The absolute value is inserted since switching is always lossy. If the motor current is negative, the loss will occur in the reverse diode. Parameters other than the current are constant, and will not affect the averaging process. Averaging this energy over a long time interval (such as one second), the total result is determined by the average of the absolute value of the current. The switching loss can be estimated as

$$P_{switch} = \frac{V_{off}|200\cos(\omega_{out}t)|t_{switch}f_{switch}}{2} = \frac{(325\,V)(400/\pi\,A)(500\,ns)(12\,kHz)}{2} \tag{13.53}$$

$$= 124\,W$$

The turn-on snubber should reduce losses, especially near the peak current. In this application, it is probably desirable to use an inductor larger than the optimum implied by the peak current, so that the snubber will still be effective at lower current levels. The optimum value at peak current is

$$L_{opt(200\,A)} = \frac{(325\,V)(500\,ns)}{\sqrt{12}\,(200\,A)} = 0.23\,\mu H \tag{13.54}$$

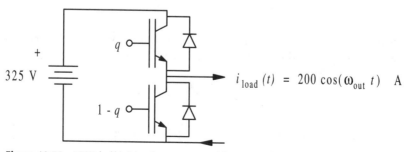

Figure 13.58   IGBT half-bridge for turn-on snubber example.

Consider the choice $L = 1 \ \mu\text{H}$—about a factor of four higher than optimum. The power loss in the semiconductor owing to turn-on is approximately

$$P_{\text{switch}} = \frac{(325 \ \text{V})^2(500 \ \text{ns})^2}{24(1 \ \mu\text{H})} f_{\text{switch}} = 13 \ \text{W} \tag{13.55}$$

The energy loss in the snubber resistor is $\frac{1}{2}Li^2$, or

$$W_{\text{r}} = \frac{1}{2}(1 \ \mu\text{H})[200 \cos(\omega_{\text{out}}t) \ \text{A}]^2 \tag{13.56}$$

The average of this loss over involves the average of the square of the current, so the resistor power loss can be found from the RMS current. The value is

$$P_{\text{r}} = \frac{1}{2}(1 \ \mu\text{H})(140 \ \text{A})^2(12 \ \text{kHz}) = 118 \ \text{W} \tag{13.57}$$

With this choice of inductor, the total loss during the turn-on transient increases slightly (less than 6%) compared to the snubberless case, but the loss in the semiconductor drops by nearly a factor of ten.

The resistor should dissipate the inductor energy quickly during the off interval. In the PWM context, one unfortunate aspect is that the off-time is shortest when the current is highest, so the resistor will need to be chosen based on the worst case. If the duty ratio does not exceed 98%, the off-time will not be less than $(1 - D)T = (0.02)/(12 \ \text{kHz}) = 1.7 \ \mu\text{s}$. A time constant of $0.83 \ \mu\text{s}$ should be sufficiently fast. With a $1 \ \mu\text{H}$ inductor, this requires $R > 1.2 \ \Omega$. The resistor rating will need to be well above 118 W.

The turn-on loss levels in this example are not small, and heat sink limitations will be a factor in circuit implementation. An important factor to remember is that switching loss— even with a snubber—is proportional to switching frequency. In many cases, the easiest way to reduce power loss is to decrease the switching frequency. The six switches in the motor drive of the Example 13.12.2 are associated with total losses of 750 W when the switching frequency is 12 kHz. If the frequency drops to 4 kHz, loss is cut by a factor of 3.

13.12.4 Combined Snubbers

The turn-off and turn-on snubbers of the preceding sections can be combined to form a single *unified snubber*, shown in Figure 13.59. The inductor avoids high peak currents during turn-on, while the capacitor prevents voltage overshoot during turn-off. To design the circuit, it is possible to consider the two portions of the snubber separately. One significant detail is that the resistor must dissipate the total snubber stored energy $\frac{1}{2}CV^2_{\text{off}} + \frac{1}{2}LI^2_{\text{on}}$, perhaps double the power when the two portions are added separately.

The unified circuit suggests an opportunity to transfer snubber energy between the inductor and capacitor. Indeed, the form provided in Figure 13.59 takes advantage of the R-L-C combination during the turn-on transient: the capacitor energy is transferred in part to the inductor. This can be used to make the snubber dissipation more uniform during the on state. It can also reduce the peak on-state current in the switch.

13.12.5 Lossless Snubbers

It seems a shame to throw away energy stored in snubber elements. In Example 13.12.2, for instance, the converter output power was about 30 kW. Total turn-on snubber losses were more than 700 W. Losses in a turn-off snubber would probably be similar, so total snubber losses approach 5% of the output. It is possible to recover the stored energy, leading to designs for *lossless snubbers* or *energy recovery snubbers*. A conceptual picture of a lossless turn-off snubber is shown in Figure 13.60. The recovery converter takes energy from the capacitor during the on interval, and returns it to the source or sends it to the load. A clever technique used in some applications is to recover the snubber energy to provide power for electronic controls. This is an efficient approach to control power that minimizes the need for external connections or power sources.

Formal converters for energy recovery are unusual in lossless snubbers because of the complexity. Charge-pump recovery circuits and resonant methods offer alternatives.

13.13 DC–DC CONVERTER DESIGN EXAMPLE

At this point in our study of power electronics, key conceptual issues, operating strategies, circuit configurations, analytical methods, and device considerations have been examined. The knowledge can be synthesized through converter designs. In this section, a boost converter design problem is covered in some detail. The design also serves to introduce engineering information as provided in manufacturers' data sheets. This is an *open-loop* converter, since it works without the control circuits and sensors needed to enforce regulation. In addition, *gate drives* will be needed to convert a given switching function into device action. Both gate drives and controls will be addressed in subsequent chapters. The dc–dc converter problem is used in the design example because it requires details of the static model, because switching loss is an important part of the issues, and because low ripple is required at both input and output. The general approach applies to an inverter, a rectifier, or to other classes of converters.

A dc–dc converter is to be built for a supercomputer disk drive application. The output is to be +24 V ± 50 mV, at loads up to 120 W. The nominal input is +12 V, but this must be tolerant enough to keep the drive running from a battery set. An input voltage range

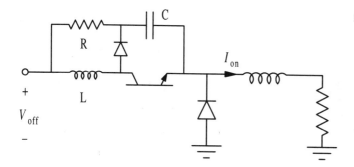

Figure 13.59   Unified snubber for buck converter.

Figure 13.60   Snubber with energy-recovery converter.

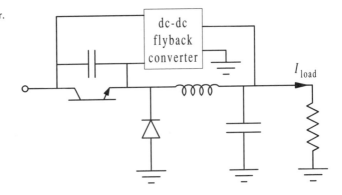

of $+12$ V $\pm$ 25% is proposed, and the customer would like to tolerate even lower input voltages if possible. The input current ripple is not as critical as output voltage ripple, but a value below 200 mA$_{\text{peak-to-peak}}$ has been suggested. The output should not deviate from the nominal 24 V even under no-load conditions. Common ground is permitted between the input and output. The operating location is climate-controlled to about 20°C, but the enclosure for the converter is slightly warmer, and it is important to ensure that the converter will not fail if the climate controller is not working. A range of 10°C to 40°C has been requested. We are asked to propose a circuit and design the components needed to implement it.

The power level and voltage requirements seem consistent with power MOSFETs. In the absence of further guidance, let us propose to use a MOSFET-based dc–dc converter switching at a frequency above 100 kHz. Final selection of the frequency will depend on the losses. With this choice, specifications are summarized in Table 13.3. A boost converter can meet the basic operating requirements, since $V_{\text{out}} > V_{\text{in}}$ under all conditions.

In a boost converter for this application, the output current will not exceed 5 A. The input current depends on load, and will be highest at the heaviest load and the lowest input voltage. Guessing an efficiency of about 85%, input of 140 W will be needed at full load. For 9 V input, this requires almost 16 A. If the input voltage is allowed to sag even lower—say to 6 V—the input current could reach 24 A. The voltage ratings in principle do not exceed 25 V, although headroom is needed for the switching trajectory. Most MOSFET rat-

TABLE 13.3
Specification Set for Section 13.13.2

Parameter	Nominal Value	Tolerance or Range
Input voltage	$+12$ V	$+9$ V to $+15$ V (lower values desired)
Output voltage	$+24$ V	± 50 mV
Output power	120 W full load	0 to 120 W
Input current	(depends on load)	± 200 mA
Switching frequency	100 kHz	Higher if possible to reduce size
Ambient temperature	25°C	10°C to 40°C
Grounding	Common input–output ground	—

Designer's™ Data Sheet
TMOS E-FET ™
Power Field Effect Transistor
N–Channel Enhancement–Mode Silicon Gate

MTP50N05E
Motorola Preferred Device

This advanced TMOS E–FET is designed to withstand high energy in the avalanche and commutation modes. The new energy efficient design also offers a drain–to–source diode with a fast recovery time. Designed for low voltage, high speed switching applications in power supplies, converters and PWM motor controls, these devices are particularly well suited for bridge circuits where diode speed and commutating safe operating areas are critical and offer additional safety margin against unexpected voltage transients.

- Avalanche Energy Specified
- Commutating Safe Operating Area (CSOA) Specified for Use in Half and Full Bridge Circuits
- Source–to–Drain Diode Recovery Time Comparable to a Discrete Fast Recovery Diode
- Diode is Characterized for Use in Bridge Circuits
- I_{DSS} and $V_{DS(on)}$ Specified at Elevated Temperature

**TMOS POWER FET
50 AMPERES
$R_{DS(on)}$ = 0.028 OHM
50 VOLTS**

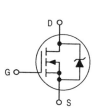

**CASE 221A–06, Style 5
TO–220AB**

MAXIMUM RATINGS (T_C = 25°C unless otherwise noted)

Rating	Symbol	Value	Unit
Drain–to–Source Voltage	V_{DSS}	50	Volts
Drain–to–Gate Voltage (R_{GS} = 1.0 MΩ)	V_{DGR}	50	Volts
Gate–to–Source Voltage — Continuous	V_{GS}	±20	Volts
Drain Current — Continuous — Continuous @ 100°C — Pulsed	I_D I_D I_{DM}	50 33A 160	Adc Apk
Total Power Dissipation Derate above 25°C	P_D	125 0.83	Watts W/°C
Operating and Storage Temperature Range	T_J, T_{stg}	−55 to 175	°C
Single Pulse Drain–to–Source Avalanche Energy — Starting T_J = 25°C (V_{DD} = 25 Vdc, V_{GS} = 10 Vpk, I_L = 50 Apk, L = 0.32 mH, R_G = 25 Ω)	E_{AS}	400	mJ
Thermal Resistance — Junction to Case — Junction to Ambient	$R_{\theta JC}$ $R_{\theta JA}$	1.2 62.5	°C/W
Maximum Lead Temperature for Soldering Purposes, 1/8″ from case for 10 seconds	T_L	260	°C

Designer's Data for "Worst Case" Conditions — The Designer's Data Sheet permits the design of most circuits entirely from the information presented. SOA Limit curves — representing boundaries on device characteristics — are given to facilitate "worst case" design.

Figure 13.61a Manufacturer Data Sheets, MTP50N05 (reprinted with permission of Motorola Corporation).

ELECTRICAL CHARACTERISTICS (T_J = 25°C unless otherwise noted)

Characteristic	Symbol	Min	Typ	Max	Unit
OFF CHARACTERISTICS					
Drain–to–Source Breakdown Voltage (V_{GS} = 0 V, I_D = 250 µAdc) Temperature Coefficient (Positive)	$V_{(BR)DSS}$	50 —	— 60	— —	Vdc mV/°C
Zero Gate Voltage Drain Current (V_{DS} = 50 Vdc, V_{GS} = 0 Vdc) (V_{DS} = 50 Vdc, V_{GS} = 0 Vdc, T_J = 150°C)	I_{DSS}	— —	— —	10 100	µAdc
Gate–Body Leakage Current (V_{GS} = ±20 Vdc, V_{DS} = 0)	I_{GSS}	—	—	100	nAdc
ON CHARACTERISTICS (1)					
Gate Threshold Voltage (V_{DS} = V_{GS}, I_D = 250 µAdc) Threshold Temperature Coefficient (Negative)	$V_{GS(th)}$	2.0 —	2.7 6.0	4.0 —	Vdc mV/°C
Static Drain–to–Source On–Resistance (V_{GS} = 10 Vdc, I_D = 25 Adc)	$R_{DS(on)}$	—	0.022	0.028	Ohm
Drain–to–Source On–Voltage (V_{GS} = 10 Vdc) (I_D = 50 Adc) (I_D = 25 Adc, T_J = 150°C)	$V_{DS(on)}$	— —	— —	1.68 1.4	Vdc
Forward Transconductance (V_{DS} ≥ 8.0 Vdc, I_D = 25 Adc)	g_{FS}	17	22	—	mhos
DYNAMIC CHARACTERISTICS					
Input Capacitance	(V_{DS} = 25 Vdc, V_{GS} = 0, f = 1.0 MHz)	C_{iss}	—	2000	3000	pF
Output Capacitance		C_{oss}	—	950	1350	
Transfer Capacitance		C_{rss}	—	290	580	
SWITCHING CHARACTERISTICS (2)					
Turn–On Delay Time	(V_{DD} = 30 Vdc, I_D = 50 Adc, V_{GS} = 10 Vdc, R_G = 9.1 Ω)	$t_{d(on)}$	—	27	50	ns
Rise Time		t_r	—	154	300	
Turn–Off Delay Time		$t_{d(off)}$	—	72	150	
Fall Time		t_f	—	99	200	
Gate Charge	(V_{DS} = 40 Vdc, I_D = 50 Adc, V_{GS} = 10 Vdc)	Q_T	—	68	75	nC
		Q_1	—	10	—	
		Q_2	—	30	—	
		Q_3	—	20	—	
SOURCE–DRAIN DIODE CHARACTERISTICS					
Forward On–Voltage	(I_S = 50 Adc, V_{GS} = 0) (I_S = 50 Adc, V_{GS} = 0, T_J = 150°C)	V_{SD}	— —	1.15 1.1	2.5 —	Vdc
Reverse Recovery Time	(I_S = 50 Adc, dI_S/dt = 100 A/µs)	t_{rr}	—	80	—	ns
INTERNAL PACKAGE INDUCTANCE					
Internal Drain Inductance (Measured from the drain lead 0.25" from package to center to die)	L_D	—	4.5	—	nH
Internal Source Inductance (Measured from the source lead 0.25" from package to source bond pad)	L_S	—	7.5	—	nH

(1) Pulse Test: Pulse Width ≤ 300 µs max, Duty Cycle = 2.0%.
(2) Switching characteristics are independent of operating junction temperature.

Figure 13.61a *Continued*

TYPICAL ELECTRICAL CHARACTERISTICS

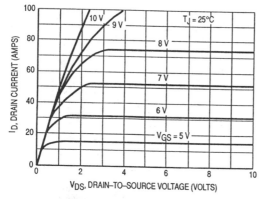

Figure 1. On–Region Characteristics

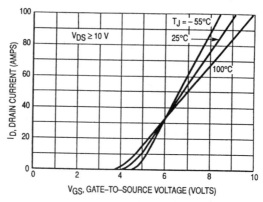

Figure 2. Transfer Characteristics

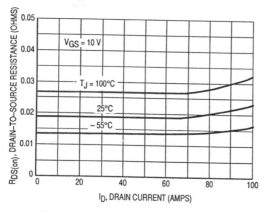

Figure 3. On–Resistance versus Drain Current

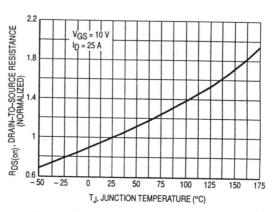

Figure 4. On–Resistance Variation With Temperature

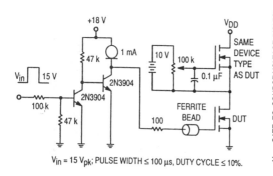

Figure 5. Gate Charge Test Circuit

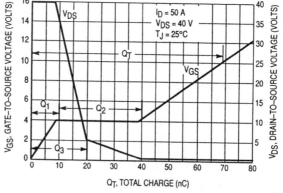

Figure 6. Gate–to–Source and Drain–to–Source Voltage versus Total Charge

Figure 13.61b Manufacturer Data Sheets, MTP50N05 (reprinted with permission of Motorola Corporation).

SAFE OPERATING AREA INFORMATION

FORWARD BIASED SAFE OPERATING AREA

The FBSOA curves define the maximum drain–to–source voltage and drain current that a device can safely handle when it is forward biased, or when it is on, or being turned on. Because these curves include the limitations of simultaneous high voltage and high current, up to the rating of the device, they are especially useful to designers of linear systems. The curves are based on a case temperature of 25°C and a maximum junction temperature of 150°C. Limitations for repetitive pulses at various case temperatures can be determined by using the thermal response curves. Motorola Application Note, AN569, "Transient Thermal Resistance–General Data and Its Use" provides detailed instructions.

SWITCHING SAFE OPERATING AREA

The switching safe operating area (SOA) of Figure 9 is the boundary that the load line may traverse without incurring damage to the MOSFET. The fundamental limits are the peak current, I_{DM} and the breakdown voltage, BV_{DSS}. The switching SOA shown in Figure 9 is applicable for both turn–on and turn–off of the devices for switching times less than one microsecond.

The power averaged over a complete switching cycle must be less than:

$$\frac{T_{J(max)} - T_C}{R_{\theta JC}}$$

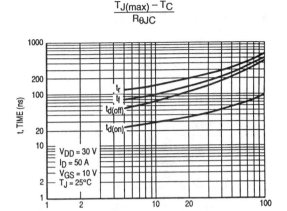

Figure 7. Resistive Switching Time Variation versus Gate Resistance

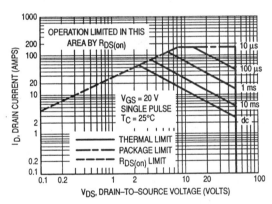

Figure 8. Maximum Rated Forward Biased Safe Operating Area

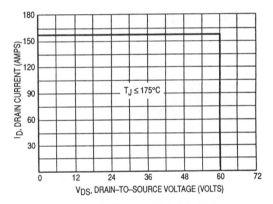

Figure 9. Maximum Rated Switching Safe Operating Area

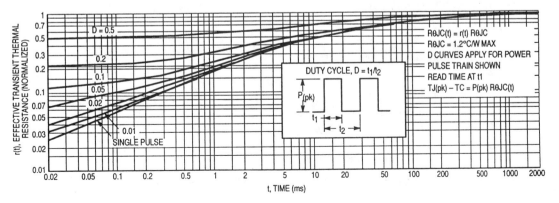

Figure 10. Thermal Response

Figure 13.61*b*   *Continued*

COMMUTATING SAFE OPERATING AREA (CSOA)

The Commutating Safe Operating Area (CSOA) of Figure 12 defines the limits of safe operation for commutated source–drain current versus re–applied drain voltage when the source–drain diode has undergone forward bias. The curve shows the limitations of I_{FM} and peak V_{DS} for a given rate of change of source current. It is applicable when waveforms similar to those of Figure 11 are present. Full or half–bridge PWM DC motor controllers are common applications requiring CSOA data.

Device stresses increase with increasing rate of change of source current so dI_S/dt is specified with a maximum value. Higher values of dI_S/dt require an appropriate derating of I_{FM}, peak V_{DS} or both. Ultimately dI_S/dt is limited primarily by device, package, and circuit impedances. Maximum device stress occurs during t_{rr} as the diode goes from conduction to reverse blocking.

$V_{DS(pk)}$ is the peak drain–to–source voltage that the device must sustain during commutation; I_{FM} is the maximum forward source–drain diode current just prior to the onset of commutation.

V_R is specified at rated $V_{(BR)DSS}$ to ensure that the CSOA stress is maximized as I_S decays from I_{RM} to zero.

R_{GS} should be minimized during commutation. T_J has only a second order effect on CSOA.

Stray inductances in Motorola's test circuit are assumed to be practical minimums. dV_{DS}/dt in excess of 10 V/ns was attained with dI_S/dt of 400 A/μs.

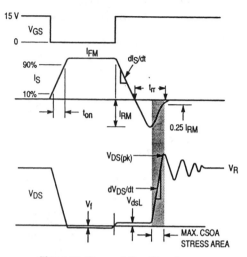

Figure 11. Commutating Waveforms

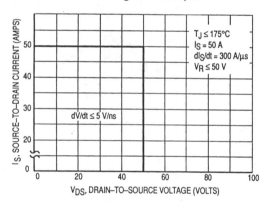

Figure 12. Commutating Safe Operating Area (CSOA)

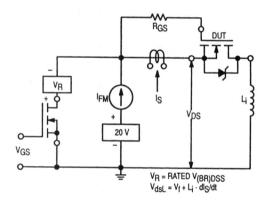

Figure 13. Commutating Safe Operating Area Test Circuit

Figure 13.61c Manufacturer Data Sheets, MTP50N05 (reprinted with permission of Motorola Corporation).

ings start at about 50 V. Data sheets for an MTP50N05 MOSFET and a Schottky diode of similar ratings are provided in Figures 13.61 and 13.62, respectively.

The data sheets provide information for static models for the two switching devices. The MOSFET has $R_{ds(on)} = 0.028$ Ω at 25°C, and can carry 40 A with low drop if its gate-source voltage is 8 V or more. One important point (shown as Figure 4 within the data sheet) is that $R_{ds(on)}$ rises significantly with junction temperature. At $T_j = 100$°C, for instance, the

MTP50N05E

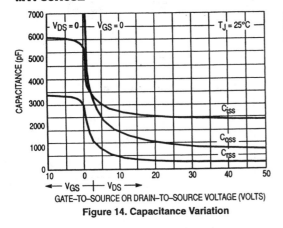

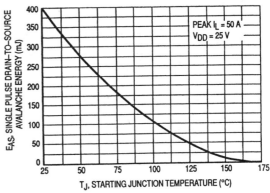

Figure 14. Capacitance Variation

Figure 15. Maximum Avalanche Energy versus Starting Junction Temperature

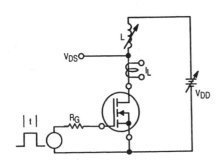

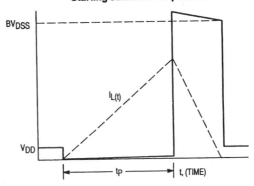

Figure 16. Unclamped Inductive Switching Test Circuit

Figure 17. Unclamped Inductive Switching Waveforms

Figure 13.61c *Continued*

resistance increases by nearly 40%. The Schottky diode is actually a pair of devices with common cathode, and both will be used in parallel to minimize forward drop at a possible peak level of 24 A. The diode forward characteristic curve (shown as Figure 1 within the data sheet) can be modelled as a 0.2 V source in series with an 0.015 Ω resistor if the junction temperature is about 100°C. The inductor and capacitor will be expected to include series resistance. A circuit diagram, including static models, is given in Figure 13.63.

The converter switches manipulate the voltage v_t and the current i_d. Assuming that the converter operates in continuous conduction mode with $I_L > 0$, the switching function expressions are

$$v_t = q_1 I_L + q_2 (V_{\text{out}} + I_L R_d + V_d),$$

$$i_d = q_2 I_L$$

(13.58)

MBR3035PT
MBR3045PT

MOTOROLA

SWITCHMODE POWER RECTIFIERS

... using the Schottky Barrier principle with a platinum barrier metal. These state-of-the-art devices have the following features:

- Dual Diode Construction — Terminals 1 and 3 May Be Connected For Parallel Operation At Full Rating
- Guardring For Stress Protection
- Low Forward Voltage
- 150°C Operating Junction Temperature
- Guaranteed Reverse Avalanche

SCHOTTKY BARRIER RECTIFIERS

30 AMPERES
35 to 45 VOLTS

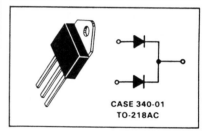

CASE 340-01
TO-218AC

1. ANODE 1
2. CATHODE(S)
3. ANODE 2
4. CATHODE(S)

RATINGS

Rating		Symbol	Maximum	Unit
Peak Repetitive Reverse Voltage Working Peak Reverse Voltage DC Blocking Voltage	MBR3035PT MBR3045PT	V_{RRM} V_{RWM} V_R	35 45	Volts
Average Rectified Forward Current Per Device (Rated V_R) T_C = 105°C Per Diode		$I_{F(AV)}$	30 15	Amps
Peak Repetitive Forward Current, Per Diode (Rated V_R, Square Wave, 20 kHz)		I_{FRM}	30	Amps
Nonrepetitive Peak Surge Current (Surge Applied at rated load conditions halfwave, single phase, 60 Hz)		I_{FSM}	200	Amps
Peak Repetitive Reverse Current, Per Diode (2.0 μs, 1.0 kHz) See Figure 6		I_{RRM}	2.0	Amps
Operating Junction Temperature		T_J	−65 to +150	°C
Storage Temperature		T_{stg}	−65 to +175	°C
Peak Surge Junction Temperature (Forward Current Applied)		$T_{J(pk)}$	175	°C
Voltage Rate of Change (Rated V_R)		dv/dt	1000	V/μs

THERMAL CHARACTERISTICS PER DIODE

	Symbol			Unit
Thermal Resistance, Junction to Case	$R_{\theta JC}$		1.4	°C/W
Thermal Resistance, Junction to Ambient	$R_{\theta JA}$		40	°C/W

ELECTRICAL CHARACTERISTICS PER DIODE

	Symbol			Unit
Instantaneous Forward Voltage (1) (i_F = 20 Amp, T_C = 125°C) (i_F = 30 Amp, T_C = 125°C) (i_F = 30 Amp, T_C = 25°C)	v_F		0.60 0.72 0.76	Volts
Instantaneous Reverse Current (1) (Rated dc Voltage, T_C = 125°C) (Rated dc Voltage, T_C = 25°C)	i_R		100 1.0	mA

(1) Pulse Test: Pulse Width = 300 μs, Duty Cycle ⩽ 2.0%
Switchmode is a trademark of Motorola Inc.

DIM	MILLIMETERS		INCHES	
	MIN	MAX	MIN	MAX
A	20.32	21.08	0.800	0.830
B	15.49	15.90	0.610	0.626
C	4.19	5.08	0.165	0.200
D	1.02	1.65	0.040	0.065
E	1.35	1.65	0.053	0.065
G	5.21	5.72	0.205	0.225
H	2.41	3.20	0.095	0.126
J	0.38	0.64	0.015	0.025
K	12.70	15.49	0.500	0.610
L	15.88	16.51	0.625	0.650
N	12.19	12.70	0.480	0.500
Q	4.04	4.22	0.159	0.166

CASE 340-01
TO-218AC

Figure 13.62 Data sheets for MBR3045PT Schottky diode (reprinted with permission of Motorola Corporation).

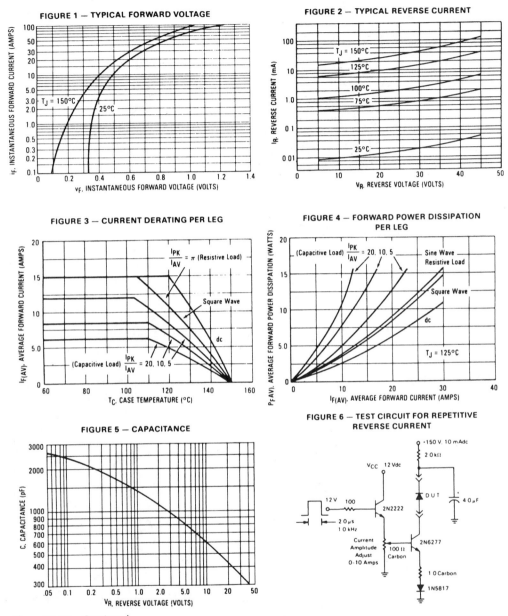

FIGURE 1 — TYPICAL FORWARD VOLTAGE

FIGURE 2 — TYPICAL REVERSE CURRENT

FIGURE 3 — CURRENT DERATING PER LEG

FIGURE 4 — FORWARD POWER DISSIPATION PER LEG

FIGURE 5 — CAPACITANCE

FIGURE 6 — TEST CIRCUIT FOR REPETITIVE REVERSE CURRENT

Figure 13.62 *Continued*

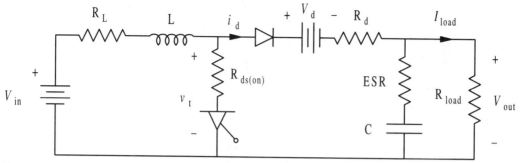

Figure 13.63   Static model of boost converter.

The average relationships require

$$\langle v_t \rangle = V_{in}\, I_L R_L = D_1 I_L R_{ds(on)} + (1 - D_1)(V_{out} + I_L R_d + V_d),$$

$$\langle i_d \rangle = I_{load} = (1 - D_1) I_L \tag{13.59}$$

This determines relationships among the input and output voltages and currents. Equation (13.59) should be used to estimate converter operation, but some of the values are unknown. Let us try to provide approximate values for them. Based on Table 11.1, the inductor will probably be wound with the equivalent of #12 AWG wire. The resistance of wire this big is 5.3 mΩ/m. Thus, R_L is likely to be only a few milliohms. Internal resistance of a backup battery or other interconnect resistance might be larger, so let $R_L = 20$ mΩ cover unexpected drops and determine the lower limit of V_{in}. Let $R_{ds(on)} = 0.040\ \Omega$ to account for the expected rise in junction temperature. At full load, $I_{load} = 5$ A. With this value, the inductor current is (5 A)/(1 − D_1), and the voltage relationship is

$$V_{in} - \frac{5\ A}{1 - D_1}(0.020\ \Omega)$$

$$= D_1 \frac{5\ A}{1 - D_1}(0.040\ \Omega) + (1 - D_1)(24.2\ V) + (5\ A)(0.015\ \Omega) \tag{13.60}$$

This can rewritten as

$$V_{in}(V) = \frac{D_1}{1 - D_1}0.2 + (1 - D_1)24.2 + 0.075 + \frac{0.10}{1 - D_1} \tag{13.61}$$

The expression for V_{in} as a function of D_1 is plotted in Figure 13.64. At full load, there is a minimum value of V_{in} below which the converter will not produce the necessary 24 V output. This minimum value is 4.38 V. Table 13.4 provides a summary of the operating results for loads above 12 W. The power loss listed in the table is static loss only. Switching losses and loss in the capacitor ESR need to be estimated later.

Consider the choice of inductor. Both ripple requirements and discontinuous mode operation are potential issues in the choice of inductor. If the ripple is set at the limit of ±200 mA,

Figure 13.64 Values of V_{in} to provide 24 V output for various duty ratios.

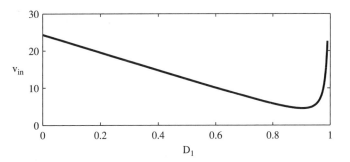

then inductor currents as low as 200 mA will maintain $I_L > 0$ and keep the converter out of discontinuous mode. With the 15 V input, 200 mA inductor current translates to about 0.125 A of load current, or 3.0 W. This is only 2.5% of full load. Since it is desired to maintain the 24 V output even with no load, a *ballast load* will be needed so that a nonzero load current is maintained when no external connection is made. To avoid a significant effect on efficiency, a ballast load equal to about 1%, or 1.2 W, is appropriate. A 470 Ω resistor is close to the correct value. With this ballast in place, the minimum load current is 0.051 A, and the minimum input current will be about 0.082 A. If the inductor is oversized to keep the current ripple below ± 0.082 A, the converter will not operate in discontinuous mode under any conditions. While this choice is somewhat arbitrary, it will tend to simplify the control later on, so let us choose L to keep the ripple below ± 0.082 A, or $\Delta i_L < 0.164$ A.

With switch #1 on, the inductor voltage is slightly less than V_{in}. The target ripple value requires

$$V_{in} = L\frac{\Delta i_L}{\Delta t}, \qquad L > \frac{V_{in}D_1 T}{0.164 \text{ A}} \tag{13.62}$$

Notice that V_{in} and D_1 are not independent in this expression. The highest value of the right-hand side will be needed to ensure that a sufficiently large inductor is selected. Equation (13.61) can be used as the basis for maximizing the product $D_1 V_{in}$. The maximum occurs when $V_{in} = 12.0$ V, which occurs with $D_1 = 0.520$ for full load. The inductor value should be at least

TABLE 13.4
Operating Parameters for Boost Converter

Input Voltage	Load Current	Duty Ratio	Input Current	Power Loss
9 V	0.5 A	0.630	1.35	0.17
	2 A	0.637	5.51	1.56
	5 A	0.651	14.32	8.84
12 V	0.5 A	0.506	1.01	0.14
	2 A	0.510	4.09	1.03
	5 A	0.520	10.42	5.09
15 V	0.5 A	0.381	0.81	0.12
	2 A	0.385	3.25	0.79
	5 A	0.393	8.23	3.51

$$L > 12 \text{ V} \frac{0.520}{0.164 \text{ A}} T, \qquad L > (38T) \text{ H} \tag{13.63}$$

given this maximum. For 100 kHz switching, this requires about 380 μH. For 200 kHz switching, 190 μH will be needed, and so on.

The inductor current can range up to 15 A under normal conditions, from Table 13.4. The stored energy $\frac{1}{2}Li^2$ therefore reaches 0.043 J with 100 kHz switching, 0.021 J with 200 kHz switching, and so on. Consider a powdered iron material, with $\mu = 25\mu_0$. With $B_{\text{sat}} = 1$ T, at 100 kHz the total core volume will need to be 2.70 cm^3 to store 0.043 J. A toroidal core with thickness 12 mm, outside diameter 50 mm, and inside diameter 25 mm will be more than large enough. For powdered iron, a switching frequency of 100 kHz should not be excessive, but higher values could increase losses significantly.

The core has reluctance $\mathcal{R} = l/(\mu A) = 2.60 \times 10^7$ H^{-1}. To provide $L = N^2/\mathcal{R} = 380$ μH, this reluctance requires 100 turns. Given #12 wire and a window fill factor of 50%, 100 turns will need a window area of at least 662 mm^2. For this core, the actual window area is 491 mm^2, so the wire will not fit. Instead, choose an inside diameter of 30 mm. The reluctance increases to 3.33×10^7 H^{-1}, and the number of turns must rise to 112. However, the turns will fit. Now, check saturation:

$$\frac{Ni}{\mathcal{R}} = \phi = BA, \qquad B < B_{\text{sat}} = 1 \text{ T} \tag{13.64}$$

For 112 turns, 15 A, $\mathcal{R} = 3.3 \times 10^7$, and $A = 120$ mm^2, the maximum flux density is 0.42 T —well below the saturation value. It appears that this core, with 112 turns of #12 wire, should meet the inductor requirements.

The capacitor must enforce 100 mV peak-to-peak output ripple. When the diode is off, the capacitor carries the full load current. The worst case condition will be the highest load and the highest transistor duty ratio. We have

$$i_{\text{load}} = C \frac{\Delta v_{\text{out}}}{D_1 T}, \qquad \Delta v_{\text{out}} < 100 \text{ mV} \tag{13.65}$$

The worst case is 9 V input and 5 A load, in which case the requirement is

$$C > (5 \text{ A}) \frac{0.651}{100 \text{ mV}} T, \qquad C > (32.6T) \text{ F} \tag{13.66}$$

For 100 kHz switching, this gives 326 μF. If a capacitor of this value has tan $\delta = 0.20$, the ESR will be (tan δ)/(ωC) = (0.20)/($2\pi f_{\text{switch}}C$). At 100 kHz, this gives ESR = 0.98 mΩ. More likely, the ESR will be dominated by lead resistance, so let us plan on 2 mΩ. The ESR jump will be I_L(ESR), with a worst-case value of (15 A)(2 mΩ) = 30 mV. The capacitor ripple therefore should be 70 mV instead of 100 mV, and $C > 466$ μF will be necessary. A commercial 470 μF device will be useful. The power loss in the ESR will not be significant in this problem. Even if the worst case 15 A flows, the loss is on the order of 100 mW.

Switching loss needs to be estimated. We might use a typical commutation parameter

$a = 2$. The MOSFET data sheets report typical rise times of 154 ns and fall times of 99 ns. Here is an estimate for the MOSFET:

$$P_{switch} \approx \frac{I_{on}V_{off}t_{switch}f_{switch}}{2} = \frac{(15 \text{ A})(24 \text{ V})(253 \text{ ns})(100 \text{ kHz})}{2} \quad (13.67)$$

This comes out to 4.6 W at the highest inductor current. The diode is faster, with a reverse recovery time of 60 ns. The data sheet does not provide reverse recovery charge, but a total switching time of 120 ns suggests 2.3 W as a reasonable worst-case estimate.

The total losses become 15.7 W under high load and low input, for efficiency of 88.4%. At nominal 12 V input, the losses can be found to be 10.1 W at the highest load, for nominal full-load efficiency of 92 %. There will be loss in the inductor, probably equal to another watt.

The MOSFET dissipation includes on-state loss and switching loss,

$$P_{MOSFET} = D_1 I_L{}^2 R_{ds(on)} + \frac{I_L(24 \text{ V})(253 \text{ ns})(100 \text{ kHz})}{2} = D_1 I_L{}^2(0.04 \text{ } \Omega) + 0.30 I_L \quad (13.68)$$

Since I_L is the load current divided by $1 - D_1$, this is worst for the highest value of D_1. This occurs under heavy load and low input. The loss reaches 10.5 W. The transistor has thermal resistance $R_{\theta ja} = 62.5$ K/W, so a heat sink will be essential. In this application, MOSFET losses tend to dominate. It would be worthwhile to identify a lower resistance part. Table 13.5 summarizes the results.

13.14 RECAP

Real power semiconductor switches exhibit an on state, and off state, and a *commutation* state representing the transition between them. In the on state, current ratings are important. Devices often have separate ratings for average current, RMS current, and peak current. These reflect different loss mechanisms and physical limitations of real devices. The *residual voltage* is the switch voltage drop during the on state, and determines the *on-state loss*. In the off state, voltage ratings are critical. The voltage rating is usually related to avalanche current or similar voltage breakdown effects. Because of this, devices usually have hard limits on the handling voltage. The *residual current* represents leakage current that contributes to *off-state loss*.

TABLE 13.5
Design Values and Results for Boost Converter

Parameter	Value	Comments
Switching frequency	100 kHz	Powdered iron core.
Inductance	380 μH	Uses a 1.2 W ballast, and avoids discontinuous mode.
Capacitance	470 μF	Needs a ripple current rating approaching 10 A.
Nominal efficiency	92%	Full load, 12 V input.
Operating range	7 V to more than 15 V	Inductor has headroom for saturation. Transistor will require extra heat sink.

Loss is generated during commutation because the voltages and currents evolve along nonzero trajectories as they make the transition back and forth from on to off. These *switching trajectories* must remain within a device's *safe operating area* to avoid violations of voltage or current ratings. In typical power converters, inductive and capacitive elements cause the trajectories to swing beyond the voltage and current source values in the circuit. This makes the safe operating area an important consideration in dynamic behavior. It also makes loss generated by commutation sensitive to load changes and to design choices.

Static models are important for analysis and design of power converters. For static models, the residual voltage and current are captured in circuit form. Static circuit models consist of restricted switches, with series voltages and series and parallel resistances to represent loss behavior. A typical example is a piecewise linear diode model, in which a real diode is represented as a series combination of an ideal diode, a voltage drop, and a resistor. Static models are effective ways to include the most important loss effects. They help reveal operating limitations and nonideal behavior of converters. It is possible to include off-state losses in static models. However, the residual current in a typical power semiconductor is many orders of magnitude lower than the on-state current, and the off-state effects are smaller than accuracy limits of the on-state models. Losses in the off state can be neglected in all but a few specialized cases.

An interesting byproduct of static modelling is the behavior of a MOSFET. A real MOSFET is a bilateral device, although the reverse-body diode limits the control capability in the reverse direction. The *synchronous rectification* technique makes use of the bilateral behavior to obtain very low residual voltages in rectifier applications. The MOSFET is turned on whenever its reverse diode attempts to turn on.

While static models address on-state and off-state loss, a separate approach is needed to analyze switching loss. The loss during a switching transition is an integral of power along the switching trajectory. The trajectory is unknown and sensitive to circuit characteristics, but it is possible to gain insight into the loss process. In this way, helpful estimates of loss can be developed. One basis for analysis is the case of *linear commutation*, in which the voltage and current make a linear transition from an initial value to a final value over a time interval t_{switch}. The time behavior of linear commutation appears as a straight line connecting the initial and final points in a current vs. voltage representation of the switching trajectory. When linear commutation is analyzed, the result show that switching loss is proportion to the *power handling value*, $I_{\text{on}}V_{\text{off}}$, multiplied by the ratio t_{switch}/T, and divided by a *commutation parameter, a*. In the linear case, $a = 6$. If instead an inductor current is to be switched, the voltage and current will change at disparate times, and the switching trajectory will appear as a rectangle with area $I_{\text{on}}V_{\text{off}}$. The loss becomes

$$P_{\text{switch}} = \frac{I_{\text{on}}V_{\text{off}}t_{\text{switch}}f_{\text{switch}}}{a} \tag{13.69}$$

with $a \approx 2$ in the inductive case. The value $a = 2$ serves as a reasonable estimate of switching loss in circuits. The time t_{switch} represents the sum of a device's turn-on and turn-off times.

In rectifiers and inverters, the switch currents and voltages are not fixed, and (13.69) is overly simple. However, power requires frequency matching. In switches with constant residual voltages, the on-state loss is the product of this voltage and the average current flow.

If a resistive element is present in the static model, an I^2R term is more appropriate. Switching loss is proportional to the difference in voltage before and after commutation, and to the difference in current before and after.

Switch losses raise the issue of heat transfer. This is a complete engineering discipline in its own right, but it is possible to establish a basic design framework for power semiconductors with simplified models. Of the three major heat transfer mechanisms—conduction, convection, and radiation—the first two involve heat flow that is proportional to a temperature difference. In terrestrial applications, radiation effects are small. The linear proportional models for conduction and convection support a definition of *thermal conductance* or *thermal resistance*. Heat energy flows between two objects or materials at a rate proportional to temperature difference and to any resistance to this power flow. The basic expression is

$$P_{\text{heat}} = \frac{\Delta T}{R_\theta} \tag{13.70}$$

where ΔT is the temperature difference and R_θ is a thermal resistance in kelvins per watt. The analysis captures the basic physical process: If a thermal insulator creates a barrier, heat flow requires a high temperature difference.

By conservation of energy, if a semiconductor device reaches a steady-state temperature, any heat generated inside it will flow to the outside world (if it did not flow out, the temperature would increase until the flow was forced to occur). Therefore any losses in a device ultimately must reach the ambient air (or water in marine applications). At the semiconductor junction where the heat is generated,

$$P_{\text{loss}}R_{\theta(\text{ja})} = T_{\text{junction}} - T_{\text{ambient}} \tag{13.71}$$

where T_{junction} is the hot-spot temperature inside the device, T_{ambient} is the temperature in the surrounding medium, and $R_{\theta(\text{ja})}$ is the thermal resistance from junction to ambient. Any semiconductor has a maximum junction temperature. A device will degrade or fail if operated at $T_{\text{junction}} > T_{\text{max}}$. Thus it is required that

$$P_{\text{loss}}R_{\theta(\text{ja})} + T_{\text{ambient}} < T_{\text{max}} \tag{13.72}$$

Power semiconductors have various maximum temperatures. The typical temperature limit for an SCR is 125°C, while that for a MOSFET or IGBT is about 150°C. If a power supply is enclosed in another instrument or exposed to a warm environment, ambient temperatures of 50°C or more might be encountered. Many designers specify a *maximum temperature rise* to help account for ambient effects. For example, if a designer attempts to keep the temperature rise $T_{\text{junction}} - T_{\text{ambient}}$ under 80°C, the product $P_{\text{loss}}R_{\theta(\text{ja})}$ must be less than 80 K. For a typical maximum temperature of 150°C, a specified 80 K rise supports ambient temperatures up to 70°C.

Power semiconductor packages are effective in transferring heat from the junction to the package surface (to the case) but are less effective in transferring heat to the surrounding air. A typical device in a TO-220 package, for instance, has a thermal resistance from junction to case $R_{\theta(\text{jc})}$ of about 1 K/W. This means that 1 W of power loss will give rise to only 1 K difference between the outside case temperature and the inside junction temperature. Power loss up to 80 W could be handled without violating an 80 K rise limit—if heat can be removed perfectly from the case. The same package has typical junction to ambient resistance of about 50 K/W. With no heat sink present, if the part dissipates 2 W, the junction will be 100 K hotter than the ambient. Power dissipation of 2 W without a heat sink is typical for this particular package.

A heat sink reduces the thermal resistance from case to ambient by providing a more effective path for heat flow. The value $R_{\theta(\text{ca})}$ adds in series to the package thermal resistance $R_{\theta(\text{jc})}$ to form the overall junction to ambient resistance. A heat sink enhances convective transfer to the air by adding surface area and extended fins. Small heat sinks suitable for direct attachment to a TO-220 package will drop the overall thermal resistance to 20 K/W. Large heat sinks with external fans are usually needed to achieve thermal resistances on the order of 1 K/W from case to ambient.

The physics of P-N junctions produce dynamic effects as well as static drops and switching losses. The dynamic effects can be interpreted through depletion and diffusion capacitance values. The *depletion capacitance* is associated with the charge depletion region near a forward-biased junction. This capacitance can be substantial, because the depletion region is narrow and the charges at its edges are substantial. To turn off a junction, the charge on the depletion capacitance must be removed. This gives rise to a *reverse recovery process* seen in diodes and other junction devices. Power diodes have *reverse recovery time*, t_{rr}, and must give up a *reverse recovery charge* during this time before turn-off is complete. The energy levels contribute to switching loss. Times are highly process dependent. There are *ultrafast* devices with recovery times of a few tens of nanoseconds, and more conventional rectifiers with recovery times of several microseconds.

The *diffusion capacitance* is associated with charge present at the ends of the recombination region formed within a reverse-biased junction. The value is considerably less than the depletion capacitance, since the charges are farther apart. However, the charge must be removed and a depletion region must be established to turn on a junction. This represents a *forward recovery* process, and is associated with a *forward recovery time*, t_{fr}. Many manufacturers do not specify t_{fr} since it is so much faster than the reverse recovery.

P-N junctions, often with a small *intrinsic region* in between (P-i-N diodes), make excellent diodes. Devices with current ratings up to about 10 kA and voltage ratings up to 10 kV are readily available. Devices rated for 600 V and 20 A are common and inexpensive. In low-voltage applications, junction diodes compete with *Schottky diodes*, formed as a junction between a metal and a semiconductor. Schottky diodes have lower forward drops than junction diodes, but their blocking capability is relatively limited, and off-state current is high. Even so, Schottky diodes are common in power supplies and dc–dc converters rated at 5 V or less.

Thyristors are P-N-P-N multilayer devices with a latching characteristic. The SCR is one of the most familiar examples, but there are several other classes of thyristors. The SCR can be understood with a two-transistor model with a cascaded PNP and NPN connection. One of the transistor base terminals serves as the device gate. If a brief gate pulse is applied, internal gain will generate enough base current to maintain an on condition until current is removed. The transistors do not need much gain (the only requirement is that the product $\beta_{\text{PNP}}\beta_{\text{NPN}}$ exceed 1). This is important, since transistor gain is low at high current levels. SCRs achieve nearly the same current and voltage ratings as diodes. However, the multiple junctions make them relatively slow. They are highly suitable for line-frequency applications such as rectifiers and cycloconverters, but are more difficult to use for inverters and high-frequency applications. The *gate turn-off SCR* (GTO) overcomes some of the control limitations, and is used in very high power inverters. The *triac* acts as a reverse parallel SCR connection with a common gate. Its main applications are in ac regulators. Many manufacturers are attempting to apply triacs to electronic circuit breakers and other ac applications of power electronics.

Bipolar junction transistors (BJTs) are becoming uncommon for power electronics applications because of improvements in MOSFETs, IGBTs, and other devices types. They were widely used in motor drives and inverters until recently. Even though they are less typical, they provide further insight into switch dynamics. In a BJT, turn-on requires injection of charge into the base region, followed by modulation of collector flow. The devices are inherently slower than diodes, since multiple processes must take place to turn them on. The turn-off process is a significant concern. If a BJT is driven very hard, the base-collector junction becomes forward biased, and excess charge is stored. The transistor cannot turn off until this charge is removed. Consequently, low forward drops produce long *storage times* that add to other turn-off processes. Storage can extend turn-off in typical devices by several microseconds.

These effects offer some insight into how a BJT should be operated. If the device is supplied with a high initial base current pulse, turn-on speed is enhanced. If an extra diode or similar circuitry is provided, forward bias of the base-collector junction can be avoided, and much of the storage time can be avoided as well. A large negative current pulse drawn from the base will speed the recovery processes, especially in the base-emitter junction. The most significant concern relative to diodes and thyristors is that the BJT is nonlatching: a highly reliable base drive is necessary to ensure that the device will not operate in its linear region. In this region, the forward voltage drop becomes substantial, and high losses can occur.

The linear region represents a potential reliability issue that can be addressed as part of the base drive design. Another failure mode is more subtle: The BJT has a negative temperature coefficient of resistivity. This means that any local crowding of current will produce local heating and reduced resistance. It is possible that a runaway process called *second breakdown* can occur if the current begins to crowd into small channels. The safe operating area for BJTs must be limited to avoid the second breakdown effect. A more practical limitation is that power BJT gains are low, with $\beta < 10$ typically. A Darlington connection can avoid this low-gain problem, at the cost of low speed because of the extra junctions.

The power MOSFET is not a junction device, and its dynamic behavior is different, and faster, than that of BJTs or thyristors. In a MOSFET, an electric field modulates the charge distribution in a semiconducting *channel*. Power devices usually operate in *enhancement mode*, meaning that gate electric field draws charge into the channel and creates a low-resistance path. There are also *depletion mode* devices, in which the channel is present with no extra field applied. A negative field can remove charge and shut the device off. In either case, the doped regions form parasitic elements, most notably a bipolar transistor. In practical devices, metallization is used to short the base-emitter junction of this parasitic BJT. This avoids unwanted current flow, but introduces an inherent *reverse-body diode*. For power conversion, the reverse diode is not usually troublesome. Indeed, it is a necessary part for inverter action.

A typical power MOSFET turns on when about 6 to 10 V is imposed between gate and source. This voltage level is convenient for analog circuits and op-amps, so MOSFET switching is relatively straightforward. In fact, the major element of dynamic behavior concerns charging and discharging the gate-source capacitance. The channel cannot conduct until the field is set up and charge is brought in.

The nature of a MOSFET channel is such that it acts as a resistive element. There are no junctions to require forward bias. At low current levels, the resistive behavior is an important advantage, and low forward voltages can be achieved. At high currents, the forward voltage across an FET is substantially higher than the drop in a BJT or IGBT of similar ratings. This is because the semiconductor material is used less effectively in a MOSFET geometry. Nevertheless, the MOSFET has become the best choice for applications below 200 V and 50 A. An important advantage of the channel behavior is that the temperature coefficient of resistance is positive: As current flow increases, resistance and voltage drops increase as well. This helps when multiple devices are connected in parallel. A device that attempts to take on extra current will increase in resistance. The increase diverts current to other devices. Because of this property, real power MOSFETs are built as parallel connections of thousands of individual cells.

The IGBT is growing in popularity, especially for inverter applications. This device can be understood as similar to a Darlington connection of a MOSFET and a BJT. The combination has the convenient gate properties of a MOSFET, but provides much higher current density than a MOSFET of similar size. The temperature coefficient is less predictable than for MOSFETs or BJTs, so the devices are more challenging than MOSFETs to connect in parallel. The Darlington arrangement has implications for speed, just as it did in the BJT case. During turn-off, typical IGBTs exhibit *current tailing*—a slow sequence during the latter part of the turn-off process. This effect is analogous to the storage time in a BJT. The effect can be reduced through processing, at the cost of increased forward voltage drop.

Snubbers are circuits that supplement power semiconductors to alter the switching trajectory. A *passive snubber* includes inductance to prevent fast current changes, a capacitor to reduce effects of fast voltage changes, a resistor to discharge energy stored in these elements, and one or more diodes to help direct the action. Since voltage and current swings are reduced with a snubber, the switching loss reduces significantly with a snubber in place. There is extra loss in the snubber resistor.

It is possible to identify an *optimum snubber* that minimizes total loss under a specific loading condition. The optimum is only approximate, since linear transitions are used to compute it. However, the concept is useful for setting a range of values for the parts. The capacitor value for this case is

$$C_{\text{opt}} = \frac{I_{\text{on}} t_{\text{fall}}}{\sqrt{12} \ V_{\text{off}}} \tag{13.73}$$

and the inductor value is

$$L_{\text{opt}} = \frac{V_{\text{off}} t_{\text{fall(voltage)}}}{\sqrt{12} \ I_{\text{on}}} \tag{13.74}$$

In practice, values higher than the optimum settings are usually used. This reduces semiconductor switching loss and improves reliability, even though there is some additional loss in the snubber resistor.

There are several techniques for producing *lossless* or *energy recovery* snubbers. In large systems, a good conceptual approach to the lossless snubber problem is to set up a conventional snubber, but substitute a dc–dc converter for the resistor. That way, discharge energy from the storage elements can be sent to perform useful work.

An extended design example considered a dc–dc boost converter for 12 V to 24 V conversion at 120 W. It was found that typical efficiency beyond 90% could be achieved at full load under nominal conditions. Power semiconductor data sheets include information about voltage and current residuals, switching times, thermal effects, and safe operating area. Most of these must factor into a useful design.

PROBLEMS

1. A six-pulse rectifier is constructed with SCRs. The load can be modelled as a 20 A_{dc} source for phase delay angles up to 75°. The SCRs can be modelled as 1.2 V forward drops in series with 0.01 Ω and FCBB switches. What is the total loss as a function of phase angle through 75°?

2. A boost converter operates at 20 kHz, and converts power from a solar panel to a battery bus. The panel has nominal voltage of 16 V and current up to 3 A. The battery bus is 48 V. The transistor can be modelled as a 0.5 V constant forward drop, while the diode is a 1.0 V drop. Which device dissipates more power? What is the converter efficiency?

3. In a clamped forward converter, the transistor forward voltage jumps to V_{clamp} as soon as turn-off begins. In a general design, the clamp voltage is chosen to be V_{in}/D to support flux resetting. What is the commutation parameter as a function of duty ratio in this application?

4. A buck-boost converter switches at 50 kHz. Its transistor can be modelled as 0.1 Ω when on. Its diode acts as a constant 1.0 V drop. The input voltage is 12.0 V. The load is a 4 Ω resistor. What is the highest possible output voltage magnitude, and at what duty ratio does it occur? What if an expensive diode with 0.5 V forward drop is used instead?

5. A new type of Schottky diode is optimized for low-voltage applications. It has typical forward voltage drop of 0.2 V, but its leakage current can be as high as 80 mA. It is being considered for a 12 V to 2.5 V converter with 50 W load. Compare the on-state and off-state losses. With this high leakage, can the off-state loss be neglected?

6. Figure 13.14 shows the forward characteristic of a 2N6508 SCR. Develop an accurate static model for the part.

7. Figure 13.65 shows an turn-on turn-off sequence from a power converter.

   a. Plot the switching trajectory based on these time waveforms.
   b. What is the commutation parameter for this device?

8. A buck converter performs 12 V to 5 V conversion at power levels up to 25 W. The switching frequency is 120 kHz. The transistor is a MOSFET with $R_{ds(on)} = 0.18$ Ω. The diode can be modelled as an ideal diode in series with 0.7 V and 0.02 Ω. The inductor has been chosen for 5% peak-to-peak current ripple, and the capacitor reduces output ripple to 1% peak-to-peak.

   a. What duty ratio will be used to give 5.0 V at full load?

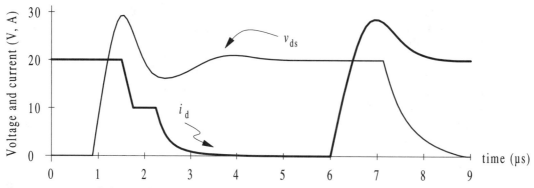

Figure 13.65 Time behavior of a power converter.

b. Estimate the losses in the switching devices, including all effects. The devices have turn-on and turn-off times of about 80 ns.

c. The ambient temperature can be up to 40°C, and the junction temperature should be limited to 130°C to provide some headroom. What is the maximum value of $R_{\theta(ja)}$ that can be tolerated for each device?

9. An IGBT carries 200 A when on, and blocks 360 V when off. This device loses about 0.03 mJ/A during turn-on and 0.11 mJ/A during turn-off. The forward drop is 2.2 V at 200 A, and the leakage current is 3 mA at 360 V. The thermal resistance from junction to the heat sink surface is 0.12 K/W. The switching frequency is 12 kHz, and the device on average has a duty ratio of 50%. What is the power loss in the part? For $T_{junction} < 150°C$ at 30°C ambient, what is the maximum value of $R_{\theta(ca)}$?

10. A MOSFET is used in a boost-buck converter. The on-state loss is 4.2 W, and the switching loss is 2.8 W. The device is in a TO-220 package. It is desired to avoid a heat sink, and a parallel connection has been suggested. If a second identical MOSFET is added in parallel, what is the loss in each part? Will a heat sink be necessary? Assume that $R_{\theta(ja)} = 45$ K/W.

11. A bridge rectifier has 230 V_{RMS} 50 Hz input. The diodes can be modelled in the on state as 0.9 V in series with 0.012 Ω. The load is a classical RC arrangement, at a power level of 500 W. The capacitor is large enough to maintain output ripple of 10 V peak-to-peak. What is the loss in the diodes?

12. A boost converter has 5 V input and 12 V output at 60 W. The diode is ideal. The transistor requires 1 μs to turn on or off. The switching frequency is 100 kHz. The storage components are large.

a. Plot the current and voltage waveforms associated with the transistor. (*Hint*: KVL and KCL must be satisfied, and an ideal diode carries current only when it is on.)

b. What is the switching loss in the transistor?

13. During a diode's reverse recovery interval, the forward voltage is relatively low. The device cannot block until charge is removed. One could argue that the reverse recovery current causes little loss in the diode. However, this current must flow elsewhere. Consider a buck converter with a diode that is ideal except for a reverse recovery time of 100 ns and charge of 10 μC. The input voltage is 100 V and the output is 12 V. The load is 120 W. How will diode reverse recovery affect the transistor loss? The transistor's switching time is 100 ns.

14. Six IGBTs form a PWM inverter with three-phase output for a motor drive. The dc input is 400 V. The motor is rated at 380 V line-to-line at 50 Hz. It is running right now at 40 Hz under constant volts per hertz control. The RMS output current is 80 A at 0.7 power factor. The IGBTs have forward drop of 2.0 V, and lose 4 mJ for each switching cycle. They have built-in reverse diodes with the same drop.

    a. What is the on-state loss in each IGBT?

    b. For 3 kHz switching, what is the switching loss in each IGBT?

    c. At 6 kHz switching, what is the device loss? Why is it not just double the 3 kHz value?

15. A MOSFET is used in a boost-buck converter. The input current is 8 A. The converter has 12 V input and -12 V output. The MOSFET switches in 50 ns and has $R_{ds(on)} = 0.1\ \Omega$. The part is in a TO-220 package with $R_{\theta(ja)} = 50$ K/W and $T_{junction} < 150°C$.

    a. Compute the loss in the MOSFET.

    b. Design a turn-off snubber for this arrangement. Try to minimize total loss.

    c. Can a snubber be designed to reduce switching loss enough to avoid a heat sink?

16. Design turn-off snubbers for the IGBTs of Problem 14.

17. A flyback converter has an input range of 3 V to 8 V, and produces 5 V output at up to 25 W. The switching frequency is 80 kHz.

    a. What is the maximum forward drop in the transistor that still supports the 3 V input?

    b. Select L and C to provide 2% peak-to-peak output ripple.

    c. Estimate the losses in each switching device, given turn-on and turn-off times of 100 ns.

    d. Design a snubber to ensure power loss below 2 W in each switching device.

18. In a clamped forward converter, the clamp enforces a voltage of $3v_{in}$ across the transistor during switching. The current is 10 A, and the transistor switches in 200 ns. The switching frequency is 50 kHz, and the input voltage is 24 V.

    a. Estimate the switching loss.

    b. Can the switching loss be cut in half with a snubber? If so, develop a snubber design.

19. A small buck converter has 24 V input and 12 V output. The transistor has $R_{ds(on)} = 0.1\ \Omega$. The diode is modelled as a 1.0 V drop. Each device has $R_{\theta(ja)} = 40$ K/W and $T_{junction} < 150°C$. You may suggest a switching frequency.

a. What is the maximum load possible without a heat sink at 20°C ambient?

b. The value of $R_{ds(on)}$ increases linearly with temperature up to 0.4 Ω at 150°C. How should the maximum load be derated with temperature to avoid the heat sink at higher ambient temperatures?

c. If a heat sink provides $R_{\theta(ja)} = 20$ K/W, how much does the load power increase?

20. A customer needs an 18 kHz source at 200 V_{RMS} for a special lighting application. The waveform should have THD below 5%. The input source is 240 V_{RMS} at 60 Hz. Present a design to meet the needs.

21. An SCR bridge rectifier is to be designed to take in 230 V_{RMS} 60 Hz and provide 150 V_{dc} output for an industrial process. The nominal load is 18 kW.

a. Propose a complete design. What phase delay angle will be used?

b. The SCRs have switching times of 3 μs. What is the switching loss for a single part?

c. What is the power converter's efficiency if the load is highly inductive?

22. A power converter is needed for 5 V input and 3.3 V output at power levels up to 10 W. MOSFETs with $R_{ds(on)}$ as low as 0.015 Ω are available. You may select a conventional or Schottky diode. Design this converter.

a. What duty ratio will be needed at full load? At 50% load? At 10% load?

b. Suggest a switching frequency. Select energy storage components to keep output ripple below 1% peak-to-peak.

c. What is the full load efficiency?

23. A 100 kW inverter is to be designed for an electric vehicle application. The battery potential is 360 V nominal. The motor planned for this application has a typical power factor of 0.85.

a. Suggest device types. Draw the circuit.

b. The desired switching frequency is 6 kHz (higher frequencies cannot always be heard, and an overly silent car will be hazardous to the blind or to children). Estimate the losses in the inverter at peak output load. What are the losses at 10% load?

c. What thermal resistance will be needed to keep the semiconductor temperature rise below 80°C?

REFERENCES

B. J. Baliga, *Power Semiconductor Devices*. Boston: PWS, 1996.

R. B. Bird, W. E. Stewart, E. N. Lightfoot, *Transport Phenomena*. New York: Wiley, 1960.

A. Blicher, *Field-Effect and Bipolar Power Transistor Physics*. New York: Academic Press, 1981.

D. A. Grant, J. Gowar, *Power MOSFETs: Theory and Applications*. New York: Wiley, 1989.

A. Jaecklin, *Power Semiconductor Devices and Circuits*. New York: Plenum Press, 1992.

International Rectifier, *HEXFET Power MOSFET Designer's Manual*. El Segundo, CA: International Rectifier, 1993.

International Rectifier, *IGBT Designer's Manual*. El Segundo, CA: International Rectifier, 1994.

Motorola, *TMOS Power MOSFET Transistor Device Data, DL 135/D*. Phoenix: Motorola, 1994.

Motorola, *Thyristor Device Data, DL 137/D*. Phoenix: Motorola, 1993.

B. E. Taylor, *Power MOSFET Design*. New York: Wiley, 1993.

B. W. Williams, *Power Electronics*. New York: Wiley, 1987.

CHAPTER 14

Figure 14.1 This drive board operates all six switches in a three-phase inverter.

INTERFACING WITH POWER SEMICONDUCTORS

14.1 INTRODUCTION

The dynamic behavior of power semiconductors is usually a major factor in determining power converter performance. Speed limitations can lead to limits on duty ratios and switching frequencies. Switching loss can be high in fast converters. The previous chapter, however, did not touch on the implementation of the switching process. How can the highest possible speed be assured? What are the design considerations associated with switch operation? The question of implementing the required switch action is still open.

This chapter addresses fundamental questions of how to operate a power semiconductor switch. The basic concern is how to translate the information in a switching function into the actual device function. The discussion begins with *gate drives*, the key circuits for this translation. The term is generic, and includes fast amplifiers to drive MOSFETs, high-current circuits to operate BJTs, pulse circuits for SCRs, driver boards for inverters like the one shown in Figure 14.1, and related hardware.

Switch isolation is a crucial but often neglected issue, discussed here in the middle portion of the chapter. Many switch matrix arrangements introduced so far require semiconductors that avoid common connections or grounds. Isolation is a challenging complication in converter design. It tends to be a limiting factor in the performance of many dc–dc converters. In certain cases, isolation can be avoided through the use of complementary switches. This practice brings its own problems, which are covered briefly.

Later in the chapter, issues associated with sensing for power electronic switches are presented. For control or protection purposes, accurate information about switch voltage and current is often needed. Methods such as differential amplifiers, current transformers, and Hall-effect devices are described. Gate drives and sensing components often consume significant power levels. Throughout the chapter, some convenient methods for analyzing the overhead power consumption associated with switch implementation are presented.

In the broadest sense, a gate drive and power semiconductor combine to form a complete two-port function, shown in Figure 14.2. A low-power switching function from an analog or digital circuit is presented to the gate port. When the switching function is high, the output port acts as a short circuit, and when it is low, the output acts as an open circuit. There

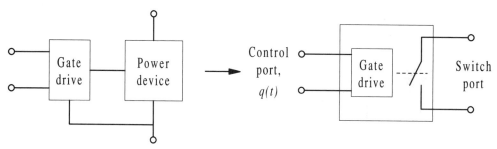

Figure 14.2 Ideal two-port switch and gate control.

is no electrical connection between the two ports, so the switch can be connected arbitrarily within a matrix. The chapter discusses the considerations in going from a real three-terminal semiconductor to the idealized two-port in Figure 14.2.

14.2 GATE DRIVES

14.2.1 Overview

All devices except the diode require correct external gate signals for turn-on, turn-off, or both. For thyristors, the signal can be a pulse, provided it is long enough to allow the various junctions to become properly biased and to allow the anode current to rise above the holding current. For transistors, the gate must be held high throughout the on state. For GTOs, turn-off pulses are needed in addition to the turn-on signals. Dynamic considerations mean that gate drives must be fast as well as efficient and reliable. In some converters, the gate drive is the limiting factor in speed. In such cases, a gate drive improvement can reduce switching losses by 30% or more.

Naturally, designers prefer systems in which gate drive implementation is simple. Power MOSFETs, with their voltage-controlled gate behavior, are convenient in this sense. Newer devices such as IGBTs and MCTs use MOSFET gate behavior to extend this convenience to higher power levels. Power BJTs have lost favor because current-controlled gate drives tend to be complicated.

14.2.2 Voltage-Controlled Gates

The relative simplicity of voltage-controlled gates makes them a helpful starting point for discussion. Consider the simplified MOSFET model in Figure 14.3. To first order, the switch is on whenever V_{gs} is above the threshold voltage V_{th}, and off when $V_{gs} < V_{th}$. In practice, V_{th} is the minimum value for current flow. A higher voltage—usually about $2V_{th}$—is more common in converter applications. Typically, this translates to 8 to 10 V for a standard power MOSFET, or 5 V for a "logic level" device. The value of V_{gs} during the off state is less critical. It should be low enough not to permit significant current flow even if external noise interferes with the gate signal.

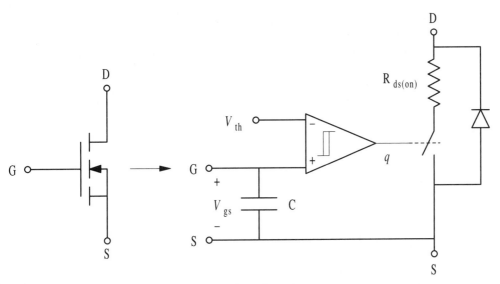

Figure 14.3 Simplified MOSFET model for gate drive discussion.

To make the transistor change state, the gate drive must charge the capacitor to sufficiently high voltage for turn-on, then remove the charge for turn-off. The charge–discharge
process is a time limiting step. It can also be a reliability limiter. For example, if the charge
process only brings the voltage up to V_{th}, the FET will operate in its active region. Losses
will be very high; a failure is likely to follow. Consider the following example.

Example 14.2.1 A power MOSFET is modelled according to Figure 14.4. The device has
$V_{th} = 4$ V and $C_{iss} = 2000$ pF. Notice that the comparator level is set to $2V_{th}$ to ensure a

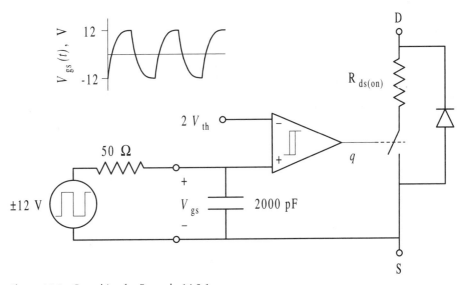

Figure 14.4 Gate drive for Example 14.2.1.

low-resistance on state. Gate oxide dielectric breakdown limits the voltage $|V_{gs}|$ to 20 V maximum. A laboratory function generator is set to provide a 12 V square wave. It has output resistance of 50 Ω, as shown in Figure 14.4. How quickly will the switch operate when commanded to do so?

The square wave drives a simple RC pair with a time constant of 100 ns. To carry high current, the gate-source voltage should probably be at least twice the threshold value, or 8 V. At turn-on, the capacitor voltage must rise from -12 V to $+8$ V in response to the square wave's rising edge. The capacitor voltage must change by 20 V compared to the total change of 24 V, or 83.3% of the square wave step. This much change requires about 1.8 time constants, or 180 ns. The turn-off behavior is similar. The switch requires 180 ns to operate. This will translate to a delay in switch operation. The time also affects the switching trajectory, and could lead to significant losses.

A typical device with characteristics along the lines of Example 14.2.1, such as the IRF640, claims a turn-on time of about 90 ns. The function generator in the example is not able to support this switching speed. The example shows that the simple voltage characteristic of the FET gate does not lead to good performance by itself. A low-impedance gate drive is required to operate the device quickly.

Low-impedance gate drives are necessary, but create problems of their own. As the impedance drops lower and lower, the gate drive becomes a switching power converter in its own right. The gate drive switches then require gate drives, which in turn require gate drives, and we are left with an unsolvable problem. Real gate drives are limited to impedances on the order of five or ten ohms. Lower values are used in some cases, especially when gate capacitance is 5000 pF or more. Higher values can be used if switching speed is not critical.

Since the function of a gate drive for a voltage-controlled device is capacitor charging and discharging, the drive need not carry a steady-state current. Current flows from the gate drive only in short pulses. However, a problem can occur if the gate drive impedance is too low. With 12 V input and 1 Ω of output resistance, a gate drive could supply at least 12 A momentarily into a MOSFET gate. Since the gate is not designed to carry load current, such a high pulse can damage the gate leads or metallization. Some manufacturers specify a *maximum gate current* to help designers avoid this problem.

The gate capacitance of a real MOSFET is nonlinear, with strong voltage variation. As shown in Figure 14.5, the capacitance actually has three components: the gate-source capacitance C_{gs}, the gate-drain capacitance C_{gd}, and the drain-source capacitance C_{ds}. In most manufacturer specification lists, these are given in terms of *common source* measurement

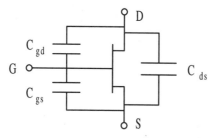

Figure 14.5 Capacitance elements associated with MOSFET.

values. Table 14.1 lists the symbols. The gate drive must charge both C_{gs} and C_{gd}. The C_{gd} portion produces interesting effects in some converter arrangements: Since switch action causes the drain voltage to swing quickly between low and high voltage values, it is possible that a substantial current $i_{gd} = C_{gd}(dv_{gd}/dt)$ will flow. The gate drive must provide enough charge to account for this current as well as the current needed to charge C_{gs}. For this reason, many manufacturers specify a *total gate charge*, Q_{gate}, as a broader indication of gate drive requirements.

Figure 14.6 illustrates several different circuits commonly used for MOSFET and IGBT gate drives. The top of the figure [circuits (a) and (b)] shows circuits based on conventional TTL and CMOS technology. These circuits are simple and convenient, although their output impedance is usually too high to support the best possible switch performance. Circuit (a) has the significant disadvantage that the bipolar device dissipates power continuously when the FET is turned off.

Circuit (c) in Figure 14.6 shows a complementary emitter follower gate drive. This circuit supports high switching performance. Transformer coupling, as shown in circuit (d), can be used when isolation is required. Since the transformer cannot support a dc component, circuit (d) has an upper limit on the switch duty ratio.

Today, it is possible to obtain ICs specifically designed as power electronic gate drives. A digital clock driver chip, such as the DS0026 shown in circuit (e) of Figure 14.6, is very effective at providing the short current pulses needed to power a gate. Newer ICs based on CMOS complementary source followers, such as the driver shown in circuit (f), have excellent characteristics for driving power MOSFETs.

Notice that all the circuits in Figure 14.6 have been arranged so that a TTL input signal (i.e. a switching function from a TTL level circuit) can operate the main switching device. Most of the circuits can also be operated with a 12 V signal from analog control hardware. The common design practice of supporting control from 5 V logic, 12 V analog, or similar voltage levels, provides flexibility in generating and controlling switching functions.

Example 14.2.2   A TTL inverter drives an external transistor, arranged as in circuit (a) of Figure 14.6. This circuit is to drive a power MOSFET with total gate capacitance of 800 pF. The TTL chip operates from +5 V, but a source of +12 V is also available for the gate drive. What resistor should be used to ensure switching speeds faster than 120 ns? How much power is lost in this resistor during the off state of the FET?

In the simple open-collector arrangement, turn-on of the FET is governed by the pull-up resistor, which connects the +12 V source to the FET gate whenever the NPN transistor is off. When the BJT is turned on, it should discharge the gate quickly. However, it will con-

TABLE 14.1
Specifications for Capacitance of a Power MOSFET

Specified Value	Symbol	In Terms of MOSFET Capacitances
Common-source input capacitance	C_{iss}	$C_{gs} + C_{gd}$
Common-source output capacitance	C_{oss}	$C_{ds} + C_{gd}$
Common-source reverse transfer capacitance	C_{rss}	C_{gd}

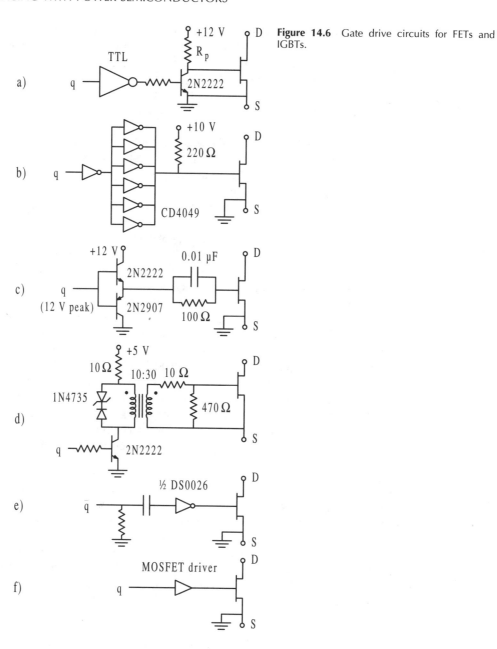

Figure 14.6 Gate drive circuits for FETs and IGBTs.

tinue to carry a current $I_c = (12 \text{ V})/R_p$, and the resistor will dissipate $V^2/R_p = (144/R_p)$ W. Given a switching time of 120 ns, the RC time constant for the gate drive should be roughly half as much, or about 60 ns. With capacitance of 800 pF, this suggests $R_p < 75$ Ω. To provide some extra headroom to maximize speed, let us use a resistor value of about 50 Ω. At low duty ratios, this resistor will dissipate $(144 \text{ V}^2)/(50 \text{ Ω}) = 2.9$ W. If the FET is used in a 30 W converter, the gate drive loss alone will drop the system efficiency by almost ten points.

As shown in Figure 14.7, CMOS gate drive ICs normally require a substantial bypass capacitor so that the high charging current spikes are not drawn from the input source. In effect, these circuits serve as charge pumps, transferring charge from the bypass capacitor to the gate capacitor. If the bypass capacitor is much larger than C_{iss}, it should be capable of providing all of the necessary gate charge.

Example 14.2.3 An MTP50N06 power MOSFET is rated as having total gate charge Q_{gate} not exceeding 100 nC, and a typical C_{iss} value of 3000 pF. The turn-on and turn-off times are rated at 200 ns with a 10 Ω gate drive impedance. Discuss the implications for the design of a CMOS gate drive, and suggest a bypass capacitor value for the drive.

For an ideal linear capacitor, the charge is the product CV. For a +12 V gate signal, the charge on a 3000 pF capacitor is 36 nC. The actual rating of 100 nC reflects the nonlinear behavior of the actual device. The bypass capacitor for the CMOS part will need to store much more than 100 nC, because the capacitor voltage must not fall much during the turn-on process. Given a factor of ten, if $CV = 1000$ nC and $V = 12$ V, the capacitance should be about 83 nF. In principle, then, a 0.1 μF capacitor is adequate for bypassing. Usually, values on the order of a few microfarads are used instead to prevent the voltage drop from affecting the 12 V source. The CMOS drive's output impedance should be about 10 Ω or less to ensure fast switching.

The gate drive circuits examined so far do not take precautions to minimize energy loss. The gate charge is removed strictly by dissipation, and an energy equal to at least $\frac{1}{2}CV^2$ will be consumed during each switching cycle. Consider the preceding example. If charge of 100 nC is taken from a capacitor at about 12 V, an energy $\int i(t)V\,dt = Q_{gate}V$ is lost during each cycle. This 1200 nJ loss must be multiplied by the switching frequency. At 100 kHz, the gate drive consumes at least 0.12 W. While this is not insignificant, it is far less than the 3 W loss encountered in the simple open-collector gate drive. The value

$$P_{gate} \approx Q_{gate}V_{supply}f_{switch} \tag{14.1}$$

provides an estimate of gate drive power. The estimate is reasonable if the bypass capacitor is large.

14.2.3 Current-Controlled Gates

Bipolar transistors need only a fraction of a volt on the base terminal to switch. The base behavior is better characterized by a current control. A gate drive for a power BJT (a *base*

Figure 14.7 CMOS gate drive with bypass capacitor.

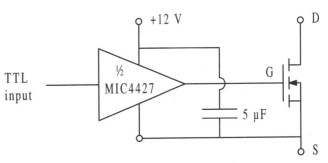

drive) must supply the current necessary to maintain the transistor in its on state. In the active region, the base current $I_b = I_c/\beta$. For on-state switching, this is a minimum value, and often $I_b > I_c/\beta$ to ensure operation close to device saturation. As mentioned in Chapter 13, β can be low for power BJTs. Values of $\beta = 5$ to 10 will be taken here as representative. Most power devices with higher gains are really Darlington pairs.

In addition to a continuous on-state current of at least I_c/β, a practical base drive should supply a brief overdrive pulse at turn-on to set up the base region quickly. This pulse is often chosen to be nearly equal to the on-state value of I_c, although base current rating limits must be considered. A brief negative current applied during turn-off helps remove the stored charge to minimize the storage time. Figure 14.8 summarizes the waveform used for high-performance BJT switching. The high currents implied by these considerations are difficult to achieve with low-power components.

Low gate drive impedance is important for FET switching, but much more critical for BJT switching. For example, a part with $\beta = 5$, intended to carry 20 A in its on state, will require at least 4 A of continuous base drive with short-term overdrive pulses as high as 20 A. The base-emitter voltage will not exceed 1.5 V. These numbers correspond to a base drive impedance of (1.5 V)/(20 A) = 0.075 Ω. Even without the overdrive pulses, the effective impedance is only (1.5 V)/(4 A) = 0.375 Ω. The necessary current and impedance values cannot be achieved with TTL, CMOS, or typical analog circuit functions.

The base-drive problem is usually addressed in one of three ways:

- Use a Darlington connection to provide a higher effective value of β. The input transistor of the Darlington pair provides the necessary base current for the output switch. This connection is much slower than a single device, particularly at turn-off. Figure 14.9 shows power Darlington and double Darlington connections with some of the auxiliary parts needed to speed up the turn-off process.

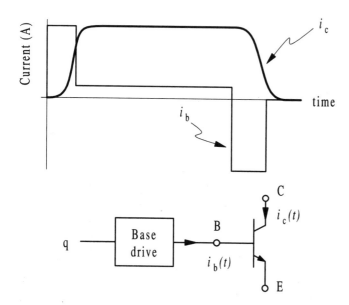

Figure 14.8   Base drive current waveform for high performance.

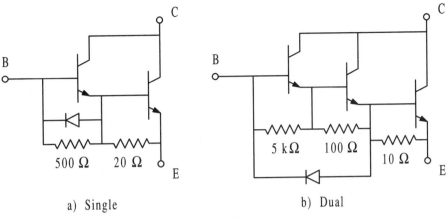

a) Single   b) Dual

Figure 14.9 Darlington connections for BJT base drives.

- Use a complementary emitter follower or similar amplifier circuit to drive the base aggressively. A speed-up capacitor is used, as illustrated in Figure 14.10, to provide the high short-term pulses needed for rapid switching.
- Use a transformer as an impedance buffer to connect an analog drive circuit to a base. A 10 : 1 transformer ratio, for instance, permits a 15 V circuit capable of 0.4 A to drive a base at 1.5 V and 4 A.

Figure 14.10 Complementary emitter follower base drive.

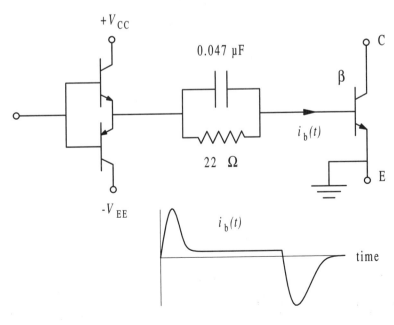

One variation on the base drive circuit combines two of these ideas in a *proportional current* driver, illustrated in Figure 14.11. A complementary emitter follower provides the short pulses for turn-on and turn-off, and the transformer provides base current during the on state. The turns ratio is chosen as $\beta : 1$, so that the transformer current on the base side will be I_c/β under all conditions.

Base drive circuits tend to consume more power than gate drives for voltage-controlled switches of similar ratings. A base current of 4 A at base-emitter voltage of 1.5 V represents continuous power consumption of 6 W at the base terminal. A complementary emitter-follower drive without a transformer consumes much more power, because it must draw the power from a source at higher voltage. Consider that if a 5 V source is available for the base drive, a continuous draw of 4 A represents total consumption of 20 W instead of just 6 W.

Example 14.2.4   A proportional base drive operates a power BJT for a PWM inverter application, switching at 3 kHz. In addition, a complementary emitter follower gate drive is used to provide the turn-on and turn-off pulses. The follower is intended to supply up to 1 A continuously for reliability, and delivers ± 10 A pulses during turn-on and turn-off. The pulses are shown in Figure 14.12. If the follower operates from a 5 V source, estimate the power it consumes.

The gate current pulses look very much like the upper portion of a sine wave. The energy lost during each switching cycle in the follower will be the sum of the energy in the turn-on pulse plus the energy in the turn-off pulse plus energy lost during the continuous 1 A conduction time. During turn-on, the base voltage rises quickly to about 1.5 V. The turn-on pulse energy is given by

$$W_{\text{pulse}} = \int_{\text{pulse time}} i_b(t)(5 \text{ V})\, dt \approx (5 \text{ V}) \int_0^{2\ \mu s} (10 \text{ A})\sin(\omega_{\text{pulse}}t)\, dt \qquad (14.2)$$

where ω_{pulse} is determined from the pulse time. Since the pulse time is 2 μs, a sine wave with period 4 μs serves as a good model. Then $\omega_{\text{pulse}} = 2\pi/T = 1.57 \times 10^6$ rad/s, and the pulse energy can be found as 63.7 μJ. The turn-off pulse is similar, except that it is delivered over only a 1.5 V potential from base to emitter. The energy lost is

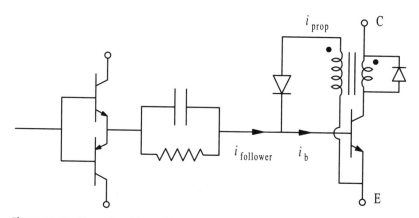

Figure 14.11   Proportional base drive circuit.

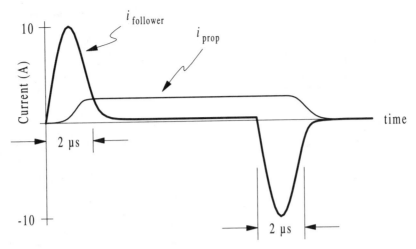

Figure 14.12   Base drive waveforms for Example 14.2.4.

$$W_{\text{pulse}} \approx 1.5 \int_0^{2 \ \mu s} 10 \sin(1.57 \times 10^6 \ t) \ dt = 19.1 \ \mu J \tag{14.3}$$

Both losses are positive, since the energy is not recovered. The sum is 108 μJ. During the rest of the on interval, power is lost at the rate of $(1.5 \ V)(1 \ A) = 1.5 \ W$. Since the converter is a PWM inverter, the overall average duty ratio should be 50%. The total loss over a long time interval should be the pulse energy loss times the switching frequency, plus the average duty ratio times the continuous 1.5 W loss. The total is

$$P_{\text{gate}} = (108 \ \mu J)f_{\text{switch}} + 0.75 \ W = 0.94 \ W \tag{14.4}$$

The loss is higher than the CMOS gate drive loss in the earlier FET example, but much lower than the loss without the proportional drive transformer.

14.2.4 Pulsed Gate Drives

The two-transistor model of the SCR is repeated in Figure 14.13. It shows that the gate connection is made to a BJT base terminal. Therefore, thyristor gates are controlled by current, although the latching behavior means that only a pulse is required. For turn-on, the gate current requirement is on the order of the holding current. Since the commutation process in a thyristor requires several microseconds, the gate pulse must continue to supply current throughout the turn-on process.

The gate current required for turn-on is highly dependent on temperature, reflecting the strong temperature dependence of the transistor gains within an SCR. Data sheets usually provide a room-temperature value for gate current, and also a worst-case value based on the minimum rated ambient temperature. The gate current must be supplied at a voltage sufficient to forward-bias the base-emitter junction of the NPN transistor.

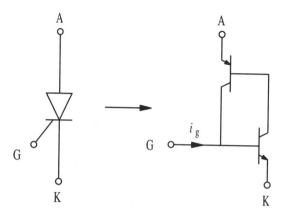

Figure 14.13 Two-transistor model of the SCR.

Gate drives can be implemented as simple switch circuits capable of providing the necessary current and voltage. It is also common to use a low-power *auxiliary thyristor* to switch the gate current of a larger device. Alternatively, a pulse transformer can be used to supply the gate current. Generic examples of these three approaches are shown in Figure 14.14. In each case, a gate-cathode resistor is usually used to improve noise immunity and to help prevent fast voltage changes from supplying enough capacitive current to trigger the gate.

In the case of the simple switch circuit of Figure 14.14a, it is usually straightforward to select most components. For example, if the available gate drive supply is at +5 V dc, a gate drive requirement of 100 mA can be ensured by setting $R_g \leq 50 \ \Omega$. A gate-cathode resistor greater than 500 Ω will have little effect on the turn-on process. The voltage is enough for forward base-emitter bias in small SCRs. Large devices have substantial capacitive and inductive parasitics, and a higher gate drive supply is often necessary to ensure forward bias of the base-emitter junction throughout the turn-on process. A gate drive supply of 20 V will work in almost any situation, given a short gate pulse. One drawback of any gate drive method is that the short-term drive current requirement is far outside the range of simple components such as TTL devices. An emitter-follower arrangement with pull-up, shown in Figure 14.15, can serve as an interconnection between conventional TTL and an SCR that requires a 100 mA gate current pulse. The TTL device might be a monostable multivibrator set to produce a short pulse.

Example 14.2.5 A 2N6508 SCR is rated for 600 V blocking and 25 A_{RMS} forward current. The holding current is 35 mA. At 25°C, the gate current requirement is 25 mA. It increases to 75 mA at −40°C. The gate-cathode voltage needed to ensure forward bias does not exceed 5 V for this relatively small device. The gate-controlled turn-on time is 1.5 μs. Design a simple switched gate drive for this part, based on a 5 V logic supply. Estimate the power consumed in the gate drive in a 60 Hz controlled rectifier application.

Although 25 mA gate current is sufficient for turn-on in a wide range of application, let us use a value of 150 mA value to ensure correct operation under low-temperature conditions and to help provide the fastest possible turn-on. With the 5 V supply, a gate-cathode voltage drop of 1 V requires the gate resistance to drop 4 V. At 150 mA, the gate resistance should be about (4 V)/(0.150 A) = 26.7 Ω. A 22 Ω resistor will support the worst-case gate current flow when resistor tolerance is taken into account. This produces a nominal gate cur-

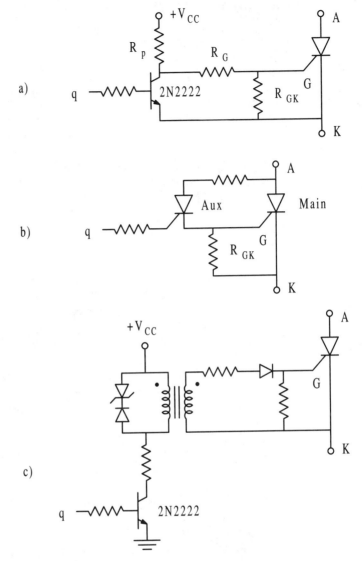

Figure 14.14 Gate drive configurations for an SCR.

a)

b)

c)

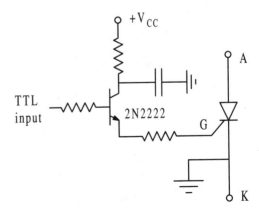

Figure 14.15 Gate drive from TTL circuit, based on emitter follower.

rent of 180 mA. To assure turn-on, let us use a 2 μs gate pulse width. Each time the device is switched, a current of 180 mA flows through a 5 V drop for 2 μs. The energy used to control the gate is approximately (5 V)(0.180 A)(2 μs) = 1.80 μJ. With a switching frequency of 60 Hz, the gate drive power is only about 110 μW.

The minuscule gate energy required by an SCR is one of its greatest advantages. In many cases, it is possible to avoid imposing current pulses on the gate drive supply. Figure 14.16 shows an example in which an *RC* combination provides more than enough energy for a 1.80 μJ pulse, while drawing minimal current from the 5 V supply. The pull-up resistor in the circuit draws more power than the gate drive itself.

These numbers reflect SCRs intended for line-frequency applications at currents below about 20 A. Larger devices require more gate voltage, especially when significant currents result from high *dv/dt* transients across the various stray capacitances. To turn on high-power SCRs quickly, the gate drive current should be able to peak at about 1 A. Pulse widths could be as long as 100 μs for the highest-power devices. The total energy needed to drive a gate in this case is still low, and gate power is a fraction of a watt. The high momentary values are a design challenge, however.

One helpful way to overcome the gate drive concerns in high-power SCRs is to use an auxiliary device to control the gate. An example appears in Figure 14.17. Since the gate drive supply in effect is the anode input voltage, there is ample voltage to operate the gate of the main SCR. The auxiliary SCR needs a rating of only an amp or two, and its gate drive requirement will be just a few milliamps. Capacitors can be used between the auxiliary and main devices to avoid the need for continuous application of gate current.

Since the duty ratio needed for an SCR gate drive current is never more than a few percent, it is reasonable to use pulse transformers for the gate drive. Pulse transformers are small magnetic transformers designed for brief energy bursts. After the pulse, a pulse transformer is often permitted to saturate, since energy transfer is needed for such a short period. The pulse-transformer gate drive has key advantages in supporting isolation and in providing a turns ratio for an extra degree of freedom.

The key parameter in designing a pulse transformer is the volt · second product that it must provide prior to saturation. Below the volt · second limit, the device acts like a conventional transformer, and provides an output current proportional to the input current. In Example 14.2.5, the gate circuit required a volt · second product on the order of

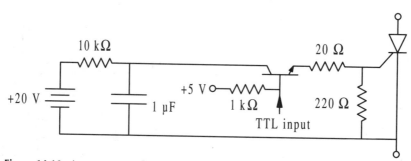

Figure 14.16 Low-current SCR gate drive circuit.

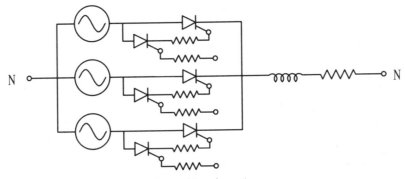

Figure 14.17 Auxiliary SCR as a gate drive alternative.

(5 V)(2 μs) = 10 μWb of flux. Let us analyze a candidate core for a similar example, and set up a possible gate drive circuit.

Example 14.2.6 An SCR is to be triggered through a pulse transformer, designed to provide up to 20 V for at least 2 μs. The input side of the transformer has access to a +20 V peak supply from a full-wave rectifier prior to its filter. The SCR requires 100 mA of gate current. A ferrite toroid with $\mu = 5000\mu_0$, magnetic path length of 50 mm, and magnetic area of 25 mm^2 has been suggested for the pulse transformer core. Can this core do the job? Plot flux density vs. time for a 60 Hz application. The circuit is shown in Figure 14.18.

When the transistor in Figure 14.18 turns on, the pulse transformer should carry the necessary gate current for at least 2 μs before saturating. Once saturation has occurred, the transistor can be shut off. Then the Zener diode will provide a return current path to allow

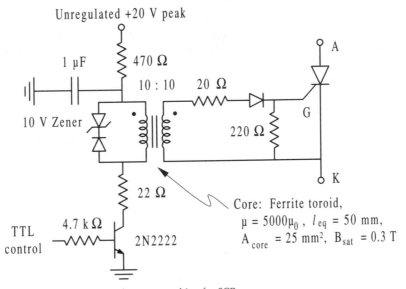

Figure 14.18 Pulse transformer gate drive for SCR.

the flux to return to zero. The saturation limit requires

$$\int v \, dt = \lambda < NB_{sat}A_{core}, \qquad (20 \text{ V})(2 \ \mu s) < N(0.3 \text{ T})(25 \times 10^{-6} \text{ m}^2) \qquad (14.5)$$

since ferrite material saturates at about 0.3 T. The inequality is satisfied when $N \geq 6$ turns, so it seems likely that this core is adequate. If 10 turns are used on each side, there should be ample time to transfer current. The 22 Ω resistor on the primary side helps to avoid excessive current flow. When the transistor turns on, about 1 A will flow into the primary winding and out of the secondary. The secondary voltage will build up to just about 20 V, and the gate will trigger quickly. After 2 μs, the transistor can be shut off. At that point, the magnetizing inductance of the transformer will force the primary current to continue flowing. The Zener diode will clamp the primary voltage to -10 V, and the flux will drop until the core is reset. Figure 14.19 shows the flux density as a function of time, given the 2 μs pulse width. The flux density builds up linearly because of the approximately constant input voltage, and falls linearly at about half the rate. The core is reset after a total of about 6 μs. This is far faster than one 60 Hz period.

14.2.5 Other Thyristors

The GTO adds a turn-off consideration to its gate drive. Recall that a GTO is prepared with high gain on the NPN transistor and low gain on the PNP device in the two-transistor SCR model. Turn-off can be achieved by drawing out a gate current sufficient to shut down the NPN device. The NPN device is effectively a power BJT; even with "high gain" its value of β probably does not exceed 5 or 10. Thus GTO turn-off requires a negative gate current at about 20% of the anode current. Even though this is a momentary requirement, the current can be extreme (tens of amperes for a 100 A GTO). A GTO gate drive therefore often resembles a BJT base drive more than an SCR pulsed gate drive.

The triac, as a device similar to a reverse-parallel dual SCR combination, requires a pulsed drive current for its gate. One significant difference is that either a positive or negative pulse can trigger a triac, in most cases without regard to the applied voltage polarity. This permits ac trigger circuits to be used.

In Figure 14.20, a triac ac circuit with an adjustable RC time delay for the gate drive

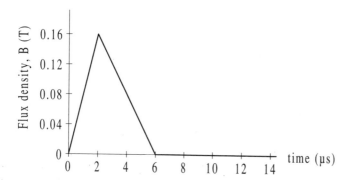

Figure 14.19 Flux density vs. time for Example 14.2.6.

is shown. This circuit is used for dimming incandescent lamps. It operates as follows: The triac is initially off. As the ac voltage rises above zero, the capacitor begins to charge at a rate dependent on $R_t + R_{load}$. When v_c exceeds the Zener voltage V_Z, current flows into the gate and turns on the triac. The Zener pair clamps v_c until the input voltage drops below V_Z. The gate current then ceases to flow, and the triac will turn off when its terminal current reaches zero. The process will repeat for both the positive and negative half-cycles.

14.3 ISOLATION

Except for the transformer-coupled gate drive circuits, each of the drive circuits in Section 14.2 shares a common node between the input switching function and the output switching device. In some circuits, this common node is not an issue. The dc–dc converters shown in Figure 14.21 have a common-source connection topology for the MOSFET. If a control circuit uses the low terminal of the supply as ground, the common-source configuration is appropriate. The boost converter in its basic form does not cause trouble. The flyback circuit is modified slightly from the usual configuration by moving the transistor to another location in the primary loop. In the push-pull circuit, both switches are in common-source configuration. These circuits can be constructed without isolation for their gate drives. Few other converters share this property.

A buck converter connects the transistor drain to the input. The source connects to the diode, and the source voltage is a square wave relative to the circuit common. A half-bridge inverter has both common-source and common-drain connections. In a phase-controlled rectifier, the cathodes of the SCRs connect to the load rather than to the common node. Each of these circuits requires extra complexity to transfer the information from a switching function into gate drive signals for a real device. In general, gate drives must be isolated from ground and from each other.

Gate drive isolation considerations include the following issues:

- The reference node for a gate drive is often not the same as ground potential or circuit common. The gate drive might need to operate up to hundreds of volts off ground.

Figure 14.20   Triac circuit, with ac gate drive.

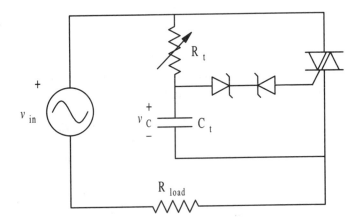

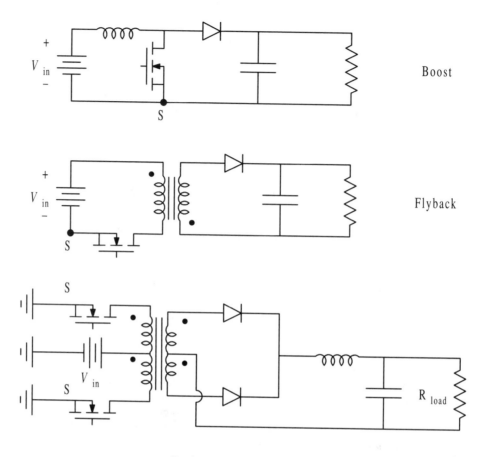

Boost

Flyback

Push-pull forward

Figure 14.21 dc–dc converters with common-source MOSFETs.

- The reference node for a gate drive is often not a constant voltage node. In buck converters, controlled rectifiers, and most other converter types, at least some of the gate drives are connected to rapidly varying nodes.

- Switches often must be isolated from each other as well as from ground. A hex-bridge inverter has four different reference nodes for the various gate drives.

A buck converter, as in Figure 14.22, illustrates another problem: the *high-side switch*. To turn the MOSFET on, a voltage $V_{gs} \approx 10$ V is required. When the device is on, its residual voltage $V_{ds} \approx 0$ V. Therefore, the on state requires a gate voltage about 10 V higher than the drain voltage. Since the drain voltage is V_{in} in this converter, the gate voltage should be $V_{in} +$ 10 V, but no voltage higher than V_{in} is available in the circuit.

Conceptually, the simplest solution to all these problems is an isolated gate drive cir-

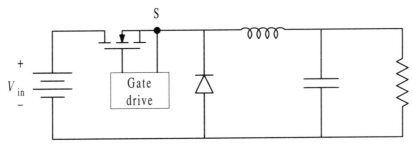

Figure 14.22   Buck converter with high-side switch.

cuit. Such a circuit can take in a switching function with an arbitrary reference point, then produce the gate drive signal necessary to operate the switching device directly. In practice, the isolation problem is difficult. The gate drive circuit needs energy to function. It needs to convey the switching function information accurately and reliably. Isolated gate drives remain a major challenge for power electronics engineers. High-side switching is a slightly different problem, since one side of the switching device has a constant voltage connection. Even here, an isolated drive would avoid problems.

The design of isolated gate drives must address two issues: transfer of information about the switching function, and transfer of energy for the gate drive. Transformer-coupled gate drives address both issues, except that there are limitations imposed on duty ratios since a transformer cannot handle dc signals. Optical coupling is helpful for information transfer. Charge pumps can be helpful for high-side switching. Table 14.2 lists a few of the alternatives.

Transformer coupling is often the preferred method if available. It does not work well, however, for PWM inverters because of the limit on duty ratio. Optical coupling can provide high performance, but a separate energy source is necessary. Commercial inverters often use optocouplers in combination with a multioutput flyback converter to drive the various switches. One interesting product is a gate drive that uses solar cells for energy transfer. The control circuit drives an LED to illuminate the cells. The power levels with this approach are too low to support fast switching, but the system is simple and functional when speed is not an issue. Charge pump methods can produce an auxiliary gate drive supply higher than V_{in}. This auxiliary voltage can be used to operate a gate drive circuit optimized for high-side switches.

The widespread use of optocoupling for gate drive isolation warrants a more detailed

TABLE 14.2
Approaches to Gate Drive Isolation

Method	Information Transfer	Energy Transfer	Comments
Transformer-coupled	Direct transfer of signal	Direct magnetic transfer	Upper limit on duty ratio
Optocoupled	Digital optocoupler	Separate dc–dc converter	Good performance, but complicated
Photovoltaic	Optocoupler	LED sends energy to a photodiode set	Low power. Does not support fast switching.
Charge-coupled	Level-shift circuit	Charge pump	Common for high-side switching.

evaluation. An example of an optically isolated gate drive for a PWM inverter is shown in Figure 14.23. Conventional optocouplers with an LED input and phototransistor output tend to be slow, so digital optocouplers are preferred. This is appropriate since only the timing information is needed for a discrete-valued switching function. The optocoupler output can be connected to a complementary emitter follower to provide a low-impedance gate drive.

Energy for the optically coupled gate drive is needed in the form of a voltage source, perhaps rated at +12 V for a MOSFET or IGBT. A low-power flyback converter can provide this source. One concern: There is stray capacitance between the windings of the flyback inductor. This capacitance will cause problems under fast switching conditions, since it represents an additional current to be supplied. The problem is more severe if multioutput flyback arrangements are used.

Example 14.3.1   A PWM inverter for an ac motor drive is built with IGBTs. The motor itself has isolated windings, so the low side of the input dc source is used as the ground reference. Notice that the circuit, shown in Figure 14.24, requires only three isolated gate drives. The devices switch in 200 ns. The high-side gate drives are optically isolated, and use a multioutput flyback converter to provide three separate +10 V sources. There is stray capacitance of 100 pF between the primary and secondary of the coupled inductor. Comment on the effects of the capacitance.

Observe what happens during converter operation, considering only the left-hand switches 1,1 and 2,1. When 2,1 is on, switch 1,1 must block 400 V. Its gate drive operates near ground potential, and maintains the off state. When it is time for 1,1 to turn on, the control applies current to the LED. The optocoupler output changes state, and the gate begins to charge. As 1,1 starts to commutate, the voltage on its emitter begins to climb from 0 to 400 V. This imposes a positive dv/dt on the stray 100 pF capacitor, and current will flow

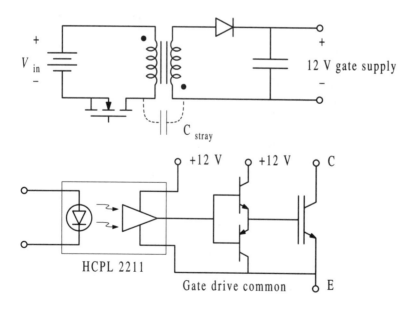

Figure 14.23 Optically isolated MOSFET gate drive.

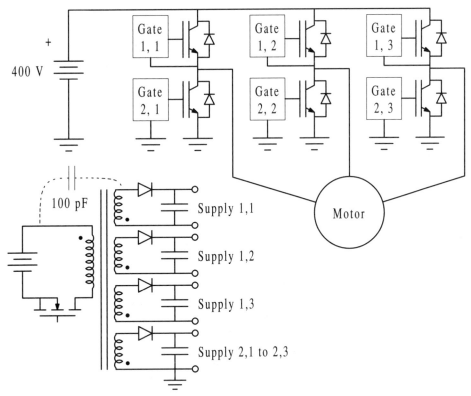

Figure 14.24 Hex-bridge inverter with optically isolated gate drives.

from the secondary to the primary of the flyback converter. The value of this current is

$$i = C\frac{dv}{dt} = (100 \text{ pF})\frac{400 \text{ V}}{200 \text{ ns}} = 0.2 \text{ A} \qquad (14.6)$$

The gate drive must provide an extra 200 mA during turn-on to charge up the stray capacitance. The turn-on process will exhibit problems if the gate drive impedance is too high to provide this current.

The stray capacitance effect is a hidden limitation of transformer-coupled circuits as well as flyback converters. Some converter packages can exacerbate the problem. MOSFETs and IGBTs with high ratings are usually packaged as groups of dies on a single substrate. The substrate is bonded to a base plate for heat sinking. It is not unusual to encounter capacitances between a terminal and this base plate on the order of 1000 pF. A general circuit model for capacitance behavior of a MOSFET in a high-side application is shown in Figure 14.25. The drain-side baseplate capacitance C_{dp} does not affect the dynamics, since its voltage is fixed at V_{in}. During turn-on, the source voltage makes the transition from 0 to V_{in}. The gate voltage must change from 0 to $V_{in} + 10$ V, and the drain-source voltage must drop from V_{in} to 0. Assuming the voltages change in time t_{on}, the various currents are

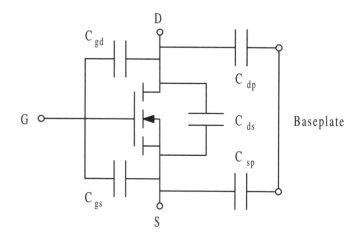

Figure 14.25 High-current MOSFET pack with capacitances.

$$i_{gd} = C_{gd}\frac{dv_{gd}}{dt} = C_{gd}\frac{-V_{in}}{t_{on}}$$

$$i_{gs} = C_{gs}\frac{dv_{gs}}{dt} = C_{gs}\frac{(10 \text{ V})}{t_{on}}, \tag{14.7}$$

$$i_{Cds} = C_{ds}\frac{-V_{in}}{dt}, \qquad i_{dp} = 0, \qquad i_{sp} = C_{sp}\frac{V_{in}}{t_{on}}$$

When the input voltage is switched in 200 ns, the capacitances to the base plate C_{dp} and C_{sp} will become additional drain-source terms when the device is connected.

In SCR-based inverters, an excessively small pulse transformer can cause commutation problems. Consider an SCR inverter switching 600 V and 500 A in 2 μs. At this current level, the SCR's package inductance and capacitance will create problems. For example, 500 A switched in a 100 nH inductor over a 2 μs interval will induce $L(di/dt) = 25$ V on the inductor. This will be enough to interfere with the gate drive in many situations. Switching 600 V in 2 μs will drive a current of 300 mA into a 1000 pF capacitor. High dv/dt and di/dt values in an SCR often make operation unreliable—one reason why limits on these derivatives are specified by most manufacturers.

Integrated circuits for high-side switching are becoming available, in part to support buck converter designs for the automotive industry. Converters operating at 12 V to 48 V do not require isolation for safety purposes, but a high-side switch is needed for many of the desired dc–dc converter circuits.

14.4 P-CHANNEL APPLICATIONS AND SHOOT-THROUGH

One alternative to high-side switching, particularly at the low voltage levels found in automobiles and portable electronics, is to use P-channel switching devices. A buck converter with a P-type MOSFET is shown in Figure 14.26. The connection is a common-source configuration, although the source is connected to V_{in}. The voltage V_{gs} must be set below the

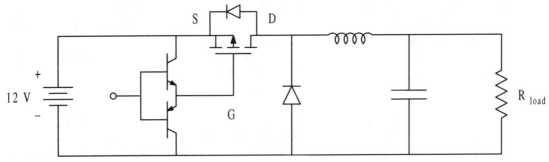

Figure 14.26   Buck converter with P-channel switch and gate drive.

(negative) threshold voltage for the device to switch it on. For example, if the gate signal is connected to the common node, $V_{gs} = -12$ V, and the device turns on. If the gate is connected to V_{in} instead, then $V_{gs} \approx 0$ V, and the transistor turns off. This behavior is convenient, and there is nothing new added to the gate drive requirements. A complementary emitter follower will still perform the job without problems. The drawback of P-channel devices is that they are more expensive for a given power rating, and tend to have lower ratings in general, than N-channel devices. This is a limiting factor in cost-sensitive applications such as the automobile industry.

The use of P-channel devices is of considerable interest in low-voltage inverters or forward converters. Consider the half-bridge in Figure 14.27. This circuit greatly simplifies the isolated gate drive problem for an inverter. When the shared gate terminal is low, the P device will be on and the N device will be off. When the shared terminal is high, the opposite occurs. Thus a single switching function $q_N(t)$ automatically provides $q_P + q_N = 1$. No gate protection or special circuitry will be needed as long as the input voltage is within the gate voltage limits of the devices. For conventional devices, input values between about 10 V and

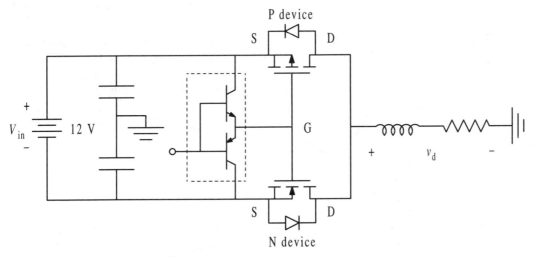

Figure 14.27   Complementary half-bridge inverter.

20 V can be handled directly in this manner. The range is suitable for small motor drives, automotive systems, and even class-D audio amplifiers. The circuit arrangement is the same as the output configuration of a CMOS digital circuit.

Unfortunately, the system issues are not this simple, and the complementary pair can be difficult to use. To turn on the N device and turn off the P device, the upper transistor in the gate drive provides the current. The P-channel device gate capacitance discharges, and the capacitance on the N device begins to charge. Meanwhile, the voltage v_d begins to swing, causing additional current to flow in the drain-source capacitance of each device. This additional current tends to counteract the gate drive action in both devices.

It is essentially impossible that the two devices be perfectly matched. Even with perfect matching, their behavior will not be in perfect complement. When the shared gate reaches +5 V, for example, the P device will have $V_{gs} = -7$ V, and the N device will have $V_{gs} = 5$ V. Both devices will be on. When this occurs, a low-resistance path appears across the input source, and high current flows through the two switches until V_{gs} on the P device becomes high enough to turn it off. This *shoot-through current* is a major limiting factor in the application of complementary devices.

Shoot-through can produce extreme losses in a low-power inverter. A typical application might allow shoot-through during only 1% of each cycle, but the current could be a factor of ten higher than the load current during that time. This will increase the switching loss by a factor of ten over the level expected ideally. Shoot-through must be avoided in real converters. The avoidance of shoot-through is often addressed in gate drive design. It is desired to provide a brief *dead time*, delaying the switching function signals to the individual switches just long enough to avoid having the devices on together. Too much dead time will force current into the reverse diodes of the MOSFETs or IGBTs. Too little will permit shoot-through loss.

Shoot-through is usually avoided in one of two ways:

1. Create independent switching functions, carefully adjusted to be nearly in complement except for the dead time, then apply them through independent gate drives. This method is direct, but it removes most of the advantages of using a complementary switch pair.
2. Use passive components to create asymmetry in the switching process. The asymmetry should add the dead time needed to minimize loss.

The second approach is not hard to implement. An example of this type of circuit is shown in Figure 14.28.

14.5 SENSORS FOR POWER ELECTRONIC SWITCHES

14.5.1 Resistive Sensing

Both converter operation and protection concerns often require sensing of currents, voltages, or other variables in power semiconductors. A gate drive might monitor device current to help time the pulses or adjust dead time. It might monitor switch voltage to confirm a successful turn-on process. In the case of resonant switch operation, precise monitoring is use-

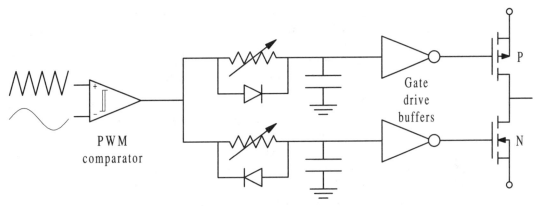

Figure 14.28 Gate drive arrangement with dead time for complementary MOSFETs.

ful in obtaining the highest possible efficiency. Currents or voltages are needed for control purposes as well.

Isolation considerations or high-side switches can make sensing difficult. If there are circuits associated with sensors, they might introduce unwanted coupling paths. Sensors often require a separate power source. In addition, many of the sensing issues in a switch require extreme dynamic range. For instance, the voltage across a switch can change by three orders of magnitude during the commutation process. A voltage sensor must be able to handle the full range if it is expected to provide useful information about on-state voltage drop. The sensors must not compromise safety, efficiency, or reliability.

A wide variety of methods can be found for power semiconductor sensing. The simplest resistive methods are shown in Figure 14.29. A pair of voltage dividers provides a differential signal proportional to V_{ds} in Figure 14.29a. In Figure 14.29b, a series resistor provides a voltage proportional to current. Another voltage divider pair allows the current to be monitored when the MOSFET is off ground potential. The differential divider approach is simple, and is common in dc–dc converters. The advantages are easy design, the ability to

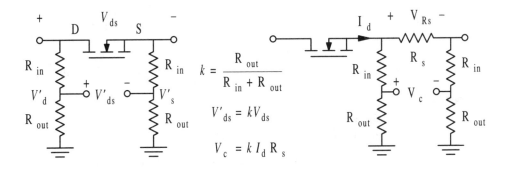

a) Drain-source voltage   b) Drain current

Figure 14.29 Resistive sensing methods for MOSFET.

support almost any voltage level, and consistent measurements. Disadvantages include the resistive connections across an isolation barrier, power loss in the resistors, and dynamic range problems.

The design of a differential divider set must take into account the desired measurement range, resistive loss, and accuracy. A design example follows.

Example 14.5.1   A resistive sensor is to be designed to measure the voltage across a high-side switch in a 200 V to 5 V forward converter. The converter is rated at 100 W, and it is desired to consume no more than $\frac{1}{2}$ W in the dividers. The control circuitry operates from the 5 V output, and cannot handle common-mode potential beyond 5 V. Suggest divider values. What effect will resistor tolerance have? What voltage will the divider provide when the switch is off (blocking 200 V) and when it is on (with voltage of 0.5 V)?

The basic configuration is the same as in Figure 14.29a. Each of the two dividers will be exposed to voltages up to 200 V. The output of each will be $R_{out}/(R_{in} + R_{out})$ times the high-side voltage. The common-mode requirement means that neither output terminal should exceed 5 V. To meet this, the divider ratio should be no more than (5 V)/(200 V) = 0.025 V/V. The power consumed in each divider is the high-side voltage squared, divided by $R_{in} + R_{out}$. Since the total should not exceed 0.5 W, each divider should not consume more than 0.25 W. The requirements are

$$\frac{(200 \text{ V})^2}{R_{in} + R_{out}} \leq 0.25 \text{ W}, \qquad R_{in} + R_{out} \geq 160 \text{ k}\Omega,$$

$$\frac{R_{out}}{R_{in} + R_{out}} \leq 0.025, \qquad 39 R_{out} \leq R_{in} \tag{14.8}$$

One possible solution is $R_{in} = 160 \text{ k}\Omega$ and $R_{out} = 4 \text{ k}\Omega$. If these are used, the voltage $V_{ds}' = V_d' - V_s' = 0.0244(V_{ds})$. When $V_{ds} = 200$ V, then V_d' will equal 4.88 V and V_s will be zero. These values are within the allowed range, and the divider output will be $V_{ds}' = 4.88$ V. When the switch is on, V_d will remain at 200 V, while V_s will swing up to 199.5 V Both V_d' and V_s' will be a bit under 5 V. The value $V_{ds} = 0.5$ V, and the divider output will be $V_{ds}' = 0.0122$ V. The divider output must be sensed to about ± 1 mV to make the V_{ds}' information useful.

What about resistor tolerance? If the resistors are all 1% tolerance types, the divider ratios will be uncertain to about $\pm 1\%$. This means the ratio will be 0.02440 $\pm 1\%$. Since V_d is fixed at 200 V, the voltage $V_d' = 4.880$ V $\pm 1\%$, or 4.880 V ± 49 mV. When the transistor is off, the voltage $V_{ds}' = 4.880$ V ± 49 mV since the value V_s is zero. When the transistor is on, the voltages are $V_d' = 4.880$ V ± 49 mV, and $V_s' = 4.868$ V ± 49 mV. The differential value V_{ds}' is $V_d' - V_s' = 12$ mV ± 98 mV. The output value is 12 mV $\pm 817\%$, and has no meaning. The resistor tolerances will need to be on the order of 0.01% if we expect to obtain accurate information about V_{ds} during the on state.

Divider tolerance is an extreme issue when it is desired to measure current. The current measuring resistor, R_s in Figure 14.29, should be as small as possible to avoid unnecessary voltage drop and loss. For example, if the MOSFET carries 10 A when on, a 40 mΩ sense resistor will drop 0.4 V and consume 4 W of power. A 10 mΩ resistor gives better results, but the differential voltage is very low. With 10 A current, the voltage V_{Rs} is 0.1 V with this value. With 1 A current, the voltage is only 10 mV. The divider tolerance problem is similar to that in the example. Even 0.01% resistors will support a current measurement that is only about 10% accurate.

The tolerance issue limits the suitability of voltage dividers for measuring small voltage differences. A divider is useful for detecting the voltage to ground at a specific location, but is much less so for isolated measurements. If only a rough measurement is needed to check switch operation, dividers can be helpful. In some critical applications the cost of high-precision resistors is justified by the need for simple sensing.

14.5.2 Integrating Sensing Functions with the Gate Drive

Protection functions benefit from sensing that is connected directly to the gate drive. This avoids isolation problems, since the gate drive is always connected directly to the switching device. Many integrated power electronic modules now being introduced have sensing and protection functions built side-by-side or even on the same chip as the semiconductor switch.

Integrated sensing brings a wealth of future possibilities:

- On-chip temperature sensing can adjust drive signals or even shut off a switch that is overheating.
- Local sensing of drain-source or collector-emitter voltages is easy. The information can be used to evaluate a device's performance or its environment. For example, an external short circuit will cause an immediate rise in the on-state voltage drop. The gate drive can be designed to shut down if the forward voltage drop exceeds a safe limit.
- Local sensing of current is also feasible. A gate drive circuit could be made to automatically adjust its operation based on the on-state current.

Figure 14.30 shows a few configurations for integrated sensors. The voltage and current sensors also can be used in nonisolated converters. In every case, the double-divider configuration is avoided.

One interesting example of an integrated sensor is the current-sense power MOSFET. You might recall that a power MOSFET is a large parallel interconnection of individual FET cells. If a few of these cells are left with their sources unconnected to the parallel set, they can provide a current divider function to make measurements more convenient. An equivalent circuit of such an arrangement is shown in Figure 14.31. There might be several thousand FET cells on the chip, all tightly matched, since they are processed together. If a few are dedicated to the divider, a current ratio of perhaps 1000 : 1 can be designed. The on-state resistance of the entire chip is low by virtue of the parallel connection. The on-state resistance of the current divider portion should be 1000 times the $R_{ds(on)}$ value of the complete chip.

Consider the implications. If the main FET is carrying a substantial current, such as 10 A, then the current divider FET will carry 10 mA when its source terminal is connected to the main source terminal. A current-to-voltage converter or similar sensing circuit can be placed in this source connection to provide convenient current sensing. This arrangement, along with a simple resistive connection, is shown in Figure 14.32. The limitation is that the

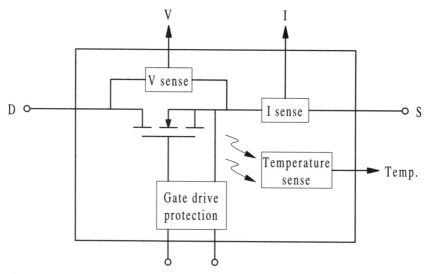

Figure 14.30 Sensor configurations for integration or for nonisolated switches.

current divider source terminal must be at the same potential as the main source terminal to provide accurate current division. For integrated sensing, this is not a problem, since the source terminal would normally be used as a reference node by the gate drive.

Example 14.5.2 A current sense FET provides a ratio of 950 : 1 between the main FET and the current divider FET. The designer wishes to avoid an op-amp in the isolated circuit, and instead places a resistance of 10 Ω between the two sets of source leads. The main MOSFET has an on-state resistance of 0.06 Ω. Of this amount, 0.02 Ω is resistance in the wire bonds of the device. Draw an equivalent circuit. What voltage is expected at the divider source terminal when the FET current is 20 A?

The value $R_{ds(on)}$ is the resistance between the drain and source terminals. The actual channel resistance is only a portion of this value. In the current-sense FET application, the drains and sources should all be connected in parallel. The current divides through the chan-

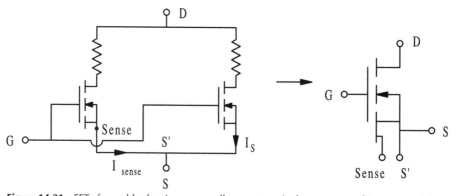

Figure 14.31 FETs formed by leaving some cell source terminals unconnected in a power MOSFET.

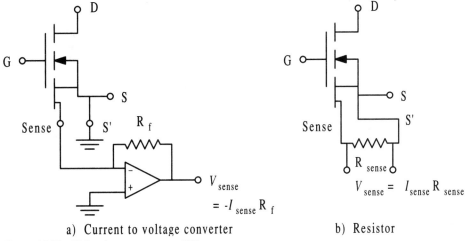

a) Current to voltage converter   b) Resistor

Figure 14.32   Using the current-sense FET.

nels, so we need the actual on-state channel resistance. For the main FET, this is 0.04 Ω ($R_{ds(on)}$ less the internal resistance in the wire bonds). The channel resistance of the current-divider FET is expected to have a value of (0.04 Ω)950 = 38 Ω. Its drain is already connected in parallel, and the source connection is made through the external 10 Ω resistor. The equivalent circuit, showing the various resistances, is shown in Figure 14.33. The current should divide as (0.04 Ω)/(48 Ω + 0.04 Ω). With 20 A total current, the current-divider FET should carry 16.7 mA. The sense voltage will be 0.167 V.

Figure 14.33   Equivalent circuit for current-sense FET.

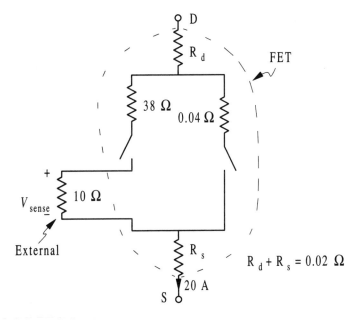

14.5.3 Nonelectrical Sensing

Gate drive isolation remains a barrier to good sensing. Integrated sensors can be used for protection, and can be used for control if the switch is at or near circuit common. In many cases, this *grounded sensing* method is not an option. In inverters for uninterruptible applications or ac motor drives, in most types of forward converters, in utility applications, and in other areas, the sensing techniques described thus far have limited usefulness. It is fundamentally difficult to measure small differences at rapidly swinging nodes in a network.

There are nonelectrical alternatives that support accurate sensing without compromising isolation. One example is a magnetic transformer, which can certainly be used to transfer ac information about voltage or current by means of magnetic flux. Another example is any of the various types of optocouplers. A thermocouple can sense temperature on a base plate or near a connection. Devices that use the Hall effect can sense dc current flow. These approaches are vital in almost all types of switching power converters.

Transformers for ac sensing can be designed much like any transformer. For sensing purposes, the terms *potential transformer* and *current transformer* (CT) are used to identify transformers specifically designed for voltage or current sensing, respectively. CTs are common in power converters. Although they cannot transfer dc information, a little creativity can often extract the necessary information from an ac flow.

Current transformers for power electronics often use just one or two turns on the high-current side, and perhaps a few dozen windings on the secondary. Figure 14.34 shows a CT installed to monitor the current in a high-side switch of a full-bridge inverter. If the turns ratio is 1 : 100, then the secondary should produce 100 mA with 10 A flowing in the primary, and so on. The secondary current can be measured with a resistor or a current-to-voltage converter.

Consider the current that will flow through the CT in Figure 14.34. It is a PWM waveform that follows both the switching function and the load. The CT will sense only the ac component. The current waveform and CT output waveform are compared in Figure 14.35. The CT output removes the dc component. However, there is enough information in the cir-

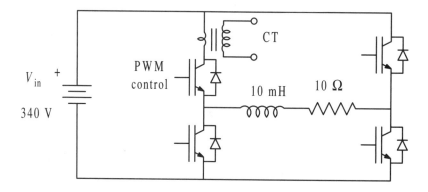

$$f_{switch} = 25\, f_{out}$$

Figure 14.34   Full-bridge inverter with 1 : 100 current transformer in one leg.

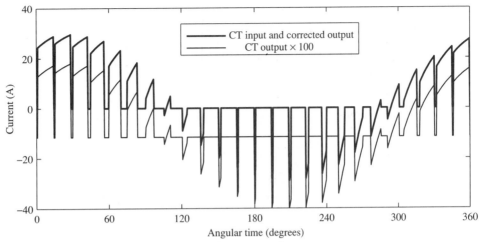

Figure 14.35 CT input, output, and corrected output waveforms.

cuit to recover the dc term: Since the current is exactly zero whenever the switch is off, a correction current can be added to the CT waveform so $i_{CT}(t) = 0$ during off intervals. When this correction is added, the exact current waveform is recovered.

One crucial distinction between a CT and other transformers is the dc rating. The current in Figure 14.35 has a significant dc component, and a transformer designed only for ac could saturate if the dc current is too high. It is vital that a CT be designed so the maximum allowed dc current will not saturate it. This is one reason for the low number of turns.

Example 14.5.3 A toroid with inside diameter of 2 cm, outside diameter of 3 cm, and thickness of 6 mm is made of ferrite with $\mu = 1000\mu_0$. It is wound with 100 turns of #28 AWG wire, then mounted on a single #10 wire expected to carry up to 50 A peak for a dc–dc conversion application. The current is a square wave with duty ratio of about 25%. Will this toroid meet the requirements? If so, what voltage will appear across a 1 Ω resistor used to terminate the secondary of the CT?

The core should avoid saturation because of dc current. In this case, the dc component is about $(0.25)(50 \text{ A}) = 12.5$ A. Of course, the safest practice would be to design for the full 50 A. The primary is a single turn. For ferrite with $B_{sat} = 0.3$ T,

$$\frac{Ni}{\Re A} < B_{sat}, \qquad i < (0.3 \text{ T})\frac{l}{\mu} \tag{14.9}$$

The magnetic path is a circle of diameter 2.5 cm and length 7.85 cm. Thus $i_{dc} < 18.7$ A. The core will work, provided the duty ratio does not increase much. The input current is a square wave, and the secondary current will be smaller by a factor of 100. The output current (and voltage on the 1 Ω resistor) is shown in Figure 14.36. The peak-to-peak value is 0.5 A, but the average is zero.

One way to avoid the dc limitations of a CT is to use a Hall-effect sensor. A sensor of this type uses a small sample of metal or semiconductor to sense a magnetic field directly.

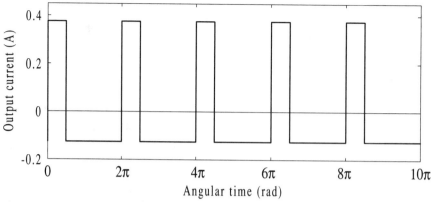

Figure 14.36 Current waveforms in CT for Example 14.5.3.

The Hall effect causes electrons or holes to be deflected by any flux as they travel through the material, so when the material is biased with a voltage, a third terminal will produce a potential proportional to the local flux. Since a current-carrying wire creates a flux around it, a Hall sensor gives a direct noncontact indication of the current.

A typical Hall-effect device is shown in Figure 14.37. A gapped core helps to direct flux to the chip. The chip has three terminals. The bias terminals are connected to a fixed supply, such as $+12$ V. The third terminal produces a voltage equal to half the supply when no current flows, and linearly proportional to the current for nonzero values. There is a maximum current that gives an output close to the supply value. Thus, with a fixed V_{bias} the sensor output is

$$V_{\text{out}} = \frac{V_{\text{bias}}}{2}\left(\frac{i}{I_{\text{max}}} + 1\right) \tag{14.10}$$

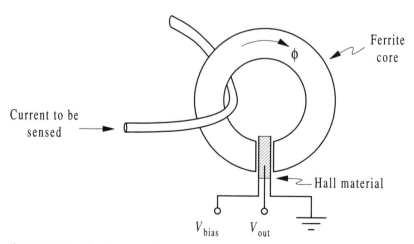

Figure 14.37 Hall-effect current sensor.

Hall devices come in maximum current ranges from about 30 A up to hundreds of amps. They are easy to use and provide good accuracy for currents ranging up to within about 20% of I_{max}.

Hall effect sensors have been used successfully in a wide variety of applications. They are the most popular type of current sensor in ac drives, and can be found in almost any industrial backup supply. The most important drawback of a Hall sensor is temperature sensitivity. Both the gain $1/I_{max}$ and the offset (the output voltage when $i = 0$) change with temperature. In a precision application, the temperature issue limits the accuracy of Hall devices.

Hall devices support some unusual sensing schemes. From equation (14.10), the output voltage is proportional to both a voltage and a current. If the bias voltage is made proportional to a voltage of interest in a circuit, the Hall device output is a direct measurement of power. A Hall sensor therefore can be used to measure power waveforms as well as current waveforms. If this sensor is further coupled to an integrator, a direct energy sensor is created.

Optocouplers can send either analog or digital information across an isolation barrier. Their biggest advantage is extreme isolation. Commercial couplers exist that can support 5 kV between the input and output. If fiber-optic cable versions are used, the isolation level can be almost any value. We saw a digital optocoupler in conjunction with gate drives. Analog couplers, such as the transistor-output version shown in Figure 14.38, tend to be slower, but still offer high isolation potentials.

The transfer function of an optocoupler is nonlinear. The LED is brighter as its current goes up, although not in a linear fashion. The LED output is, in effect, injected into the transistor base. Since β is not constant, the output collector current will not be a linear function of the light level. This problem can be avoided with a conventional technique: Bias the LED current to an intermediate level, then allow only a small-signal swing of current around the bias level. This approach produces excellent linearity from i_{LED} to i_c. The LED current can be obtained from a divider, or produced by an op-amp or other circuit connected directly to the switch. A voltage such as a switch V_{ds} value generates a current, then the output collector current allows convenient measurement of the voltage. There is no need for a precise system to measure small differences.

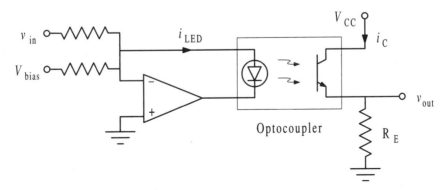

Figure 14.38 Analog optocoupler in a measurement application.

14.6 RECAP

Gate drive circuits provide the primary interconnection between a switching function and a power semiconductor device. Gate drives have considerable effect on the performance of power semiconductor devices. Good designs are required to provide the best possible performance from a given power electronic system. An ideal gate drive operates a switch as quickly as possible, uses no energy, and includes isolation and protection functions.

For voltage-controlled switches such as MOSFETs and IGBTs, the gate drive must act to charge the input capacitance quickly. Even though no continuous current is needed to maintain such a switch in its on state, the charging and discharging processes consume energy and require a circuit capable of high momentary current. Low-impedance gate drives work well in operating voltage-controlled switches. Gate drive impedances below 10 Ω are common in practice. Complementary emitter followers or equivalent CMOS circuits are usually used when high performance is required.

In general, MOSFETs and IGBTs have similar requirements for gate drives, although an IGBT of a given rating has lower input capacitance than a similar MOSFET. A MOSFET rated for 400 V and 10 A has typical input capacitance of about 1600 pF, and requires about 50 nC of gate charge and V_{gs} = 10 V to turn on. An IGBT with 600 V, 10 A ratings has about 1/3 as much capacitance and requires only about 20 nC of gate charge to switch. The nonlinear behavior of the capacitance often makes charge a more useful specification.

The energy consumed in a gate drive for a voltage-controlled switch can be estimated as the total charge required to switch, times the supply voltage from which the charge is drawn. The power is this energy times the switching frequency, so the gate drive power is

$$P_{\text{gate}} \approx Q_{\text{gate}} V_{\text{supply}} f_{\text{switch}} \tag{14.11}$$

For the BJT, a high-current base drive is required for switching. Power BJTs have low gain, with $\beta \sim 5$ not unusual. This means i_b is substantial. For fast turn-on and turn-off, an additional current pulse, possibly as high as the collector current, is applied at turn-on and at turn-off. Darlington combinations can be used to increase the effective value of β, but only at the expense of lower switching speed and extra forward drop. Proportional base drives use a current transformer to supply $i_b = i_c/\beta$. With a proportional drive, the gate drive circuit only needs the supply a large pulse of current for turn-on and turn-off.

Base drive power consumption is usually at least several watts for a power BJT operating in a circuit at 20 A or more. The power can be estimated by multiplying the base current by the gate drive supply voltage value, then by the switch duty ratio. Losses caused by the turn-on and turn-off pulses can be estimated by assuming that the pulses are portions of a sine wave.

SCRs, triacs, GTOs, and other thyristors require a brief current pulse to trigger their latching switch action. The energy levels are much lower than those of BJT base inputs. A common technique is to provide a supply level of up to about 20 V, then impose this potential on the SCR gate terminal through a 20 Ω resistor. The 1 A pulse that results will help speed the turn-on process, and the 20 V level will ensure that stray inductive and capacitive voltages within the switch will not cause reverse bias on the gate-cathode junction, even for large devices.

Pulse transformers are commonly used for SCR gate drives, since they provide isolation in addition to gate triggering. A pulse transformer is a transformer designed to handle a specific volt·second value. When the volt·second limit is exceeded, the transformer is permitted to saturate. Small toroids can meet SCR gate triggering requirements.

A GTO requires a large negative turn-off current pulse in addition to a small turn-on pulse. The negative pulse can have a magnitude on the order of 20% of the anode current—reflecting the model of the GTO as a pair of power transistors. The turn-off current is often handled separately from the turn-on pulse. It might use a pulse transformer or a high-current complementary emitter follower.

A triac can be triggered with either polarity of gate current. This is an important advantage in ac applications, since it facilitates the design of simple circuits for symmetric operation during positive and negative half-cycles.

Isolation is a critical issue in most gate drive circuits, since a power semiconductor switch is often not connected to circuit ground. Not only is off-ground operation essential, but operation must be reliable when the off-ground potential varies quickly. Stray capacitance can produce high $C(dv/dt)$ currents during switching, and these currents can affect gate drive operation. In converters such as three-phase hex-bridge inverters, the several gate drives must be mutually isolated as well.

Magnetic, optical, and charge pump techniques are common in creating isolated gate drives. A gate drive requires both a signal (the switching function) and energy. In many industrial converters, a multioutput forward or flyback converter provides several separate supplies for use with gate drives. With this approach, optocouplers can be used to transfer information, while the flyback converter provides energy. Photovoltaic energy transfer has also been used, although the power level is too low to support high-performance switching.

In many simple converters such as buck or half-bridge arrangements, there is a *high-side switch*, connected to the input rail. When this switch is a MOSFET or IGBT, a voltage about 10 to 15 V above the input rail will be necessary for successful switching. This extra voltage can be created with charge-pump methods instead of with a separate dc–dc converter. Charge-pump gate drives are popular in the automotive industry for the control of buck converters.

One alternative to high-side switching is to use P-channel or complementary devices in place of conventional N-channel transistors. This is not common because P-channel switches tend to be more expensive for a given rating than N-channel devices. However, the practice can simplify system design, especially in 5 V and 12 V applications. In inverters or other circuits that use a complementary pair, a common gate signal can be used to control both switches. The complementary devices will exhibit complementary response to the gate signal, so that $q_P + q_N = 1$ with minimal effort. An unfortunate drawback is that commutation overlap might occur. When this happens, a low-resistance path is presented to the input source, and a high *shoot-through current* can flow. Complementary switches require a *dead time* to allow current to commutate between the two switches without KVL problems.

Gate drives often make use of sensing, or their energy source might serve to power a sensor on an isolated switch. Simple voltage sensing methods, such as voltage dividers, are not effective for making differential measurements of small voltages in an isolated circuit. The error caused by the common-mode offset is too high for reasonable resistor tolerance levels. Instead, optocouplers, transformers, or noncontact sensors such as Hall devices are used to transfer information from a switch to a control circuit.

Many switching devices are being built with integrated sensing. Some IGBT modules incorporate temperature sensing and collector-emitter voltage sensing, and shut the device off if the values are outside acceptable limits. In MOSFETs, some of the parallel cells can be dedicated to a current divider. This supports current measurements with simple circuits, and permits accurate monitoring of the switch current without much power loss.

Current transformers are common for providing isolated information about current in a converter. Since portions of a switch current waveform (the off portions) are known in advance, the inability of a CT to transfer dc information can be avoided in many applications. The CT is connected in series with the switch, and its output waveform is shifted up so that the off-state current is zero.

Optocouplers can be used to transfer analog information, but they require a bias current and use a small-signal swing around that current for linearity. Discrete-valued information can be transferred with digital optocouplers.

PROBLEMS

1. A power MOSFET requires a gate charge of 90 nC for turn-on. If the gate drive can produce up to 2 A, how long is the turn-on time?

2. It is often desirable to use a TTL circuit for direct gate drive of a power MOSFET. An open-collector output can be used to provide the 10 V signal necessary for turn-on. It is unlikely that a pull-up resistor smaller than about 500 Ω will be feasible. How long will this type of gate drive require to turn on a MOSFET with $C_{iss} = 800$ pF?

3. Gate drive power for two MOSFETs is to be compared. The first device requires $Q_{gate} = 40$ nC for complete turn-on. A 12 V gate drive source with resistance of 50 Ω provides gate current of approximately 200 mA. The second device requires $Q_{gate} = 200$ nC. Its gate drive uses 12 V and 10 Ω, with gate current of about 1 A. How much power does each drive consume, given 200 kHz switching?

4. A low-resistance MOSFET has $C_{iss} = 4000$ pF and $V_{th} = 4$ V. Suggest a gate drive arrangement. Estimate the turn-on time and gate drive power consumption for your choice.

5. A complementary emitter follower is to be used for a MOSFET gate drive. The follower draws power from a +12 V supply. The MOSFET is a large device, with $C_{iss} = 15\ 000$ pF. The follower uses a speed-up capacitor, in parallel with a 20 Ω resistor, to connect to the gate terminal. The follower's output transistors gives it an effective output resistance of 1 Ω in addition to the 20 Ω part.

   a. Draw the circuit as described.
   b. How long will the gate require to reach $V_{gs} > 6$ V if no speed-up capacitor is used?
   c. How long will the gate require to reach $V_{gs} > 6$ V if the speed-up capacitor is 0.1 μF?

6. A MOSFET with gate drive is illustrated in Figure 14.39. Assume that the input capacitance is the major determinant of speed, and that the device is fully on when $V_{gs} > 6$ V. It is fully off when $V_{gs} < 4$ V. How long will the turn-on and turn-off processes require in this converter? If the gate drive input is a square wave with exactly 50% duty ratio, what is the duty ratio for the converter, reflecting the effects of the gate drive?

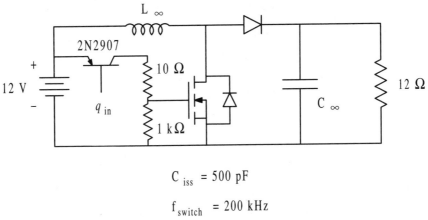

$$C_{iss} = 500 \text{ pF}$$

$$f_{switch} = 200 \text{ kHz}$$

Figure 14.39 Gate drive and converter for Problem 6.

7. The ratio I_c/I_b in a BJT switching application is often called the *forced beta* value, since the base current is set externally. In the circuit of Figure 14.40, the device manufacturer recommends a forced beta value of 8 to 10. Estimate the value of R_{base} that will provide this level of base current. The inductance is large.

8. A base drive for a BJT supplies peak current of 40 A, continuous on-state current of 6 A, and peak negative current of -25 A, similar to the shape in Figure 14.10. The base drive operates from ± 10 V sources. If the duty ratio is 50% and the turn-on and turn-off processes each require 4 μs, plot base drive power as a function of frequency between 1 kHz and 50 kHz.

9. One drawback of a proportional base drive is the turn-off process. The external base circuit must divert all the current from the transformer. In addition the base circuit must provide extra current sinking so that the net base current is close to $-I_c$. Consider a BJT intended to carry 75 A when on. The proportional drive transformer uses a 5 : 1 turns ratio. If the base drive uses ± 5 V sources, how much power is consumed in the gate drive during the turn-off process if the negative current flow lasts 3 μs? The switching frequency is 15 kHz.

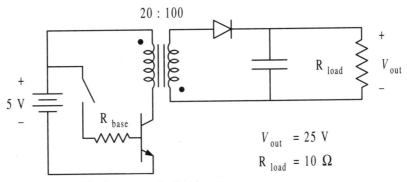

$$V_{out} = 25 \text{ V}$$

$$R_{load} = 10 \text{ } \Omega$$

Figure 14.40 Flyback converter for Problem 7.

10. An SCR gate drive is shown in Figure 14.41. The 2N2222 is driven by a pulse 20 μs in duration. The SCR gate-cathode junction can be modelled as a 1 V drop. The SCR itself has an anode-cathode drop of 1.5 V when on. Estimate the gate drive power for a switching frequency of 1.2 kHz. If the average SCR current is 20 A, estimate the ratio of gate drive power to device on-state loss.

11. Consider the triac gate drive and circuit shown previously in Figure 14.20. The ac input is a 120 V_{RMS} 60 Hz source. The Zener voltages are 5.1 V. The timing components are set to provide an RC time constant of about 1 ms. Determine the phase delay angle if the load resistance is much smaller than the timing resistance. What is the phase delay angle if the RC time constant changes to 2 ms?

12. Gate drive timing considerations are often set by the external circuit. An SCR-based midpoint rectifier with three-phase input of 230 $V_{RMS(line-to-line)}$ at 50 Hz supplies a load of 4 Ω in series with 100 mH. The gate drives use pulse transformers, and meet the "20 V, 20 Ω" empirical rule.

  a. Draw the circuit.

  b. Consider the initial turn-on (from $i_L = 0$). If the SCR holding current is 50 mA, how long will it take the current to ramp to this level when an SCR is turned on?

  c. Based on part (b), what duration of gate pulse will be necessary to ensure reliable start-up of this midpoint rectifier?

13. An SCR based on a full 100 mm silicon wafer can handle 5000 A_{RMS} and blocks up to 5000 V. Its gate requirements are 300 mA and up to 2.5 V. The device requires up to 500 μs for reverse recovery, and requires a similar time for turn-on if inductance does not limit di/dt. Suggest a gate drive circuit based on discharge of a capacitor. Determine the gate drive power consumption per device in a 60 Hz full-bridge rectifier application.

14. IGBTs are being used for a motor drive for a gantry crane. The input power source is 600 $V_{RMS(line-to-line)}$ at 50 Hz source. This is rectified into a dc bus, then inverted with the IGBTs for the motor drive. The IGBTs require about 500 ns to switch. They require total gate charge of 500 nC at 15 V from gate to emitter to ensure turn-on under worst-case conditions. The gate drive power supply comes from a flyback converter with 500 pF of stray coupling between primary and secondary.

  a. Assume the gate charge action is well represented by a constant capacitance. Recommend a gate drive impedance to charge this capacitance in well under 500 ns.

  b. How much current will flow in the stray capacitance of the flyback converter when a high-side device turns on? Compare this to the expected gate drive current.

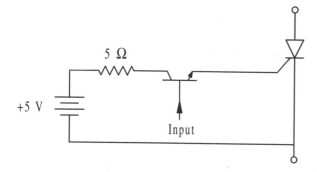

Figure 14.41   SCR gate drive for Problem 10.

15. An experimental dc–dc converter operates a MOSFET at 5 MHz. The MOSFET is expected to switch about 48 V in 20 ns in a high-side arrangement. The MOSFET input capacitance $C_{\text{iss}} = 200$ pF. A 9 V battery is used for the gate drive, but the layout creates about 200 pF of coupling between the MOSFET source and the circuit common node.

    a. Estimate the power consumed by the gate drive, ignoring the effects of stray coupling.

    b. Assuming that the charging process for the stray coupling capacitance ultimately produces power loss, how much extra loss does the stray element represent?

16. A half-bridge circuit based on complementary MOSFETs operates at 200 kHz. The devices have $R_{\text{ds(on)}} = 0.04\ \Omega$. The input source is 400 V.

    a. If the gate drive process allows 10 ns of shoot-through, estimate the power consumed in the shoot-through process. You may ignore stray inductances.

    b. If 10 ns of shoot-through occurs, but stray inductance limits the current rise to 400 A, estimate shoot-through losses.

17. A dual resistive divider supports measurement of V_{ds} in a 48 V to 5 V buck converter. A measurement accuracy of ± 0.5 V is desired for a protection function. All measurement circuitry can handle up to 5 V. Recommend a divider circuit, and specify resistor tolerance levels to meet the requirements.

18. For a vehicle application, a resistive shunt with $R = 0.2$ mΩ, rated up to 500 A, is used for battery current sensing. The batteries operate at up to 400 V, so it is desirable to use $100 : 1$ dividers for a safe measurement of the current. The resistors in the divider network should consume no more than 0.25 W total. Design a divider arrangement for this situation. What accuracy can be expected for the current measurement if 0.1% resistors are used in the divider?

19. A Hall-effect sensor draws power for a 12 V source. At 25°C, its output is 6.00 V with no current flow, 2.4 V with -100 A flow, and 9.6 V with $+100$ A flow. The output voltage is a linear function of the current within 0.5%.

    a. When the output reading is 4.00 V, what is the nominal current, and what is the tolerance associated with this current value?

    b. The Hall voltage has a temperature coefficient. If this coefficient is $+25$ mV/K, what current will provide output of 6.00 V at 100°C?

20. A ferrite toroid with $\mu = 5000\mu_0$ is to be used for a $200 : 1$ current transformer. If the toroid has outside diameter of 10 mm, inside diameter of 5 mm, and thickness of 5 mm, what is the highest dc current that can be tolerated?

REFERENCES

B. J. Baliga, *Power Semiconductor Devices*. Boston: PWS, 1996.

R. Chokhawala, J. Catt, B. Pelly, "Gate drive considerations for IGBT modules," in *IGBT Designer's Manual*. El Segundo, CA: International Rectifier, catalog IGBT-3, 1994.

D. A. Grant, J. Gowar, *Power MOSFETs: Theory and Applications*. New York: Wiley, 1989.

J. Kassakian, M. Schlecht, G. Verghese, *Principles of Power Electronics*. Reading, MA: Addison-Wesley, 1991.

K.-W. Ma, Y.-S. Lee, "Technique for sensing inductor and dc output currents of PWM dc-dc converter," *IEEE Trans. Power Electronics*, vol. 9, No. 3, pp. 346–354, May 1994.

Motorola, Inc., "Gate drive requirements," in *TMOS Power MOSFET Device Data*. Phoenix: Motorola, catalog DL135/D, Rev. 5, 1995.

Motorola, Inc., "Thyristor drives and triggering," in *Thyristor Device Data*. Phoenix: Motorola, catalog DL137/D, Rev. 5, 1993.

PART IV

CONTROL ASPECTS

CHAPTER 15

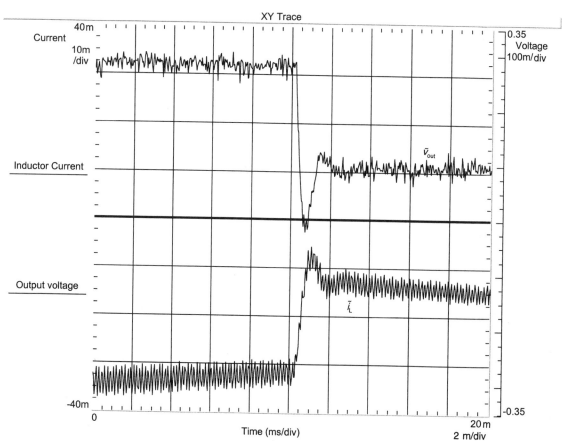

Figure 15.1 Transient response of a flyback converter to a 10% load change.

OVERVIEW OF FEEDBACK CONTROL FOR CONVERTERS

15.1 INTRODUCTION

The definition of power electronics emphasizes *control* of energy flow. One of the most significant advantages of switching conversion is that it supports automated control action. This is essential for proper operation of almost any power electronic system. Without control, a converter is constrained by its environment. Many loads require very tight regulation of their supply voltage. Control is much more than selecting the correct switching functions and a reliable gate drive. The switching functions must adjust to maintain precise operation, and adjustment must be performed continuously whenever the converter operates.

With *feedback control*, the converter output and other internal voltages and currents are measured, then this information is used to adjust operation to obtain a desired result. Feedback control is considered essential in high-performance power electronics. A related concept, termed *feedforward control*, uses information about the input waveform or the system behavior to help determine the correct system operation. Many power electronic systems use combinations of feedforward and feedback to minimize error between the actual and desired behavior. The usual purpose is good regulation: the desire to hold a constant output under changing conditions. A second purpose is *dynamic response* suggested by the transient behavior shown in Figure 15.1: It is desired that a power supply act quickly to correct errors.

In this chapter, basic control concepts are introduced in the context of power electronic systems. The regulation problem is revisited to demonstrate the need for feedback. The discussion requires key definitions and concepts from feedback system theory. These are presented with a few of the special concerns unique to power conversion. Important concepts such as stability are examined. The basic feedback process is discussed.

Later in the chapter, some general modelling approaches for control are given. Feedback implementation is described for dc–dc converters and rectifiers. Models for control design are covered further in Chapter 16.

15.2 THE REGULATION AND CONTROL PROBLEM

15.2.1 Introduction

Real power converters do not provide perfect regulation unless an external control adjusts their operation. Most converters produce an output dependent on the input, and do not provide inherent line regulation. The effects of ESR, the voltage drops on semiconductor switches, and even wire resistance will make operation load dependent. Switch residual voltages depend on temperature. Components change over long periods of time.

To achieve good regulation, the phase angle of a rectifier must be adjusted to cancel the effect of line variation. The duty ratio of a dc–dc converter must be altered to account for switch voltage drops. The modulating function of an inverter must change to compensate for temperature effects in switches and other parts. Some earlier examples have addressed the need for control. An ideal buck converter, for instance, achieves perfect line regulation if the duty ratio is adjusted as the inverse of the input, such that $D = V_{out}/V_{in}$. An active control circuit will be needed to sense the input and adjust duty ratio accordingly. Load regulation effects in a nonideal converter can be taken into account by monitoring the actual output voltage and then adjusting D as needed. How can a converter be made to function as an ideal source in the face of uncertain conditions? What are the limitations involved, and the ultimate level of performance that can be achieved? How should control functions be implemented?

15.2.2 Defining the Regulation Problem

Good performance means good regulation and fast dynamics. The design objective might be to produce a fixed dc output. It might be to generate a well-defined sine wave independent of the dc input energy source. It might be to make operation insensitive to temperature extremes or to any number of other factors. Regulation and insensitivity must be addressed in two very different contexts:

Definition: Static regulation *is concerned with the effects of changes on the steady-state operation of a converter. Static regulation can be expressed in terms of partial derivatives or ratios, as discussed in Chapter 3.*

Definition: Dynamic regulation, *or* tracking, *is the ability of a system to maintain the desired operation even during rapid changes. The term also refers to the ability to follow time-varying outputs.*

Equally important is the converter's reaction to large changes or to problems. For example, any converter must cope with a starting transient when the energy source is first connected. We want the converter to reach the desired steady-state operation without difficulty. In practice, it is common to see momentary short circuit loads during installation, and mistakes such as polarity mismatches or confusion between input and output are common.

The user of a power electronic circuit usually has little interest in worrying about regulation or large-signal operation. The converter must be automated enough to cope with expected problem situations without requiring the end user to be trained in power electronics. There are two ways to cope with the issues:

1. Design a system that is inherently insensitive to changing conditions. In a simple buck dc–dc converter, for example, the load regulation is perfect for an ideal converter. This insensitivity is very helpful for static regulation.
2. Provide the system with information about operating conditions, and then design controls to adjust as needed.

The most common approach to the second method is to monitor the output, and make adjustments to keep the output very close to the desired value.

15.2.3 The Control Problem

Control in a power converter is performed only by altering switch action. This fundamental limitation is a special characteristic of power electronics. The control problem in power electronics is part of the software problem introduced in Chapter 2: How should a switch matrix operate to provide the desired function? Regulation is a primary control concern, but does not give the entire picture. The control must ensure *stability* so that a converter will return to the desired operating point after a disturbance.

> **Definition:** *A system is said to be* stable *if it returns to the original operating conditions after being disturbed or altered.* Stability *can be defined in terms of small disturbances, large ones, or specific types of changes in the load or other conditions. A variety of specific mathematical definitions of stability are used in the literature.*

Stability is needed for small, fast disturbances such as noise pulses, for large disturbances such as loss of load or startup, and for continuous small disturbances such as ripple. The control should provide *robustness*, meaning that uncertainty in component values will not alter the basic action. The control must be fast enough to be of use with rapidly changing loads.

In practice, it is common to separate large-signal and small-signal aspects of the control problem. A startup circuit ensures success as a converter begins to operate. Small-signal feedback controls provide stability when noise is imposed. Overcurrent detectors shut off the circuit if short circuits develop. Crowbar circuits divert current if the output voltage rises too high. Short-term large-signal effects are often outside the scope of regulation specifications. A likely future trend is to combine large-signal and small-signal controls to build converters with broad application.

15.3 REVIEW OF FEEDBACK CONTROL PRINCIPLES

15.3.1 Open-Loop and Closed-Loop Control

A mathematical model of a system is the starting point for control. It is natural for a circuit designer to focus on the loop and node equations that describe the system of interest. For more general systems, equations of force, motion, and energy will be used to develop the model. In power electronics, the equations are those of the circuits, combined with the switching functions. The *input–output* behavior of the system can be defined as the output behavior in response to these control inputs.

Open-loop control is the simplest way to operate a system. In this case, the control inputs are independent variables, and are not adjusted. For example, a dc–dc converter can be operated open-loop, with switching functions having fixed duty ratio. An open-loop controlled rectifier, such as the one in Figure 15.2, will produce an average output proportional to the cosine of the phase angle and to the peak of the sinusoidal input. The limitation of open-loop control is that no corrective action can take place. In the rectifier, the output will match a specific desired value only if α is set correctly. However, even if α is correct for a particular set of conditions, a change in source voltage will alter the output.

Open-loop control does not cancel the effects of changes, but still can be very useful. A spacecraft deep in interplanetary space can save fuel by allowing itself to fall freely through local gravitational fields. A six-legged robot can walk over solid terrain with simple foot motions without regard to surface roughness or rocks. An open-loop rectifier gives a useful dc output if some variation can be tolerated. An ac induction motor can drive a load at nearly fixed speed without the complication of sensors and controls.

Although open-loop control is simple, and indeed often preferred if it can meet the needs of a user, it is not sufficient in general for power converter regulation or for many other everyday systems. If an automobile reaches a curve in the road, open-loop steering control is a problem. If a drifting spacecraft begins to spin, open-loop control will not keep its radio antenna pointed at the earth. If a robot with open-loop control topples over, it will not take action to right itself.

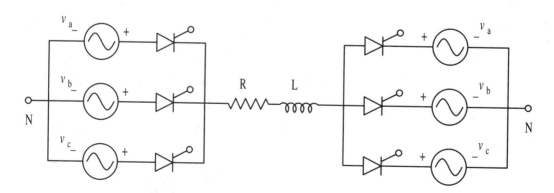

Fixed delay angle α Delay angle $\alpha + 180°$

Figure 15.2 Open-loop controlled rectifier.

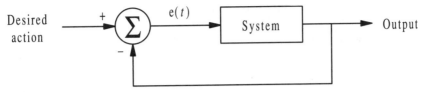

Figure 15.3 Closed-loop system based on output feedback.

Closed-loop or feedback control makes a measurement of the output or some other variable in a system. The measurement can be compared to the desired action. An *error signal* e(t) is developed, and the control input is altered to account for this error. The concept is given in Figure 15.3. Converter regulation is an excellent example. Other examples include heating systems that operate with a thermostat, driver action in an automobile, and corrective movements made by a baseball player attempting to catch a fly ball.

Consider the controlled rectifier: The output average voltage is a linear function of cos α. If α increases (from $0°$), the output decreases and vice versa. If the output average value is monitored, the controller can increase or decrease α to correct any error relative to a desired reference value V_{ref}. A system that uses an error signal to make this adjustment is shown in Figure 15.4. Just about any converter with a direct relationship between the control input and the output can use a feedback loop similar to the one in this figure.

An engineer seeks a *control parameter* that can serve as the command input for a system.

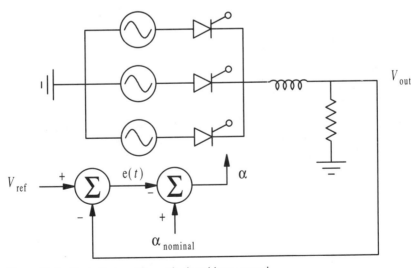

Figure 15.4 Controlled rectifier with closed-loop control.

> **Definition:** A control parameter *is a value that provides a way to alter the output or other desired variable. In power electronics, the conventional control parameters include duty ratio and phase. In resonant converters, on-time or off-time might serve as control parameters instead.*

Sometimes control parameters appear as functions. In a rectifier, the function $\cos \alpha$ is simpler to use in a control loop than the phase angle itself. In a PWM inverter, the modulating function can be treated as the control parameter.

Closed-loop control brings with it a paradox: The error signal is intended to be zero. If control action is successful in making $e(t) \rightarrow 0$, then no error signal will be available to drive the control parameter. Two methods are commonly used to avoid this *zero-error paradox*. The first is *high-gain control*, in which an amplifier is applied to the error signal. In this way, even very small errors generate strong control action. The second is *integral control*, in which the signal $\int e(t)\, dt$ is used. This integral produces nonzero output even when $e(t)$ reaches zero.

High-gain control can produce instability. Consider again a controlled rectifier, in which the difference between the actual average output and a desired reference value is used to adjust $\cos \alpha$. The representation is

$$\cos \alpha = k(V_{\text{ref}} - V_{\text{out}}) \tag{15.1}$$

where k is the feedback gain. If the output is a bit too low, $\cos \alpha$ will increase to raise it. The trouble is that $\cos \alpha$ is limited to the range -1 to 1. A high-gain value, such as 1000, will drive the control to the maximum value of $\cos \alpha$ even with an error of only 1 mV. This increase in $\cos \alpha$ quickly produces a negative error, and $\cos \alpha$ is driven to -1. The system will *chatter* back and forth between the extreme values $\cos \alpha = 1$ and $\cos \alpha = -1$. In any real system, the feedback gain should not be set above some maximum value. Integral control also can create instability, especially if fast disturbances occur. The challenges to the control designer are to choose gain values, to select the control method, and to add such elements as might be necessary to give good performance without instability.

15.3.2 Block Diagrams

Once a system is expressed mathematically, it can be represented in block diagram form. A typical example is a dc motor. A circuit model for a dc motor is given in Figure 15.5. The circuit has the equation

$$v_a(t) = i_a R_a + L_a \frac{di_a}{dt} + k_t \omega \tag{15.2}$$

From this expression and from the figure, the voltage v_a serves as the control parameter, while i_a is a dependent variable. The speed ω or the current i_a can be taken as outputs.

Figure 15.5 Circuit model for dc motor.

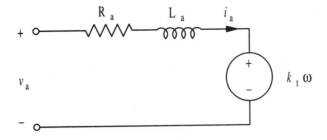

Equation (15.2) can be rewritten as

$$\frac{di_a}{dt} = \frac{v_a - i_a R_a - k_t \omega}{L_a} \tag{15.3}$$

with solution

$$i_a(t) = \frac{1}{L}\int [v_a(t) - i_a R_a - k_t \omega]\, dt \tag{15.4}$$

Figure 15.6 shows a block diagram form of (15.4). Addition is represented by the summation symbol within a circle. A constant value within a rectangle represents scalar gain. Mathematical operations such as integration or differentiation are shown explicitly within a block. The diagram demonstrates that there is current and speed feedback in the motor behavior, with feedback gains R_a and k_t, respectively. In a dc motor, the shaft output torque is $T_e = k_t i_a$. Newton's Laws relate the net torque to the angular acceleration, and

$$T_e - T_{\text{load}} = J\frac{d\omega}{dt} \tag{15.5}$$

where J is the moment of inertia. Figure 15.7 shows a more complete block diagram, in which the motor drives a mechanical load and the speed is the output.

The block diagram of Figure 15.7 shows feedback loops, but no control has been applied to the input voltage. Therefore, the diagram represents the open-loop dynamics of a dc motor. For closed-loop control, speed or current will be measured, and an error signal will be used to adjust v_a. In the diagram, higher v_a will result in higher i_a and T_e. Assuming that

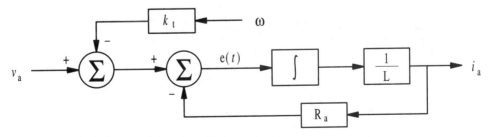

Figure 15.6 Block diagram of dc motor with i_a as output.

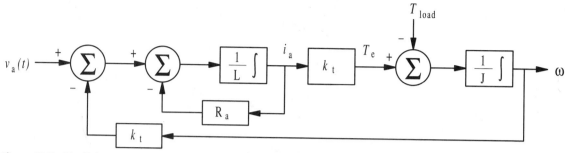

Figure 15.7   Block diagram of dc motor with speed as output.

the motor load moves faster if more torque is applied, higher v_a results in higher speed. A speed control can be developed, based on feedback with

$$v_a = k_\omega(\omega_{\text{ref}} - \omega) \tag{15.6}$$

If speed is below the reference value, positive v_a will increase the torque and speed up the motor. If speed is too high, negative v_a will slow the motor down. This speed control system, with a tachometer for feedback, is shown in Figure 15.8.

The differential equations of a system lend themselves to block diagram format. The equations have solutions expressed as integrals of the independent and dependent variables. Dependent variables will appear as feedback loops. Parameters will appear as gains. Non-linear blocks such as limiters, multipliers for dependent variables, or switches are common in power electronics.

15.3.3 System Gain

If the system is described by a set of linear differential equations, it is a linear system. Solutions of the form $c_1 e^{st} + c_2$ would be expected for each equation in a first-order set. In this case, Laplace transforms can be used in place of the equations themselves. The differentia-

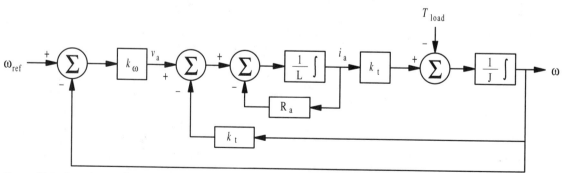

Figure 15.8   Speed control system for dc motor.

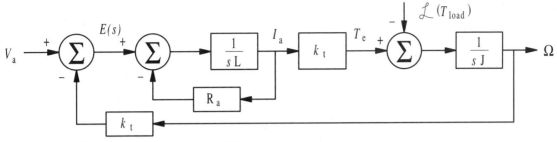

Figure 15.9 Laplace transform representation of the open-loop dc motor.

tion process is equivalent to multiplication by s, and integration is equivalent to multiplication by $1/s$. In the dc motor example, the Laplace transform of the block diagram representation is given in Figure 15.9. A gain function $G(s) = \Omega(s)/V_a(s)$ can be defined, where $\Omega(s)$ and $V_a(s)$, respectively, are the Laplace transforms of the output speed and armature voltage time functions. This represents the open-loop system gain or *open-loop transfer function*, and the block diagram of Figure 15.7 can be replaced with a single gain $G(s)$.

A feedback control loop in a linear system can also be represented with a Laplace transform. For the unloaded dc motor, the system with speed control loop can be represented as a combination of an open-loop gain $G(s)$ and a feedback loop gain $H(s)$, as shown in Figure 15.10. This provides a relationship between the input speed reference $\Omega_{ref}(s)$ and the output speed $\Omega(s)$. Consider that

$$[\Omega_{ref}(s) - H(s)\Omega(s)]\, G(s) = \Omega(s) \tag{15.7}$$

Solving for $\Omega(s)$, the result is the *closed-loop transfer function*

$$\Omega(s) = \frac{G(s)}{1 + G(s)H(s)}\Omega_{ref}(s) \tag{15.8}$$

The symbol $K(s)$ will be used for this function, so that $\Omega(s) = K(s)\Omega_{ref}(s)$. The error function $E(s) = \Omega_{ref}(s) - \Omega(s)H(s)$ is given by

$$E(s) = \frac{1}{1 + G(s)H(s)}\Omega_{ref}(s) \tag{15.9}$$

Notice that the error can be made small with a large value for $H(s)$. This is exactly the principle of high-gain feedback. It is often convenient to move the feedback loop gain $H(s)$

Figure 15.10 Transfer function representation with feedback.

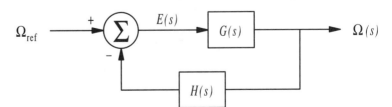

through the summing block in Figure 15.10 to form an *equivalent open-circuit transfer function*, $G(s)H(s)$. This function represents the signal to be subtracted as feedback, and is very useful for stability analysis. The equivalent feedback is unity in this case.

If the open-loop and feedback-loop gains $G(s)$ and $H(s)$ are positive, the open-loop gain is always higher than the closed-loop gain. The effect is that more *control energy* will be necessary for a closed-loop system than for the open-loop case. In many systems—particularly power converters—there are physical limits on the input signals. Thus physical limitations and stability constraints limit the practical value of $H(s)$. Even moderate values of feedback gain offer performance improvements, however.

Example 15.3.1   An electronic amplifier is to be used to generate the input for a dc motor control. The amplifier is not very precise. Its gain is 100 ± 10% over the full input range. The output cannot exceed 100 V, and the available input does not exceed 10 V. In the open-loop case, a voltage reference signal between 0 and 1 V is applied. It is desired to build a closed-loop control for this amplifier to improve precision. Suggest an approach.

Here is a simple solution: Connect a voltage divider to the amplifier output, and use negative feedback in an attempt to correct the input. Both the open-loop and feedback loop gains will be constants, and the Laplace transform of a constant is simply the constant itself. The open-loop gain is approximately $G(s) = 100$. Since the output does not exceed 100 V and the input does not exceed 10 V, the closed-loop gain should not be less than 10 to allow full usage of the amplifier range. (A lower value makes it impossible to obtain 100 V at the output.) This requires

$$\frac{G}{1 + GH} \geq 10, \qquad \frac{100}{1 + 100H} \geq 10, \qquad H \leq 0.09 \qquad (15.10)$$

If we choose $H(s) = 0.09$, as in Figure 15.11, the system gain will be approximately 10. What about error? In the open-loop case, the gain could be anywhere between 90 and 110. In the closed-loop case, $H(s) = 0.09$ in equation (15.10) gives system gain between 9.89 and 10.09, for a value of 10 ± 1.1%. The gain error has been reduced by a factor of 9 even with a very small feedback gain. The original low-quality amplifier can be used to produce high-quality output through the use of feedback. As long as the feedback divider is precise, the precision of the overall system improves dramatically.

In power electronics, the control input must be distinguished from the input energy source. The function $K(s)$ is sometimes called the *control-to-output transfer function* to avoid confusion.

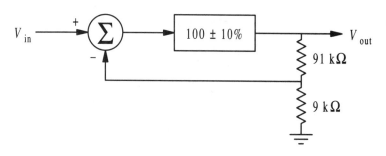

Figure 15.11   Feedback amplifier.

The Laplace transform approach to analysis is valuable for control design, although the concept is valid mainly for linear systems. One additional advantage is that the transfer functions are relatively easy to measure. A unit impulse function has a Laplace transform of 1. The open-loop response to an impulse is $G(s)$, while the closed-loop response is $K(s)$. A step input has a transform of $1/s$, and gives the integral of $G(s)$ or $K(s)$. Response measurements can help identify linear approximations to a nonlinear system. In Chapter 16, other methods for constructing linear models will be introduced.

15.3.4 Transient Response

The *transient response* of a system specifies the time behavior following a disturbance. In power electronics, the transient response can be expressed as the time to return to a small output error following a significant disturbance. It can also be expressed as the peak error value during a transient or as a combination of time and error. The transient response of any power electronic system is limited in part by the switching rate: When a disturbance occurs, no action can be taken until the next time a switch operates. In the worst case, switch action can impose a delay of up to one cycle before any control action occurs.

Transient response of a linear closed-loop system is analyzed conveniently by considering a frequency-domain representation of $K(s)$. The complex frequency $j\omega$ is substituted for s, then the magnitude and phase of $K(j\omega)$ are plotted as a function of ω. Figure 15.12 provides an example of such a representation, called a *Bode plot* or a *Bode diagram*. The gain magnitude is usually expressed in decibels (a voltage or current ratio in decibels is twenty times the common logarithm of the ratio).

The frequency domain approach makes it easy to represent the gain values, to determine the relative effect of noise at various frequencies, and to determine the range of frequencies that can be corrected effectively by the control. The concept of *system bandwidth* is an example of the utility of this approach.

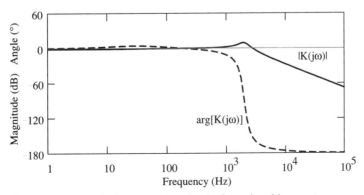

Figure 15.12 Sample frequency domain plot for a closed-loop system.

> *Definition:* System bandwidth *represents the highest frequency of disturbance (in the load or source, for power converters) at which there is enough overall gain to correct the error. The standard definition is the point at which the closed-loop gain magnitude is down 3 dB from its low-frequency value.*

In a converter, the input source might experience a disturbance, such as line frequency harmonics or external noise. The *source-to-output transfer function* $V_{out}(s)/V_{in}(s)$ should be low at the disturbance frequencies to reject source noise. This transfer function is called the *audio susceptibility* (based on older terminology from systems engineering), since it provides a measure of the effect of ac noise at the output. Ideally, the audio susceptibility should be low. Ratios below -40 dB are obtained in many power supplies. Audio susceptibility in effect provides a transient measure corresponding to line regulation.

Load disturbances can be represented with the transfer function $V_{load}(s)/I_{load}(s)$. This is the system output impedance—ideally 0 for voltage source behavior and ∞ for current source behavior. This provides a transient measure corresponding to load regulation. Load disturbances or output noise will have minimal effect on the output voltage if the output impedance is much less than the load impedance.

15.3.5 Stability

When a system is stable, its steady-state operation is consistent and predictable. For a given input, the system will reach a steady-state *operating point*, defined by the voltages, currents, and energy levels after transients have died out. Short disturbances or impulse inputs might alter the output temporarily, but the system will return to the operating point. In linear systems, the return to the operating point takes the form of an exponential decay. The point to which a stable system converges is termed an *equilibrium point*.

Stability can be analyzed through the closed-loop transfer function $K(s)$ or through the open-loop transfer function $G(s)H(s)$. The function $K(s)$ can be written as a ratio of polynomials in s, $P(s)/Q(s)$, and each of these polynomials can be factored. The roots of the numerator represent *zeroes* of the transfer function, while roots of the denominator represent *poles* or *singularities* of the function. The basis of the Laplace transform is that the solution of a linear system can be written as a sum of exponential functions. Consider a system exposed to a noise impulse. The system is stable if its response to the impulse decays over time. This requires each of the exponential terms to have a negative exponent. The requirement that the response must decay away over time is equivalent to requiring that all roots of the denominator polynomial $Q(s)$ have negative real parts. Imaginary parts represent oscillatory behavior, while negative real parts represent exponential damping.

Stability can also be determined by analysis of feedback behavior. A system will be unstable if a disturbance grows with time. This occurs if there is positive feedback gain of the disturbance: A small disturbance will produce a higher output, and the positive feedback will reinforce the disturbance. The output will grow larger with time. To avoid this effect, two alternatives are possible:

1. Ensure that a disturbance is attenuated instead of amplified. Then the output will be smaller than the input.

2. Ensure that the feedback is negative, so that the disturbance is cancelled and not reinforced.

Negative feedback is assumed in the closed-loop transfer function, as illustrated in Figure 15.10. However, if the transfer function $G(s)H(s)$ introduces a phase shift of 180°, the feedback taken from the output will become positive. The frequency domain rule for stability can be stated as the *Nyquist criterion*. One statement of this criterion is the gain-phase rule:

Empirical Rule: *To ensure stability for a control system, the open-loop gain* $|G(s)H(s)|$ *must be less than 0 dB whenever the phase reaches or exceeds 180°.*

This rule is accurate for systems in which an increase in feedback gain eventually creates instability—the usual behavior in power conversion systems. One of the most powerful attributes of the Nyquist criterion, and this frequency domain statement of it, is that the concept of *relative stability* can be defined. We can ask, for example, how close the phase angle is to 180° when the gain drops to 0 dB. Or we might ask the value of gain at the point when the phase reaches 180°. Figure 15.13 shows a few Bode plots of the function $G(s)H(s)$ for various system characteristics. Three of the plots illustrate stable systems, while the fourth system is unstable. The system illustrated in Figure 15.13c is stable, based on the Nyquist criterion; however, the phase angle is just a bit less than 180° when the gain drops to 0 dB. In system (c), small component errors, minor changes, or an extra time delay could shift the phase characteristic slightly, and the system would become unstable. Three terms are used to describe relative stability.

Definition: *The* crossover frequency *of a system is the frequency at which the open-loop gain is 0 dB.*

Definition: *The* phase margin *of a system is the difference angle* $180° - \phi_c$, *where* ϕ_c *is the phase of the transfer function* $G(s)H(s)$ *at the crossover frequency.*

Definition: *The* gain margin *of a system is inverse of the open-loop gain magnitude at the frequency where the phase angle is 180°. Thus if the gain is* -16 *dB at an angle of 180°, the gain margin is 16 dB.*

Control engineers choose values of gain or phase margin to ensure good operation of a system. The values are linked to specific time responses, such as overshoot and settling time. Many designers prefer phase margin of at least 60°. Gain margin is used less often, but values of 12 dB or more are typical targets.

In second-order systems, the damping parameter ξ (see Section 8.2) is closely related to the phase margin. Margin of 45°, for example, corresponds to ξ equal to about 40% of the resonant frequency, while 60° margin yields about 60%. A system with low damping has substantial overshoot in response to a step function. Damping at 60%, for instance, is asso-

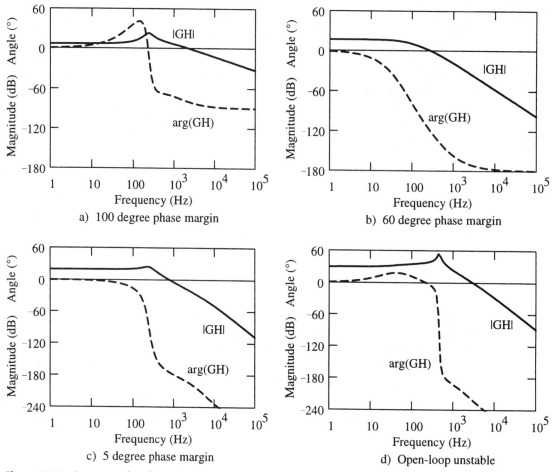

Figure 15.13   Four examples of open-loop responses.

ciated with overshoot of about 10%, while 40% damping produces about 25% overshoot. More damping can be provided if overshoot is a significant issue (as is often the case in power electronics), but the time required to reach the equilibrium point will lengthen with more damping.

Example 15.3.2   A dc motor closed-loop control is shown in Figure 15.14. It is used to operate a load at approximately constant speed. The load is governed by Newton's Second Law for rotational systems,

$$T_e - T_{\text{load}} = J\frac{d\omega}{dt} \tag{15.11}$$

where T_e is the motor output torque $k_t i_a$, J is the moment of inertia, and T_{load} is a mechanical load torque described by the linear expression $T_{\text{load}} = c\omega$. The various values for the mo-

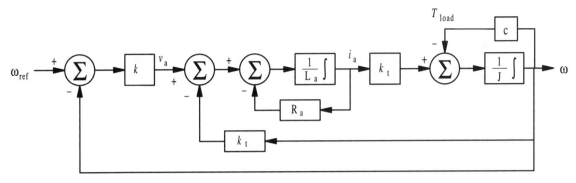

Figure 15.14 Block diagram for Example 15.3.2.

tor are $R_a = 1\ \Omega$, $L_a = 20$ mH, $k_t = 1$ N·m/A, $c = 0.02$ N·m·s, and $J = 0.04$ kg·m^2. The tachometer gain (for speed feedback) is 0.01 V · s/rad. Given a gain k, with $v_a = k(\omega_{ref} - \omega)$ for a value $\omega_{ref} = 100$ rad/s, find the open-loop transfer function and plot a Bode diagram for it. Are there any stability limitations on the value of k? If so, suggest a value of k to provide phase margin of 45°. What is the steady-state speed?

The various loops in the block diagram can be reduced through algebra to give an overall relationship between ω_{ref} and the output speed, ω. For instance, the torque $T_e = k_t i_a$, so the last loop with the load gives the relationship

$$\omega = \frac{k_t i_a - c\omega}{sJ} \tag{15.12}$$

The loop for armature current i_a has the relationship

$$i_a = \frac{v_a - k_t\omega - R_a i_a}{sL_a} \tag{15.13}$$

Equation (15.12) allows i_a to be written as

$$i_a = \omega\frac{c + sJ}{k_t} \tag{15.14}$$

When this is substituted for i_a in equation (15.13), the result is

$$\omega\frac{c + sJ}{k_t} = \frac{v_a - k_t\omega - R_a\omega(c + sJ)/k_t}{sL_a} \tag{15.15}$$

In the open-loop case, with $v_a = k\omega_{ref}$, the transfer function ω/ω_{ref} reduces to

$$G(s)H(s) = \frac{\omega}{\omega_{ref}} = \frac{kk_t}{s^2JL_a + sR_aJ + scL_a + R_ac + k_t^2} \tag{15.16}$$

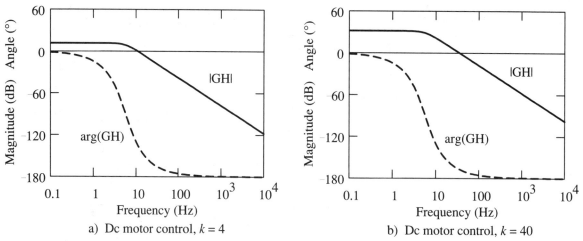

Figure 15.15 Bode diagram corresponding to equation (15.16) for two gain settings.

In the complete closed-loop system, the ratio $\omega/\omega_{\text{ref}}$ reduces to

$$K(s) = \frac{\omega}{\omega_{\text{ref}}} = \frac{kk_t}{s^2JL_a + sR_aJ + scL_a + R_ac + kk_t + k_t^2} \tag{15.17}$$

Stability requires that the denominator polynomial of $K(s)$ has roots with strictly negative real parts. With the parameters given in the problem statement, this is true for any positive value of k, so there are no direct stability limitations. The Bode diagrams for $k = 4$ and $k = 40$ are given in Figure 15.15. Even though any value of k will work in principle, there is a phase margin issue that might affect performance. With $k = 4$, there is phase margin of about 45°, while with $k = 40$, the phase margin is less than 15°. Since the details of the model and behavior are not known, the higher phase margin will probably be needed for design.

The steady-state speed can be found by setting $s = 0$ in (15.17). This is because s represents a frequency, and the steady-state value is equivalent to the response at dc. With these parameters,

$$\omega = \frac{k\omega_{\text{ref}}}{1.02 + k} \tag{15.18}$$

If $k = 4$, the speed will be just about 20% below ω_{ref}. If $k = 40$, the speed will be about 2.5% less than ω_{ref}. In the limit of high gain ($k \rightarrow \infty$), the steady speed will match ω_{ref}.

15.4 CONVERTER MODELS FOR FEEDBACK

15.4.1 Basic Converter Dynamics

It would be helpful to study the open-loop behavior of converters before trying to design closed-loop controls. Let us consider samples of a few converter types, and evaluate their open-loop performance with step-change tests. Figure 15.16 shows two dc–dc converters—a buck and a flyback—designed for similar loads and ripple levels. Three open-loop tests will be performed, beginning from steady-state periodic conditions with 40% duty ratio:

1. Test the control-to-output behavior by altering the switching function for 60% duty ratio.

2. Test the input–output behavior by imposing a 5 V step increase in the source voltage.

3. Test the output behavior with a 25% decrease in the load resistance.

Results of these three tests for both converters are illustrated in Figure 15.17. None of the tests suggests unstable behavior. Except for ripple, there is little to suggest that the systems are nonlinear. The responses are very much like those of a conventional second-order linear system. The buck converter responds more quickly than the flyback because its output capacitor can be much smaller for a specified ripple level.

Figure 15.18 shows a rectifier and a voltage-sourced inverter for open-loop testing. The test sequence is similar:

1. Test the control effect with a 30° step in the phase delay for the rectifier and the relative phase delay for the inverter.

2. Test the input effect with a 25% step in the input voltage amplitudes.

3. Test the output effect with a 25% decrease in load resistance.

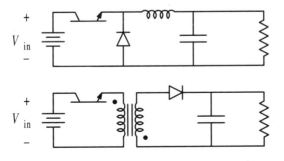

Figure 15.16 dc–dc converters for open-loop tests.

Figure 15.17 Open-loop transient response of dc–dc converters.

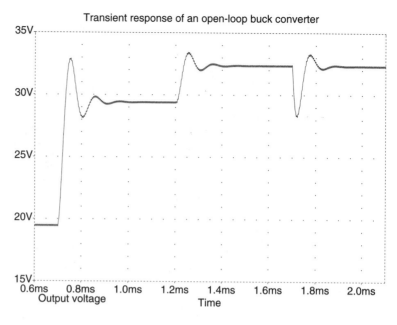

Transient response of an open-loop buck converter

Output voltage

Time

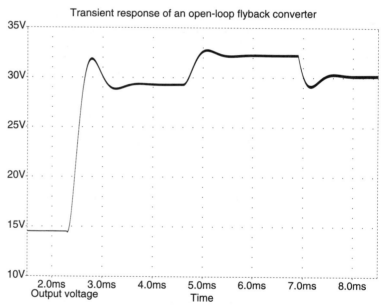

Transient response of an open-loop flyback converter

Output voltage

Time

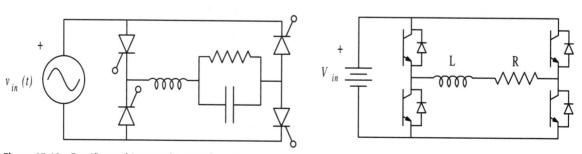

Figure 15.18 Rectifier and inverter for open-loop tests.

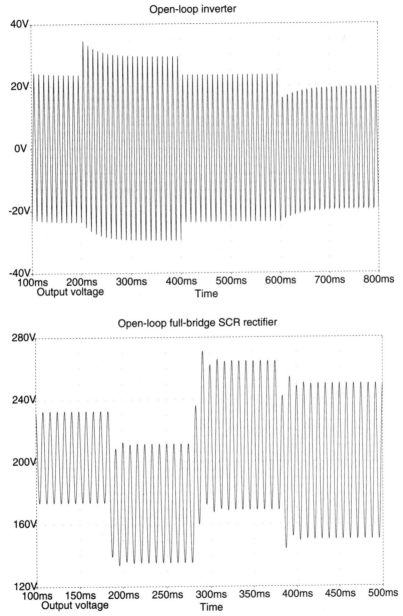

Figure 15.19 Open-loop test results for rectifier and inverter.

Results are provided in Figure 15.19. These circuits are not quite as well behaved, mostly because the filtering properties are less ideal given the low switching frequency. Even so, the transient behavior is stable and does not exhibit especially unusual dynamics.

The transient behavior of a power converter often is tied to its filter design. If the interface components filter out most effects of switching, the dominant dynamics are those of a relatively simple system.

15.4.2 Fast Switching

A converter can be made to switch very rapidly—often much faster than load dynamics. In general, the behavior of a system is governed by the slowest dynamics if there is wide separation between the relevant time scales. For example, the switching process in a 100 kHz dc–dc converter will not have much effect on a load at rates up to a few kilohertz. If a PWM inverter switching at 5 kHz is used to control an ac motor at frequencies up to a few hundred hertz, commutation and fast details will not have much impact on operation.

The principle of *fast switching* is the basis for one class of converter models. A PWM inverter, for instance, can be modelled in terms of its modulating function, given that $v_{out}(t) \approx m(t)V_{in}$. An ideal buck converter can be modelled as $V_{out} \approx DV_{in}$. A related technique, system averaging, will be addressed in Chapter 16. Fast-switching converter models replace the switch with a continuous element, or just a gain value. This supports conventional control design methods. As a result, fast-switching models are common for control analysis and design. Some converters have linear fast-switching models, although in general the models are nonlinear.

15.4.3 Piecewise-Linear Models

A linear system can be examined in time domain as well as in frequency domain. The differential equations can be split into a system of first-order equations. Given a set of m *state variables* (normally continuous variables that determine the stored energy, such as inductor currents and capacitor voltages), $\mathbf{x}(t)$, and three independent input functions $\mathbf{u}(t)$, the differential equations can be written

$$
\begin{bmatrix}
\dot{x}_1(t) \\
\dot{x}_2(t) \\
\dot{x}_3(t) \\
\cdot \\
\cdot \\
\cdot \\
\dot{x}_m(t)
\end{bmatrix}
=
\begin{bmatrix}
a_{1,1} & a_{1,2} & a_{1,3} & \cdots & a_{1,m} \\
a_{2,1} & a_{2,2} & a_{2,3} & \cdots & a_{2,m} \\
a_{3,1} & a_{3,2} & a_{3,3} & \cdots & a_{3,m} \\
\cdot & \cdot & \cdot & & \cdot \\
\cdot & \cdot & \cdot & & \cdot \\
\cdot & \cdot & \cdot & & \cdot \\
a_{m,1} & a_{m,2} & a_{m,3} & \cdots & a_{m,m}
\end{bmatrix}
\begin{bmatrix}
x_1(t) \\
x_2(t) \\
x_3(t) \\
\cdot \\
\cdot \\
\cdot \\
x_m(t)
\end{bmatrix}
+
\begin{bmatrix}
b_{1,1} & b_{1,2} & b_{1,3} \\
b_{2,1} & b_{2,2} & b_{2,3} \\
b_{3,1} & b_{3,2} & b_{3,3} \\
\cdot & \cdot & \cdot \\
\cdot & \cdot & \cdot \\
\cdot & \cdot & \cdot \\
b_{m,1} & b_{m,2} & b_{m,3}
\end{bmatrix}
\begin{bmatrix}
u_1(t) \\
u_2(t) \\
u_3(t)
\end{bmatrix}
$$

$$(15.19)$$

The values $a_{1,1}$, $a_{1,2}$, $b_{1,1}$, and so on are constants. The number of input functions need not match the number of state variables (three input functions might reflect a three-phase input source, for example). It is convenient to rewrite equation (15.19) in a matrix shorthand,

$$\dot{\mathbf{x}}(t) = \mathbf{A}\mathbf{x}(t) + \mathbf{B}\mathbf{u}(t) \qquad (15.20)$$

Here $\mathbf{x}(t)$ is a vector of m variables and $\mathbf{A}$ is a constant $m \times m$ matrix. For p inputs, B is a constant $m \times p$ matrix, and $\mathbf{u}(t)$ is a vector of p independent inputs. The equation also requires an initial condition, such as $\mathbf{x}(0) = \mathbf{x}_0$.

In the converters examined so far, each of the possible configurations is a linear circuit. Each configuration has its own set of differential equations. As switches operate, they alter the values in the $\mathbf{A}$ and $\mathbf{B}$ matrices. The overall action can be analyzed with a step-wise approach: The initial condition and initial configuration are solved to give the values of state variables until a *final condition* at which a switch changes state. The final condition becomes the initial condition for the next configuration, and so on until a full switching period has been studied. Control circuits can be added to the circuit model to complete the picture. This modelling approach develops a power electronic system as a *piecewise-linear system*.

A piecewise-linear model can be written conveniently in terms of switching functions. Consider a converter with N possible configurations. There is a vector of N switching functions, and an individual configuration n is associated with a value of this vector, q_n. There might also be a zero configuration, in which all switches are off. The differential equation is

$$\dot{\mathbf{x}}(t) = \mathbf{A}_0\mathbf{x}(t) + \mathbf{B}_0\mathbf{u}(t) + \sum_{n=1}^{N} q_n(\mathbf{x},t)[\mathbf{A}_n\mathbf{x}(t) + \mathbf{B}_n\mathbf{u}(t)] \qquad (15.21)$$

Keep in mind that each value $\mathbf{A}_n$ and $\mathbf{B}_n$ is a complete matrix, possibly taking on different element values as n changes. Notice that the switching functions can depend both on time and on the values of states. Control action will alter the switching behavior based on state values. Equation (15.21) is called the *network equation* for a power electronic system. Control methodology for piecewise-linear systems continues to develop.

15.4.4 Discrete-Time Models

For piecewise-linear systems, the transition in a given time interval from initial condition to final condition follows an exponential function that can be computed. The converter action can be characterized by considering only the values at the switching instants rather than the simpler transition process during each interval. Such a characterization tracks the converter behavior at multiples of the switching period rather than continuously. This naturally leads to a *discrete-time model*, in which the states at time t_k, combined with the switch action, determine the states at the next switching point t_{k+1}.

A discrete-time model has the advantage of getting to the heart of converter operation—the values at the moment of switching. There is a large body of literature on discrete-time control. Many of the techniques have been applied to power converters and to their control. Most of the stability and design concerns have direct counterparts in continuous systems. The drawback of discrete-time models is that most converters do not really function this way. It is true that switch action is a discrete process. However, the switching instants are an analog process: Switches can turn on and off at almost any time. This aspect has been an impediment to widespread use of discrete-time control techniques for power converters.

15.5 VOLTAGE-MODE AND CURRENT-MODE CONTROLS FOR DC–DC CONVERTERS

15.5.1 Voltage-Mode Control

The open-loop action of converters, combined with fast-switching models, provides a good basis for control function. As a typical example, consider a buck converter. With static models included for the switching devices,

$$V_{out} = D(V_{in} - I_L R_{ds(on)} + V_{diode}) - V_{diode} \qquad (15.22)$$

[the same as equation (13.9)]. This is not a complete solution for V_{out} since I_L depends on V_{out} and on the load. Even so, the behavior is clear: the output changes from a minimum when $D = 0$ to a maximum when $D = 1$. If D increases or decreases, the output increases or decreases with it. The duty ratio can be taken as a control parameter, and the output voltage can be sensed and used for feedback. Although the relationship between source and load might not be known, feedback should help give the desired output.

Figure 15.20 shows a block diagram of a buck converter and a possible controller for it. The output voltage is compared to a reference value. This error signal is amplified to provide a duty ratio command. The input V_{in} represents an additional gain. Then

$$V_{out} = [k_d(V_{ref} - V_{out})]V_{in},$$
$$V_{out} = \frac{k_d V_{in}}{1 + k_d V_{in}} V_{ref} \qquad (15.23)$$

If $k_d V_{in} >> 1$ (true, for instance, if $k_d = 10$ and $V_{in} = 12$ V), (15.23) shows that $V_{out} \approx V_{ref}$. The control will work if V_{in} is high enough to support the desired operating point with $D < 1$. In steady state, the output error $V_{out} - V_{ref}$ is given by

$$V_{out} - V_{ref} = \frac{-1}{1 + k_d V_{in}} V_{ref} \qquad (15.24)$$

The output is always slightly less than the reference voltage. In principle, the error can be made as small as possible. If $k_d V_{in} > 100$, the error will be less than 1% of the reference value. A feedback loop that includes integral control can eliminate the error entirely.

The block diagram suggests a stable system. In practice, however, the gain k_d must be limited. When k_d is set high, tiny variations in V_{out} will cause large fluctuations in the duty

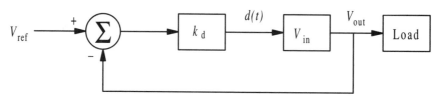

Figure 15.20 Block diagram for output voltage control of buck converter.

ratio. The output is accompanied by ripple, and possibly by noise picked up in the sensing process. Ripple of 50 mV$_{\text{peak-to-peak}}$ and gain $k_d = 20$ will be enough to cause duty ratio swings from 0 to 1 even in normal operation unless extra filtering is provided between the output and the duty ratio control. This does not correspond to instability in the sense of a growing disturbance, but the unpredictable variation in d must be avoided. It is logical to provide a feedback block with a low-pass characteristic to prevent ripple from affecting behavior in this way. Unfortunately, the extra phase delay introduced by this filter tends to be destabilizing.

The block diagram in Figure 15.20 does not take advantage of knowledge about the open-loop behavior that can be obtained from measurement or simulation. Recall that the response of a buck converter is very much like that of an underdamped second-order system. Therefore, an approximate transfer function is that of a second-order system,

$$G(s) \approx \frac{\omega_r^2 V_{\text{in}}}{s^2 + 2\xi\omega_r s + \omega_r^2} \tag{15.25}$$

where ω_r is the resonant frequency $1/\sqrt{(LC)}$ and ξ is a damping ratio. A low-pass filter should be added to prevent ripple from influencing the feedback signal. This can be generated with a simple RC divider, shown in Figure 15.21, that has the response

$$H(s) = \frac{1}{s\tau + 1}, \qquad \tau = RC \tag{15.26}$$

A more complete block diagram is shown in Figure 15.22. The closed-loop transfer function has three poles in its denominator. A gain choice on the order of $k_d = 1$, along with a low-pass filter with a time constant of about 20 μs, should work well for a converter switching at about 100 kHz.

The voltage feedback arrangement is known as *voltage-mode control* when applied to dc–dc converters. Voltage mode control is widely used. Its key advantage is that it is easy to implement: An error signal equal to $k_d(V_{\text{ref}} - V_{\text{out}})$ is compared to a triangle or sawtooth waveform just as in conventional PWM. The duty ratio then is proportional to the error signal, and the converter control is complete. The fact that many dc–dc converters tend to act much like second-order systems is especially helpful. Techniques for second-order systems are well developed.

Voltage mode control has disadvantages. Since V_{in} is a significant parameter in the loop gain, any change in V_{in} will alter the gain and change the system dynamics. A sinusoidal disturbance at the source will be hard to correct, since the disturbance is delayed in

Figure 15.21   *RC divider for low-pass action.*

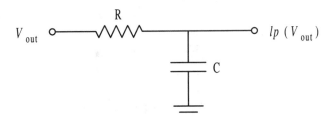

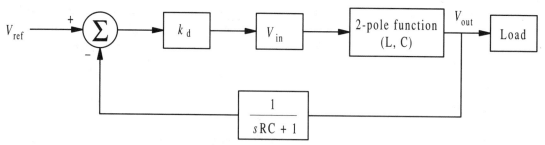

Figure 15.22 Complete block diagram of buck converter.

phase by the inductor and capacitor prior to the output. The central issue is that voltage mode control cannot correct any disturbance or change until it is detected at the output. However, even given its disadvantages, a properly designed voltage mode controller is simple, works well, and has good immunity to disturbances at the reference input.

Example 15.5.1 A voltage-mode control is desired for the buck converter shown in Figure 15.23. First, choose switching devices and inductance and capacitance so the converter provides an output of 12 V with less than 2% output ripple for loads between 100 W and 500 W. Next, identify a second-order system model based on fast switching. Propose feedback gain R_f/R_i to keep the total load and line regulation under 1%.

A load of 500 W implies load current of about 42 A. Since the input voltage does not exceed 30 V, it should be possible to use power MOSFETs for this application. A device rated 50 V and 50 A (such as the MTP50N05) has $R_{ds(on)} = 0.028 \ \Omega$. When a MOSFET is used, a switching frequency of 100 kHz should be reasonable. Given a diode drop of 1 V, the open-loop converter relationship is

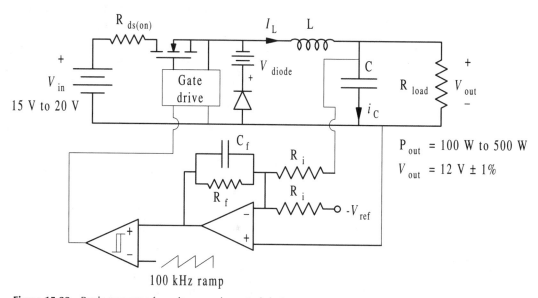

Figure 15.23 Buck converter for voltage-mode control design.

$$V_{\text{out}} = D_1(V_{\text{in}} - I_L R_{\text{ds(on)}} + V_{\text{diode}}) - V_{\text{diode}} \tag{15.27}$$

from (15.22). The lowest duty ratio occurs at the highest input voltage and the lightest load. For a 100 W load, the inductor current is 8.33 A. For this load, a 20 V input gives $D_1 = 0.626$, while the 15 V input gives $D_1 = 0.824$. For the 500 W load, the current is 41.7 A. The duty ratio with 20 V input is 0.655, while the duty ratio with 15 V input is 0.876. The complete duty ratio range is therefore 0.626 to 0.876, and depends on both the input voltage and the load. The closed-loop control must adjust it.

The critical inductance for this converter occurs at the lightest load, when $\Delta i_L = 16.7$ A. During the transistor off-time, the inductor voltage is 12 V. Therefore L_{crit} can be determined by

$$v_L = L \frac{\Delta i_L}{\Delta t}, \qquad 12 \text{ V} = L_{\text{crit}} \frac{16.7 \text{ A}}{D_2 T} \tag{15.28}$$

With the duty ratio range given above, and a switching period of 10 μs, the critical inductance is 2.7 μH. To avoid discontinuous mode and to minimize flux change, let us choose an inductance of ten times this value, or 27 μH. The capacitor sees a triangular current, with a peak-to-peak value of 1.67 A with this inductor choice. To keep the output voltage ripple within limits, the capacitor voltage must not change by more than 0.24 V while the capacitor current is positive. The current is positive half of each cycle. Thus

$$i_C = C \frac{dv_{\text{out}}}{dt}, \qquad \frac{1}{C} \int_{i_C > 0} i_C \, dt < 0.24 \text{ V} \tag{15.29}$$

Since the current is triangular with a peak of (1.67 A)/2, the integral is the area $\frac{1}{2}(T/2)$ (0.833 A) = 2.08 μA·s. The capacitor should be at least 8.7 μF to meet the requirements. We might choose 10 μF to account for ESR drop.

The second-order converter model can be obtained by assuming fast switching. If the switching is very rapid, the output RLC combination sees a voltage $d_1 V_{\text{in}}$. The inductor voltage is this value less v_{out}. With $v_C = v_{\text{out}}$, the circuit loop and node equations give

$$L \frac{di_L}{dt} = d_1(V_{\text{in}} - i_L R_{\text{ds(on)}} + V_{\text{diode}}) - V_{\text{diode}} - v_C,$$

$$C \frac{dv_C}{dt} = i_L - \frac{v_C}{R_{\text{load}}} \tag{15.30}$$

With Laplace transforms, the derivatives become multiplication by the operator s, and the expressions reduce to the single form

$$\frac{R_{\text{load}}(d_1 V_{\text{in}} + d_1 V_d - V_d)}{s^2 LCR_{\text{load}} + sL + s d_1 CR_{\text{load}} R_{\text{ds(on)}} + R_{\text{load}} + d_1 R_{\text{ds(on)}}} \tag{15.31}$$

For small V_d and $R_{\text{ds(on)}}$, this simplifies to the second-order form

$$\frac{d_1 R_{\text{load}} V_{\text{in}}}{s^2 LCR_{\text{load}} + sL + R_{\text{load}}} \tag{15.32}$$

Now, the output voltage will need to be filtered, and there should be a gain such that $d_1 = k_p(V_{ref} - v_{out})$. In the open-loop case, the gain $G(s)H(s)$ becomes

$$G(s)H(s) = \left(\frac{d_1 R_{load} V_{in} k_p}{s^2 LCR_{load} + sL + R_{load}} \right) \left(\frac{1}{sR_f C_f + 1} \right), \qquad k_p = \frac{R_f}{R_i} \qquad (15.33)$$

To meet the regulation requirements, there must be sufficient gain so a 1% change in V_{out} will cause the duty ratio to swing over its full range. A gain of just $k_d = 2$ will change D by 0.24 for an input change of 0.12 V. Let us use a gain of 5. One problem is the potential effect of output ripple. The capacitor C_f provides a low-pass function to help avoid trouble. A Bode plot of $G(s)H(s)$ was used to check values for $R_f C_f$. A time constant of 8 ms provides stable operation and reasonable performance.

The circuit response is shown in Figure 15.24 for both line and load disturbances. At time $t = 1000$ μs, the input steps from 15 V to 18 V. At time $t = 1800$ μs, the nominal load current steps from 20 A to 24 A. These 20% steps bring about short-term changes on the order of 10%, but the steady-state result is very close to 12.00 V.

15.5.2 Current-Mode Control

It is possible to improve on the performance of a voltage-mode control by including information about the source. Since the source acts as an independent input not subject to control, source sensing is an example of *feedforward* control, in which an input is fed into the

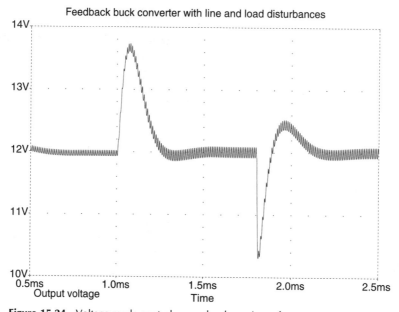

Figure 15.24 Voltage-mode control example: dynamic performance.

system to allow direct compensation. In a buck converter, this is equivalent to using the product DV_{in} as the control parameter rather than the duty ratio alone.

A direct feedforward technique would provide the same error loop as in Figure 15.22, except that a gain $1/V_{in}$ is inserted between the error amplifier and the duty ratio input. It is possible to implement this approach with a minor change to the PWM system: If the amplitude of the PWM carrier ramp is proportional to V_{in}, the switching function will reflect a specific value of DV_{in} instead of just the duty ratio. This *input voltage feedforward* technique is sometimes used by itself. More commonly, feedback control blocks with the same source-dependent action are used.

Feedforward gives the controller information about converter action prior to the energy storage elements at the output. In most converters, there is a feedback signal available earlier in the converter path that offers similar information: the inductor current. The inductor current in all the major dc–dc converter types gives an indication of the input energy without the extra delay in the output filter. Inductor current can be used in one of two ways:

1. It can be used as a feedback variable. A general representation is an error signal given by

$$e(t) = k_v(V_{ref} - V_{out}) + k_i(I_{ref} - I_L) \tag{15.34}$$

2. It can be used to modify the PWM process.

Converter processes that make use of inductor current sensing are termed *current-mode controls*. This control approach brings some additional advantages to a converter: Since the current is being measured directly, it is a simple matter to add overcurrent protection. It is equally easy to operate several converters in parallel. If they share current references, they will divide the load current evenly. In industry, current-mode control in some form is often used for high-performance applications. Most implementations use the second process, in which the current alters the PWM action.

The trouble with the general form in equation (15.34) is that the current reference I_{ref} is not usually available. The current is a function of the unknown load. The concept of *two-loop control*, shown in Figure 15.25, offers a way around this problem. The principle of the two-loop arrangement is that if the output voltage is too low, the converter will need to increase the inductor current to raise it. The voltage error can be used as a *virtual current reference*. Since the current experiences less delay than the voltage, the two-loop approach tends to have better dynamics than voltage-mode control alone. In particular, the additional current loop compensates for line changes just like a source feedforward. In a buck converter, for instance, any increase in source voltage will cause the inductor voltage and current to

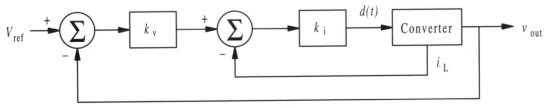

Figure 15.25   Cascaded two-loop control to avoid current reference.

rise before there is a detectable change at the output. The current loop will act to alter the duty ratio with the output voltage error unchanged.

The use of current to modify the PWM process is based on the fact that the inductor current is a triangular waveform just like a PWM carrier. It is possible to substitute the current for the carrier waveform if a triggering clock is provided separately. The process is illustrated in Figure 15.26. In this arrangement, a short pulse from the clock causes the main switch to turn on by setting the latch. The inductor current begins to ramp up. The comparator tests the relationship between the voltage error signal $k_v(V_{ref} - V_{out})$ and the current. The latch is reset and the switch turns off when the two signals match. This method is also called *peak current-mode control* since the switch turn-off point determines the peak inductor current. The voltage error signal is used as a reference for the maximum inductor current. It is also possible to process the current to provide a signal more representative of its average, representing *average current-mode control*. Other variations are possible, but the basic action still uses the current to form the PWM carrier.

Current-mode control has good operational properties within certain limits. An increase in reference voltage will increase the duty ratio and raise the current. An increase in input source voltage will decrease the duty ratio without altering the output current. The method is almost as easy to implement as voltage mode control.

One intriguing limitation of the PWM current-mode approach is a limit on the duty ratio. If D exceeds 50%, instability occurs. This can be seen in Figure 15.27. The figure shows an inductor current, compared to a fixed output voltage error signal. The voltage is high enough to command a duty ratio beyond 50%. Imagine what would happen if a disturbance current Δi is added beginning at time t_0 because of a small change in the load. The current will ramp up, and will reach the error signal sooner than in the preceding cycle. In the buck converter, the on-state slope is $m_{on} = (V_{in} - V_{out})/L$ and the off-state slope is $m_{off} = -V_{out}/L$. At the moment of turn-off, there is a change in the on-time $\Delta(DT)$ such that $m_{on} = \Delta i/\Delta(DT)$. The current then ramps down, and finishes the cycle at time $t_1 = t_0 + T$ with a value $i_0 - \Delta i_1$. The change Δi_1 is such that $m_{off} = \Delta i_1/\Delta(DT)$. The ratio of the two current changes can be found as

$$\frac{\Delta i_1}{\Delta i} = -\frac{m_{off}}{m_{on}} \tag{15.35}$$

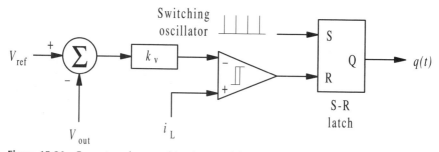

Figure 15.26 Current-mode control implemented through the PWM process.

where the negative sign reflects the fact that m_{off} is negative. For stability, the disturbance must not grow with time. The ratio $m_{\text{off}}/m_{\text{on}}$ was $-D/(1-D)$ before the disturbance. If D exceeds 0.5, then the ratio exceeds one, and the magnitude of Δi_1 will exceed Δi. The disturbance will grow, and an instability occurs.

The instability can be avoided by altering the slopes. One approach is to use a ramp signal in place of the error voltage signal. In this case, a ramp oscillator output is subtracted from the error voltage. The result is compared to the current waveform. This control signal is called a *stabilizing ramp*. The effect, shown in Figure 15.28, is to increase the duty ratio at which the instability occurs. If $-m_{\text{r}}$ is the ramp slope, the condition to be met is

$$\frac{m_{\text{off}} - m_{\text{r}}}{m_{\text{on}} + m_{\text{r}}} < 1 \qquad (15.36)$$

The choice $m_{\text{r}} = m_{\text{off}}$ (a higher ramp slope than shown in Figure 15.28) will guarantee stability for all possible duty ratio values, and is sometimes called the *optimum stabilizing ramp*. A stabilizing ramp is a bit inconvenient, especially since the slopes depend on inductance and voltage values. However, many manufacturers provide integrated circuits that support peak current-mode control with a stabilizing ramp.

15.5.3 Large-Signal Issues in Voltage-Mode and Current-Mode Control

The buck converter is a forgiving system for the design of controllers. The output is a direct function of the duty ratio, the energy delivered to the load increases whenever the active switch is on. Other converters do not share this characteristic. In a boost converter, for instance, energy is delivered to the load indirectly, while the active switch is off. A real boost converter also has two values of duty ratio at which a particular output voltage is achieved.

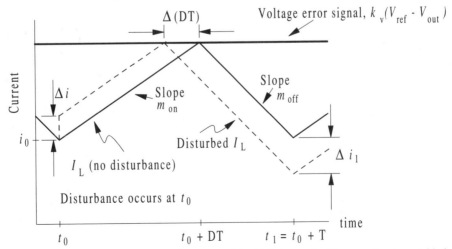

Figure 15.27 Current waveform and error signal for high duty ratio. The operation is unstable because $\Delta i_1 > \Delta i$.

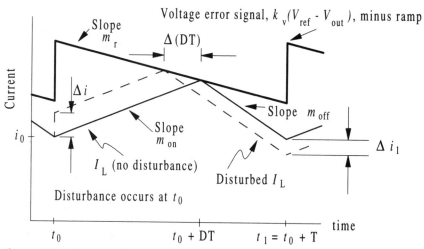

Figure 15.28 Current-mode control with a stabilizing ramp on the voltage error signal.

The desired value is approximately $D = 1 - V_{in}/V_{out}$. The second value is close to 1, and produces high internal losses. There is a duty ratio at which the output is maximized. Furthermore, the relationship between the output and input is nonlinear, since $V_{out} = V_{in}/(1 - D)$ in the ideal case.

The implications of these differences can be observed by testing a control function. Imagine a simple voltage error loop. If the error signal is positive, the active switch turns on. The inductor current rises, but the output voltage falls since the diode is off. The voltage error increases, and keeps the active switch on. The output continues to fall. The end result is that the active switch is held on indefinitely while the output voltage decays to zero.

The problem is an extra time lag in the conversion process. An initial attempt to increase the output voltage must increase i_L. As i_L rises, the output voltage will fall. Thus the initial response of the converter is the opposite of what is desired, although the output will increase on average if the duty ratio rises. This behavior is called *nonminimum phase response* in controls, because there is an extra delay.

Figure 15.29 shows the output voltage as a function of duty ratio for a boost converter with static loss models for switches and other components included. In the figure, the characteristic is increasing for duty ratios below about 0.85, but it is seen that there are two duty ratios for each output voltage when $V_{out} \geq V_{in}$. A voltage-mode control loop will drive the system to the undesired high duty ratio solution. The system has a *large-signal stability problem* since there is more than one operating point associated with a given value of V_{out}. The nonminimum phase behavior and the large-signal stability problem both reflect an effect of the nonlinear system action on converter control.

Large-signal instability in the boost converter can be avoided entirely by ensuring that D is limited. In Figure 15.29, for example, a hard limit that keeps $D < 0.85$ in this converter will make a voltage-mode control stable. In fact, a simple duty ratio limit along these lines allows both voltage-mode control and current-mode control to be used with a boost converter without further complication. Similar effects appear in buck-boost and other indirect converters, and a duty ratio limit is good practice in almost any conversion system.

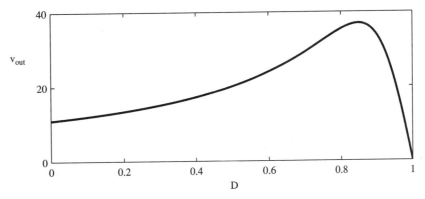

Figure 15.29 Output voltage vs. duty ratio for nonideal boost converter.

Example 15.5.2 The boost converter shown in Figure 15.30 has been designed for 100 kHz operation. The capacitor provides about 1% ripple, while the inductor enforces current ripple of about 20%. The control elements filter the output voltage, then use an error signal $k_v(50 - v_{out})$ as the basis for a current-mode control design. Discuss a current-mode control for this application. Show the performance of a candidate design.

The converter is intended to boost 20 V to 50 V, so a duty ratio above 50% will be necessary. A stabilizing ramp will be required to support high duty values. When the diode is on, the inductor current falls at a rate $v_L/L = (30\text{ V})/(50\ \mu\text{H}) = 0.6$ A/μs. A stabilizing ramp that falls at 1 V/μs should work, if the current sensor gain is 1 Ω. With a frequency of 100 kHz, the ramp will need a peak value of 10 V to achieve this slope. Since D falls between 0 and 100%, a ramp with 10 V peak in effect represents a gain of 0.1 (it will require

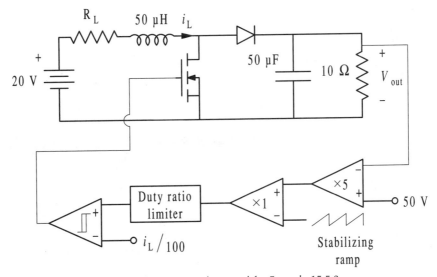

Figure 15.30 Converter with current-mode control for Example 15.5.2.

an error voltage of 10 V to swing the duty ratio from 0 to 1). The fast switching model for this converter gives the open-loop transfer function

$$G(s) = \frac{v_{\text{out}}}{d_2} = \left(\frac{V_{\text{in}}R_{\text{load}}}{s^2 LCR_{\text{load}} + sL + d_2^2 R_{\text{load}}} \right) \frac{1}{s\tau + 1} \tag{15.37}$$

where τ is the low-pass filter time constant and d_2 is the diode duty ratio. This transfer function represents a nonlinear system, since the duty ratio appears in it. However, it can still be used to estimate stability limitations. Feedback gain values $k_v = 5$ and $k_i = 0.01$ produce stable action for a filter time constant of 4 ms. In addition, a hard duty ratio limit is imposed: The comparator output is logically *and*ed with a square wave of 85% duty. This ensures that the active switch will always be off at least 15% of each cycle.

The performance is illustrated in Figure 15.31. At time $t = 3000$ μs, a 20% increase in V_{in} is imposed. The dynamic effect is small, reflecting the current-mode action. At time $t = 6000$ μs, the load resistance drops by 10%. The converter recovers within 2 ms.

15.6 COMPARATOR-BASED CONTROLS FOR RECTIFIER SYSTEMS

Rectifiers tend to be much slower systems than dc–dc converters, and conventional control techniques are not always appropriate. An alternative is to control the switch delay directly, based on the desired output. Rectifiers often operate open loop because their dynamics are so constrained by the low switching frequency. When closed-loop controls are used, they tend to address specific current or voltage limits. While each switch is on, the controller can

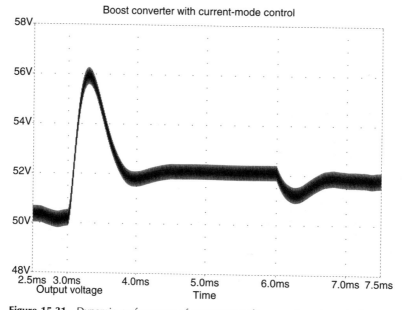

Figure 15.31   Dynamic performance of converter under current-mode control.

determine whether limits are being violated, and can take action at the next switching time. We have already examined open-loop behavior of rectifier systems. In this section, a comparator method for closed-loop control will be introduced.

The controlled two-pulse bridge rectifier in Figure 15.32 serves as the basis for the initial discussion. For convenience, assume that the inductor is large enough to maintain nonzero current flow. If the source voltage has a peak value of V_0, the average output voltage is

$$\langle v_{\text{out}} \rangle = \frac{2V_0}{\pi} \cos \alpha \tag{15.38}$$

This is nonlinear in the control parameter α, but would be linear if somehow $\kappa = \cos \alpha$ could be used as the control parameter instead.

A possible control design methodology would be as follows:

1. Sense and feed back the output voltage through a low-pass filter (to remove the ripple).
2. Design a conventional closed-loop control that develops a value for κ. Ensure that $-1 < \kappa < 1$.
3. Compute $\alpha = \cos^{-1} \kappa$, then control the turn-on delay accordingly.

The inverse cosine step might seem complicated, but there is an elegant way to go directly from the value of κ to the delay time. Figure 15.33 shows waveforms relevant to the process for the choice $\alpha = 30°$. In the figure, the source voltage is shown as $v_{\text{in}}(t) = V_0 \sin(\omega t)$. The integral of the input is $-\omega V_0 \cos(\omega t)$. If the frequency is nearly constant (as will be the case for utility line potential), the integral can be scaled and inverted to give $v_{\text{int}}(t) = (2V_0/\pi)\cos(\omega t)$. This cosine function can be used to give the value of α: The value of α determines a time t_{on} such that $\alpha = \omega t_{\text{on}}$. Define the desired average output as a reference potential $V_{\text{ref}} = (2V_0/\pi)\cos \alpha$. Then $V_{\text{ref}} = v_{\text{int}}(t_{\text{on}})$. That is, the time when V_{ref} matches $v_{\text{int}}(t)$ is exactly the time to turn on the switch.

The inverse cosine therefore can be implemented with a comparator. A reference potential is compared to the integral of the source voltage. When the two match, a short pulse is generated to fire the SCR. To control all four switches, a rectified image of the integral can be used. The comparator process is a direct way to fire SCRs in single-phase circuits. The process adjusts for variation in V_0. It also supports conventional feedback design. For

Figure 15.32 Controlled bridge rectifier.

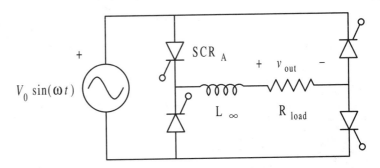

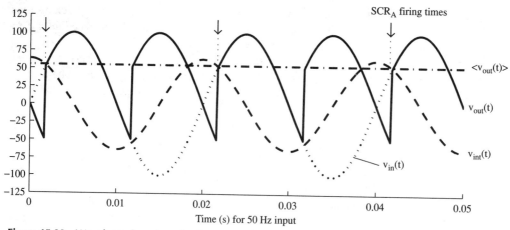

Figure 15.33 Waveforms for automatic phase control in a bridge rectifier.

example, V_{ref} can be obtained from an error signal instead of directly. With a simple feedback gain, the switch is fired when

$$v_{int}(t) = k(V_{ref} - V_{out(ave)}) \qquad (15.39)$$

Figure 15.34 shows a block diagram to implement the SCR control process.

In industrial applications, six-pulse rectifiers are common. The circuit is reviewed in Figure 15.35. The integral process cannot be used directly for this circuit, because the firing point is shifted by 60° relative to the single-phase case. The important waveforms are shown in Figure 15.36. The respective source voltages are $v_a(t) = V_0 \cos(\omega t)$, $v_b(t) = V_0 \cos(\omega t - 2\pi/3)$, and $v_c(t) = V_0 \cos(\omega t + 2\pi/3)$. For phase a, the switch turn-on point is $\omega t_{on} = -\pi/3$ when $\alpha = 0$, and in general $\alpha = \omega t_{on} + \pi/3$. Therefore, we need

$$V_{ref} = V_0 \cos \alpha = V_0 \cos\left(\omega t_{on} + \frac{\pi}{3}\right) \qquad (15.40)$$

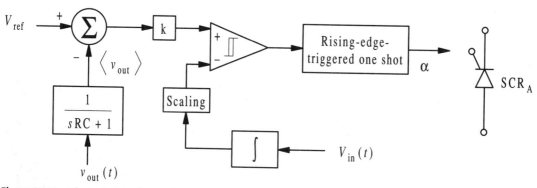

Figure 15.34 Block diagram for rectifier controller.

Figure 15.35 Six-pulse rectifier for industrial applications.

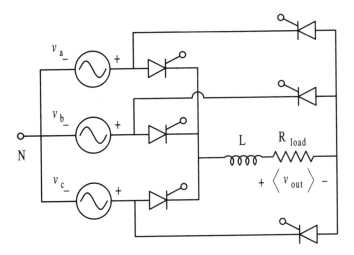

A cosine function delayed by $\pi/3$ or 60° is needed for the comparison. Such a function can be generated with an op-amp circuit by recognizing the relationship

$$V_0 \cos\left(\omega t + \frac{\pi}{3}\right) = \frac{1}{2}V_0 \cos(\omega t) - \frac{\sqrt{3}}{2}V_0 \sin(\omega t) \qquad (15.41)$$

This is not hard to implement since the integral of $v_a(t)$ is proportional to $\sin(\omega t)$. [It is also tempting to use the voltage $-v_b(t) = V_0 \cos(\omega t + \pi/3)$ in the comparison process. This does not work as well as (15.41) because of possible voltage magnitude imbalance.]

The result from (15.41) can be scaled to give an "integral" waveform for control. A reference value that matches the desired average output can be defined. The switch for phase a is fired when the reference matches the integral signal. Firing signals for phases b and c can be generated with separate comparators, but using the same reference voltage.

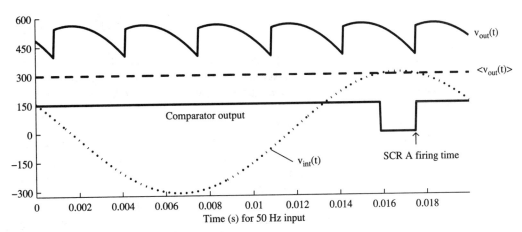

Figure 15.36 Waveforms for rectifier control.

The comparator process for controlled rectifiers is very common, and works well in practice. A disadvantage is the general reliance on nonzero load current. It is not at all unusual for rectifiers to operate in discontinuous mode. A rectifier in discontinuous mode has a higher average output than would be expected with nonzero current. With feedback, discontinuous mode can be handled without much extra trouble. If the output voltage is too high, the phase angle will be further delayed. Whether the rectifier operates in continuous or discontinuous mode, added phase delay will produce a lower average output.

15.7 PROPORTIONAL AND PROPORTIONAL-INTEGRAL CONTROL APPLICATIONS

The example systems in this chapter so far have used *proportional control*, meaning that the control parameter is strictly proportional to the error signal. One attribute of a proportional control is that the error signal cannot be driven to zero. In steady state, a small error signal must remain to provide the input for the control parameter. The error can be reduced with high gain, but this is not always an option, as we have seen.

One way around the steady-state error issue is to apply an integration to the error signal. The signal $\int e(t)\, dt$ will rise when the error is positive, fall when it is negative, and remain constant even when the error is zero. If an integrator drives a control parameter, the system output will change until the error is exactly zero. Notice that this effect is independent of gain. Even an integrator with very low gain will provide zero error in steady state. In a system with control parameter κ, the idea is to use the signal

$$\kappa = k_i \int (V_{\text{ref}} - v_{\text{out}})\, dt + k_p(V_{\text{ref}} - v_{\text{out}}) \tag{15.42}$$

This is termed *proportional-integral* (PI) control because of the respective integration and proportional gain terms applied to the error signal $V_{\text{ref}} - v_{\text{out}}$. More generally, a derivative term can be added as well to implement *proportional-integral-differential* (PID) control,

$$\kappa = k_i \int (V_{\text{ref}} - v_{\text{out}})\, dt + k_p(V_{\text{ref}} - v_{\text{out}}) + k_d \frac{d}{dt}(V_{\text{ref}} - v_{\text{out}}) \tag{15.43}$$

The PI control approach is one of the most widely used methods, especially for systems with only a few state variables. In power electronics, the PI approach is very common; however, the PID approach is rare because commutation spikes and switching functions have derivatives with extreme values.

A natural question is "Why not use the integration alone and leave out proportional gain entirely?" The integrator prevents steady-state error, and will cause a change in the control parameter any time a change is seen at the output. The trouble with an integrator alone is the action in response to large disturbances. For example, if the output drops suddenly (owing to a load change or to some temporary problem), the integrator will begin accumulating error. It will tend to drive the output too high to compensate for the accumulated error. In addition, the integrator will hold the output too high until the accumulated error is cancelled out. In a 5 V converter, for instance, an interval of 1 ms spent at 2 V will cause an error accumulation of 3 mV·s. This error will need to be cancelled out, perhaps by operating at 6 V for 3 ms. In any case, the output will overshoot as it recovers to 5.0 V. This effect is called *integrator windup*. One way to reduce windup problems is to keep the integral gain low and provide a relatively high proportional gain to alter the control parameter immediately when a large disturbance occurs. Another way to reduce windup is to clamp the integrator and prevent an unlimited buildup of V·s product.

PI control is straightforward to implement. A simple op-amp circuit that performs an inverted PI operation is shown in Figure 15.37. In this circuit, the proportional gain $k_p = -R_2/R_1$. The integral gain $k_i = -1/(R_1 C)$. While there are many ways to address the windup problem, the anode-to-anode Zener clamp shown in the figure is one of the simplest. The Zener voltage can be chosen to give the desired range of integral operation.

A PI controller adds an extra pole (and zero) to the system transfer function. Consider a system with feedback function $H(s)$ and open-loop gain $G(s)$. A PI controller will multiply $G(s)$ by $k_i + k_p/s$. The closed-loop transfer function becomes

$$V_{out}(s) = \frac{(k_i + k_p s)G(s)}{s + k_p sG(s)H(s) + k_i G(s)H(s)} V_{in}(s) \qquad (15.44)$$

The zero occurs at the value $s = -k_i/k_p$. The extra pole is present, although its value is less obvious, and depends on the form of $G(s)$ and $H(s)$. Since both k_p and k_i can be selected by the designer, the value of the pole can be set by the designer. The design of PI controls is an example of *pole placement*. Recall that a pole is a root of the denominator polynomial. There are many design practices for pole-placement methods. Poles can be selected to optimize specific performance parameters, to maximize system bandwidth, to minimize overshoot, or to meet other specifications. A conventional objective is as follows:

Empirical Rule: *While there are many alternatives for pole-placement design, a good general purpose rule is to produce poles with roughly equal (and high) magnitudes, spread along an arc in the left-half complex s plane. Imaginary parts should not be larger than real parts for any pole.*

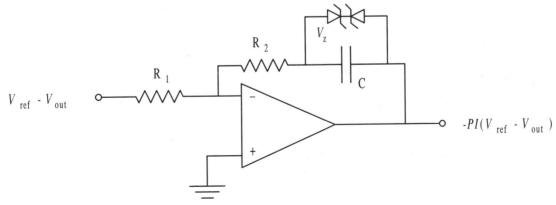

Figure 15.37 Op-amp circuit to implement PI control.

PI control and pole-placement approaches are discussed in depth in most practical texts on control systems.

Figure 15.38 shows a boost converter with two-loop PI current-mode control. The PI function of the output voltage error signal serves as a virtual current reference for the inner current loop. This converter has a very important advantage: The regulation properties are perfect, at least in principle. In practice, regulation is limited only by noise and by issues such as temperature variation of the reference voltage and of resistors. The dynamic properties are good when a stabilizing ramp is used and the gains are chosen for stable operation with reasonable phase margin. This circuit example represents a high-performance converter.

15.8 RECAP

Feedback control is necessary for any power converter that must adjust for external change. It is only through control that excellent regulation can be achieved. Control design attempts to address two major operating concerns. *Static regulation* represents the steady-state effects of changes on the output, measured with partial derivatives or ratios. A control system might also address *dynamic regulation*, which can represent either a converter's ability to maintain the desired operation during rapid changes or a converter's ability to follow a time-varying command. In addition to providing good regulation, a control design must provide *stability*—the property of returning to a desired operating point after a disturbance. In power electronics, many types of stability are important. Small-signal stability represents the ability to recover from noise spikes or small variations in source or load. Large-signal stability represents the ability to recover from extreme changes such as loss of load or momentary short circuits, and also the ability to reach a desired operating condition when the converter first starts.

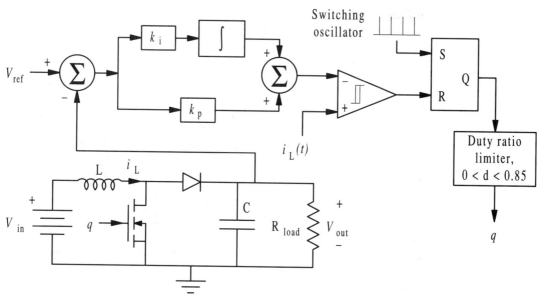

Figure 15.38 Boost converter with two-loop current-mode control. The outer loop uses PI control.

Open-loop control methods, or methods that include feedforward, are sometimes used in place of the feedback approach. Open-loop systems offer certain advantages, including simplicity, higher system gain compared to feedback, lower energy requirements, and no stability problems beyond those inherent in the base system. However, open-loop systems do not react to changes. Only feedback control allows a system to adjust for uncertainty and ensures good static and dynamic regulation under unexpected conditions. Open-loop control is viable in situations that do not require high accuracy, as in the case of many practical rectifiers and ac regulators.

In closed-loop control, information from the output or from various *state variables* (most commonly capacitor voltages and inductor currents) is processed with a reference value to produce an error signal e(*t*). Some function is applied to the error signal, and the result is used as the input to a *control parameter* in the system. In power electronics, the usual control parameters are duty ratio and phase. A typical example of a control process uses inductor current and filtered output voltage to set the phase angle of a controlled rectifier,

$$\cos \alpha = k_i \left[k_v \int V_{\text{ref}} - lp(v_{\text{out}}) \, dt - i_L \right] \qquad (15.45)$$

where $lp(v)$ represents a low-pass filter function. Control of this type achieves good regulation either because of the integral term (which can have a nonzero value even with no error) or because gain values such as k_i are large. Gains in general must be limited to avoid instability, or often to avoid driving control parameters to hard physical limits.

An electronic control system has associated differential equations. In the case of linear systems, Laplace transform methods substitute polynomial functions of the complex frequency parameter s for the differential equations. In this case, a system is associated with a gain function, denoted $G(s)$. In conventional control notation, a feedback function $H(s)$ returns information to generate the error signal. Without feedback, the open-loop gain is just $G(s)$. When the feedback loop is closed, the overall system gain becomes

$$V_{\text{out}}(s) = \frac{G(s)}{1 + G(s)H(s)} V_{\text{ref}}(s) \qquad (15.46)$$

In a power electronic system, the reference voltage must be distinguished from the source input voltage V_{in}. Equation (15.46) represents the system's *control-to-output* transfer function, $K(s)$. The complex frequency $j\omega$ can be substituted for s to provide a frequency domain transfer function $K(j\omega)$.

The magnitude and phase of $K(j\omega)$ are useful for the analysis of dynamic regulation. *System bandwidth*, the highest frequency at which the system performs a useful control function, can be studied directly from the transfer function magnitude. The conventional definition of system bandwidth is the frequency at which the transfer function magnitude is down 3 dB from its low-frequency value, provided the system is stable up to this frequency. Stability can be studied in frequency domain with the open-loop transfer function $G(s)H(s)$ by means of the Nyquist criterion. One statement of this constraint is that feedback that tends to reinforce a disturbance rather than attenuate it will create instability. In magnitude and phase terms, the following rule applies.

> **Empirical Rule:** *To ensure stability for a control system, it is necessary that the open-loop transfer function gain magnitude* $|G(s)H(s)|$ *be less than 0 dB whenever the phase reaches or exceeds 180°.*

The 180° phase value represents the value at which negative feedback turns into positive feedback. If positive feedback is present with a gain above 0 dB, the feedback signal will add to a disturbance and cause its effects to grow. The Nyquist criterion supports definitions of relative stability. The *crossover frequency* is the frequency at which the gain becomes 0 dB. Let the phase at the crossover frequency be ϕ_c. Designers are interested in sufficient *phase margin*, defined as the extra angle $180° - \phi_c$ that is present at the crossover frequency. Nonzero phase margin provides some robustness in case the design values are not exactly accurate. Phase margins of 45° or more represent good practice. Sometimes, the concept of *gain margin* (the extra gain that could be tolerated while keeping the overall gain below 0 dB when ϕ_c reaches 180°) is used in place of phase margin as a relative stability measure.

Stability and performance can be studied through the polynomial functions of s that appear in a transfer function. The denominator polynomial of $K(s)$ is especially important. The roots of this polynomial correspond to poles of the $K(s)$ function. These poles should have negative real parts to ensure stability. The numerator polynomial is less critical since its roots are associated with zeroes of $K(s)$. However, a control system with negative real parts for its zeroes performs better and is easier to design than one with positive real parts for zeroes.

All power electronic circuits are nonlinear, so Laplace transforms and frequency domain methods do not apply in a strict sense. However, when ripple is ignored, the response of many types of converters shows relatively simple behavior. Simple buck, boost, and flyback circuits behave like underdamped second-order systems when a step change in duty ratio is imposed on them. This reflects the presence of two energy storage elements, and suggests that there are converter models that avoid nonlinear switching behavior and provide a useful basis for control design.

Three different types of approximate converter models were introduced. In *fast-switching* models, the switching process is assumed to be far faster than any other system dynamics. Continuous parameters, such as duty ratio, can be substituted for switching functions in this case. The switch action has the same effect as a gain, and equations well suited to control analysis and design result. Fast-switching models are not necessarily linear, but they avoid the discontinuities implied by switch action.

A more general model is based on the *piecewise-linear* approach. This approach takes advantage of the fact that any power converter has well-defined configurations, and acts by switching among these configurations. If the switching process is ideal, the state variables within a converter do not change abruptly as the configurations change. A solution can be pieced together by examining the configurations separately. Since each configuration is usually a linear circuit, a piecewise-linear solution approach works well. The study of control techniques for piecewise-linear systems is an active research area.

The piecewise-linear model helps motivate *discrete-time* models, in which converter action is examined at specific evenly spaced sample times instead of continuously. Control system design is well developed for discrete-time models. The main limitation of the discrete-time approach is that few power electronic systems are truly periodic. When feedback control is added, at least one switching time each cycle becomes an unpredictable function of the state variables and external conditions. In practice, the discrete-time model has been limited to analysis and to converters controlled by clocked microprocessors.

In control design for dc–dc converters, distinctions are made between voltage feedback and current feedback processes. In *voltage-mode control*, an error signal is generated from the output voltage. This error signal is compared to a carrier ramp to create a pulse-width modulated switching function. The error signal can be produced with *proportional control* or with *proportional-integral* (PI) *control*. The general form is

$$e(t) = k_i \int (V_{ref} - v_{out}) \, dt + k_p(V_{ref} - v_{out}) \tag{15.47}$$

In proportional control, the value $k_i = 0$ in this expression. Voltage-mode PI control has excellent static regulation properties and tends to be stable for reasonable values of gain. Dynamic performance is limited because a voltage-mode controller cannot react to a disturbance until the effects have appeared at the converter output. In many applications, the advantages of voltage-mode control outweigh these limits on dynamic performance. The approach is very common for dc power supplies.

When better dynamic performance is desired, feedback from inductor currents can augment the control approach. Assuming only proportional gain for simplicity in the notation, a general *current-mode control* error signal can be identified as

$$e(t) = k_v(V_{ref} - v_{out}) + k_i(I_{ref} - i_L) \tag{15.48}$$

This error signal can be compared to a PWM ramp to generate a switching function. However, an alternative technique is more common. In (15.48), the value of I_{ref} is likely to be unknown, since it depends on the load. A process known as *two-loop control* uses a voltage error signal in place of the current reference, and

$$e(t) = k_i[k_v(V_{ref} - v_{out}) - i_L] \tag{15.49}$$

The inductor current is a triangular function of time. Because of this, equation (15.49) can be implemented by comparing the voltage error signal to the inductor current, instead of with a separate PWM ramp.

Current-mode control improves the dynamic behavior of a converter because the current changes immediately in the event of a source change. However, a direct comparator approach introduces a large-signal stability concern: If the duty ratio tries to rise above 50%, small disturbances will tend to grow from cycle to cycle, and the converter becomes unstable. It is possible to avoid this large-signal problem by adding a *stabilizing ramp* to the current signal. With proper choice of the ramp slope, the converter can be stable for duty ratios right up to 100%.

Most converters have nonlinear fast-switching models. In these cases, a large-signal instability mode might be present. For example, a real boost converter has a maximum output voltage, and has two possible duty ratios for any boosted output level below this maximum. One simple way to avoid large-signal instability and prevent the high-loss high-duty ratio solution is to limit the duty ratio to less than the value associated with the highest output. When a duty ratio limit is imposed, voltage-mode and current-mode controls can be used successfully.

Fast-switching models do not apply to most practical rectifiers, since the switching frequency is tied to an ac line input. Open-loop control for rectifiers sets a phase-delay angle to provide a desired average output. However, only closed-loop control can actively adjust to enforce current or voltage limits or maintain a specific output voltage. Closed-loop control can be implemented by using cos α as a control parameter. The sinusoidal input waveform can be processed to provide a comparison function. This function maps the cos α term into a firing time for an SCR. For example, if $\alpha = \omega t_{on}$, the necessary comparison function is $V_0 \cos(\omega t)$. The SCR is fired when the error signal crosses this function.

A control system can involve *proportional control*, in which the error signal is proportional to the difference between the desired result and the actual result. In proportional control of an output voltage, the signal is

$$e(t) = k_p(V_{ref} - V_{out}) \tag{15.50}$$

This is easy to implement. In steady state, a small output error is present under this type of control, because a nonzero signal is needed to drive the control parameter. Usually, high gain is desirable to minimize the output steady-state error.

An integral term can be added to create *proportional-integral* (PI) *control*. The integral can maintain a nonzero error signal even when the output matches the reference exactly. Thus PI control supports zero output error in steady state. It can be further supplemented with a derivative term to create proportional-integral-differential (PID) control, although this is rare in power electronics because of high switching derivatives. The general PID form is

$$e(t) = k_p(V_{ref} - V_{out}) + k_i \int (V_{ref} - V_{out}) \, dt + k_d \frac{d}{dt}(V_{ref} - V_{out}) \tag{15.51}$$

PI and PID control are among the most widely used general methods. The PI form is common in power supplies and other general purpose power conversion applications. The gain values in PI control can be selected to modify the poles of $K(s)$, and PI gives good design freedom for systems with just one or two state variables.

PROBLEMS

1. Stability properties do not always require mathematical analysis. Consider the basic dc–dc converters (buck, boost, buck-boost, and boost-buck). Given a resistive load and open-loop control with a fixed duty ratio, are these converters *stable* (do they return to the operating point) if they are disturbed by momentarily short-circuiting the output?

2. List the control parameters for the following desired actions (*Example*: automobile travel direction—the parameter is steering wheel position angle).

   a. Automobile travel direction.

   b. Inverter output voltage magnitude.

   c. Output current from buck converter.

   d. High-voltage dc line power flow magnitude.

   e. Electric heater temperature.

   f. Ac regulator output power.

3. Draw a block diagram for PI control of the torque output of a dc motor.

4. Consider a PI control for dc motor torque (Problem 3). Given the motor parameters $k_t = 1$ N·m/A, $J = 0.04$ kg·m^2, $R_a = 1$ Ω, $L_a = 20$ mH, is the system stable for all values of feedback gain k_p and k_i? You may assume that a load is present, and that it requires a torque that is a linear function of speed. If not, what values support stable operation?

5. Example 15.3.1 explored the use of feedback to improve the performance of a low-quality amplifier. Now, consider the effects of the feedback divider. If the feedback ratio has possible error of $\pm 1\%$, what is the tolerance on overall system gain?

6. An open-loop system has the Laplace transfer function

$$\frac{v_{out}}{v_{in}} = \frac{1}{s(s + 10000)} \tag{15.52}$$

It is desired to use a closed-loop system with proportional gain k_p. Plot the Bode diagram for the open-loop system for a few values of k_p. Will the closed-loop system be stable? Is there a limit on the gain value?

7. An amplifier has gain of approximately 10 000, but the gain value has substantial tolerance ($\pm 50\%$). Use feedback with this amplifier to produce overall gain of 100. This problem represents typical behavior of an operational amplifier.

   a. What is the tolerance in this gain?

   b. Now consider that the amplifier also has a low-pass characteristic with a corner frequency of 100 Hz. This means it can be represented with the model shown in Figure 15.39. With feedback in place for system gain of 100, find the transfer function $K(s)$ for the system. Plot the magnitude and phase as a function of frequency.

   c. What is the phase margin in part (b)?

8. The system with frequency characteristic shown in Figure 15.13a ($100°$ phase margin) has the transfer function

$$G(s)H(s) = \frac{14\ 600(s + 2200)(s + 800)}{(s + 5000)(s^2 + 600s + 2.25 \times 10^6)} \tag{15.53}$$

   What is the time response of the closed-loop system to an input step of 1 V?

9. A system has the transfer function

$$G(s) = \frac{s + 5}{s^2 + 2s - 3} \tag{15.54}$$

   Proportional feedback is to be used for control, with $H(s) = 1$. Generate the Bode diagram for $G(s)H(s)$. Is the system stable with this feedback? What about other values of gain?

10. A system has the closed-loop transfer function

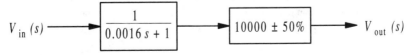

$$V_{in}(s) \longrightarrow \boxed{\frac{1}{0.0016\ s + 1}} \longrightarrow \boxed{10000 \pm 50\%} \longrightarrow V_{out}(s)$$

Figure 15.39 Amplifier model for Problem 6.

$$K(s) = \frac{s + 4000}{s^2 + ks + 20\ 000} \tag{15.55}$$

In the complex plane, plot the poles of the function, parametrized by values of k. Comment on the nature of the results. Is there an especially good value of k from a control standpoint?

11. A dc–dc buck converter uses current mode control. The design is intended for 24 V to 8 V conversion at power levels up to 80 W. The switching frequency is 150 kHz, and the inductor is five times the critical value at 100% output power. A capacitor and filter set at the output limit the peak-to-peak ripple to 100 mV. The inductor current signal is sensed through a 20 mΩ resistor, then amplified. At the maximum possible current level, the output of this amplifier is 5.00 V. The voltage loop is a proportional gain loop, with a gain of 10.

    a. Draw the circuit and its control.

    b. Assume the circuit uses a reference potential of exactly 8 V. It has been running for a while at full output power, and is in steady state. Plot the inductor current waveform under this condition. Overlay the output of the voltage loop to confirm the switching times. What is the average output?

    c. While the converter is running with full power, midway through a switching period the input voltage suddenly falls to 20 V. Make an approximate plot of the change in inductor current and output voltage over the subsequent 25 μs. Is recovery apparent after this brief interval?

12. It is sometimes asserted that PID control is not used in power electronics because the derivative information is redundant, not because of noise. Consider a dc–dc boost converter in which $L = 5L_{crit}$ and $C = 100C_{crit}$. The input is 20 V, the load is 200 W, and the nominal duty ratio is 58%. Plot the inductor current and capacitor voltage, the integral of these, and the derivative of these. Consider how the transistor gate signal might be used to generate derivative information.

13. A single-phase bridge rectifier uses comparator control with current feedback to implement a battery charger. An interface inductor helps smooth the current into the battery. The basic technique is a proportional control involving the current.

    a. Draw the circuit, given a peak input voltage of 16.5 V, input frequency of 60 Hz, and a nominal battery voltage of 12 V. Also, draw the sinusoid to be used for the comparison process.

    b. Select a value of inductor. At time $t = 0$, the system is turned on. The reference current level is set to 10.0 A. The proportional gain is 5. For your choice of inductor, plot the current as it evolves over the first 0.1 s.

    c. For a reference current of 10.0 A, what is the actual steady-state average output current?

14. In a single-phase rectifier, the comparator control method used a scaled integral of the voltage. In the three-phase case, a waveform with 60° shift was required. What is the reference waveform needed for the general case of m phases?

15. In a certain dc motor application, position control is desired. In this case, a PI loop produces an error signal computed from a reference position and from measured position. Draw a block diagram for this arrangement. Given the dc motor of Example 15.3.2, will such a control be stable for all values of gain?

16. When a low-pass filter is inserted in a feedback loop, does the overall system become more stable or less stable, as reflected in the phase margin? For a system with crossover frequency 50 kHz and phase margin 45°, what will the crossover frequency and phase margin become if a low-pass filter with corner frequency 1 kHz is used in the feedback loop?

17. In a buck converter, any level of feedback gain can be used in principle. In practice, chattering that results from the limits on duty ratio becomes a problem at high gain values. Consider a buck converter for 200 V to 120 V conversion at 2 kW. The switching frequency is 20 kHz, and the time constant L/R_{load} is 10 ms. Compare action, through simulation or through computation and plots, for proportional control with gains of 1 and of 1000, respectively. The arrangement is $d(t) = k_p(V_{ref} - v_{out})$. Both choices are stable in some mathematical sense, but are they practical?

18. Prepare a simulation of the open-loop response of a SEPIC converter to changes in line and load. Does the result look like that of a second-order system?

19. A PWM inverter uses current and voltage feedback to set the modulating function. The converter is a half-bridge circuit with a capacitive divider. The voltage feedback is taken from one of the capacitors, while current is measured at the output inductor. The reference signal is $I_{ref} = 20\cos(\omega t)$, in units of amperes. The input bus voltage is 400 V. The load is modelled as a 10 Ω resistor.

   a. Draw this system, and show a block diagram of the control.

   b. Will the concept work for reasonable choices of switching frequency and other component values? Will the converter provide a sinusoidal output current?

REFERENCES

N. M. Abdel-Rahim, J. E. Quaicoe, "Analysis and design of a multiple feedback loop control strategy for single-phase voltage-source UPS inverters," *IEEE Trans. Power Electronics*, vol. 11, no. 4, pp. 532–541, July 1996.

A. R. Brown, R. D. Middlebrook, "Sampled-data modeling of switching regulators," in *IEEE Power Electronics Specialists Conf. Rec.*, 1981, pp. 349–369.

C. W. Deisch, "Simple switching control method changes power converter into a current source," in *IEEE Power Electronics Specialists Conf. Rec.*, 1979, pp. 300–306.

S. B. Dewan, G. R. Slemon, A. Straughen, *Power Semiconductor Drives.* New York: Wiley, 1984.

R. C. Dorf, *Modern Control Systems, 5th ed.* Reading, MA: Addison-Wesley, 1989.

G. F. Franklin, J. D. Powell, A. Emami-Naeini, *Feedback Control of Dynamic Systems, 3rd ed.* Reading, MA: Addison-Wesley, 1994.

B. Friedland, *Control System Design.* New York: McGraw-Hill, 1986.

J. G. Kassakian, M. F. Schlecht, G. C. Verghese, *Principles of Power Electronics.* Reading, MA: Addison Wesley, 1991.

R. D. Middlebrook, "Topics in multiple-loop regulators and current-mode programming," *IEEE Power Electronics Specialists Conf. Rec.*, 1985, p. 716.

G. C. Verghese, M. Ilic, J. H. Lang, "Modeling and control challenges in power electronics," in *Proc. 25th Conf. Decision and Control*, 1986, p. 39.

CHAPTER 16

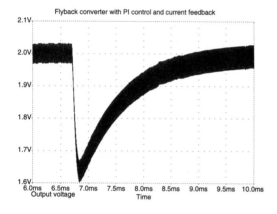

Output voltage

Figure 16.1 Transient response of flyback converter.

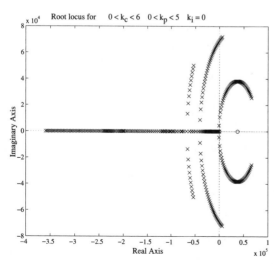

APPROXIMATE METHODS FOR CONTROL DESIGN

16.1 INTRODUCTION

Feedback control is essential for precise, repeatable operation of a power converter. Approaches such as voltage-mode and current-mode control show how feedback control might be implemented. An integral term can be added in a control loop to eliminate steady-state error. Comparators allow PWM and phase modulation techniques to be implemented directly. The missing information at this point concerns design methodology. Can concepts taken from linear control system theory be applied to large-signal nonlinear systems? How does the switch action in a converter complicate the control design process?

Power electronics engineers usually use approximate models to support the control design process. In Chapter 16, we consider *averaged models* used for this purpose. Later, in Chapter 17, *geometric methods* are introduced that support direct control designs for power converters. In the case of approximate models, one common procedure is to build an averaged model that can follow the "bulk" behavior of a converter without the need for detailed modelling of switches. This averaged model is *linearized* according to conventional methods for nonlinear systems. Linearized models have a very important property: They support control design methods based on linear systems and Laplace transforms. One such method, the root locus approach, is used in Figure 16.1 to produce a fast flyback converter.

Chapter 16 develops averaging methods, then creates small-signal linearized models based on averaged models. Design methods for linearized models are the major focus of the chapter. For dc–dc converters, the concept of *compensation* is introduced, and examples are provided for voltage-mode control. Averaged models are not necessarily linear, and some engineers advocate their direct use for nonlinear controls. This issue is discussed briefly at the end of the chapter.

16.2 AVERAGING METHODS AND MODELS

Averaging has been one of our most important tools for power converter design. In dc–dc converters, the functional relationships between sources, outputs, and control parameters were explored through the use of averaging. In PWM inverters, the modulating function represented a moving average of the switch action. The output, on average, tracked the modulating function. In phase-modulated ac–ac converters, the average output behavior was a sinusoidal function of a modulating function.

Average behavior gives information about the dc or low-frequency action of a converter, ignoring ripple, commutation, and other fast effects. Even though ripple is not captured in the average output, an average can still be useful in determining both static regulation and transient response. If an averaged model can be developed for a converter, it should be able to track large-scale changes in voltages and currents as the source, load, or control inputs change, or when component tolerances and similar issues are taken into account. Ripple can be added in at the end of the modelling process.

16.2.1 Formulation of Averaged Models

In Chapter 15, the *infinite switching frequency model* was introduced. In this model, duty ratios take the place of switching functions. Averaging provides a more complete way to establish this type of model. The sequence of configurations in a converter adds and removes energy to produce well-defined average results at the input or output. There ought to be some *average configuration* that generates the same average input and output.

The network equation is a basis for the averaging process itself. It is probably easier to follow the treatment for a specific circuit, so let us begin with the buck converter of Figure 16.2. The converter has two state variables—the inductor current and capacitor voltage—and two configurations if discontinuous mode is avoided. Define configuration #1 for the transistor on and #2 for the diode on. The state variables act according to

$$\dot{\mathbf{x}} = A_1\mathbf{x} + B_1\mathbf{u}, \quad \text{\#1 on}$$
$$\dot{\mathbf{x}} = A_2\mathbf{x} + B_2\mathbf{u}, \quad \text{\#2 on} \tag{16.1}$$

where $\mathbf{x}$ is the state variable vector of inductor current and capacitor voltage. Consider the state variable values at the beginning of a switching period, $\mathbf{x}(0)$. Start in configuration #1. If the derivatives are approximately constant (true if the ripple is triangular or the switching frequency is high), the states will change linearly at the rate $\dot{\mathbf{x}}(0)$ until time $t = D_1T$. The value for the $\mathbf{x}$ vector would be

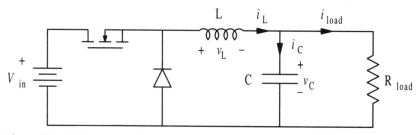

Figure 16.2 Buck converter for averaging process.

$$\mathbf{x}(\Delta t) \approx \mathbf{x}(0) + \dot{\mathbf{x}}(0)\Delta t, \qquad \Delta t = D_1 T \tag{16.2}$$

Substituting from (16.1),

$$\mathbf{x}(D_1 T) \approx \mathbf{x}(0) + (A_1 \mathbf{x} + B_1 \mathbf{u})D_1 T \tag{16.3}$$

The value $\mathbf{x}(D_1 T)$ from configuration #1 serves as the initial condition for configuration #2. At time $t = T$, triangular ripple or fast switching means

$$\mathbf{x}(T) \approx \mathbf{x}(D_1 T) + \dot{\mathbf{x}}(D_1 T)\Delta t, \qquad \Delta t = D_2 T \tag{16.4}$$

Substitution from (16.1) for $\dot{\mathbf{x}}(D_1 T)$ yields

$$\begin{aligned}
\mathbf{x}(T) &\approx \mathbf{x}(D_1 T) + (A_2 \mathbf{x} + B_2 \mathbf{u})D_2 T \\
&\approx \mathbf{x}(0) + (A_1 \mathbf{x} + B_1 \mathbf{u})D_1 T + (A_2 \mathbf{x} + B_2 \mathbf{u})D_2 T \\
&\approx \mathbf{x}(0) + [(D_1 A_1 + D_2 A_2)\mathbf{x} + (D_1 B_1 + D_2 B_2)\mathbf{u}]T
\end{aligned} \tag{16.5}$$

The utility of this result is perhaps unclear, but if average matrices $\overline{A} = D_1 A_1 + D_2 A_2$ and $\overline{B} = D_1 B_1 + D_2 B_2$ are defined, an important simplification is possible. The final result of equation (16.5) can be expressed as

$$\mathbf{x}(T) \approx \mathbf{x}(0) + T(\overline{A}\mathbf{x} + \overline{B}\mathbf{u}) \tag{16.6}$$

This is exactly the result that would have been produced for the *averaged system*,

$$\dot{\mathbf{x}} = \overline{A}\mathbf{x} + \overline{B}\mathbf{u} \tag{16.7}$$

Notice also that the $\overline{A}$ and $\overline{B}$ matrices are averages of the configurations, weighted by the fraction of the cycle spent in each configuration.

Equation (16.7) is the same as the infinite switching frequency model, and gives exact results when the switching period is much shorter than any of the circuit time constants. The equation is termed the *state-space average* since it represents an average of the configurations. Strictly speaking, the state variables in (16.7) are only approximations to the original variables. No switching is reflected in the model, so the states do not exhibit ripple. In a formal sense, we have found a *new system* of equations that models the behavior of new state variables $\mathbf{y}(t)$ driven by new inputs $v(t)$, such that

$$\dot{\mathbf{y}} = \overline{A}\mathbf{y} + \overline{B}v \tag{16.8}$$

This is a continuous system corresponding to the original piecewise system. In the infinite frequency limit, the values $\mathbf{y}(t)$ match the original states $\mathbf{x}(t)$, and the new inputs $v(t)$ should match $u(t)$. However, the results are useful even when the frequency is not infinite. It has been proved in the literature that the new states $\mathbf{y}(t)$ actually do track the average behavior of $\mathbf{x}(t)$, provided that $v(t)$ is defined to track the average behavior of $u(t)$. At low switching frequency, there is steady-state error in the tracking process for certain kinds of converters, but the basic averaging process still works.

The advantage of the state-space average is that the process of building a model is direct. The circuit equations are written for each state, then a weighted average is created with the appropriate duty ratios. Another way to generate an averaged model is with switching functions. It is sufficient to write the circuit equations in terms of switching functions, then substitute duty ratios for switching functions just as was done in steady-state analysis.

Let us consider the buck converter in a little more depth. The inductor current derivative is proportional to v_L. This voltage is $V_{in} - v_C$ when the transistor is on and $-v_C$ when the diode is on. The capacitor current determines the voltage derivative, and is $i_L - i_{load}$ in both configurations. With switching functions, we can write these KVL and KCL expressions as

$$L\frac{di_L}{dt} = q_1(t)V_{in} - v_C(t)$$

$$C\frac{dv_C}{dt} = i_L(t) - i_{load}(t)$$

(16.9)

Since the inductor and capacitor energies are determined by i_L and v_C, respectively, these are the state variables. The differential equations, written only in terms of the state variables, are

$$L\frac{di_L(t)}{dt} = q_1(t)V_{in} - v_C(t)$$

(16.10)

$$C\frac{dv_C(t)}{dt} = i_L(t) - \frac{v_C(t)}{R_{load}}$$

Divide the current equation by L and the voltage equation by C. This supports a matrix equation,

$$\begin{bmatrix} \dfrac{di_L}{dt} \\ \dfrac{dv_C}{dt} \end{bmatrix} = \begin{bmatrix} 0 & \dfrac{-1}{L} \\ \dfrac{1}{C} & -\dfrac{1}{R_{load}C} \end{bmatrix} \begin{bmatrix} i_L \\ v_C \end{bmatrix} + \begin{bmatrix} q_1(t)\dfrac{V_{in}}{L} \\ 0 \end{bmatrix}$$

(16.11)

Equation (16.11) is strictly a notational change that uses matrix multiplication to write the differential equations of the system. In this buck converter, we see that

$$A_1 = A_2 = \bar{A} = \begin{bmatrix} 0 & -\dfrac{1}{L} \\ \dfrac{1}{C} & -\dfrac{1}{R_{load}C} \end{bmatrix}$$

(16.12)

The product Bu changes as the configurations alternate, and we have

$$B_1u = \begin{bmatrix} \dfrac{V_{in}}{L} \\ 0 \end{bmatrix}, \qquad B_2u = \begin{bmatrix} 0 \\ 0 \end{bmatrix}$$

(16.13)

The averaged result $\bar{B}v$ becomes

$$\bar{B}v = \begin{bmatrix} \dfrac{D_1 V_{\text{in}}}{L} \\ 0 \end{bmatrix} \tag{16.14}$$

The averaged differential equations can be written in matrix form; however, let us write them out explicitly for now. Given an *average inductor current variable*, $\bar{i}_L$ and an *average capacitor voltage*, v_C, the averaged system has the equations

$$L\frac{d\bar{i}_L(t)}{dt} = D_1 V_{\text{in}} - \bar{v}_C(t)$$

$$C\frac{d\bar{v}_C(t)}{dt} = \bar{i}_L(t) - \frac{\bar{v}_C(t)}{R_{\text{load}}} \tag{16.15}$$

If the duty ratio is fixed, this represents a linear second-order system. In Chapter 15, it was noted that the open-loop behavior of a buck converter is much like that of a second-order system. Equation (16.15) helps explain why this is true.

Consider the boost converter in Figure 16.3, with its two configurations (discontinuous conduction mode is ignored here). The inductor voltage and capacitor current, from KVL and KCL, become

$$L\frac{di_L}{dt} = V_{\text{in}} - q_2(t)v_C(t)$$

$$C\frac{dv_C}{dt} = q_2(t)i_L(t) - i_{\text{load}}(t) \tag{16.16}$$

The differential equations, in the state variables alone, are

$$L\frac{di_L(t)}{dt} = V_{\text{in}} - q_2(t)v_C(t)$$

$$C\frac{dv_C(t)}{dt} = q_2(t)i_L(t) - \frac{v_C(t)}{R_{\text{load}}} \tag{16.17}$$

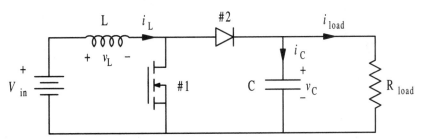

Figure 16.3   Boost converter for averaged modelling.

The A matrices are

$$A_1 = \begin{bmatrix} 0 & 0 \\ 0 & \dfrac{-1}{R_{\text{load}}C} \end{bmatrix}, \quad A_2 = \begin{bmatrix} 0 & \dfrac{-1}{L} \\ \dfrac{1}{C} & \dfrac{-1}{R_{\text{load}}C} \end{bmatrix}, \quad \bar{A} = \begin{bmatrix} 0 & \dfrac{-D_2}{L} \\ \dfrac{D_2}{C} & \dfrac{-1}{R_{\text{load}}C} \end{bmatrix} \qquad (16.18)$$

The Bu matrices are

$$B_1 u = B_2 u = \bar{B} v = \begin{bmatrix} \dfrac{V_{\text{in}}}{L} \\ 0 \end{bmatrix} \qquad (16.19)$$

and the averaged differential equations are

$$L\frac{d\bar{i}_L(t)}{dt} = V_{\text{in}} - D_2\bar{v}_C(t)$$

$$C\frac{d\bar{v}_C(t)}{dt} = D_2\bar{i}_L(t) - \frac{\bar{v}_C(t)}{R_{\text{load}}} \qquad (16.20)$$

Other dc–dc converters can be analyzed along the same lines.

The real value of averaging is that it supports even complicated switching converters. Consider a more complete model of a buck converter, with stray resistances and static switch models, given in Figure 16.4. Although the equations are more complicated, the averaging process can be used effectively.

Example 16.2.1 A buck converter, with stray resistances and static switch models, is shown in Figure 16.4. Find A_1, B_1, A_2, B_2, and the averaged model for this converter. Then substitute values of $V_{\text{in}} = 20$ V, $R_{\text{ds(on)}} = 0.2$ Ω, $V_d = 0.8$ V, $R_d = 0.02$ Ω, $R_L = 0.1$ Ω, $R_{\text{ESR}} = 0.1$ Ω, $L = 100$ μH, $C = 100$ μF, and a 1 Ω load. Given 200 kHz switching at 25% duty ratio, comment on start-up simulation comparisons between the complete piecewise linear system and the averaged system.

The KVL and KCL relationships yield inductor voltage and capacitor current. The output voltage is not quite the same as the capacitor voltage because of the ESR, but it is a function of the capacitor voltage, with $v_C = v_{\text{out}} - i_C R_{\text{ESR}}$. The differential equations are

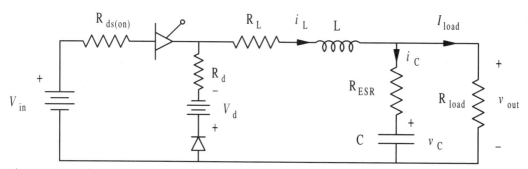

Figure 16.4 Buck converter for averaging analysis.

$$L\frac{di_L}{dt} = q_1(V_{in} - i_L R_{ds(on)}) + q_2(-V_d - i_L R_d) - i_L R_L - v_{out}$$

$$(16.21)$$

$$C\frac{dv_C}{dt} = i_L - \frac{v_{out}}{R_{load}}$$

The voltage v_{out} can be eliminated with some algebraic effort. The equations become

$$\frac{di_L}{dt} = \frac{1}{L}\left[q_1(V_{in} - i_L R_{ds(on)}) + q_2(-V_d - i_L R_d) - i_L R_L\right.$$

$$\left. - v_C\frac{R_{load}}{R_{load} + R_{ESR}} - i_L\frac{R_{load}R_{ESR}}{R_{load} + R_{ESR}}\right]$$

$$(16.22)$$

$$\frac{dv_C}{dt} = \frac{1}{C}\left[i_L\frac{R_{load}}{R_{load} + R_{ESR}} - \frac{v_C}{R_{load} + R_{ESR}}\right]$$

In matrix form, we have

$$\begin{bmatrix} \dfrac{di_L}{dt} \\ \dfrac{dv_C}{dt} \end{bmatrix} = \begin{bmatrix} \dfrac{-q_1 R_{ds(on)} - q_2 R_d - R_L - c_d R_{ESR}}{L} & \dfrac{-c_d}{L} \\ \dfrac{c_d}{C} & \dfrac{-c_d}{R_{load}C} \end{bmatrix}\begin{bmatrix} i_L \\ v_C \end{bmatrix} + \begin{bmatrix} \dfrac{q_1 V_{in} - q_2 V_d}{L} \\ 0 \end{bmatrix}$$

$$(16.23)$$

where c_d is the divider constant $c_d = R_{load}/(R_{load} + R_{ESR})$. The various matrices are

$$A_1 = \begin{bmatrix} \dfrac{-R_{ds(on)} - R_L - c_d R_{ESR}}{L} & -\dfrac{c_d}{L} \\ \dfrac{c_d}{C} & -\dfrac{c_d}{R_{load}C} \end{bmatrix}, \qquad B_1\boldsymbol{u} = \begin{bmatrix} \dfrac{V_{in}}{L} \\ 0 \end{bmatrix},$$

$$(16.24)$$

$$A_2 = \begin{bmatrix} \dfrac{-R_d - R_L - c_d R_{ESR}}{L} & -\dfrac{c_d}{L} \\ \dfrac{c_d}{C} & -\dfrac{c_d}{R_{load}C} \end{bmatrix}, \qquad B_2\boldsymbol{u} = \begin{bmatrix} \dfrac{-V_d}{L} \\ 0 \end{bmatrix}$$

The average matrix $\bar{A} = D_1 A_1 + D_2 A_2$. In continuous conduction mode, $D_1 = 1 - D_2$, and

$$\bar{A} = \begin{bmatrix} \dfrac{-D_1 R_{ds(on)} - (1 - D_1)R_d - R_L - c_d R_{ESR}}{L} & -\dfrac{c_d}{L} \\ \dfrac{c_d}{C} & -\dfrac{c_d}{R_{load}C} \end{bmatrix},$$

$$(16.25)$$

$$\bar{B}\boldsymbol{u} = \begin{bmatrix} \dfrac{D_1 V_{in} - (1 - D_1)V_d}{L} \end{bmatrix}$$

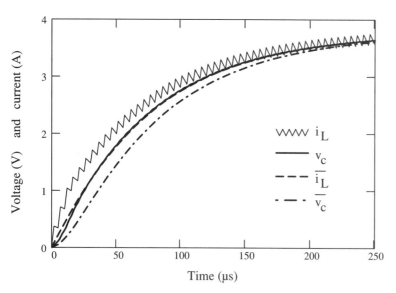

Figure 16.5 Simulation results for Example 16.2.1.

The $\dot{\mathbf{x}} = \mathbf{A}\mathbf{x} + \mathbf{B}\mathbf{u}$ expressions lend themselves well to simulation. In Figure 16.5, the original system and the averaged system are simulated using a fourth-order Runge–Kutta solution constructed in Mathcad (see Appendix C for details). The figure shows the simulated solution of the original A_1 and A_2 equations, starting from zero, through forty cycles. These results step through the two configurations, and compute eight points per switching cycle. The simulation of the averaged model is much faster since switching is not present. In the figure, the averaged model is computed once every other switching cycle.

The current ripple is evident in the original system simulation. Capacitor voltage ripple is also present, but it is harder to see in the simulation results because it is relatively small. The averaged system results do not show ripple, but they do show time behavior similar to the general behavior of the original waveforms. One discrepancy is a general delay between the original system and the averaged system. It is almost as if the averaged system starts a little later than the original system. This is a matter of initial conditions. The averaged variables really should not start out at zero, but rather should start out to match the average initial behavior. However, the discrepancy is not critical: The two sets of waveforms have the same time constants, and are close together after forty cycles of operation.

Example 16.2.2 Formulate the averaged model differential equations for the boost-buck converter of Figure 16.6.

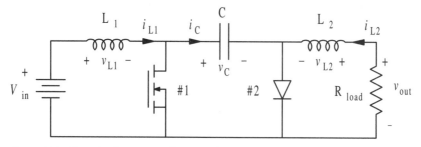

Figure 16.6 Boost-buck converter for Example 16.2.2.

The boost-buck converter has three energy storage elements, so three state variables i_{L1}, i_{L2}, and v_C are needed. The KVL and KCL expressions in terms of the inductor voltages and capacitor current are

$$v_{L1} = V_{in} - q_2 v_C$$
$$i_C = q_1(-i_{L2}) + q_2 i_{L1} \qquad (16.26)$$
$$v_{L2} = -v_{out} + q_1 v_C$$

In continuous conduction mode, $q_1 + q_2 = 1$. Also, $v_{out} = i_{L2} R_{load}$. When equation (16.26) is written entirely in terms of state variables, the equations are

$$L_1 \frac{di_{L1}}{dt} = V_{in} - q_2 v_C$$

$$C \frac{dv_C}{dt} = -q_1 i_{L2} + q_2 i_{L1} \qquad (16.27)$$

$$L_2 \frac{di_{L2}}{dt} = q_1 v_C - i_{L2} R_{load}$$

The averaged model is obtained by substituting D_1 for q_1 and $1 - D_1$ for q_2. The model for the averaged variables is

$$L_1 \frac{d\bar{i}_{L1}}{dt} = V_{in} - (1 - D_1)\bar{v}_C$$

$$C \frac{d\bar{v}_C}{dt} = -D_1 \bar{i}_{L2} + (1 - D_1)\bar{i}_{L1} \qquad (16.28)$$

$$L_2 \frac{d\bar{i}_{L2}}{dt} = D_1 \bar{v}_C - \bar{i}_{L2} R_{load}$$

16.2.2 Averaged Circuit Models

One way to help make averaged models more useful is to generate *averaged circuits* that correspond to the model equations. Look at the circuit in Figure 16.7. The transformer is ideal. The differential equations for the circuit are

Figure 16.7 Averaged circuit model for buck converter.

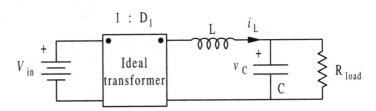

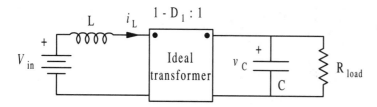

Figure 16.8 Averaged circuit corresponding to a boost dc–dc converter.

$$L\frac{di_L(t)}{dt} = D_1 V_{in} - v_C(t)$$

$$C\frac{dv_C(t)}{dt} = i_L(t) - \frac{v_C(t)}{R_{load}}$$

(16.29)

These are exactly the same as in (16.15). Thus Figure 16.7 is an averaged circuit model for the buck converter.

Figure 16.8 shows the averaged circuit for a basic boost converter. As in the buck case, the switching action acts on average like an ideal transformer. Figure 16.9 shows averaged circuits for buck-boost and boost-buck converters. Figure 16.10 shows a low-frequency circuit model corresponding to a half-bridge PWM inverter. In the inverter case, the switch action is represented with an equivalent source instead of a transformer (the transformer would have required a time-varying turns ratio).

The averaged equivalent circuits are structurally similar. It is natural to ask whether all averaged circuits substitute a transformer for the switch pair. In general, this is not the case. Figure 16.11 shows an averaged circuit that corresponds to the buck converter with internal resistances from Example 16.2.1. Notice that the static models for the switches appear as scaled parameters in the circuit, to match the dynamic equations.

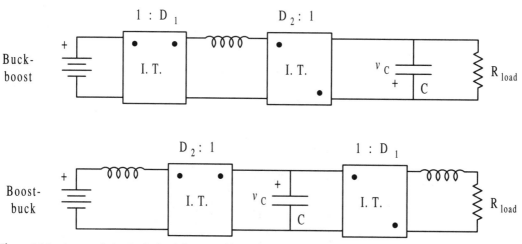

Figure 16.9 Averaged circuits for buck-boost and boost-buck circuits.

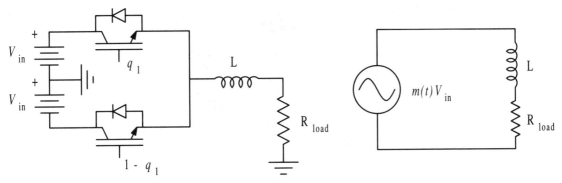

Figure 16.10 Low-frequency equivalent circuit for half-bridge PWM inverter.

16.3 SMALL-SIGNAL ANALYSIS AND LINEARIZATION

16.3.1 The Need for Small-Signal Models

In any controlled converter, the duty ratios are time-varying quantities. The switching functions are not merely replaced with a fixed duty ratio parameter, but instead by time functions $d(t)$. This has at least two significant implications:

- Since the duty ratio can vary, it is not strictly an average of the associated switching function.
- In most of the averaged models, the duty ratio appears with a state variable in a product term. Since the duty ratio is the control parameter, the control has a multiplicative nonlinearity.

The first of these brings into question the validity of the averaging process. The trouble can be avoided by considering a moving average. Thus a time-varying duty ratio does not introduce fundamental problems.

The second issue makes averaged models nonlinear. Nonlinear control approaches sacrifice many important tools of linear control, including Laplace transforms and frequency domain representations. Design guidelines such as phase margin are defined for linear sys-

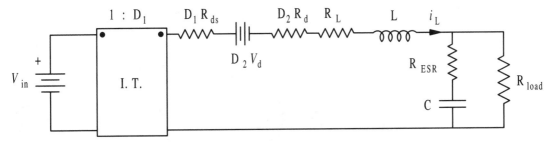

Figure 16.11 Averaged circuit model for Example 16.2.1.

tems. Transfer functions with poles and zeroes can be interpreted for linear systems. To use these tools, it will be necessary to have a linear approximate model for the converter.

16.3.2 Obtaining Models

For dc–dc converters, there is usually a well-defined nominal output. Small-signal analysis can be used to find a model for small deviations around the nominal point. A small-signal model only helps over a limited range of operation, but it should give insight into the dynamic properties of the averaged variables in a converter. The small-signal approach is conventional:

1. Replace the control parameters and state variables with *perturbed* quantities (represented here with the tilde character). In each case, the variable is defined as a nominal value plus a small time-varying component. For instance, let $v_C(t) = V_0 + \tilde{v}_C(t)$, let $d(t) = D_{nom} + \tilde{d}(t)$, and so on.
2. Rewrite the equations in terms of perturbed quantities. Since the perturbations are small, products of perturbations will be neglected.
3. Treat the nominal values as constant parameters.

One unconventional aspect is that the process will be applied to the averaged model rather than to the converter itself. After all, a switching function cannot be considered to have small deviation. This means that an accurate small-signal model requires an accurate averaged model.

Let us explore the linearization technique through the averaged equations for a boost converter, written in terms of the control parameter d_1 as

$$L \frac{d\bar{i}_L(t)}{dt} = V_{in} - (1 - d_1)\bar{v}_C(t)$$

$$C \frac{d\bar{v}_C(t)}{dt} = (1 - d_1)\bar{i}_L(t) - \frac{v_C(t)}{R_{load}} \tag{16.30}$$

First, define each state variable, the control parameter d_1, and the input voltage as a sum of a constant nominal value and a perturbation function. The nominal values $I_{L(nom)}$, $V_{C(nom)}$, $D_{1(nom)}$, and V_{in} have zero time derivatives by definition. Equation (16.30) becomes

$$L \frac{d\tilde{i}_L}{dt} = V_{in} + \tilde{v}_{in} - (1 - D_{1(nom)} - \tilde{d}_1)(V_{C(nom)} + \tilde{v}_C)$$

$$C \frac{d\tilde{v}_C}{dt} = (1 - D_{1(nom)} - \tilde{d}_1)(I_{L(nom)} + \tilde{i}_L) - \frac{V_{C(nom)} + \tilde{v}_C}{R_{load}} \tag{16.31}$$

In (16.31), products such as $\tilde{d}\tilde{V}$ and $\tilde{d}\tilde{i}$ are to be neglected. Notice as well the following relationships for this boost converter:

- The nominal inductor current is the same as the nominal input current, so $I_{L(\text{nom})} = I_{\text{in}}$.
- The nominal capacitor voltage is the nominal output voltage, $V_{C(\text{nom})} = V_{\text{out}}$.

Writing out (16.31) with this in mind, the result is

$$L\frac{d\tilde{i}_L}{dt} = V_{\text{in}} + \tilde{v}_{\text{in}} - (1 - D_{1(\text{nom})})V_{\text{out}} + \tilde{d}_1 V_{\text{out}} - (1 - D_{1(\text{nom})})\tilde{v}_C$$

$$C\frac{d\tilde{v}_C}{dt} = (1 - D_{1(\text{nom})})I_{\text{in}} - \tilde{d}_1 I_{\text{in}} + (1 - D_{1(\text{nom})})\tilde{i}_L - \frac{V_{\text{out}}}{R_{\text{load}}} - \frac{\tilde{v}_C}{R_{\text{load}}}$$

(16.32)

Since there are no losses, the average relationships with no perturbations require $(1 - D_{1(\text{nom})})V_{\text{out}} = V_{\text{in}}$ and $(1 - D_{1(\text{nom})})I_{\text{in}} = I_{\text{out}}$. Given that $I_{\text{out}} = V_{\text{out}}/R_{\text{load}}$, several terms cancel to give

$$L\frac{d\tilde{i}_L}{dt} = \tilde{v}_{\text{in}} + \tilde{d}_1 V_{\text{out}} - (1 - D_{1(\text{nom})})\tilde{v}_C$$

$$C\frac{d\tilde{v}_C}{dt} = -\tilde{d}_1 I_{\text{in}} + (1 - D_{1(\text{nom})})\tilde{i}_L - \frac{\tilde{v}_C}{R_{\text{load}}}$$

(16.33)

This last set of equations is linear given constant nominal values. The linearization process can be applied to any of the averaged models we have developed. In this case, the result models the open-loop behavior of the converter.

An equivalent circuit that corresponds to equation (16.33) is shown in Figure 16.12. The circuit is indeed linear, since the transformer turns ratio is fixed. The duty ratio perturbation appears in two linear controlled sources, each with a source gain dependent on the nominal operating point. Notice in the figure that the specific numerical value of the operating point does not alter the form of the circuit. For 10% duty ratio, the equivalent circuit is well defined, although it will have different values compared to the circuit for 90% duty or some other value.

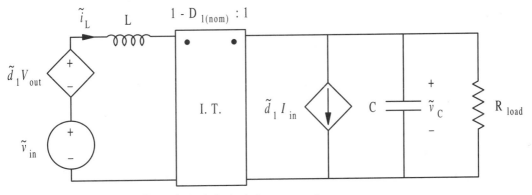

Figure 16.12   Equivalent small-signal circuit for basic boost converter.

TABLE 16.1
Boost Converter Nominal Operating Points for D_1 = 20%, 50%, and 70%

D_1	$1 - D_1$	V_{out} (V)	I_{out} (A)	I_{in} (A)
0.20	0.80	15.0	1.25	1.56
0.50	0.50	24.0	2.00	4.00
0.70	0.30	40.0	3.33	11.11

Example 16.3.1   A boost converter has a 12 V input and a 12 Ω load. Draw small-signal equivalent circuits for several duty ratios.

Let us find the circuits for duty ratio values of 20%, 50%, and 70% to give good variety. Table 16.1 shows the operating point values. The corresponding small-signal equivalents are shown in Figure 16.13.

16.3.3 Generalizing the Process

When a perturbation is added to each of the variables, the results include a dc expression that represents the average periodic steady-state solution, and a perturbed expression that represents

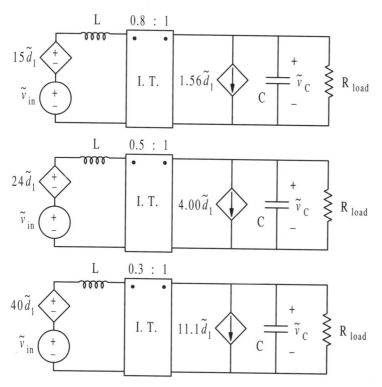

Figure 16.13   Small-signal equivalents for the boost converter at various duty ratios.

the small-signal dynamics. Let us consider, in a general sense, how these two separate results emerge from the analysis. In many converters, the nonlinearities are products of duty ratios and state variables. In this case, the averaged model can be written in the matrix form

$$\dot{\mathbf{x}} = A\mathbf{x} + Bu + C\mathbf{x}d + Ed \tag{16.34}$$

with A, B, C, and E taken as constant matrices from the averaged model. Now let $\mathbf{x} = \mathbf{x}_{\text{nom}} + \tilde{\mathbf{x}}$, $u = u_{\text{nom}} + \tilde{u}$, and $d = d_{\text{nom}} + \tilde{d}$. Then

$$\begin{aligned}
\frac{d}{dt}(\mathbf{x}_{\text{nom}} + \tilde{\mathbf{x}}) &= A\mathbf{x}_{\text{nom}} + A\tilde{\mathbf{x}} + Bu_{\text{nom}} + B\tilde{u} \\
&+ C\mathbf{x}_{\text{nom}}d_{\text{nom}} + C\mathbf{x}_{\text{nom}}\tilde{d} + C\tilde{\mathbf{x}}d_{\text{nom}} + C\tilde{\mathbf{x}}\tilde{d} + Ed_{\text{nom}} + E\tilde{d}
\end{aligned} \tag{16.35}$$

To analyze this, recognize that the $\tilde{\mathbf{x}}\tilde{d}$ term should be neglected. The dc and perturbed terms must separately satisfy the equation, since by definition the dc terms contain none of the ac information and the perturbation terms avoid dc. For the dc terms,

$$0 = A\mathbf{x}_{\text{nom}} + Bu_{\text{nom}} + C\mathbf{x}_{\text{nom}}d_{\text{nom}} + Ed_{\text{nom}} \tag{16.36}$$

This represents the average periodic steady-state solution corresponding to the analysis in Chapter 4. The perturbation terms must satisfy the expression

$$\frac{d\tilde{\mathbf{x}}}{dt} = A\tilde{\mathbf{x}} + B\tilde{u} + C\mathbf{x}_{\text{nom}}\tilde{d} + C\tilde{\mathbf{x}}d_{\text{nom}} + E\tilde{d} \tag{16.37}$$

This is the small-signal linearized model for a converter with an averaged model given by (16.34).

There are only a few possibilities more general than (16.34). In some converters, the input function u is multiplied by the duty ratios. The converter output voltage or current might be related linearly to the state variables, instead of actually being one of the $x(t)$ values (an example is the case when an output capacitor's ESR makes $v_{\text{out}} \neq v_C$). These extensions would add more terms to (16.34), but both the dc and small-signal solutions would appear through perturbation analysis.

Example 16.3.2 Consider the buck converter with stray resistance, as analyzed previously in Example 16.2.1. Write the averaged differential equations, then linearize them.

The matrices for the averaged equations for this circuit were given in Equation (16.25). The averaged equations can be written out as

$$\begin{aligned}
L\frac{d\bar{i}_L}{dt} &= [-D_1 R_{\text{ds(on)}} - (1 - D_1)R_d - R_L - c_d R_{\text{ESR}}]\bar{i}_L \\
&- c_d \bar{v}_C + D_1 V_{\text{in}} - (1 - D_1)V_d
\end{aligned} \tag{16.38}$$

$$C\frac{d\bar{v}_C}{dt} = c_d \bar{i}_L - \frac{c_d}{R_{\text{load}}}\bar{v}_C$$

For linearization, the state variables, the duty ratio, and the input voltage need to be redefined in terms of a nominal value and a perturbation. The initial result is

$$L\frac{d\tilde{i}_L}{dt} = [-D_{1(\text{nom})}R_{\text{ds(on)}} - \tilde{d}_1 R_{\text{ds(on)}} - (1 - D_{1(\text{nom})} - \tilde{d}_1)R_d - R_L - c_d R_{\text{ESR}}]I_{L(\text{nom})}$$
$$+ [-D_{1(\text{nom})}R_{\text{ds(on)}} - (1 - D_{1(\text{nom})})R_d - R_L - c_d R_{\text{ESR}}]\tilde{i}_L$$
$$- c_d(V_{c(\text{nom})} + \tilde{v}_C) + D_{1(\text{nom})}V_{\text{in}} + \tilde{d}_1 V_{\text{in}} + D_{1(\text{nom})}\tilde{v}_{\text{in}} \tag{16.39}$$
$$- (1 - D_{1(\text{nom})} - \tilde{d}_1)V_d$$

$$C\frac{d\tilde{v}_C}{dt} = c_d I_{L(\text{nom})} + c_d\tilde{i}_L - \frac{c_d}{R_{\text{load}}}V_{C(\text{nom})} - \frac{c_d}{R_{\text{load}}}\tilde{v}_C$$

The linearized expression separates out the perturbation terms, and

$$L\frac{d\tilde{i}_L}{dt} = [-\tilde{d}_1 R_{\text{ds(on)}} + \tilde{d}_1 R_d]I_{L(\text{nom})}$$
$$+ [-D_{1(\text{nom})}R_{\text{ds(on)}} - (1 - D_{1(\text{nom})})R_d - R_L - c_d R_{\text{ESR}}]\tilde{i}_L$$
$$- c_d\tilde{v}_C + \tilde{d}_1 V_{\text{in}} + D_{1(\text{nom})}\tilde{v}_{\text{in}} + \tilde{d}_1 V_d \tag{16.40}$$

$$C\frac{d\tilde{v}_C}{dt} = c_d\tilde{i}_L - \frac{c_d}{R_{\text{load}}}\tilde{v}_C$$

This is a linear expression. It can be used for analysis, simulation, or control design. An equivalent circuit that represents it is shown in Figure 16.14. Recall that c_d was the ratio $R_{\text{load}}/(R_{\text{ESR}} + R_{\text{load}})$. In the equivalent circuit, the physical ESR resistor and load resistor have been shown instead. In the equivalent circuit, every part after the switch matrix is left unchanged. This is the case for extra linear elements connected outside the switch matrix for filtering or to form a more complete load model.

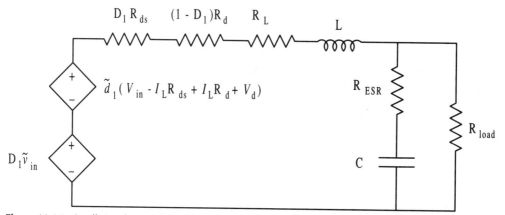

Figure 16.14 Small-signal equivalent circuit for buck converter with internal resistance.

16.4 CONTROL AND CONTROL DESIGN BASED ON LINEARIZATION

16.4.1 Transfer Functions

The small-signal linear models developed thus far represent open-loop converters. We can use them to develop open-loop transfer functions $G(s)$ between the control parameter and the output. The basic buck converter has the averaged model presented earlier. If perturbed variables are substituted into the averaged model, the small-signal model can be found as

$$L\frac{d}{dt}\tilde{i} = D_{1(\text{nom})}\tilde{v}_{\text{in}} + \tilde{d}V_{\text{in}} - \tilde{v}_C$$

$$C\frac{d}{dt}\tilde{v}_C = \tilde{i}_L - \frac{\tilde{v}_C}{R_{\text{load}}}$$

(16.41)

A block diagram form of this system of equations is given in Figure 16.15. To simplify the notation, let i, v, d, and v_{in} represent, respectively, the Laplace transforms of the perturbed variables. The differentiation process is equivalent to multiplication by s. The Laplace form of equation (16.41) is

$$sLi = D_{1(\text{nom})}v_{\text{in}} + dV_{\text{in}} - v$$

$$sCv = i - \frac{v}{R_{\text{load}}}$$

(16.42)

Since the output of the converter is v, let us eliminate i from the system of equations to give a transfer function for the output voltage. After some arithmetic,

$$v = \frac{R_{\text{load}}}{s^2 LCR_{\text{load}} + sL + R_{\text{load}}}(D_{1(\text{nom})}v_{\text{in}} + dV_{\text{in}})$$

(16.43)

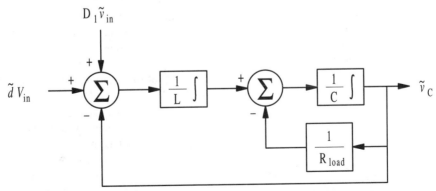

Figure 16.15   Block diagram for buck converter small-signal model.

Two transfer functions can now be written. The *control-to-output transfer function* is the ratio v/d if v_{in} is zero, and the *source-to-output transfer function* is the ratio v/v_{in} when d is zero. These are

$$\frac{v}{d} = \frac{R_{load}V_{in}}{s^2 LCR_{load} + sL + R_{load}} \qquad (16.44)$$

and

$$\frac{v}{v_{in}} = \frac{R_{load}D_{1(nom)}}{s^2 LCR_{load} + sL + R_{load}} \qquad (16.45)$$

The control-to-output transfer function represents the open-loop system transfer function, and will be used shortly for frequency-domain control design. The source-to-output transfer function should be examined in the final closed-loop system to show the effects of line disturbances. If a simple proportional control

$$d = k_p(V_{ref} - v_{out}) \qquad (16.46)$$

is used, a perturbation $\tilde{d}$ can be substituted into (16.41). The reference voltage should be set to the perturbed value $V_{ref} = V_r + \tilde{v}_r$ to represent control action. The duty ratio perturbation is $\tilde{d} = k_p(\tilde{v}_r - \tilde{v}_C)$. Its Laplace transform is $d = k_p(v_r - v)$. This makes the feedback function $H(s) = k_p$. The overall transfer function becomes $K(s) = G(s)/[1 + G(s)H(s)]$. After simplification, this gives

$$v = \frac{R_{load}D_{1(nom)}v_{in} + R_{load}V_{in}k_p v_r}{s^2 LCR_{load} + sL + R_{load}k_p V_{in} + R_{load}} \qquad (16.47)$$

This shows that the gain k_p can influence the transient response to disturbances in the source voltage or in the control input. Equation (16.47) therefore represents the dynamic line regulation behavior of the buck converter and also the dynamic control response.

It is important to remember that (16.47), as well as earlier expressions, is based on the converter's averaged model. Therefore, the expressions are valid only well below the switching frequency. In the Bode plots that follow, the response is plotted through 100 kHz. The responses will be valid for switching frequencies above 200 kHz. For slower switching, any response information above $f_{switch}/2$ is inaccurate because of the model limitations.

Equation (16.47) can be used to analyze the response to line disturbances alone if v_r is set to zero. Figure 16.16 shows a Bode plot of equation (16.47) for $v_r = 0$, $V_{in} = 48$ V, nominal duty ratio of 25%, and parameters $L = 100$ μH, $C = 100$ μF, and $R_{load} = 4$ Ω. Response plots for gain values $k_p = 0.1$, 1.0, and 10.0 are shown. The response shows peaking behavior associated with an equivalent resonant frequency. The frequency can be altered by adjusting the gain value. The plot shows that higher gains tend to reduce the effect of line perturbations. With a gain $k_p = 1.0$, the peak value of system gain is about -20 dB. This means that a disturbance at the source will appear a factor of ten lower at the output. From the plot, input noise at 120 Hz (which would reflect ripple in the original rectified ac line of a dc power supply) will appear at the output attenuated by a factor of about 10 if the pro-

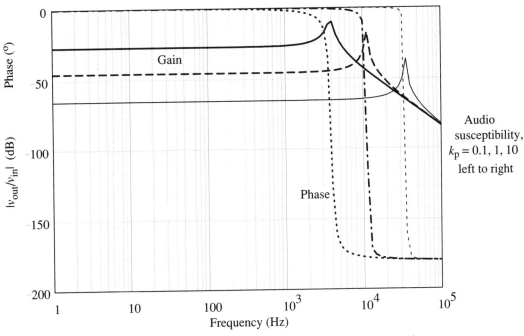

Figure 16.16 Bode plot of dynamic line response for buck converter for various gain values.

portional gain is 0.1, a factor of 100 if the gain is 1.0, and a factor of 1000 if the gain is 10.0.

The dynamic control response can be studied from the open-loop transfer function $G(s)H(s)$. Equation (16.44) provides $G(s)$, and $H(s)$ is the feedback gain k_p. Figure 16.17 shows a Bode plot for $G(s)H(s)$, with the same parameters $R_{load} = 4\ \Omega$, $L = 100\ \mu H$, and so on. The response peaks at about 1600 Hz—exactly coinciding with the resonant frequency $1/\sqrt{(LC)}$. The phase response is independent of gain, and shows a resonant transition from $+0°$ to $-180°$ at the resonant frequency. For gain of 1.0, the crossover frequency is just over 10 kHz.

The analysis so far does not address dynamic load regulation. The load resistor is fixed. Since it is a parameter rather than a state variable or input, it is hard to resolve the issue with a "perturbed resistance." In laboratory tests, a good way to evaluate dynamic load regulation is to inject a small-signal current back into the output port. Figure 16.18 shows a circuit configuration for such a test. The current adds a perturbation $\tilde{i}_{load}$ to the equation for capacitor voltage. Without repeating the details, the closed loop response expression is found to be

$$v = \frac{R_{load}(D_{1(nom)}v_{in} + V_{in}k_p v_r + sLi_{load})}{s^2 LCR_{load} + sL + R_{load} + k_p R_{load}V_{in}} \tag{16.48}$$

The dynamic load response is v/i_{load}. It is useful to normalize this ratio—the output impedance—with respect to R_{load}. This ratio, $v/(i_{load}R_{load})$, can be expressed in decibels. A value

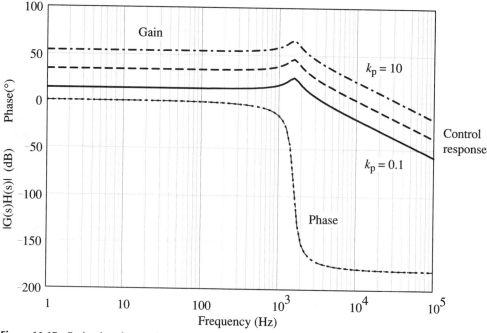

Figure 16.17   Bode plot of control response for buck converter under proportional control.

of 0 dB means that the impedance matches R_{load}. A value of -40 dB means that the output impedance is 1% of R_{load}, and so on. Figure 16.19 shows this ratio for the same parameters as above. The graph shows that a gain of 10.0 produces a peak output impedance ratio of about -20 dB. The higher the gain, the lower the value of $|Z_{out}|$ in the range of interest. As in the case of line regulation, the gain alters the resonant frequency effect.

Now consider closed-loop control. When PI control is used, we have

$$d = k_p(V_{ref} - v_{out}) + k_i \int (V_{ref} - v_{out}) \, dt \qquad (16.49)$$

In the case $v_{out} = v_C$, the perturbation forms $v_C = V_{C(nom)} + \tilde{v}_C$, $d = D_{nom} + \tilde{d}$, and $V_{ref} = V_r + \tilde{v}_r$ can be substituted. The dc portion of the PI expression sets the steady-state duty ratio,

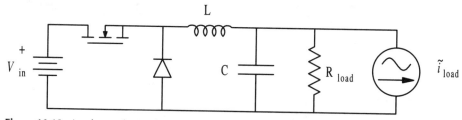

Figure 16.18   Load perturbation based on small-signal current source.

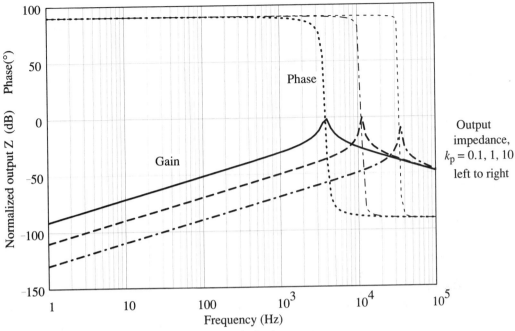

Figure 16.19 Bode plot of dynamic load behavior for buck converter.

$$D_{\text{nom}} = k_p(V_{\text{ref}} - V_{C(\text{nom})}) + k_i \int (V_{\text{ref}} - V_{C(\text{nom})}) \, dt \qquad (16.50)$$

If (16.50) has a solution, it requires $V_{C(\text{nom})} = V_{\text{ref}}$ because of the integral term. The small-signal portion is

$$\tilde{d} = k_p(\tilde{v}_r - \tilde{v}_C) + k_i \int \tilde{v}_r - \tilde{v}_C \, dt \qquad (16.51)$$

In terms of Laplace transforms,

$$d = k_p(\mathrm{v}_r - v) + \frac{k_i}{s}(v_r - v) \qquad (16.52)$$

A system block diagram with PI control is shown in Figure 16.20. The Laplace expression in equation (16.52) can be substituted for d in equation (16.48) to give the PI version of the closed-loop transfer function. After some algebra, the result is

$$v = \frac{sD_{\text{nom}}R_{\text{load}}v_{\text{in}} - s^2 LR_{\text{load}}i_{\text{load}}}{s^3 LCR_{\text{load}} + s^2 L + sR_{\text{load}} + sk_p V_{\text{in}}R_{\text{load}} + k_i V_{\text{in}}R_{\text{load}}} \qquad (16.53)$$

With two gain values, there is more freedom for control design.

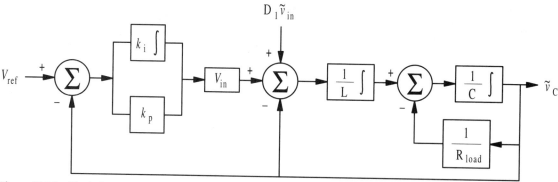

Figure 16.20   Block diagram for small-signal buck converter model with PI control.

16.4.2 Control Design—Introduction

The open-loop and closed-loop transfer functions support two different approaches to control design. In the open-loop case, a feedback function $H(s)$ is selected. The open-loop function $G(s)H(s)$ must have sufficient phase margin to ensure closed-loop stability. This transfer function is the basis of frequency domain design. In the closed-loop case, the denominator polynomial of the transfer function $K(s)$ can be used as the basis of pole-placement design. Frequency domain design tends to be popular in power electronics because the transfer functions $G(s)$ are often simpler than the closed-loop functions. Both methods are treated in depth in good textbooks on control, and the discussion here is intended as an introduction. The example that follows analyzes a frequency domain control approach.

Example 16.4.1   Design a buck dc–dc converter for 250 kHz switching and maximum 1% output ripple. The input voltage is nominally 16 V. The output is 3.3 V. The load ranges from 1 W to 20 W. A PI loop is to be added to control this converter. Develop a small-signal model for this design, with its control. Produce a Bode plot that reflect the open-loop transfer function $G(s)H(s)$. Select PI gains to provide a phase margin of at least 45° and system bandwidth of 10 kHz or more. For this choice of gains, test the closed-loop response to a load transient.

Since the PI loop can compensate for small dc errors, we will base the design on a lossless converter model, except for estimated 1 V forward voltage drops in the semiconductors. The design follows our previous work, and will be summarized rather than detailed. Values are found to be switching period $T = 4$ μs, nominal duty ratio $D_{1(\text{nom})} = (V_{\text{out}} + 1)/V_{\text{in}} = 4.3/16 = 0.269$, $D_{2(\text{nom})} = 0.731$, and $L_{\text{crit}} = 21$ μH. Let us set $L = 25$ μH to allow for tolerance and small changes while maintaining continuous mode operation. To ensure 1% ripple (peak-to-peak), the capacitance should be 10 μF. However, a higher value will be needed to account for ESR jump. A standard value of 22 μF might be a good choice. For a 20 W load, $R_{\text{load}} = 0.54$ Ω. For a 1 W load, $R_{\text{load}} = 10.9$ Ω.

The averaged differential equations, after the small-signal approximation, become

$$sLi = dV_{\text{in(nom)}} + D_{1(\text{nom})}v_{\text{in}} - v$$

$$sCv = i_L + i_{\text{load}} - \frac{v}{R_{\text{load}}}$$

$$(16.54)$$

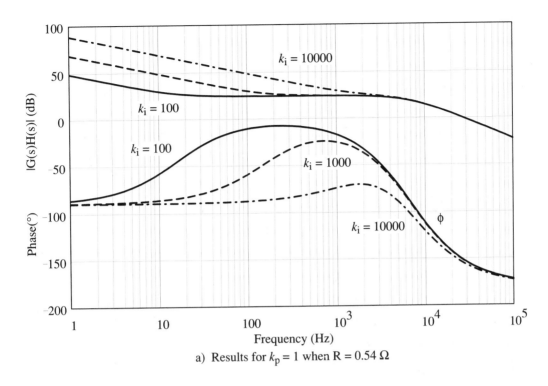

a) Results for $k_p = 1$ when R = 0.54 Ω

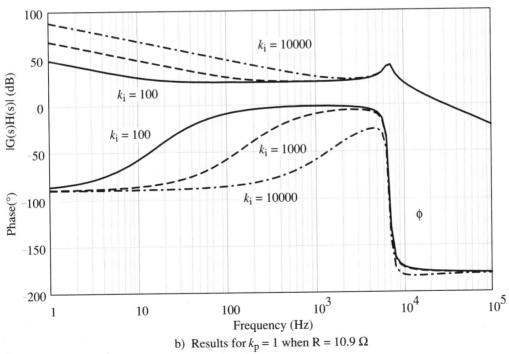

b) Results for $k_p = 1$ when R = 10.9 Ω

Figure 16.21 Open-loop function $G(s)H(s)$, plotted for Example 16.4.1.

With PI feedback, the function $G(s)H(s)$ is

$$\frac{v}{d} = \frac{R_{\text{load}}V_{\text{in}}}{s^2LCR_{\text{load}} + sL + R_{\text{load}}}\left(k_p + \frac{k_i}{s}\right) \tag{16.55}$$

The Bode plot based on (16.55) is given in Figure 16.21 for $k_p = 1$ and a few values of k_i. The upper graph in Figure 16.21a shows the response for $R_{\text{load}} = 0.54\ \Omega$, while the trace in Figure 16.21b shows the response for $R_{\text{load}} = 10.9\ \Omega$. The integrator gain k_i/s produces unlimited gain for dc frequencies; This explains how an integrator avoids any steady-state error. At frequency k_i rad/s, the integrator gain reaches 0 dB. Above this frequency, its effect vanishes as the proportional gain dominates. For frequencies well below k_i, the integrator produces a phase shift of $-90°$. Above k_i, the second-order converter model drives the phase toward $-180°$. The figure shows that at light loads, the converter is underdamped and exhibits peak response corresponding to the L-C resonance point. At heavy loads, the converter does not exhibit a peaking in its response because of the damping effect of the load.

It is desired to obtain bandwidth of at least 10 kHz with phase margin of 45° or more. Over the entire load range, the proportional gain must be at least 0.2 to keep the bandwidth at 10 kHz or more. At the heavy load, this gain is associated with almost 60° of phase margin if gain $k_i = 100$ is selected. However, at light load the phase margin is inadequate. With a PI loop as the only control structure, it is not possible to meet the requirements completely. The transfer function $G(s)H(s)$ is illustrated in Figure 16.22 for this case at the two extremes of load. The figure shows that peaking behavior at light load degrades the phase margin in that case. Figure 16.23 shows transient response of the converter with $k_p = 0.2$ and $k_i = 1000$. The load is initially 0.6 Ω. The figure illustrates a 2 V increase in source voltage, imposed at a time of 2500 μs. A load increase from 0.6 Ω to 0.54 Ω is imposed at a time of 6000 μs. The action is stable, and the converter returns to 3.3 V output after a time in the face of a disturbance.

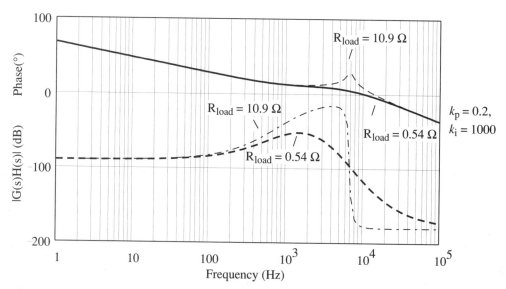

Figure 16.22 Bode plot for $k_p = 0.2$ and $k_i = 1000$ for PI buck converter.

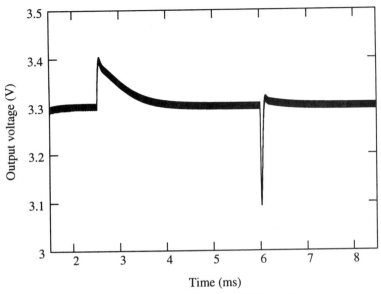

Figure 16.23 Time response of buck converter with PI control.

In the example, the proportional gain had to be limited to keep the phase margin as high as possible. The crossover frequency was 10 kHz—a factor of ten below the switching frequency. Higher values of k_p move the magnitude curve up proportionally, and will generate higher crossover frequencies. At higher gain, fast disturbances sometimes cause aliasing or other unusual effects related to switch action. The common practice is to seek crossover frequencies a factor of 8 to 10 or more below the switching frequency.

Gain values must be selected with proper attention to the frequency domain response. Each extra pole causes the gain to fall off at the rate of 20 dB/decade, and ultimately adds $-90°$ to the phase. For PI control, one alternative is to choose the gain k_p to achieve the desired crossover frequency. The gain k_i is chosen well below this crossover frequency so the phase margin will not be compromised. In the example problem, integrator gains up to about 1000 meet these requirements, and should be successful.

A circuit that implements proportional control for a dc–dc converter, complete with duty ratio limiter, is given in Figure 16.24. This op-amp circuit provides a direct gain equal to (R_f/R_{in}) multiplied by the error value $V_{ref} - V_{out}$. An accurate reference is provided by the left Zener diode. In addition, the circuit provides low-pass filter action with a corner frequency corresponding to the time constant $R_f C_f$. An offset voltage, appearing on the node of resistor R_2, sets a midrange duty ratio when the error is zero. This decreases the steady-state error, since the duty ratio has a nonzero nominal value. Finally, the right Zener diode prevents the gain circuit from driving above the upper limit of the triangle waveform. This is a direct way to limit the duty ratio and avoid the large-signal instability in boost or buck-boost converters. For PI control, a separate op-amp stage can perform integration prior to the Zener duty-ratio limit diode.

The general design process is as follows:

1. Determine whether proportional or PI control will meet the requirements. PI control avoids steady-state error, but complicates the dynamic response.

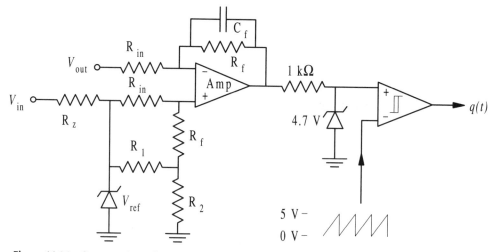

Figure 16.24 Op-amp circuit for proportional control with filter and duty-ratio limiter.

2. Plot $G(s)H(s)$ for the converter.
3. Set the proportional gain to generate the correct crossover frequency.
4. Set the integral gain. The gain must be low enough so the proportional term dominates near the crossover frequency. This avoids imposing an extra 90° phase shift in addition to the natural two-pole 180° shift in the converter.
5. Check the net new $G(s)H(s)$ curve to ensure stability and reasonable phase margin.

The integrator represents a 90° phase shift for low frequencies. It also causes the gain to increase by 20 dB/decade as the y-axis is approached. PI control cannot meet every need, but the design process is straightforward, and a working solution can be developed if one exists. The process is identical for other types of dc–dc converters, provided a duty ratio limit is enforced.

16.4.3 Compensation and Filtering

There are two troubles with the analysis so far. First, since the average model only makes sense for changes slower than the switching frequency, low-pass filters might be necessary to remove switching frequency ripple from the feedback variables. This is not reflected accurately in the Bode plots. If information is present in the feedback variables near the switching frequency, it can mix with the switching function to produce sum and difference terms. A low-pass filter serves an *antialias* function to help avoid this. The second problem is phase margin. The Bode plots in Example 16.4.1 show low phase margin—less than 2° at light load! In principle, the phase delay with these PI controls never exceeds 180°, and instability should never occur. In practice, however, the model is only approximate. In addition to the stray resistors that were left out of the circuit, there will be time delays in op-amps or

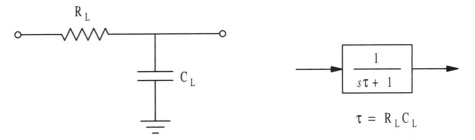

Figure 16.25 Single-pole low-pass filter and transform block.

the switch commutation process. These can add enough delay (which can be measured as additional phase lag) to push the angle beyond $180°$ and cause instability.

Filters can be addressed by adding low-pass transfer blocks to the feedback paths. An *RC* low-pass filter is shown with its Laplace transform block in Figure 16.25. The block adds an extra pole, and will further delay the phase. In practical converters, it is not unusual to set the low-pass filter pole well below the switching frequency to eliminate ripple. This makes the low-pass filter pole dominate the transfer function. Two-pole or higher-order filters can be used, but each extra pole requires additional components and delays the phase more. It is unusual to encounter multipole filters in power converter feedback applications.

How can the phase margin be improved without complicating the control? This question is especially important when filter stages impose additional phase lags. The process of *compensation* seeks to add simple circuitry to alter the phase behavior and improve the margins. Two widely used methods are *phase-lead* compensation and *phase-lag* compensation. *RC* circuits for lead and lag compensation are shown in Figure 16.26.

The objective of phase-lead compensation is to increase the phase margin directly by adding positive phase in the transfer function. The phase-lead circuit also increases gain and increases the crossover frequency. Component values must be chosen so that the phase margin improves even at the new, higher, crossover frequency. The objective of phase-lag compensation is to improve phase margin with a decrease in crossover frequency. The circuit also causes an increase in system phase lag. Component values must be chosen so the frequency effect is more important than the phase delay effect.

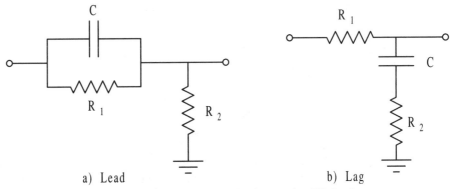

a) Lead   b) Lag

Figure 16.26 Phase-lead and phase-lag compensators implemented as *RC* circuits.

A general compensator of the lead-lag type (covering these two circuits and other forms) has the Laplace transform

$$G(s) = k_{ll} \frac{s + \omega_z}{s + \omega_p} \tag{16.56}$$

where k_{ll} is the compensator gain, and ω_z and ω_p are, respectively, radian frequencies associated with a zero and a pole in the compensator. For a phase-lead compensator, $|\omega_z| < |\omega_p|$. Both the pole and the zero must appear in the right half-plane. For a phase-lag compensator, $|\omega_p| < |\omega_z|$. The phase-lead compensator in Figure 16.26 has the parameters

$$RC \text{ phase lead:} \quad \omega_z = \frac{1}{R_1 C}, \quad \omega_p = \frac{R_1 + R_2}{R_1 R_2 C}, \quad k_{ll} = 1 \tag{16.57}$$

and the lag circuit has

$$RC \text{ phase lag:} \quad \omega_z = \frac{1}{R_2 C}, \quad \omega_p = \frac{1}{(R_1 + R_2)C}, \quad k_{ll} = \frac{R_2}{R_1 + R_2} \tag{16.58}$$

For both of these circuits, the phase change peaks at the geometric mean frequency $\omega_m = \sqrt{(\omega_z \omega_p)}$. This maximum phase change, θ_m, is such that

$$\sin \theta_m = \frac{\omega_p - \omega_z}{\omega_p + \omega_z} \tag{16.59}$$

This ratio can be made high by setting $|\omega_p| \gg |\omega_z|$. However, there is a practical limit to the phase value. Instead of an ideal 90° maximum, for instance, the value $\omega_p = 10\omega_z$ gives a phase lead of only 54.9°. A ratio $\omega_p/\omega_p = 57.7$ is needed to produce a 75° change. In a few applications, compensator networks are cascaded to increase the phase change.

Figure 16.27 shows a Bode plot associated with a phase-lead circuit. Notice the gain, rising by 20 dB per decade for frequencies between ω_z and ω_p. The high-frequency gain value is ω_p/ω_z (the value in dB is 20 $\log(\omega_p/\omega_z)$ by definition). At frequency ω_m, the gain in dB is half this much, or 10 $\log(\omega_p/\omega_z)$. This extra gain will raise the crossover frequency when the compensation circuit is used in a feedback loop. Usually, a designer chooses ω_m to match the new crossover frequency, and selects values of ω_p and ω_z to provide the desired phase margin.

Example 16.4.2   A feedback control system has phase margin of 10°, and a crossover frequency of 100 kHz. At the crossover frequency, the gain is dropping off at 20 dB/decade, while the phase is dropping at about 30° per decade. Design a phase-lead compensator to improve the phase margin to 45°.

Since the uncompensated phase margin is 10°, at least 35° of extra phase lead is needed. However, since the crossover frequency will rise with compensation, even more phase lead will be necessary. Let us plan on a doubling of the frequency, plus some headroom. Phase lead of 50° might be a good starting point. Thus ω_m should correspond to 200 kHz, or $\omega_m = 1.26 \times 10^6$ rad/s. From equation (16.59), the ratio $(\omega_p - \omega_z)/(\omega_p + \omega_z) = \sin 50° = 0.766$. These constraints, solved simultaneously, yield $\omega_p = 3.45 \times 10^6$ rad/s and $\omega_z = 4.57 \times 10^5$.

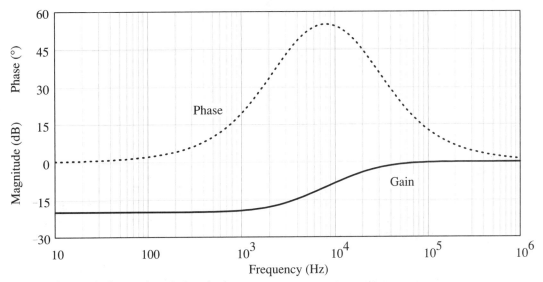

Figure 16.27 Bode diagram for *RC* phase-lead compensator.

The extra gain in dB is $10 \log(\omega_p/\omega_z) = 8.78$ dB. In the uncompensated system, a gain of -8.78 dB corresponds to 275 kHz. Consequently, the new crossover frequency will be 275 kHz. At this frequency, the system phase margin is $-3.17°$ (assuming the total 30° change over the decade follows a log-linear characteristic). With compensation, the net margin is $-3.17° + 50° = 46.8°$. The compensator has improved the phase margin considerably, while more than doubling the system bandwidth. Figure 16.28 shows the uncompensated and compensated Bode plot for this system.

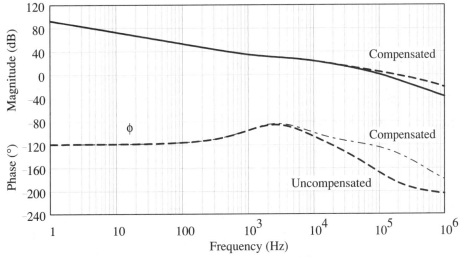

Figure 16.28 Uncompensated and compensated characteristics for Example 16.4.2.

Figure 16.29 shows the Bode plot for a phase-lag network. This network is used much differently from a phase-lead circuit: Its objective is to improve gain margin through its attenuation effect. In this case, the geometric mean frequency $\sqrt{(\omega_p \omega_z)}$ is placed well below the target crossover frequency. This way, there will be little extra phase lag at crossover, but the frequency itself will change significantly. In Example 16.4.2, for instance, the phase margin could have been set to 45° by dropping the gain by 24 dB. This would have dropped the crossover frequency to about 6300 Hz, and would increase the phase margin to 10° + 36° = 46°. The loss in bandwidth is substantial, however.

The choice of whether to use lead or lag compensation is coupled to the application requirements. Phase-lead compensation usually improves system bandwidth, and speeds up the response to transient changes. This implies that operation might become more sensitive to noise as well. This would be an appropriate choice for a high-performance dc–dc converter, especially if fast-changing loads are expected. Phase-lag compensation drops system bandwidth, but increases noise immunity. If a converter is not exposed to fast line or load changes, phase-lag compensation is usually preferred.

16.4.4 Compensated Feedback Examples

Let us revisit some of the control designs, and see how performance changes when compensation is added. Two examples will be presented for phase-lead compensation. Rectifiers are often intended for applications with slow transients, and a phase-lag compensator is developed for one such case.

Example 16.4.3   Example 15.5.1 developed a buck converter for 18 V to 12 V conversion over the range of 100 W to 500 W. In that example, a switching frequency of 100 kHz was

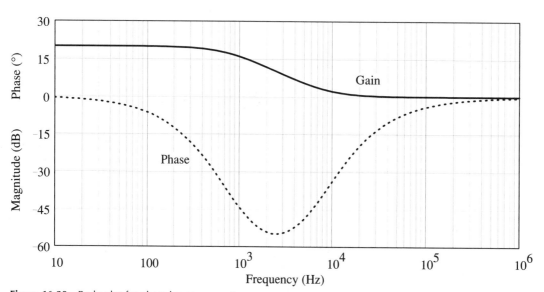

Figure 16.29   Bode plot for phase-lag compensator.

proposed. The inductor and capacitor were set at 27 μH and 10 μF. For feedback control, a low-pass filter was inserted, then a proportional gain was selected to maintain about 1% output regulation. The low-pass filter time constant was set to 8 ms. This time constant is slow enough to dominate the system performance. It is desired to improve system response while increasing the phase margin. Design a phase-lead compensator for this converter. Decrease the filter time constant if possible. Show the Bode plot for the $G(s)H(s)$ of the compensated system. Determine the time response to line and load transients.

The uncompensated Bode plot for this converter is shown in Figure 16.30. The crossover frequency is 2.5 kHz for maximum load and 5.0 kHz for minimum load. The phase margin is 32° for the maximum load, and higher for lighter loads. The slope break in the magnitude plots at 20 Hz reflects the slow filter time constant. The filter time constant could be reduced with the filter time constant remaining as the dominant pole. Let us try to reduce the filter time constant by a factor of 2 to 4 ms. With phase-lead compensation, the crossover frequency will increase because of the extra gain and the filter change. Let us specify a ratio of 16 : 1 for the pole-zero frequency ratio. After a couple of iterations, a crossover frequency of 16 kHz seems to be a reasonable value for $\sqrt{(f_p f_z)}$. With $f_p/f_z = 16$, then $f_p = 64$ kHz and $f_z = 4000$ Hz. The radian values are 4.02×10^5 rad/s and 2.51×10^4 rad/s, respectively. The time constants are 2.5 μs and 40 μs, respectively. Figure 16.31 shows a phase-lead compensation circuit that meets these characteristics.

The Bode plot for the converter small-signal model, with compensation, is shown in Figure 16.32 for the maximum and minimum load. The new crossover point is 5 kHz for the maximum load and 16 kHz for the minimum load. The phase margin improves to 60° for the maximum load, but decreases to 20° for the minimum load because of the shape near the converter's resonant point. If the minimum load is a true worst-case condition, this small margin might work. Figure 16.33 shows the transient response of the compensated circuit,

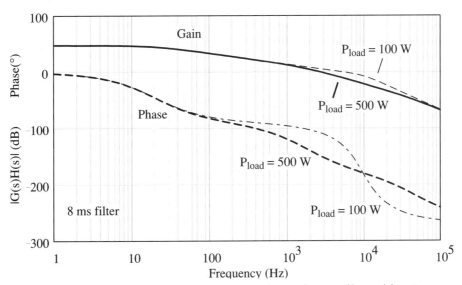

Figure 16.30 Bode plot for 18 V to 12 V converter with 8 ms low-pass filter and $k_p = 5$.

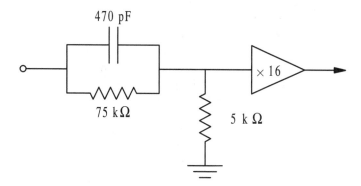

Figure 16.31   Phase-lead compensator with a zero at 6.3 kHz and a pole at 63 kHz,

with a 240 W load. The filter time constant has been further reduced to 0.4 ms. The response is substantially faster than without compensation.

Example 16.4.4   Add phase-lead compensation to the converter of Example 16.4.1 to speed its transient response.

Compensation can be combined with PI control. In this converter, the switching frequency is 250 kHz. The inductor is 25 μH, and the output capacitor is 22 μF. This is a 16 V to 3.3 V converter for 1 W to 20 W output. The original design had $k_p = 0.2$ and $k_i = 1000$. The crossover frequency was about 10 kHz. The phase margin was very low for the light load condition. For compensation, let us guess a new crossover frequency of 20 kHz. If the zero is set to correspond to 5 kHz, the pole should be set to 80 kHz to make the maximum phase correspond to the expected crossover frequency.

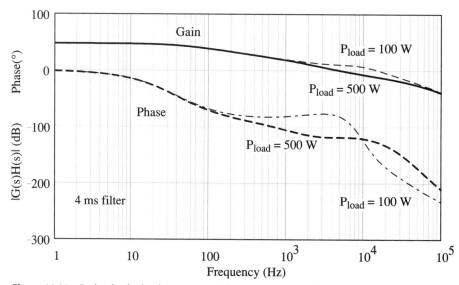

Figure 16.32   Bode plot for buck converter with compensation and with $k_p = 5$.

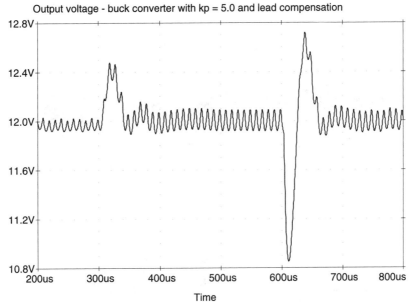

Figure 16.33   Transient response of buck converter with proportional control, low-pass feedback filter, and phase-lead compensation.

The Bode plot with this phase-lead compensation network is shown in Figure 16.34. The phase margin is very much improved. Even though the integral gain is still 1000, the phase margin is 60° at the light load condition. The transient response is shown in Figure 16.35. It is considerably faster. The converter recovers from a 10% load step in less than

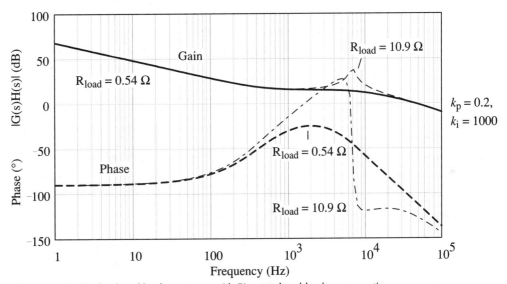

Figure 16.34   Bode plot of buck converter with PI control and lead compensation.

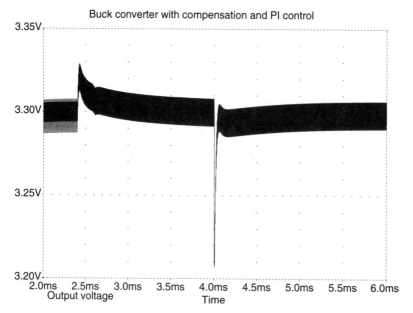

Figure 16.35 Transient response of lead compensated buck converter under PI control.

150 μs. Equally important, the converter is able to react to the load change in a direct fashion. Without compensation, the load change caused a 200 mV transient at the moment of the load change. With compensation, the transient is 100 mV. The smaller swing and better transient performance will be beneficial in high-performance applications such as low-voltage analog or digital circuits.

Example 16.4.5   A six-pulse rectifier is used for a chemical process. The input is stepped down to 20 V_{ll}, 60 Hz, three phase. The nominal output is 12.6 V at 4500 A. The rectifier has a 50 μH output inductor to keep the output current approximately constant. In the application, noise immunity is considered very important. At such a high current, momentary glitches in the SCRs would be destructive. It has been proposed to use PI control and a comparator process to set the phase angle. This is needed because the voltage must follow the process requirements, while the current is adjusted by an operator to achieve the desired action. To keep noise immunity high, phase-lag compensation has been suggested. Examine this application and propose actual controller values. With these values, what is the transient response?

A potential of 20 V_{ll} indicates the line-to-line RMS value. For a six-pulse rectifier, the peak voltage is $\sqrt{2}V_{ll} = 28.3$ V. The average output is $(6V_0/\pi)\sin(\pi/6)\cos \alpha$, or $(27.0 \text{ V})\cos \alpha$. Since the current is 4500 A at 12.6 V, the output resistance is 0.0028 Ω. The averaged model gives

$$(27.0 \text{ V})\cos \alpha = L\frac{di}{dt} + iR \qquad (16.60)$$

Let $k = \cos \alpha$. The output voltage is iR. The Laplace transform gives

$$i = \frac{27.0 \; k/R}{s \; \tau + 1}, \qquad v_{out} = \frac{27.0 \; k}{s \; \tau + 1}, \qquad \tau = \frac{L}{R} \qquad (16.61)$$

Here the circuit time constant L/R is 0.0179 s. A possible Bode plot, based on PI control, is shown in Figure 16.36. A low-pass filter has been added in the feedback loop to help avoid harmonics, but the gain values have been set aggressively. In this case, the crossover frequency is actually higher than the switching frequency because of high gain. To improve gain margin, a phase-lag compensator can be used to move the crossover frequency to a lower value. Figure 16.37 shows a Bode plot for a candidate design. This design will perform the necessary function, while avoiding effects of line harmonics.

16.4.5 Challenges for Control Design

The challenge in many converter control designs is that the small-signal model depends on operating point. In these examples, the model and the frequency domain behavior depend on the load, on the nominal duty ratio, and on other circuit parameters. It is desirable to perform worst-case design, but this is challenging when the model itself is variable. Converters with nonlinear models such as boost, buck-boost, and boost-buck circuits always require extra effort, since the duty ratio appears in the control transfer function.

Small-signal designs provide a systematic way to attack these challenges. Although the models are approximate, they provide valuable insight into controller performance. It can be argued, however, that nonlinear control methods would be more direct and perhaps are more appropriate for power electronics. It is certainly true that all the systems of interest in conversion have switching nonlinearities. It is also true that there is active interest in the con-

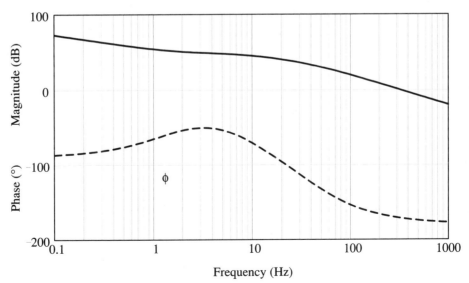

Figure 16.36   Possible Bode plot for rectifier control.

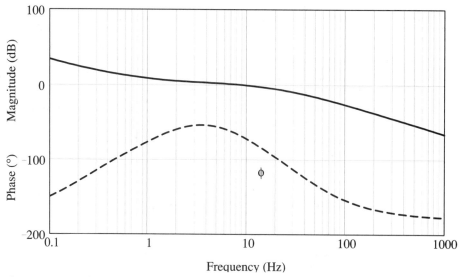

Figure 16.37 Bode plot for rectifier control with phase-lag compensation.

trol engineering community in a wide variety of methods for nonlinear systems. Switching controls, such as the sliding mode approach, have become popular.

Even so, power electronic systems involve special challenges. The nonlinearities take the form of discontinuities in derivatives. These switching discontinuities are outside the scope of most mathematical theory for nonlinear control systems. Only recently has some of the theory begun to meet the characteristics of power converters. We can expect important new developments in nonlinear controls for power electronics.

In Chapter 17, one class of nonlinear control is introduced. This class of *geometric control* provides a direct way to address large-signal requirements in power converters. The discussion should be viewed as a way to supplement small-signal design. The reader needs to be aware that there is no consensus on design methods for control of power converters—because only a few possible methods have been explored in detail. There are ample opportunities for new ideas in this area.

16.5 RECAP

Conventional methods for control design address linear systems. In most control practice, the use of conventional methods requires a *linearization* step. In linearization, a model is created for the system under study, based on small variations around a nominal operating point. Strictly speaking, power electronic systems do not involve small variations. They cannot be linearized directly. However, it is possible to make use of *averaging* in a formal way to develop approximate converter models. These *averaged models* meet the requirements of the linearization process, and serve as the basis for *small-signal models*.

Averaged models are the same as *infinite switching frequency* models introduced in Chapter 15. The models can be developed through a step-by-step analysis of circuit configurations. In general, the end result substitutes duty ratios for switching functions. In this way, an *averaged system* of differential equations can be created for a converter. The variables in these equations track the average behavior of currents and voltages in the network, but do not consider ripple or other fast effects. The averaged system equations can be written in terms of circuit models. Circuit models based on averaging are useful for simulation and for a variety of design problems.

The averaged models obtained for most converters are nonlinear. Typically, they involve products of duty ratios and state variables. Since the duty ratios serve as control parameters, the introduction of state feedback will generate product terms in the equations. This is a dilemma, especially since most converter models are relatively simple: They involve only two or three state variables, and the averaged circuit models often have only a few nodes. To support conventional control design methods, linearization can be applied to averaged models.

One way to form a linearized model is to replace state variables and inputs with a *perturbation* expression. This expression replaces a variable with a dc term plus an ac small-signal perturbation. Thus a capacitor voltage $v_C(t)$ becomes $V_{C(\text{nom})} + \tilde{v}_C(t)$. When the substitution is complete, the expressions include a set of dc terms, a linear set of perturbation terms, and a few products of perturbations. Since perturbations are small by definition, products of two perturbation terms are neglected in the models. The dc terms can be solved separately from the ac portion. The remaining expressions form a linear system, and represent a small-signal averaged model for the underlying power converter. In general, the model includes operating point information written as parameters. Thus the nominal duty ratio $D_{1(\text{nom})}$ might represent the gain of a controlled source in the small-signal model. There will be a different set of parameters for each operating point.

Once a small-signal model is available, it can be analyzed through Laplace transform techniques or through methods of time-domain linear systems. Block diagrams can be drawn. Various feedback techniques can be analyzed. The Laplace transform leads to transfer functions that represent frequency domain response of the converter. Frequency domain methods are popular for control design in power electronic systems.

Important transfer functions include the *control-to-output* transfer function, usually v_{out}/d. This is the system's *open-loop transfer function, G(s),* and supports frequency domain control design methodology. The *source-to-output transfer function*, $v_{\text{out}}/v_{\text{in}}$, also called the *audio susceptibility*, measures the effects of small line disturbances or noise on the converter output. The *output impedance* transfer function $v_{\text{out}}/i_{\text{out}}$ measures the effects of load changes. If a resistive model is used for the converter load, a separate *perturbation source* can be introduced for i_{out} to help identify this transfer function. Each transfer function is the ratio of the associated Laplace transform terms, with others set null.

The *equivalent open-loop transfer function*, $G(s)H(s)$, indicates the effects of a candidate feedback design. The Nyquist criterion can be used to evaluate closed-loop stability through examination of this transfer function. In control practice, the function is often termed the *open-loop transfer function*, in place of $G(s)$. The *closed-loop transfer function, K(s) = G(s)/(1 + G(s)H(s))* is the end product of control. The denominator of this function must have all its roots in the complex left half-plane to ensure stability. This fact is the basis for *root locus* feedback design, in which the root locations are examined as functions of gains or other system parameters.

Feedback design examples were developed from the open-loop transfer function based on conventional frequency domain techniques. With these methods, it is possible to evaluate different gain settings or other effects quickly, and make graphical adjustments to address performance requirements. An especially important graphical method in the frequency domain is *compensation*. The term has general meaning in control systems. Some of the most common approaches include *phase-lead compensation* and *phase-lag compensation*. In both cases, a pole-zero pair is generated, and the combination serves as a feedback transfer function. Phase-lead compensation seeks to improve phase margin by adding phase lead at the crossover frequency. It usually extends the system bandwidth, and is a useful approach for speeding up the transient response. Phase-lag compensation seeks to improve gain margin by rolling the gain off and reducing the crossover frequency. This approach decreases system bandwidth, and is an excellent tool for improving noise immunity.

While both phase-lead and phase-lag methods apply to power conversion, the operating point dependency can become a challenge. Extra phase margin at one operating point does not guarantee extra phase margin at all points. The designer must check for worst-case conditions if a robust control design is required. It is possible to create local instability if the control design is not checked at all relevant points.

Nonlinear methods suitable for power electronic systems are beginning to appear. They can be expected to influence control methodology in the near future. Small-signal methods offer good insight into operation and limitations, but by definition cannot address large transients or fault conditions in converters.

PROBLEMS

1. A flyback converter uses a 30 : 1 turns ratio in an application for 400 V to 12 V conversion. Find an averaged model for such a converter, and draw an averaged circuit model.

2. Simulate the open-loop response of a boost-buck converter to an input transient. The converter has $C = 100 \ \mu F$, and runs with a switching frequency of 50 kHz. Both inductors are 400 μH. The input voltage is nominally 20 V, and the output is -12 V. The load draws 100 W. How will the converter react to a 10% increase in input voltage?

3. A boost converter operates at 12 kHz. The input is 15 V at up to 5 A, and the output is a 72 V battery bus. The inductor has a value of 500 μH, and a 5 μF capacitor interfaces to the batteries. The inductor's internal resistance is 0.2 Ω. The transistor is a MOSFET with $R_{ds(on)} = 0.2 \ \Omega$. The diode can be modelled as a 0.5 V drop in series with 0.02 Ω.

   a. Draw the circuit.
   b. Develop the averaged model for this circuit. What is the nominal operating point? Since the output is a battery, should the capacitor voltage be included as a state variable?
   c. A duty ratio limit will be needed to prevent large-signal instability. Recommend a limit value.

4. Create an averaged model of a SEPIC converter. Ignore internal resistances and residual voltages. What is the small-signal model?

5. Develop an averaged model for a flyback converter, then prepare a small-signal model. The turns ratio on the coupled inductor is $a : 1$. Develop and draw an averaged circuit corresponding to the flyback small-signal model.

6. It is desired to test the tradeoffs between inductor and capacitor values in a buck converter. Consider a buck converter, with a conventional L-C pair at the output, with a resistive load. Feedback control uses proportional gain. The load is 5 V at up to 100 W, and the input is 12 V. Switching occurs at 100 kHz.

   a. Create a small-signal averaged model for the case $C = 0$. With $L = 100L_{crit}$, provide a Bode diagram for the system. What is the bandwidth for a given gain?

   b. Create a small-signal averaged model for the case $L = L_{crit}$, and C selected for $\pm 1\%$ ripple. Provide a Bode diagram. What is the bandwidth for a given gain?

   c. Consider an intermediate case with $L = 10L_{crit}$, and C selected for $\pm 1\%$ ripple. Again provide a model, a Bode diagram, and the system bandwidth.

   d. Compare these cases. Are they similar from a control perspective, or does a particular choice improve the overall system dynamics?

7. It was pointed out in Chapter 15 that a large-signal instability can occur in many converters if duty ratio is not limited. This is hard to capture with a small-signal model. Consider a dc–dc boost converter. The converter is initially "off," meaning that the transistor has been off for a long time and $v_C = V_{in}$, $i_L = V_{in}/R_{load}$. At time $t = 0$, proportional control is initiated, with $d = k_p(V_{ref} - v_C)$ (subject to the limitation that $0 \leq d \leq 1$). The values are $V_{ref} = 24$ V and $k_p = 10$. Is this stable for some choice of load resistance and other parameters? What is the behavior? Is the small-signal averaged model stable for the same case?

8. Compensation circuits can be cascaded if extra phase shifts are needed. Provide Bode diagrams for these cases:

   a. Two phase-lead compensators in cascade, with matched poles and zeroes.

   b. Two phase-lag compensators in cascade, with matched poles and zeroes.

   c. A phase-lead circuit cascaded with a phase-lag circuit. The poles of the lead circuit are zeroes of the lag circuit, and vice versa.

9. A certain system has a crossover frequency of 20 kHz and phase margin of 10°. The gain changes by 40 dB/decade, and the phase rate of change near f_c is 90° per decade.

   a. Can a phase-lead compensator be designed to improve phase margin to 60°? If so, what is the new crossover frequency?

   b. Consider a cascade of two phase-lead circuits instead. Can the phase margin be raised to 60° or better? What is the expected new crossover frequency?

10. A buck converter uses a capacitor with significant ESR. In fact the ESR value is half the value of the load resistor. The specific application is a 5 V to 2.5 V converter at 80 W. The inductor is set to about $10L_{crit}$, given a switching frequency of 100 kHz. The capacitor value is 1000 μF. Develop a small-signal averaged model. What are the pole and zero values for the open-loop case? How does the ESR affect the model? Provide a Bode diagram for the model.

11. In an ideal boost converter, the models suggest that infinite output is possible. We would prefer to capture the true behavior—a maximum output value limited by internal losses. Develop an averaged model for a boost converter with a nonzero value of on-state resistance for the transistor. Does this averaged model (which is still a large-signal model) show that a maximum output limit exists?

12. A boost converter is to be designed for 5 V to 24 V conversion. The switching frequency is 120 kHz. The load can vary from 0 to 20 W. The line can vary by ± 4 V. Output ripple is to be less than $\pm 1\%$. Line and load regulation should be 0.5% or better.

   a. Choose L and C to meet the output ripple requirements.

   b. Add PI control and a duty ratio limiter to the converter (for tight regulation and large-signal stability). Can you find proportional and integral gain values to provide phase margin of 45° or better?

   c. For your choice of PI gain values, simulate the behavior in response to a load decrease from 100% to 50%. Is the system stable?

13. For ac systems, an approach analogous to averaging is called the *describing function approach*. In averaging, we focus on the dc behavior of a system. With describing functions, we focus on some other Fourier component. Consider an ac–ac converter with 50 Hz input and 350 Hz switching. Try to develop a model for the 300 Hz output component.

14. Design an inverter that takes 12 V dc input and provides $170\cos(377t)$ V output, with less than 3% error. The load can range from 0 to 300 W. Suggest a closed-loop control to achieve this function.

15. Figure 16.1 shows the response of a specific flyback converter design to a 10% load increase. The second part of the figure shows a root locus for the particular case. The application was 360 V to 12 V conversion, with a switching frequency of 150 kHz, an output capacitor of 30 μF, and a 30 Ω load. The inductor, measured from the output side, is 250 μH. Generate an averaged model for this case, then prepare a Bode diagram of the linearized system. Can a stable closed-loop control be prepared with proportional gain alone, with just output voltage feedback?

16. Design a system for 12 V to 5 V conversion at 0 to 20 W. Feedback control should be provided to support 0.1% load regulation. The output ripple should not exceed 100 mV peak to peak.

17. Design a system for 48 V to -12 V conversion at a nominal power level of 40 W. The load can vary between 10 W and 40 W. The line tolerance is $\pm 25\%$. The desired line and load regulation should not exceed 1%.

REFERENCES

G. F. Franklin, J. D. Powell, A. Emami-Naeini, *Feedback Control of Dynamic Systems, 3rd ed.* Reading, MA: Addison-Wesley, 1994.

B. Friedland, *Control System Design.* New York: McGraw-Hill, 1986.

J. G. Kassakian, M. F. Schlecht, G. C. Verghese, *Principles of Power Electronics.* Reading, MA: Addison Wesley, 1991.

B. Y. Lau, R. D. Middlebrook, "Small-signal frequency response theory for piecewise-constant two-switched-network dc-to-dc converter systems," in *IEEE Power Electronics Specialists Conf. Rec.*, 1986, pp. 186–200.

B. Lehman, R. M. Bass, "Extensions of averaging theory for power electronic systems," *IEEE Trans. Power Electronics*, vol. 11, no. 4, pp. 542–553, July 1996.

R. D. Middlebrook, "Small-signal modeling of pulse-width modulated switched-mode power converters," *Proc. IEEE*, vol. 76, no. 4, pp. 343–354, April 1988.

R. D. Middlebrook, "Modeling current-programmed buck and boost regulators," *IEEE Trans. Power Electronics*, vol. 4, no. 1, pp. 36–52, January 1989.

R. D. Middlebrook, S. Ćuk, "A general unified approach to modelling switching-converter power stages," in *IEEE Power Electronics Specialists Conf. Rec.*, 1976, pp. 18–34.

D. M. Mitchell, *Dc–Dc Switching Regulator Analysis*. New York: McGraw-Hill, 1988.

N. Mohan, T. Undeland, W. P. Robbins, *Power Electronics, Second Edition*. New York: Wiley, 1995.

R. B. Ridley, B. H. Cho, F. C. Lee, "Analysis and interpretation of loop gains of multiloop-controlled switching regulators," *IEEE Trans. Power Electronics*, vol. 3, no. 4, pp. 489–498, October 1988.

G. W. Wester, R. D. Middlebrook, "Low frequency characterization of dc–dc converters," *IEEE Trans. Aerospace Electronic Systems*, vol. AES-9, no. 3, pp. 376–385, May 1973.

CHAPTER 17

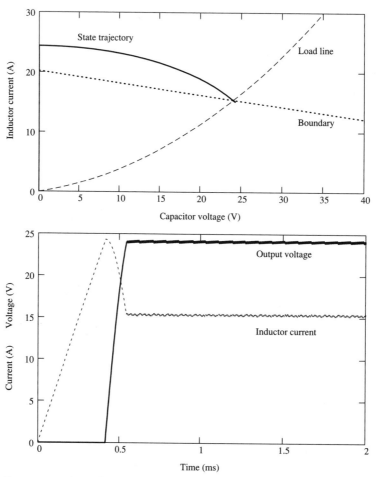

Figure 17.1 Phase-plane and time domain action of a high-performance boundary controller for a buck-boost dc–dc converter.

GEOMETRIC CONTROL FOR POWER CONVERTERS

17.1 INTRODUCTION

In the control systems literature, the term *geometric control* relates to *sliding mode control*, a method used for control of nonlinear systems by imposing specific state action through switching. The term *boundary control* usually refers to the idea of using different control strategies or gain settings near a particular operating trajectory, or *boundary*. In this chapter, a general approach to geometric control, appropriate for power electronics, is developed. Since a power converter can be controlled only by switch action, it is reasonable to define the action itself in terms of the variables in a converter. In power electronics, geometric control naturally refers to the way in which switch action interacts with the state trajectories. Switches open as currents cross zero, for example, or close when a reference voltage crosses a triangle oscillator waveform. In this general view, the switching boundaries themselves are adjusted to manipulate the operation of a converter. Therefore, the term *boundary control* is used here to indicate system control by means of switching boundaries. The concept can be very straightforward, especially when graphical tools are used to study it. It offers a powerful alternative for control of power electronics. In many cases, very fast performance can be obtained, as in Figure 17.1.

Boundary control is a direct large-signal control approach, meaning that it addresses the complete operation of a converter and does not separate start-up, steady-state, and protection modes. One basic example can be demonstrated with a buck converter. If the transistor is turned on whenever the output voltage falls too low, and then turned off when it becomes too high, the converter will be constrained to operate close to a fixed output voltage. The intended output voltage defines a boundary, and switch action is taken when the actual output voltage crosses this boundary.

In the first part of the chapter, boundary control is introduced, and conventional examples are illustrated. The buck converter controlled by an output boundary can be termed a *hysteresis control* approach. The hysteresis method has been used for ac motor drive inverters, dc–dc converters, and power factor correction circuits. Its properties are identified and evaluated.

The middle sections of the chapter describe *fixed boundary control*, a generalization of hysteresis control. A properly designed fixed boundary control has excellent steady-state and dynamic properties, and changes quickly from any starting point to the desired operating value. Fixed boundaries include sliding mode control techniques as a subset.

Fixed boundary controllers have some important operating advantages. They are stable even for extreme disturbances, and can be chosen to directly guarantee ripple specifications or other important operating issues in a converter. Implementation is easy. One property that many designers find undesirable is that the switching frequency is not fixed, but varies with converter conditions. The idea of *moving boundaries* can be used to represent a PWM process, or to enforce a fixed switching frequency. This is discussed briefly.

17.2 HYSTERESIS CONTROL

17.2.1 Definition and Basic Behavior

Many familiar control systems perform an abrupt change when the output value crosses a threshold. A thermostat, for instance, switches on a heating system when the temperature falls below a specified level, and switches it off above a second, higher level. Figure 17.2 shows this simple behavior. As long as the high-temperature setting turns off the system and the low-temperature setting turns it on, the room temperature, starting from any point, will move into the desired temperature range and then stay there. The high and low settings can be associated with *switching boundaries* that determine the control action.

> **Definition:** A switching boundary *is a geometric representation of switch action in a controller. Switch action takes place when the state trajectory crosses the boundary.*

When both a high and low boundary are used, the space in between the two boundaries defines a *dead band*, in which no control action occurs. The dead band is an important design requirement for a switching controller if there is just one state variable. Imagine a system in which both the on and off levels are the same. Then the system might *chatter*, try-

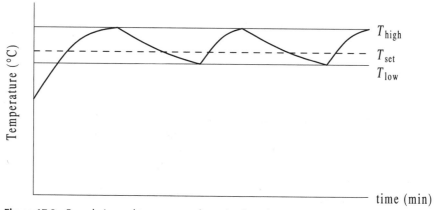

Figure 17.2   Boundaries and temperature dynamics for a heating system.

ing to switch on and off infinity fast in an attempt to keep the operating state at exactly the desired value. In a more complicated system, with two or more states, the extra dynamics of the other states might prevent chattering. In a power converter, however, the excessively fast switching associated with chattering is destructive, and it is essential to avoid the behavior. Therefore, a dead band is a typical feature of a power electronic boundary control.

A dead band implies irreversibility: The turn-on command and turn-off command occur under different conditions. The irreversibility represents *hysteresis*, analogous to that of the B-H loop of a magnetic material. When a simple boundary control like that of a thermostat is applied to a power converter, the process is termed *hysteresis control*.

> **Definition:** *In* hysteresis control, *switching boundaries are defined in terms of a single state variable or the system output. Two boundaries with a small separation control the switch turn-on and turn-off action.*

This control approach is the same as *bang–bang* control: A selection is made between two different control actions, based on a single measurement. The control is either full on or full off. The hysteresis process is extremely effective for certain types of converters. In some cases, the dead band directly determines the output ripple as well.

17.2.2 Hysteresis Control in dc–dc Converters

Hysteresis action for a buck converter is shown in a time plot in Figure 17.3. Regardless of where the voltage starts, switch action picks up as soon as a boundary is encountered. In the figure, converter start-up is illustrated. The transistor turns on initially because the voltage is below the turn-on boundary. The voltage rises from 0 to 5.05 V at a rate limited only by the inductor and capacitor values. The transistor then turns off, and remains off until the voltage falls to 4.95 V. The specific switching times and state values are determined solely by

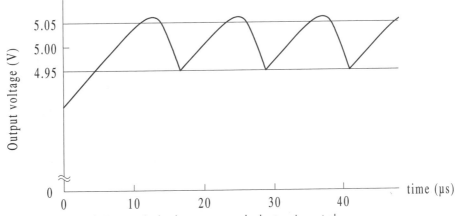

Figure 17.3 Time behavior of a buck converter under hysteresis control.

the action of a linear circuit. For example, the capacitor voltage must increase as long as the current $i_C > 0$. This current, $i_C = i_L - v_C/R_{load}$, is positive for a short time after the transistor turns off, so the output voltage must overshoot, and will rise somewhat beyond 5.05 V.

Notice that once the voltage is between the boundaries, switch action will keep it in the vicinity of the boundaries under all conditions. The general operation becomes independent of line, load, and even the inductor and capacitor values. The specific action, such as the degree of overshoot and the switching frequency, depends on the L and C values as well as on the load. Hysteresis control in principle eliminates output variations other than ripple. This property is called *robustness*: The system will stay close to the desired 5 V output even if component values are uncertain or the load changes dramatically. Most boundary controls are robust to system uncertainty—a very desirable attribute for a practical controller. Hysteresis control also provides immediate response to disturbances.

An important drawback is that hysteresis control does not work with all types of systems. If a control of this type, based on the output voltage, is tested with a boost converter, the results are disastrous. Starting from zero, the transistor turns on—then stays on indefinitely, since no energy flows to raise the output voltage. This was also true for simple proportional control if the duty ratio was not limited. In either case, why do such simple controls provide excellent results for a buck converter, yet completely fail to work with the boost circuit? The answer can be found in phase space.

Buck and boost circuits are shown for reference in Figure 17.4. Each converter has the usual circuit configurations, with the transistor on or off. In each configuration, there is a single stable equilibrium point that will be reached if the configuration is held for a long time. Table 17.1 summarizes these points. When a switch is thrown, the state variables change along trajectories toward the corresponding equilibrium point. Figure 17.5 shows the two

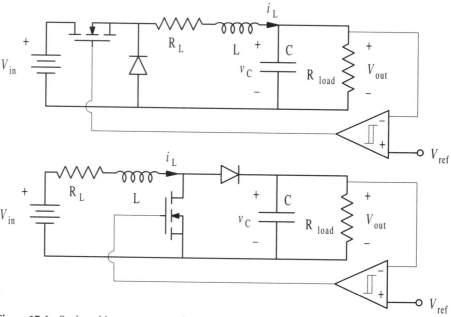

Figure 17.4 Buck and boost converters for analysis of boundary control.

TABLE 17.1
Equilibrium Points of the Direct dc–dc Converters

Converter and Configuration	Equilibrium Point	Comments
Buck, transistor on	$v_C = V_{in}$, $i_L = V_{in}/R_{load}$	Resistive load assumed
Buck, transistor off	$v_C = 0$, $i_L = 0$	Same whether diode is on or off
Boost, transistor on	$v_C = 0$, $i_L = V_{in}/R_L$	Current limited only by stray resistance of inductor
Boost, transistor off	$v_C = V_{in}$, $i_L = V_{in}/R_{load}$	Resistive load assumed

sets of trajectories for a buck converter in phase space. The inductor current is on the vertical axis, and the capacitor voltage is on the horizontal axis. Each curve represents the path taken from a given starting point.

Figure 17.6 shows the corresponding shapes for a boost converter. The transistor-on state will drive the inductor current to an extremely high value, too high for the equilibrium point to be plotted on a scale of realistic operating current values. Notice in the figure that the on-state trajectories approach a high point on the y-axis.

One interesting property of any real power converter is that the operating point we intend for the system is not the same as any equilibrium point of its configurations. After all, switching would not be necessary if the operating point matched one of the equilibrium points. Nevertheless, the states can only follow trajectories like those in the figures. Switch action pieces together short segments of the families of trajectories to coax the circuit toward the desired operating point.

There is a family of possible operating points defined by fixed duty ratios. For the buck converter, this family falls along a line joining the points $(v_C, i_L) = (0,0)$ and $(V_{in}, V_{in}/R)$. At $D = 0.5$, the point is $(0.5V_{in}, 0.5V_{in}/R)$, and so on. The values $D = 0$ and $D = 1$ correspond to the equilibrium points. Figure 17.7 shows the family of operating points taken at various

Figure 17.5 Equilibrium points and trajectory families for buck converter.

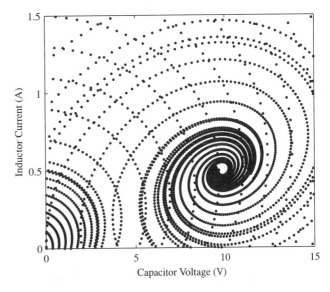

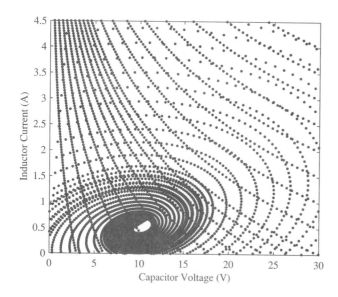

Figure 17.6 Equilibrium points and trajectory families for boost converter.

duty ratios in each of the two direct converters. The boost converter shows a large curve that starts at $(V_{in}, V_{in}/R)$, swings out to higher voltages, then eventually contacts the x-axis at a current of V_{in}/R_L. The family of points can be termed a *load line*, since the points are those that are physically consistent with the load resistance at various values of D.

In phase space, output-based hysteresis control is very easy to represent: The hysteresis boundary is a vertical line at the desired V_{out} (or, more strictly, a pair of lines at $V_{in} - \Delta V/2$ and $V_{in} + \Delta V/2$). This boundary splits the phase plane into two half-planes. Whenever the state trajectory tries to cross a boundary, switching occurs. For the buck converter, this is summarized in Figure 17.8. Whenever the state is to the left of the boundaries, the tran-

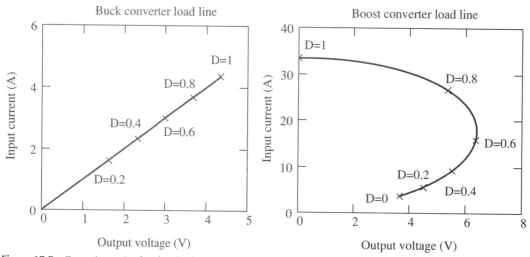

Figure 17.7 Operating point families for buck and boost converters.

Figure 17.8 State evolution for buck converter under hysteresis control.

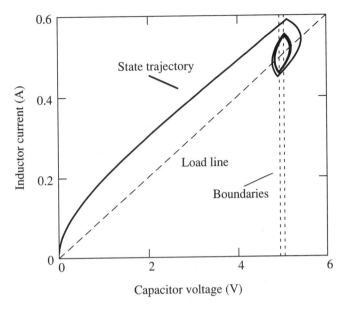

sistor is on. When the state is to the right of the boundaries, the transistor turns off. The figure shows how the converter evolves from the origin to the final operating point. Since the boundary keeps the equilibrium points separate, there is no chance for converter action to somehow be pulled toward the wrong operating point.

The boost converter is very different. A vertical boundary can be drawn at some voltage greater than V_{in} on Figure 17.7, but what should be done with the switch? If the state is to the left of the boundary, turning the transistor on could move the system toward the point $(0, V_{in}/R_L)$, a high-loss point with no output. The control would never recover. A vertical boundary is not a feasible choice for this converter, since it is not guaranteed to establish a unique operating point. Looking ahead for a moment, imagine that Figures 17.5 and 17.6 are used with nonvertical boundaries. If the slope of the boundary is used as a design parameter, many new possibilities for control open up.

Hysteresis control can be used in a converter that meets certain conditions. These include:

- When the hysteresis control is represented as a boundary in phase space, this boundary must keep the equilibrium points of the configurations separated.
- The switches must act in opposition to the natural evolution toward an equilibrium point. In the buck converter, for example, if the trajectories are moving toward (0,0), the transistor should turn on to force movement in a different direction. This way, the system cannot "lock up" at the wrong point.
- The switch action must have a direct influence on the state variable. For instance, the variable should rise when the switch is on and fall when the switch is off.
- The boundary must pass through the desired operating point to ensure that switch action can drive the system to that point.
- A dead band should be provided to avoid chattering.

In a buck converter, boundaries based on either the output voltage or the inductor current will meet these requirements. In a boost converter, the input current can be used. The output current in a half-bridge inverter will work, and the inductor current in a buck-boost converter can also provide a successful hysteresis control.

The switching frequency in a dc–dc converter controlled by hysteresis is determined entirely by the way in which the trajectories interact with the boundary. However, it is fairly simple to determine the steady-state duty ratio: It must be V_{out}/V_{in} in a buck converter in which $L > L_{crit}$. An example might help to show some of the behavior.

Example 17.2.1   An ideal 24 V to 5 V buck converter uses hysteresis control with a 50 mV dead band. The inductor value is 100 μH and the capacitor value is 2 μF. What are the switching frequency, duty ratio, and output ripple with a 1.25 Ω load? What happens if the load changes abruptly to 0.25 Ω?

Even with the hysteresis control, basic operating principles of the buck converter do not change. The operation must reach periodic steady state at some frequency. If the average output is to be 5 V, the duty ratio for the transistor must be 5/24 in steady state. The inductor current will be a triangle waveform. The capacitor voltage will follow a parabolic shape (as in Figure 3.21). Given that the capacitor voltage is nearly constant, the steady-state analysis can follow conventional lines. With the transistor on, the inductor relationship gives the change in current,

$$v_{L(on)} = L\frac{di}{dt}, \qquad V_{in} - V_{out} = L\frac{\Delta i}{DT} \tag{17.1}$$

Here, both T and Δi are unknown. We obtain

$$\Delta i = \frac{DT(V_{in} - V_{out})}{L} \tag{17.2}$$

The capacitor current is $i_C = i_L - i_{load}$. The output voltage change is determined from

$$\frac{1}{C}\int_{t_p} i_C \, dt = \Delta v_C \tag{17.3}$$

where $t_p = T/2$ is the time interval during which $i_C > 0$. Since the inductor current is a triangular waveform, as in Figure 17.9, the integral is the triangular area

$$\Delta v_C = \frac{1}{2C}t_p i_{C(max)} \tag{17.4}$$

The maximum capacitor current is equal to $\Delta i_L/2$, and $t_p = T/2$, so the change in voltage is

$$\Delta v_C = \frac{1}{2C}\frac{T}{2}\frac{DT(V_{in} - V_{out})}{2L} = \frac{DT^2(V_{in} - V_{out})}{8LC} \tag{17.5}$$

The remaining unknowns are the period T and the actual change in voltage Δv_C. Another equation is needed. This equation is the full differential equation of the converter: During

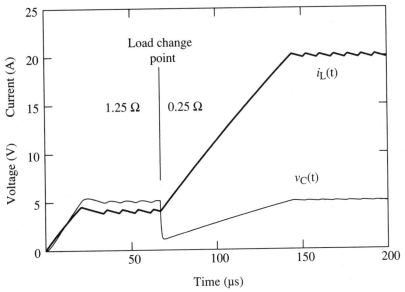

Figure 17.9 Inductor current and capacitor voltage for Example 17.2.1.

the on-time of the transistor, the capacitor voltage makes the transition from the lower boundary to the upper boundary. The change is governed by the equation of a second-order system with input V_{in}.

The boundary result in Figure 17.8 shows that voltage ripple under hysteresis control can be greater than the dead band. Equation 17.5 can be used to find the actual ripple—once the period is known. It is perhaps more direct to obtain the ripple result from a computer animation. Table 17.2 lists a BASIC program that uses the forward Euler method to determine and plot time or state evolution for this converter. At light loads (above about 20 Ω), discontinuous mode will affect D, but the hysteresis converter will still maintain approximate 5 V output. Figure 17.10 shows the phase plane behavior for the 1.25 Ω and 0.25 Ω loads. The behavior with a load of 20 Ω is also shown for comparison. When the simulation results were printed, the output voltage for the 1.25 Ω load ranged from 4.958 V to 5.141 V, for total ripple of 183 mV peak-to-peak. The switching frequency was 112 kHz. The corresponding results for the 0.25 Ω load were a range from 4.973 V to 5.040 V, ripple of 67 mV peak-to-peak, and a switching frequency of 122 kHz.

The hysteresis control in the example always maintains the converter output close to 5 V. The trajectory shapes, which depend on L, C, and R_{load}, determine the degree of overshoot, the switching frequency, and the actual ripple value. Values of L and C can be selected to provide a desired trajectory shape, and thereby alter the behavior. For example, higher capacitance values tend to make the trajectories more nearly vertical in the vicinity of the operating point (a change in the inductor current has less effect on v_C if C is large).

TABLE 17.2
BASIC Program to Determine the Operation of a Hysteresis Control

```
'Hysteresis control, Example 17.2.1.   BASIC program (for QBASIC and other PC versions)

L = 100E-06: C = 2E-06: Vin = 24: Vout = 5          'Circuit parameters
Voff = 5.025: Von = 4.975                           'Boundaries

PRINT "Do you want to plot in time domain or state domain"; : INPUT x$ 'Find out what to plot.
PRINT "Load resistance"; : INPUT Rload              'Load resistance is an input.

DEF fndil (il, vc) = (q1 * Vin − vc) / L            'Define derivatives for IL and
DEF fndvc (il, vc) = (il − vc / Rload) / C          ' VC for numerical method.

'Set up the plot scales. The values are chosen for EGA/VGA.
ym = 350 / (1.5 * Vout / Rload): yb = 175 + ym * Vout/Rload    'ym is linear scale, yb is vertical offset,
vm = 350 / (1.5 * Vout): vb = 175 + vm * Vout        'vm and vb are scale and offset for y-axis volts
xm = 640 / (Vin): xb = xm * Vout − 320: IF xb < 0 THEN xb = 0   'Scale, offset for x-axis voltage.

t = 0: vc = 0: il = vc / Rload                      'Initial conditions.
delt = 1E-07: tlast = .0002: tscale = 640/tlast     'Time step, ending time, time scale.

SCREEN 12: CLS: LINE (0, 0) − (0, yb): LINE (0, yb) − (640, yb)    'Set graphics mode and draw axes.

IF x$ = "state" THEN                'If a state domain plot is needed, draw the load line and boundaries.
      LINE (−xb, yb) − (xm * Vin − xb, yb − ym * Vin / Rload)    'Load line.
      LINE (xm * Voff − xb, 0) − (xm * Voff − xb, yb)    'Off boundary.
      LINE (xm * Von − xb, 0) − (xm * Von − xb, yb)      'On boundary.
      LINE (−xb, yb) − (−xb, yb)                   'Starting point.
END IF
IF x$ = "time" THEN                'If a time domain plot is needed, draw the boundaries but no load line.
      LINE (0, vb − vm * Voff) − (640, vb − vm * Voff)
      LINE (0, vb − vm * Von) − (640, vb − vm * Von)
END IF
        'Here is the simulation loop. The idea is that Xnew = Xold + DeltaT*(dX/dt).
WHILE t < tlast
tnew = t + delt                    'Update time.
IF vc > Voff THEN q1 = 0           'Which boundary is active?
IF vc < Von THEN q1 = 1            'Transistor on or off?
'Compute and plot as appropriate.
IF x$ = "state" THEN LINE −(vc * xm − xb, yb − ym * il)
ilnew = il + delt * fndil(il, vc): IF ilnew < 0 THEN ilnew = 0
vcnew = vc + delt * fndvc(il, vc)
IF x$ = "time" THEN
      LINE (t * tscale, yb − il * ym) − (tnew * tscale, yb − ilnew * ym)
      LINE (t * tscale, vb − vc * vm) − (tnew * tscale, vb − vcnew * vm)
END IF
vc = vcnew: il = ilnew: t = tnew   'Update the values.
WEND                               'End of loop.
END
```

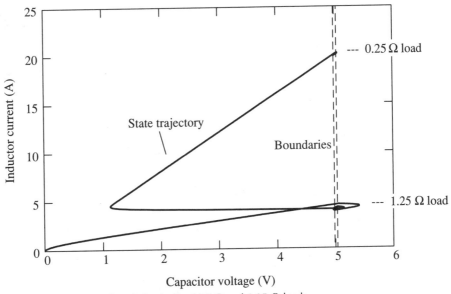

Figure 17.10   Phase plane behavior for 1.25 Ω and 0.25 Ω loads.

Figure 17.11 shows results for a few choices of L and C for a 24 V to 5 V converter with a 1.25 Ω load. Some results might be surprising. For example, when the capacitance is increased, the ripple actually goes up because the switching frequency decreases and there is more energy change during the on and off states. Higher inductor values can reduce both the frequency and the ripple, since there is less change in the input current when L is large. If the inductor is made very large, the ripple approaches 50 mV, and is determined solely by the boundaries.

Hysteresis control can involve variables other than the output voltage of a buck converter, as mentioned above. The inductor current value can be controlled, for example, by using horizontal boundaries on a phase plot. The problem with this approach is that the control setting becomes load dependent. For a 5 V converter with 1.25 Ω load, the current setting should be 4 A. For a 0.25 Ω resistor, it should be 20 A, and so on. The performance is qualitatively different. Results for the 24 V to 5 V converter with $L = 100\ \mu H$, $C = 2\ \mu F$, and a 1.25 Ω load are shown in Figure 17.12. The current reaches the boundary and then "slides" along it to the operating point. Many boundary controls have the attributes of sliding mode control, in which the system dynamics are effectively governed by the boundary rather than the trajectories. In Figure 17.12, the output voltage shows essentially no overshoot, and the output ripple is determined almost entirely by the dead band.

Current-based hysteresis control can be implemented along the lines of the two-loop method discussed in Chapter 16. Just as in two-loop control, the error between the actual and desired voltage represents an additional current that must be delivered to the load. A PI block can use the voltage error signal to provide a current value for a hysteresis control. A system based on this configuration is shown in Figure 17.13. Like a two-loop converter, it can be difficult to set proportional and integral gains in this system, since gain settings de-

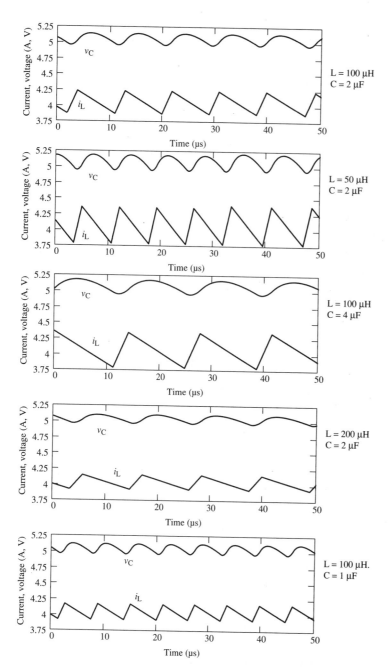

Figure 17.11 Behavior near operating point for 24 V to 5 V converter, 1.25 Ω load, with various L and C values.

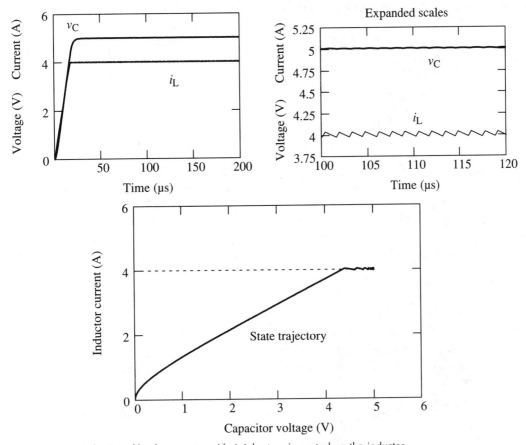

Figure 17.12 Behavior of buck converter with 4 A hysteresis control on the inductor.

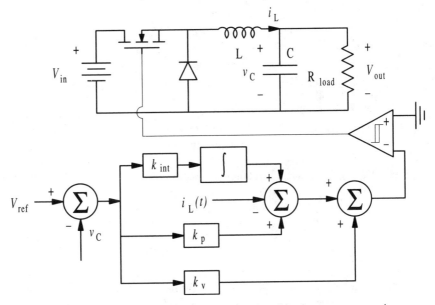

Figure 17.13 Block diagram of PI system for current-based buck converter control.

pend on the value of R_{load}. A relatively low integral gain will ensure that the system reaches the correct operating point without causing extra overshoot. A low proportional gain also seems to work well in this application.

17.2.3 Power Factor Corrector

The power factor correction boost converter introduced in Section 6.5 is an excellent example of an opportunity for hysteresis control. In the PFC boost converter, the *shape* of the desired current waveform should be a full-wave rectified sinusoid, identical in shape to the input voltage waveform after full-wave rectification. The magnitude of the current can be set with an outer PI loop that measures the difference between the actual output voltage and its desired value. Once the current waveform is determined, the transistor can be turned on when the current is too low, and shut off when it is too high.

The boost converter's input current can be controlled through hysteresis because it meets the requirements:

- The boundary is horizontal in the (v_C, i_L) plane. A horizontal line drawn through a realistic operating point in Figure 17.7 separates the two equilibrium points.
- The switch action is opposite to the trajectory action. If the transistor is on, the converter moves toward the high-current point $(0, V_{in}/R_L)$, but the switch will shut off when the current becomes too high.
- The transistor has a direct and immediate effect on the inductor current, unlike the capacitor voltage effect, which is indirect.
- A dead band can be produced simply by adding or subtracting a small offset from the desired current waveform.

A block diagram of this PFC boost conversion concept is shown in Figure 17.14, as it might be used in a dc link power supply application. Figure 17.15 shows the voltage and actual current in a sample converter drawing 50 W from a 120 V, 60 Hz source. The current tracks a rectified sinusoid. The use of circuits like this is growing rapidly. Hysteresis control simplifies the design and keeps the system operation consistent even over wide load variations.

Phase-plane representation of a PFC boost application is not as straightforward as in more conventional dc–dc conversion applications. The output voltage is to be held constant, but the input current varies sinusoidally from 0 to a peak value. Since the load line shows possible operating points for a given input, the load line in a PFC application becomes time varying. The desired operation is along a line segment rather than at a specific point since constant v_C with varying i_L is desired.

17.2.4 Inverters

Inverters, especially those for ac drives, are often operated in a hysteresis mode. In an ac drive, the control computations can be used to identify a reference value for motor current. This reference value, combined with a small dead band, can serve as the basis for hysteresis control.

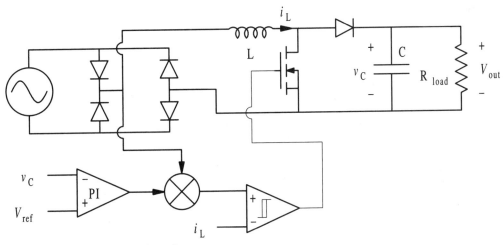

Figure 17.14   Hysteresis-based PFC boost converter.

Consider the half-bridge inverter in Figure 17.16. This circuit will meet all the requirements for hysteresis control if the upper switch is turned on when the inductor current is low and the lower switch is turned on when it is high. If a microcomputer or similar processor engine has provided a desired value of current, it is a simple matter to add a dead band and control the two transistors in a hysteresis mode. As in the case of the PFC boost converter, the desired current varies with time, but the basic hysteresis action still works.

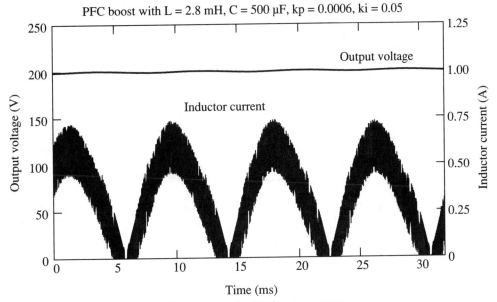

Figure 17.15   Output voltage and input current for PFC boost converter.

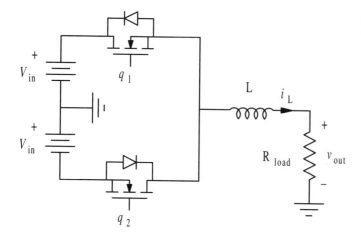

Figure 17.16   Half-bridge inverter for hysteresis control.

Example 17.2.2   A half-bridge inverter uses ± 100 V dc supplies. The output inductor value is 10 mH. The load resistance is 10 Ω. It is desired to impose a current of 5 A_{RMS} at 50 Hz on the load resistor, with a 500 mA dead band. Start from 0 A, and show the output current that results from this operating strategy.

The desired current is $5\sqrt{2}\sin(100\pi t)$ A (since it starts from 0). To provide the dead band, let us choose a strategy in which the upper switch turns on when $i(t) <$

TABLE 17.3
BASIC Program for Example 17.2.2

```
'BASIC program for Example 17.2.2; Hysteresis control of inverter.

'Establish the L-R circuit equation for di/dt.
Rload = 10: L = .01: Vin = 100: deli = .5: pi = 4 * ATN(1)      'Parameters.
DEF fni (t) = 5 * SQR(2) * SIN(100 * pi * t)                    'Desired current
DEF fndi (il) = ((q1 * Vin − q2 * Vin) − il * Rload) / L        'di/dt

q1 = 0: q2 = 0: il = 0                                          'Initial values
delt = .00002: tlast = 2 / 50: tscale = 640 / tlast            'Time and scaling

yscale = 30: yoff = 240                                         'Y axis scale and offset

SCREEN 12: CLS ': LINE (0, yoff) − (640, yoff): LINE (0, 0) − (0, 350)   'Draw axes
LINE (0, yoff − yscale * il) − (0, yoff − yscale * il)          'Initialize plot

'Begin main loop.
WHILE t < tlast
t = t + delt                                                   'Update time
IF il < fni(t) − deli / 2 THEN q1 = 1: q2 = 0                   'Check boundaries
IF il > fni(t) + deli / 2 THEN q2 = 1: q1 = 0

il = il + delt * fndi(il)                                       'Update current and plot
LINE −(t * tscale, yoff − yscale * il)                         ' based on Euler method.
WEND
END
```

$5\sqrt{2}\sin(100\pi t) - 0.25$ A, and the lower switch turns on when $i(t) > 5\sqrt{2}\sin(100\pi t) + 0.25$ A. To avoid KVL violations, when one switch is required to turn on, the other must shut off. A BASIC program that evaluates this circuit is given in Table 17.3. The current waveform, given in Figure 17.17, gives the desired result: moderate ripple around the reference current waveform.

The half-bridge inverter exhibits some basic design constraints. To follow a specified current, for instance, the *L-R* circuit must be fast enough to meet the desired maximum value of *di/dt*. The current provided in Example 17.2.2 has a maximum *di/dt* value of about 2200 A/s. With 100 V, the inductor permits about 10000 A/s, ignoring the voltage across the resistor. If the inductor were larger by a factor of five, the inverter would not be able to follow the desired current change. This limitation would affect operation regardless of the control technique, but the issues are more directly evident for a hysteresis control.

It is also possible to control an inverter directly with a boundary. The inductor current and capacitor voltage in the half bridge will trace out an ellipse in state space when the inverter operates correctly. This ellipse defines the desired operation. It can be used as the basis for a control boundary if attention is given to the polarities. The references at the end of this chapter include an elliptical boundary inverter.

17.2.5 Design Approaches

Hysteresis control is relatively straightforward for the converters discussed here. In most cases, the dead band is determined directly by the boundaries. In general, the procedure follows along these lines:

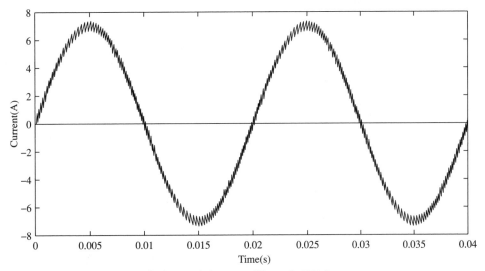

Figure 17.17 Output current for hysteresis inverter of Example 17.2.2.

1. Identify the basic expectations for the converter.
2. Choose a target switching frequency and nominal load.
3. Choose L and C values, as in Chapter 4, to meet the desired ripple levels.
4. For a dc–dc converter, determine the expected Δi_L and Δv_C values. For current, select a dead band consistent with the expected Δi_L. For voltage in a buck converter, the voltage dead band needs to produce the desired Δi_L value. It can be determined by solving for the trajectories or by iterating through a few quick attempts.
5. For an inverter, ensure that the maximum derivative determined by $v_L = L(di/dt)$ is high enough to follow the desired current change. If the current is $I_0 \sin(\omega t)$, then $v_L/L > \omega I_0$. The value of v_L is not the input voltage, since $v_L = V_{in} - i(t)R_{load}$. Given the exponential behavior, it is probably sufficient to choose $v_L \approx V_{in}e^{-1}$. Thus $V_{in}/L > 2.8\omega I_0$ should provide a workable design.
6. Select the boundaries to meet the key boundary control requirements: The equilibrium points must be kept separate, the switch action must oppose the natural evolution toward the equilibrium points, and the boundary must intersect the load line at the desired operating point.

A major advantage of a hysteresis control is that stray components and other unknowns will not affect the operation very much, so the design procedure does not require detailed analysis. A second advantage is that if the requirements for the boundaries are met, there is no need for a separate start-up control.

Example 17.2.3   It is desired to operate a 12 V to 48 V boost converter for a telephone application with hysteresis control. The expected load is 10 W, although the load could vary from 0 to as much as 25 W. Suggest a possible design, and plot its time and state performance when 12 V is applied.

Since this is a boost converter, voltage-based hysteresis control will not work. Since the load is unknown, current-based hysteresis control will require a PI loop. The nominal duty ratio is 3/4 for 12 V to 48 V conversion, and the output current will not exceed (25 W)/(48 V) = 0.52 A. Assuming 90% efficiency, the input current should not exceed 2.5 A. We can use this as an upper limit on the PI control to keep the transistor-on equilibrium point separate from the transistor-off point. Let us suggest an output ripple of 250 mV— about 0.5%. At this power level, a switching frequency of 200 kHz should be reasonable. Under nominal operation, the output current is about 0.2 A. The transistor has $D = 0.75$, and should be on for about 3.75 μs at nominal load. The ripple expressions give

$$i_C = C\frac{dv}{dt}, \qquad 0.2 \text{ A} = C\frac{0.25 \text{ V}}{3.75 \text{ }\mu s}, \qquad C = 3 \text{ }\mu\text{F} \qquad (17.6)$$

Hysteresis control has no problems if operation extends into discontinuous mode, so presumably a small inductor can be used. Under nominal conditions, the input current is (10 W)/(12 V), or about 0.8 A. To avoid excessive input current ripple, let us suggest $\Delta i_L = 0.16$ A (i.e., $\pm 10\%$) for the dead band. Then

$$v_L = L\frac{di}{dt}, \qquad 12 \text{ V} = L\frac{0.16 \text{ A}}{3.75 \text{ }\mu s}, \qquad L = 280 \text{ }\mu\text{H} \qquad (17.7)$$

The only remaining issue is the choice of PI loop gains to set the reference current. This current could be as low as zero or as high as 2.5 A, but should not be allowed to be any higher. The intended voltage is 48 V, and analog control circuits probably will not handle more than about 10 V. Therefore a proportional gain not more than 0.2 seems appropriate. Some testing shows good overall results with a proportional gain of 0.1 and an integral gain of 150. One caveat: The output voltage sensed for the control loop should be low-pass filtered. If it is not, the output voltage ripple will interact with the boundary, possibly in an undesired fashion. Time domain and state domain plots of the simulated performance are shown in Figure 17.18.

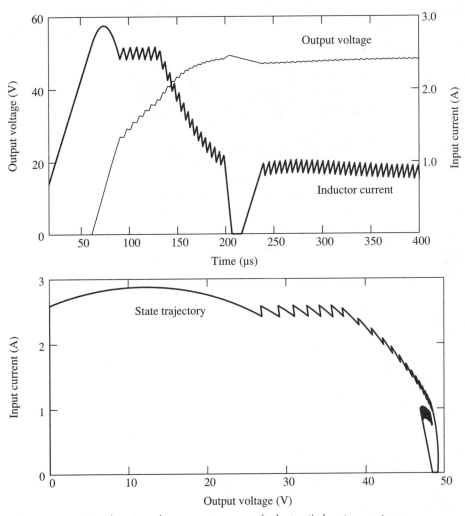

Figure 17.18   Time domain and state-space response for hysteretic boost converter.

17.3 GENERAL BOUNDARY CONTROL

17.3.1 Behavior Near a Boundary

Hysteresis control is a special case in which a boundary is selected based on a single state variable. In general, combinations of states and any appropriate type of boundary can be used for control. Alternative choices offer certain advantages. In a simple two-state dc–dc converter, for example, the control boundary could be any one of a family of lines. Perhaps curved boundaries might have merit as well. In a converter with three states, the boundary becomes a plane in state space.

Boundary control in a broader sense uses one or more structures in state space, with one less dimension than that of the space. When more than two state variables are involved, the boundary becomes a *switching surface* (a general structure in n-space with dimension $n - 1$). In many cases, the boundary might represent a combination of individual sections. Example 17.2.3 combined a fixed current limit of 2.5 A with an adjustable boundary controlled with a PI loop. Hard limits on current or voltage are easy to represent as boundary controls.

Any boundary that might be chosen interacts with the trajectories associated with individual converter configurations and with equilibrium points of the configuration. Since switching occurs when a boundary is crossed, it is helpful to consider the trajectory behavior at a boundary. There are only three possibilities, shown in Figure 17.19:

1. On one side of the boundary, trajectories approach it. On the other side, trajectories lead away from it. This gives rise to *refractive* behavior, in which the states evolve in a new direction when the boundary is crossed and switching occurs.

2. Trajectories on both sides of the boundary approach it. In this case, switch action immediately redirects the system back without allowing the boundary to be crossed. This can be termed *reflective* behavior.

3. Trajectories on both sides of the boundary lead away from it. In this case, switch action cannot force the states to evolve toward the boundary. The behavior might be called *rejective*.

In boundary control, stability means that states tend to get closer and closer to the desired operating point as time moves along.

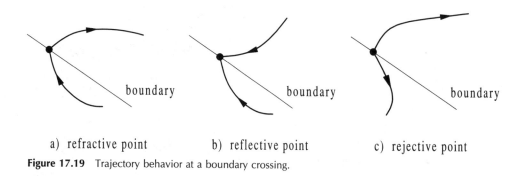

a) refractive point   b) reflective point   c) rejective point

Figure 17.19 Trajectory behavior at a boundary crossing.

17.3.2 Possible Behavior

When the boundary exhibits refractive behavior, the system operation follows along the lines of Figure 17.19. The trajectories form a loop that tends to get smaller over time. In a real converter, switching frequency limits or a dead band will ultimately lead to a stable loop. The steady-state operation becomes a *limit cycle* in state space.

> **Definition:** *A* limit cycle *is a closed loop in state space that describes the steady-state operation of a periodic system.*

Ripple is related to the size of the limit cycle. The limit cycle (and the ripple) can always be made smaller by allowing faster switching. In the infinite frequency limit, the limit cycle reduces to the desired operating point. As long as the boundary passes through the desired point, the system will move to a limit cycle near that point if it is stable.

In a refractive case, the dead band determines the nature of the limit cycle, but not directly. A small dead band might still produce high ripple if the trajectories are highly curved near the operating point. In practice, it may not be necessary to add an explicit dead band in a converter with refractive operation. This is because speed limitations of comparators and op-amps will provide enough dead band to keep the system functioning properly.

The same is not true in a reflective case. When the trajectories are reflective, a dead band is required because the switching frequency tends immediately to infinity. In the reflective case, the system in effect reverses direction when a boundary is encountered. If the system starts on the "switch on" side of the boundary, the switch will chatter as soon as the boundary is encountered. The system is then constrained to act only along the boundary. Since the boundary does not include any equilibrium points, the system operating point will move along the boundary. This defines a *sliding mode*, as used in the control literature.

Sliding mode control is a very powerful technique. Since the boundary determines the operation of the system when a sliding mode occurs, the system dimension can be reduced by one order. For example, consider a buck converter controlled by a linear boundary described by the line $k_v(v_C - V_{out}) + k_i(i_L - V_{out}/R_{load})$. If the converter exhibits reflective behavior along this boundary, the inductor current and capacitor voltage are not independent. The system acts as a first-order linear system, and its operating point will decay exponentially to the point $(V_{out}, V_{out}/R_{load})$. Sliding mode control has an additional advantage for ripple: The dead band determines the ripple, and the system is constrained to operate inside the dead band. From a design standpoint, these advantages normally make a reflective boundary the preferred arrangement.

Rejective boundaries are undesirable. Since all trajectories move away from them, rejective boundaries cannot be used to draw the system to a desired operating point. Indeed, if the operating point happens to sit on a rejective boundary, there is no way to reach the point. Therefore, it is important to choose boundaries to avoid rejective characteristics.

17.3.3 Choosing a Boundary

The ideal boundary provides stability, good large-signal operation, and fast dynamics. Typically, a good choice for a boundary has a large sliding mode region (a reflective region) around the operating point. The behavior can be evaluated by defining normal and tangential vectors based on the candidate boundary, as shown in Figure 17.20. Each trajectory also can be associated with a velocity vector. The general direction of trajectories can be found with a dot product between the velocity vectors and the normal vector. Reflective behavior requires that the dot product be negative on the side of the boundary defined by a positive normal vector, and that the dot product be positive on the other side. This is a vector-based description of the property that both sets of trajectories are directed toward the boundary.

The velocity vector is simply a vector form of the derivative $\dot{\mathbf{x}}$, taken from the network equation of a converter system

$$\dot{\mathbf{x}} = f(\mathbf{x}, u, q) \tag{17.8}$$

By setting each switching function q_i to a value of 1 in turn, velocity vectors can be found for each configuration. In the cases considered here, the vector is two-dimensional, and just two or three configurations must be considered. A linear boundary for control must pass through the desired operating point, given by (v_{C0}, i_{L0}) for a converter with a single inductor and capacitor. This line has equation

$$k_v(v_C - v_{C0}) + k_i(i_L - i_{L0}) = 0 \tag{17.9}$$

and slope $-k_v/k_i$. If the state space is associated with unit vectors $\mathbf{a_v}$ and $\mathbf{a_i}$, the unit normal vector $\mathbf{n}$ to the boundary is given by

$$\mathbf{n} = \frac{k_v}{\sqrt{k_v^2 + k_i^2}} \mathbf{a_v} + \frac{k_i}{\sqrt{k_v^2 + k_i^2}} \mathbf{a_i} \tag{17.10}$$

The velocity vector has the value

$$\dot{\mathbf{x}} = \dot{v}_C \mathbf{a_v} + \dot{i}_L \mathbf{a_i} \tag{17.11}$$

Of interest is the normal component of the velocity vector relative to the boundary, given by

$$\dot{\mathbf{x}} \cdot \mathbf{n} = \frac{k_v \dot{v}_C}{\sqrt{k_v^2 + k_i^2}} + \frac{k_i \dot{i}_L}{\sqrt{k_v^2 + k_i^2}} \tag{17.12}$$

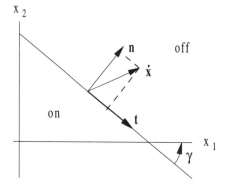

Figure 17.20   Boundary with normal and tangential vectors.

A final issue to be established is the side of the boundary associated with the on and off configurations, respectively. If there is a single active switch which should be off above or to the right of the boundary (that is, on the boundary side associated with a positive normal direction), then reflective action requires

$$\dot{\mathbf{x}}_{\text{off}} \cdot \mathbf{n} = \frac{k_v \dot{v}_{C(\text{off})}}{\sqrt{k_v^2 + k_i^2}} + \frac{k_i \dot{i}_{L(\text{off})}}{\sqrt{k_v^2 + k_i^2}} < 0 \tag{17.13}$$

$$\dot{\mathbf{x}}_{\text{on}} \cdot \mathbf{n} = \frac{k_v \dot{v}_{C(\text{on})}}{\sqrt{k_v^2 + k_i^2}} + \frac{k_i \dot{i}_{L(\text{on})}}{\sqrt{k_v^2 + k_i^2}} > 0$$

This can be simplified to give

$$k_v \dot{v}_{C(\text{off})} + k_i \dot{i}_{L(\text{off})} < 0 \tag{17.14}$$

$$k_v \dot{v}_{C(\text{on})} + k_i \dot{i}_{L(\text{on})} > 0$$

and the gain values k_v and k_i can be selected in an attempt to meet the requirements for reflective action around the point (v_{C0}, i_{L0}). The gain values can also be selected to meet dynamic objectives: Along the boundary, the inductor current is given by

$$i_L(t) = \frac{-k_v v_C(t) + k_v v_{C0} + k_i i_{L0}}{k_i} \tag{17.15}$$

and the inductor current can be eliminated as an independent state variable.

Example 17.3.1   A buck-boost converter is to provide 12 V to -24 V conversion. The inductor value is 200 μH and the capacitor is 100 μF. The load is a 4.8 Ω resistor. The inductor's internal resistance is 20 mΩ. Boundary control is to be used, with sliding action near the desired operating point. It is also desirable that the sliding action provide fast dynamics to move the converter toward the desired operation. Design such a control.

The voltages and currents can be defined as shown in Figure 17.21. With these selections, the switching function relationships are

$$v_t = q_1 V_{\text{in}} - q_2 v_C, \qquad i_d = q_2 i_L \tag{17.16}$$

$$\langle v_t \rangle = i_L R_L = D_1 V_{\text{in}} - D_2 V_{\text{out}}, \qquad \langle i_d \rangle = \frac{v_C}{R_{\text{load}}} = D_2 I_L$$

Assuming that continuous mode applies, so that $D_1 + D_2 = 1$, the desired operating point gives $v_C = 24$ V and $i_L = 15.4$ A. The duty ratio $D_1 = 0.675$ to produce this operating point. The boundary will be given by

$$k_v(v_C - 24) + k_i(i_L - 15.4) = 0 \tag{17.17}$$

When the differential expressions $v_L = L(di_L/dt)$ and $i_C = C(dv_C/dt)$ are substituted into equation (17.16), the differential equations are obtained, and

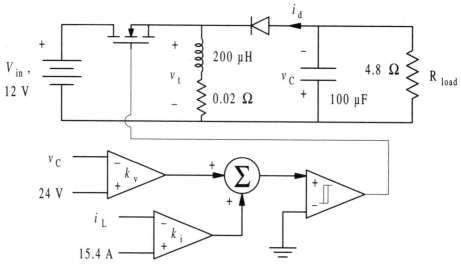

Figure 17.21 Buck-boost converter for Example 17.3.1.

$$\dot{v}_C = \frac{q_2 i_L}{C} - \frac{v_C}{R_{\text{load}}C} \tag{17.18}$$

$$i_L = \frac{q_1 V_{\text{in}} - q_2 v_C - i_L R_L}{L}$$

For the off configuration, reflective (sliding mode) action requires

$$k_v\left(\frac{i_L}{C} - \frac{v_C}{R_{\text{load}}C}\right) + k_i\left(\frac{-v_C - i_L R_L}{L}\right) < 0 \tag{17.19}$$

For the on configuration, the requirement is

$$k_v\left(\frac{-v_C}{R_{\text{load}}C}\right) + k_i\left(\frac{V_{\text{in}} - i_L R_L}{L}\right) > 0 \tag{17.20}$$

When the circuit parameters are used, near the operating point (24 V, 15.4 A) the specific requirements are

$$1040k_v < 1215k_i, \qquad 58\,460k_i > 50\,000k_v \tag{17.21}$$

Both of these are met at the operating point if $k_i/k_v > 0.86$. For good performance, it would be helpful to have sliding mode action over a wider range around the operating point, instead of simply at the point itself. During converter start-up, for instance, the capacitor voltage will be less than 24 V, and a sliding mode would prevent voltage overshoot. Let us require a sliding mode whenever $v_C > 6$ V. Then the required ratio is $k_i/k_v > 4.68$.

Figure 17.22 shows time domain and state domain behavior for the choice $k_i/k_v = 5$, starting from zero volts and amps. The current overshoots to the point where the boundary intersects the current axis, then the voltage comes up rapidly to 24 V. The behavior at other slope settings can be observed with the BASIC program given in Table 17.4. Some testing shows that the converter does not function properly if $k_i/k_v < 0.86$, as expected. The fastest performance is obtained with $k_i/k_v = 2.75$, because this value stores just enough extra energy in the inductor to bring the capacitor up to 24 V immediately.

All of the sliding-mode configurations produce chattering, and a dead band must be included to avoid excessive switching frequency. In the computer program, chattering is avoided by enforcing a minimum off time or minimum on time of 2 μs for the transistor.

One of the key properties of boundary control is stability. The operation of the converter in Example 17.3.1 is independent of its starting point. For any initial capacitor volt-

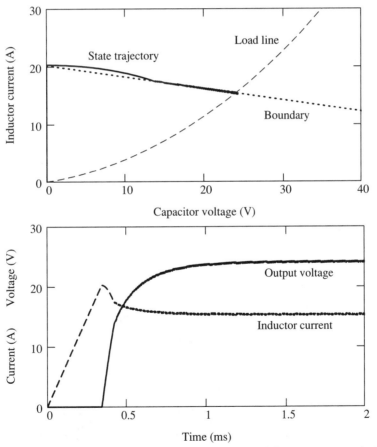

Figure 17.22   State plane and time domain action of buck-boost converter under boundary control with $k_i/k_v = 5$.

TABLE 17.4
BASIC Simulation Program for Example 17.3.1

```
'Animation of buck-boost converter under boundary control, Example 17.3.1
'Some suggested slope values: 12, 5, 2.7, 0.87, 0.82, 0.7

'First define the values and parameters, then get the plot type and gain ratio.
L = .0002: C = .0001: Vin = 12: Vout = 24: RL = .02: Rload = 4.8: kv = 1: Iref = 15.395: Vref = Vout
PRINT "Do you want to plot in time domain or state domain"; : INPUT x$
PRINT "Boundary slope, ki/kv"; : INPUT ki

'Define the derivatives for current and voltage.
DEF fndil (il, vc) = (q1 * Vin − il * RL − q2 * vc) / L
DEF fndvc (il, vc) = (q2 * il − vc / Rload) / C

'Set up the plot scales.
zoom = .5                  'Set zoom > 1 for closeup view
yscale = zoom * 350 / Iref : yoffset = 300 + yscale * Iref
vscale = zoom * 350 / Vref : voffset = 300 + vscale * Vref
xscale = zoom * 640 / Vref : xoff = xscale * Vref − 320: IF xoff < 0 THEN xoff = 0
t = 0: vc = 0: il = 0 : delt = .000002: tlast = .00005 * 100: tscale = 640 / tlast 'Initial conditions and times.
SCREEN 12: CLS: LINE (0, 0)-(0, yoffset): LINE (0, yoffset)-(640, yoffset) 'Set graphics and draw axes.

'If a state domain plot is needed, draw the load line and boundary.
IF x$ = "state" THEN
        LINE (−xoff, yoffset) − (−xoff, yoffset)
        FOR D = 0 TO .8 STEP .01                                    'Load line
        FOR j = 1 TO 4: vop = −(iop*RL − D*Vin) / (1 − D): iop = (vop/Rload) / (1 − D): NEXT j
        LINE −(xscale * vop − xoff, yoffset − yscale * iop)
        NEXT D
        LINE (−xoff, yoffset − yscale*(Vref*kv/ki + Iref)) − (xscale*(Iref*ki/kv + 24)-xoff, yoffset) 'Boundary
END IF

'If a time domain plot is needed, draw the boundaries but no load line.
IF x$ = "time" THEN
        LINE(0,voffset − vscale*Vref) − (640,voffset − vscale*Vref): LINE(0,yoffset − yscale*Iref) − (640,yoffset − yscale*Iref)
END IF

'Here is the simulation loop. The idea is that Xnew ~ Xold + DeltaT*(dX/dt).
WHILE t < tlast
tnew = t + delt
Ibound = ki * (il − Iref) + kv * (vc − Vref)
IF Ibound < 0 THEN q1 = 1: q2 = 0
IF Ibound > 0 THEN q1 = 0: q2 = 1

'Compute and plot as appropriate.
IF x$ = "state" THEN LINE −(vc * xscale − xoff, yoffset − yscale * il)
ilnew = il + delt * fndil(il, vc): IF ilnew < 0 THEN ilnew = 0
vcnew = vc + delt * fndvc(il, vc)
IF x$ = "time" THEN
        LINE (t * tscale, yoffset − il * yscale) − (tnew * tscale, yoffset − ilnew * yscale)
        LINE (t * tscale, voffset − vc * vscale) − (tnew * tscale, voffset − vcnew * vscale)
END IF
vc = vcnew: il = ilnew: t = tnew: intcont = intnew      'Update the values.
WEND                                                    'End of loop.
END
```

age and inductor current, the converter will evolve toward small ripple around the desired operating point if conditions for sliding mode are met. The converter recovers from pulse disturbances or from abrupt steps in the state values, no matter how large. This implies large-signal stability. The linear boundary of Example 17.3.1 was not necessarily chosen for robustness, however. The reference current value depends on both the load resistance and the input voltage. If either the input voltage or load power changes, the desired operating point must change accordingly.

In the example, the robustness problem can be illustrated by altering the input voltage. If the input is changed to ten volts instead of twelve volts, for instance, a new load line appears. The boundary $k_i(i_L - 15.4 \text{ A}) + k_v(v_C - 24 \text{ V}) = 0$ will not intersect this load line at (24 V, 15.4 A). In steady state, the new operating point will move to the left, depending on the ratio k_i/k_v. If $k_i/k_v = 2.7$, the new intersection point will be close to (22.6 V, 15.9 A). The line regulation can be defined as the relative change, or

$$\text{line reg.} = \frac{24 \text{ V} - 22.6 \text{ V}}{12 \text{ V} - 10 \text{ V}} = 0.7 \tag{17.22}$$

This poor regulation level (70%) reflects the fact that the current reference level is no longer consistent with the load. To make the boundary control practical, a PI loop can be used in place of a fixed current reference, so the operating point is uniquely defined by the desired voltage and the actual load.

17.4 OTHER CLASSES OF BOUNDARIES

In the previous section, boundaries were fixed structures (usually linear) that constrained converter operation. Control based on fixed boundaries can be termed *state-dependent control*, since the action of switches depends on the voltage and current values rather than on time. For instance, the switching function q for a boundary-controlled converter can be represented in terms of *Heaviside's step function*, $u(x)$, with $u(x) = 1$ when $x > 0$ and $u(x) = 0$ when $x \leq 0$. For a linear boundary, the control is

$$q(\mathbf{x}) = u[k_i(I_{\text{ref}} - i_L) + k_v(V_{\text{ref}} - v_C)] \tag{17.23}$$

State-dependent action is simple in principle, but the time behavior becomes difficult to regulate. For instance, the switching frequency cannot be predicted for such a control unless the details of the circuit and load are known. In many applications, this is not a problem. However, most power electronics designers prefer to have limitations on the switching frequency or even to impose a strict fixed frequency.

Pulse-width modulation is the conventional approach for fixed frequency control in a power converter. Closed-loop PWM has an interesting boundary interpretation: The triangle carrier waveform can be represented as a moving boundary, with a slope dependent on feedback gains for voltage and current. With a symmetric triangle waveform, the boundary moves back and forth in state space. With a ramp, the boundary moves in one direction, then jumps abruptly to its starting position. In most closed-loop converters, switch action takes place when the state values cross the moving boundary. The action can be represented as

$$q(\mathbf{x},t) = u[k_i(I_{ref} - i_L) + k_v(V_{ref} - v_C) - \text{tri}(t,T)] \qquad (17.24)$$

where $\text{tri}(t,T)$ represents the triangular or ramp-based PWM carrier function with period T.

The basic requirements of boundary control can be combined with motion to obtain a desired converter action. Consider a boost dc–dc converter. If a linear boundary is selected with a slope that keeps the equilibrium points separated, then even a moving boundary will meet some of the basic requirements of boundary control. Equation (17.23) shows that feedback error between the actual and desired output voltage or current will be needed to ensure that the converter ultimately is driven to the intended operating point.

Moving boundary controls lend themselves to animation studies of control action. Figure 17.23 shows one frame of an animation for a boost converter. The boundary in this case is formed from a triangular carrier, and moves back and forth across the desired operating point. A conventional PWM method adds a latch instead of reversing the boundary motion. The PWM oscillator sets the latch, then the latch is reset when the moving boundary is crossed. If the latch-setting operation is represented as a logical function $clk(t)$, the switch action can be written as a logical combination of (17.23) (controlling switch turn-off) and $clk(t)$ (controlling switch turn-on).

Moving boundary control differs in requirements from the static case. Since switching frequency control is an important motivating issue for moving boundaries, sliding mode behavior is unwanted in this case. Sliding modes would result in fast switching, decoupled from the boundary motion. In fact, it is known that if the state trajectory moves faster then the boundary and catches up with it, chaotic behavior can occur when the state reaches the boundary.

17.5 RECAP

Switching boundaries offer a direct way to command power electronic circuits to track a desired voltage or current. Switch action takes place when a state trajectory crosses a boundary. Therefore control based on boundaries constrains the trajectories and can be used to direct them to the desired operating condition. Boundaries are represented as surfaces in state space. This geometric interpretation is useful for design and analysis, although the technique is not limited to systems of just two or three state variables.

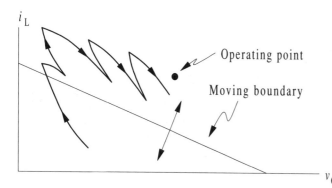

Figure 17.23 Animation frame from state behavior of boost converter.

Hysteresis control is one of the most common boundary control techniques. In this case, a single state variable is used to define switch action. The switch operates when the state crosses a predetermined threshold. Usually both a lower and upper threshold are provided so that a *dead band* exists in which no control action is taken. The dead band avoids *chattering*, the condition in which infinitely fast switching occurs as the state moves back and forth across a single threshold.

In converters with a buck structure, including buck dc–dc converters and voltage-sourced inverters, hysteresis based on the output voltage is a simple way to enforce a specific operating condition or output waveform. Output voltage hysteresis for a buck converter is stable. It exhibits *robust* behavior: The output is maintained even under extreme changes in load, line, or component values. As long as the input source can serve the power needs of the converter, hysteresis control will ensure the desired output.

Hysteresis control is common in power-factor correction (PFC) circuits, in which a sinusoidal or rectified sinusoidal input current is required. The input current in a boost converter, for example, can be compared to a desired waveform. The converter's active switch is turned on to raise the current and turned off to decrease it.

Although it is simple and direct, hysteresis control does not work for many situations. If the method is used based on the output voltage of a boost or flyback converter, it fails to drive the converter to the correct operating point. More general switching boundaries, which take into account the action of many state variables, can be used for these converters.

The design of boundary controls in general requires knowledge of the *equilibrium points* of the possible circuit configurations. Switch action should prevent the converter from reaching a stable equilibrium point, since it might remain there instead of moving toward the desired output. The range of possible operating points of a converter, parametrized by duty ratio for example, defines a *load line*. The control boundary should intersect the load line at the desired operating point. If this intersection point is correct, and subject to certain conditions on the boundary slope, boundary control action will produce large-signal stability. Since the operating point depends on the load and on the input, it is usually necessary to use feedback control such as from a PI loop to drive the intersection to the correct location.

Successful hysteresis or boundary control requires conditions on the state-space geometry. In particular

- The boundary or switching surface must be positioned to separate equilibrium points of circuit configurations.
- The switch action must be selected to drive the states away from the equilibrium points in each configuration.
- The boundary must pass through the desired operating point.
- A dead band or timing limit is needed if chattering is a possibility.

Hysteresis control works best if switch action has a direct influence on the state variable. That is, the state should move in a consistent direction dependent on the switch state. This is not a necessary condition, but it is helpful for good performance.

Boundary control works well when it involves a sliding mode: a region near the operating point over which the trajectories are forced toward the boundary immediately after the switch action. Sliding mode control exhibits chattering, which must be avoided with a dead band. However, it forces the states to evolve along the boundary, and usually helps avoid overshoot or oscillatory behavior during a transient. Sliding modes can be generated by proper choice of boundary slope. If the system trajectories always have a velocity component directed toward the boundary, a sliding mode can appear.

In an inverter, a change in configuration will cause the output voltage to rise or fall, presumably to follow a desired output time waveform. The conditions on the boundaries themselves are not sufficient to ensure correct operation. In addition, the converter must be able to ramp the output current or voltage quickly enough to follow the intended signal. The derivatives are usually limited by internal filter components such as inductors rather than by the control. Thus a minimum di/dt or dv/dt specification must be met by the energy storage components.

A switching boundary can exhibit three major classes of behavior:

1. State trajectories on both sides of the boundary approach it. This is the basis for a sliding mode, and is termed a *reflective* boundary.

2. Trajectories on one side approach the boundary, while on the other side they move away from it. This *refractive* behavior usually produces loops or arcs in state behavior along the boundary.

3. If trajectories are directed away from the boundary on both sides, the behavior is *rejective*. An operating point in a rejective region cannot be reached with a boundary control. The boundary slope or other parameters must be adjusted to avoid rejective behavior near the operating point.

Refractive boundaries generally lead to stable *limit cycle* action, in which the converter retraces the state behavior around a closed loop.

Trajectory directions can be compared to the boundary by testing the state velocities, taken from the *network equation*

$$\dot{\mathbf{x}} = f(\mathbf{x}, u, q) \tag{17.25}$$

against the normal vector associated with the boundary. The dot product $\dot{\mathbf{x}} \cdot \mathbf{n}$ can be used to evaluate the directions.

More generally, boundary control can be represented as a unit step function applied to a function of the state variables. For a linear *fixed boundary*, the Heaviside step function $u(x)$ represents the switching function in terms of the state variables,

$$q(\mathbf{x}) = u[\mathbf{k} \cdot (\mathbf{x}_{\text{ref}} - \mathbf{x})] \tag{17.26}$$

where $\mathbf{k}$ is a gain vector for the states and $\mathbf{x}_{\text{ref}}$ is the reference value vector representing the desired operating point. The step function has the value 1 when its input is positive, and 0 when the input is null or negative. It is also possible to use curved boundaries.

The action of most PWM controllers lends itself to a slightly different boundary representation. In most PWM applications of power electronics, switch action is determined by the interaction of the states with a time-based carrier waveform. For the typical triangular carrier $\text{tri}(t, T)$, switch action can be represented as

$$q(\mathbf{x},t) = u[\mathbf{k} \cdot (\mathbf{x}_{ref} - \mathbf{x}) - \text{tri}(t,T)] \tag{17.27}$$

Geometrically, this appears as a *moving boundary* in state space. Switch action occurs as state trajectories cross this moving surface. Moving boundaries are usually not designed for sliding mode action, since the purpose of the motion is to constrain the switching frequency and to define a periodic steady-state behavior over a definite time interval. In many implementations, a latch resets the boundary after switch action occurs, so that multiple crossings and extra switch actions are avoided.

PROBLEMS

1. A 12 V to 5 V buck dc–dc converter is to operate with hysteresis control. Rated load is 100 W. Suggest inductor, capacitor, and switching boundaries (constant voltage) to give ripple below 0.1 V peak-to-peak with a switching frequency close to 100 kHz.

2. A boost converter is to be tested for the possibility of hysteresis control. The input is 5 V and the output is 12 V. The power level is 20 W. Are there any values of inductor and capacitor that will allow an output voltage hysteresis control to function successfully?

3. A current-based hysteresis control is to be used for a boost dc–dc converter. The input is 5 V, the load is 4 Ω, and the current setting is 7.2 A. Will this function successfully? If so, select an inductor and capacitor to give output ripple of about 0.2 V peak-to-peak and a switching frequency of 80 kHz.

4. A hysteretic control for a buck converter has an upper boundary at 12.1 V and a lower boundary at 11.9 V. The converter has an inductor of 20 μH and a capacitor of 20 μF. The input voltage is about 50 V, and the rated output power is 24 W. What are the steady-state output ripple and switching frequency? If the load decreases to 6 W, what are the new values of ripple voltage and switching frequency?

5. It is proposed to stabilize a boost converter hysteresis control by imposing a second limitation on inductor current: No matter what the hysteresis boundary, the transistor is turned off if the current tries to rise above I_{max}. Consider a 20 V to 80 V boost converter with nominal load of 100 W.

   a. Does this boundary arrangement meet requirements for hysteresis control? Assume that a dead band is provided and that the value of I_{max} is well below the equilibrium current with the transistor on.
   b. Will this be stable given any inductor and capacitor values? If not, are there L and C values for which it could work?

6. The boost converter from Problem 5 operates with a voltage boundary at 80 V $\pm$ 1 V, and a current limit at 20 A. The inductor is 100 μH and the capacitor is 100 μF. Simulate the action starting from zero initial conditions.

7. A boost converter has $L = 400$ μH and $C = 50$ μF. The input is 20 V, and the load is 20 Ω. The transistor has $R_{ds(on)} = 0.04$ Ω. The diode forward drop is 1.0 V.

   a. Find the equilibrium points in the various configurations.

b. Will a boundary given by $15i_L + 4v_C - 155 = 0$ (with i_L in amps and v_C in volts) be stable? With such a boundary, what is the final operating point?

8. A buck converter operates from 400 V input. The inductance is 40 μH and the output capacitor is 50 μF. The load is a 10 Ω resistor, and a 100 V output is desired. A hysteresis control, based on voltage sensing only, is set up with a dead band 1 V wide. What is the output ripple, and at what switching frequency will this converter operate?

9. Set up a simulation to show the dynamic response for a boundary control in a buck-boost circuit. In this circuit, $L = 25$ μH and $C = 500$ μF. The nominal input and output voltages are 12 V and -48 V, respectively. The load draws 50 W at nominal output. The inductor has parasitic resistance of 0.01 Ω. A boundary control based on the line $0 = 20(v_C - v_{C(nom)}) + 1(i_L - i_{L(nom)})$ has been proposed.

    a. Be careful with polarities. Should the boundary be defined in terms of $+v_C$ or $-v_C$? Is the nominal voltage to be added or subtracted to obtain proper operation? (*Hint*: Plot the behavior in the *i-v* plane.)

    b. Simulate control action from start-up. Is the controller stable? If so, what are the peak-to-peak output ripple and the switching frequency?

    c. Simulate from start-up for 10 ms. Then double the load resistance. How quickly does the converter recover? Is the new ripple value higher or lower?

10. It is desired to provide a boost converter with very good, stable, boundary control. Both output voltage and inductor current will be part of the error signal. The input voltage is 5 V, the load is 4 Ω, and the desired output is 12 V. The target switching frequency is 80 kHz, and output ripple should be less than 200 mV peak-to-peak. Select L and C values, and voltage and current gains for good operation. Simulate the performance for your choices.

11. In a buck converter with output filter capacitor, the output ripple is larger than the dead band under voltage hysteresis control because of the shapes of the trajectories. Is this true if no capacitor is present? Analyze the case of no capacitor, and find relationships between L, the switching frequency, the dead band width, and the output ripple. The input is 48 V, and the intended output is 12 V at 300 W.

12. An inverter operates according to a circular boundary in state space. If the system is inside the circle, one switch action will take place. Outside the circle, the opposite action will occur. The dc input is 50 V, and the circle has been selected to be consistent with the load magnitude. Will this process meet the requirements of boundary control?

13. For the converter shown in Figure 17.24,

    a. Write the state variable expressions for each configuration. Include the configuration in which both switches are on.

    b. What are the equilibrium points?

    c. What are the velocities $\dot{x}$ near the point $i_{L1} = 5$ A, $i_{L2} = -10$ A, and $v_C = 15$ V?

    d. Simulate operation for a boundary with gains $k_{iL1} = k_{iL2} = k_v = 1$, given a desired nominal duty ratio $D_1 = 0.25$. Is control action stable with this boundary?

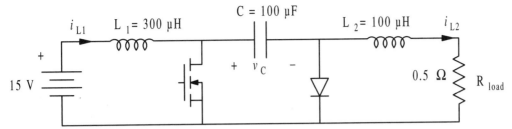

Figure 17.24 Boost-buck converter for Problem 13.

REFERENCES

R. M. Bass, P.T. Krein, "Switching boundary geometry and the control of single-phase inverters," in *Conf. Rec., IEEE Industry Applications Soc. Ann. Meet.* 1989, pp. 1052–1056.

W. W. Burns III, T. G. Wilson, "State trajectories used to observe and control dc-to-dc converters," *IEEE Trans. Aerospace Electronic Systems*, vol. AES-12, pp. 706–717, 1976.

M. Carpita, M. Marchesoni, "Experimental study of a power conditioning system using sliding mode control," *IEEE Trans. Power Electronics*, vol. 11, no. 5, pp. 731–742, September 1996.

R. A. DeCarlo, S. H. Zak, G. P. Matthews, "Variable structure control of nonlinear multivariable systems: a tutorial," *Proc. IEEE*, vol. 76, no. 3, pp. 212–232, March 1988.

R. W. Erickson, S. Ćuk, R. D. Middlebrook, "Large-signal modelling and analysis of switching regulators," in *IEEE Power Electronics Specialists Conf. Rec.*, 1982, pp. 240–250.

P. T. Krein, R. M. Bass, "Geometric formulation, classification and methods for power electronic systems," *IEEE Power Electronics Specialists Conf. Rec.*, 1990.

R. Munzert, P. T. Krein, "Issues in boundary control," in *IEEE Power Electronics Specialists Conf. Rec.*, 1996, pp. 810–816.

H. Sira-Ramirez, "A geometric approach to pulse-width-modulated control design," in *Proc. 26th Conf. Decision Control*, 1987, pp. 1771–1776.

APPENDIXES

APPENDIX A: SOME USEFUL TRIGONOMETRIC IDENTITIES

$$\sin^2 A + \cos^2 A = 1 \tag{A.1}$$

$$\sin(A \pm B) = \sin A \cos B \pm \cos A \sin B \tag{A.2}$$

$$\cos(A \pm B) = \cos A \cos B \mp \sin A \sin B \tag{A.3}$$

$$2 \sin A \cos B = \sin(A + B) + \sin(A - B) \tag{A.4}$$

$$2 \cos A \cos B = \cos(A + B) + \cos(A - B) \tag{A.5}$$

$$2 \sin A \sin B = \cos(A - B) - \cos(A + B) \tag{A.6}$$

$$\sin A \pm \sin B = 2 \sin\left(\frac{A \pm B}{2}\right)\cos\left(\frac{A \mp B}{2}\right) \tag{A.7}$$

$$\cos A + \cos B = 2 \cos\left(\frac{A + B}{2}\right)\cos\left(\frac{A - B}{2}\right) \tag{A.8}$$

$$\cos A - \cos B = 2 \sin\left(\frac{A + B}{2}\right)\sin\left(\frac{B - A}{2}\right) \tag{A.9}$$

$$\sin^2 A - \sin^2 B = \sin(A + B)\sin(A - B) \tag{A.10}$$

$$\cos^2 A - \cos^2 B = \sin(A + B)\sin(B - A) \tag{A.11}$$

$$\cos^2 A - \sin^2 B = \cos^2 B - \sin^2 A = \cos(A + B)\cos(A - B) \tag{A.12}$$

$$\sin^2 A = \tfrac{1}{2}[1 - \cos(2A)], \quad \cos^2 A = \tfrac{1}{2}[1 + \cos(2A)] \tag{A.13}$$

$$\cos(2A) = \cos^2 A - \sin^2 A = 2 \cos^2 A - 1 \tag{A.14}$$

$$\sin(2A) = 2 \sin A \cos A \tag{A.15}$$

$$\cos A = \sin\left(A + \frac{\pi}{2}\right), \quad \sin A = \cos\left(A - \frac{\pi}{2}\right), \quad \cos A = \cos(-A) \tag{A.16}$$

$$\sin A = -\sin(-A), \quad \sec A = \frac{1}{\cos A}, \quad \csc A = \frac{1}{\sin}A, \quad \tan A = \frac{\sin A}{\cos A} \tag{A.17}$$

$$\sec^2 A - \tan^2 A = 1, \quad e^{jx} = \cos x + j\sin x \tag{A.18}$$

$$\cos\left(\frac{\pi}{6}\right) = \frac{\sqrt{3}}{2}, \quad \cos\left(\frac{\pi}{3}\right) = \frac{1}{2}, \quad \cos\left(\frac{\pi}{4}\right) = \frac{\sqrt{2}}{2}, \quad \sin\left(\frac{\pi}{6}\right) = \frac{1}{2}, \quad \sin\left(\frac{\pi}{3}\right) = \frac{\sqrt{3}}{2} \tag{A.19}$$

$$\cos(x\cos A) = J_0(x) - 2J_2(x)\cos(2A) + 2J_4(x)\cos(4A) - \cdots \tag{A.20}$$

$$\cos(x\sin A) = J_0(x) + 2J_2(x)\cos(2A) + 2J_4(x)\cos(4A) + \cdots \tag{A.21}$$

$$\sin(x\cos A) = 2J_1(x)\cos A - 2J_3(x)\cos(3A) + 2J_5(x)\cos(5A) - \cdots \tag{A.22}$$

$$p\cos A + q\sin A = r\cos(A - \phi) \quad \text{where } r = \sqrt{p^2 + q^2}, \quad \phi = \tan^{-1}\left(\frac{q}{p}\right), \tag{A.23}$$

$$\text{also } \sin\phi = \frac{q}{r}, \quad \cos\phi = \frac{p}{r}$$

$$\cos A + \cos\left(A - \frac{2\pi}{3}\right) + \cos\left(A + \frac{2\pi}{3}\right) = 0 \tag{A.24}$$

APPENDIX B: MEASUREMENT SYSTEMS

Power electronics engineers encounter a variety of different units. Older units such as horsepower and pound-foot are common in motor drive applications. Cgs units such as gauss and oersted are common in magnetics. Customary units such as the amp-hour and kilowatt-hour are common in utility applications. This appendix provides conversions among a variety of units, and also identifies the systems. Emphasis is placed on SI units. A table of SI constants is also provided.

Customary Unit	System and Application	SI Unit and Symbol	Conversion Factor: Multiply Customary by Factor to Obtain SI Equivalent
Area		*square meter*	
square inch	English, used for U.S. parts	m^2	6.4516×10^{-4} m^2/in^2
cm^2	Cgs unit, magnetics	m^2	10^{-4} m^2/cm^2
Energy		*joule* (same as watt-second)	
calorie	Cgs unit, heat	J	4.184 J/cal
electron volt, eV	Customary, semiconductors.	J	1.602×10^{-19} J/eV
kilowatt-hour, kW · hr	Customary, power	J	3.6×10^6 J/(kW·hr)
BTU	English, mechanical energy	J	1054 J/BTU
Force		*newton* (same as $kg \cdot m/s^2$)	
pound-force	English, weight	N	4.448 N/lbf
kg-force	Customary, weight	N	9.807 N/kgf
dyne	Cgs, force	N	10^{-5} N/dyne
Length		*meter*	
inch	English, U.S. parts	m	0.0254 m/in
mil	English, 0.001 in, U.S. tolerances.	m	2.54×10^{-5} m/mil
angstrom	Cgs, semiconductors	m	10^{-10} m/Å
Magnetics			
gauss, G	Cgs, flux density	tesla, T	10^{-4} T/G
gilbert, Gi	Cgs, MMF	amp-turn, A	0.7958 A/Gi
maxwell, Mx	Cgs, flux	weber, Wb	10^{-8} Wb/Mx
oersted, Oe	Cgs, magnetic field intensity	amp-turn/meter, A/m	79.58 A/m/Oe
Miscellaneous			
amp-hour, A·hr	Customary, charge in batteries	coulomb, C	3600 C/(A·hr)
pound-mass	English, mass	kilogram, kg	0.4536 kg/lbm
horsepower, hp	English, power	watt, W	746 W/hp
pound-foot	English, torque	N·m	1.356 N·m/(lb·ft)
mile per hour, mph	English, velocity	m/s	0.447 m/s/mph
cubic inch	English, volume	m^3	1.639×10^{-5} m^3/in^3

The SI system of units is the standard for virtually all measurements, and is based on the older MKS (meter-kilogram-second) system. An excellent source of information is *IEEE Standard 268-1992*, "IEEE's American National Standard for Metric Practices." In the SI formulation, only a few units are truly fundamental in the physical sense. These include the meter, kilogram, second, ampere, radian, kelvin, and candela. Table A.1 lists some of the major relationships, while Table A.2 lists a few physical constants.

TABLE B.1
Derived Units for Electromagnetics

Quantity	Unit, Symbol	Base Unit
Energy	joule, J	$kg \cdot m^2/s^2$
Charge	coulomb, C	$A \cdot s$
Voltage	volt, V	$J/C = (kg \cdot m^2)/(A \cdot s^3)$
Magnetic flux	weber, Wb	$V \cdot s = (kg \cdot m^2)/(A \cdot s^2)$
Magnetic flux density	tesla, T	$V \cdot s/m^2 = kg/(A \cdot s^2)$
Magnetic field intensity	amp-turn/meter, A/m	A/m
Resistance	ohm, Ω	$V/A = (kg \cdot m^2)/(A^2 \cdot s^3)$
Inductance	henry, H	$V \cdot s/A = kg \cdot m^2/(A^2 \cdot s^2)$
Capacitance	farad, F	$C/V = (A^2 \cdot s^4)/(kg \cdot m^2)$
Power	watt, W	$V \cdot A = J/s = kg \cdot m^2/s^3$

TABLE B.2
Some Important Constants, Given in SI Units

Constant	Symbol	Value (SI)
Permittivity of free space	ϵ_0	8.854 pF/m
Permeability of free space	μ_0	$4\pi \times 10^{-7}$ H/m
Acceleration due to earth gravity	g	9.80665 m/s^2
Electron charge	e	1.6022×10^{-19} C
Speed of light	c	2.99792458×10^8 m/s
Planck constant	h	6.626×10^{-34} J $\cdot$ s
Boltzmann constant	k	1.381×10^{-23} J/K
Stefan–Boltzmann constant	σ	5.671×10^{-8} W/(m$^2 \cdot$ K^4)

APPENDIX C: COMPUTER ANALYSIS EXAMPLES

C.1 MATHEMATICA[1] LISTINGS

Solution of Equation (1.2) in Mathematica

This is a very direct solution of a first-order differential equation. The initial condition is made clear from switch action: The current starts at zero, coincident with the voltage zero crossing. The solution is valid until the current reaches zero.

Variables: w—radian frequency 2 Pi f
L—inductance
r—resistance
v0—peak voltage
t0—time of voltage zero crossing

In[1] :=
t0 = −Pi/(2*w)

Out[1]=

$$\frac{-Pi}{2\ w}$$

In[2]:=
DSolve[{v0 Cos[w t] == L i′[t] + r i[t],i[t0]==0},i[t],t]

---

[1]S. Wolfram, *Mathematica*. Reading, MA: Addison-Wesley, 1988. *Mathematica* is a registered trademark of Wolfram Research, Inc. These listings should function with version 2.1 and higher.

Out[2]=

$$\left\{\left\{i[t] \to \frac{E^{-((r\,t)/L)-(Pi\,r)/(2\,L\,w)}L\,v0\,w}{r^2 + L^2 w^2} + \frac{r\,v0\,Cos[t\,w] + L\,v0\,w\,Sin[t\,w]}{r^2 + L^2 w^2}\right\}\right\}$$

The above result Out[2] is the solution given in equation (1.3)

Solution of Equation (1.4) with Mathematica

This is an average integral, on an angular time scale.

In[1]:=
voutave=1/(2 Pi) Integrate[vpeak Cos[theta],{theta,−Pi/3,2 Pi/3}]

Out[1]=

$$\frac{Sqrt[3]\ vpeak}{2\ Pi}$$

In[2]:=
N[%]

Out[2]=
0.275664 vpeak

Equation (2.15) Illustrated with Mathematica

Fourier coefficients of full-wave function

In[1]:=
a[n_]=4/Pi Integrate[v0 Cos[theta] Cos[2 n theta],{theta,0,Pi/2}]

Out[1]=

$$\frac{-4\ v0\ Cos[n\ Pi]}{(-1 + 4n^2)\ Pi}$$

Equation (2.29) Solved with Mathematica

Fourier coefficients of switching function.
First, obtain an and bn. Next, find cn.

In[1]:=
an=2/per Integrate[1 Cos[n 2 Pi/per t],{t,t0−d per/2,t0+d per/2}]

Out[1]=

$$\left(\frac{-\left(2 \text{ per } \text{Sin}\left[\frac{n \text{ Pi } (-(d \text{ per} + 2 \text{ t0}))}{\text{per}}\right]\right)}{2 \text{ n Pi}} + \frac{2 \text{ per } \text{Sin}\left[\frac{n \text{ Pi } (d \text{ per} + 2 \text{ t0})}{\text{per}}\right]}{2 \text{ n Pi}}\right)\Big/\text{per}$$

In[2]:=
bn=2/per Integrate[1 Sin[n 2 Pi/per t],{t,t0−d per/2,t0+d per/2}]

Out[2]=

$$\frac{\dfrac{2 \text{ per } \text{Cos}\left[\dfrac{n \text{ Pi } (-(d \text{ per}) + 2 \text{ t0})}{\text{per}}\right]}{2 \text{ n Pi}} - \dfrac{2 \text{ per } \text{Cos}\left[\dfrac{n \text{ Pi } (d \text{ per} + 2 \text{ t0})}{\text{per}}\right]}{2 \text{ n Pi}}}{\text{per}}$$

In[3]:=
Simplify[an ^2 + bn ^2]

Out[3]=

$$\frac{4 \text{ Sin}^2[d \text{ n Pi}]}{n\text{Pi}^2}$$

In[4]:= (This is simulated, since the program will not reduce square roots).
Sqrt[%]

Out[4]=

$$\left[\frac{2 \text{ Sin}[d \text{ n Pi}]}{n \text{ Pi}}\right]$$

Equation (5.29) with Mathematica

Average output of midpoint rectifier.

In[1]:=
voutave=m/(2 Pi) Integrate[v0 Cos[theta],{theta,−Pi/m +a,Pi/m +a}]

Out[1]=

$$\frac{m \left(-\left(v0 \text{ Sin}\left[a - \frac{\text{Pi}}{m}\right]\right) + v0 \text{ Sin}\left[a + \frac{\text{Pi}}{m}\right]\right)}{2 \text{ Pi}}$$

In[2]:=
Simplify[%]

Out[2]=

$$\frac{m \ v0 \ Cos[a] \ Sin\left[\dfrac{Pi}{m}\right]}{Pi}$$

Equation (7.11) with Mathematica

RMS values from ac regulator.

In[1]:=
voutRMS=Sqrt[1/Pi Integrate[v0^2 (Sin[theta])^2,{theta,a,Pi}]]

Out[2]=

$$Sqrt\left[\frac{\dfrac{Pi \ v0}{2} - \dfrac{v0^2(2a^2 - Sin[2 \ a])}{4}}{Pi}\right]$$

With a little more reduction, this matches the equation.

Example 5.2.1 Analyzed with Mathematica

Classical rectifier ripple analysis.

In[1]:=
v=10; vp=10*Sqrt[2]; vin=vp*Cos[w*t]; w=120*Pi;

In[2]=
avel=w/Pi *Integrate[vin,{t,−1/240,1/240}]

Out[2]=

$$\frac{20 \ Sqrt[2]}{Pi}$$

In[3]:=
N[%]

Out[3]=
9.00316

In[4]:=
6/%

Out[4]=
0.666432

In[5]:=
r=%3^2/6

Out[5]=
13.5095

In[6]:=
DSolve[{vin−l*i′[t]−i[t]*r==0,i[−1/240]==i[1/240]},i[t],t]

$$\{\{i[t] \rightarrow (0.74048\ E^{0.0562895/l}\ l) \ /$$

$$((-1. + E^{0.112579/l})\ E^{(13.5095\ t)/l}$$

$$(0.0126741 + 9.8696\ l^2\ )) +$$

$$(0.112579\ E^{(13.5095\ t)/l}\ l\ Cos[120\ Pi\ t] +$$

$$E^{(13.5095\ t)/l}\ l\ Pi^2\ Sin[120\ Pi\ t]) \ /$$

$$(6\ Sqrt[2]\ E^{(13.5095\ t)/l}\ l\ (0.0126741 + l^2\ Pi^2\ ))\}\}$$

In[7]:=
Plot3D[i[t],{t,−1/240,1/240},{l,0.2,5}]

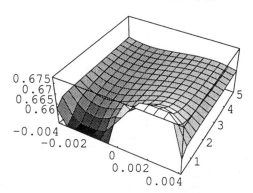

**Plot[{i[t]/.l → 0.5,i[t]/.l → 0.6,i[t]/.l → 0.7},
{t,−1/240,1/240}]**

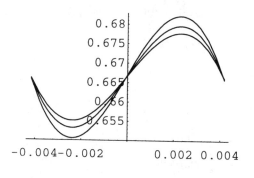

Example 8.5.1 with Mathematica

In[1]:=
wr=1/Sqrt[Lt Cd];zc=Sqrt[Lt/Cd];iout=vout/rload

Out[1]=
vout/rload

In[2]:=
iL[t_]=vin/zc Sin[wr t] + iout; vC[t_]=vin−vin Cos[wr t]

Out[2]=

$$\text{vin} - \text{vin Cos}\left[\frac{t}{\text{Sqrt[Cd Lt]}}\right]$$

In[3]:=
toff=(ArcSin[−iout zc/vin] + 2 Pi)/wr

Out[3]=

$$\text{Sqrt[Cd Lt]}\left(2\text{ Pi} - \text{ArcSin}\left[\frac{\text{Sqrt}\left[\frac{\text{Lt}}{\text{Cd}}\right]\text{vout}}{\text{rload vin}}\right]\right)$$

In[4]:=
vCoff=vC[toff]

Out[4]=

$$\text{vin} - \text{vin Sqrt}\left[1 - \frac{\text{Lt vout}^2}{\text{Cd rload}^2 \text{ vin}^2}\right]$$

In[5]:=
vave=fswitch (Integrate[vC[t],{t,0,toff}]+ 1/2 Cd vCoff ∧ 2/iout)

Out[5]=

$$\text{fswitch}\left(\frac{\dfrac{\text{Sqrt}\left[\frac{\text{Lt}}{\text{Cd}}\right]\text{Sqrt[Cd Lt] vout}}{\text{rload}} + \text{Cd rload}\left(\text{vin} - \text{vin Sqrt}\left[1 - \frac{\text{Lt vout}^2}{\text{Cd rload}^2 \text{ vin}^2}\right]\right)^2}{2\text{ vout}}\right.$$
$$\left. + \text{Sqrt[Cd Lt] vin}\left(2\text{ Pi} - \text{ArcSin}\left[\frac{\text{Sqrt}\left[\frac{\text{Lt}}{\text{Cd}}\right]\text{vout}}{\text{rload vin}}\right]\right)\right)$$

Figure 7.32 with Mathematica

Power output from ac regulator.

In[1]:=
f=60; w=2*Pi*f; m=2; v0=1; r=1;l= r*xi/w; tstart=(−Pi/m+a)/w;

In[2]:=
iall=DSolve[{v0*Cos[w t]==l*i'[t]+r*i[t],i[tstart]==0},i[t],t]

Out[2]=
$$\{\{i[t]\rightarrow-((E^{(2\ a\ -\ Pi)/(2\ xi)\ -\ (120\ Pi\ t)/xi}$$

$$(-(xi\ Cos[a])\ +\ Sin[a]))\ /\ (1\ +\ xi^2\ ))\ +$$

$$(Cos[120\ Pi\ t]\ +\ xi\ Sin[120\ Pi\ t])\ /\ (1\ +\ xi^2\ )\}\}$$

In[3]:=
iall2[t_,xi_,a_]=Simplify[iall[[1]][[1,2]]]

Out[3]=

$$-\left(\frac{E^{(2\ a-Pi-240\ Pi\ t)/(2\ xi)}(-(xi\ Cos[a])\ +\ Sin[a])}{1\ +\ xi^2}\right)+\frac{Cos[120\ Pi\ t]\ +\ xi\ Sin[120\ Pi\ t]}{1\ +\ xi^2}$$

In[4]:=
iall0[t_,xi_,a_]:=If[N[iall2[t,xi,a]] > 0,iall2[t,xi,a],0]

In[5]:=
tend=tstart+1/(f m); ts[a_] = (− Pi/m + a)/w; te[a_] = ts[a] + 1/(fm)

Out[5]=
Plot[iall0[t,1,Pi/2],{t,tstart/.a → Pi/2,tend/.a → Pi/2}]
-Graphics-

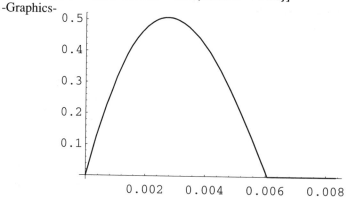

In[6]:=
irms2[xi_,a_]:=120*NIntegrate[(iall0[t,xi,a])^2,
    {t,ts[a],te[a]}];

In[7]:=
Plot[{irms2[.075,a*Pi/180.],irms2[.6,a*Pi/180.], irms2[1., a*Pi/180.], irms2[1.5,
a*Pi/180.]},{a,5,150}]

In[8]=
Show[%,Ticks → {Range[0,180,20], Range[0,0.5,0.1]}]

-Graphics-

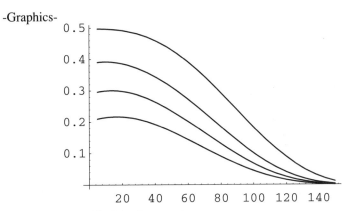

Figure 8.4 with Mathematica

In[1]:=
DSolve[{vin==vc″[t]+1/q vc′[t]+vc[t],vc[0]==0,vc′[0]==0},vc[t],t]

Out[1]=

$$\{\{vc[t] \rightarrow vin + E^{((-1 + Sqrt[1 - 4\,q^2])\,t)/(2\,q)}\left(\frac{-vin}{2} - \frac{vin}{2\,Sqrt[1 - 4\,q^2]}\right) + E^{((-1 - Sqrt[1 - 4\,q^2])\,t)/(2\,q)}\left(\frac{-vin}{2} + \frac{vin}{2\,Sqrt[1 - 4\,q^2]}\right)\}\}$$

In[2]:=
v[t_,q_]=%[[1]][[1,2]];v[t_,1/2]=DSolve[{vin==vc″[t]+2vc′[t]+vc[t],vc[0]==0,vc′[0]
==0},vc[t],t][[1]][[1,2]]

Out[2]=

$$vin - \frac{vin}{E^t} - \frac{t\,vin}{E^t}$$

In[3]:=
Solve[{1/(l*c)==1,Sqrt[l/c]==q},c,l]

Out[3]=

$$\{\{c \rightarrow -(-\tfrac{1}{q})\}, \{c \rightarrow -\tfrac{1}{q}\}\}$$

In[4]:=
i[t_,q_]=1/q * D[v[t,q],t]

Out[4]=

$$\left(\left(E^{((-1 + \text{Sqrt}[1 - 4\,q^2])\,t)/(2\,q)}\,(-1 + \text{Sqrt}[1 - 4\,q^2])\left(\frac{-\text{vin}}{2} - \frac{\text{vin}}{2\,\text{Sqrt}[1 - 4\,q^2]}\right)\right) / (2\,q) +$$

$$\left(E^{((-1 - \text{Sqrt}[1 - 4\,q^2])\,t)/(2\,q)}\,(-1 - \text{Sqrt}[1 - 4\,q^2])\left(\frac{-\text{vin}}{2} + \frac{\text{vin}}{2\,\text{Sqrt}[1 - 4\,q^2]}\right)\right) / (2\,q)\right) / q$$

In[5]:=
Plot[{v[t,0.08],v[t,0.2],v[t,1/2],v[t,2],v[t,8]},{t,0,20}, PlotRange → All]

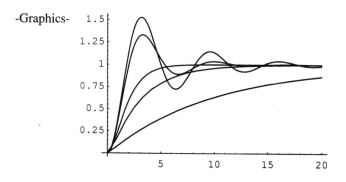

-Graphics-

C.2 MATHCAD LISTINGS

Example 2.4.5

$V0 := 1.2$ $w := 120 \cdot \pi$ Illustrates the application of the Heaviside step function as a way to generate switching functions.

$Va(t) := V0 \cdot \cos(w \cdot t)$ $Vb(t) := V0 \cdot \cos\left(w \cdot t - 2 \cdot \dfrac{\pi}{3}\right)$ $Vc(t) := V0 \cdot \cos\left(w \cdot t + 2 \cdot \dfrac{\pi}{3}\right)$

$qa(t) := \Phi\left(Va(t) - \dfrac{V0}{2}\right)$ $qb(t) := \Phi\left(Vb(t) - \dfrac{V0}{2}\right)$ $qc(t) := \Phi\left(Vc(t) - \dfrac{V0}{2}\right)$

$i := 0..600$ $t_i := \dfrac{1.2}{60} \cdot \dfrac{i}{500}$ $vout(t) := Va(t) \cdot qa(t) + Vb(t) \cdot qb(t) + Vc(t) \cdot qc(t)$

Example 2.4.6

$V0 := 1.2$ $w := 120 \cdot \pi$

$Va(t) := V0 \cdot \cos(w \cdot t)$ $Vb(t) := V0 \cdot \cos\left(w \cdot t - 2 \cdot \dfrac{\pi}{3}\right)$ $Vc(t) := V0 \cdot \cos\left(w \cdot t + 2 \cdot \dfrac{\pi}{3}\right)$

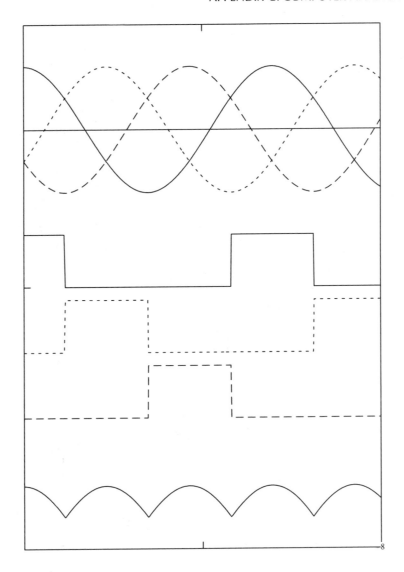

$$qa(t) := \Phi\left(|Va(t)| - \frac{V0}{2} \cdot \sqrt{3}\right) \quad qb(t) := \Phi\left(|Vb(t)| - \frac{V0}{2} \cdot \sqrt{3}\right) \quad qc(t) := \Phi\left(|Vc(t)| - \frac{V0}{2} \cdot \sqrt{3}\right)$$

$$i := 0..600 \qquad t_i := \frac{1.2}{60} \cdot \frac{i}{500} \qquad vout(t) := Va(t) \cdot qa(t) + Vb(t) \cdot qb(t) + Vc(t) \cdot qc(t)$$

Section 2.8

MathCad version of Fourier component computation program. This code gives a list of Fourier coefficients of a piecewise-sinusoidal function.
P. T. Krein, January 1997, copyright © 1997, P. T. Krein. All rights reserved.

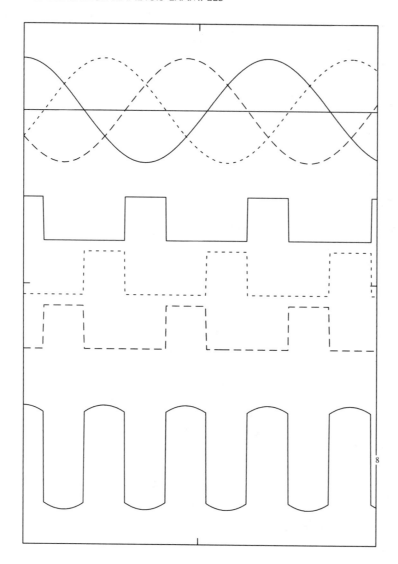

One interesting fact is that a version of the FFT can also be used to produce Fourier coefficients, provided that the waveform is defined for a number of points equal to an integer power of 2 (like 256, 512, 1024, and so on). The value, except for dc, is 2*FFT(data). Here, instead, the integral is used explicitly.

An example of a six-segment ac–ac converter is used to fill out the terms here
TOL := 0.00001
How many component values to compute and list?   ncomp := 20   nn := 1..ncomp
How many pieces ("segments") are needed to define a full period?   nseg := 6
What is the function frequency?   freq := 60
What frequency should be used for the angular time scale?   fscale := 60
t = 0 is the time associated with 0 degrees. t = 1/fscale = 0.01667 is the time at 360 degrees.
For this choice of angular time scale, a full waveform period corresponds to $360 \cdot$ fscale/freq = 360 degrees.

For each segment, Mathcad needs the starting angle, ending angle, and parameters to define the cosine waveform during the segment. The parameters and amplitude, frequency, and phase angle. Phase angle of zero shows a cosine input. Phase angle of $-90°$ indicates a sine, and so on. These data are entered into a matrix with nseg rows.

Start, end, amplitude, frequency, phase

$$\text{wavedat} := \begin{bmatrix} -30 & 30 & 1 & 60 & 0 \\ 30 & 90 & 1 & 60 & -120 \\ 90 & 150 & 1 & 60 & 120 \\ 150 & 210 & 1 & 60 & 0 \\ 210 & 270 & 1 & 60 & -120 \\ 270 & 330 & 1 & 60 & 120 \end{bmatrix}$$

$$\text{per} := \frac{1}{\text{freq}}$$

$$\omega\text{for} := \frac{2 \cdot \pi}{\text{per}}$$

$$f_{nn} := nn \cdot \text{freq}$$

From the raw data, produce time and function data in preparation for the integration step. Augment the results to be sure a full period is covered.

$$\text{nseg3} := \text{nseg} + 2 \qquad ii := 0..\text{nseg} - 1 \qquad jj := 0..\text{nseg3} - 1$$

$$\text{tstart}_{ii} := \text{wavedat}_{ii,0} \cdot \frac{\pi}{180} \cdot \frac{1}{2 \cdot \pi \cdot \text{fscale}} \qquad \text{tend}_{ii} := \text{wavedat}_{ii,1} \cdot \frac{\pi}{180} \cdot \frac{1}{2 \cdot \pi \cdot \text{fscale}}$$

$$\text{amp}_{ii} := \text{wavedat}_{ii,2} \qquad \omega_{ii} := 2 \cdot \pi \cdot \text{wavedat}_{ii,3} \qquad \phi_{ii} := \frac{\pi}{180} \cdot \text{wavedat}_{ii,4}$$

$$\text{tstart}_{\text{nseg}} := \text{wavedat}_{0,0} \cdot \frac{\pi}{180} \cdot \frac{1}{2 \cdot \pi \cdot \text{fscale}} + \text{per} \qquad \text{tend}_{\text{nseg}} := \text{wavedat}_{0,1} \cdot \frac{\pi}{180} \cdot \frac{1}{2 \cdot \pi \cdot \text{fscale}} + \text{per}$$

$$\text{amp}_{\text{nseg}} := \text{wavedat}_{0,2} \qquad \omega_{\text{nseg}} := 2 \cdot \pi \cdot \text{wavedat}_{0,3} \qquad \phi_{\text{nseg}} := \frac{\pi}{180} \cdot \text{wavedat}_{0,4} - 2 \cdot \pi \cdot \frac{\text{fscale}}{\text{freq}}$$

$$\text{tstart}_{\text{nseg}+1} := \text{wavedat}_{1,0} \cdot \frac{\pi}{180} \cdot \frac{1}{2 \cdot \pi \cdot \text{fscale}} + \text{per}$$

$$\text{tend}_{\text{nseg}+1} := \text{wavedat}_{1,1} \cdot \frac{\pi}{180} \cdot \frac{1}{2 \cdot \pi \cdot \text{fscale}} + \text{per}$$

$$\text{amp}_{\text{nseg}+1} := \text{wavedat}_{1,2} \qquad \omega_{\text{nseg}+1} := 2 \cdot \pi \cdot \text{wavedat}_{1,3}$$

$$\phi_{\text{nseg}+1} := \frac{\pi}{180} \cdot \text{wavedat}_{1,4} - 2 \cdot \pi \cdot \frac{\text{fscale}}{\text{freq}}$$

Plot it and see if the waveform is correct. This gives an approximate plot for nplot points.

$$\text{nplot} := 500 \quad iii := 0..\text{nplot} \quad t_{iii} := \frac{\text{per}}{\text{nplot}} \cdot iii$$

$$v_{iii} := \sum_{jj} \text{if}(t_{iii} \geq \text{tstart}_{jj}, \text{if}(t_{iii} < \text{tend}_{jj}, \text{amp}_{jj} \cdot \cos(\omega_{jj} \cdot t_{iii} + \phi_{jj}), 0), 0)$$

In contrast to the plot, the Fourier component computation is exact in form, since it is done as a true integral. (The actual accuracy is limited because of the internal algorithms that Mathcad uses for integration.)

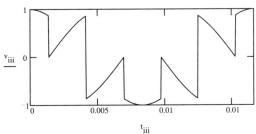

$$\frac{v_{iii}}{}$$

$$t_{iii}$$

$$a_0 := \frac{1}{per} \cdot \sum_{ii} \int_{tstart_{ii}}^{tend_{ii}} amp_{ii} \cdot \cos(\omega_{ii} \cdot t + \phi_{ii})\, dt \qquad\qquad \text{The average.}$$

$$RMS := \sqrt{\frac{1}{per} \cdot \sum_{ii} \int_{tstart_{ii}}^{tend_{ii}} (amp_{ii} \cdot \cos(\omega_{ii} \cdot t + \phi_{ii}))^2\, dt} \qquad \begin{array}{l}\text{Exact RMS } \textit{IF}\text{ no segments}\\ \text{overlap.}\end{array}$$

$$a_{nn} := \frac{2}{per} \cdot \sum_{ii} \int_{tstart_{ii}}^{tend_{ii}} amp_{ii} \cdot \cos(\omega_{ii} \cdot t + \phi_{ii}) \cdot \cos(nn \cdot \omega for \cdot t)\, dt \qquad \text{Fourier cosine terms.}$$

$$b_{nn} := \frac{2}{per} \cdot \sum_{ii} \int_{tstart_{ii}}^{tend_{ii}} amp_{ii} \cdot \cos(\omega_{ii} \cdot t + \phi_{ii}) \cdot \sin(nn \cdot \omega for \cdot t)\, dt \qquad \text{Fourier sine terms.}$$

Now translate to c_n, θ_n form. $c_{nn} := \sqrt{(a_{nn})^2 + (b_{nn})^2}$ $\theta_{nn} := angle(a_{nn}, b_{nn}) \cdot \dfrac{180}{\pi}$

Fix numerical quirk in θ computation. $\theta_{nn} := if(\theta_{nn} > 359.999, \theta_{nn} - 360, \theta_{nn})$
$\theta_{nn} := if(c_{nn} < 10^{-8}, 0, \theta_{nn})$

THD, ignoring a_0 and assuming that   An approximate RMS that
the RMS value is accurate: works with overlap:

$$RMSac := \sqrt{RMS^2 - (a_0)^2} \qquad\qquad RMSapp := \sqrt{\sum_{nnn=0}^{ncomp} \frac{(c_{nnn})^2}{2}}$$

$$THD := \sqrt{\frac{2 \cdot RMSac^2 - (c_1)^2}{(c_1)^2}} \qquad\qquad RMSapp = 0.69155$$

Now, the results! $a_0 = 0$ $a_0 \cdot \pi = 0$
RMS = 0.70711   THD = 0.67983

nn	f_{nn}	a_{nn}	$a_{nn} \cdot \pi$	b_{nn}	$b_{nn} \cdot \pi$	c_{nn}	θ_{nn}
1	60	0.82699	2.59808	0	0	0.82699	0
2	120	0	0	0	0	0	0
3	180	0	0	0	0	0	0
4	240	0	0	0	0	0	0
5	300	0.4135	1.29904	0	0	0.4135	0
6	360	0	0	0	0	0	0
7	420	-0.20675	-0.64952	0	0	0.20675	180
8	480	0	0	0	0	0	0
9	540	-1.22345·10^{-7}	-3.84359·10^{-7}	0	0	1.22345·10^{-7}	180
10	600	0	0	0	0	0	0
11	660	-0.1654	-0.51962	0	0	0.1654	180
12	720	0	0	0	0	0	0
13	780	0.11814	0.37115	0	0	0.11814	0
14	840	0	0	0	0	0	0
15	900	0	0	0	0	0	0
16	960	0	0	0	0	0	0
17	1020	0.10337	0.32476	0	0	0.10337	0
18	1080	0	0	0	0	0	0
19	1140	-0.0827	-0.25981	0	0	0.0827	180
20	1200	0	0	0	0	0	0

Example 3.5.1

$k := 0..50 \qquad L := .025 \qquad R := 2 \qquad Z_k := R + \sqrt{-1} \cdot 240 \cdot k \cdot \pi \cdot L$

$Za(Z) := |Z| + \sqrt{-1} \cdot arg(Z) \cdot \dfrac{180}{\pi} \qquad vk_k := 4 \cdot \dfrac{40}{\sqrt{2} \cdot \pi} \cdot \dfrac{(-1)^{(k+1)}}{(2 \cdot k - 1) \cdot (2 \cdot k + 1)}$

$ik_k := \dfrac{vk_k}{Z_k} \qquad Zm_k := Za(Z_k) \qquad im_k := Za(ik_k)$

$j := 1..3$

$i := 0..50$

$vk_1 \cdot \dfrac{\sqrt{2}}{40} = 0.424413$

$t_i := \dfrac{i}{50} \cdot 2 \cdot \pi$

$vt_i := 2 \cdot \dfrac{40}{\pi} + \displaystyle\sum_j vk_j \cdot \cos(j \cdot t_i) \cdot \sqrt{2} \qquad it_i := \dfrac{40}{\pi} + \displaystyle\sum_j (Re(ik_j) \cdot \cos(j \cdot t_i) \cdot \sqrt{2}$

$+ Im(ik_j) \cdot \sin(j \cdot t_i) \cdot \sqrt{2})$

$\max(it) = 13.60622$
$\min(it) = 11.823084$
$\max(it) - \min(it) = 1.783135$

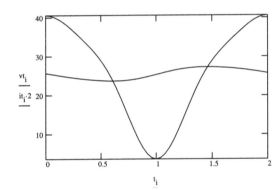

Example 5.2.1

$w := 120 \cdot \pi \qquad V0 := 10 \cdot \sqrt{2} \qquad r := 13.5 \qquad C := 6200 \cdot 10^{-6}$

$vin(t) := |V0 \cdot \cos(w \cdot t)| \qquad wtoff := atan\left(\dfrac{1}{w \cdot r \cdot C}\right) \qquad V1 := V0 \cdot \cos(wtoff)$

$il(t) := \left(\dfrac{d}{dt} vin(t)\right) \cdot C + \dfrac{vin(t)}{r} \qquad toff := \dfrac{wtoff}{w}$

$v2(t) := V1 \cdot e^{-\frac{mod\left(t, \frac{\pi}{w}\right) - toff}{r \cdot C}} \cdot \Phi\left(mod\left(t, \dfrac{\pi}{w}\right) - toff\right) \qquad v2(t) := V0 \cdot e_{-\frac{t}{r \cdot C}}$

$$\text{vout(t)} := \text{if}(\text{vin(t)} > \text{v1(t)}, \text{vin(t)}, \text{v1(t)}) \qquad x := \frac{\pi}{w} \cdot 0.9$$

$$\text{iin(t)} := \text{if}(\text{vin(t)} > \text{v1(t)}, \text{il(t)}, 0) \qquad \text{ton} := \text{root}(\text{vin(x)} - \text{v1(x)}, x)$$

$$i := 0..540 \qquad \text{ton} \cdot 1000 = 7.248$$

$$t_i := \frac{i}{180} \cdot \frac{\pi}{w} \qquad \text{vave} := \frac{w}{\pi} \cdot \int_0^{1/120} \text{vout(t)} \, dt \qquad \text{vave} = 13.58$$

$$\text{il(ton)} = 14.107 \qquad \frac{\text{vave}^2}{r} = 13.661 \qquad \frac{\text{vave}}{r} = 1.006 \qquad \text{vout(ton)} = 12.976$$

$$\text{vrms} := \sqrt{\frac{w}{\pi} \cdot \int_0^{1/120} \text{vout(t)}^2 \, dt} \qquad \text{irms} := \sqrt{\frac{w}{\pi} \cdot \int_0^{1/120} \text{iin(t)}^2 \, dt} \qquad \text{vrms} = 13.584$$

$$\text{vrip} := V0 - \text{vout(ton)} \qquad \text{irms} = 3.09 \qquad \text{Sin} := \text{irms} \cdot 10 \qquad \text{pout} := \frac{\text{vrms}^2}{r} \qquad \text{vrip} = 1.167$$

$$\text{pf} := \frac{\text{pout}}{\text{Sin}} \qquad \text{pout} = 13.669 \qquad \frac{\text{vrip}}{\text{vave}} = 0.086 \qquad \text{Sin} = 30.879 \qquad \text{pf} = 0.442$$

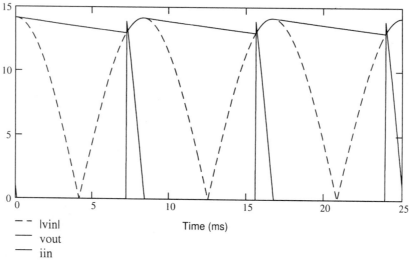

- - - |vin|
— vout
— iin

Time (ms)

Example 5.2.2

$$w := 120 \cdot \pi \qquad V0 := 10 \cdot \sqrt{2} \qquad r := 6 \qquad C := 33000 \cdot 10^{-6}$$

$$\text{vin(t)} := |V0 \cdot \cos(w \cdot t)| \qquad \text{wtoff} := \text{atan}\left(\frac{1}{w \cdot r \cdot C}\right) \qquad V1 := V0 \cdot \cos(\text{wtoff})$$

$$\text{il(t)} := \left(\frac{d}{dt}\text{vin(t)}\right) \cdot C + \frac{\text{vin(t)}}{r} \qquad \text{toff} := \frac{\text{wtoff}}{w}$$

$$\text{v1(t)} := V1 \cdot e^{-\frac{\text{mod}\left(t, \frac{\pi}{w}\right) - \text{toff}}{r \cdot C}} \cdot \Phi\left(\text{mod}\left(t, \frac{\pi}{w}\right) - \text{toff}\right) \qquad \text{v2(t)} := V0 \cdot e^{-\frac{t}{r \cdot C}}$$

$$x := \frac{\pi}{w} \cdot 0.95 \qquad \text{vout(t)} := \text{if}(\text{vin(t)} > \text{v1(t)}, \text{vin(t)}, \text{v1(t)})$$

$t_{on} := \text{root}(\text{vin}(x) - v1(x), x)$ $\text{iin}(t) := \text{if}(\text{vin}(t) > v1(t), \text{il}(t), 0)$ $t_{on} \cdot 1000 = 7.655$

$i := 0..540$ $\text{vout}(0) = 12.142$

$t_i := \dfrac{i}{180} \cdot \dfrac{\pi}{w}$ $\text{vave} := \dfrac{w}{\pi} \cdot \displaystyle\int_0^{1/120} \text{vout}(t)\, dt$ $\text{vave} = 11.918$

$\text{il}(t_{on}) = 46.437$ $\dfrac{\text{vave}^2}{r} = 23.672$ $\dfrac{\text{vave}}{r} = 1.986$ $\text{vout}(t_{on}) = 11.683$

$\text{vrms} := \sqrt{\dfrac{w}{\pi} \cdot \displaystyle\int_0^{1/120} \text{vout}(t)^2\, dt}$ $\text{irms} := \sqrt{\dfrac{w}{\pi} \cdot \displaystyle\int_{t_{on}}^{1/120 + \text{toff}} \text{il}(t)^2\, dt}$ $\text{vrms} = 11.918$

$\text{vrip} := \text{vout}(0) - \text{vout}(t_{on})$ $\text{irms} = 7.856$ $\text{Sin} := \text{irms} \cdot 10$

$\text{pout} := \dfrac{\text{vrms}^2}{r}$ $\text{vrip} = 0.46$ $\text{pf} := \dfrac{\text{pout}}{\text{Sin}}$

$\text{pout} = 23.674$ $\dfrac{\text{vrip}}{\text{vave}} = 0.039$ $\text{Sin} = 78.562$ $\text{pf} = 0.301$

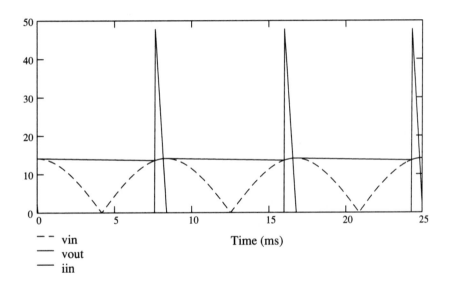

- - vin
— vout
— iin

Time (ms)

Example 6.3.2

$R := 2$ $L := .003$ $k := 35$ $d := 30 \cdot \dfrac{\pi}{180}$ $\text{Vin} := 144$ $w := 120 \cdot \pi$

$n := 1..k$ $tt := \dfrac{d}{w}$ $i := 0..500$ $t_i := \dfrac{1}{30} \cdot \dfrac{i}{500}$

$V(t) := \dfrac{4 \cdot \text{Vin}}{\pi} \cdot \displaystyle\sum_n \dfrac{\sin\left(n \cdot \dfrac{\pi}{2}\right)}{n} \cdot \cos\left(\dfrac{n \cdot d}{2}\right) \cdot \cos\left(n \cdot w \cdot t - n \cdot \dfrac{d}{2}\right)$

$I(t) := \displaystyle\sum_n \dfrac{4 \cdot \text{Vin}}{\pi} \cdot \dfrac{\sin\left(\dfrac{n \cdot \pi}{2}\right)}{n \cdot \sqrt{R^2 + n^2 \cdot w^2 \cdot L2}} \cdot \cos\left(n \cdot \dfrac{d}{2}\right) \cdot \cos\left(n \cdot w \cdot t - n \cdot \dfrac{d}{2} - \text{atan}\left(n \cdot w \cdot \dfrac{L}{R}\right)\right)$

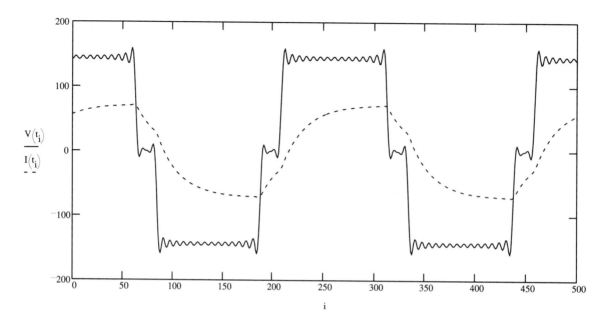

Section 6.4

PWM data creation program   TOL := 10^{-6}
Inputs are the modulation index 0<=m<=1,
and the frequency ratio k.

$$k := 25 \qquad m := 0.85 \qquad i := 0..4 \cdot k + 2 \qquad nf := 1024$$

$$\text{tri}(t,k) := \left| \frac{\left| \mod\left(t, \frac{nf}{k}\right) \right|}{\frac{nf}{k}} - 0.5 \right| \cdot 2 - 0.5 \qquad \text{ref}(t,m) := m \cdot \cos\left(t \cdot 2 \cdot \frac{\pi}{nf}\right)$$

$$\text{trip}(t,k) := \left| \frac{\left| \mod\left(t, \frac{nf}{k}\right) \right|}{\frac{nf}{k}} - 0.5 \right| \cdot 2 \qquad \text{trin}(t,k) := \left| \frac{\left| \mod\left(t, \frac{nf}{k}\right) \right|}{\frac{nf}{k}} - 0.5 \right| \cdot 2 - 1$$

This version uses a fcarrier/fmod ratio of   breaks $(t,k,m) := \text{root}(\text{tri}(t,k) - \text{ref}(t,m),t)$
k, and a modulation index of m. Two
waveforms are generated. The first switches $$t_i := i \cdot \frac{360}{2 \cdot k} - 1 \qquad\qquad j := 0..nf$$
from zero to rail, while the second switches
from rail to rail. The Fourier Transform is $$ii := 0..4 \cdot k - 1$$
used to find the series coefficients that
represent each waveform.

Difference for finding q →
Prepare matrix in proper format

$$d(t,k,m) := \Phi\left(\frac{ref(t,m)}{2} - tri(t,k)\right)$$

$$s(t,k,m) := 2 \cdot d(t,k,m) - 1$$

$$s3lev(t,k,m) := \Phi(ref(t,m) - trip(t,k)) \cdot \Phi(ref(t,m)) - \Phi(trin(t,k) - ref(t,m)) \cdot \Phi(-ref(t,m))$$

$$s32lev(t,k,m) := \Phi(ref(t,m) - 0.5) \cdot d(t,k,m) - \Phi(0.5 - ref(t,m)) \cdot (1 - d(t,k,m))$$

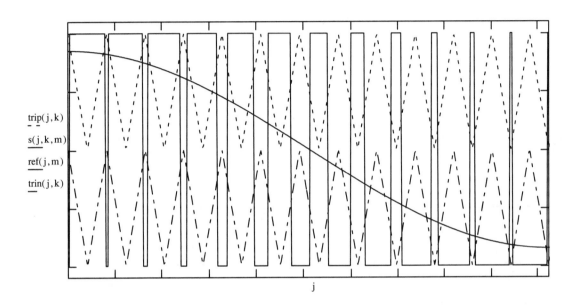

$$\frac{trip(j,k)}{}$$
$$\underline{s(j,k,m)}$$
$$ref(j,m)$$
$$\underline{trin(j,k)}$$

$$j$$

$$kk := 0..400 \qquad iii := 0..nf - 1$$

$$s1_{iii} := s(iii,k,m) \qquad s3_{iii} := s3lev(iii,k,m) \qquad g1 := 2 \cdot FFT(s1) \qquad g3 := 2 \cdot FFT(s3)$$

$$g1_1 = 0.874033 + 3.90625 \cdot 10^{-3}i \qquad g3_0 = 0$$

$$g1_0 = 0 \qquad g3_1 = 0.843705$$

$$jjj := 1..\frac{nf}{2} \qquad g3_0 := g1_0 \qquad g3_k = -0.446003$$

$$g3i_{jjj} := \frac{g3_{jjj} \cdot 340}{2 \cdot (4.5 + 120 \cdot \pi \cdot jjj \cdot .00742i)} \qquad g1_{25} = -0.740357 + 3.90625 \cdot 10^{-3}i$$

$$g2i_{jjj} := \frac{g1_{jjj} \cdot 340}{2 \cdot (4.5 + 120 \cdot \pi \cdot jjj \cdot .00742i)} \qquad s3i := IFFT(g3i) \qquad s2i := IFFT(g2i)$$

$$m = 0.85 \qquad k = 25$$

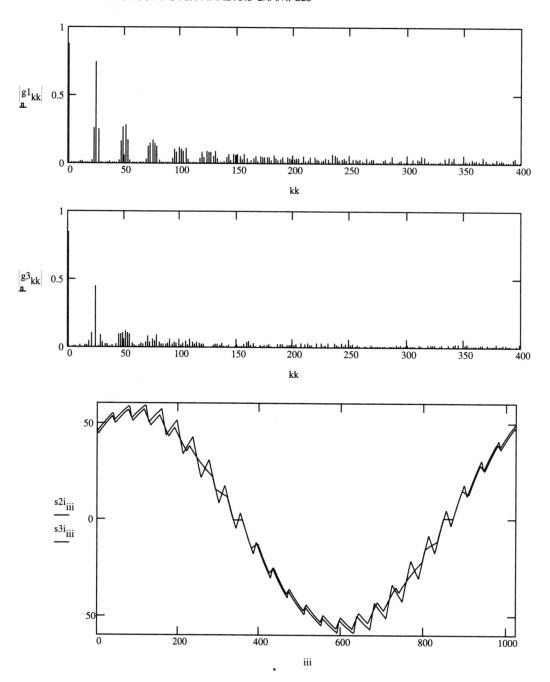

$$ssi_{iii} := d(iii, k, m) \cdot \left| s3i_{iii} \right| \cdot 2$$

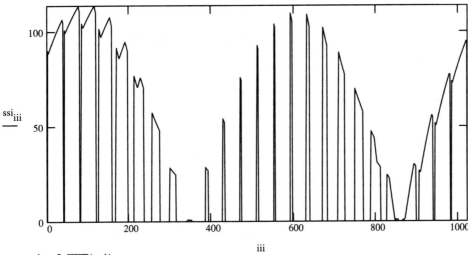

$$gsi := 2 \cdot FFT(ssi)$$

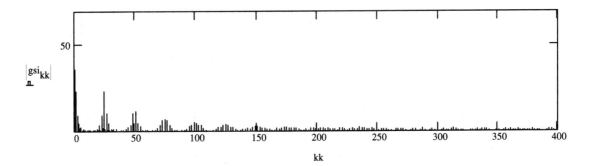

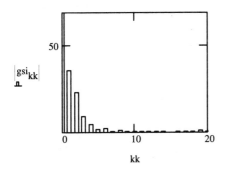

$$Pl := \sum_{iii} d(iii, k, m) \cdot \left| s3i_{iii} \right| \cdot 2$$

$$\frac{Pl}{512} = 68.829$$

Example 6.8.2

$n := 1..9 \qquad w := 120 \cdot \pi \qquad del := 0 \qquad vin := 132$

$R := 7.2 \qquad L := .1 \qquad C := 70 \cdot 10^{-6} \qquad td := \dfrac{del \cdot \dfrac{\pi}{180}}{w} \qquad j := \sqrt{-1}$

$Z_n := R + j \cdot n \cdot w \cdot L + \dfrac{1}{j \cdot n \cdot w \cdot C} \qquad Z_3 = 7.2 + 100.466i$

$q1(t) := 0.5 + \dfrac{2}{\pi} \cdot \displaystyle\sum_n \dfrac{\sin\left(n \cdot \dfrac{\pi}{2}\right)}{n} \cdot \cos(n \cdot w \cdot t)$

$q2(t) := 0.5 + \dfrac{2}{\pi} \cdot \displaystyle\sum_n \dfrac{\sin\left(n \cdot \dfrac{\pi}{2}\right)}{n} \cdot \cos(n \cdot w \cdot (t - td))$

$vd(t) := vin \cdot (\Phi(\cos(w \cdot t)) + \Phi(\cos(w \cdot (t - td))) - 1) \qquad i := 0..200 \qquad t_i := \dfrac{i}{200} \cdot \dfrac{1}{60}$

$$vout(t) := vin \cdot R \cdot \left[\left[\left[0.5 + \left[\dfrac{2}{\pi} \cdot \sum_n \dfrac{\sin\left(n \cdot \dfrac{\pi}{2}\right)}{n} \cdot \dfrac{\cos(n \cdot w \cdot t - \arg(Z_n))}{|Z_n|}\right]\right] \right.\right.$$
$$\left.\left. + \left[0.5 + \dfrac{2}{\pi} \cdot \sum_n \dfrac{\sin\left(n \cdot \dfrac{\pi}{2}\right)}{n} \cdot \dfrac{\cos[n \cdot w \cdot (t - td) - \arg(Z_n)]}{|Z_n|}\right] - 1\right]\right]$$

$voutrms := \sqrt{\dfrac{w}{2 \cdot \pi} \cdot \displaystyle\int_0^{2 \cdot \pi/w} vout(t)^2 dt} \qquad v31 := 118.8372 \qquad v1 := 118.7982 \qquad v231 := 3.0493$

$voutrms = 118.837 \qquad THD := \sqrt{\dfrac{v31^2 - v1^2}{v1^2}} \qquad THD = 0.0256$

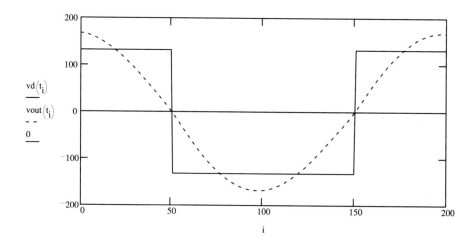

Example 7.3.3

$$D := \frac{1}{2} \qquad fout := 20 \qquad fin := 50 \qquad f := 110 \qquad T := \frac{1}{f} \qquad \omega := 2 \cdot \frac{\pi}{T} \qquad tp := 0$$

$$Tout := \frac{1}{fout} \qquad \text{*f is the switching frequency, 1/T}$$

$$q(t,t0) := if\left(mod\left(t - t0 + 5 \cdot T + D \cdot \frac{T}{2}, T\right) < D \cdot T, 1, 0\right)$$

$$vin(t,p) := cos\left(2 \cdot fin \cdot \pi \cdot t - p \cdot \frac{\pi}{180}\right) \qquad vout(t) := q(t,tp) \cdot vin(t,0) + q\left(t, \frac{T}{2} + tp\right) \cdot vin(t,180)$$

$$imax := 512 \qquad i := 0..imax - 1 \qquad k := 0..\frac{imax}{2} \qquad k := 0..20$$

$$t_i := Tout \cdot \frac{i}{imax} \qquad vouti_i := vout(t_i) \qquad four := FFT(vouti) \cdot 2$$

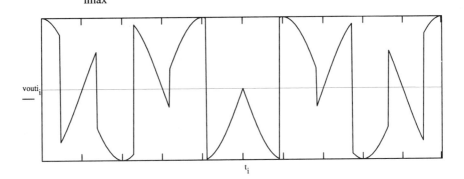

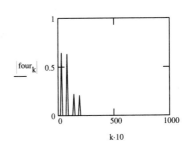

k·fout	$|four_k|$	$|four_k \cdot \pi|$
0	0.00057	0.00178
20	0.00081	0.00254
40	0.00364	0.01142
60	0.64328	2.02091
80	0.00137	0.0043
100	0.00062	0.00194
120	0.00062	0.00194
140	0.00135	0.00424
160	0.63011	1.97954
180	0.00375	0.01178
200	0.00081	0.00256
220	0.00057	0.00178
240	0.0008	0.00252
260	0.00353	0.01108
280	0.21901	0.68805
300	0.00139	0.00436
320	0.00062	0.00195
340	0.00061	0.00193
360	0.00133	0.00417
380	0.20583	0.64664
400	0.00387	0.01215

Example 7.3.4

$$D := \frac{1}{3} \qquad \text{fout} := 5 \qquad f := 60 + \text{four} \qquad T := \frac{1}{f} \qquad \omega := 2 \cdot \frac{\pi}{T} \qquad \text{Tout} := \frac{1}{\text{fout}}$$

$$q(t,t0) := if\left(\text{mod}\left(t - t0 + 5 \cdot T + D \cdot \frac{T}{2}, T \right) < D \cdot T, 1, 0 \right)$$

$$vin(t,p) := 4160 \cdot \sqrt{2} \cdot \cos\left(120 \cdot \pi \cdot t - p \cdot \frac{\pi}{180} \right)$$

$$vout(t) := q(t,0) \cdot q\left(t, -\frac{T}{6} \right) \cdot vin(t,0) + q(t,0) \cdot q\left(t, \frac{T}{6} \right) \cdot vin(t,60)$$

$$+ q\left(t, \frac{T}{3} \right) \cdot q\left(t, \frac{T}{6} \right) \cdot vin(t,120) + q\left(t, \frac{T}{3} \right) \cdot q\left(t, \frac{T}{2} \right) \cdot vin(t,180)$$

$$+ q\left(t, 2 \cdot \frac{T}{3} \right) \cdot q\left(t, \frac{T}{2} \right) \cdot vin(t,240) + q\left(t, 2 \cdot \frac{T}{3} \right) \cdot q\left(t, -\frac{T}{6} \right) \cdot vin(t,300)$$

$$\text{imax} := 512 \qquad i := 0..\text{imax} - 1 \qquad t_i := 0.2 \cdot \frac{i}{\text{imax}} \qquad vouti_i := vout(t_i) \qquad \text{four} := FFT(vouti)$$

$$k := 1..\frac{\text{imax}}{2} \qquad g_k := \frac{\text{four}_k}{k} \qquad g_0 := 0 \qquad voutf = IFFT(g)$$

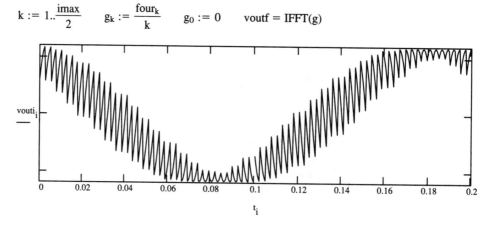

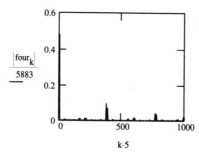

Example 7.3.5

$$D := \frac{1}{3} \qquad \text{fout} := 60 \qquad f := 60 + \text{fout} \qquad T := \frac{1}{f} \qquad \omega := 2 \cdot \frac{\pi}{T} \qquad \text{Tout} := \frac{1}{\text{fout}}$$

$$q(t,t0) := if\left(mod\left(t - t0 + 5 \cdot T + D \cdot \frac{T}{2}, T\right) < D \cdot T, 1, 0\right)$$

$$vin(t,p) := 4160 \cdot \sqrt{2} \cdot cos\left(120 \cdot \pi \cdot t - p \cdot \frac{\pi}{180}\right)$$

$$vout(t) := q(t,0) \cdot vin(t,0) + q\left(t,\frac{T}{3}\right) \cdot vin(t,120) + q\left(t,2 \cdot \frac{T}{3}\right) \cdot vin(t,240)$$

$$imax := 512 \qquad i := 0..imax - 1 \qquad t_i := Tout \cdot \frac{i}{imax} \qquad vouti_i := vout(t_i)$$

$$four := FFT(vouti) \cdot 2 \qquad k := 0..\frac{imax}{2}$$

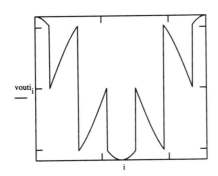

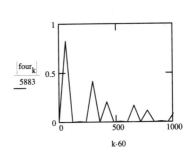

Example 7.4.1

$$w := \pi \quad T := 2 \cdot \frac{\pi}{w}$$

$$a(k,t) := acos(k \cdot cos(w \cdot t)) \cdot \frac{180}{\pi}$$

$$imax := 1024$$

$$i := 0..imax - 1 \quad t_i := 5 \cdot T \cdot \frac{i}{imax}$$

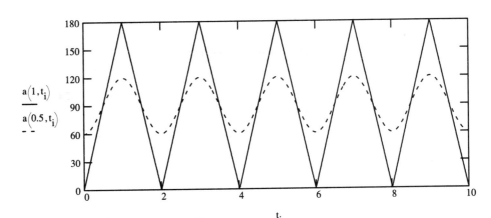

$$aa(t) := \frac{\pi}{2} - 0.2 \cdot \cos(w \cdot t)$$

$$aout(t) := \cos\left(a(0.5,t) \cdot \frac{\pi}{180}\right)$$

$$aout1(t) := \cos(aa(t)) \qquad aaa_i := aout1(t_i)$$

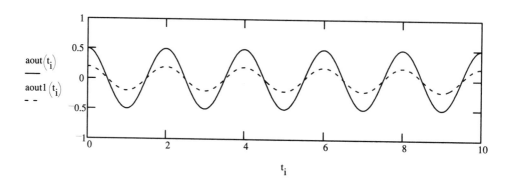

aout(t_i)
—
aout1 (t_i)
- -

t_i

Figure 7.18

$$q(t,D,T,t_0) := if\left[mod\left[\left[(t - t_0) + 10 \cdot T + D \cdot \frac{T}{2}\right],T\right] < D \cdot T,1,0\right]$$

$$ka := 1 \qquad T := \frac{1}{60} \qquad \omega := 2 \cdot \frac{\pi}{T} \qquad p := 3$$

$$t1(t) := Re\left(\frac{-acos(ka \cdot \cos(40 \cdot \pi \cdot t))}{\omega}\right) \qquad t0(t) := Re\left(\frac{acos(ka \cdot \cos(40 \cdot \pi \cdot t))}{\omega}\right)$$

$$iout(t) := \cos(40 \cdot \pi \cdot t) \qquad q3(t,j) := q\left(t,\frac{1}{p},T,\frac{T}{p} \cdot j + t0(t)\right) \qquad q2(t,j) := q\left(t,\frac{1}{p},T,\frac{T}{p} \cdot j + t1(t)\right)$$

$$v3(t,j) := \cos\left(\omega \cdot t - \frac{2 \cdot \pi}{p} \cdot j\right)$$

$$i := 0..1023 \qquad j := 0..p - 1$$

$$t_i := \frac{i}{1024} \cdot T \cdot 6 \qquad tt0_i := t0(t_i) \qquad vout(t) := \sum_j q3(t,j) \cdot v3(t,j) \qquad vout1(t) := \sum_j q2(t,j) \cdot v3(t,j)$$

$$q_i := q\left(t_i,\frac{1}{2},T,tt0_i\right) \qquad vi_i := vout(t_i) \qquad vi2_i := vout1(t_i) \qquad k := 0..24 \qquad fourier := 2 \cdot FFT(vi)$$

$$fourier_0 := \frac{fourier_0}{2}$$

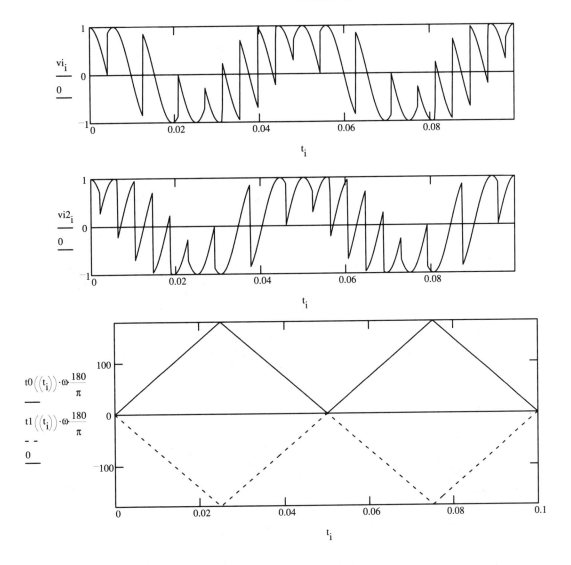

Example 9.3.1

$$i(L,t) := -\frac{1700}{\exp\left(10 \cdot \dfrac{t}{L}\right) \cdot (100 + 14400 \cdot L^2 \cdot \pi^2)}$$

$$+ 170 \cdot \frac{10 \cdot \cos(120 \cdot \pi \cdot t) + 120 \cdot L \cdot \pi \cdot \sin(120 \cdot \pi \cdot t)}{100 + 14400 \cdot L^2 \cdot \pi^2}$$

$$ii := 0..100 \qquad j := 0..5 \qquad t_{ii} := ii \cdot \frac{\frac{1}{180}}{100} \qquad L_j := j \cdot .003 + .003 \qquad vii_{ii} := 17 \cdot \cos(120 \cdot \pi \cdot t_{ii})$$

$$ip(L,t) := if(i(L,T) > 0, i(L,t), 0) \qquad iii_{j,ii} := ip(L_j, t_{ii}) \qquad iii_{j,ii+100} := iii_{j,ii} + iii_{j,100}$$

$$iii_{j,ii+200} := iii_{j,ii} + iii_{j,100} \cdot 2 \qquad jj := 0..300 \qquad vii_{ii+100} := vii_{ii} \qquad vii_{ii+200} := vii_{ii}$$

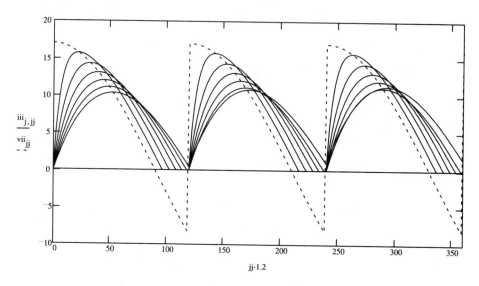

$$jj \cdot 1.2$$

Example 10.5.5

$$j := \sqrt{-1} \qquad w := 120 \cdot \pi \qquad vin := 120 \qquad iout := \frac{80}{\pi} \cdot \sin\!\left(\frac{\pi}{3}\right) \cdot \frac{1}{\sqrt{2}}$$

$$jj := 0..100 \qquad i := 2..100$$

$$Cin := 0 \qquad C7 := 59 \cdot 10^{-6} \qquad C5 := 10 \cdot 10^{-6} \qquad C19 := 2 \cdot 10^{-6}$$

$$L7 := \frac{1}{49 \cdot w^2 \cdot C7} \qquad L5 := \frac{1}{25 \cdot w^2 \cdot C5} \qquad L19 := \frac{1}{361 \cdot w^2 \cdot C19}$$

$$Ceq := Cin + C7 \cdot \frac{49}{48} + C5 \cdot \frac{25}{24} \qquad n_i := i \qquad Ls := .0016 \qquad Zs := 0.08 + j \cdot w \cdot Ls$$

$$L5 = 0.02814 \qquad Ceq = 7.06458 \cdot 10^{-5} \qquad iin := iout + j \cdot w \cdot Ceq \qquad vs := vin - iin \cdot Zs$$

Given

$$vin - iin \cdot Zs = vs \qquad iin = vs \cdot j \cdot w \cdot Ceq + out \qquad a := Find(iin, vs)$$

$$L7 = 0.00243 \qquad Iin := a_0 \qquad Vs := a_1 \qquad Iin = 15.85549 + 3.21379i$$

$$Vs = 120.67007 - 9.82091i \qquad df := \cos(\arg(Vs) - \arg(Iin)) \qquad df = 0.96073$$

$$Y1(n) := \frac{1}{(0.08 + j \cdot n \cdot w \cdot Ls)} \qquad In(n) := \frac{40}{\pi \cdot \sqrt{2}} \cdot \frac{\sin\!\left(n \cdot \frac{\pi}{3}\right)}{n} \cdot (1 - \cos(n \cdot \pi))$$

$$\text{Int}(t) := \sum_{jj} \text{In}(jj) \cdot \cos(jj \cdot w \cdot t) \cdot \sqrt{2}$$

$$\text{Ytot}(n) := \text{Y1}(n) + j \cdot n \cdot w \cdot \text{Cin} + \cfrac{1}{j \cdot n \cdot w \cdot \text{L7} + \cfrac{1}{j \cdot n \cdot w \cdot \text{C7}} + .0001}$$

$$+ \cfrac{1}{j \cdot n \cdot w \cdot \text{L5} + \cfrac{1}{j \cdot n \cdot w \cdot \text{C5}}} \qquad \text{Iin}(n) := \text{In}(n) \cdot \frac{\text{Y1}(n)}{\text{Ytot}(n)}$$

$$\text{iiin}_i := \text{In}(n_i) \qquad \text{iiin}_0 = 0 \qquad \text{iiin}_1 := a_0$$

$$\text{Ithd} := \sqrt{\frac{\sum_i \text{In}(i)^2}{\text{In}(1)^2}} \qquad \text{Iinthd} := \sqrt{\frac{\sum_i (|\text{Iin}(i)|)^2}{(|a_0|)^2}} \qquad \text{Ithd} = 0.30538 \qquad \text{Iinthd} = 0.0922$$

$$\text{Iint}(t) := \sqrt{2} \cdot \sum_{jj} \text{Re}(\text{iiin}_{jj}) \cdot \cos(jj \cdot w \cdot t) + \text{Im}(\text{iin}_{jj}) \cdot \sin(jj \cdot w \cdot t)$$

$$\text{ioutt}(t) := \text{if}(\cos(w \cdot t) > 0.5, 20, 0) + \text{if}(\cos(w \cdot t) < -0.5, -20, 0)$$

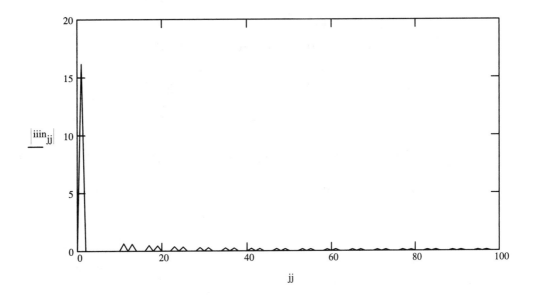

Example 16.2.1

This solves a vector equation of the form x dot = Ax + Bu for a piecewise-linear system. It is assumed that there are only two pieces in this version. The method is the fourth-order Runge–Kutta approach.

$$d := 0.25 \qquad \text{fswitch} = 200000 \qquad T := \frac{1}{\text{fswitch}} \qquad \text{iL} := 0 \qquad \text{vC} := 0$$

$$\text{Vin} := 20 \qquad \text{Rds} := 0.2 \qquad \text{Vd} := 0.8 \qquad \text{Rd} := 0.02 \qquad \text{RL} := 0.1$$

$$\text{RESR} := 0.1 \qquad \text{Rload} := 1 \qquad L := 10^{-4} \qquad C := 10^{-5}$$

$$Rp := \frac{Rload}{(Rload + RESR)} \qquad D := 0.25 \qquad \text{Variables: } x_0 = iL, \ x_1 = vC.$$

The averages are y_0 and y_1, respectively.

$$A1 := \begin{bmatrix} \dfrac{-Rds - RL - Rp \cdot RESR}{L} & -\dfrac{Rp}{L} \\ \dfrac{Rp}{C} & -\dfrac{Rp}{Rload \cdot C} \end{bmatrix} \qquad B1u := \begin{pmatrix} \dfrac{Vin}{L} \\ 0 \end{pmatrix} \qquad xdot1(x,t) := A1 \cdot \begin{pmatrix} x_0 \\ x_1 \end{pmatrix} + B1u$$

$$A2 := \begin{bmatrix} \dfrac{-Rd - RL - Rp \cdot RESR}{L} & -\dfrac{Rp}{L} \\ \dfrac{Rp}{C} & -\dfrac{Rp}{Rload \cdot C} \end{bmatrix} \qquad B2u := \begin{pmatrix} -\dfrac{Vd}{L} \\ 0 \end{pmatrix} \qquad xdot2(x,t) := A2 \cdot \begin{pmatrix} x_0 \\ x_1 \end{pmatrix} + B2u$$

$$Abar := A1 \cdot D + A2 \cdot (1 - D) \qquad Bubar := B1u \cdot D + B2u \cdot (1 - D)$$

$$ydot(y,t) := Abar \cdot \begin{pmatrix} y_0 \\ y_1 \end{pmatrix} + Bubar$$

Runge–Kutta method (defines four functions, f1, f2, f3, f4)

$$f1(xdot,x,t,dt) := xdot(x,t)$$

$$f2(xdot,x,t,dt) := xdot(x + .5 \cdot dt \cdot f1(xdot,x,t,dt), t + .5 \cdot dt)$$

$$f3(xdot,x,t,dt) := xdot(x + .5 \cdot dt \cdot f2(xdot,x,t,dt), t + .5 \cdot dt)$$

$$f4(xdot,x,t,dt) := xdot(x + dt \cdot f3(xdot,x,t,dt), t + dt)$$

$$f0(xdot,x,t,dt) := \frac{dt}{6} \cdot (f1(xdot,x,t,dt,) + 2 \cdot f2(xdot,x,t,dt) + 2 \cdot f3(xdot,x,t,dt) + f4(xdot,x,t,dt))$$

Set up the simulation. Initial and final times, time step, number of points to compute:

$$t0 := 0 \qquad tf := 50 \cdot T \qquad nx := 400 \qquad dt := \frac{tf - t0}{nx} \qquad i := 1..nx \qquad t_0 := 0$$

$$t_i := t_{i-1} + dt \qquad kx := 0..nx \qquad x^{\langle 0 \rangle} := \begin{pmatrix} 0 \\ 0 \end{pmatrix} \qquad y^{\langle 0 \rangle} := \begin{pmatrix} 0 \\ 0 \end{pmatrix}$$

$$ny := 40 \qquad dty := \frac{tf - t0}{ny} \qquad iy := 1..ny \qquad ty_0 := 0 \qquad ky := 0..ny$$

Start-up. Begin from 0 initial conditions.

$$ty_{iy} := ty_{iy-1} + dty$$

$$x^{\langle i \rangle} := x^{\langle i-1 \rangle} + if[mod[(t_{i-1}),T] \leq D \cdot T, f0(xdot1, x^{\langle i-1 \rangle}, t, dt), f0(xdot2, x^{\langle i-1 \rangle}, t, dt)]$$

Actual variables.

$$y^{\langle iy \rangle} := y^{\langle iy-1 \rangle} + f0(ydot, y^{\langle iy-1 \rangle}, ty, dty)$$

Averaged variables.
Plot results:

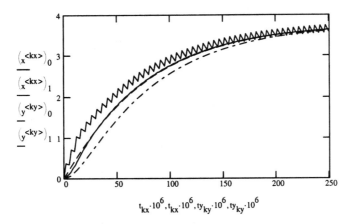

$$t_{kx} \cdot 10^6, t_{kx} \cdot 10^6, ty_{ky} \cdot 10^6, ty_{ky} \cdot 10^6$$

Example 17.3.1

$$Vin := 12 \qquad RL := 0.02 \qquad Rload := 4.8$$

$$Vout(D,Rload) := D \cdot \frac{Vin \cdot (1 - D)}{\left[\dfrac{-RL}{Rload} - (1 - D)^2\right]} \qquad IL(D,Rload) := \frac{-Vout(D,Rload)}{Rload \cdot (1 - D)}$$

$$i := 0..700 \qquad d_i := \frac{i}{1000} \qquad V_i := -Vout(d_i,Rload) \qquad I_i := IL(d_i,Rload) \qquad I_{1000} := 600$$

$$D := 0.66 \qquad TOL := 0.0001$$

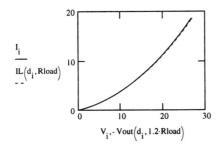

$$\frac{I_i}{IL(d_i,Rload)}$$

$$V_i, -Vout(d_i, 1.2 \cdot Rload)$$

Given
$$Vout(D,Rload) = -24$$
$$a := Find(D) \qquad a \cdot 100 = 67.5219$$
$$IL(a,Rload) = 15.395$$

Three-Phase Paper Generation

$$T := 12 \qquad \omega := 2 \cdot \frac{\pi}{T} \qquad p := 3 \qquad i := 0..1023$$

$$t_0 := 0 \qquad k := 0..96 \qquad k1 := 0..48 \qquad t_i := \frac{i}{1024} \cdot T \cdot 4$$

$$x_{k1 \cdot 2} := k1 \qquad y_{k1 \cdot 4} := -1 \qquad y_{k1 \cdot 4+2} := 1$$

$$v3(t,j) := \cos\left(\omega \cdot t - \frac{2 \cdot \pi}{p} \cdot j\right) \qquad x_{k1 \cdot 2+1} := k1 \qquad y_{k1 \cdot 4+1} := 1 \qquad y_{k1 \cdot 4+3} := -1$$

$j := 0..p - 1$

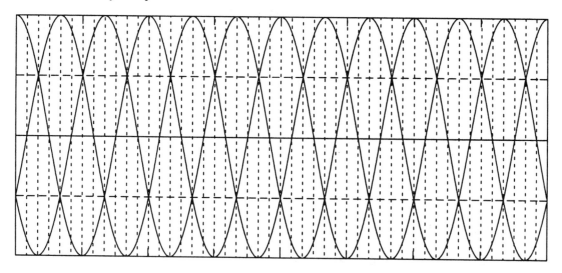

C.3 SPICE LISTINGS

These listings are compatible with the student version of PSPICE

Diode Circuit, Example 2.7.1

```
.opt acct opts nomod nopage it14=20 it15=10000
.tran 20us 50ms 0 20us

*Sinusoidal input (at 60 Hz) into
*4 diodes in a bridge.

*The bridge supplies a resistive string with a dc bias voltage.

*A fifth diode is in the output string.

*vout(t) is from node 12 to node 0

vin 1 2 sin (0 40 60 0 0)
d1 1 10 dio
d2 2 10 dio
d3 0 1 dio
d4 0 2 dio
vdc 10 11 dc 10
r1 11 12 1k
d5 12 13 dio
```

```
vdc2 13 14 dc 10
r2 14 0 100
rout 12 0 1k
.model dio D(Is=10n Rs=0.01)
.probe
.end
```

Diode Bridge Example 5.5.1

```
.opt acct nomod nopage reltol=.001 itl4=20 itl5=10000
.width out=80
.temp 30
.tran 1ms 200ms 0 0.2m uic
vin 99 0 sin(0 22.0 60 0 0)
d11 99 1 dio
d12 2 99 dio
d21 0 1 dio
d22 2 0 dio
lout 1 3 0.0463 ic=0
cout 3 2 1000uF ic=0
rout 3 2 6
*d1 98 0 dio
.model dio D(Is=20p Rs=0.1 N=1)
.probe
.end
```

VSI into RL and RLC loads. Text example 6.3.2 P. Krein

```
.opt acct nomod nopage reltol=.001 itl5=0
.width out=80
.temp 30
.tran 10us 200ms 0 100us uic
vvsip 20 0 pulse(0 144 −.004166667 0 0 .00833333 .01666667)
vvsin 21 20 pulse(0 144 −.002777777 0 0 .00833333 .01666667)
vvsi0 22 21 dc −144
l1 22 23 3mh ic=0
r1 23 0 2
l2 22 33 3mh ic=0
c2 33 34 2345uf ic=0
r2 34 0 2
l3 22 43 30mh ic=50a
c3 43 44 234.5uf ic=0
r3 44 0 2
```

```
.probe
.end
```

VSI into 277 V. Text example 6.8.1 P. Krein

```
.opt acct nomod nopage reltol=.001 itl5=0
.width out=80
.temp 30
.tran 10us 200ms 0 100us uic
vac 10 0 sin(0 391.737 60 −.004166667 0)
vvsip 20 0 pulse(0 500 −.002703333 0 0 .0037356 .01666667)
vvsin 21 20 pulse(0 −500 .00563 0 0 .0037356 .01666667)
l1 21 22 132.629uh ic=2500
r1 22 10 .002
.probe
.end
```

Figure 8.15—Resonant BUCK CONVERTER for Chapter 8 (ZCS or ZVS)

```
.opt acct nomod nopage reltol=.001 itl5=0
.width out=80
.temp 30
.tran 100ps 100ns 0 200ps uic
vg 101 2 pulse(0 12 3.5ns 0 0 11ns 25ns)
*rg 101 104 5
vcc 3 0 dc 12
lin 3 31 50nH ic=0A
rin 31 32 0.001
*cds 32 2 200pF ic=0V
*lout 2 1 250uH ic=1A
.model dioi D(Is=0.1p Rs=0.01)
dout 0 2 dioi
cd 2 0 80pF ic=0V
*resr 1 201 0.35
*cout 1 0 10uf ic=0V
*rload 1 0 1
iload 2 0 dc 0.3
m1 32 101 2 2 mosfast
.model mosfast NMOS(Level=3 Gamma=0 Delta=0 Kappa=0 Vmax=0 Xj=0
+     Tox=20n Uo=600 Phi=.6 Rs=1m Kp=20 u W=.3 L=0.2u Vto=2.5 Rd=1m
+     Rds=500K Cbd=3p Pb=.8 Mj=.5 Fc=.5 Cgso=8p Cgdo=1.5p Rg=15
+     Is=2E−16 N=1 Tt=1n)
```

```
*.print tran i(lin),v(101,2),v(2),v(1),i(lout)
.probe
.end
```

Figure 15.24—FEEDBACK BUCK CONVERTER

```
.opt acct nomod nopage reltol=.001 itl5=0 itl4=40
.width out=80
.temp 30
.tran 0.1us 1000us 0u 0.5u uic
*vg 101 102 pulse(0 12 0 0 0 16us 50 us)
*rg 101 104 5
*vcc 3 0 dc 24
vcc 3 0 pwl(0 20 1002u 20 1003u 24 3000u 24)
s1 2 0 21 10 spwr
l1 3 2 50uH ic=12.5A
.model dio D(Is=10n Rs=0.01)
dout 2 1 dio
cout 1 0 50uf ic=50V
rload 1 0 10
sload 1 11 22 0 spwr
rload2 11 0 100
egain 20 0 value={V(30) * (15 − 0.3*V(1) − 0.02*I(L1))}
rf 20 21 10k
cf 21 0 0.1uF ic=0V
vramp 10 0 pulse(0 5 0 9.9u 0 0 10us)
es 22 0 value={(time−.0018)*1e5}
vr 23 0 pwl(0 20 1800u 20 1805u 24 3000u 24)
rr 23 0 1k
vlim 30 0 pulse(0 1 0 0 0 8u 10u)
.model spwr vswitch(ron=0.028 roff=10E7 von=0.5 voff=0)
.probe
.end
```

Figure 15.31—CURRENT-MODE FEEDBACK BOOST CONVERTER

```
.opt acct nomod nopage reltol=.001 itl5=0 itl4=100 abstol=1n numdgt=5
.width out=80
.temp 30
.tran 1us 9000us 2500u 0.25u uic
*vcc 100 0 dc 20
```

```
vi 100 103 dc 0
rl 103 3 0.1
vcc 100 0 pwl(0 20 3002u 20 3003u 24 9000u 24)
s1 2 0 25 40 spwr
cl 2 0 100pf ic=0.625V
ll 3 2 50uH ic=12.5A
hl 40 0 vi 0.1
.model dio D(Is=10n Rs=0.01)
dout 2 1 dio
cout 1 0 50uf ic=50V
rload 1 0 10
sload 1 11 26 0 spwr
rload2 11 0 100
vswl 26 0 pwl(0 0 6000u 0 6001u 1 9000u 1)
rd 26 0 1k
egain 20 0 value={0.3 * (50 − V(1) )+5}
rf 20 21 10k
cf 21 0 0.1uF ic=2V
estab 25 0 value={v(30)*(v(21)−v(12))}
vramp2 12 0 pulse(0 5 0 9.8u 0.2u 0 10u)
rdummy 12 0 1k
vlim 30 0 pulse(0 1 0 0 0 8u 10u)
rdummy2 30 0 1k
.model spwr vswitch(ron=0.05 roff=1E6 von=0.05 voff=−0.05)
.print tran v(1)
.probe
.end
```

Figure 16.1—Flyback Converter with PI Control and Current Feedback

```
.opt acct nomod nopage reltol=.001 itl5=0 itl4=40 numdgt=5
.width out=80
.temp 30
.tran 1us 1500us 100u 0.25u uic
*vg 101 102 pulse(0 12 0 0 0 16us 50us)
*rg 101 104 5
*vcc 3 0 dc 16
vcc 3 0 dc 360
s1 3 2 32 10 spwr
rsnub 3 103 5000
csnub 103 2 100pF ic=0V
dsnub 3 103 dio
vi1 2 102 dc 0
ll 102 0 0.36H ic=0.027A
vi2 0 104 dc 0
```

```
l2 104 4 400uH ic=0A
k1 l1 l2 0.99
.model dio D(Is=10n Rs=0.01)
dout 4 1 dio
cout 1 0 30uf ic=12V
rload 1 0 30
sload 1 11 43 0 spwr
lload2 11 12 600uH ic=0A
rload2 12 0 150
egain 20 0 value={V(1) − 12}
eamp 21 0 0 23 10000
rin 20 23 1000
rf 23 25 1500
cf 25 21 0.4uf ic=−0.5V
econt 32 0 value={v(21)−3*(30*i(vi1)+i(vi2))}
*rcomp 21 31 30k
*ccomp 31 33 1uF ic=0V
*rcomp2 33 0 2000
*ecomp 32 0 value={(v(31)−10*i(vi1))*v(45)}
vramp 10 0 pulse(0 1 0 6.6u 0.02u 0.047u 6.667us)
rramp 10 0 1k
vr 43 0 pwl(0 0 6700u 0 6705u 1 10000u 1)
rr 43 0 1k
vlim 45 0 pulse(0 1 0 0.1u 0.1u 5u 6.667u)
rvlim 45 0 1k
.model spwr vswitch(ron=0.028 roff=10E7 von=0.05 voff=0)
.probe
.print tran v(1)
.end
```

C4 OTHER PROGRAM LISTINGS

Fig 16.1—This Matlab File Produces the Root Locus for Figure 16.1

```
for kc=0:2:6
   for kp=0:.1:5
     ki=0;
     %Define circuit parameters
     L=.0004;C=30e−6;R=30;
     D=.5;I=.8;Vin=12;Vout=−12;

     %Define state equations
     a=[−kc*(Vin−Vout)/L      (1−D)/L+kp/L*(Vin−Vout)      ki*(Vin−Vout)/L
        (D−1)/C−kc*I/C   −1/(R*C)+kp*I/C        ki*I/C
              0                   1               0];
```

```
        b=[kp/L*(Vin−Vout)
        kp*I/C
          1];

        c=[0 1 0];
        d=[0];

        %Find poles and zeros
        pzmap(a,b,c,d);
        if kp==0 & kc==0 & ki==0
          [p,z]=pzmap(a,b,c,d)
        hold
        end

    end
  end
  hold
```

Figure 17.24—This BASIC Program Simulates a PFC Boost Converter

```
'Initializing values
increment = 0: t = 0: vc = 200: il = 0: kerr = 0: kp = 0.01: ki = 0.01:
Vref = 200

'Define circuit element values
Rload = 200: L = 2.55e−2: C = 7.5e−3

'Define simulation parameters
tlast = 0.06: delt = 1e−7: po = 500

'Get data file name from user
DialogSheets("in_name").Show
in_file = DialogSheets("in_name").EditBoxes(1).Text
Open in_file For Output As #1

'Begin simulation loop
While t < tlast
t = t + delt

'Rectified signal and PI control calculation
Vins = 170 * Abs(Sin(377 * t))
erro = ki * (Vref − vc)
kerr = kerr + erro * delt
Ireff = Vins * (kerr + kp * (Vref − vc))
```

```
'Switching control
If il <= Ireff − 0.1 Then q1 = 1: q2 = 0
If il > Ireff + 0.1 Then q1 = 0: q2 = 1

'Use forward Euler equations to solve circuit
dil_dt = (q1 * Vins + q2 * (Vins − vc)) / L
il = il + delt * dil_dt
If il < 0 Then il = 0: q2 = 0
dvc_dt = (−vc / Rload + q2 * il) / C
vc = vc + delt * dvc_dt

'Print data to a file
If increment = po Then
    Print #1, t; ";"; vc / 200; ";"; il
    increment = 0
End If
increment = increment + 1

Wend

Close #1

End Sub
```

A FORTRAN program to find Fourier components of a piecewise-sinusoidal function. This code can be obtained at ftp://power.ece.uiuc.edu/

```
      PROGRAM FOURCOM
      REAL T1(201),T2(201),V(201),AK(201),PHI(201)
      COMMON T1,T2,V,AK,PHI
      CHARACTER*60,NAME
      INTEGER*2,IMON,IDAY,IYR
      DATA VERS/2.2/
C
C         ......FOURCOM PROGRAM. VERSION 2.0, 9/22/87.
C     ......VERSION 2.01, 11/11/87. VERSION 2.02 (IBM-PC) 10/13/88.
C     ......VERSION 2.03, 9/15/89, ADDS FULL POLAR OUTPUT.
C     ......VERSION 2.04, 9/18/89, FIXES RMS BUG.
C     ......VERSION 2.05, 9/28/89, ADDS I/O ERROR HANDLING.
C     ......VERSION 2.06, 9/21/90, FIXES −180° PROBLEM.
C     ......VERSION 2.1, 10/22/90, ADDS T.H.D., AND ALTERNATE TIME SCALE.
C     ......VERSION 2.2, 12/31/96, IMPROVES DOCUMENTATION.
C     ......by Philip T. Krein, University of Illinois, Urbana.
C     ......COPYRIGHT © 1987–1997 PHILIP T. KREIN.
C     ......ALL RIGHTS RESERVED, EXCEPT AS STATED BELOW.
C     ......
C     ...... PERMISSION IS GRANTED TO COPY AND USE THIS PROGRAM IN
C     ......AN EDUCATIONAL SETTING, FREE OF CHARGE,
```

```
C    ......PROVIDED THAT THIS COPYRIGHT NOTICE
C    ......IS INCLUDED IN ALL COPIES, AND ACKNOWLEDGEMENT OF THE
C    ......AUTHOR IS MADE.
C    ......
C    ...... FOURCOM IS A PROGRAM TO COMPUTE THE
C    ......FOURIER COEFFICIENTS OF A PIECEWISE-SINUSOIDAL
C    ......WAVEFORM. THE WAVEFORM IS DEFINED IN "SEGMENTS" THAT
C    ......DEFINE THE VARIOUS SINUSOIDAL PIECES. AN ANGULAR TIME
C    ......SCALE THETA = OMEGA*TIME IS USED THROUGHOUT, WITH VALUES
C    ......GIVEN IN DEGREES. EACH SEGMENT NEEDS THE FOLLOWING DATA:
C    ......STARTING POINT, ENDING POINT, AND
C    ......THE PARAMETERS OF THE SINUSOIDAL
C    ......FUNCTION THAT IS IN EFFECT DURING THE SEGMENT. THESE
C    ......DATA APPEAR IN THE CODE AS:
C    ......THETA1 (THE STARTING ANGLE), THETA2 (THE ENDING ANGLE),
C    ......V (THE AMPLITUDE OF THE SEGMENT SINUSOID),
C    ......AK (K IS THE NUMBER, WHICH, WHEN MULTIPLIED BY
C    ......THE FREQUENCY OMEGA IN THE FOURIER SERIES, GIVES THE
C    ......FREQUENCY OF THE COSINE IN THE SEGMENT), AND
C    ......PHI (PHASE SHIFT OF THE SINUSOID IN THE SEGMENT).
C    ......PHI IS ZERO FOR THE COSINE.
C    ......
C    ...... THE PROGRAM WORKS BY INTEGRATING THE WAVEFORM
C    ......PIECE-BY-PIECE (I.E. SEGMENT BY SEGMENT), AND THEN
C    ......ADDING THE RESULTS. THIS MEANS THAT WAVEFORMS
C    ......REPRESENTED BY SUMS OF PIECEWISE SINUSOIDS, SUCH AS
C    ......BRIDGE OUTPUT WAVEFORMS, CAN ALSO BE ANALYZED.
C
C    ...... BEGIN ACTUAL CODE. COMMENTS DESCRIBE EACH SECTION.
C    ......PLEASE NOTICE THAT THERE ARE A FEW LITTLE QUIRKY THINGS
C    ......TO ADD RARELY-USED FEATURES WITHOUT AN EXCESSIVE SET OF
C    ......PROMPTS. THESE QUIRKS ARE NOTED WITH THE "!" SIGN.
C
C    ......STATEMENT FUNCTIONS FOR AVERAGE AND RMS. THE RMS
C    ......CALCULATION ONLY WORKS PROPERLY IF NO PIECES OVERLAP,
C    ......SINCE THE SQUARE IS TAKEN BEFORE THE SUM.
C
     CA0(J)=(SIN(AK(J)*T2(J)+PHI(J))-SIN(AK(J)*T1(J)+PHI(J)))/2./AK(J)
     CRMS(J)=
     &   ((SIN(2.*AK(J)*T2(J)+2.*PHI(J))-SIN(2.*AK(J)*T1(J)+2.*PHI(J)))
     & /AK(J)/2. +T2(J)-T1(J))/4.
C
C    ......CONSTANTS FOR TRIGNONOMETRIC CONVERSION.
     PI=3.141592654
     DEGRAD=PI/180.
     RADDEG=180./PI
```

```
C
C     ......INPUT INFORMATION:
C     ......   NCOMP — THE NUMBER OF FOURIER COMPONENTS TO BE
C     ......       CALCULATED. THERE IS NO SPECIAL UPPER LIMIT,
C     ......       AND HUNDREDS OF COMPONENTS CAN BE COMPUTED.
C     ......!!   IF NCOMP IS GIVEN AS NEGATIVE, THE PROGRAM
C     ......       LOOKS INTO FILE UNIT #11 FOR INPUT,
C     ......       INSTEAD OF THE DEFAULT UNIT #5.
C     ......   FREQ — THE FUNCTION FREQUENCY IN HERTZ. THIS IS
C     ......       USED TO ESTABLISH AN ANGULAR TIME SCALE, IN WHICH
C     ......       1/FREQ CORRESPONDS TO 360 DEGREES.
C     ......   TSCALE — STARTING WITH VERSION 2.07, A DIFFERENT
C     ......       FREQUENCY CAN BE USED FOR THE ANGULAR TIME SCALE,
C     ......       IF DESIRED. IF THIS FEATURE IS USED, A FULL PERIOD
C     ......       WILL NOT CORRESPOND TO 1/TSCALE = 360 DEGREES.
C     ......   NSEG — THE NUMBER OF PIECEWISE-SINUSOIDAL "SEGMENTS"
C     ......       REQUIRED TO DEFINE ONE FULL FUNCTION PERIOD. THE
C     ......       NATURE OF THE ALGORITHM IMPLICITLY ASSUMES A VALUE
C     ......       OF ZERO FOR ANY PORTION OF THE PERIOD LEFT UNDEFINED.
C     ......   FOR EACH OF THE NSEG DEFINED SEGMENTS, THE PROGRAM
C     ......       REQUESTS:
C     ......       ANGULAR TIME BOUNDARIES (START AND END), GIVEN
C     ......       IN DEGREES, PER THE DEFINED ANGULAR SCALE.
C     ......       THE AMPLITUDE, FREQUENCY, AND PHASE OF THE
C     ......         PIECEWISE SINUSOIDAL SEGMENT. (NOTICE THAT
C     ......         AN AMPLITUDE COMBINED WITH FREQUENCY OF 0 AND
C     ......         PHASE OF 0 CAN BE USED TO DEFINE A DC SEGMENT.)
C     ......THE INFORMATION IS SENT OUT TO UNIT #12 AS IT IS INPUT.
C     ......AVERAGE AND RMS VALUES ARE COMPUTED AS THE SEGMENT
C     ......INFORMATION COMES IN.
C
C     ......   SEND PROMPTS TO THE CONSOLE.
      WRITE (*,*) 'FOURIER COMPONENTS OF PIECEWISE-SINUSOIDAL FUNCTION,'
      WRITE (*,670) VERS
  670 FORMAT(1X,'VERSION ',F4.1,', DECEMBER 1996, P.T. KREIN.')
      WRITE (*,680)
  680 FORMAT(/1X,'Please type in your name: '\)
      READ (*,'(A)') NAME
      WRITE (*,690)
  690 FORMAT(/1X,'How many Fourier components should be computed? '\)
      READ (*,*) NCOMP
      IF (NCOMP.GT.200)
     &WRITE(*,*) 'That''s a lot of components! Stop me if it is wrong.'
C
C     ......!!IF NCOMP<0, PULL DATA FROM A FILE, FORTRAN UNIT #11.
C
```

```
          IFILE=5
          IF (NCOMP.LT.0) THEN
              INDP=1
              NCOMP=-NCOMP
              IFILE=11
              GOTO 10
   5          IFILE=5
              INDP=0
   10         CONTINUE
          ENDIF
C
C     ......READY FOR FUNCTION ANGULAR TIME SCALE DEFINITION.
C
          IF (INDP.NE.1) INDP=0
          IF (INDP.NE.1) WRITE(*,695)
   695    FORMAT(/1X,'What is the function frequency (Hz)? '\)
          READ (IFILE,*) FREQ
          TSCALE=FREQ
C
C     ......START AN OUTPUT FILE TO STORE INFORMATION AND RESULTS.
C     ......PUT IN THE DATE (BASED ON MICROSOFT'S COMPILER FORMAT).
C
          IF (INDP.NE.1) WRITE(*,*)
   &          'Trying to open the file DATOUT.DAT for output.'
          OPEN(UNIT=12,FILE='DATOUT.DAT',MODE='READWRITE',ERR=200)
          IF (INDP.NE.1) WRITE(*,*)
   &          'Successful open. Output file is DATOUT.DAT.'
   20     CALL GETDAT(IYR,IMON,IDAY)
C
C     ......   FIRST REFERENCE TO UNIT #12
C
   30     WRITE(12,700,ERR=210) VERS,NAME,IMON,IDAY,IYR
          WRITE(12,710,ERR=210) FREQ
   700    FORMAT(1X,'FOURIER COMPONENTS OF PIECEWISE-SINUSOIDAL
   &     FUNCTION.'
   &     /' VERSION 'F4.1//1X,A60,5X,I2,'/',I2.2,'/',I4)
   710    FORMAT(1X/3X,'FUNCTION FREQUENCY (HERTZ): ',10X,F14.6)
          IF (INDP.NE.1) THEN
              WRITE(*,711) FREQ
   711    FORMAT(/1X,'If you would prefer to use some frequency other '
   &     'than ',F12.3/' for the angular time scale, enter it now '
   &     '(else type 0): '\)
          READ(IFILE,*) TSCALE
          IF(TSCALE.EQ.0.) TSCALE=FREQ
          WRITE(*,712)
   712    FORMAT(/1X,
```

```
     &    ' Based on your value, I have defined a time scale in degrees.')
          WRITE(*,*) 'T=0 IS TIME OF ZERO DEGREES. T=',
     &       1./TSCALE,' IS TIME OF 360 DEGREES.'
          IF (TSCALE.NE.FREQ) WRITE(*,*) ' In your case, a full ',
     &       'period corresponds to ',360.*TSCALE/FREQ,' degrees.'
C
C    ......FUNCTION SEGMENT DATA INPUT SECTION.
C    ......CALCULATE DC AND RMS VALUES AS THE DATA ARE ENTERED.
C    ......ECHO THE DATA TO UNIT #12.
C
          WRITE (*,715)
715       FORMAT(/1X,
     &       'How many pieces are needed to define your function? '\)
          ENDIF
          READ (IFILE,*) NSEG
          IF (INDP.NE.1) THEN
             WRITE(*,718)
718       FORMAT(/1X,'For each function piece, '
     &       'input start time, end time (in degrees), and ')
          WRITE (*,*)
     &'       the amplitude, frequency, and phase angle within the piece.'
          ENDIF
          WRITE(12,719,ERR=210) TSCALE
719       FORMAT(3X,'ANGULAR TIME SCALE FREQUENCY (HERTZ): ',F14.6/)
          WRITE(12,720)
720       FORMAT(1X,'USER INPUT DATA:'/1X,'START ANGLE',2X,'END ANGLE',
     &       2X,'AMPLITUDE',2X,'SEGMENT FREQUENCY',2X,'SEGMENT PHASE')
          A0=0
          ARMS=0
          VMAX=0.
          DO 40 I=1,NSEG
35        IF (INDP.NE.1) WRITE(*,725) I
725       FORMAT(1X,'piece ',I3, '? '\)
          READ(IFILE,*,IOSTAT=IOCHK) T1(I),T2(I),V(I),AK(I),PHI(I)
          IF(IOCHK.GT.0) THEN
             WRITE(*,*) 'Error in input. Try again.'
             GOTO 35
          ENDIF
          WRITE(12,730) T1(I),T2(I),V(I),AK(I),PHI(I)
730       FORMAT(1X,F11.5,1X,F11.5,1X,F9.4,1X,F17.6,1X,F13.6)
C
C    ......CONVERT ANGULAR TIME SCALE DATA TO ABSOLUTE TIME DATA.
C
          T1(I)=DEGRAD*T1(I)*FREQ/TSCALE
          T2(I)=DEGRAD*T2(I)*FREQ/TSCALE
          AK(I)=AK(I)/FREQ
```

```
                   PHI(I)=DEGRAD*PHI(I)
                   IF (V(I).EQ.0.) GOTO 40
                   VMAX=MAX(V(I),VMAX)
                     IF (AK(I).NE.0.) THEN
                       A0=A0+CA0(I)*V(I)
                       ARMS=ARMS+CRMS(I)*V(I)**2
                     ELSE
                       A0=A0+(T2(I)−T1(I))*V(I)*COS(PHI(I))/2.
                       ARMS=ARMS+(T2(I)−T1(I))*V(I)**2/2.
                     END IF
       40    CONTINUE
                   IF (INDP.NE.1) WRITE(*,*) 'DATA COMPLETE. PROGRAM RUNNING.'
C
C    ......END OF INPUT SECTION. NOW, CALCULATE AND
C    ......TABULATE THE SERIES, ONE COMPONENT AT A TIME.
C    ......SIX DIGIT ACCURACY LIMITS ARE ASSUMED.
C
      ARMS=SQRT(ARMS/PI)
      WRITE (12,840) ARMS
      IF (ABS(A0).LT.1E-6*VMAX) A0=0.
      DO 100 I=1,NCOMP
          AN=0.
          BN=0.
          DO 50 J=1,NSEG
             IF (V(J).EQ.0) GOTO 50
             AN=AN+CAN(I,J)*V(J)
             BN=BN+CBN(I,J)*V(J)
       50    CONTINUE
C
C    ......CALCULATE T.H.D. BASED ON THE N=1 COMPONENT, AND
C    ......EXCLUDING THE DC COMPONENT.
C
                   IF (I.EQ.1) THEN
                     A1=((AN/PI)**2+(BN/PI)**2)/2
                     IF (A1.LT.0.0001) GOTO 60
                     THD=SQRT((ARMS**2-A1-(A0/PI)**2)/A1)
                     WRITE(12,850) THD*100
       60          WRITE(12,820) A0/PI,A0
                   ENDIF
                   IF (ABS(AN).LT.1E-6*VMAX) AN=0.
                   IF (ABS(BN).LT.1E-6*VMAX) BN=0.
      100   WRITE (12,800) I,FREQ*FLOAT(I),AN/PI,BN/PI,AN,BN,
      &     SQRT(AN**2+BN**2)/PI,ANGLE(BN,AN)
      800   FORMAT (1X,I3,F8.1,6F11.5)
      820   FORMAT(/3X,'N',4X,'FREQ',7X,'A',10X,'B',9X,'A*PI',7X,
      &     'B*PI',8X,'C ',7X,'THETA'
      $     /3X,'0',5X,'0.',1X,F11.5,11X,F11.5)
```

```
840   FORMAT (//1X,'RMS VALUE OF FUNCTION: ',1PG14.5,
&         '(accurate only if segments never overlap)')
850   FORMAT (1X,'THD (%) FOR FUNCTION: ',1PG14.5,
&         '(computed from RMS value, C1, and C0.)')
      IF (INDP.NE.1) THEN
         WRITE(*,860)
860   FORMAT (/1X,
&         'END PROGRAM — DATA WRITTEN TO OUTPUT FILE NAMED
&      ABOVE.'/)
      WRITE(*,*) 'THE FILE CAN NOW BE TYPED OR PRINTED.'
      WRITE(*,*) 'BE SURE TO USE THE CORRECT FILE NAME.'
      ENDIF
      STOP
C
C     ......I/O ERROR HANDLING.
C
200   OPEN(12,FILE='C:DATOUT.DAT',MODE='READWRITE',ERR=202)
      IF (INDP.NE.1) WRITE(*,*)
&'First attempt failed. Output file changed to C:DATOUT.DAT'
      GOTO20
201   OPEN(12,FILE='CON:',ERR=202)
      WRITE(*,*) 'Could not open output file.'
      WRITE(*,*) 'Will send output to screen instead.'
      GOTO 20
202   WRITE(*,*) 'ERROR WHILE TRYING TO OPEN OUTPUT FILE.'
      WRITE(*,*) 'PLEASE PROVIDE A NEW FILE NAME AT THE PROMPT.'
      OPEN(12,FILE=' ',MODE='READWRITE',ERR=201)
      GOTO 20
210   CLOSE(12)
      GOTO 202
      END
C
C     ......INTEGRATION FUNCTION FOR THE COSINE (A SUB N) TERMS.
C
      FUNCTION CAN(N,J)
      REAL T1(201),T2(201),V(201),AK(201),PHI(201)
      COMMON T1,T2,V,AK,PHI
      SINNK(T,A)=SIN(N*T+A*AK(J)*T+A*PHI(J))/(N+A*AK(J))/2.
      CAN=SINNK(T2(J),1.)-SINNK(T1(J),1.)
      IF(REAL(N).EQ.AK(J)) THEN
         CAN=CAN+(T2(J)-T1(J))*COS(PHI(J))/2.
         RETURN
      END IF
      CAN=CAN+SINNK(T2(J),-1.)-SINNK(T1(J),-1.)
      RETURN
      END
```

```
C
C      ......INTEGRATION FUNCTION FOR THE SINE (B SUB N) TERMS.
C
       FUNCTION CBN(N,J)
       REAL T1(201),T2(201),V(201),AK(201),PHI(201)
       COMMON T1,T2,V,AK,PHI
       COSNK(T,A)=-COS(N*T+A*AK(J)*T+A*PHI(J))/(N+A*AK(J))/2.
       CBN=COSNK(T2(J),1.)-COSNK(T1(J),1.)
       IF (REAL(N).EQ.AK(J)) THEN
           CBN=CBN-(T2(J)-T1(J))*SIN(PHI(J))/2.
       RETURN
       END IF
       CBN=CBN+COSNK(T2(J),-1.)-COSNK(T1(J),-1.)
       RETURN
       END
C
C      ......FUNCTION TO CALCULATE ANGLE OF FOURIER COMPONENT.
C
       FUNCTION ANGLE(B,A)
       REAL B,A,RADDEG,PI
       ANGLE=0.
       IF (B.EQ.0.) THEN
       IF (A.GE.0.) RETURN
       ANGLE=180.
       RETURN
       ENDIF
       IF (A.EQ.0.) THEN
       ANGLE=SIGN(90.,-B)
       RETURN
       ENDIF
       PI=3.141592654
       RADDEG=180./PI
       ANGLE=RADDEG*ATAN2(B,A)
       RETURN
       END
```

APPENDIX D: REFERENCE MATERIALS

D.1 FOURIER SERIES OF SELECTED WAVEFORMS

General periodic waveform; a function $f(t)$ such that $f(t + T) = f(t)$

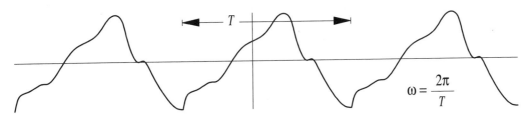

$$\omega = \frac{2\pi}{T}$$

Figure D.1 A periodic function with period T.

$$f(t) = a_0 + \sum_{n=1}^{\infty} a_n\cos(n\omega t) + b_n\sin(n\omega t), \qquad \omega = \frac{2\pi}{T} \tag{D.1}$$

$$a_0 = \frac{1}{T}\int_{\tau}^{\tau+T} f(t)dt$$

$$a_n = \frac{2}{T}\int_{\tau}^{\tau+T} f(t)\,\cos(n\omega t)dt \tag{D.2}$$

$$b_n = \frac{2}{T}\int_{\tau}^{\tau+T} f(t)\,\sin(n\omega t)dt$$

Even function; a function $f(t)$ such that $f(-t) = f(t)$ (same symmetry as cosine)

$$a_0 = \frac{1}{T} \int_{\tau}^{\tau+T} f(t)dt$$

$$a_n = \frac{2}{T} \int_{\tau}^{\tau+T} f(t) \cos(n\omega t)dt \qquad \text{(D.3)}$$

$$b_n = 0$$

Odd function; a function $f(t)$ such that $f(-t) = -f(t)$ (same symmetry as sine)

$$a_n = 0 \qquad \qquad \text{(D.4)}$$

$$b_n = \frac{2}{T} \int_{\tau}^{\tau+T} f(t) \sin(n\omega t)dt$$

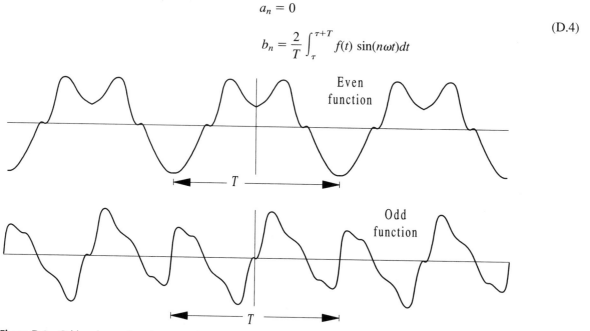

Figure D.2   Odd and even function examples.

Switching function $q(t)$

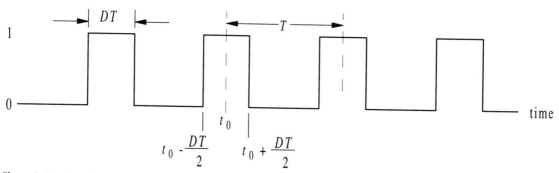

Figure D.3   General periodic switching function $q(t)$.

$$q(t) = D + \frac{2}{\pi} \sum_{n=1}^{\infty} \frac{\sin(n\pi D)}{n} \cos(n\omega t - n\phi_0) \tag{D.5}$$

$$q_{RMS} = \sqrt{D} \tag{D.6}$$

Square wave $sq(t)$ (the signum function applied to a cosine waveform)

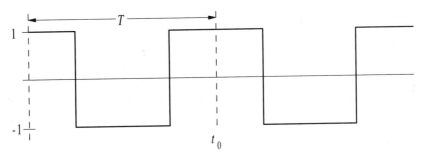

Figure D.4   Square wave function sq(t).

$$sq(t) = \frac{4}{\pi} \sum_{n=1}^{\infty} \frac{\sin(n\pi/2)}{n} \cos(n\omega t - n\phi_0)$$

$$= \frac{4}{\pi} \sum_{n=1,3,5,\cdots}^{\infty} \frac{(-1)^{\frac{n-1}{2}}}{n} \cos(n\omega t - n\phi_0) \tag{D.7}$$

$$sq_{RMS} = 1, \qquad sq_{THD} = \sqrt{\frac{\pi^2}{8} - 1} \tag{D.8}$$

Triangle wave $tri(t)$

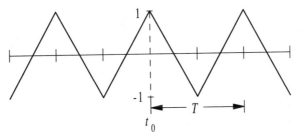

Figure D.5   Symmetric triangle waveform.

$$tri(t) = \frac{4}{\pi^2} \sum_{n=1}^{\infty} \frac{1 - \cos(n\pi)}{n^2} \cos(n\omega t - n\phi_0)$$

$$= \frac{8}{\pi^2} \sum_{n=1,3,5,\cdots}^{\infty} \frac{1}{n^2} \cos(n\omega t - n\phi_0) \tag{D.9}$$

$$tri_{RMS} = \frac{\sqrt{3}}{3}, \qquad tri_{THD} = \sqrt{\frac{\pi^4}{96} - 1} \tag{D.10}$$

Full wave rectifier function $rect(t)$

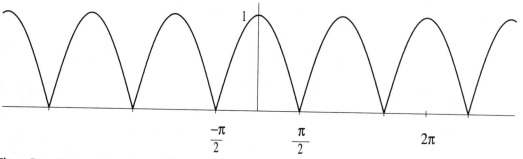

Figure D.6 Output voltage from a full-wave rectifier.

$$rec(t) = \frac{2}{\pi} + \frac{4}{\pi} \sum_{n=1}^{\infty} \frac{\cos(n\pi)}{1 - 4n^2} \cos(2n\omega_{in}t) \tag{D.11}$$

$$rect_{RMS} = \frac{\sqrt{2}}{2} \tag{D.12}$$

Quasi-square wave (voltage-sourced inverter output) $vsi(t)$

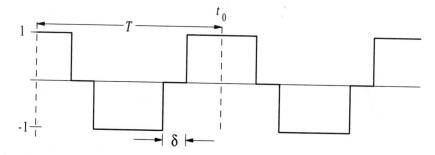

Figure D.7 Voltage-sourced inverter output.

$$vsi(t) = \frac{4}{\pi} \sum_{n=1}^{\infty} \frac{\sin(n\pi/2)\cos(n\delta/2)}{n} \cos(n\omega t - n\phi_0) \tag{D.13}$$

$$vsi_{RMS} = \sqrt{2d} \quad \text{with} \quad d = \frac{1}{2} - \frac{\delta}{2\pi} \tag{D.14}$$

m-pulse rectifier output *mrect(t)*

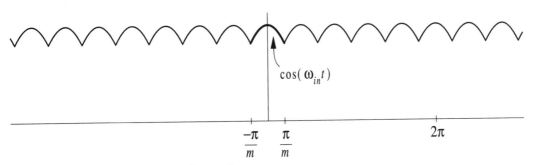

Figure D.8   Rectifier output waveform for m-pulse system.

$$mrect(t) = \frac{m}{\pi} \sin\frac{\pi}{m} + \frac{m}{\pi} \sin\frac{\pi}{m} \sum_{n=1}^{\infty} \left[\frac{\cos(n\pi)}{nm + 1} - \frac{\cos(n\pi)}{nm - 1} \right] \cos(nm\omega_{in}t) \quad (D.15)$$

$$mrect_{RMS} = \sqrt{\frac{1}{2} + \frac{m}{4\pi} \sin\left(\frac{2\pi}{m}\right)} \quad (D.16)$$

D.2 POLYPHASE GRAPH PAPER

The grids on the following pages are set up with three-phase and six-phase sinusoidal wave-forms. These grids may be copied freely. They are ideal for problems involving two-pulse, three-pulse, and six-pulse rectifiers and current-sourced inverters.

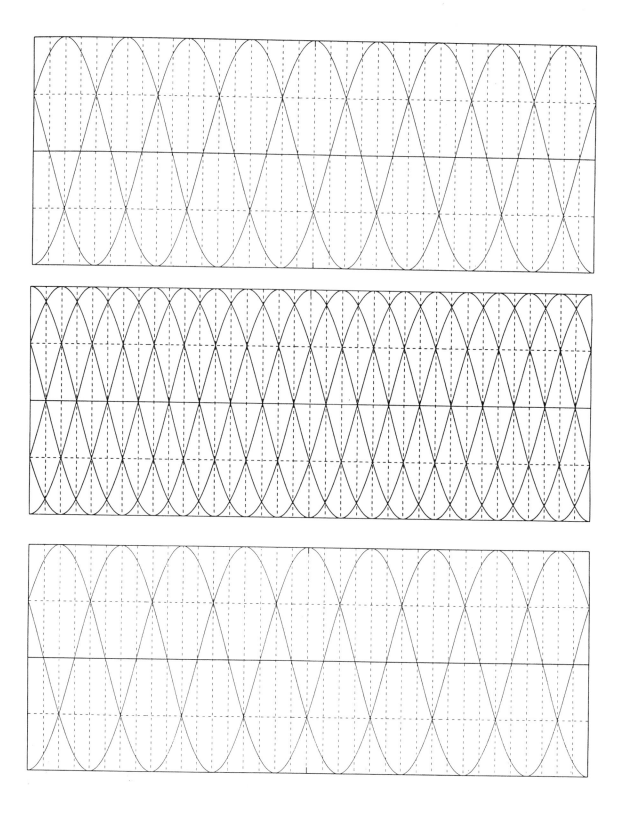

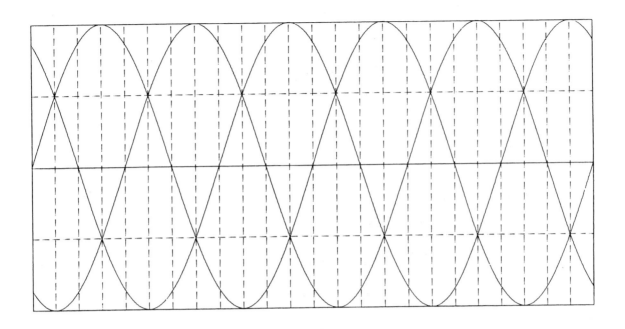

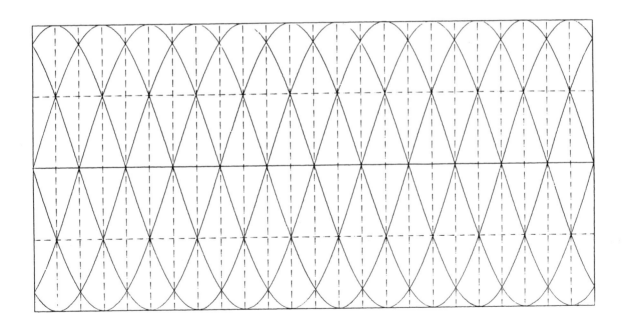

INDEX

Absolute phase, 211
ac current source,
    as a load model, 208
    as an input, 229
    equivalent source, 227, 240
    interface for, 369, 375
ac drives, 33
    converters for, 232–34, 676
    current sensors in, 565
ac link circuit, 142
ac motor drives, 22, 207, 232–34, 242
    ac-ac conversion applications, 257
    dc link conversion applications, 268
    frequency requirements, 216, 233–34
    isolated gate drive issues, 552
    sensing issues, 562
ac motors, 11, 242
    as an ac load, 208
    control of, 15–18, 21, 34, 233–34
    current control for, 232, 663
    inverters for, 206, 209, 224, 240
    motor-generator conversion, 10, 18
    time scales, 594
    torque characteristics, 235
    with rotor converters, 253–54
ac regulators, 250, 270–78
    analysis, 271–73
    and in-rush control, 275
    applications, 273–75
    commutation failure in, 343–45
    discontinuous mode analysis, 325, 343–45
    integral cycle control alternative, 276
    load effects, 271
    low inductance requirement, 343, 345
    phase delay control, 271, 277
ac voltage source,
    as input source, 102, 163, 164
    as load model, 242
    as PFC input, 229
    interfaces, 369–71, 375
    KVL issue in classical rectifier, 166
    timing from, 15
ac-ac conversion,
    ac regulators, 270–76
    bidirectional switch requirement, 57, 249
    cycloconverters, 13, 260–65
    dc link, 266–70
    frequency matching condition, 251
    integral cycle control, 276
    linear phase modulation, 259

negative converter, 262
nonlinear phase modulation, 260–66
positive converter, 262
PWM control, 266–67
slow-switching converter (SSFC), 252–55
subharmonics, 253
unrestricted frequency converter (UFC), 255–57
wanted component, 67, 91
ac-ac matrix, 249
ac-dc converter. *See* Controlled rectifier
Active ac load, 208, 210
Active state, 489
Adjustable speed drives, 234. *See also* ac motor drives
Adjustment of switch action, 77
Aerospace, 19, 21
Air gap,
    design examples, 436–40
    distributed, 432, 446
    energy storage, 409, 428, 432
    for linear inductance, 431–33, 445–46
    for maximum energy, 428, 445
    in magnetic circuit, 413–15, 419
    volume, 428, 434, 436
Aliasing, 645–46
Aluminum,
    conductor, 358
    electrolytic capacitors, 382, 391
    in ferromagnetic alloys, 412
    resistivity, 399
    temperature coefficient of resistivity, 399
    thermal property, 474
AM (amplitude modulation), 16
Ambient environment, 475
Ambient temperature, 476–78, 543
    in examples, 502, 509
American wire gauge. *See* AWG
Amp-turn limit,
    effect of air gap, 432
    impact on stored inductive energy, 429
    imposed by magnetic saturation, 428, 441, 445
    imposed by wire size, 441
Ampère's Law, 410
    loops, 416, 419, 422, 429
    magnetic form, 411–12
    MMF loop law, 413–14, 444
Ampere-turn limit. *See* Amp-turn limit
Analogue,
    electrical heat flow, 474–75

six-phase bridge, 185
Angular time, definition and change of variables, 66
Animation, 672, 688, 690
Anti-alias, 646
Antiresonance, 292–93
Antiresonant filters, 294–95, 315
Apparent power, 108, 370
Appliances, 21
    efficiency improvements, 10, 23
    power supplies, 19, 119
    triacs in, 54, 271
    variable speed, 15, 271
Approximations,
    average variables, 623
    for classical rectifiers, 171
    for magnetic analysis, 410
    for nonlinear systems, 585
Armature current, 589
Audio amplifier, 8, 319
Audio susceptibility, 586, 657
Automotive industry, high-side switching, 554, 556, 567
Auxiliary thyristor, 544
Avalanche current, 452, 521
Average behavior,
    in discontinuous mode, 327, 338
    models for, 622, 623, 657
Average configuration, 622
Average current rating, 452
Average current-mode control, 602
Average matrix, 627
Average power, 26–27, 59, 129, 284, 368, 369
    balance in discontinuous mode, 331
    balance in storage devices, 30–31, 136
    computation from components, 104, 190, 237, 276, 459
    ESR loss, 394
    Fourier Series computation, 71–74, 79
    frequency matching condition, 250–51
    heat transfer, 478
    in transfer source, 132, 140
    switching loss, 463
Averaged circuit, 629–31, 657, 658
Averaged models, 621–29, 657
    for control design, 637–38
    linearization, 631–35
    rectifiers, 654
AWG, definition and table, 396–97

757